STRUCTURAL ANALYSIS

STRUCTURAL ANALYSIS

Using Classical and Matrix Methods

Third Edition

James K. Nelson, Jr.
Western Michigan University

Jack C. McCormac
Clemson University

JOHN WILEY & SONS, INC.

ACQUISITIONS EDITOR Wayne Anderson
ASSOCIATE EDITOR Jenny Welter
MARKETING MANAGER Katherine Hepburn
SENIOR PRODUCTION EDITOR Caroline Sieg
NEW MEDIA EDITOR Martin Batey
SENIOR DESIGNER Madelyn Lesure
ILLUSTRATION EDITOR Sandra Rigby

About the Cover:
New building for the College of Engineering and Applied Sciences, Western Michigan University, Kalamazoo, MI.

Cover photo by Neil Rankin Photography.

This book was set in Gill Sans by Techbooks and printed and bound by Hamilton Printing Company. The cover was printed by Phoenix Color.

This book is printed on acid free paper.

Library of Congress Cataloging-in-Publication Data

Nelson, James K.
Structural analysis: Using classical and matrix methods / James K. Nelson, Jack C. McCormac.
p. cm.

McCormac's name appears first on previous ed.
Includes index.
ISBN 0-471-40273-7 (cloth : alk. paper)
1. Structural analysis (Engineering) I. McCormac, Jack C. II. Title.
TA645 .M38 2002
624.1'71—dc21

2002028821

Printed in the United States of America

10 9 8 7 6 5 4 3 2 1

This Book Is Dedicated to Our Families,

Whose Support and Encouragement Is Always Appreciated

Preface

The purpose of this book is to help students develop a fundamental understanding of the principles of structural analysis in the modern engineering office. An equally important purpose is to develop in the students an understanding of the behavior of structural systems under load. Some of the more significant changes in this edition of the textbook are:

- The educational version of SAP2000 is included on the CD-ROM. Data files for the computer examples are now included on the CD-ROM and not shown in the book. SAP2000 is a commercially available computer program for structural analysis. The CD-ROM includes Manuals in PDF form to help you use SAP2000. The Manual files are accessible in the default directory of the install: i.e. C:\ComputersandStructures\SAP2000student.
- SABLE (**S**tructural **A**nalysis and **B**ehavior for **L**earning **E**ngineering) software is available for download from the website at www.wiley.com/college/nelson. The program SABLE was developed by the authors to introduce students to software for structural analysis and to provide a tool to obtain a better understanding of structural behavior.
- Some methods of analysis that are no longer used in common practice have been eliminated.
- Examples have been improved and expanded to enhance clarity. Furthermore, matrix notation and methods of equation solving are used frequently throughout the text. These methods lend themselves to solution using hand-held calculators and computational software.
- Some chapters have been extensively rewritten to reflect current thinking and needs, in particular, Chapters 4, 5, 11 and 13.
- Chapters 2 and 3 have been updated to include the provisions of ASCE 7-98, the current standard for structural loads. In addition, reference is made to the IBC 2000 building code in these chapters. IBC 2000 is a model building code being adopted by many jurisdictions.
- Matrix methods have been expanded to include a finite element formulation.

There are many excellent computer-based computational tools available for use in performing engineering calculations. These include Mathcad® and Matlab®. Many of the examples in this book have been verified with Mathcad® and the data files for these example problems are included on the CD-ROM inside the front cover of the book. Most of the homework problems at the end of each chapter have been worked in the same manner. To minimize the mundane burden of the multitude of hand calculations necessary to perform structural analysis, students are encouraged to use these available computational tools. By doing so, they can spend more time studying the solution procedure and the behavior of the structure. Little is learned by performing mind-numbing calculations, but much is learned by studying resulting response.

The Solutions Manual for in this textbook is available to instructors who have adopted the book for their courses. Please visit the companion web site at **www.wiley.com/college/nelson** to request access.

The authors thank the following persons who gave much time to review this edition.

- John W. van de Lindt of Michigan Technological University
- R. Craig Henderson of Tennessee Technological University
- Paul R. Heyliger of Colorado State University
- Norman C. Cluley of California State Polytechnic University, Pomona

The comments and suggestions offered by these individuals are appreciated by the authors and improved the quality of the textbook.

The authors also wish to acknowledge with much gratitude the many people who assisted them in the development of this textbook. These people include: Dr. David Clarke of Clemson University who provided the numerous pictures of railroad bridges and trains; Dr. Peter Sparks of Clemson University for providing information for the discussion of wind loads; Dr. Scott Schiff of Clemson University for reviewing the chapter on shearing force and bending moment diagrams; and Christopher Strang, Jessica Cummings, Julie Babb, Elizabeth Haseldon, and Jennifer Aulick of Clemson University for helping work many of the homework problems contained in this book and for providing comments from the perspective of the student.

James K. Nelson, Jr.
Jack C. McCormac

Table of Contents

CHAPTER 3
System Loading and Behavior 36

CHAPTER 4
Reactions 52

CHAPTER 5
Shearing Force and Bending Moment 90

CHAPTER 6
Introduction to Plane Trusses 120

PART ONE

STATICALLY DETERMINATE STRUCTURES

Chapter 1

Introduction

1.1 STRUCTURAL ANALYSIS AND DESIGN

The application of loads to a structure causes the structure to deform. Because of the deformation, various forces are produced in the components that comprise the structure. Calculating the magnitude of these forces, and the deformations that caused them is referred to as *structural analysis,* which is an extremely important topic to society. Indeed, almost every branch of technology becomes involved at some time or another with questions concerning the strength and deformation of structural systems.

Structural design includes the arrangement and proportioning of structures and their parts so they will satisfactorily support the loads to which they may be subjected. More specifically, structural design involves the following: the general layout of the structural system; studies of alternative structural configurations that may provide feasible solutions; consideration of loading conditions; preliminary structural analyses and design of the possible solutions; the selection of a solution; and the final structural analysis and design of the structure. Structural design also includes preparation of design drawings.

This book is devoted to structural analysis, with only occasional remarks concerning the other phases of structural design. Structural analysis can be so interesting to engineers that they become completely attached to it and have the feeling that they want to become 100% involved in the subject. Although analyzing and predicting the behavior of structures and their parts is an extremely important part of structural design, it is only one of several important and interrelated steps. Consequently, it is rather unusual for an engineer to be employed solely as a structural analyst. An engineer, in almost all probability, will be involved in several or all phases of structural design.

It is said that Robert Louis Stevenson studied structural engineering for a time, but he apparently found the "science of stresses and strains" too dull for his lively imagination. He went on to study law for a while before devoting the rest of his life to writing prose and poetry[1]. Most of us who have read *Treasure Island, Kidnapped,* or his other works would agree that the world is a better place because of his decision. Nevertheless,

[1]Proceedings of the First United States Conference on Prestressed Concrete (Cambridge, Mass.: Massachusetts Institute of Technology, 1951), 1.

White Bird Canyon Bridge, White Bird, Idaho (Courtesy of the American Institute of Steel Construction, Inc.)

there are a great number of us who regard structural analysis and design as extremely interesting topics. In fact some of us have found it so interesting that we have gone on to practice in the field of structural engineering. The authors hope that this book will inspire more engineers to do the same.

1.2 HISTORY OF STRUCTURAL ANALYSIS

Structural analysis as we know it today evolved over several thousand years. During this time many types of structures such as beams, arches, trusses, and frames were used in construction for hundreds or even thousands of years before satisfactory methods of analysis were developed for them. Though ancient engineers showed some understanding of structural behavior (as evidenced by their successful construction of great bridges, cathedrals, sailing vessels, and so on), real progress with the theory of structural analysis occurred only in the last 150 years.

The Egyptians and other ancient builders surely had some kinds of empirical rules drawn from previous experiences for determining sizes of structural members. There is, however, no evidence that they had developed any theory of structural analysis. The Egyptian Imhotep who built the great step pyramid of Sakkara in about 3000 B.C. sometimes is referred to as the world's first structural engineer.

Although the Greeks built some magnificent structures, their contributions to structural theory were few and far between. Pythagoras (about 582–500 B.C.), who is said to have originated the word *mathematics,* is famous for the right angle theorem that bear his name. This theorem actually was known by the Sumerians in about 2000 B.C. Further, Archimedes (287–212 B.C.) developed some fundamental principles of statics and introduced the term *center of gravity.*

The Romans were excellent builders and very competent in using certain structural forms such as semicircular masonry arches. But, as did the Greeks, they too had little knowledge of structural analysis and made even less scientific progress in structural theory. They probably designed most of their beautiful buildings from an artistic viewpoint. Perhaps their great bridges and aqueducts were proportioned with some rules of thumb; however if these methods of design resulted in proportions that were insufficient, the structures collapsed and no historical records were kept. Only their successes endured.

One of the greatest and most noteworthy contributions to structural analysis, as well as to all other scientific fields, was the development of the Hindu–Arabic system of numbers. Unknown Hindu mathematicians in the 2nd or 3rd century B.C. originated a numbering system of one to nine. In about 600 A.D. the Hindus invented the symbol *sunya* (meaning empty), which we call zero. The Mayan Indians of Central America, however, had apparently developed the concept of zero about 300 years earlier.[2]

In the 8th century A.D. the Arabs learned this numbering system from the scientific writings of the Hindus. In the following century, a Persian mathematician wrote a book that included the system. His book later was translated into Latin and brought to Europe.[3] In around 1000 A.D., Pope Sylvester II decreed that the Hindu–Arabic numbers were to be used by Christians.

Before real advances could be made with structural analysis, it was necessary for the science of mechanics of materials to be developed. By the middle of the 19th century, much progress had been made in this area. A French physicist Charles Augustin de Coloumb (1736–1806) and a French engineer–mathematician Claude Louis Marie Henri Navier (1785–1836), building upon the work of numerous other investigations over hundreds of years, are said to have founded the science of mechanics of materials. Of particular significance was a textbook published by Navier in 1826, in which he discussed the strengths and deflections of beams, columns, arches, suspension bridges, and other structures.

Andrea Palladio (1508–1580), an Italian architect, is thought to have been the first person to use modern trusses. He may have revived some ancient types of Roman structures and their empirical rules for proportioning them. It was actually 1847, however, before the first rational method of analyzing jointed trusses was introduced by Squire Whipple (1804–1888). His was the first significant American contribution to structural theory. Whipple's analysis of trusses often is said to have signalled the beginning of modern structural analysis. Since that time there has been an almost continuous series of important developments in the subject.

Several excellent methods for calculating deflections were published in the 1860s and 1870s, which further accelerated the development of structural analysis. Among the important investigators and their accomplishments were James Clerk Maxwell (1831–1879) of Scotland, for the reciprocal deflection theorem in 1864; Otto Mohr (1835–1918) of Germany, for the method of elastic weights presented in 1870; Carlo Alberto Castigliano (1847–1884) of Italy, for the least-work theorem in 1873; and Charles E. Greene (1842–1903) of the United States, for the moment-area theorems in 1873.

The advent of railroads gave a great deal of impetus to the development of structural analysis. It was suddenly necessary to build long-span bridges capable of carrying very heavy moving loads. As a result, the computation of stresses and strains became increasingly important.

Further, fatigue and impact stresses became serious matters. Furthermore, up until this time there had not been a great deal of need to analyze statically indeterminate structures, however, continuous span railroad bridges created the need to do so.

One method for analyzing continuous statically indeterminate beams—the three-moment theorem—was introduced in 1857 by the Frenchman B. P. E. Clapeyron (1799–1864), and was used for analyzing many railroad bridges. In the decades that followed, many other

[2]*The World Book Encyclopedia* (Chicago, IL, 1993, Book N–O), pg. 617.

[3]*Ibid.*

advances were made in indeterminate structural analysis that were based upon the recently developed deflection methods.

Otto Mohr, who worked with railroads, is said to have reworked into practical, usable forms many of the theoretical developments up to his time. Particularly notable in this regard was his 1874 publication of the method of consistent distortions for analyzing statically indeterminate structures.

In the United States, two great developments in statically indeterminate structure analysis were made by G. A. Maney (1888–1947) and Hardy Cross (1885–1959). In 1915 Maney presented the slope deflection method, whereas Cross introduced moment distribution in 1924.

In the first half of the 20th century, many complex structural problems were expressed in mathematical form, but sufficient computing power was not available for practically solving the resulting equations. This situation continued in the 1940s, when much work was done with matrices for analyzing aircraft structures. Fortunately, the development of digital computers made the use of equations practical for these and many other types of structures, including high-rise buildings.

Pacific Gas and Electric Company headquarters, San Francisco (Courtesy of Bethlehem Steel Corporation)

Some particularly important historical references on the development of structural analysis include those by Kinney[4], Timoshenko[5], and Westergaard.[6] They document the slow but steady development of the fundamental principles involved. It seems ironic that the college student of today can learn in a few months the theories and principles of structural analysis that took many scholars several thousand years to develop.

[4]J. S. Kinney, *Indeterminate Structural Analysis* (Reading, Mass.: Addison-Wesley, 1957), 1–16.

[5]S. P. Timoshenko, *History of Strength of Materials* (New York: McGraw-Hill, 1953), 1–439.

[6]H. M. Westergaard, "One Hundred Fifty Years Advance in Structural Analysis," (*ASCE-94,* 1930), 226–240.

Cold-storage warehouse, Grand Junction, Colorado (Courtesy of the American Institute of Steel Construction, Inc.)

1.3 BASIC PRINCIPLES OF STRUCTURAL ANALYSIS

Structural engineering embraces an extensive variety of structural systems. When speaking of structures, people typically think of buildings and bridges. There are, however, many other types of systems with which structural engineers deal, including sports and entertainment stadiums, radio and television towers, arches, storage tanks, aircraft and space structures, concrete pavements, and fabric air-filled structures. These structures can vary in size from a single member, as in the case of a light pole, to structures such as the Sears Tower in Chicago, which is over 1450 ft tall, or the Humber Estuary Bridge in England, which has a suspended span that is over 4626 ft long. In 1999 the Akashi–Kaikyo suspension bridge in Japan was completed with its suspended span of approximately 6530 ft.

To be able to analyze this wide range of sizes and types of structures, a structural engineer must have a solid understanding of the basic principles that apply to all structural systems. It is unwise to learn how to analyze a particular structure, or even a few different types of structures. Rather, it is more important to learn the fundamental principles that apply to all structural systems, regardless of their type or use. One never knows what types of problems the future holds or what type of structural system may be conceived for a particular application, but a firm understanding of basic principles will help us to analyze new structures with confidence.

The fundamental principles used in structural analysis are Sir Isaac Newton's laws of inertia and motion, which are:

1. A body will exist in a state of rest or in a state of uniform motion in a straight line unless it is forced to change that state by forces imposed on it.
2. The rate of change of momentum of a body is equal to the net applied force.
3. For every action there is an equal and opposite reaction.

These laws of motion can be expressed by the equation

$$\sum F = ma \qquad \textbf{Eq. 1.1}$$

In this equation, $\sum F$ is the summation of all the forces that are acting on the body, m is the mass of the body, and a is its acceleration.

In this textbook, we will be dealing with a particular type of equilibrium called *static equilibrium,* in which the system is not accelerating. The equation of equilibrium thus becomes

$$\sum F = 0 \qquad \textbf{Eq. 1.2}$$

These structures either are not moving, as is the case for most civil engineering structures, or are moving with constant velocity, such as space vehicles in orbit. Using the principle of static equilibrium, we will study the forces that act on structures and methods to determine the response of structures to these forces. By response, we mean the displacement of the system and the forces that occur in each component of the system. This emphasis should provide readers with a solid foundation for advanced study, and we hope convince them that structural theory is not difficult and that it is not necessary to memorize special cases.

1.4 STRUCTURAL COMPONENTS AND SYSTEMS

All structural systems are composed of components. The following are considered to be the primary components in a structure:

- *Ties:* those members that are subjected to axial tension forces only. Load is applied to ties only at the ends. Ties cannot resist flexural forces.
- *Struts:* those members that are subjected to axial compression forces only. Like a tie, a strut can be loaded only at its ends and cannot resist flexural forces.
- *Beams* and *Girders:* those members that are primarily subjected to flexural forces. They usually are thought of as being horizontal members that are primarily subjected to gravity forces; but there are frequent exceptions (e.g., rafters).
- *Columns:* those members that are primarily subjected to axial compression forces. A column may be subjected to flexural forces also. Columns usually are thought of as being vertical members, but they may be inclined.
- *Diaphragms:* structural components that are flat plates. Diaphragms generally have very high in-plane stiffness. They are commonly used for floors and shear-resisting walls. Diaphragms usually span beams or columns. They may be stiffened with ribs to better resist out-of-plane forces.

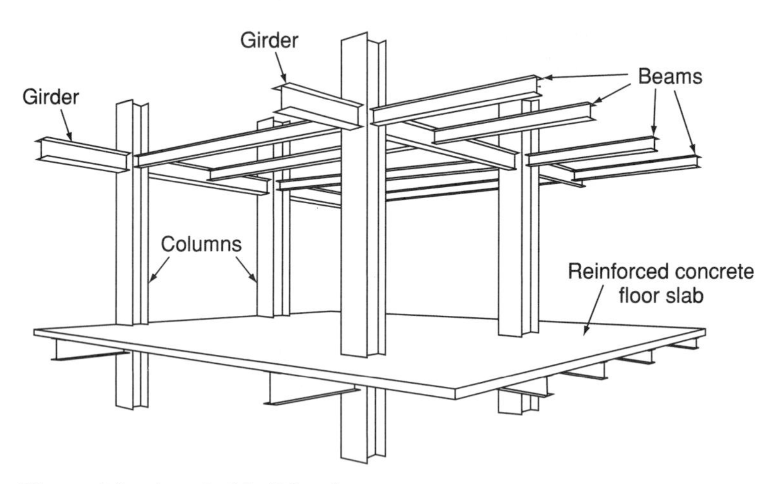

Figure 1.1 A typical building frame

Figure 1.2 Some components of a railroad bridge truss

Structural components are assembled to form structural systems. In this textbook, we will be dealing with typical framed structures. A building frame is shown in Figure 1.1. In this figure, a girder is considered to be a large beam with smaller beams framing into it.

A *truss* is a special type of structural frame. It is composed entirely of struts and ties. That is to say, all of its components are connected in such a manner that they are subjected only to axial forces. All of the external loads acting on trusses are assumed to act at the joints and not directly on the components, where they might cause bending in the truss members. An older type of bridge structure consisting of two trusses is shown in Figure 1.2. In this figure, the *top and bottom chords* and the *diagonals* are the primary load carrying components of trusses. *Floor beams* frame into the joints of the truss and support the roadway, or roadbed in the case of a railroad bridge.

There are other types of structural systems. These include fabric structures (e.g., tents and outdoor arenas) and curved shell structures (e.g., dams or sports arenas). The analysis of these types of structures requires advanced principles of structural mechanics and is beyond the scope of this book.

1.5 STRUCTURAL FORCES

A structural system is acted upon by forces. Under the influence of these forces, the entire structure is assumed herein to be in a state of static equilibrium and, as a consequence, each component of the structure also is in a state of static equilibrium. The forces that act on a structure include the applied loads and the resulting reaction forces.

Las Vegas Convention Center (Courtesy of Bethlehem Steel Corporation)

The applied loads are the known loads that act on a structure. They can be the result of the structure's own weight, occupancy loads, environmental loads, and so on. The reactions are the forces that the supports exert on a structure. They are considered to be part of the external forces applied and are in equilibrium with the other external loads on the structure.

To introduce loads and reactions, three structural systems are shown in Figure 1.3. A railroad bridge with a train crossing is shown in Figure 1.3(a). Loads such as these are moving loads. During the passage of the train, the loads acting on the structure are continually changing. In Chapter 9 we will discuss moving loads and where to place the loads on the structure to cause the maximum forces. The people in the ballpark

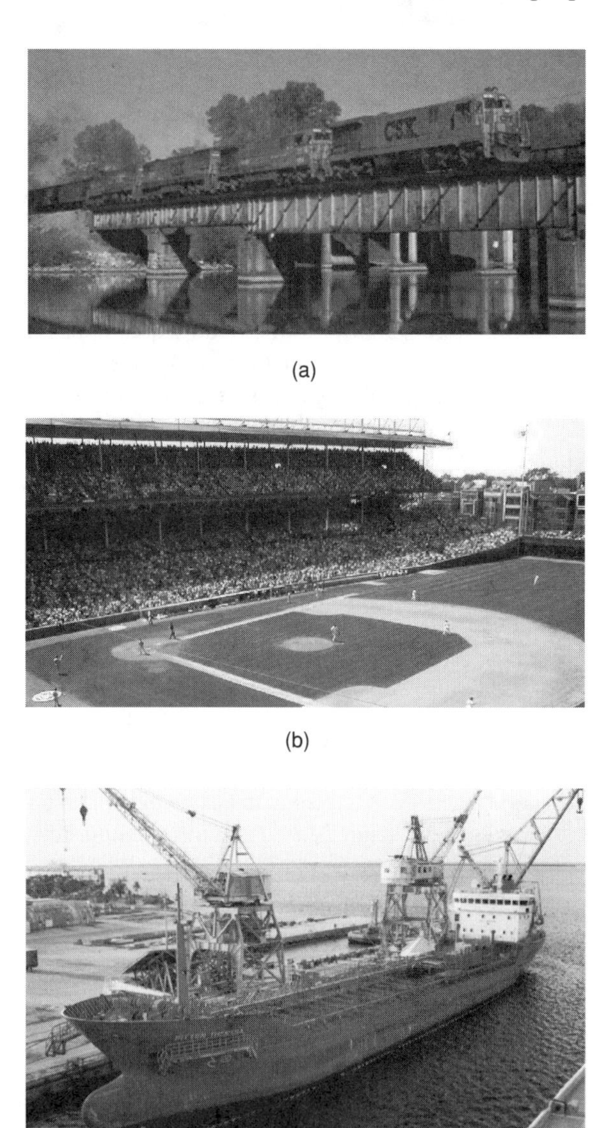

(a)

(b)

(c)

Figure 1.3 Some types of structures and loads

shown in Figure 1.3(b) are a typical live load with which structural engineers deal. Normally this type of load is treated as a uniformly distributed load. The people are a gravity load and, as such, act downward. A discussion of loads acting on buildings is presented in Chapters 2 and 3. In Figure 1.3(c), the ship is carrying cargo that can be represented as either distributed or concentrated loads. The structural support, or reaction, is the hydrostatic force of the water, which is distributed uniformly along the bottom of the ship. A detailed discussion of reactions and their computation is presented in Chapter 4.

1.6 STRUCTURAL IDEALIZATION (LINE DIAGRAMS)

To calculate the forces in the various parts of a structure with reasonable simplicity and accuracy, it is necessary to represent the structure in a simple manner that is conducive to analysis. Structural components have width and thickness. Concentrated forces rarely act at a single point; rather, they are distributed over small areas. If these characteristics are taken into consideration in detail, however, an analysis of the structure will be very difficult, if not impossible to perform.

The process of replacing an actual structure with a simple system conducive to analysis is called *structural idealization.* Most often, lines that are located along the centerlines of the components represent the structural components. The sketch of a structure idealized in this manner usually is called a *line diagram.* The preparation of line diagrams is shown in Figure 1.4. A railroad girder bridge is shown in part (a). The deck girders are supported on several piers, which are in turn supported on the river floor. A common representation of this bridge for analysis is shown in Figure 1.4(b). The piers are assumed to be immovable supports. The girder rests on rollers that rest on the supports. This is a typical sketch of a structural system for purposes of analysis, and is often used in this textbook.

Sometimes the idealization of a structure involves assumptions about the structures behavior. As an example, consider the bolted steel truss of Figure 1.5(a). The joints in

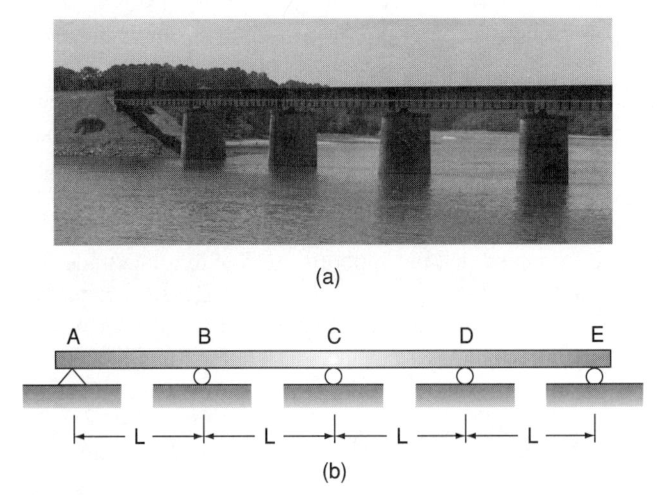

Figure 1.4 Replacing a structure and its forces with a line diagram

Figure 1.5 A line diagram for a portion of a steel truss

trusses often are made with large connection or gusset plates and, as such, can transfer moments to the ends of the members. However, experience has shown that the stresses caused by the axial forces in the members greatly exceed the stresses caused by flexural forces. Consequently, for purposes of analysis we can assume that the truss consists of a set of pin-connected lines, as shown in Figure 1.5(b).

Access Bridge, Renton, Washington (Courtesy of Bethlehem Steel Corporation)

Although the use of simple line diagrams for analyzing structures will not result in perfect analyses, the results usually are quite acceptable. Sometimes, though, there may be some doubt in the mind of the analyst as to the exact line diagram or model to be used for analyzing a particular structure. For instance, should beam lengths be clear spans between supports, or should they equal the distances center to center of those supports? Should it be assumed that the supports are free to rotate under loads, are fixed against rotation, or do they fall somewhere in between? Because of many questions such as these, it may be necessary to consider different models and perform the analysis for each one to determine the worst cases.

1.7 CALCULATION ACCURACY

A most important point that many students with their superb pocket calculators and personal computers have difficulty understanding is that structural analysis is not an exact science for which answers can confidently be calculated to eight or more significant digits. Computations to only three places probably are far more accurate than the estimates of material strengths and magnitudes of loads used for structural analysis and design. The common materials dealt with in structures (wood, steel, concrete, and a few others) have ultimate strengths that can only be estimated. The loads applied to structures may be known within a few hundred pounds or no better than a few thousand pounds. It therefore seems inconsistent to require force computations to more than three or four significant figures.

Hungry Horse Dam and Reservoir, Rocky Mountains, in northwest Montana (Courtesy of the Montana Travel Promotion Division)

Several partly true assumptions will be made about the construction of trusses such as: truss members are connected with frictionless pins; or the deformation of truss members under load is so slight as to cause no effect on member forces. These deviations from actual conditions emphasize that it is of little advantage to carry the results of structural analysis to many significant figures. Furthermore, calculations to more than four significant figures may be harmfully misleading; it may be giving a fictitious sense of precision.

1.8 CHECKS ON PROBLEMS

A definite advantage of structural analysis is the possibility of making either mathematical checks on the analysis by some method other than the one initially used, or by the same method from some other position on the structure. You should be able in nearly every situation to determine if your work has been done correctly.

All of us, unfortunately, have the weakness of making exasperating mistakes, and the best that can be done is to keep them to the absolute minimum. The application of the simple arithmetical checks suggested in the following chapters will eliminate many of these costly blunders. The best structural designer is not necessarily the one who makes the fewest mistakes initially, but probably is the one who discovers the largest percentage of his or her mistakes and corrects them.

Oxford Valley Mall, Langehorne, Pennsylvania (Courtesy of Bethlehem Steel Corporation)

1.9 IMPACT OF COMPUTERS ON STRUCTURAL ANALYSIS

The availability of personal computers has drastically changed the way in which structures are analyzed and designed. In nearly every engineering school and office, computers are used to address structural problems. Though in the past computers were used much more for analysis than for design, the situation is rapidly changing as more and more design software is developed and sold commercially.

Though computers do increase productivity by expediting the work and reducing mathematical errors, they undoubtedly tend to reduce the analyst's "feel" for structures because less thought is given to behavior when an analytical model is developed and used. This can be a serious problem, particularly for young engineers with little previous experience. Unless engineers develop this feel, computer usage may occasionally result in large mistakes. That is, completely unrealistic answers may be obtained and yet not be recognized as such.

Nevertheless, the authors believe that students should know how to perform structural analysis using computers. In the front of this book is a CD-ROM that contains a structural analysis program. The program is the student version of SAP2000, which is a commercially available computer program used extensively in industry. Many examples in this book have been worked with SAP2000; files containing the input data for those problems are contained in the examples directory on the CD-ROM. A second program called SABLE is available on the web. This program enables rapid analysis of simple structures and also provides visual means for the student to begin understanding the behavior of structural systems. Both computer programs operate on IBM compatible PCs using the Windows® operating system.

Chapter 2

Structural Loads

2.1 INTRODUCTION

This textbook discusses a large number of structures (beams, frames, trusses, etc.) that have all sorts of loads acting on them. In fact, the reader may very well wonder "Where in the world did they get these loads?" That very important and logical question is answered in this and the next chapter.

Structural engineers today generally use computer software in their work. Although the typical software will enable them to quickly analyze and design structures after the loads are established, it will provide little help in selecting the loads. Perhaps the most critical—and difficult—task faced by structural engineers is the accurate estimation of the magnitude and character of the loads that may be applied to a structure during its lifetime. No loads that can reasonably be expected to occur can be overlooked.

In this chapter, various types of loads are introduced and specifications are presented with which the individual magnitudes of the loads may be estimated. Our objective is to be able to answer questions such as the following: How heavy could the snow load be on a structure in Minneapolis? What maximum wind force might be expected on a hotel in Miami? How large a rain load is probable for a flat roof in Houston?

The methods used for estimating loads are constantly being refined and may involve some very complicated formulas. Do not be concerned about committing such expressions to memory. Rather, learn the types of loads that can be applied to a structure and where to obtain information for estimating the magnitude of these loads.

2.2 SPECIFICATIONS AND BUILDING CODES

The design of most structures is controlled by specifications. Even if they are not so controlled, the designer probably will refer to them as a guide. No matter how many structures a person has designed, it is impossible for him or her to have encountered every situation. By referring to specifications, an engineer is making use of the best available material on the subject. Engineering specifications that are developed by various organizations present the best opinion of those organizations as to what represents good practice.

Municipal and state governments concerned with the safety of the public have established building codes with which they control the construction of various structures within their jurisdiction. These codes, which actually are laws or ordinances, specify design loads, design stresses, construction types, material quality, and other factors. They

The world's largest radio telescope, Green Bank, West Virginia (Courtesy of Lincoln Electric Company)

can vary considerably from city to city, which can cause some confusion among architects and engineers.

The determination of the magnitude of loads is only a part of determining the structural loads. The structural engineer must be able to determine which loads can be reasonably expected to act concurrently on a structure. For example, would a highway bridge completely covered with ice and snow be simultaneously subjected to fast moving lines of heavily loaded trucks in every lane and a 90-mile-per-hour lateral wind? Instead, is some lesser combination of these loads more reasonable and realistic? The topic of concurrent loads is addressed initially in Chapter 3 along with a related problem; the placement of loads on a structure so as to create the most severe conditions.

Quite a few organizations publish recommended practices for regional or national use. Their specifications are not legally enforceable, however, unless they are embodied in local building codes or made part of a particular contract. Among these organizations are the American Society of Civil Engineers (ASCE)[1], the American Association of State Highway and Transportation Officials (AASHTO)[2], and the American Railway Engineering Association (AREA).[3]

Recently, the International Code Council has developed the *International Building Code®*.[4] This code was developed to meet the need for a modern building code for building systems that emphasize performance. The *International Building Code* (IBC-2000) is intended to serve as a set of model code regulations to safeguard the public in all communities.

Readers should note that logical and clearly written codes are quite helpful to structural engineers. Furthermore, fewer structural failures occur in areas with good codes that

[1]American Society of Civil Engineers, *ASCE 7-98, Minimum Design Loads for Buildings and Other Structures,* Reston, Virginia 20191-4400, 2000.

[2]American Association of State Highway and Transportation Officials, *AASHTO LRFD Bridge Design Specifications,* (Washington, D.C. 1994).

[3]American Railway Engineering Association, *Specifications for Steel Railway Bridges,* (Washington, D.C. 1994).

[4]*2000 International Building Code,* (Falls Church, Virginia 22041-3401, International Code Council, Inc., 2000).

are strictly enforced. The specifications published by the organizations mentioned are frequently used to estimate the maximum loads to which buildings, bridges, and some other structures may be subjected during their estimated lifetimes.

Some people feel that specifications prevent engineers from thinking for themselves—and there may be some basis for the criticism. The pundits say that the ancient engineers who built the great pyramids, the Parthenon, and the great Roman bridges were controlled by few specifications. This statement is certainly true. On the other hand, only a few score of these great projects were built over many centuries, and they were apparently built without regard to the cost of material, labor, or human life. They probably were built by intuition and by certain rules-of-thumb developed by observing the minimum size or strength of members that would fail only under given conditions. Quite likely, there were numerous failures not recorded in history. Only the successes endured.

Today, at any one time hundreds of projects are being constructed in the United States that rival in importance and magnitude the famous structures of the past. Building codes and specifications are prepared by experts with knowledge in particular topics to provide guidance to engineers and a minimum standard of acceptable practice for design in a particular region. The result is that there are fewer disastrous failures and the public is better protected. *The important thing to remember about specifications is that their purpose is not to restrict engineers. Rather, their purpose is to protect the public.* Yet, no matter how many specifications are written, they cannot address every possible situation. Consequently, no matter which code or specification is or is not being used, the ultimate responsibility for the design of a safe structure lies with the structural engineer.

Specifications on many occasions clearly prescribe the loads for which structures are to be designed. Despite the availability of this information, however, the designer's ingenuity and knowledge of the situation often are needed to predict the loads a particular structure will have to support in years to come. For example, over the past several decades, insufficient estimates of future traffic loads by bridge designers have resulted in a great amount of replacement with wider and stronger structures.

This chapter introduces the basic types of loads with which the structural engineer needs to be familiar. Its purpose is to help the reader develop an understanding of structural loads and their behavior, and to provide a foundation for estimating their magnitudes. It should not be regarded, however, as a complete essay on the subject of the loads that might be applied to any and every type of structure the engineer may design.

Since building loads are the most common type encounted by designers, they are the loads most frequently referred to in this text. The basic document currently being used by a large number of structural designers for estimating the loads to be applied to buildings is the ASCE 7 specification. It is often referred to in this chapter and has been incorporated into many model building codes. This specification was originally prepared and published by the American National Standards Institute and is referred to as the ANSI 58.1 Standard. It has gone through several revisions. In 1988, it was taken over by ASCE and renamed ANSI/ASCE 7. Much information in this book is based on the 1998 edition of this specification, which now is called ASCE 7-98. Reference is also made to the *International Building Code*.

When studying the information provided in this chapter, or when reviewing any standard providing design loads, the reader is cautioned that minimum design load standards are presented. An engineer should always view minimum design standards with some skepticism. The design standards are excellent and well prepared for most situations. However, there may be a building configuration, or a building use, for which the specified design loads are not adequate. A structural engineer should evaluate the minimum specified design loads to determine whether they are adequate for the structural system being designed.

South Fork Feather River Bridge in Northern California, being erected by use of a 1626-ft-long cableway strung from 210-ft-high masts anchored on each side of the canyon (Courtesy Bethlehem Steel Corporation)

2.3 TYPES OF STRUCTURAL LOADS

Structural loads usually are categorized by means of their character and duration. Loads commonly applied to buildings are categorized as follows:

- *Dead loads:* those loads of constant magnitude that remain in one position. They include the weight of the structure under consideration, as well as any fixtures that are permanently attached to it.
- *Live loads:* those loads that can change in magnitude and position. They include occupancy loads, warehouse materials, construction loads, overhead service cranes, and equipment operating loads. In general, live loads are caused by gravity.
- *Environmental loads:* those loads caused by the environment in which the structure is located. For buildings, the environmental loads are caused by rain, snow, wind, temperature, and earthquakes. Strictly speaking, these are also live loads, but they are the result of the environment in which the structure is located.

With the exception of the earthquake loads, each of these types of loads is discussed in considerable detail in the sections that follow. Earthquake, or seismic, loads are only briefly discussed, as those loads are a subject requiring advanced study.

2.4 DEAD LOADS

The dead loads that must be supported by a particular structure include all of the loads that are permanently attached to that structure. They include the weight of the structural frame and also the weight of the walls, roofs, ceilings, stairways, and so on.

Permanently attached equipment, described as "fixed service equipment" in ASCE 7-98, also is included in the dead load applied to the building. This equipment will include ventilating and air-conditioning systems, plumbing fixtures, electrical cables, support racks, etc. Depending upon the use of the structure, kitchen equipment

TABLE 2.1 WEIGHT OF SOME COMMON BUILDING MATERIALS

Building Material	Unit Weight	Building Material	Unit Weight
Reinforced Concrete	150 pcf	2 × 12 @ 16-in. Double Wood Floor	7 psf
Acoustical Ceiling Tile	1 psf	Linoleum or/ Asphalt Tile	1 psf
Suspended Ceiling	2 psf	Hardwood Flooring (7/8-in.)	4 psf
Plaster on Concrete	5 psf	1-in. Cement on Stone-Concrete Fill	32 psf
Asphalt Shingles	2 psf	Movable Steel Partitions	4 psf
3-Ply Ready Roofing	1 psf	Wood Studs w/1/2-in. gypsum	8 psf
Mechanical Duct Allowance	4 psf	Clay Brick, 4 in. wythe	39 psf

such as ovens and dishwashers, laundry equipment such as washers and dryers, or suspended walkways could be included in the dead load.

The dead loads acting on the structure are determined by reviewing the architectural, mechanical, and electrical drawings for the building. From these drawings, the structural engineer can estimate the size of the frame necessary for the building layout and the equipment and finish details indicated. Standard handbooks and manufacturers' specifications can be used to determine the weight of floor and ceiling finishes, equipment, and fixtures. The approximate weights of some common materials used for walls, floors, and ceilings are shown in Table 2.1.

The estimates of building weight or other structural dead loads may have to be revised one or more times during the analysis–design process. Before a structure can be designed, it must be analyzed. Among the loads used for the first analysis are the estimates of the weights of the components of the frame, which then is designed using the results of that analysis. Its weight can then be computed and compared with the initial weight estimate used for analysis. If there is a significant difference, the structure should be reanalyzed using a revised frame weight estimate. This cycle is repeated as many times as is necessary.

2.5 LIVE LOADS

Live loads are those loads that can vary in magnitude and position with time. They are caused by the building being occupied, used, and maintained. Most of the loads applied to a building that are not dead loads are live loads. Environmental loads, which are actually live loads by our usual definition, are listed separately in ASCE 7-98 and IBC-2000. Although environmental loads do vary with time, they are not all caused by gravity or operating conditions, as is typical with other live loads.

Some typical live loads that act on building structures are presented in Table 2.2. The loads shown in the table were taken from Table 4-1 in ASCE 7-98 and Table 1607.1 in IBC-2000. They are acting downward and are distributed uniformly over the entire floor or roof.

TABLE 2.2 SOME TYPICAL UNIFORMLY DISTRIBUTED LIVE LOADS

Area Utilization	Live Load	Area Utilization	Live Load
Lobbies of Assembly Areas	100 psf	Classrooms in Schools	40 psf
Dance Hall and Ballrooms	100 psf	Upper Floor Corridors in Schools	80 psf
Library Reading Rooms	60 psf	Stairs and Exitways	100 psf
Library Stack Rooms	150 psf	Heavy Storage Warehouse	250 psf
Light Manufacturing	125 psf	Retail Stores—First Floor	100 psf
Offices in Office Buildings	50 psf	Retail Stores—Upper Floors	75 psf
Residential Dwelling Areas	40 psf	Walkways and Elevated Platforms	60 psf

TABLE 2.3 TYPICAL CONCENTRATED LIVE LOADS

Area or Structural Component	Concentrated Live Load
Elevator Machine Room Grating on 4-in.^{2}	300 lbs
Office Floors	2000 lbs
Center of Stair Tread on 4-in.^{2}	300 lbs
Sidewalks	8000 lbs
Accessible Ceilings	200 lbs

Many building specifications provide concentrated loads to be considered in design. This is the situation in Section 4.3 of the ASCE 7-98 and Section 1607.4 of IBC-2000. These specifications state that the designer must consider the effect of certain concentrated loads as an alternative to the previously discussed uniform loads. The intent, of course, is that the loading used for design be the one that causes the most severe stresses.

Presented in Table 4-1 of ASCE 7-98 and Table 1607.1 of IBC-2000 are the minimum concentrated loads to be considered. Some typical values from these tables are shown in Table 2.3. The appropriate loads are to be positioned on a particular floor or roof so as to cause the greatest stresses (a topic to be discussed in more detail in Chapters 3 and 9). Unless otherwise specified, each of the concentrated loads is assumed to be uniformly distributed over an area 2.5 ft sq (6.25 ft^2).

When estimating the magnitudes of the live loads that may be applied to a particular structure during its lifetime, engineers need to consider the future utilization of that structure. For example, modern office buildings often are constructed with large open spaces that may later be divided into offices and other work areas by means of partitions. These partitions may be moved, removed, or added to during the life of the structure. Building codes typically require that partition loads be considered if the floor live load is less than 80 psf, even if partitions are not shown on the drawings. A rather common practice of structural designers is to increase the otherwise specified floor design live loads of office buildings by 20 psf to estimate the effect of the impossible-to-predict partition configurations of the future. This is the minimum partition load specified in Section 1607.5 of IBC-2000.

The method used to establish the magnitude of the ASCE 7-98 live loads is a rather complicated process that is described in the Commentary of that specification. Among the factors contributing to a particular specified value are the mean expected load, its variation over time, the magnitude of short duration transient loads, and the reference time period, which is typically assumed to be 50 years.

To convince the reader that the specified loads are reasonable, a brief examination of one of the specified values is considered. The example used here is the 100-psf live load specified by ASCE 7-98 for the lobbies of theaters and for assembly areas. Determine if such a load is reasonable for a crowd of people standing quite close together. Assume that the area in question is full of average adult males each weighing 165 pounds and each occupying an area 20 in. by 12 in., or 1.67 sq ft. The average load applied equals $165/1.67 = 98.8$ psf. As such, the 100-psf live load specified seems reasonable. It actually is on the conservative side, as it would be rather difficult to have men standing that close together over a floor area that is either small or large.

2.6 LIVE LOAD IMPACT FACTORS

Impact loads are caused by the vibration and sudden stopping or dropping of moving or movable loads. It is obvious that a crate dropped on the floor of a warehouse or a truck bouncing on the uneven pavement of a bridge causes greater forces than would occur if

TABLE 2.4 LIVE LOAD IMPACT FACTORS

Equipment or Component	Impact Factor
Elevator Machinery	100%
Motor-Driven Machinery	20%
Reciprocating Machinery	50%
Hangers for Floors and Balconies	33%

the loads were applied gently and gradually. Impact loads are equal to the difference between the magnitude of the loads actually caused and the magnitude of the loads had they been dead loads. In other words, impact loads result from the dynamic effects of a load as it is applied to a structure. With static loads, these effects are short–lived and do not necessitate dynamic structural analysis. They do cause, however, an increase in structural stress that must be considered. Impact loads usually are specified as percentage increases of the basic live load. Table 2.4 shows the impact percentages for buildings given in Section 4.7.2 of ASCE 7-98 and Section 1607.8.2 of IBC-2000

2.7 LIVE LOADS ON ROOFS

The live loads that act on roofs are handled in most building codes in a little different manner than are the other building live loads. The pitch of the roof (the ratio of the rise of the roof to its span) affects the amount of load that realistically can be placed upon it. As the pitch increases, the amount of load that can be placed on the roof before it begins to slide off decreases. Furthermore, it is less likely that the entire roof area will be loaded at any one time since the area of the roof that contributes to the load acting on a supporting component increases.

The largest roof live loads usually are caused by repair and maintenance operations that probably do not occur simultaneously over the entire roof. This is not true of the environmental rain and snow loads, however, which are considered in Sections 2.8 and 2.10 of this chapter.

In the equations presented in this section, the term *tributary area* is used. This term, discussed in detail in Chapter 3, is defined as the loaded area of a structure that directly contributes to the load applied to a particular member. The tributary area for a member is assumed to extend from the member in question halfway to the adjacent members in each direction. When a building is being analyzed, it is customary for the analyst to assume that the load supported by a member is the load that is applied to its tributary area. The tributary area for a column is shown in Figure 2.1.

The basic minimum roof live load to be used in design is 20 psf. This value is specified in Section 4.9 of ASCE 7-98 and Section 1607.11.2.1 of IBC-2000. Depending on the size of the tributary area and the rise of the roof, this value may be reduced. The actual value to be used is determined with the expression:

$$L_r = 20R_1R_2$$
$$12 < L_r \leq 20 \qquad \textbf{Eq. 2.1}$$

The term L_r represents the roof live load in psf of horizontal projection, whereas R_1 and R_2 are reduction factors. R_1 is used to account for the size of the tributary area A_t and R_2 is included to estimate the effect of the rise of the roof. The greater the tributary area (or the greater the rise of the roof) the smaller will be the applicable reduction factor and the smaller the roof live load. The maximum roof live load is 20 psf and the minimum is

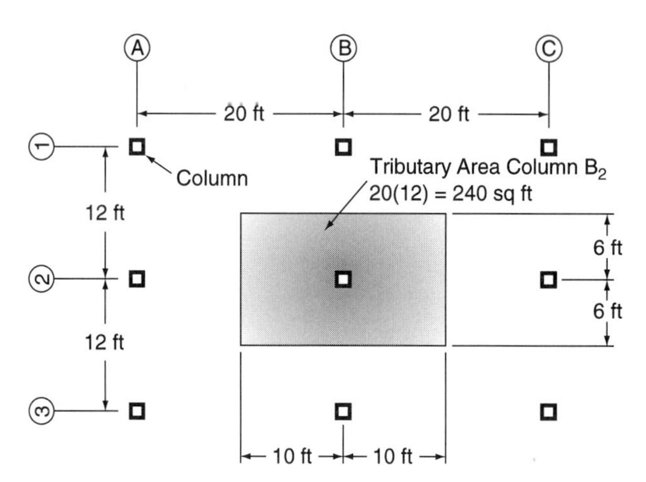

Figure 2.1 Tributary area for a column

12 psf. Expressions for computing R_1 and R_2 are

$$R_1 = \begin{cases} 1.0 & A_t \leq 200 \text{ ft}^2 \\ 1.2 - 0.001A_t & 200 \text{ ft}^2 < A_t < 600 \text{ ft}^2 \\ 0.6 & A_t \geq 600 \text{ ft}^2 \end{cases} \quad \textbf{Eq. 2.2}$$

$$R_2 = \begin{cases} 1.0 & F \leq 12 \\ 1.2 - 0.05F & 4 < F < 12 \\ 0.6 & F \geq 12 \end{cases} \quad \textbf{Eq. 2.3}$$

The term F represents the number of inches of rise of the roof per foot of span. If the roof is a dome or an arch, the term F is the rise-to-span ratio of the structure multiplied by 32.

2.8 RAIN LOADS

It has been claimed that almost 50% of the lawsuits faced by building designers are concerned with roofing systems.[5] *Ponding,* a problem with many flat roofs, is one of the common subjects of such litigation. If water accumulates more rapidly on a roof than it runs off, ponding results because the increased load causes the roof to deflect into a dish shape that can hold more water, which causes greater deflections, and so on. This process continues until equilibrium is reached or collapse occurs. Through proper selection of loads and good design providing adequate roof stiffness, we try to avoid the latter situation. Many useful references on the subject of ponding are available.[6,7,8,9]

[5] Gary Van Ryzin, "Roof Design: Avoid Ponding by Sloping to Drain," *Civil Engineering* (New York: ASCE, January 1980), 77–81.

[6] F. J. Marino, "Ponding of Two-Way Roof System," *Engineering Journal,* AISC, 3rd quarter, no. 3 (1966), 93–100.

[7] L. B. Burgett, "Fast Check for Ponding," *Engineering Journal,* AISC, 10, 1st quarter, no. 1 (1973), 26–28.

[8] J. Chinn, "Failure of Simply-Supported Flat Roofs by Ponding of Rain," *Engineering Journal,* AISC, 2nd quarter, no. 2 (1965), 38–41.

[9] J. L. Ruddy, "Ponding of Concrete Deck Floors," *Engineering Journal*, AISC, 23, 3rd quarter, no. 2 (1986), 107–115.

During a rainstorm, water accumulates on a roof for two reasons. First, when rain falls, time is required for the rain to run off the roof. Therefore, some water will accumulate. Second, roof drains may not be even with the roof surface and they may become clogged. Generally, roofs with slopes of 0.25 in. per ft or greater are not susceptible to ponding, unless the roof drains become clogged, thus enabling deep ponds to form.

In addition to ponding, another problem may occur for very large flat roofs (with perhaps an acre or more of surface area). During heavy rainstorms strong winds frequently occur. If there is a great deal of water on the roof, a strong wind may very well push a large quantity of the water toward one end. The result can be a dangerous water depth influencing the load on that end of the roof. For such situations, *scuppers* are sometimes used. Scuppers are large holes or tubes in walls or parapets that enable water above a certain depth to quickly drain off the roof.

Generally, two different drainage systems are provided for roofs. These normally are referred to as the primary and secondary drains. Usually the primary system will collect the rainwater through surface drains on the roof and direct it to storm sewers. The secondary system consists of scuppers or other openings or pipes through the walls that permit the rainwater to run over the sides of the building. The inlets of the secondary drains are normally located at elevations above the inlets to the primary drains.

The secondary drainage system is used to provide adequate drainage of the roof in the event that the primary system becomes clogged or disabled in some manner. The rainwater design load therefore is based on the amount of water that can accumulate before the secondary drainage system becomes effective.

The determination of the water that can accumulate on a roof before runoff during a rainstorm will depend upon local conditions and the elevation of secondary drains. Section 8 of ASCE 7-98 specifies that the rain load (in psf) on an undeflected roof can be computed from

$$R = 5.2(d_s + d_h) \qquad \textbf{Eq. 2.4}$$

This is the same equation as that in IBC-2000. The term d_s is the depth of water (in inches) on the undeflected roof up to the inlet of the secondary drainage system when the primary drainage system is blocked. This is the static head, which can be determined from the drawings of the roof system. The term d_h is the additional depth of water on the undeflected roof above the inlet of the secondary drainage system at its design flow. This is the hydraulic head. It is dependent upon the capacity of the drains installed and the rate at which rain falls.

From Section 8.3 of the ASCE 7-98 Commentary, the flow rate (in gallons per minute) that a particular drain must accommodate can be computed from

$$Q = 0.0104Ai \qquad \textbf{Eq. 2.5}$$

The term A is the area of the roof (in square feet) that is served by a particular drain, and i is the rainfall intensity (in inches per hour). The rainfall intensity is specified by the code that has jurisdiction in a particular area. After the flow quantity is determined, the hydraulic head can be determined from Table 2.5 (ASCE 7-98, Table C8-1) for the type of drainage system being used. If the secondary drainage system is simply runoff over the edge of the roof, the hydraulic head will equal zero.

Example 2.1 illustrates the calculation of the design rainwater load for a roof with scuppers using the ASCE 7-98 specification.

TABLE 2.5 FLOW RATE, Q, IN GALLONS PER MINUTE OF VARIOUS DRAINAGE SYSTEMS AND HYDRAULIC HEADS

	Hydraulic Head d_h (in.)									
Drainage System	1	2	2.5	3	3.5	4	4.5	5	7	8
4-in. diameter drain	80	170	180							
6-in. diameter drain	100	190	270	380	540					
8-in. diameter drain	125	230	340	560	850	1100	1170			
6-in. wide channel scupper	18	50		90		140		194	321	393
24-in. wide channel scupper	72	200		360		560		776	1284	1572
6-in. wide, 4-in. high, closed scupper	18	50		90		140		177	231	253
24-in. wide, 4-in. high, closed scupper	72	200		360		560		708	924	1012
6-in. wide, 6-in. high, closed scupper	18	50		90		140		194	303	343
24-in. wide, 6-in. high, closed scupper	72	200		360		560		776	1212	1372

Note: Interpolation is appropriate, including between scupper widths. Closed scuppers are four sided and channel scuppers are open topped.

EXAMPLE 2.1

A roof measuring 240 feet by 160 feet has 6-in. wide channel-shaped scuppers serving as secondary drains. The scuppers are 4 inches above the roof surface and are spaced 20 feet apart along the two long sides of the building. The design rainfall for this location is 3 inches per hour. What is the design roof rain load?

Solution. The area served by each scupper is

$$A = 20(80) = 1600 \text{ ft}^2 \qquad A = 20 \times 80 = 1600 \text{ ft}$$

The runoff quantity for each scupper is

$$Q = 0.0104Ai = 0.0104(1600)(3) = 49.92 \text{ gpm}$$

Referring to Table 2.5, observe that the hydraulic head at this flow rate for the scupper used is 2 in. The design roof rain load, then, is

$$R = 5.2(d_s + d_h) = 5.2(4 + 2) = 31.2 \text{ psf} \quad \blacksquare$$

2.9 WIND LOADS

A survey of engineering literature for the past 150 years reveals many references to structural failures caused by wind. Perhaps the most infamous of these have been bridge failures such as those of the Tay Bridge in Scotland in 1879 (which caused the deaths of 75 persons) and the Tacoma Narrows Bridge (Tacoma, Washington) in 1940. However, there were some disastrous building failures due to wind during the same period, such as the Union Carbide Building in Toronto in 1958. It is important to realize that a large percentage of building failures due to wind have occurred during their construction.[10]

Considerable research has been conducted in recent years on the subject of wind loads. Nevertheless, a great deal more study is needed, as the estimation of wind forces can by no means be classified as an exact science. The magnitude and duration of wind loads vary with geographical locations, the height of the structure above ground, the type of terrain around the structure, the proximity of other buildings, and the character of the wind itself.

Fortunately, structural engineers usually do not need to determine the exact forces at every location in a building as a function of time. The approach taken by most building

[10]Wind Forces on Structures, Task Committee on Wind Forces. Committee on Loads and Stresses, Structural Division, ASCE, Final Report, *Transactions ASCE* 126, Part II (1961): 1124–1125.

codes for a regularly shaped building is to determine a set of distributed loads that encompass or envelop the actual loads expected. These distributed loads are applied to the building in a series of wind-loading conditions.

Section 6 of the ASCE 7-98 specification provides a rather lengthy procedure for estimating the wind pressures applied to buildings. The procedure involves several factors with which we attempt to account for the terrain around the building, the importance of the building regarding human life and welfare, and of course, the wind speed at the building site. Though use of the equations is rather complex, the work can be greatly simplified with the tables presented in the specification. The reader is cautioned, however, that the tables presented are for buildings of regular shapes. If a building having an irregular or unusual geometry is being considered, wind tunnel studies may be necessary. Guidelines for conducting such studies are provided in Section 6.6 of the ASCE 7-98 specification.

The determination of wind loads acting on a building is often the subject of an entire course. The discussion about wind forces that follows is intended to be only an introduction to the subject and does not consider wind speed-up caused by topographic features. Furthermore, it addresses only the wind forces applied to the main wind-force resisting system of low-rise buildings with roof slopes of less than 10 degrees. It also applies to wind forces applied to components of the building and *cladding* of these buildings. Cladding is the exterior covering of the structural parts of the building. A simplified and conservative design procedure is presented in Section 6.4.2 of ASCE 7-98.

As we begin the discussion, some necessary definitions are

- *Low-Rise Building:* An enclosed or partially enclosed building in which the mean roof height is less than 60 ft and the height is not greater that the least horizontal dimension.
- *Open Building:* A building in which the total area of the openings in each wall is equal to at least 80% of the total area of the wall.
- *Partially Enclosed Building:* A building in which the total area of the openings in a wall that receives positive external pressure exceeds (a) the sum of the areas of the openings in the rest of the building envelope by more than 10%, and (b) 4 sf, or 1%, of the area of that wall, whichever is smaller, and the percentage of openings in the rest of the building does not exceed 20%.
- *Enclosed Building:* Any building except open or partially enclosed buildings.
- *Building Envelope:* The cladding, roofing, exterior walls, doors, windows, and other components that enclose the building.
- *Main Wind-Force Resisting System:* An assemblage of structural elements assigned to provide support and stability for the overall structure. The wind-force resisting system usually receives wind load from more than one surface.

There are six steps for determining the basic wind loads acting on structural systems. These steps are

1. Determine the design wind speed,
2. Determine the importance factor,
3. Determine the worst exposure category at the site,
4. Compute the velocity pressure,
5. Compute the wind pressure on the main wind-force resisting system,
6. Compute the wind pressure on the components and cladding.

Each of these steps is discussed in the following sections. During the discussion, reference is made to the figures and tables in Appendix C of this book.

TABLE 2.6 IMPORTANCE FACTOR I FOR WIND LOADS

Building Use Category	Non-Hurricane Prone Regions and Hurricane-Prone Regions with V = 85 − 100 mph and Alaska	Hurricane Prone Regions with V > 100 mph
I	0.87	0.77
II	1.00	1.0
III	1.15	1.15
IV	1.15	1.15

Source: ASCE 7-98

2.9.1 Determine the Design Wind Speed, V

The basic wind speed to be used in design is determined from Figure C.1, which is Figure 6.1 in ASCE 7-98. The wind is assumed to approach the building from any direction. The wind speeds obtained from this table are not to be used in mountainous areas, gorges, and other regions where unusual wind conditions may exist.

2.9.2 Determine the Importance Factor, I

The importance factor is intended to bring into the calculation of wind forces a measure of the consequences of failure. Critical buildings, such as schools and hospitals, will have a higher importance factor and therefore higher design wind forces. Building whose failure will have little consequence on human life, such as farm buildings, will have a lower importance factor and therefore lower design wind forces.

The classification of buildings for wind, snow, and earthquake loads is shown in Table C.1. This table is taken from Table 1-1 in ASCE 7-98. When computing wind loads, the importance factor associated with each classification is shown in Table 2.6.

2.9.3 Determine the Worst Exposure Category at the Site

When computing the design wind forces, the terrain and obstructions to wind at the location of the building must be considered. For every direction of the wind evaluated, an exposure category must be determined that takes into consideration the terrain and obstructions for the building being designed. The exposure categories as presented in ASCE 7-98 in Section 6.5.6.1 are listed in Table 2.7

The exposure category for the building is used when computing the velocity pressure. If the building being designed lies in a region between two exposure categories, the exposure category causing the highest wind forces is used.

2.9.4 Compute the Velocity Pressure

The velocity pressure, in units of pounds per square foot, is computed from the equation

$$q_h = 0.00256 K_d K_z V^2 I \qquad \textbf{Eq. 2.6}$$

In Equation 2.6, V is the basic wind speed and I is the importance factor. The term K_d is a directionality factor that is equal to 0.85 when the loading combinations specified in Section 2 of ASCE 7-98 are used in design. Otherwise, the factor is equal to unity. Load combinations are discussed in Chapter 3.

The term K_z is the velocity pressure exposure coefficient and is determined from Table 6-5 in ASCE 7-98 (Table C.2). The height used when determining K_z for low-rise

TABLE 2.7 EXPOSURE CATEGORIES FOR WIND FORCES

Exposure Category	Description
A	Large city centers where at least 50% of the buildings are taller than 70 ft. Use of this category is limited to those areas for which terrain representative of Exposure A prevails in the upwind direction for a distance of at least 0.5 miles or 10 times the height of the building or other structure, whichever is greater.
B	Urban and suburban areas, wooded areas, and other terrain with numerous closely spaced obstructions having the size of single family dwellings or larger. Use of this exposure category is limited to those areas for which terrain that is representative of Exposure Category B prevails for at least 1500 ft or 10 times the height of the building or other structure, whichever is greater.
C	Open terrain with scattered obstructions that are generally less than 30 ft high. This category includes flat open country, grasslands, and shorelines in hurricane-prone areas.
D	Flat, unobstructed areas exposed to wind flowing over open water excluding the shoreline in hurricane-prone areas. Shorelines that are included in Exposure Category D include inland waterways, the Great Lakes, and the coastal areas of California, Oregon, Washington, and Alaska. This exposure category extends inland from the shore a distance of 1500 ft or 10 times the height of the building or other structure, whichever is greater.

Source: ASCE 7-98

buildings is the mean roof height. If topographic features such as ridges are present, the velocity pressure will need to be modified as discussed in Section 6.5.7 of ASCE 7-98.

2.9.5 Pressure of the Main Wind-Force Resisting System

The design pressure for wind loads acting on the main wind-force resisting system of a low-rise building is specified in Section 6.5.12.2.2 of ASCE 7-98. That pressure can be calculated from

$$p = q_h[(GC_{pf}) - (GC_{pi})] \quad \textbf{Eq. 2.7}$$

Jacobs Field, the home of the Cleveland Indians (Courtesy of The Lincoln Electric Company)

In this equation, the term (GC_p) is the external pressure coefficient. It can be found from Figure C.2, which is Figure 6-4 in ASCE 7-98. The term (GC_{pi}) is the internal pressure coefficient. This coefficient can be found from Table C.3, which is Table 6-7 in ASCE 7-98. The velocity pressure, q_h, used in the equation is the pressure evaluated at the mean roof height. According to Section 6.5.6.2.2 of ASCE 7-98, the velocity pressure to be used for the design of low-rise buildings is the highest for any wind direction at the location of the building when the pressure coefficients previously specified are used.

The minimum design wind loads are stated in Section 6.1.4 of ASCE 7-98. The design wind load for the main wind-force resisting system for an enclosed or partially enclosed building, as specified in Section 6.1.4.1, shall not be less than 10 psf multiplied by the projected area of the structure normal to the assumed direction of the wind.

2.9.6 Pressure on the Components and Cladding

The design pressure, in units of pounds per square foot, for wind loads acting on the components and cladding of a low-rise building is specified in Section 6.5.12.4.1 of ASCE 7-98. That pressure can be calculated from

$$p = q_h[(GC_p) - (GC_{pi})] \qquad \textbf{Eq. 2.8}$$

In this equation, the term (GC_p) is the external pressure coefficient, which is found using Figures C.3 to C.5. These are Figures 6-5 to 6-7, respectively, in ASCE 7-98. The term (GC_{pi}) is the internal pressure coefficient and is found on Table C.3, which is Table 6-7 in ASCE 7-98. The velocity pressure q_h, used in the equation, is evaluated at the mean roof height using the exposure, category causing the highest wind loads for any wind direction at the location of the building. The procedure that we have just discussed is applied to a typical low-rise building in Example 2.2.

EXAMPLE 2.2

Determine the wind loads acting on the building shown in the figure. These building frames are located 20 feet on-center. This building is located along the beach at Galveston on the Texas Gulf Coast. It is oriented so that the ridgeline is parallel with the beach. The primary use of the building will be hotel rooms; there are no areas in which more than 300 people can congregate.

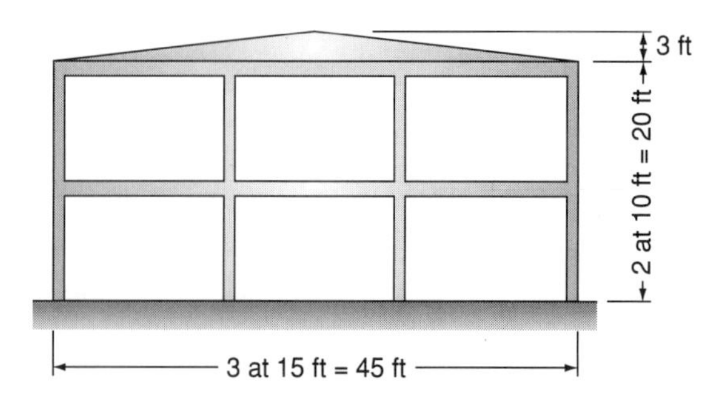

Solution. Figure C.1 indicates that the basic wind speed for Galveston is 140 mph. Because this building is located on the hurricane-prone coastline, it is an Exposure Category C as described in Table 2.7. Because of the occupancy of the building, it is a Category II structure as described Table C.1, therefore the importance factor, I, is 1.0 as determined from Table 2.6.

We can now compute the velocity pressure acting on this building. The velocity pressure is computed using Equation 2.6. The directionality factor, K_d, is equal to 0.85 because the load combinations in ASCE-7 will be used for design. The velocity pressure coefficient, as obtained from Table C.2 for a mean roof height of 21.5 ft, is 0.912.

$$q_h = 0.00256 K_d K_z V^2 I$$
$$q_h = 0.00256(0.85)(0.912)(140)^2(1.0) = 38.9 \text{ psf}$$

We next need to compute the wind pressure acting on the main wind-force resisting system. To perform this calculation, we will need the external pressure coefficient. It can be obtained from Figure C.2. In this example, we will compute the pressure on the roof surface away from the edge of the roof on the windward side of the building. For this case, external pressure coefficient is -1.07. We also need to know the internal pressure coefficient, which can be found on Table C.3. That factor is equal to ± 0.18. The design pressures used to determine forces in the main wind-force resisting system are then:

$$p = q_h[(GC_{pf}) - (GC_{pi})] = 38.9[-1.07 \pm 0.18] = \begin{cases} -34.6 \text{ psf} \\ -48.6 \text{ psf} \end{cases}$$

Pressures acting on the other parts of the structural system resisting the wind force would be calculated in a similar manner.

Lastly, we will compute the wind pressures acting on the roof cladding. We will compute this pressure acting on the central part of the roof on the windward side of the roof. The effective wind area is going to be greater than 100 sf so we will use a value of -0.8 for external pressure coefficient as determined from Figure C.3b. The internal pressure coefficient is the same as we previously determined. The pressure acting on the roof cladding, then, is

$$p = q_h[(GC_p) - (GC_{pi})] = 38.9[-0.8 \pm 0.18] = \begin{cases} -24.1 \text{ psf} \\ -38.1 \text{ psf} \end{cases}$$

Pressures acting on the other parts of the components and cladding would be calculated in a similar manner. ■

2.10 SNOW LOADS

In the colder states, snow and ice loads are often quite important. One inch of snow is equivalent to approximately 0.5 psf, but it may be higher at lower elevations where snow is denser. For roof design, snow loads ranging from 10 to 40 psf are usually specified. The magnitude depends primarily on the slope of the roof, and to a lesser degree on the character of the roof surface. The larger values are used for flat roofs and the smaller values for sloped roofs. Snow tends to slide off sloped roofs, particularly those with metal or slate surfaces. A load of approximately 10 psf might be used for roofs with 45° slopes while a 40-psf load might be used for flat roofs. Studies of snowfall records in areas with severe winters may indicate the occurrence of snow loads much greater than 40 psf, with values as high as 100 psf in northern Maine.

TABLE 2.8 EXPOSURE COEFFICIENTS FOR SNOW LOADS

Terrain Category (See ASCE 7-98, Section 6.5.6.1)	Exposure of the Roof		
	Fully Exposed	Partially Exposed	Sheltered
A: Large city center	N/A	1.1	1.3
B: Urban and suburban areas	0.9	1.0	1.2
C: Open terrain with scattered obstructions	0.9	1.0	1.1
D: Unobstructed areas with wind over open water	0.8	0.9	1.0
Above the tree line in windswept mountainous areas	0.7	0.8	N/A
Alaska in areas with trees not within 2 miles of the site	0.7	0.8	N/A

Snow is a variable load that may cover an entire roof or only part of it. There may be drifts against walls or buildup in valleys or between parapets. Snow may slide off one roof onto a lower one. The snow may blow off one side of a sloping roof or it may crust over and remain in position even during very heavy winds.

The snow loads that are applied to a structure are dependent upon many factors, including geographic location, the pitch of the roof, sheltering, and the shape of the roof. The discussion that follows is intended to provide only an introduction to the determination of snow loads on buildings. When estimating these loads, consult ASCE 7-98 for information that is more complete.

According to Section 7.3 of ASCE 7-98, the basic snow load to be applied to structures with flat roofs in the contiguous United States can be obtained from the expression

$$p_f = 0.7C_eC_tIp_g \qquad \textbf{Eq. 2.9}$$

This expression is for unobstructed flat roofs with slopes equal to or less than 5° (a 1 in./ft slope is equal to 4.76°). In the equation, C_e is the exposure index. It is intended to account for the snow that can be blown from the roof because of the surrounding locality. The exposure coefficient is lowest for highly exposed areas and is highest when there is considerable sheltering. Values of C_e are presented in Table 2.8 (Table 7-2 in ASCE 7-98).

The terrain category and roof exposure condition chosen must be representative of the anticipated conditions during the life of the structure. In this table the following definitions are used for the exposure of the roof:

- *Partially exposed:* All roofs except as described as fully exposed or sheltered.
- *Fully exposed:* Roofs exposed on all sides with no shelter afforded by terrain, higher structures, or trees. Roofs that contain several large pieces of mechanical equipment or other obstructions are not in this category.
- *Sheltered:* Roofs located tight in among conifers that qualify as obstructions.

TABLE 2.9 THERMAL FACTOR FOR SNOW LOADS

Representative Anticipated Winter Thermal Conditions	C_t
All structures except as indicated below	1.0
Structures kept just above freezing with cold ventilated roofs in which the thermal resistance between ventilated and heated space exceeds 25°Fhft²/BTU	1.1
Unheated structures and structures intentionally kept below freezing	1.2
Continuously heated greenhouses with a roof having a thermal resistance of less than 2.0°Fhft²/BTU.	0.85

TABLE 2.10 IMPORTANCE FACTOR I FOR SNOW LOADS

Building Use Category	I
I	0.8
II	1.0
III	1.1
IV	1.2

Obstructions within a distance of $10h_o$ provide "shelter." The term h_o is the height of the obstruction above the roof level. If the only obstructions are a few deciduous trees that are leafless in winter, the "fully exposed" category should be used except for terrain category A. Please note that these are heights above the roof. The height used to establish the Terrain Category in Section 6.5.3 of ASCE 7-98 is the height above the ground.

The term C_t in Equation 2.9 is the thermal index. Values of this coefficient are shown in Table 2.9 (Table 7-3 in ASCE 7-98). As shown in the table, the coefficient is equal to 1.0 for heated structures, 1.1 for structures that are minimally heated to keep them from freezing, and 1.2 for unheated structures.

The values of the importance factors I for snow loads are shown in Table 2.10 (Table 7-4 in ASCE 7-98). The categories of building use are the same as those used for the computation of wind loads and are shown in Table C.1 in Appendix C.

The last term in Equation 2.9, p_g, is the ground snow load in pounds per square foot. Typical ground snow loads for the United States are shown in Figure C.6 in Appendix C. These values are dependent upon the climatic conditions at each site. If data are available that show local conditions are more severe than the values given in the figure, the local conditions should always be used.

The minimum value of p_f is $p_g(I)$ in areas where the ground snow load is less than or equal to 20 psf. In other areas, the minimum value of p_f is 20(I) psf.

Example 2.3, which follows, illustrates the calculation of the design snow load for a building in Chicago according to the ASCE 7-98 specification.

EXAMPLE 2.3

A shopping center is being designed for a location in Chicago. The building will be located in a residential area with minimal obstructions from surrounding buildings and the terrain. It will contain large department stores and enclosed public areas in which more that 300 people can congregate. The roof will be flat, but to provide for proper drainage it will have a slope equal to 0.5 in./ft. What is the roof snow load that should be used for design?

Solution. Because the slope of the roof is less than 5°, it can be designed as a flat roof. From Figure C.6 the ground snow load, p_g, for Chicago is 25 psf. The exposure factor, C_e, can be taken to equal 0.9 because there are minimal obstructions, though not necessarily an absence of all obstructions. Furthermore, because the building will be located in a residential area, it is unlikely there will be any obstructions to wind blowing across the roof. The thermal factor C_t is 1.0 because this will have to be a heated structure. Lastly, the importance factor I to be used is 1.1 because more than 300 people can congregate in one area. The design snow load to be used, then, is

$$p_f = 0.7C_eC_tIp_g = 0.7(0.9)(1.0)(1.1)(25) = 17.3\,\text{psf}$$

or

$$p_f = 20(1.1) = 22\,\text{psf} \qquad \text{CONTROLS} \blacksquare$$

Another expression is given in Section 7.4 of ASCE 7-98 for estimating the snow load on sloping roofs. It involves the multiplication of the snow load for flat roofs by a roof slope factor, C_s. Values of C_s and illustrations are given for warm roofs, cold roofs, and for other situations in Sections 7.4.1 through 7.4.4 of ASCE 7-98.

2.11 OTHER LOADS

There are quite a few other kinds of loads that the designer may occasionally face. These include the following

2.11.1 Traffic Loads on Bridges

Bridge structures are subject to a series of concentrated loads of varying magnitude caused by groups of truck or train wheels. Such loads are discussed in detail in Sections 9.11 and 9.12 of this text.

2.11.2 Seismic Loads

Many areas of the world fall into "earthquake territory," and in those areas it is necessary to consider seismic forces in the design of all types of structures. Through the centuries there have been catastrophic failures of buildings, bridges, and other structures during earthquakes. It has been estimated that as many as 50,000 people died in the 1988 earthquake in Armenia.[11] The 1989 Loma Prieta and 1994 Northridge earthquakes in California caused many billions of dollars of property damage as well as considerable loss of life.

Recent earthquakes have clearly shown that the average building or bridge that has not been designed for earthquake forces can be destroyed by a relatively moderate earthquake. Most structures can be economically designed and constructed to withstand the forces caused during most earthquakes. On the other hand, the cost of providing seismic resistance to existing structures (called retrofitting) can be extremely high.

Some engineers seem to think that the seismic loads needed in design are merely percentage increases of the wind loads. This assumption is incorrect, however. Seismic loads are different in their action and are not proportional to the exposed area of the building, but rather are proportional to the distribution of the mass of the building above the particular level being considered.

Another factor to be considered in seismic design is the soil condition. Almost all of the structural damage and loss of life in the Loma Prieta earthquake occurred in areas that have soft clay soils. Apparently, these soils amplified the motions of the underlying rock.[12]

It is well to understand that earthquakes add load to structures in an indirect fashion. The ground is displaced, and because the structures are connected to the ground, they also vibrate and are displaced. Consequently, various deformations and stresses are caused throughout the structures.

From this information it is clear that no external forces are applied above ground by earthquakes to structures. Procedures for estimating seismic forces such as the ones presented in Section 9 of ASCE 7-98 are very complicated. As such, they usually are

[11] V. Fairweather, "The Next Earthquake," *Civil Engineering* (New York: ASCE, March 1990), 54–57.

[12] *Ibid.*

addressed in advanced structural analysis courses such as structural dynamics or earthquake resistance design courses.

2.11.3 Ice Loads

Ice has the potential of causing the application of extraordinarily large forces to structural members. Ice can emanate from two sources: (1) surface ice on frozen lakes, rivers, and other bodies of water; and (2) atmospheric ice (freezing rain and sleet). The latter can form even in warmer climates.

In colder climates, ice loads often will greatly affect the design of marine structures. One such situation occurs when ice loads are applied to bridge piers. For such situations it is necessary to consider dynamic pressures caused by moving sheets of ice, pressures caused by ice jams, and uplift or vertical loads in water of varying levels causing the adherence of ice.

The breaking up of ice and its movement during spring floods can significantly affect the design of bridge piers. During the breakup, tremendous chunks of ice may be heaved upward, and when the jam breaks up, the chunks may rush downstream, striking and grinding against the piers. Furthermore, the wedging of pieces of ice between two piers can be extremely serious. It is thus necessary to either keep the piers out of the dangerous areas of the stream or to protect them in some way. Section 3 of the AASHTO specifications provides formulas for estimating the dynamic forces caused by moving ice.

Bridges and towers—any structure, for that matter—are sometimes covered with layers of ice from 1 to 2 in. thick. The weight of the ice runs up to about 10 psf. A factor that influences wind loads is the increased surface area of the ice-coated members.

Atmospheric icing is discussed in Section 10 of ASCE 7-98. Detailed information for estimating the thickness and weight of ice accumulations is provided. This ice typically accumulates on structural members as shown on Figure 10.1 in ASCE 7-98. The thickness to which the ice accumulates must be determined from historic data for the site. It can also be determined from a meteorological investigation of the conditions at the site. The extent of ice accumulation is such a localized phenomenon that general tables cannot be reasonably prepared.

2.11.4 Miscellaneous Loads

Some of the many other loads with which the structural designer will have to contend are

- *soil pressures:* such as the exertion of lateral earth pressures on walls or upward pressures on foundations;
- *hydrostatic pressures:* such as water pressure on dams, inertia forces of large bodies of water during earthquakes, and uplift pressures on tanks and basement structures;
- *flooding:* caused by heavy rain or melting snow and ice;
- *blast loads:* caused by explosions, sonic booms, and military weapons;
- *thermal forces:* due to changes in temperature resulting in structural deformations and resulting structural forces;
- *centrifugal forces:* such as those caused on curved bridges by trucks and trains, or similar effects on roller coasters, etc.;
- and *longitudinal loads:* caused by stopping trucks or trains on bridges, ships running into docks, and the movement of traveling cranes that are supported by building frames.

2.12 PROBLEMS FOR SOLUTION

Use the basic building layout shown in the figure for the solution of Problems 2.1 through 2.5. Determine the requested loads for your area of residence, or for an area specified by your instructor. Use the building code specified by your instructor. If no building code is specified, use ASCE 7-98. You do not need to determine the weight of the beams, girders, and columns for these problems.

Basic building layout

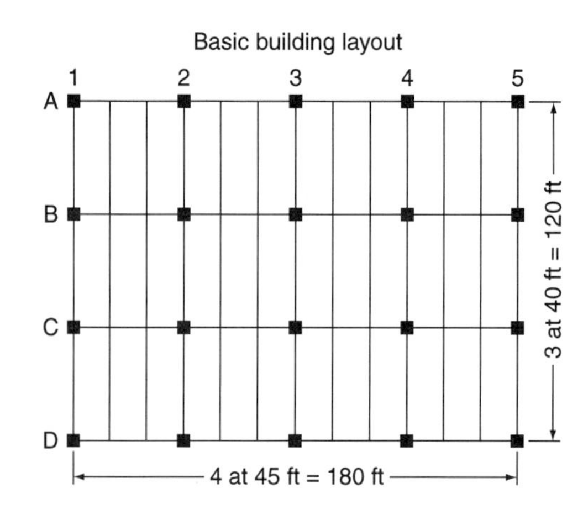

2.1 The roof of the building is flat. It is composed of 4-ply felt and gravel on 3 in. of reinforced concrete. The ceiling beneath the roof is a suspended steel channel system. Determine
a. The roof dead load
b. The live load in psf to be applied to Column B2

2.2 The roof of the building is flat. It is composed of 2 in. of reinforced concrete on 18-gauge metal decking. A single-ply waterproof sheet will be used. The ceiling beneath the roof is unfinished, but allowance for mechanical ducts should be provided. Determine
a. The roof dead load (*Ans.* 32.7 psf)
b. The live load in psf to be applied to Column B1 (*Ans.* 12 psf)

2.3 Determine the dead load and live load for the second floor of a library in which any area can be used for stacks. Assume that there will be a steel channel ceiling system and asphalt tile on the floors. The floors are 6-in. reinforced concrete. Allowance should be provided for mechanical ducts.

2.4 Determine the dead load and live load for a floor in a light manufacturing warehouse/office complex in which any area can be used for storage. Assume that there will be no ceiling or floor finish and that the floors are 4-in. reinforced concrete. Allowance for mechanical ducts should be provided. (*Ans.* D = 54 psf, L = 125 psf)

2.5 Determine the dead load and live load for a typical upper floor in an office building with movable partition walls. The ceiling is a steel channel system and the floors have a linoleum finish. The floors are 3-in. reinforced concrete. Allowance should be provided for mechanical ducts.

2.6 Determine the loads on an upper floor in a school with steel stud walls (1/2-in. gypsum on each side), a steel channel ceiling system, and a 3-in. reinforced concrete floor with asphalt tile covering. Allowance should be made for mechanical ducts. (*Ans.* d = Dead load is 52.5 psf, l = Live load is 40 psf in the classroom and 80 psf in the hallway)

For Problems 2.7 through 2.9, using the basic building layout shown, determine the requested wind loads acting on the main wind-force resisting system of the building by using ASCE 7-98. Assume that the long side of the building is perpendicular to the direction of the wind and there are no local changes in terrain.

2.7 Determine the windward wall of a 30-ft tall building with a flat roof. The building is an essential facility located in Chicago, Illinois, on the shore of Lake Michigan.

2.8 Determine the leeward wall if the building is a hospital, is 30 ft tall, and is located in a wooded residential suburb of Dallas, Texas. (Assume the roof is flat.).

2.9 Determine the sidewalls of a 2-story condominium located along the coast in Miami Beach, Florida. Assume each story is 10-ft high and that the roof slope is 10°.

2.10 Determine the design rain load on the roof of a building that is 250 ft wide and 500 ft long. The architect has decided to use 6-in. diameter drains spaced uniformly around the perimeter of the building at 50-ft intervals for the secondary drainage system. The drains are located 1.5 in. above the roof surface. The rainfall intensity at the location of this building is 2.5 in. per hour. (*Ans.* 15.9 psf)

For Problems 2.11 through 2.12, use the basic building layout shown on the previous page for the solution of these problems. Determine the snow load for your area of residence or for an area specified by your instructor. Use ASCE 7-98 as the basis for computation.

2.11 The roof is flat and the building is an unheated warehouse. Assume this is a Category I structure and is sheltered by conifers.

2.12 The roof is flat and the building is an office building. This building is taller than the surrounding trees, buildings, and terrain. (*Ans.* 20.2 psf)

Chapter 3

System Loading and Behavior

3.1 INTRODUCTION

In Chapter 2 we discussed different types of loads that might be applied to structural systems. Methods for estimating the individual magnitudes of the loads were presented. In that discussion, however, we did not consider whether the loads acted at the same time or at different times, nor did we address how and where to place them on the structure to cause maximum system response.

System response is a catchall phrase that really refers to a particular quantity of structural behavior. The response could be the negative bending moment in a floor beam, the displacement at a particular location in the structure, or the force at one of the structural supports. We probably know very little about how to calculate these aspects of response. Nevertheless, we will use the computer software enclosed at the back of the book to help us develop an understanding of the interaction of system loading and system response.

After the magnitudes of the loads have been computed, the next step in the analysis of a particular structure includes the placing the loads on the structure and calculating its response to those loads. When placing the loads on a structure, two distinct tasks must be performed:

1. We must decide which loads can reasonably be expected to act concurrently in time. Because different loads act on the structure at different times, several different loading conditions must be evaluated. Each of these loading conditions will cause the structural system to respond in a different manner.
2. We also need to determine where to place those loads on the structure. After loads are placed on the structure, the response of the structure is computed. If the same loads are placed on the structure in different positions, the response of the system will be different. We need to determine where to place the loads to obtain maximum response. For example, would the *bending moments* in the floor beams be greater if we placed the floor live loads on every span or on every other span?

Placing the live loads to cause the worst effects on any member of a structure is the responsibility of the structural engineer. Theoretically, his or her calculations are subject to the review of the appropriate building officials, but seldom do such individuals have the time and/or the ability to make significant reviews. Consequently, these calculations remain the responsibility of the engineer.

Martin Towers, Bethlehem, Pennsylvania (Courtesy Bethlehem Steel Corporation)

3.2 TRIBUTARY AREAS

In Section 2.7, the term *tributary area* was briefly defined. In this section, this term is discussed at greater length. In the next section, a related term, *influence area,* is introduced.

The tributary area is the loaded area of a particular structure that directly contributes to the load applied to a particular member in the structure. It is best defined as the area that is bounded by lines halfway to the next beam or to the next column. Tributary areas are shown for several beams and columns by the shaded areas in Figure 3.1. The component that the tributary area serves is indicated in black.

The tributary areas shown for the girders in Figure 3.1 are the tributary areas used in common practice. The theoretical tributary area for a typical interior girder and typical edge girder are shown in Figure 3.2.

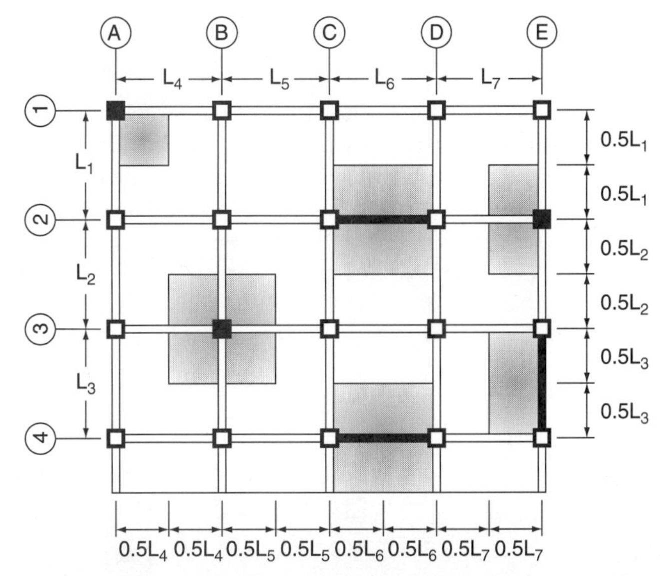

Figure 3.1 Tributary area for selected columns and girders

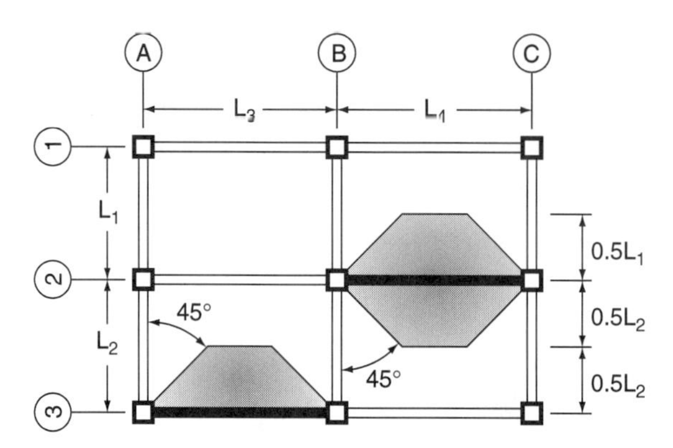

Figure 3.2 Theoretical tributary area for a beam

We see that in the middle of the beam the tributary area extends halfway to the next beam in each direction. At the ends of the beam, however, load is supported partly by the beams in the perpendicular direction. Therefore, the boundary of the tributary area will fall halfway between the two, that is, at a 45° angle. The tributary areas for beams shown in Figure 3.2 are rarely, if ever, used in practice because of the difficulty in dealing with the resulting trapezoidal load. Using the tributary area shown on Figure 3.1 instead of those shown on Figure 3.2 is conservative because there will be more load acting on the member when it is analyzed than will actually occur.

Very often floor loads are supported by floor beams as shown in Figure 3.3(a). Floor beams extend from one girder to another. Normally the floor beams are connected to the girders with connections that can be idealized as simple supports—there is no moment at the end-of-floor beams. When the floor beams are framed in this manner, the loads that act on them are shown in Figure 3.3(b). The girders support the end reactions of the beams as shown in Figure 3.3(c). The magnitudes of the individual forces can be determined from the formulas presented in Equation 3.1.

Floor Beam $\quad W = p_o s$

Girder
$$P = \begin{cases} w\left(\dfrac{L_1 + L_2}{2}\right) & \text{Interior girder} \\ w\left(\dfrac{L_1 \text{ or } L_2}{2}\right) & \text{Edge girder} \end{cases} \qquad \textbf{Eq. 3.1}$$

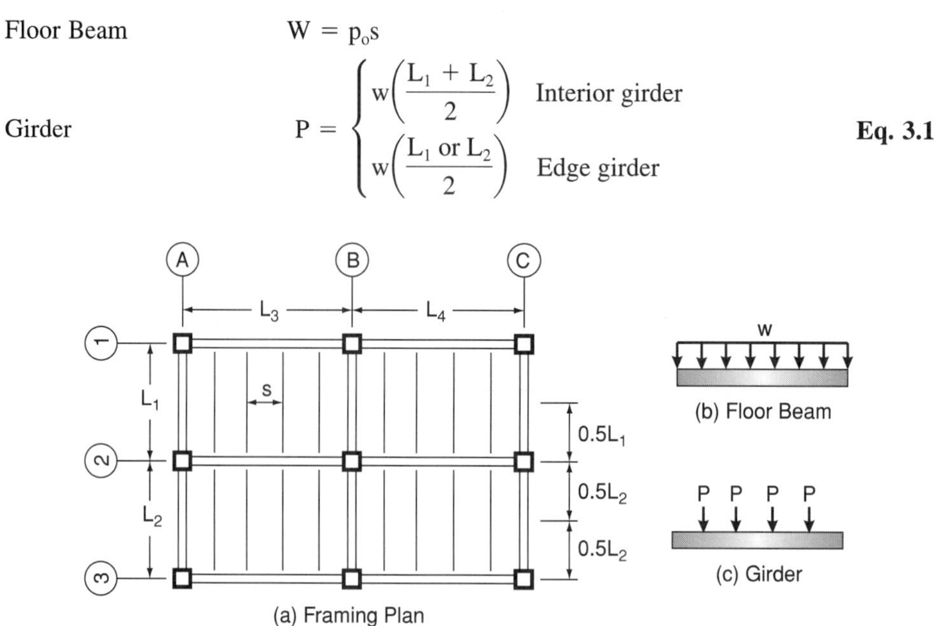

Figure 3.3 A typical floor framing system

Examples 3.1 and 3.2 illustrate the computation of loads acting on columns, beams, and girders. Before working the examples, one additional comment should be made regarding structural framing. The beams and girders in a frame can be connected to the columns either as simple supports or in a manner that allows moment to exist at the ends of the members. If moment is resisted at the ends of the beams and girders, the frame is referred to as a moment-resisting frame. If simple nonmoment resisting connections are used, diagonal bracing must be provided for stability and the frame is referred to as a braced-frame.

EXAMPLE 3.1

The building floor shown Figure 3.1 is to be designed to support a uniformly distributed load of 50 psf over its entire area. The dimensions of the floor are

L_1 to L_3 20 ft L_4 to L_7 25 ft

Using these dimensions, determine the loads transferred from this floor to: (a) interior column B3, (b) edge column E2, and (c) corner column A1.

Solution
Column B3

$$P = 50\left(\frac{L_2 + L_3}{2}\right)\left(\frac{L_4 + L_5}{2}\right) = 50\left(\frac{20 + 20}{2}\right)\left(\frac{25 + 25}{2}\right) = 25{,}000 \text{ lbs}$$

Column E2

$$P = 50\left(\frac{L_1 + L_2}{2}\right)\left(\frac{L_7}{2}\right) = 50\left(\frac{20 + 20}{2}\right)\left(\frac{25}{2}\right) = 12{,}500 \text{ lbs}$$

Column A1

$$P = 50\left(\frac{L_1}{2}\right)\left(\frac{L_4}{2}\right) = 50\left(\frac{20}{2}\right)\left(\frac{25}{2}\right) = 6250 \text{ lbs} \quad \blacksquare$$

EXAMPLE 3.2

The building floor shown in Figure 3.3 is to be designed to support a uniformly distributed load of 50 psf over its entire area. The dimensions of the floor are

L_1 and L_2 20 ft L_3 and L_4 25 ft

Using these dimensions, determine the loads on: (a) a typical floor beam, (b) interior girder 2-A-B, and (c) edge girder 1-B-C.

Solution
Floor Beam

$$w = p_0 s = 50(5) = 250 \text{ plf}$$

Girder 2-A-B

$$P = w\left(\frac{L_1 + L_2}{2}\right) = 250\left(\frac{20 + 20}{2}\right) = 5000 \text{ lbs}$$

This load is spaced at 5-ft intervals along the girder beginning 5 ft from the end.

Girder 1-B-C

$$P = w\left(\frac{L_1}{2}\right) = 250\left(\frac{20}{2}\right) = 2500 \text{ lbs}$$

This load is spaced at 5-ft intervals along the girder beginning 5 ft from the end. ■

3.3 LIVE LOAD REDUCTION

Under some circumstances, the code-specified live loads for a building can be reduced. Before considering such reductions, however, the area that directly influences the forces in a particular member needs to be defined. That area, which is referred to as the *influence area,* is illustrated in Figure 3.4 for several different beams and columns in a building structure.

In both ASCE-7 and IBC2000, the influence area is defined as:

$$A_I = K_{LL}A_T \qquad \textbf{Eq. 3.2}$$

The influence area is equal to the live load element factor, K_{LL}, times the tributary area, A_T. The live load element factor can be calculated from building geometry, or can be taken from Table 3.1.

From this figure and table you can see that the influence area for an interior column is four times as large as its tributary area, whereas that for an interior beam is twice as large as its tributary areas.

As the influence area contributing to the load on a particular member increases, the possibility of having the full design live load on the entire area at the same time decreases. Consequently, building codes usually permit some reduction of the specified values. In Section 4.8 of ASCE 7-98, and Section 1607.9.1 of IBC 2000, the following reduction

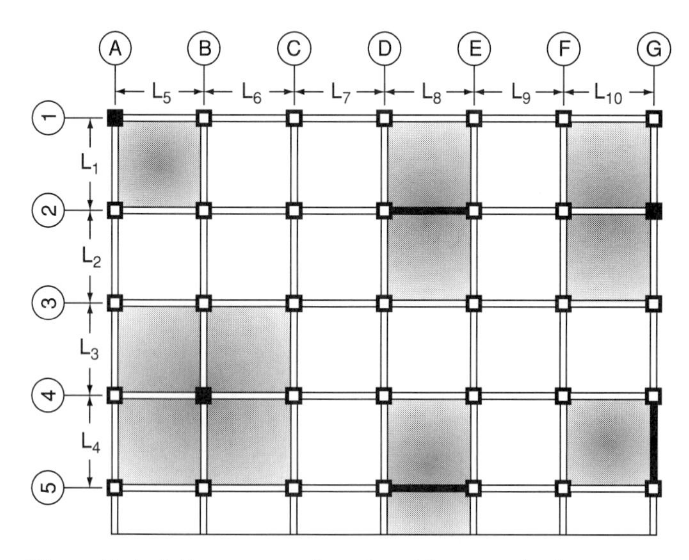

Figure 3.4 Influence areas for selected beams and columns

TABLE 3.1 LIVE LOAD ELEMENT FACTOR, K_{LL}*

Type of Element	K_{LL}
Interior column, Exterior columns without cantilever slabs.	4
Edge columns with cantilever slabs.	3
Corner columns with cantilever slabs, Edge beams without cantilever slabs, Interior beams.	2
All other beams not identified including: Edge beams with cantilever slabs, cantilever beams, two-way slabs, and members without provisions for continuous shear transfer normal to their span.	1

*IBC2000 Table 1607.9.1

factor is given

$$L = L_0\left(0.25 + \frac{15}{\sqrt{K_{LL}A_T}}\right) \qquad \textbf{Eq. 3.3}$$

In this equation, L is the reduced live load, L_0 is the code-specified live load, and the term in parentheses is the reduction factor.

From this equation, we can see that live loads will be reduced only when the influence area is greater than 400 sq ft. There are limits, though, as to how much the live load can be reduced. If the structural member supports one floor only, the live load cannot be reduced by more than 50%. For members supporting more than one floor, the live load cannot be reduced by more than 60%. Usually, beams are used to support one floor. Columns often support more than one floor, in fact, they support all of the floors above them. A plot of live-load reduction factors, as obtained from Equation 3.2, is presented in Figure 3.5 for different influence areas.

However, live-load reduction is not permitted in all cases by ASCE 7-98. If the unit live load is 100 psf or less and the loaded area is used for a place of public assembly, the reduction cannot be taken. Also, the reduction cannot be taken if the loaded area is a roof or a one-way slab and the unit live load is less than or equal to 100 psf. A one-way slab is a slab that is primarily supported on two opposite edges only.

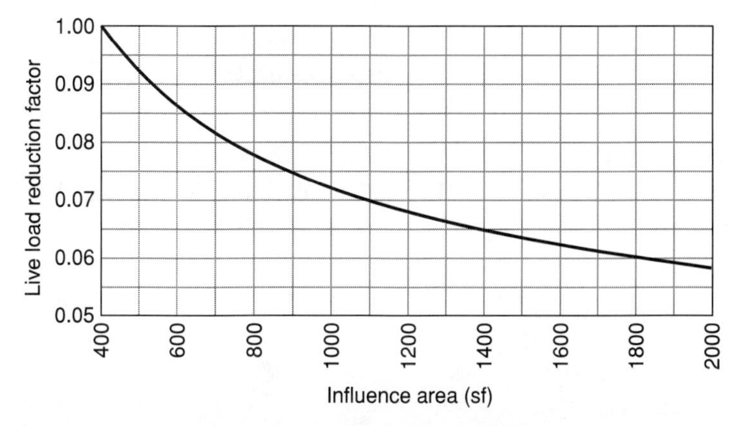

Figure 3.5 Live load reduction factor versus influence area

When the unit live load exceeds 100 psf, Equation 3.1 does not apply. In these cases, a reduction of 20% can be taken for structural members supporting more than one floor. There is no permitted reduction for members that support only one floor when the unit live load is greater than 100 psf. The basis for this 20% reduction is that higher unit live loads tend to occur in buildings used for storage and warehousing. In this type of building, several adjacent spans may be loaded concurrently, but studies have indicated that rarely is an entire floor loaded to more than 80% of its rated design load.

The provision for live load reduction and the related limitations have two significant implications on structural analysis. First, the loads used to obtain column design forces and those used to obtain floor beam design forces may be different. This situation occurs because the live load reduction factors for each are likely to be different. Also, because roofs and floors are treated differently, it does not appear that the live-load reduction is always permitted when columns support a floor and the roof. Typical of such columns are those supporting the top story in a building. When a column supports a floor and the roof, that column should be considered to support a single floor for purposes of determining the permissible live-load reduction.

3.4 LOADING CONDITIONS FOR ALLOWABLE STRESS DESIGN

Now that information has been presented concerning the calculation of the magnitude of various loads, it is necessary to consider which of those loads can act concurrently. For this discussion, the following nomenclature is used

D = dead load
L = live load
L_r = roof live load
W = wind load
E = earthquake load
R = rain load
S = snow load
F = flood load
T = self-straining force
H = load due to the weight and lateral pressure of soil and water in soil

In accordance with the guidelines of ASCE 7-98, the following different loading conditions need to be considered as a minimum in the analysis of a structural system when allowable stress design procedures are being used:

1. D
2. D + L + F + H + T + (L_r or S or R)
3. D + (W or 0.7E) + L + (L_r or S or R)
4. 0.6D + W + H
5. 0.6D + 0.7E + H

The load combinations in IBC-2000 are similar. As specified in IBC-2000, Section 1605.3.1, the basic load combinations to be considered are:

1. D
2. D + L
3. D + L + (L_r or S or R)
4. D (W or 0.7E) + L + (L_r or S or R)
5. 0.6D + W
6. 0.6D + 0.7E

When using IBC-2000, flood loads, self-straining loads, and lateral pressure loads are treated as live loads.

You will note that all of these loads, other than dead loads, will vary appreciably with time—there is not always snow on the structure and the wind is not always blowing, for example. In the second and third ASCE-7 loading conditions, the dead load has been combined with multiple variable loads, yet in the fourth and fifth loading conditions the dead load has been combined with only two variable loads. Observe also in the fourth and fifth loading combinations that the full dead load is not being considered. The two variable loads in these combinations, the wind and earthquake loads, generally have a lateral component. As such, they tend to cause the structure to overturn. A dead load, on the other hand, is a gravity load, which tends to cause the structure to right itself. Consequently, a more severe condition can occur if for some reason the full dead load were not acting.

Most likely when two or more loads are acting on a structure in addition to dead load, the loads other than dead load are not likely to achieve their absolute maximum values simultaneously. Load surveys seem to bear out this assumption. ASCE 7-98 (Section 2.4.3[1]) and IBC 2000 (Section 1605.3.1.1) permit the load effects in these loading conditions, except dead load, to be multiplied by 0.75 provided the result is not less than that produced by dead load and the load causing the greatest effect. IBC-2000 eliminates the 0.7 factor on earthquake loads when the 0.75 reduction is used. *Remember that the code is listing the minimum conditions that must be considered.* If the design engineer feels that the maximum values of two variable loads (wind and rain, for example) may occur at the same time in his or her area, it is not required that the 0.75 value be used.

These load combinations are only the recommended minimum load combinations that need to be considered. As with the determination of the loads themselves, the engineer must evaluate the structure being analyzed and determine whether these load combinations comprise all of the possible combinations for a particular structure. Under some conditions, other loads and load combinations may be appropriate.

EXAMPLE 3.3

An observation deck at an airport has girders configured as shown in the figure.

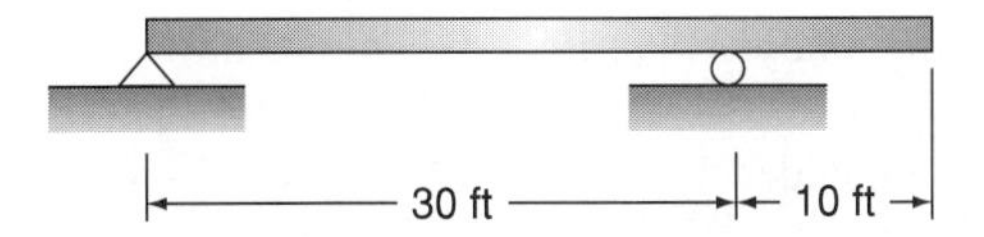

These girders are spaced 15 ft on-center. Assume that all of the loads are uniformly distributed over the deck and have been found to be as follows

Dead load: 32 psf

Live load: 100 psf

Snow load: 24 psf

Rain load: 10 psf

Using the loading conditions from ASCE-7, what are the combined loads that can reasonably be expected to act on this girder?

[1]ASCE 7-98 explicitly does not permit the reduction unless there are two or more loads acting in combination with a dead load, excluding earthquake loads. Shortly after the code was published, this provision was reported to have been removed and it will not appear in the next edition. The next edition will permit the reduction as discussed above.

Solution. Because the girders are 15 ft on-center, the tributary area for this beam has a width of 15 ft. The applicable loading combinations for this beam are

Combination 1	$15[32] = 480$ plf	Dead
Combination 2	$15(32 + 100 + 10) = 2130$ plf	Dead + Live + Rain
Combination 3	$15(32 + 100 + 24) = 2340$ plf	Dead + Live + Snow

The other load combinations either are not appropriate given the loads on the structure or are redundant. The beam should then be analyzed using these three conditions of load to determine the complete range of response. In this particular case, though, only Combination 3 will have significance from a design viewpoint because it is the largest. ■

3.5 LOADING CONDITIONS FOR STRENGTH DESIGN

A design philosophy that has become common is the strength-design procedure. With this method the estimated loads are multiplied by certain load factors that are almost always larger than 1.0 and the resulting ultimate or "factored" loads are then used for designing the structure. The structure is proportioned to have a design ultimate strength sufficient to support the ultimate loads.

The purpose of load factors is to increase the loads to account for the uncertainties involved in estimating the magnitudes of dead or live loads. For instance, how close in percent could you estimate the largest wind or snow loads that ever will be applied to the building that you are now occupying?

The load factors used for dead loads are smaller than those used for live loads because engineers can estimate more accurately the magnitudes of dead loads than they can the magnitudes of live loads. In this regard, notice that loads that remain in place for long periods will be less variable in magnitude, whereas those applied for brief periods, such as wind loads, will have larger variations.

When a structure is to be designed using strength procedures, other load combinations and load factors may apply. Although we are not directly concerned about design while performing analysis, we must be cognizant of the design method that will be used so that the analysis results will have meaning for the design engineer.

The recommended load combinations and load factors for strength design usually are presented in the specifications of the different code-writing bodies such as the International Council of Building Officials, the American Institute of Steel Construction, or the American Petroleum Institute. The load factors for use with strength design are determined statistically, and consideration is given to the type of structure upon which the loads are acting. As such, the load combinations and factors for a building will be different from those for a bridge. Both of these will likely be different from those for an offshore oil production platform. A structural analyst always should refer to the design guide or recommendation appropriate for the system being analyzed.

Following are the recommended load combinations for building structures as recommended by ASCE 7-98:

1. $1.4D$
2. $1.2(D + F + T) + 1.6(L + H) + 0.5(L_r \text{ or } S \text{ or } R)$
3. $1.2D + 1.6(L_r \text{ or } S \text{ or } R) + (0.5L \text{ or } 0.8W)$
4. $1.2D + 1.6W + 0.5L + 0.5(L_r \text{ or } S \text{ or } R)$
5. $1.2D + 1.0E + 0.5L + 0.2S$

6. 0.9D + 1.6W + 1.6H
7. 0.9D + 1.0E + 1.6H

It may appear peculiar at first that only 90% of the dead load is used in the last two load combinations. After all, the dead load always is acting and is defined quite well. However, if the nature of the other loads in the load combination is considered, using only 90% of the dead load may be a more severe condition than if the entire dead load had been applied. As discussed, wind and seismic loads cause overturning moments to occur in the structure; they try to tip the structure over. The dead load causes a righting moment, that is, a resisting moment. It tries to keep the structure from tipping over. By using reduced dead load, there is less moment resisting the overturning effects of the wind and seismic loads.

The load combinations specified in the IBC-2000 code are similar to those in ASCE-7, but there are differences. Following are the load combinations specified in IBC-2000:

1. 1.4D
2. 1.2D + 1.6L + 0.5(L_r or S or R)
3. 1.2D + 1.6(L_r or S or R) + (f_1L or 0.8W)
4. 1.2D + 1.6W + f_1L + 0.5(L_r or S or R)
5. 1.2D + 1.0E + f_1L + f_2S
6. 0.9D + (1.0E + 1.6W)

A significant difference between these load combinations and the ones recommended in ASCE-7 is the introduction of the factors f_1 and f_2. The values that these factors can assume are shown in Table 3.2. As with the other load combinations previously discussed, these load combinations are only the recommended minimum load combinations that need to be considered. The engineer must decide whether these load combinations represent all of the possible combinations for the building being analyzed.

TABLE 3.2 IBC-2000 LIVE AND SNOW LOAD COEFFICIENTS

Factor	Design Condition	Value
f_1	Floors in places of public assembly, for live loads in excess of 100 psf, and for parking garage live load	1.0
	Other live loads	0.5
f_2	Roof configurations (such as saw tooth) that do not shed snow off the structure	0.7
	Other roof configurations	0.2

EXAMPLE 3.4

Repeat Example 3.3, but this time determine the load combinations for a structure to be designed using strength design procedures.

Solution. The applicable load combinations are as follows:

Combination 1	15[1.4(32)] = 672 plf	Dead
Combination 2	15[1.2(32) + 1.6(100) + 0.5(24)] = 3,156 plf	Dead + Live + Snow
Combination 3	15[1.2(32) + 1.6(100) + 0.5(10)] = 3,051 plf	Dead + Live + Rain
Combination 4	15[1.2(32) + 1.6(24) + 0.5(100)] = 1,902 plf	Dead + Live + Snow
Combination 5	15[1.2(32) + 1.6(10) + 0.5(100)] = 1,566 plf	Dead + Live + Rain

The other load combinations either are not appropriate or cause a lower total load to be placed on the structure. ■

3.6 PLACING LOADS ON THE STRUCTURE

After the magnitudes of the various loads that may be applied to a structure have been computed, the engineer must determine where to place the loads on the structure to cause the most severe situations. There usually is little question about the position of the environmental loads, but this is not the case for the gravity live loads. The placement of these loads is discussed at length in Chapter 9 and is introduced only briefly in this section.

At some point in time, the gravity live loads may not be acting at their maximum values everywhere on the structure. But at some other point in time, they may be acting at their full magnitudes everywhere. Consequently, we need to learn where to place them for analysis so the worst possible effects can be found. The problem is complicated further because the loading pattern that causes the maximum value of one response of a structure may not be the same pattern that causes the maximum value of some other response. For example, the loading condition causing a maximum support reaction may not be the same loading condition that causes the maximum bending moment or deflection. An easy way to demonstrate this concept is through an example.

EXAMPLE 3.5

With the enclosed computer program SAP2000, determine the maximum reaction at the left support and the maximum negative moment at the right-hand support in the beam shown in the figure. The term *bending moment,* with which you are already familiar, will be discussed at length in Chapter 5. Use units of inches and pounds for this example.

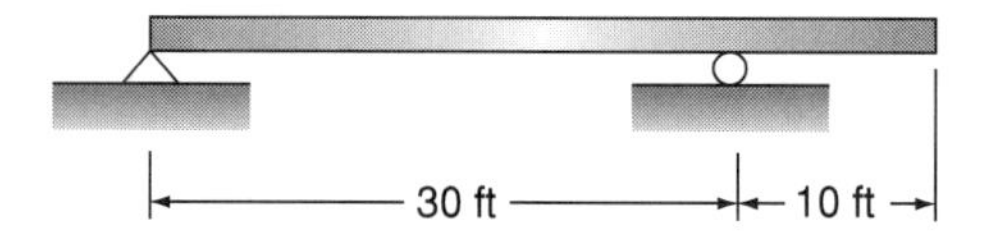

The dead load acting on the beam, including the self-weight of the beam is 480 plf. The live load on the beam, which can act anywhere, is 1500 plf. The beam is made of steel, it has a cross section area on 45 in^2 and second moment of area of 500 in^4.

Solution. The three basic loads that we will consider, and where they act on the beam, are shown in the following figure. The dead load is always acting. Therefore, it is placed on the entire beam. The live load may be acting on the entire beam, on the left span only, or on the cantilever span only. By using various combinations of these basis loads, we can achieve the possible loading conditions.

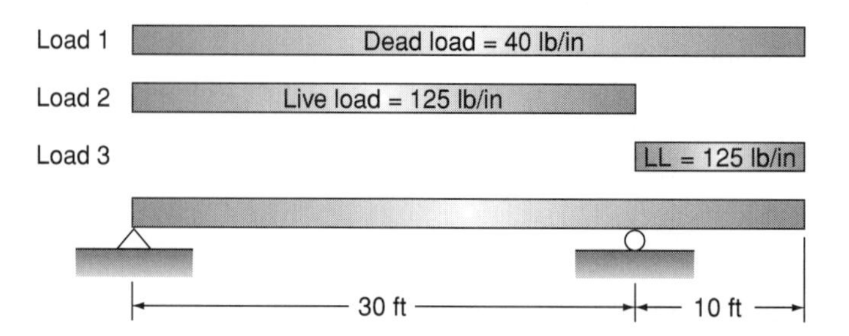

To begin the analysis, detailed information regarding the joints, members, and loading data is entered into the computer. The joints and members are numbered for identification in SABLE as shown in the figure that follows. The data file for this example is included with the program SABLE.

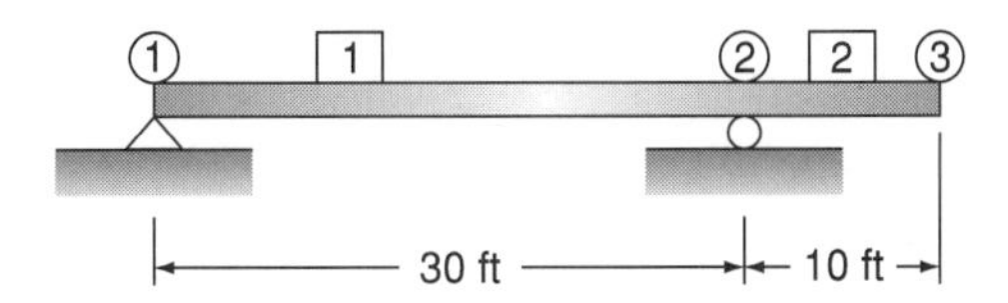

The information regarding the joints, members, and loading data is shown in Tables E3.5-1 and E3.5-2. Table E3.5-1 lists the locations of the joints and the types of restraint at each joint. Note that the left hinge support has restraint in the x and y directions, but no moment restraint is provided.

TABLE E3.5-1 JOINT LOCATION AND RESTRAINT DATA

	Location (in)		Restraints		
Joint	X	Y	Disp. X	Disp. Y	Rotn. Z
1	0	0	Y	Y	N
2	360	0	N	Y	N
3	480	0	N	N	N

In Table E3.5-2 detailed data for each of the members of the structure are provided. First, the connectivity of the members is shown, that is, the joint numbers at both ends of each member are given. Then the properties of the members (areas, moments of inertia, and moduli of elasticity) are entered. Next, the loads applied to each member are specified.

TABLE E3.5-2 BEAM PROPERTY AND LOAD DATA

	Connectivity		Properties			Load Cases		
Beam	i joint	j joint	A	I	E	W_1	W_2	W_3
1	1	2	45	500	29×10^6	−40	−125	0
2	2	3	45	500	29×10^6	−40	0	−125

The dead load on the beam is 480 plf or 40 pounds per inch. The live load is 1500 plf or 125 pounds per inch. These data are entered for the three basic load cases that we are considering. Observe that the loads have a negative sign. The negative sign indicates that the loads are acting downward. Using the load combination feature of SABLE, we will combine these individual load cases into the possible loading conditions acting on the structure.

We wish to find the maximum vertical reaction at the left support and the maximum negative moment at the right support. The forces of reaction are obtained by displaying joint forces after a static analysis has been performed. The maximum moment in the beam is obtained by plotting the shearing force and bending-moment diagrams for the beam. The results obtained from the analysis are presented in Table E3.5-3 for the basic load cases.

TABLE E3.5-3 ANALYSIS RESULTS AND DATA INTERPRETATION

	Basic load cases (lb., in.)			Combinations of basic load cases (lb, in.)			
	Load 1	Load 2	Load 3	1 only	1 + 2	1 + 3	1 + 2 + 3
Force Y Joint 1	6400	22,500	−2500	6400	28,900	3900	26,400
Moment at Joint 2	−288,000	0	−900,000	−288,000	−288,000	−1,188,000	−1,188,000

In this example, there are four possible loading combinations. These are (1) the dead load only acting on the entire beam; (2) the dead load on the entire beam and the live load on the left span; (3) the dead load on the entire beam and the live load on the cantilever span; and (4) the dead load and the live load on the entire span. The results of these loading combinations also are presented in Table E3.5-3.

Notice the range of forces in Table E3.5-3 for each of the response quantities considered. The maximum vertical upward force at the left support varies from 3900 lbs to 28,900 lbs. The largest value occurs when load cases 1 and 2 are combined. The entire live load is not acting on the beam; it is acting on the left span only.

The maximum negative moment at the right-hand support ranges from 288,000 to 1,188,000 lb-ft. Its maximum value occurs when the total load is acting on the beam (also when we have dead load on the entire beam plus live load on the cantilever span). From this simple example, notice that the placement of live loads can cause very large variations in the magnitudes of the forces in a structural system. ■

Other types of gravity loads that can vary with time are moving loads such as those that occur on bridges. A more thorough discussion about where and how to place loads that can change with time is presented in Chapter 9. For now, however, we should recognize that gravity live loads might not be acting at their full magnitude on all areas of the structure concurrently and that the maximum response does not necessarily occur in the fully loaded structure.

3.7 CONCEPT OF THE FORCE ENVELOPE

When loads are applied to a structure, the structure responds in reaction to those loads. The forces in a particular component in the system are caused by (1) the loads acting on the structure, and (2) where those loads are located. From analysis of the response to the various forces acting, we can determine the maximum and minimum forces that can exist in any component. This range of forces is called a *force envelope*.

We saw a force envelope in the last example. The reaction in the left support had a minimum value of 3900 lbs and a maximum value of 28,900 lbs. These values were determined by placing loads on the system in locations to cause the maximum effect. For this reaction, the force envelope is 3900 to 28,900 lbs. No matter where these loads are placed on the structure, the force at that reaction will always fall within this range. Force envelopes are very important to design engineers because they define the full range of response for which the structure must be designed.

Example 3.6 further illustrates the idea of force envelopes. A force envelope is prepared for the bending moment in a simple beam using the procedures learned in earlier courses.

EXAMPLE 3.6

Using the principles of mechanics, draw diagrams showing the variation of bending moment throughout the beam illustrated here for each of the two loads shown. The loads do not necessarily act simultaneously.

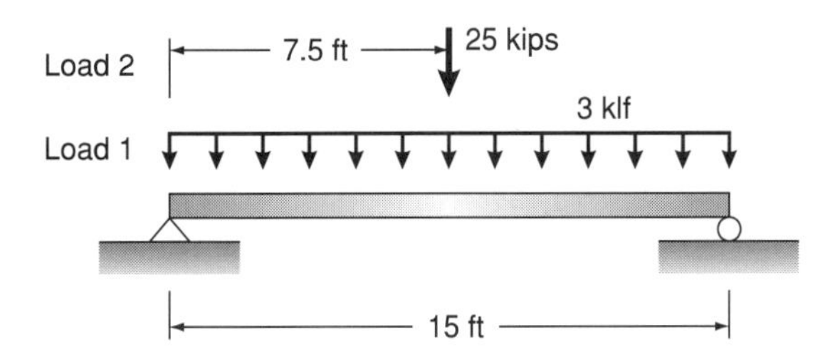

Solution. Expressions for moment are written with the origin placed at the left end of the beam. Note that due to symmetry the end reactions are each equal to one half the total load. The reactions at the left and right support caused by Load 1 are 22.5 kips each. The bending moment in the beam caused by Load 1 is

$$M_1(x) = 22.5x - 3x\left(\frac{x}{2}\right)$$

The reactions for the second load are 12.5 kips at each support. Because the load changes halfway across the beam, the moment equation changes. Two such expressions are needed for Load 2. These expressions are

$$M_2(x) = 12.5x \qquad \text{for } 0 < x \leq 7.5$$
$$M_2(x) = 12.5x - 25(x - 7.5) \qquad \text{for } 7.5 < x \leq 15$$

The preceding expressions for M_1 and M_2 are plotted in the following figure. The shaded areas shown on the figure represent the moment envelope.

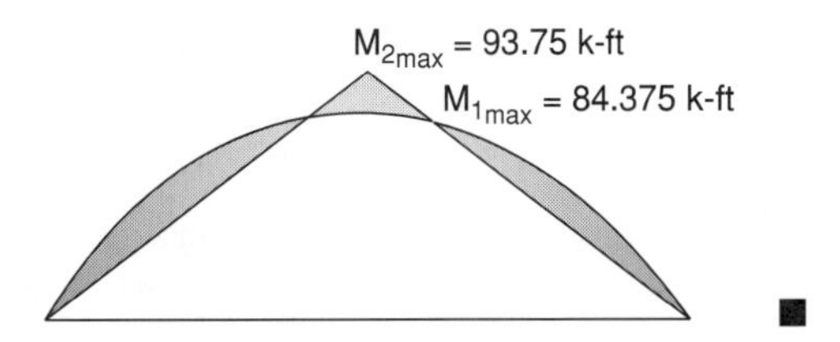

■

3.8 PROBLEMS FOR SOLUTION

For Problems 3.1 through 3.5, use the basic floor framing plan shown in the adjacent figure when solving these problems. Consider live load reduction following the provisions of ASCE 7-98 as appropriate. For purposes of these problems assume that this is an upper story in a multi-story office building, but it is not the top story.

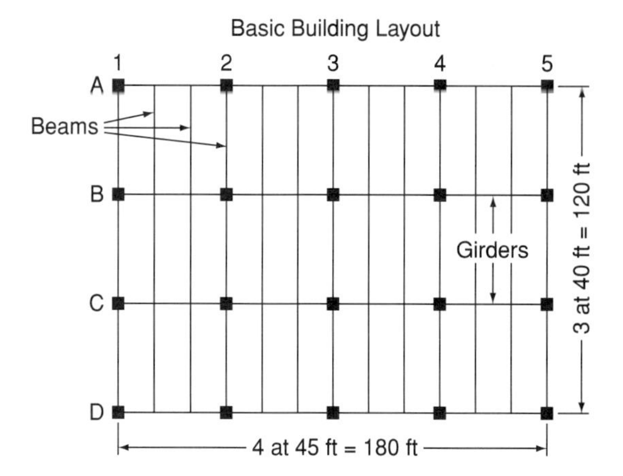

3.1 Load on Column B3 contributed by this floor if the live load is 150 psf

3.2 Load on Column A3 contributed by this floor if the live load is 75 psf (*Ans.* 33,750 lbs)

3.3 Load on Column A1 contributed by this floor if the live load is 100 psf

3.4 Load on an interior floor beam if the live load is 75 psf (*Ans.* 768.4 plf)

3.5 Load on Girder B2–B3 if the live load is 50 psf

For Problems 3.6 through 3.10, given the loads specified, compute the maximum combined load using the ASCE 7-98 load combinations for working stress design.

3.6 $D = 50$ psf, $L_r = 75$ psf, $R = 8$ psf, $S = 20$ psf (*Ans.* 125 psf)

3.7 $D = 45$ psf, $L = 60$ psf

3.8 $D = 2750$ lbs, $L = 4500$ lbs, $L_r = 1500$ lbs, $R = 1250$ lbs, $S = 1000$ lbs (*Ans.* 8750 lbs)

3.9 $D = 87$ psf, $L = 150$ psf

3.10 $D = 75$ psf, $L_r = 35$ psf, $R = 12$ psf (*Ans.* 110 psf)

For Problems 3.11 through 3.15, repeat Problems 3.6 to 3.10 using the ASCE 7-98 LRFD load combinations.

3.11 Repeat Problem 3.6

3.12 Repeat Problem 3.7 (*Ans.* 150 psf)

3.13 Repeat Problem 3.8

3.14 Repeat Problem 3.9 (*Ans.* 344.4 psf)

3.15 Repeat Problem 3.10

For Problems 3.16 through 3.20, repeat Problems 3.6 through 3.10 using the specified IBC-2000 load combinations.

3.16 Repeat Problem 3.6 using the IBC-2000 ASD load combinations

3.17 Repeat Problem 3.7 using the IBC-2000 LRFD load combinations

3.18 Repeat Problem 3.8 using the IBC-2000 ASD load combinations

3.19 Repeat Problem 3.9 using the IBC-2000 LRFD load combinations

3.20 Repeat Problem 3.10 using the IBC-2000 ASD load combinations

For Problems 3.21 through 3.25, for each of the beams shown, determine the range of the vertical reaction at Support B if the dead load is 540 plf and the live load is 960 plf. The dead load must act over the entire beam all the time. The live load can act anywhere on the beam, but does need to act on entire spans when it acts. Use SAP2000 or SABLE to solve these problems. For analysis, let $A = 50$ in.^{2}, $I = 550$ in.^{4}, and $E = 29{,}000$ ksi.

3.21 (*Ans*. 13,500 lbs to 37,500 lbs)

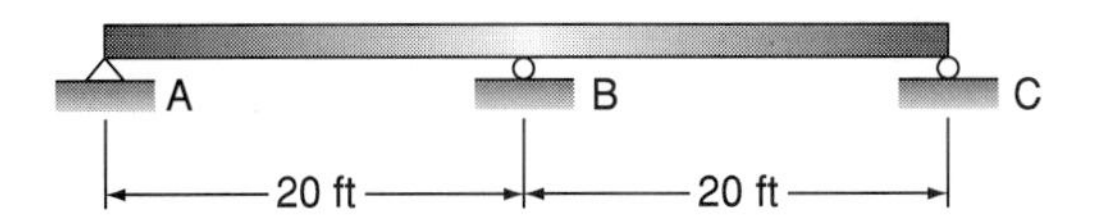

3.22

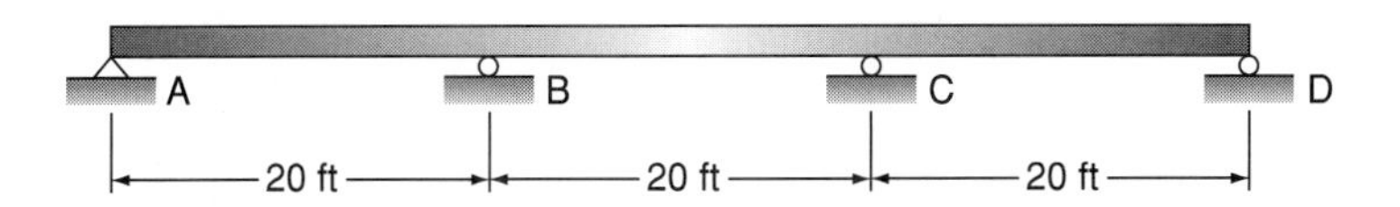

3.23 (*Ans*. 5600 lbs to 21,600 lbs)

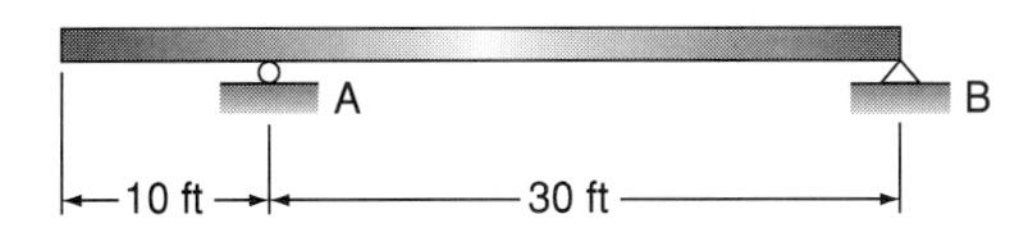

3.24

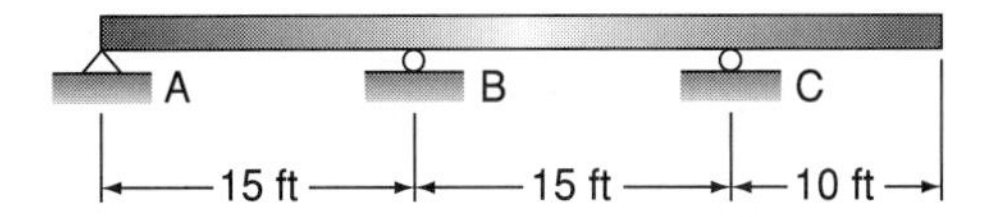

3.25 (*Ans*. 8400 lbs to 32,400 lbs)

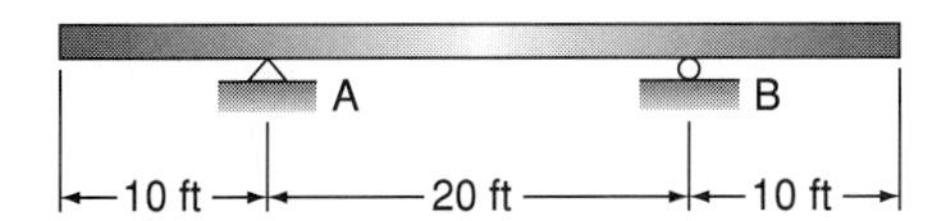

Chapter 4

Reactions

4.1 EQUILIBRIUM

A body at rest is said to be in *static equilibrium*. The resultant of the external forces acting on the body—including the supporting forces, which are called reactions—is zero. Not only must the sum of all forces (or their components) acting in any possible direction be zero, but the sum of the moments of all forces about any axis also must be equal to zero.

If a structure, or part of the structure, is to be in equilibrium under the action of a system of loads, it must satisfy the six equations of static equilibrium. Using the Cartesian x, y, and z coordinate system, the equations of static equilibrium can be written as

$$\sum F_x = 0 \qquad \sum F_y = 0 \qquad \sum F_z = 0$$
$$\sum M_x = 0 \qquad \sum M_y = 0 \qquad \sum M_z = 0 \qquad \textbf{Eq. 4.1}$$

For purposes of analysis and design, the large majority of structures can be considered as being planar structures without loss of accuracy. For these structures, which are usually assumed to be in the xy plane, the sum of the forces in the x and y directions and the sum of the moments about an axis perpendicular to the plane must be zero. The equations of equilibrium reduce to

$$\sum F_x = 0 \qquad \sum F_y = 0 \qquad \sum M_z = 0 \qquad \textbf{Eq. 4.2}$$

These equations are commonly written as

$$\sum F_H = 0 \qquad \sum F_V = 0 \qquad \sum M = 0 \qquad \textbf{Eq. 4.3}$$

These equations cannot be proved algebraically; they are merely statements of Sir Isaac Newton's observation that for every action on a body at rest there is an equal and opposite reaction. Whether the structure under consideration is a beam, a truss, a rigid frame, or some other type of assembly supported by various reactions, the equations of static equilibrium must apply if the body is to remain at rest.

The structures discussed in the first seven chapters of this textbook are called planar structures; the entire structure lies in a plane. Three-dimensional trusses, also called space trusses, are discussed in Chapter 8.

4.2 MOVING BODIES

The statement was made in the preceding section that a body at rest is in a state of static equilibrium. However, being at rest is not a necessary condition for static equilibrium. A body that is moving at constant velocity also can be in a state of static equilibrium; the net force acting on the body is equal to zero. This concept can be proven with the impulse–momentum relationship:

$$F(\Delta T) = m(\Delta v) \qquad \textbf{Eq. 4.4}$$

In this equation, F is the net force acting on the body, ΔT is the time that the force acts, Δv is the change in velocity of the body, and m is the mass of the body. If the net force is equal to zero, the left side of the equations becomes zero, which implies that the change in velocity must be equal to zero since the mass of any real body cannot be equal to zero. When the change in velocity is equal to zero, the body is not accelerating—it is moving at constant velocity. There is nothing in the relationship to imply that the body is stationary.

When a body is accelerating, there are additional forces acting that must be included in equilibrium calculations. These additional forces are the inertial forces, which are caused by the mass of the body. When the inertial forces are included and the net force acting on the body, including the inertial forces, is equal to zero, the body is said to be in a state of dynamic equilibrium. We will not investigate dynamic equilibrium in this book.

4.3 CALCULATION OF UNKNOWNS

To completely define a force, three properties of that force must be defined. These properties are its magnitude, its line of action, and the direction in which it acts along the line of action. All these properties are generally known for each of the externally applied loads. However, when dealing with structural reactions, only the point at which the reaction force acts, and perhaps its direction, are known. The magnitude of the forces of reaction, and sometimes the directions in which they act, are unknown quantities that must be determined.

The number of unknowns that can be determined using the equations of static equilibrium is limited by the number of independent equations of static equilibrium available. For any structure lying in a plane, there are only three independent equations of static equilibrium. These three equations can be used to determine at most three unknown quantities for the structure. When there are more than three unknowns to evaluate, additional equations must be used in conjunction with the equations of static equilibrium. We will see that in some instances, because of special construction features, equations of condition are available in addition to the usual equations of static equilibrium. In later chapters, we will introduce equations of compatibility of displacement.

City Park Lake Bridge, Baton Rouge, Louisiana (Courtesy of the American Institute of Steel Construction, Inc.)

4.4 TYPES OF SUPPORT

Structural frames may be supported by hinges, rollers, fixed ends, or links. These supports are discussed in the following paragraphs. Common notations for showing the different types of supports are shown in Figure 4.1.

A *hinge* or pin-type support is assumed to be connected to the structure with a frictionless pin. This type of support prevents movement in a horizontal or vertical direction, but does not prevent slight rotation about the hinge. There are two unknown forces at a hinge: the magnitude of the force required to prevent horizontal movement and the magnitude of the force required to prevent vertical movement. The support supplied at a hinge may also be referred to as an inclined force, which is the resultant of the horizontal force and the vertical force at the support. Two unknowns remain: the magnitude and direction of the inclined resultant.

A *roller* support is assumed to offer resistance to movement only in a direction perpendicular to the supporting surface beneath the roller. There is no resistance to slight

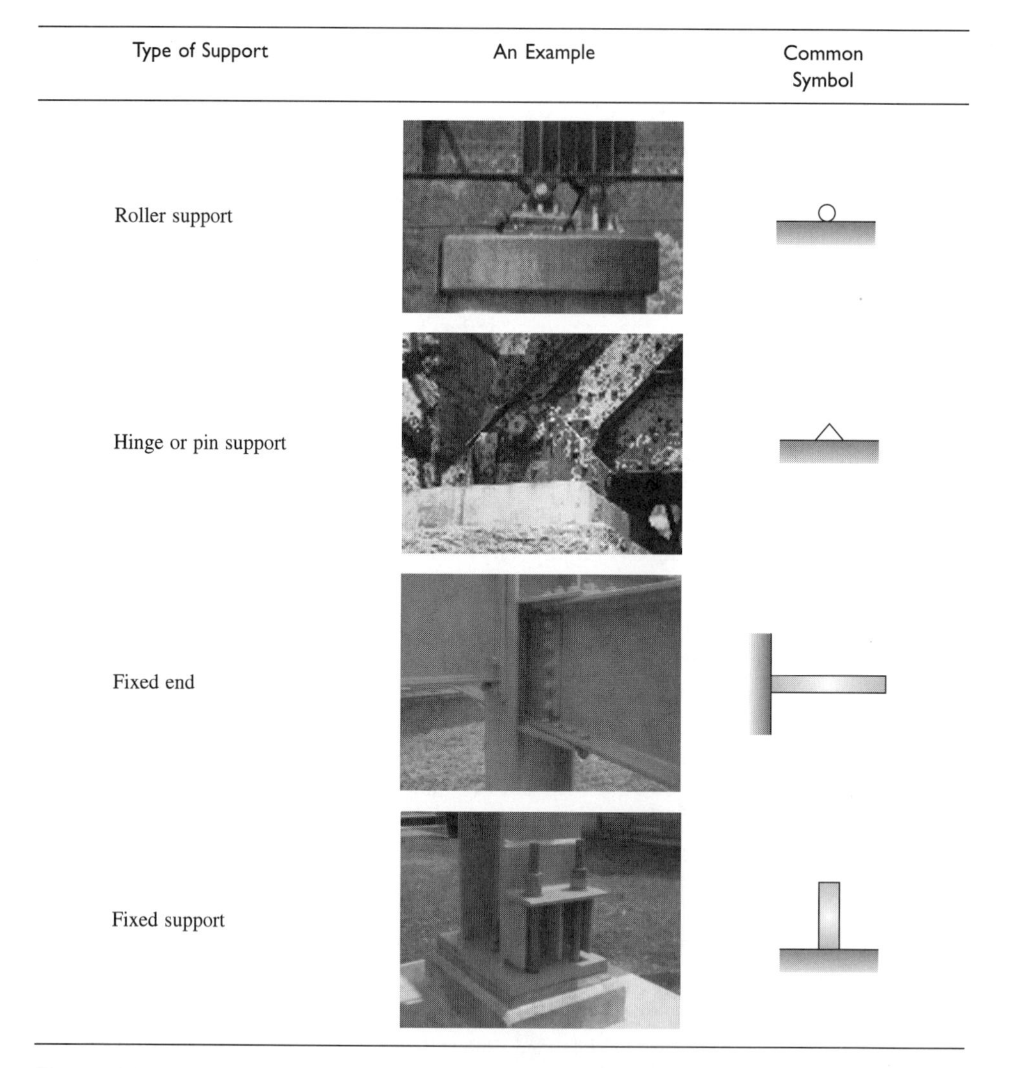

Figure 4.1 Common representation of structural supports

rotation about the roller or to movement parallel to the supporting surface. The magnitude of the force required to prevent movement perpendicular to the supporting surface is the one unknown. Rollers may be installed in such a manner that they can resist movement either toward or away from the supporting surface.

A *fixed-end* support is assumed to offer resistance to both rotation about the support and to movement vertically and horizontally. There are three unknowns: the magnitude of the force to prevent horizontal movement, the magnitude of the force to prevent vertical movement, and the magnitude of the force to prevent rotation.

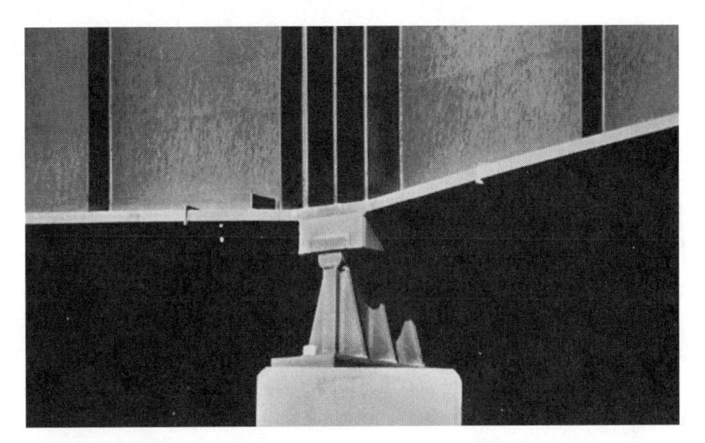

Hinged support for a bridge girder (Courtesy Bethlehem Steel Corporation)

Another type of support is a *link* support. It is similar to the roller in its action because the pins at each end are assumed to be frictionless. The line of action of the supporting force must be in the direction of the link and through the two pins. One unknown is present: the magnitude of the force in the direction of the link.

4.5 FREE-BODY DIAGRAMS

For a structure to be in equilibrium, every part of the structure also must be in equilibrium. The equations of static equilibrium are as applicable to each piece of a structure as they are to an entire structure. It is therefore possible to draw a diagram of any part of a structure, including all of the forces that are acting on that part of the structure, and apply the equations of static equilibrium to that part. Such a diagram is called a *free-body diagram*. The forces acting on the free-body are the external forces acting on that piece of the structure, including structural reactions, and the internal forces applied from the adjoining parts of the structure.

A simple beam is shown in Figure 4.2(a). This beam has two supports and is acted upon by two loads. A free-body diagram of the entire beam Figure 4.2(b) shows all of the forces of reaction. We also can cut the beam at point A and draw a free-body diagram for each of the two pieces. The result is shown in Figure 4.2(c). Observe that we now have included the internal forces at the location of the cut on the diagrams. The internal forces are the same on the two pieces, but the direction in which they act is reversed. In essence, the effects of the right side of the beam on the left side are shown on

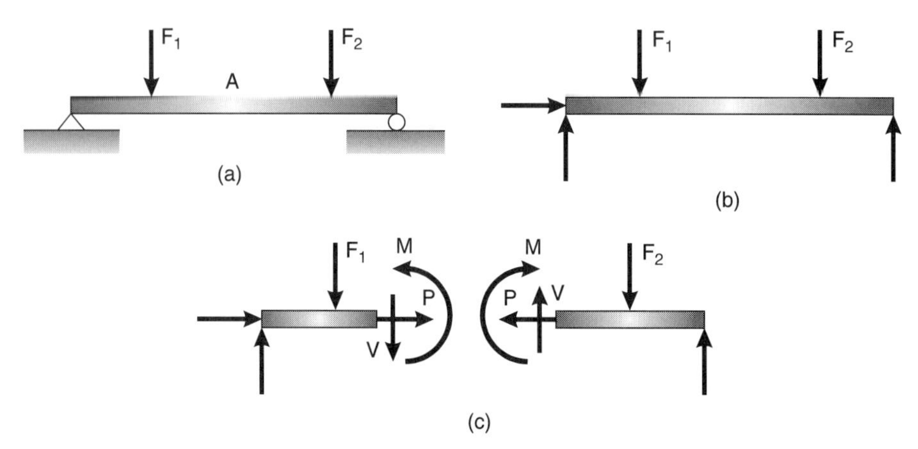

Figure 4.2 A beam and two free-body diagrams

the left free-body and vice versa. For instance, the right-hand part of the beam tends to push the left-hand free body down while the left-hand part is trying to push the right-hand free body up.

Isolating certain sections of structures and considering the forces applied to those sections is the basis of all structural analysis. It is doubtful that this procedure can be overemphasized to the reader, who will, it is hoped, discover over and over that free-body diagrams open the way to the solution of structural problems.

When drawing free-body diagrams, a good practice is to compute the horizontal and vertical components of inclined forces for use in making calculations. If this practice is not followed, the perpendicular distances from the lines of action of inclined forces to the point where moment is being taken will have to be found. The calculation of these distances is often difficult, and the possibility of making mistakes in setting up the equations is greatly increased.

4.6 SIGN CONVENTION

The particular sign convention used for tension, compression, and so forth is of little consequence as long as a consistent system is used. We use the following sign convention in computations for the purposes of this text.

1. Unknown forces of reaction are generally assumed to act upward or to the left. Unknown reaction moments are usually assumed to act clockwise.
2. A positive sign is used for *tensile* forces. This convention was selected because components in tension become longer; the change in length is positive.
3. A negative sign is used for pieces in compression. This sign convention is used because members in compression become shorter; the change in length is negative.
4. A positive sign is used for moments acting in a clockwise direction. Conversely, a negative sign is used for moments acting counterclockwise.

Again, the sign convention used is not critical. The important thing to remember is that once a sign convention is adopted in an analysis, it should be used consistently throughout the analysis to avoid confusion.

On many occasions, it is possible to determine the direction of a reaction by inspection. When such a determination is not possible, a direction is assumed and the

Mississippi River Bridge, St. Paul, Minnesota (Courtesy of Kenneth M. Wright Studios, St. Paul, Minnesota)

appropriate equations of equilibrium are written. If on solution of the equation the numerical value for the reaction is positive, then the assumed direction was correct; if negative, the correct direction is opposite that assumed.

4.7 STABILITY, DETERMINACY, AND INDETERMINACY

The discussion of supports shows there are three unknown reaction components at a fixed end, two at a hinge, and one at a roller or link. If, for a particular structure, the total number of forces of reaction is equal to the number of equations of static equilibrium available, the unknowns may be determined and the structure is then said to be statically determinate externally. Should the number of unknowns be greater than the number of equations available, the structure is statically indeterminate externally; if less, it is unstable externally. From this discussion, note that stability, determinacy, and indeterminacy are dependent upon the configuration of the structure; they are not dependent upon the loads applied to the structure. The following examples will demonstrate the application of these concepts to structural systems.

EXAMPLE 4.1

Determine the statical classification of the simply supported beam shown in the figure. The unknown forces of reaction are illustrated. There are two unknown forces of reaction at the left support because it is a pinned support. There is one unknown force of reaction at the right support because it is a roller.

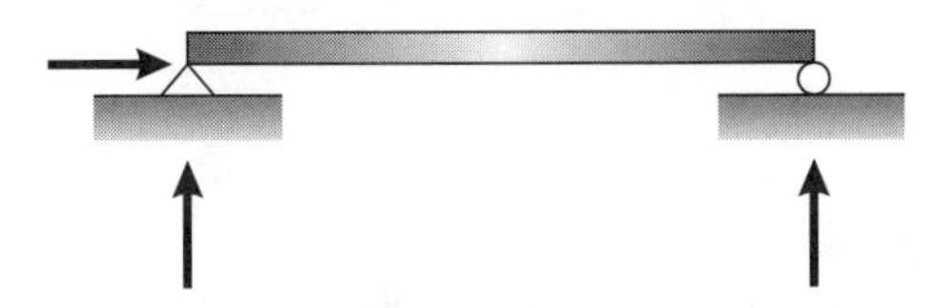

There are three applicable equations of static equilibrium: summation of forces vertically, summation of forces horizontally, and summation of moments. Because there are three unknown forces of reaction and three applicable equations of equilibrium, the beam is stable and statically determinate externally—the number of unknowns is equal to the number of equations of static equilibrium. ■

EXAMPLE 4.2

Determine the statical classification of the simply supported beam shown in the figure. The unknown forces of reaction are illustrated. There is one unknown force of reaction at each of the two supports because the supports are rollers.

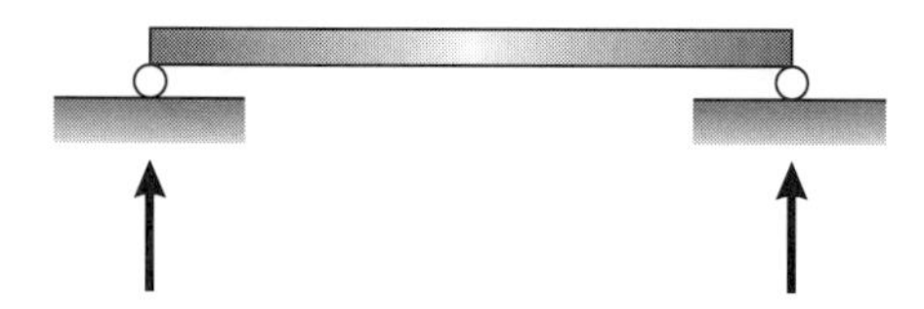

There are three applicable equations of static equilibrium: summation of forces vertically, summation of forces horizontally, and summation of moments. Because there are two unknown forces of reaction and three applicable equations of equilibrium, the beam is unstable—the number of equations of static equilibrium exceeds the number of unknowns. ■

EXAMPLE 4.3

Determine the statical classification of the continuous beam shown in the figure. The unknown forces of reaction are illustrated. There are two unknown forces of reaction at the left support because that support is pinned. There is one unknown force of reaction at each of the other supports because they are rollers.

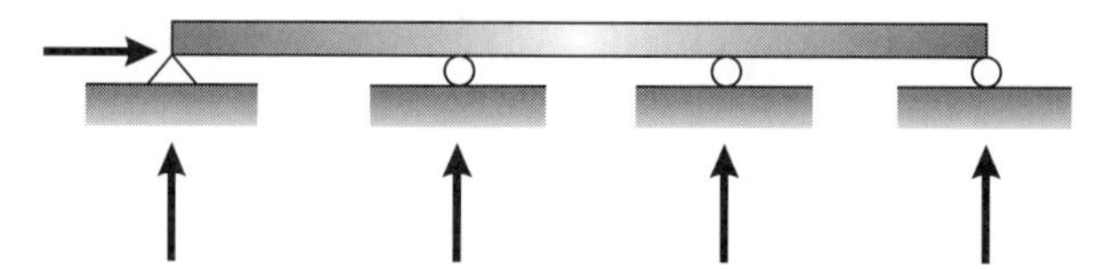

There are three applicable equations of static equilibrium: summation of forces vertically, summation of forces horizontally, and summation of moments. Because there are five unknown forces of reaction and three applicable equations of equilibrium, the beam is stable and statically indeterminate to the second degree externally—there are two more unknowns than there are equations of static equilibrium. ■

EXAMPLE 4.4

Determine the statical classification of the propped cantilever beam shown in the figure. The unknown forces of reaction are illustrated. There are three unknown forces of reaction at the left support because it is fixed. There is one unknown force of reaction at the right support because that support is a roller.

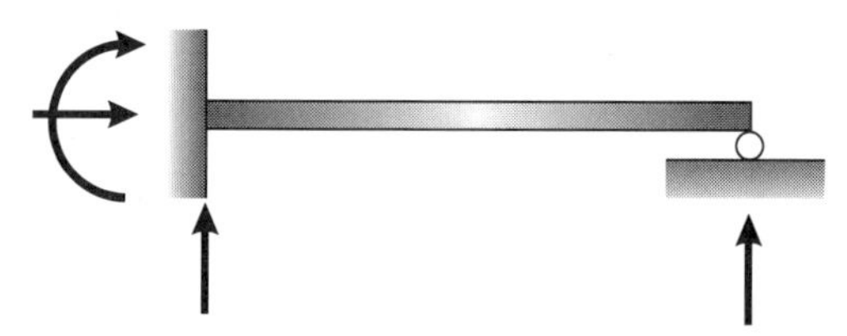

There are three applicable equations of static equilibrium: summation of forces vertically, summation of forces horizontally, and summation of moments. Because there are four unknown forces of reaction and three applicable equations of equilibrium, the beam is stable and statically indeterminate to the first degree externally—there is one more unknown than there are equations of static equilibrium. ■

The internal arrangement of some structures is such that one or more equations of condition are available. This often occurs when there are hinges or links in the structure. A special condition exists because the internal moment at the hinge, or at the ends of the link, must be zero regardless of the loading. The internal moment is zero because of the "frictionless" pin used to make the connection:—rotation cannot be transferred between the adjacent parts of the structure. A similar statement cannot be made for any continuous section of the beam. If the number of condition equations plus the three equations of static equilibrium is equal to the number of unknowns, the structure is statically determinate; if more, it is unstable; and if less, it is statically indeterminate.

EXAMPLE 4.5

Determine the statical classification of the propped cantilever beam shown in the figure. The unknown forces of reaction are illustrated. This beam is the same as in the last example except that now a link is used at the right support instead of a roller. There are three unknown forces of reaction at the left support because it is a fixed support. There is only one unknown force of reaction at the right side even though the link is pin-connected at the support. A link can only transmit force along its axis, therefore the direction in which the reaction force is acting is known.

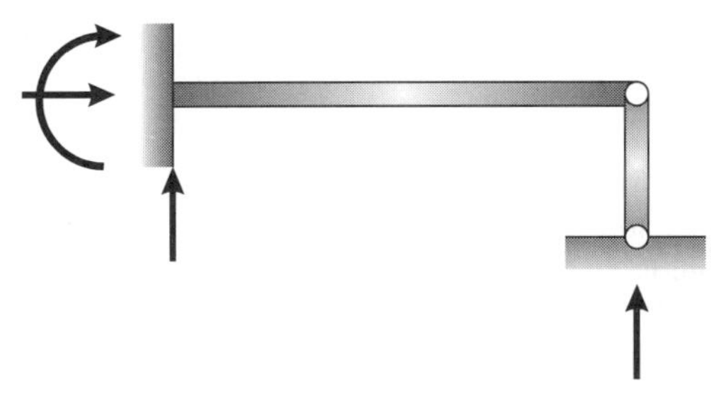

There are three applicable equations of static equilibrium: summation of forces vertically, summation of forces horizontally, and summation of moments. Because there are four unknown forces of reaction and three applicable equations of equilibrium, the beam is stable and statically indeterminate to the first degree externally—there is one more unknown force than there are equations of static equilibrium. ■

EXAMPLE 4.6

Determine the statical classification of the structural system shown in the figure.

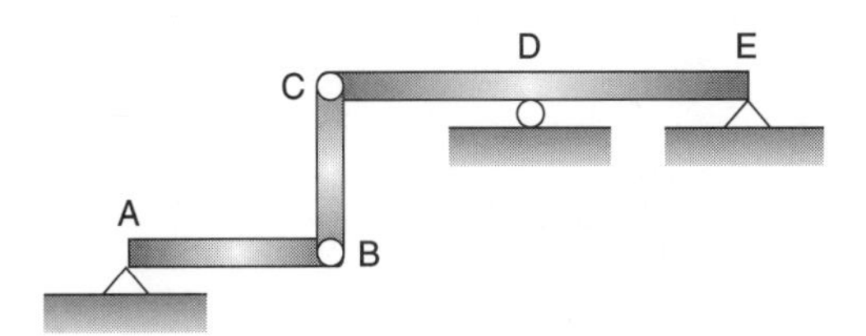

In this structural system, there are five unknown forces of reaction: 2 forces at A, 1 force at D, and 2 forces at E. On the surface, this structure appears to be statically indeterminate to the second degree because there are only three applicable equations of static equilibrium. However, because of the link connected to points B and C in the system, an equation of condition can be introduced. The pins at B and C are assumed frictionless so the moment at B and at C must be equal to zero. Given this condition, the two free-body diagrams shown here can be drawn.

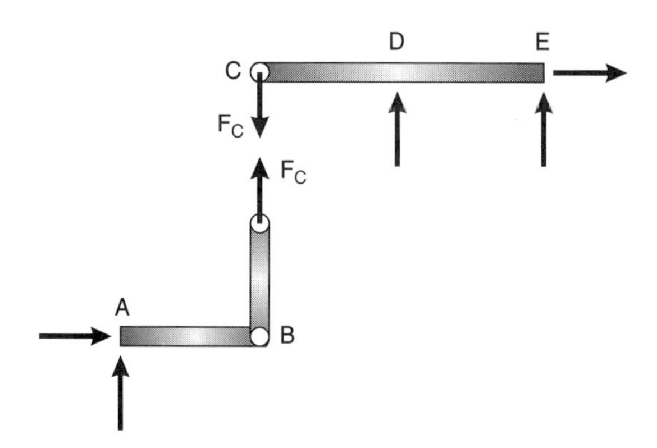

Observe that there are six unknown forces: the forces of reaction previously mentioned and the force in the pin at C. There are three applicable equations of static equilibrium for each of the free-bodies. As such, we have six equations of static equilibrium and six unknown forces. This structure is therefore stable and statically determinate. Note that the free bodies could be cut at B instead of C. ■

4.8 UNSTABLE EQUILIBRIUM AND GEOMETRIC INSTABILITY

The ability of a structure to adequately support the loads applied to it is dependent not only on the number of reaction components but also on the arrangement of those components. A structure can be unstable and yet be stable under a certain set of loads. The beam in Example 4.2 is an example of such a structure. This beam is supported on its ends with rollers and is unstable. The beam will slide laterally if any horizontal force is applied. However, the beam can support vertical loads and is stable if only vertical loads are applied. Although a structure may be stable under one arrangement of loads, if it is not stable under any other conceivable set of loads, it is unstable. This condition is sometimes referred to as *unstable equilibrium*.

It is also possible for a structure to have as many or more reaction components as there are equations available and yet still be unstable. This condition is referred to as *geometric instability*. The frame of Figure 4.3(a) has three reaction components and three equations available for their solution. However, a study of the moment at B shows that the structure is unstable. The line of action of the reaction at A passes through the reaction at B. Unless the line of action of the force P passes through the same point, the sum

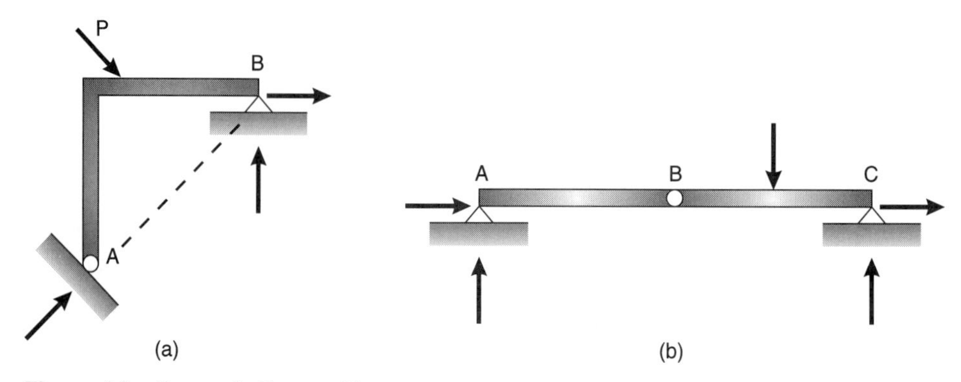

Figure 4.3 Geometrically unstable structures

of the moments about B cannot be equal to zero. There would be no resistance to rotation about B, and the frame would immediately begin to rotate. It might not collapse, but it would rotate until a stable situation were developed when the line of action of the reaction at A did not pass through B. Of prime importance to the engineer is for a structure to hold its position under load (even though it may deform). One that does not do so is unstable.

Another geometrically unstable structure is shown in Figure 4.3(b). Four equations are available to compute the four unknown forces of reaction—three equations of static equilibrium and one equation of condition. Nevertheless, rotation will instantaneously occur about the hinge at B. After a slight vertical deflection at A, the structure probably will become stable.

4.9 REACTIONS CALCULATED WITH EQUATIONS OF STATIC EQUILIBRIUM

Reaction calculations using the equations of static equilibrium are illustrated in Examples 4.7 to 4.9. When applying the equation for summation of moments, a point may usually be selected as the center of moments so that the lines of action of all but one of the unknown forces pass through that point. That unknown force can then be determined from the moment equation. The other components of reaction are found by summing forces vertically and horizontally.

The beam of Example 4.7 has three unknown reaction components: vertical and horizontal ones at A and a vertical one at B. Moments are taken about A to find the value of the vertical component at B. All of the vertical forces are equated to zero, and the vertical reaction component at A is found. A similar equation is written for the horizontal forces applied to the structure, and the horizontal reaction component at A is found to be zero.

The calculations for reactions may be checked by taking moments about another point on the structure, usually another support, as illustrated in Example 4.7. For future examples, space is not taken to show the checking calculations. *A problem, however, should be considered incomplete until a mathematical check of this nature is made.*

EXAMPLE 4.7

Compute the reaction components for the beam shown in the figure.

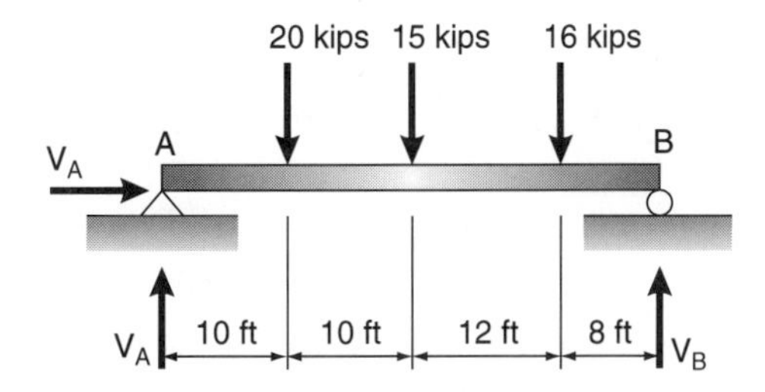

Solution. The forces of reaction and their assumed direction are shown on the figure. Begin the solution by summing forces horizontally to determine H_A.

$$\sum F_H = H_A = 0$$
$$\therefore H_A = 0$$

Next, sum moments clockwise about the left support. After doing so we obtain the equation

$$\sum M_A = 20(10) + 15(20) + 16(32) - V_B(40) = 0$$
$$\therefore V_B = 25.3 \text{ kips}$$

The result for V_B is positive so the assumed direction is correct; the reaction at B acts upward. Lastly, forces are summed vertically to compute the remaining reaction.

$$\sum F_V = V_A - 20 - 15 - 16 + V_B = 0$$
$$V_A - 20 - 15 - 16 + 25.3 = 0$$
$$\therefore V_A = 25.7 \text{ kips}$$

Again, the computed reaction at A is positive so the assumed direction is correct. We can sum moments about B to check our calculations.

$$\sum M_B = 25.7(40) - 20(30) - 15(20) - 16(8)$$
$$\sum M_B = 0$$

Because the summation of moments is equal to zero, the computed reactions are correct. ■

The beam in Example 4.8 is subjected to an inclined 50-kip load. For convenience when summing moments, this load will be replaced with its vertical and horizontal components. The structure must also support for a short distance a uniform load of 3 klf (kips or kilo-pounds per linear foot).

EXAMPLE 4.8

Find all of the reaction components in the structure shown in the figure.

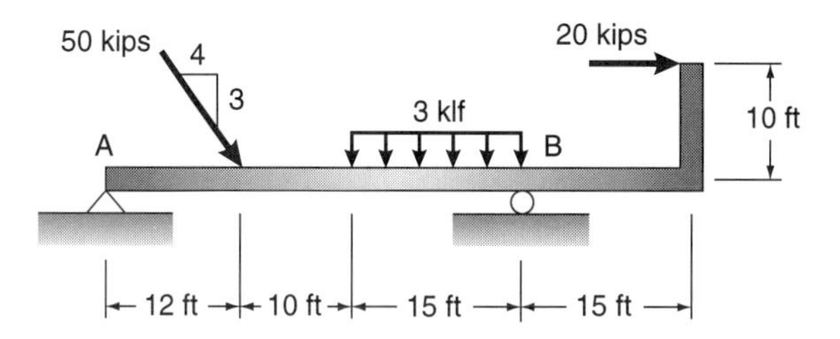

Solution. The free-body diagram for this beam is shown here. Observe that the inclined force has been replaced with its horizontal and vertical components, and the distributed load has been replaced with an equivalent concentrated load acting at its centroid.

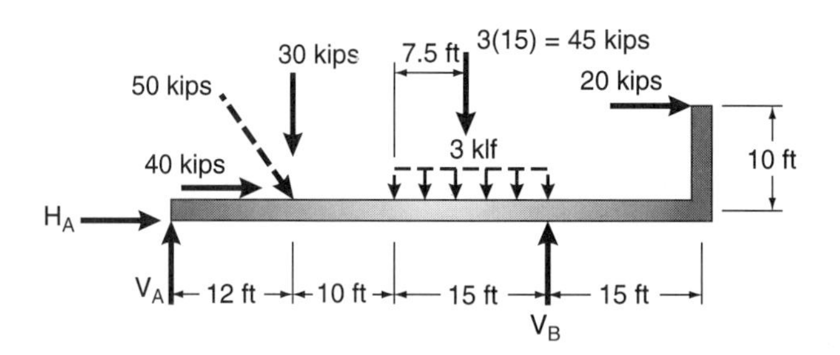

Begin by summing forces horizontally to compute the horizontal reaction at A.

$$\sum F_H = 40 + 20 + H_A = 0$$
$$\therefore H_A = -60 \text{ kips}$$

Observe that the computed value is negative so the actual direction of the reaction is opposite that assumed. Next we will sum moments clockwise about A to determine the vertical reaction at B.

$$\sum M_A = 30(12) + 45(29.5) - V_B(37) + 20(10) = 0$$
$$\therefore V_B = 51 \text{ kips}$$

Because the computed sign on V_B is positive the assumed direction is correct. We will sum forces vertically to compute the vertical reaction at A.

$$\sum F_V = V_A - 30 - 45 + V_B = 0$$
$$V_A - 30 - 45 + 51 = 0$$
$$\therefore V_A = 24 \text{ kips}$$

Again, the assumed direction of V_A is correct. ■

The roller in the frame in Example 4.9 is supported by an inclined surface. The direction of the reaction at the roller is known; it is perpendicular to the supporting surface. If the direction of the reaction is known, the relationship between the vertical component, the horizontal component, and the reaction itself is known. Here the reaction has a slope of four vertically to three horizontally (4:3), which is perpendicular to the slope of the supporting surface of three to four (3:4). Moments are taken about the left support, which gives an equation including the horizontal and vertical components of the reaction at the inclined roller. Both components are in terms of that reaction. Therefore only one unknown is present in the equation and its value is easily obtained.

EXAMPLE 4.9

Compute the reactions for the frame shown in the figure.

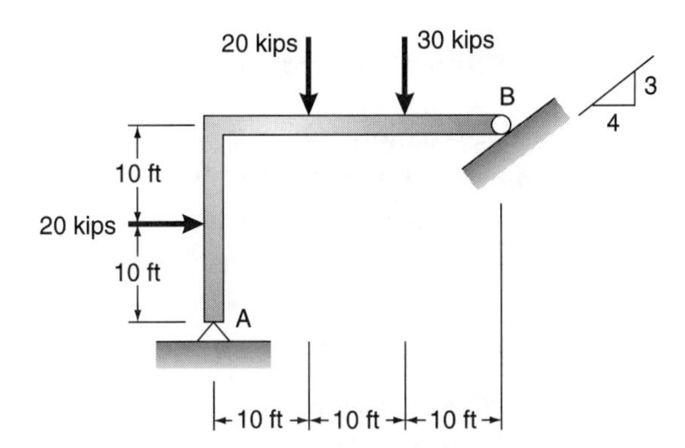

Solution. The free-body diagram for this frame is shown next. Observe that the inclined force of reaction at B has been replaced with its horizontal and vertical components. These components act at the same point as the reaction.

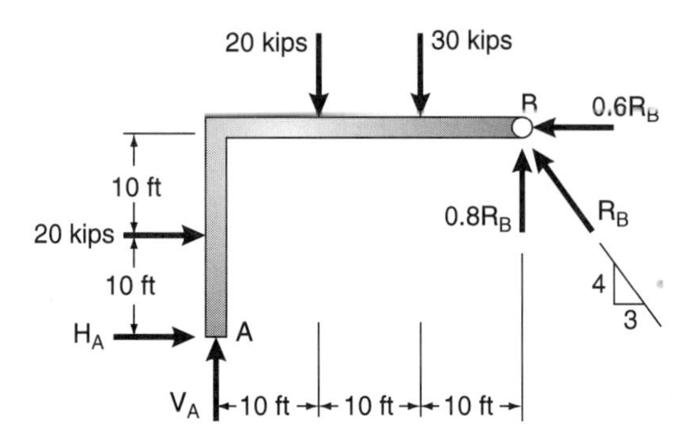

We will begin by summing moments about A to determine the reaction at B. In the calculation we will use the components of the reaction.

$$\sum M_A = 20(10) + 20(10) + 30(20) - 0.8R_B(30) - 0.6R_B(20) = 0$$
$$\therefore R_B = 27.8\,\text{kips}$$

The sign on the computed reaction at B is positive so the assumed direction is correct. Next we will sum forces vertically to obtain the vertical reaction at A.

$$\sum F_V = V_A - 20 - 30 + 0.8R_B = 0$$
$$V_A - 20 - 30 + 0.8(27.8) = 0$$
$$\therefore V_A = 27.8$$

Again, the sign on the computed reaction is positive so the assumed direction is correct. By summing the horizontal forces we can evaluate the horizontal reaction at A.

$$\sum F_H = H_A + 20 - 0.6R_B = 0$$
$$H_A + 20 - 0.6(27.8) = 0$$
$$\therefore H_A = -3.3\,\text{kips}$$

The sign on the computed reaction is negative so the force of reaction is actually acting to the left, opposite the direction indicated. ■

4.10 PRINCIPLE OF SUPERPOSITION

As we proceed with our study of structural analysis, we will encounter structures subject to large numbers of forces and to different kinds of forces (concentrated, uniform, triangular, dead, live, impact, etc.). To assist in handling such situations there is available an extremely useful tool called the *principle of superposition.* The principle of superposition can be stated as follows:

> If the structural behavior is linearly elastic, the forces acting on a structure may be separated or divided in any convenient fashion and the structure analyzed for the separate cases. The final results can then be obtained by adding algebraically the individual results.

This concept is graphically represented in Figure 4.4.

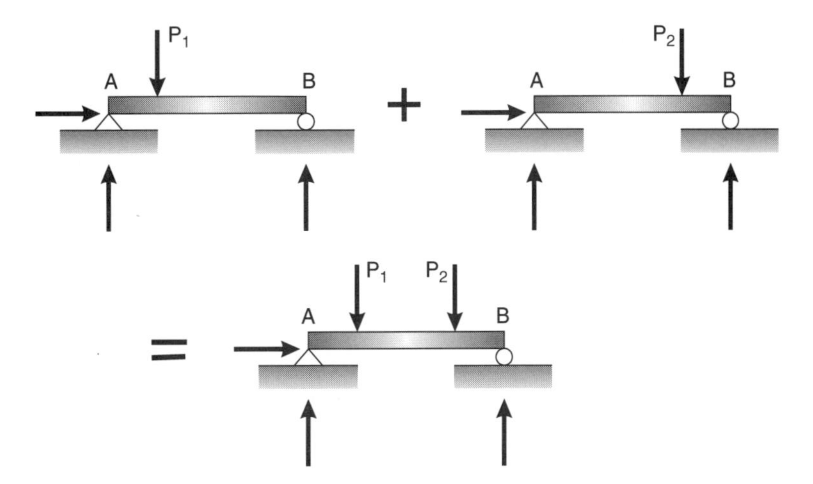

Figure 4.4 Principle of superposition

The principle of superposition is not valid in two important situations. The first occurs where the geometry of the structure is appreciably changed under the action of the loads. The second occurs where the structure consists of a material for which stresses are not directly proportional to strains. This latter situation can occur when the stress is beyond the elastic limit of the material. It can also occur when the material does not follow Hooke's law for any part of its stress-strain curve.

United Airlines hangar, San Francisco, California (Courtesy of the Lincoln Electric Company)

4.11 THE SIMPLE CANTILEVER

The simple cantilever in Example 4.10 has three unknown reaction components supporting it at the fixed end; they are the forces required to resist horizontal movement, vertical movement, and rotation. They may be determined with the equations of static equilibrium as illustrated.

EXAMPLE 4.10

Find all reaction components for the cantilever beam shown in the figure.

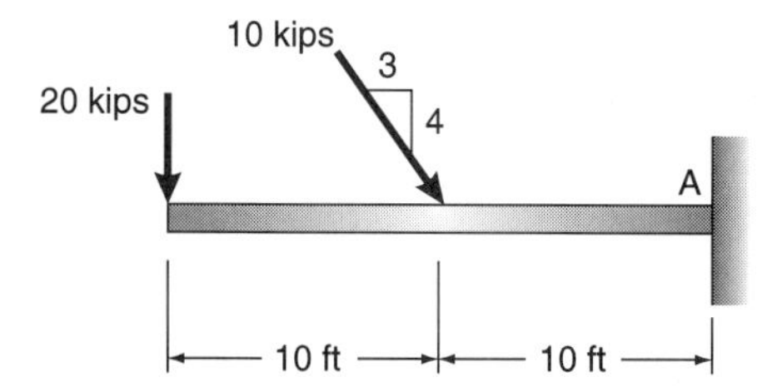

Solution. The free-body diagram used for the analysis is illustrated here.

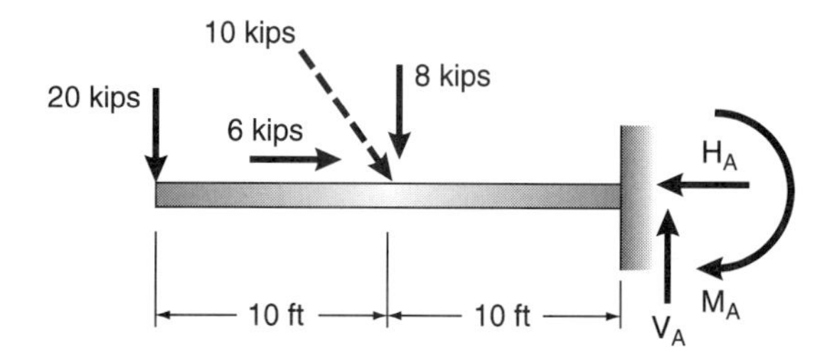

Summing forces vertically yields the vertical reaction at A

$$\sum F_V = -20 - 8 + V_A = 0$$
$$\therefore V_A = 28\,\text{kips}$$

Summing forces horizontally yields the horizontal reaction at A

$$\sum F_H = 6 - H_A = 0$$
$$\therefore H_A = 6\,\text{kips}$$

Finally, summing moments clockwise about A yields the rotational component of the reaction.

$$\sum M_A = -20(20) - 8(10) + M_A = 0$$
$$\therefore M_A = 480\,\text{k-ft} \quad \blacksquare$$

4.12 CANTILEVERED STRUCTURES

Internal forces in structures that are simply supported increase rapidly as their spans become longer. We will see that bending moment increases approximately in proportion to the square of the span length. Stronger and more expensive structures are required to resist the greater forces. For very long spans, bending moments can be so large that it becomes economical to introduce special types of structures that will reduce some of the internal bending moment. One special type of structure is a *cantilevered construction,* as illustrated in Figure 4.5(a).

A cantilevered structure as in Figure 4.5(a) can be used instead of the three simple beams shown in Figure 4.5(b). This is accomplished by making the beam continuous over the interior supports B and C, and by introducing hinges in the center span at C and D. An equation of condition (summation of moments is equal to zero) is available at each hinge. As such, there are five equations and five unknowns; the structure is statically determinate.

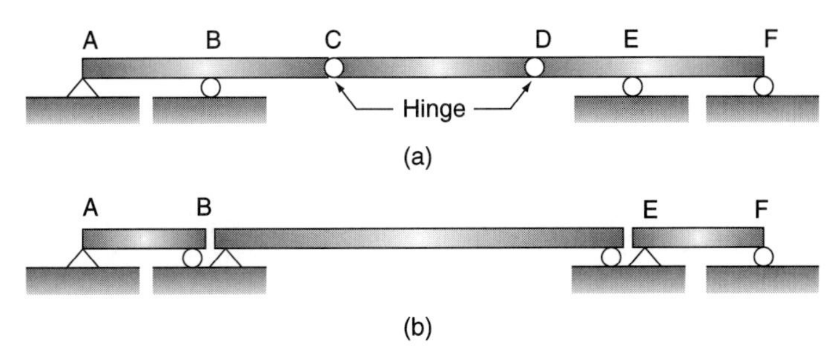

Figure 4.5 Cantilevered construction (a), versus three-beam construction (b)

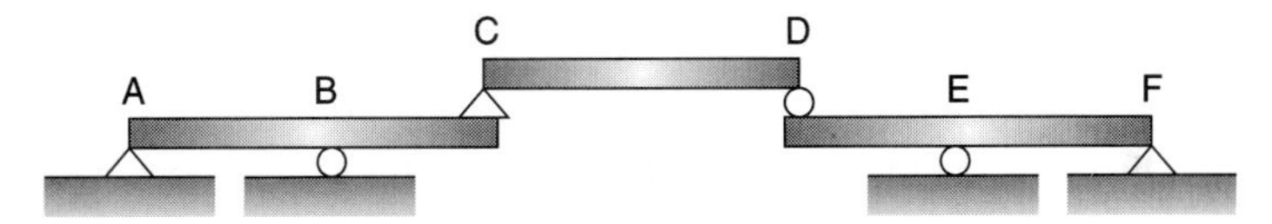

Figure 4.6

Cantilevered construction is essentially two simple beams, each with a cantilevered (overhanging) end, and a beam simply supported by the cantilevered ends (Figure 4.6).

The first step in determining the reactions for cantilevered structures is to isolate the center, simple beam and compute the forces necessary to support it at each end. Second, these forces are applied as downward loads on the respective cantilevers, and as a final step the end-beam reactions are determined individually. The next example illustrates the entire process.

EXAMPLE 4.11

Calculate all reactions for the cantilevered structure illustrated.

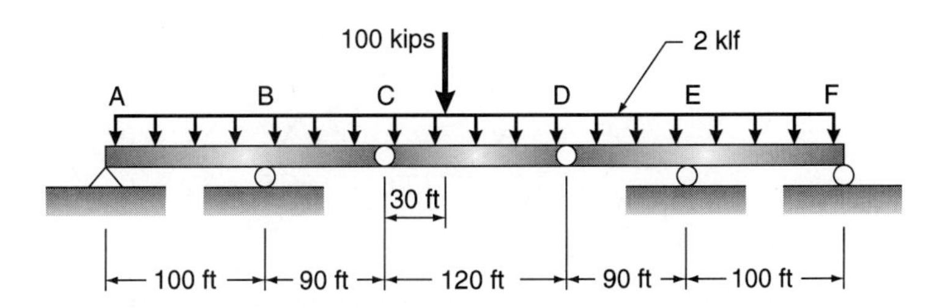

Solution. The free-body diagram that we will use is:

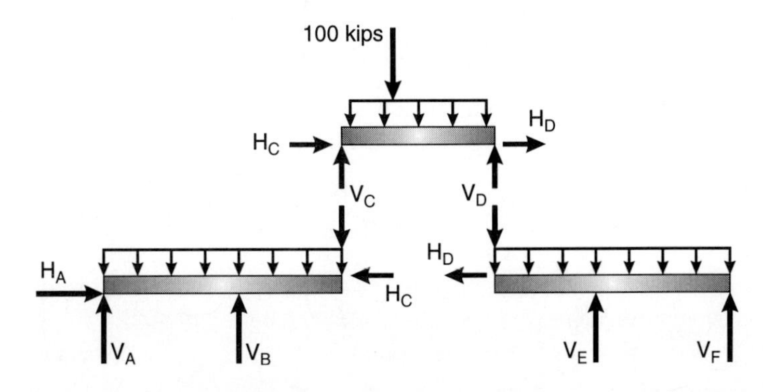

Because of the equations of condition that exist at C and D, we were able to isolate the center section from the two end sections with the forces acting at the hinge, as shown on the free-body. We can then use the free-body of the center section to compute the forces V_C and V_D,

$$\sum M_C = 100(30) + 2(120)\left(\frac{120}{2}\right) - V_D(120) = 0$$

$$\therefore V_D = 145\,\text{kips}$$

then by summing forces on the center section vertically, we can obtain V_D.

$$\sum F_V = V_C - 100 - 2(120) + V_D = 0$$

$$V_C - 100 - 2(120) + V_D = 0$$

$$\therefore V_C = 195\,\text{kips}$$

Finally, by summing forces horizontally on the center section we obtain a relationship between the horizontal forces.

$$\sum F_H = H_C + H_D = 0$$

$$\therefore H_C = -H_D$$

Now that we have computed the vertical forces acting on the center section, we will turn to the right section. First, sum moments counterclockwise about the right end.

$$\sum M_F = V_D(190) + 2(190)\left(\frac{190}{2}\right) - V_E(100) = 0$$

$$195(190) + 2(190)\left(\frac{190}{2}\right) - V_E(100) = 0$$

$$\therefore V_E = 636.5\,\text{kips}$$

We can then sum forces vertically and horizontally to obtain the other components of reaction on the right section.

$$\sum F_V = -V_D - 2(120) + V_E + V_F = 0$$

$$-145 - 2(120) + 636.5 - V_F = 0$$

$$\therefore V_F = -111.5\,\text{kips}$$

The negative sign indicates that the reaction is acting downward, opposite the direction assumed.

$$\sum F_H = H_D = 0$$

$$\therefore H_D = 0 = H_C$$

We will work with the left section to compute the remaining forces of reaction. Begin by summing moments clockwise about A.

$$\sum M_A = 2(190)\left(\frac{190}{2}\right) + V_C(190) - V_B(100) = 0$$

$$2(190)\left(\frac{190}{2}\right) + 195(190) - V_B(100) = 0$$

$$\therefore V_B = 731.5\,\text{kips}$$

The remaining vertical reaction is found by summing forces vertically.

$$\sum F_V = V_A - 2(190) - V_C + V_B = 0$$
$$V_A - 2(190) - 195 + 731.5$$
$$\therefore V_A = -156.5 \text{ kips}$$

Again, the negative sign indicates that the force of reaction is acting opposite that assumed. Finally, we compute the horizontal reaction.

$$\sum F_H = H_A - H_C = 0 = H_A - 0$$
$$\therefore H_A = 0 \quad \blacksquare$$

An examination of the reactions obtained in the previous example for the left-end and right-end sections of this beam show why cantilevered bridges often are called "seesaw" bridges. These structures are primarily supported by the first interior supports on each end, where the reactions are quite large. The end supports may very well have to provide downward reaction components. Thus, an end section of a cantilevered structure seems to act as a seesaw over the first interior support with downward loads on both sides.

4.13 ARCHES

Historically, arches were the only feasible form that could be used for large structures made up of materials with negligible tensile strength, such as bricks and stones. Masonry arches of such materials have been used for thousands of years.

In effect, an arch takes vertical loads and turns them into thrusts that run around the arch and put the elements of the arch in compression, as shown in Figure 4.7. The parts of a stone arch are called *voussoirs*. As you can see in the inset, these are stones in the shape of truncated wedges. They are pushed together in compression.

Arches are very rigid, stable structures that are not appreciably affected by movements of their foundations. It is rather interesting to note that excavations of ancient ruins show that arches are the structures that generally have survived best.

Theoretically, an arch can be designed for a single set of gravity loads so that only axial compression stresses are developed in it. Unfortunately, however, in practical structures the loads change and move so that bending stresses are developed in the arch. Nevertheless, arches generally are designed so that their predominant loading primarily causes compression stresses.

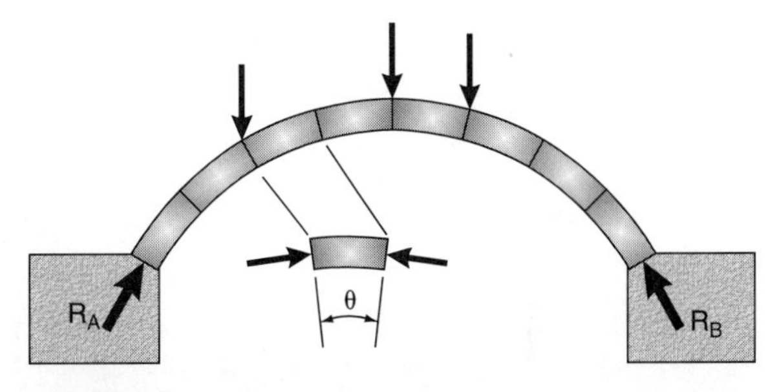

Figure 4.7 A stone arch

Arch structures have horizontal and vertical components of reaction at each support under vertical loading. The horizontal forces are produced because arches tend to become flat under load—the ends tend to move away from each other. As such, arches must be fixed against horizontal movement at their supports. There is an old saying to the effect that "an arch never sleeps." The thought is that the horizontal reactions are always present, even when no live loads are applied, because of the structure's own weight.

Structural arch construction: U.S.A.F. hangar, Edwards Air Force Base (Courtesy Bethlehem Steel Corporation)

4.14 THREE-HINGED ARCHES

Arches may be constructed with two or three hinges; very rarely are they constructed with only one hinge. Quite often in reinforced concrete construction, arches are constructed without any hinges. The three-hinged arch is discussed because it is the only statically determinate arch.

Examination of the three-hinged arch pictured in Figure 4.8 reveals two reaction components at each support, for a total of four. Three equations of static equilibrium and one condition equation are available to find the unknowns. The equation of condition is the summation of moments about the crown hinge from either the left or the right.

The arch in Example 4.12 is analyzed by taking moments about one of the supports to obtain the vertical reaction component at the other support. Because the supports are on the same level, the horizontal reaction component at the second support passes through the point where moments are being taken. The horizontal reaction components are obtained by taking moments at the crown hinge of the forces either to the left or to the right. The only unknown appearing in either equation is the horizontal reaction component on

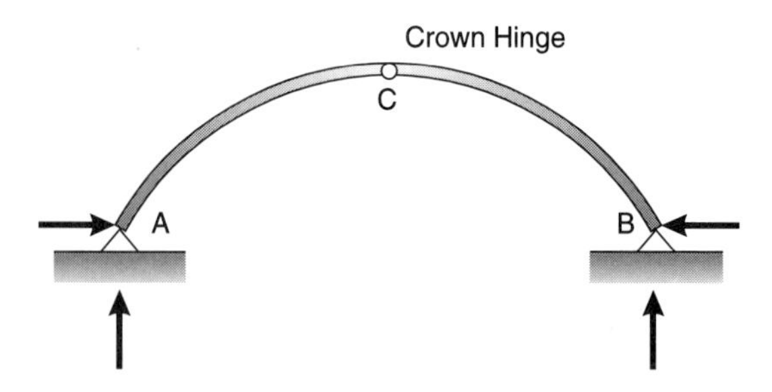

Figure 4.8 Three-hinged arch

that side, and the equation is solved for its value. Once the reactions are determined, it is easy to compute the moment and axial force in the arch at any point using equations of static equilibrium.

Laminated timber beams, Maumee, Ohio (Courtesy of Unit Structures, Inc.)

EXAMPLE 4.12

Find all reaction components for the three-hinged arch shown in the figure.

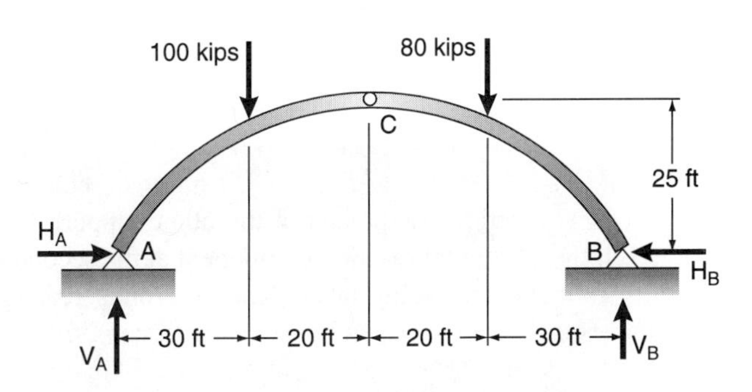

Solution. We will begin the solution by summing moments about the left support.

$$\sum M_A = 100(30) + 80(70) - V_B(100) = 0$$
$$\therefore V_B = 86\,\text{kips}$$

Next, we will sum forces vertically to compute the vertical component of reaction at the left support.

$$\sum F_V = V_A - 100 - 80 + V_B = 0$$
$$V_A - 100 - 80 + 86 = 0$$
$$\therefore V_A = 94\,\text{kips}$$

By summing forces horizontally, we find that the two horizontal components of reaction are equal to each other, but we cannot determine their magnitude.

$$\sum F_H = H_A - H_B$$
$$\therefore H_A = H_B$$

To compute the magnitude of the horizontal components of reactions, we will need to use the free-body shown in the following diagram and sum moments about C. This free-body was obtained from the equation of condition that finds moment at C is equal to zero.

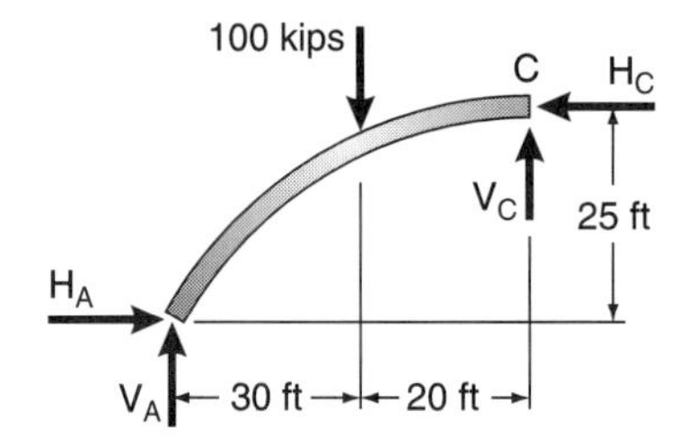

$$\sum M_C = V_A(50) - H_A(25) - 100(20) = 0$$
$$94(50) - H_A(25) - 100(20) = 0$$
$$\therefore H_A = 108\,\text{kips} = H_B \quad \blacksquare$$

The computation of reactions for the arch of Example 4.13 is slightly more complicated because the supports are not on the same level. Summing moments about one of the supports results in an equation involving both the horizontal and vertical components of reaction at the other support. Moments may be taken about the crown hinge of the forces on the same side as those two unknowns. The resulting equation contains the same two unknowns. Solving the equations simultaneously yields the magnitude of those reactions and then, by summing forces vertically and horizontally, the magnitude of the remaining reaction components can be determined.

North Dakota State Teachers College fieldhouse, Valley City, North Dakota (Courtesy of the American Institute of Timber Construction)

EXAMPLE 4.13

Determine the components of reaction for the structure shown in the figure below.

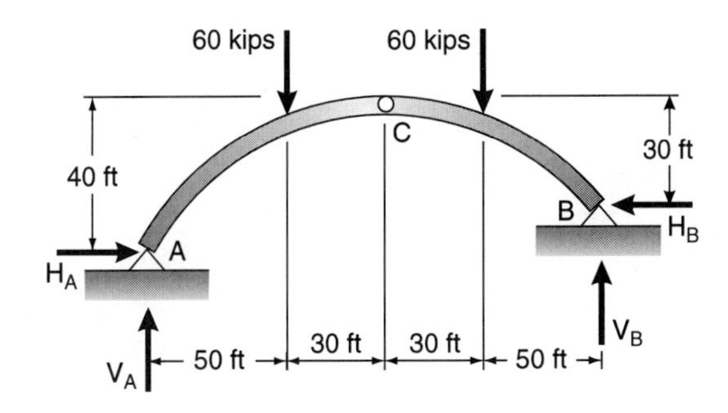

Solution. We begin by summing moments clockwise about A.

$$\sum M_A = 60(50) + 60(110) - H_B(10) - V_B(160) = 0$$
$$10H_B + 160V_B = 9600$$

We then utilize the equation of condition at C to obtain the free-body shown next. Using this free-body, we sum moments about C.

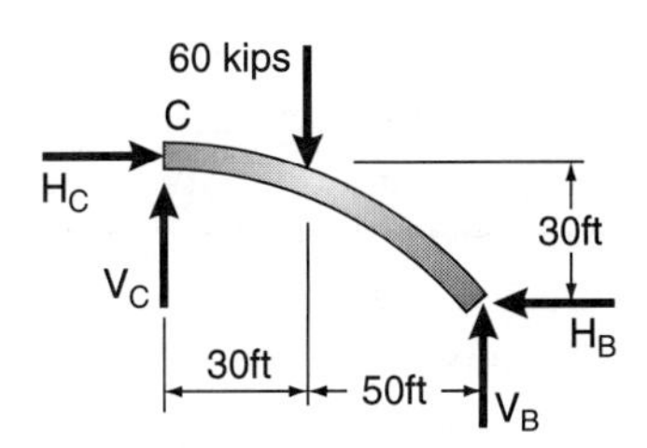

$$\sum M_C = 60(30) + H_B(30) - V_B(80) = 0$$
$$30H_B - 80V_B = -1800$$

The two simultaneous linear equations that resulted from these two summations of moment can be represented as

$$\begin{bmatrix} 10 & 160 \\ 30 & -80 \end{bmatrix} \begin{Bmatrix} H_B \\ V_B \end{Bmatrix} = \begin{Bmatrix} 9600 \\ -1800 \end{Bmatrix}$$

Upon solution of this matrix equation we find that

$$H_B = 85.7\,\text{kips}$$
$$V_B = 54.6\,\text{kips}$$

We can now sum forces vertically and horizontally to compute the remaining components of reaction.

$$\sum F_H = H_A - H_B = H_A - 85.7 = 0$$
$$\therefore\ H_A = 85.7\,\text{kips}$$

$$\sum F_V = V_A - 60 - 60 + V_B = V_A - 60 - 60 + 54.6 = 0$$
$$\therefore\ V_A = 65.4\,\text{kips}$$ ■

Glued laminated three-hinged arches, The Jai Alai Fronton, Riviera Beach, Florida (Courtesy of the Forest Products Association)

4.15 USES OF ARCHES AND CANTILEVERED STRUCTURES

Three-hinged steel arches are used for short and medium-length bridges with spans of up to approximately 600 ft. They are used for buildings when large clear spans are required underneath, as for hangars, field houses, and armories. Two-hinged steel arches generally are economical for bridges from 600 to 900 ft in length, with a few exceptional spans being over 1600 ft long. Reinforced-concrete arches without hinges are used for bridges with spans ranging from 100 to 400 ft. Cantilever-type bridges are used for spans from approximately 500 ft up to very long spans, such as the 1800-ft center span of the Quebec Bridge.

An arch is a structure that requires foundations capable of resisting the large horizontal reaction components, called thrusts, at the supports. In arches for buildings, it is possible to carry the thrusts by tying the supports together with steel rods, with steel sections, or even with specially designed floors. Arches so constructed are referred to as *tied arches*. For many locations with poor foundation conditions, and thereby the possibility of settlement, the three-hinged arch is selected over the statically indeterminate arches because the forces will not change in a three-hinged arch when settlement occurs. We will see in later chapters that foundation settlement can cause severe stress changes in statically indeterminate structures. Ease of erection is another advantage of three-hinged arches. It often is convenient to assemble and ship the two halves of a precast-concrete, structural steel, or laminated-timber arch separately and assemble them on the site.

The fact that cantilever-type construction reduces bending moments for long spans has been previously demonstrated. Arch-type construction also reduces moments, because the reactions at the supports tend to cause bending in the arch in a direction opposite to that caused by the downward loads. Because of this characteristic of having small bending moments, arches were admirably suited for masonry construction as practiced by the ancient builders.

4.16 CABLES

Cables provide perhaps the simplest means for supporting loads. They are used for supporting bridge and roof systems, as guys for derricks, radio towers, and similar type structures, as well as for many other applications. To the student, the most common use may seem to be the cable car systems at hundreds of ski slopes around the world.

Steel cables are economically manufactured from high-strength steel wire, providing perhaps the lowest cost-to-strength ratio of any common structural members. Further,

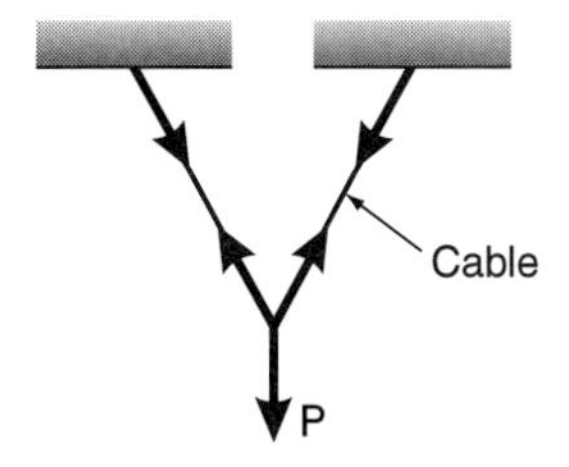

Figure 4.9 A simple cable structure

they are easily handled and positioned, even for very long spans. For the discussion to follow, the cable weight is neglected. When a cable of a given length is suspended between two supports, the shape it takes is determined by the applied loads.

The shape cables take in resisting loads is called a *funicular* curve. You may have noticed that the cable car systems in Europe often are called *funiculars*. Cables are quite flexible and support their loads in pure tension as shown in Figure 4.9. It can be seen in this figure that the load P must be balanced by the vertical components of tension in the cable; thus, the cable must have a vertical projection in order to support the load. The greater the vertical projection, the smaller will be the cable tension. If the cable moves or if other loads are applied, the cable will change shape.

The resultant tension at any point can be obtained from the following equation:

$$T = \sqrt{H^2 + V^2} \qquad \textbf{Eq. 4.5}$$

In this equation, H and V are, respectively, the horizontal and vertical components of tensile force in the cable at that point. We can see from this equation that the tension varies along the length of the cable. However, if only vertical loads are present the value of H will be constant throughout the cable and the maximum tensile force occurring can be determined by substituting the maximum value of the vertical force into the equation.

Cables are assumed to be so flexible that they cannot resist bending or compression; they act in direct tension. An equation of condition is available for analysis: the summation of moments to the left or right at any point along the cable is equal to zero. Should the position or sag of a cable at a particular point be known, the reactions at the cable ends and the sag at any other point in the cable can be determined with these equations. A numerical example follows. The weight of the cable is assumed to be negligible in this case.

Cable-stayed Sitka Harbor Bridge, Sitka, Alaska (Courtesy of the Alaska Department of Transportation)

EXAMPLE 4.14

Determine the reactions for the cable in the figure and the sag at the 40-kip load.

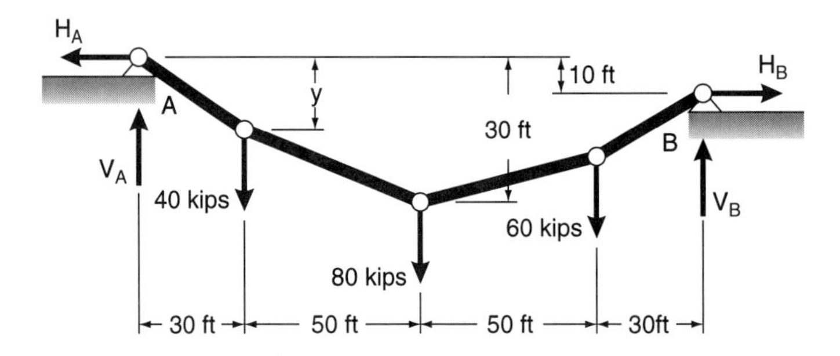

Solution. Begin the solution by summing moments clockwise about the right reaction.

$$\sum M_B = V_A(160) - H_A(10) - 40(130) - 80(80) - 60(30) = 0$$
$$160V_A - 10H_A = 13{,}400$$

Next we will utilize the equation of condition and sum moments of the forces to the left of the 80-kip load.

$$\sum M_{80} = V_A(80) - H_A(30) - 40(50) = 0$$
$$80V_A - 30H_A = 2000$$

The equations that resulted from these two moment summations can be represented in matrix form and solved for the reaction components at the left support.

$$\begin{bmatrix} 160 & -10 \\ 80 & -30 \end{bmatrix} \begin{Bmatrix} V_A \\ H_A \end{Bmatrix} = \begin{Bmatrix} 13{,}400 \\ 2{,}000 \end{Bmatrix}$$
$$\begin{bmatrix} V_A \\ H_A \end{bmatrix} = \begin{bmatrix} 160 & -10 \\ 80 & -30 \end{bmatrix}^{-1} \begin{Bmatrix} 13{,}400 \\ 2{,}000 \end{Bmatrix} = \begin{Bmatrix} 95.5 \\ 188.0 \end{Bmatrix}$$

The vertical reaction at the left support is equal to 95.5 kips and the horizontal reaction at that support is equal to 188 kips. To determine the other two reaction components we can sum forces vertically and horizontally.

$$\sum F_V = V_A - 40 - 80 - 60 + V_B = 0$$
$$95.5 - 40 - 80 - 60 + V_B = 0$$
$$\therefore V_B = 84.5 \text{ kips}$$

$$\sum F_H = -H_A + H_B = 0$$
$$-188 + H_B = 0$$
$$\therefore H_B = 188 \text{ kips}$$

Lastly, we can use the equation of condition again and determine the sag at the 40-kip load by summing moments to the left of the 40-kip load.

$$\sum M_{40} = V_A(30) - H_A y = 0$$
$$95.5(30) - 188y = 0$$
$$\therefore y = 15.24 \text{ ft.}$$

The sag, y, at the location of the 40-kip load is 15.24 ft. ■

Very often, the actual geometry of the cable structure will not be known initially. We may know only the total length of the cable and the locations at which the loads are to be applied. From this information, we must determine the geometry of the cable structure before we can determine the forces that exist in the cable. The following example demonstrates such a situation.

EXAMPLE 4.15

For the cable structure shown in the figure, determine the forces of reaction, the tension in each segment of the cable, and the sag in the cable. The total length of the cable is 65 ft.

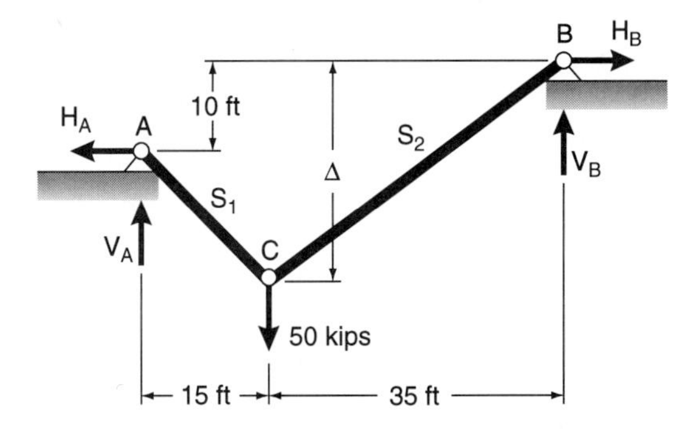

Solution. Before the forces of reaction can be computed, the sag in the cable, Δ, must be computed. Using the Pythagorean theorem, equations for the sag can be written in terms of the left and right cable segments. These equations are

$$\Delta = 10 + \sqrt{S_1^2 - 15^2} \quad \text{in terms of the left segment}$$

$$\Delta = \sqrt{S_2^2 - 35^2} \quad \text{in terms of the right segment}$$

Both of these equations can be expressed in terms of one of the segment lengths, S_1 or S_2. To do so, we must solve for the length of one segment in terms of the other. We will solve for S_1 in terms of S_2, which yields:

$$S = S_1 + S_2 = 65\,\text{ft}$$

$$\therefore S_1 = 65 - S_2$$

If we substitute this expression for S_1 into the equations for sag, we obtain the equations:

$$\Delta = 10 + \sqrt{(65 - S_2)^2 - 15^2}$$

$$\Delta = \sqrt{S_2^2 - 35^2}$$

We can iterate on S_2 until the sag we compute using both equations is the same. Upon doing so we find that

$$S_2 = 43.361\,\text{ft}$$

$$\Delta = 25.596\,\text{ft}$$

We can now solve for the components of reaction and the force in each segment of the cable. By summing moments clockwise about the left support, we obtain the equation

$$\sum M_A = H_B(10) - V_B(15 + 35) + 50(15) = 0$$

$$10H_B - 50V_B = -750$$

We obtain the following equation by summing forces horizontally:

$$\sum F_H = -H_A + H_B = 0$$
$$-H_A + H_B = 0$$

By summing forces vertically, we obtain:

$$\sum F_V = V_A + V_B - 50 = 0$$
$$V_A + V_B = 50$$

Lastly, we can use the equation of condition that exists at C. At that location on the cable the bending moment is equal to zero. As such, if we sum moments in the right segment about C we obtain the equation

$$\sum M_{C(Right)} = H_B(\Delta) - V_B(35) = 0$$
$$25.596H_B - 35V_B = 0$$

These four equations can be represented as a matrix equation of the form

$$\begin{bmatrix} 0 & 0 & 10 & -50 \\ -1 & 0 & 1 & 0 \\ 0 & 1 & 0 & 1 \\ 0 & 0 & 25.596 & -35 \end{bmatrix} \begin{Bmatrix} H_A \\ V_A \\ H_B \\ V_B \end{Bmatrix} = \begin{Bmatrix} -750 \\ 0 \\ 50 \\ 0 \end{Bmatrix}$$

From this equation, we can solve for the forces of reaction, which are:

$$\begin{Bmatrix} H_A \\ V_A \\ H_B \\ V_B \end{Bmatrix} = \begin{bmatrix} 0 & 0 & 10 & -50 \\ -1 & 0 & 1 & 0 \\ 0 & 1 & 0 & 1 \\ 0 & 0 & 25.596 & -35 \end{bmatrix}^{-1} \begin{Bmatrix} -750 \\ 0 \\ 50 \\ 0 \end{Bmatrix} = \begin{Bmatrix} 28.23 \\ 29.35 \\ 28.23 \\ 20.65 \end{Bmatrix}$$

Using the free-body diagram shown in the following figure we can compute the tension in the left cable segment.

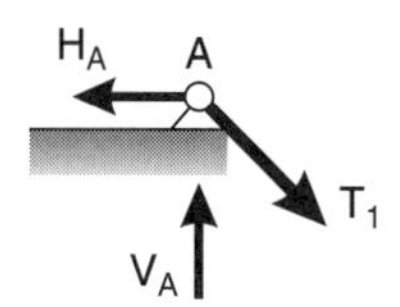

$$T_1 = \sqrt{H_A^2 + V_A^2} = \sqrt{28.23^2 + 29.35^2} = 40.73\,\text{kips}$$

Using a similar free-body diagram, we find the tension in the right cable segment to be

$$T_2 = \sqrt{H_B^2 + V_B^2} = \sqrt{28.23^2 + 20.65^2} = 34.98\,\text{kips} \quad \blacksquare$$

The distortion, or deflection, of most structures is assumed to be negligible when computing the forces produced in those structures. Such an assumption is not correct, however, for many cable structures, particularly the flat ones where a little sag can drastically affect cable tensions. This topic is not considered herein, but is described very well in a book by Firmage.[1] Flat cables cause very large horizontal reaction components and thus have very high tensile forces.

[1] D. A. Firmage, *Fundamental Theory of Structures* (New York: Wiley, 1963), 258–265.

A cable supporting a load that is uniform along its length, such as a cable loaded only by its own weight, will take the form of a catenary. On many occasions, concentrated loads are applied to cables by hangers. If they are closely spaced, the loading will approach that of a uniform load along the horizontal projection of the cable. Cables supporting the roadway of suspension bridges usually are assumed to fall into this class. The analysis of such cables is presented in Appendix A.

4.17 PROBLEMS FOR SOLUTION

For Problem 4.1, determine which of the structures shown in the accompanying illustration are statically determinate, statically indeterminate (including the degree of indeterminacy), and unstable in regards to outer forces.

4.1

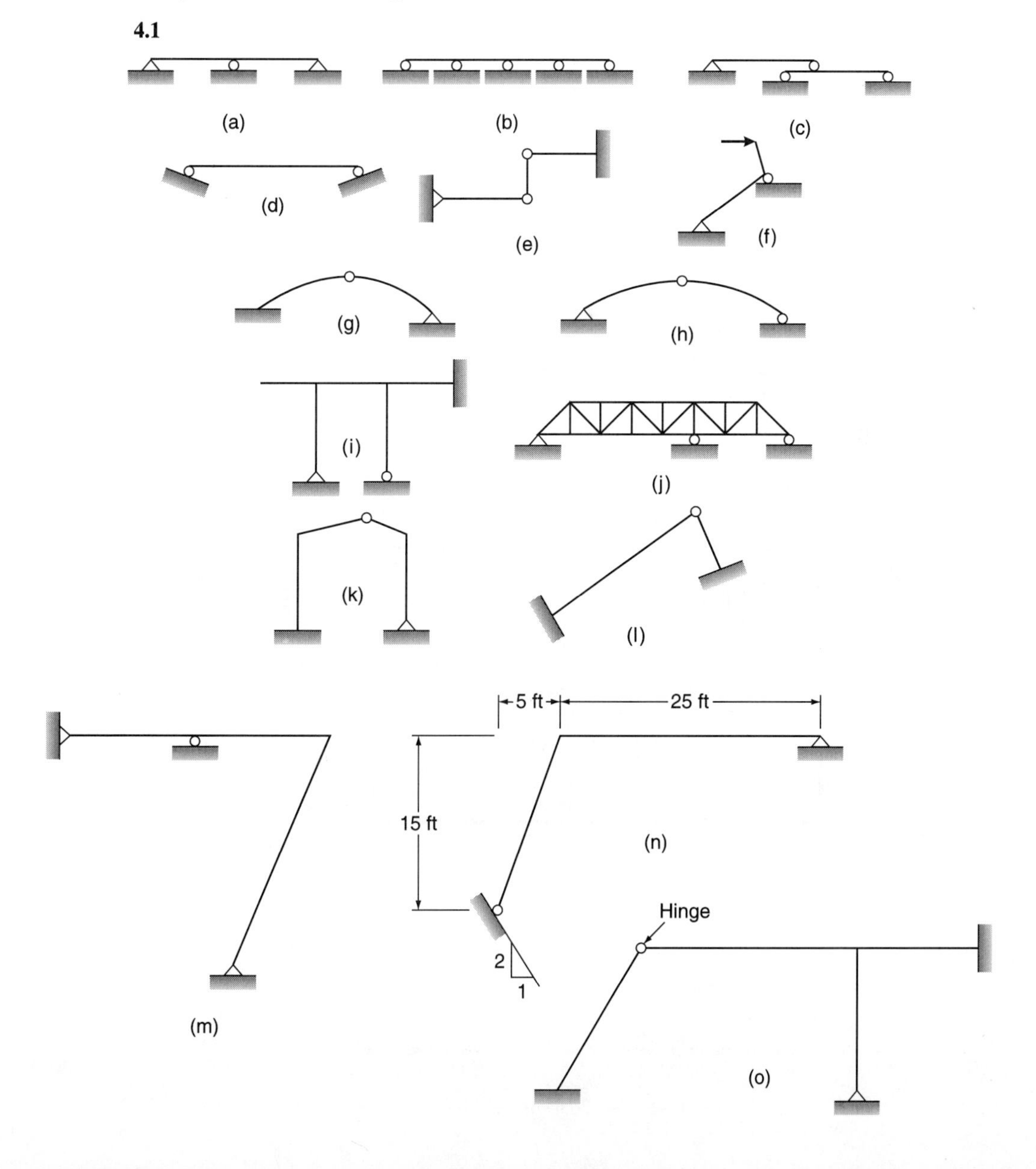

For Problems 4.2 to 4.46, compute the reactions for the structures.

4.2 (*Ans.* $V_L = 65.71$ k ↑, $V_R = 94.29$ k ↑)

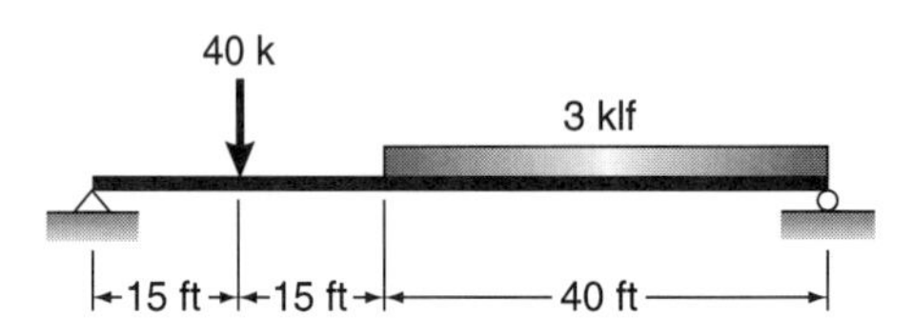

4.3

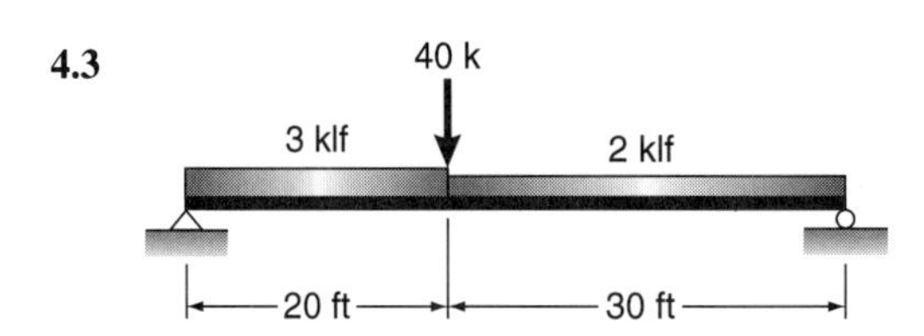

4.4 (*Ans.* $V_L = 28.33$ k ↑, $V_R = 21.67$ k ↑)

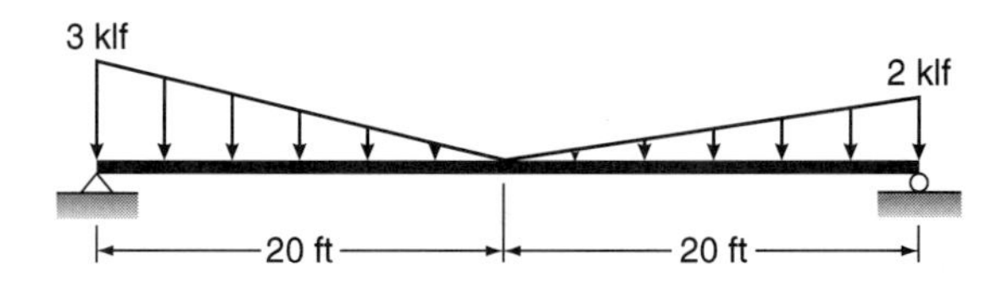

4.5

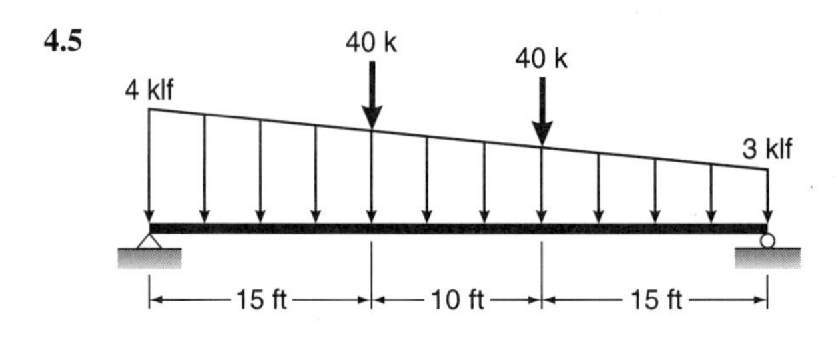

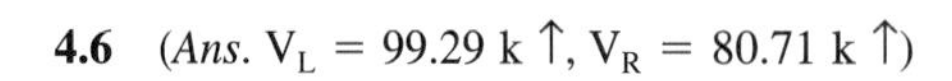

4.6 (*Ans.* $V_L = 99.29$ k ↑, $V_R = 80.71$ k ↑)

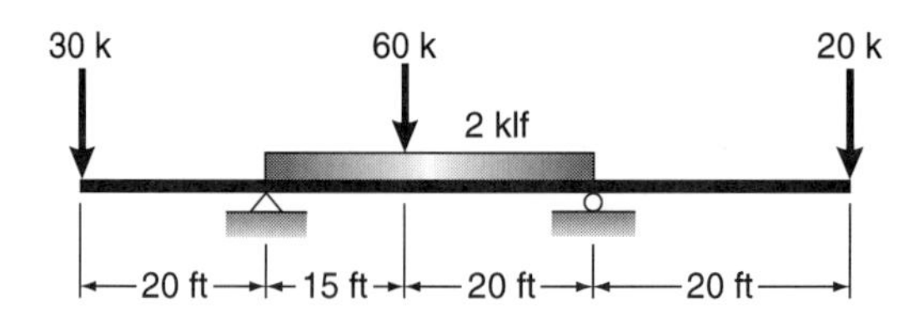

4.7

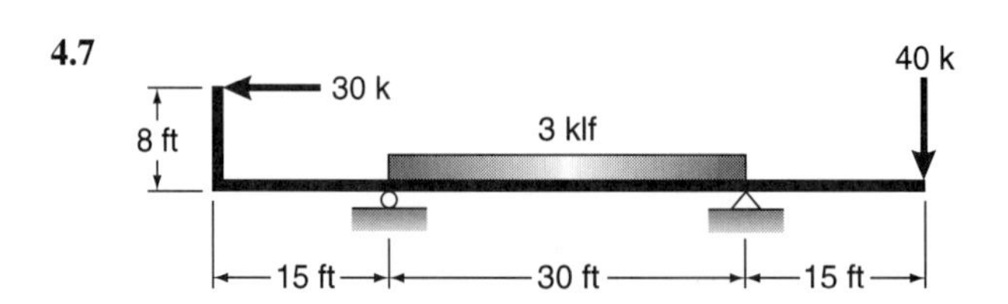

4.8 (*Ans.* $V_L = 115.82$ k ↑, $V_R = 5.10$ k ↑, $H_R = 66.97$ k →)

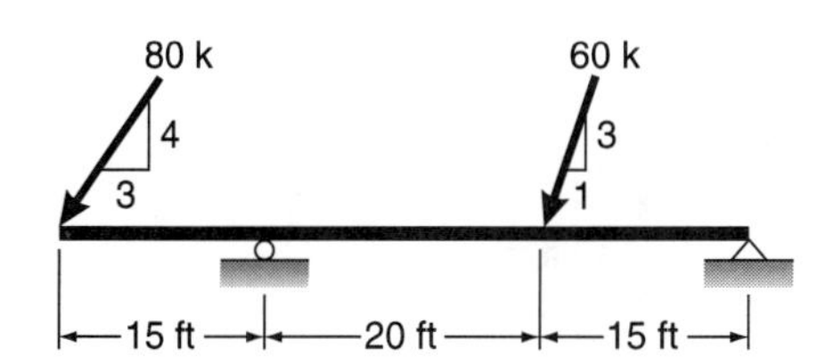

4.9

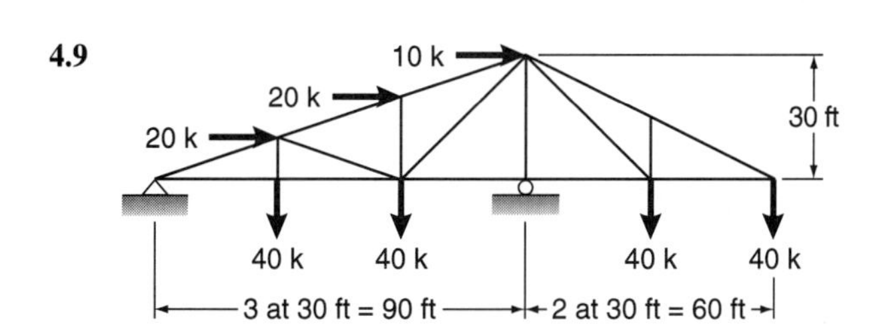

4.10 (*Ans.* $H_A = 240$ k →, $H_B = 240$ k ←, $V_B = 150$ k ↑)

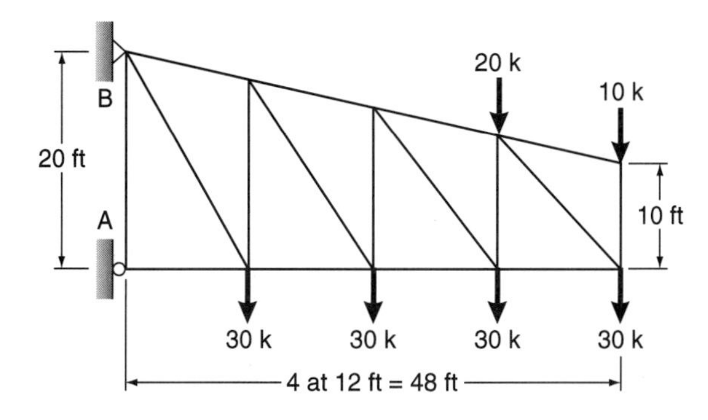

4.11

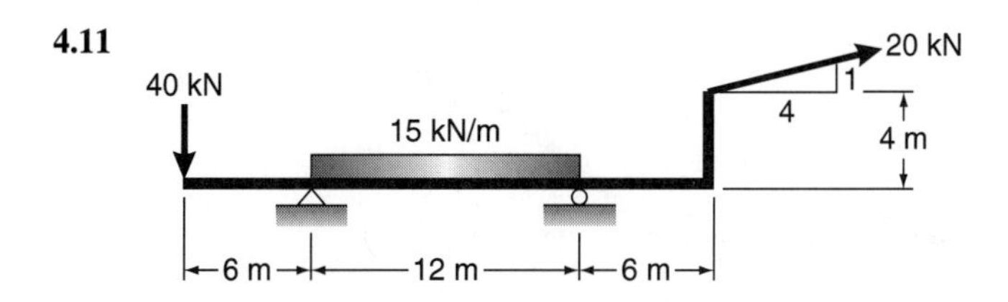

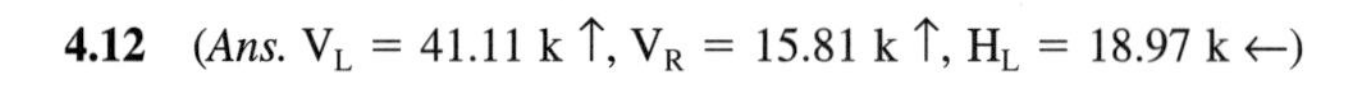
4.12 (*Ans.* $V_L = 41.11$ k ↑, $V_R = 15.81$ k ↑, $H_L = 18.97$ k ←)

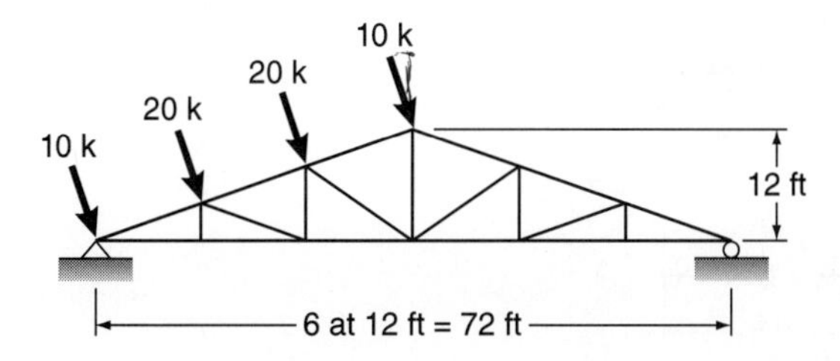

4.13

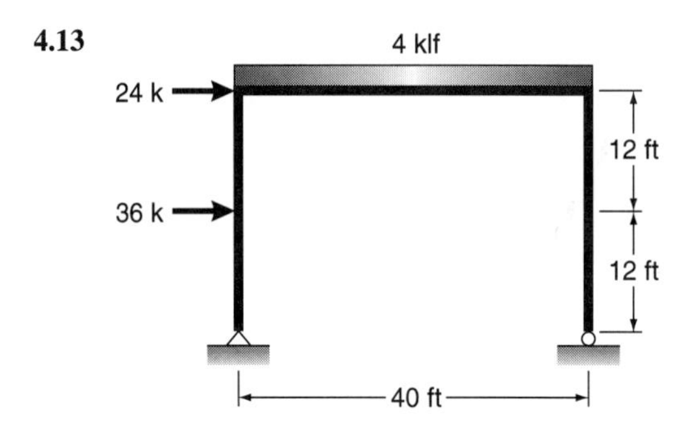

4.14 (*Ans.* $V_L = 78.75$ k ↑, $V_R = 71.25$ k ↑, $H_R = 45$ k ←)

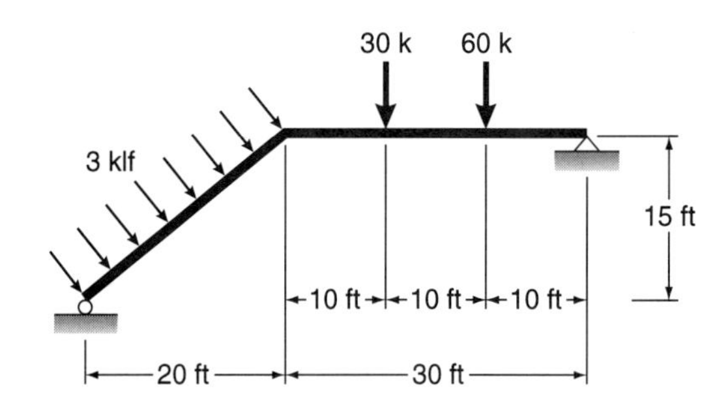

4.15

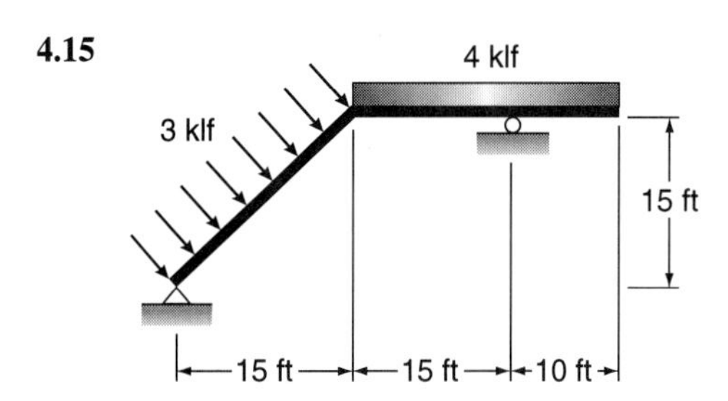

4.16 (*Ans.* $V_L = 43.64$ k ↑, $H_L = 12.27$ k →, $V_R = 16.36$ k ↑, $H_R = 12.27$ k ←)

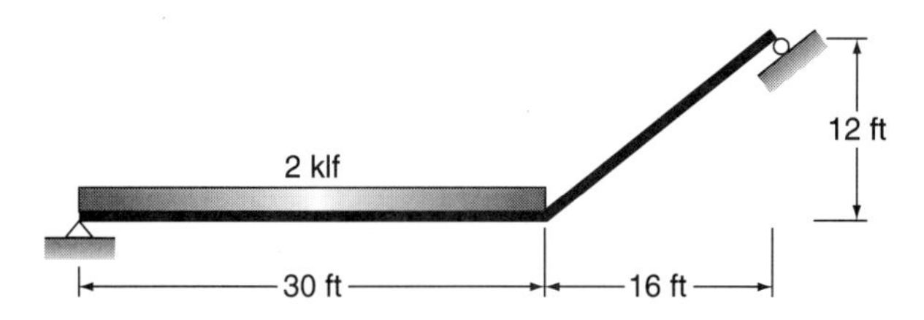

4.17

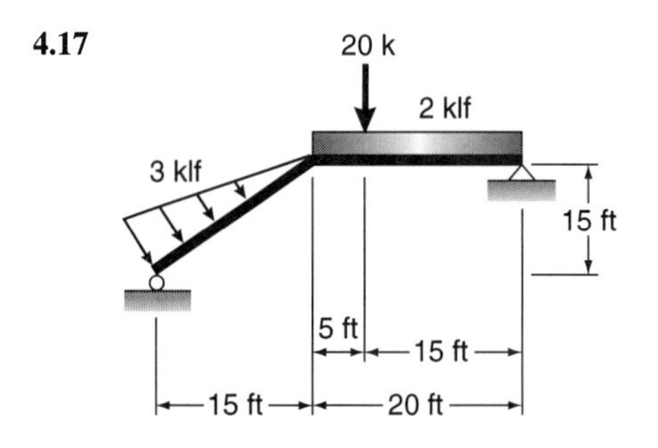

4.18 (*Ans.* $V_L = 127.09$ k ↑, $H_L = 19.07$ k ←, $V_R = 56.72$ k ↑, $H_R = 28.36$ k ←)

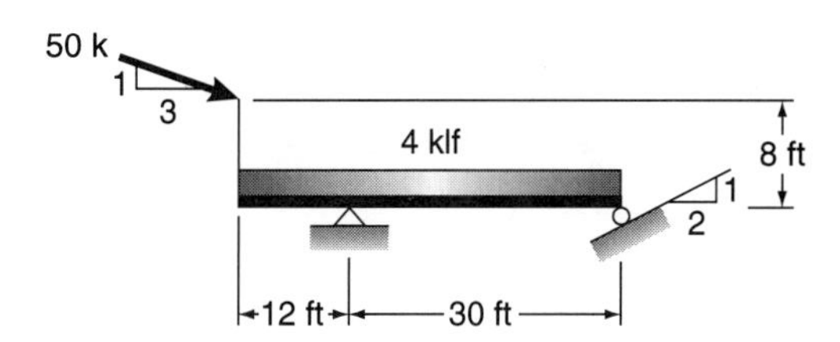

4.19

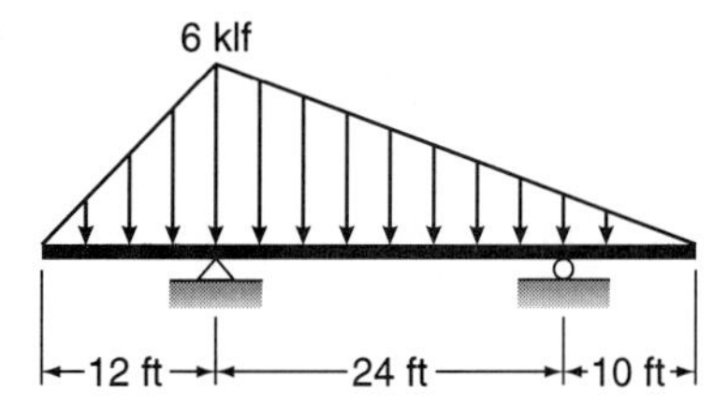

4.20 (*Ans.* $V_L = 58.32$ k ↑, $V_R = 16.67$ k ↑)

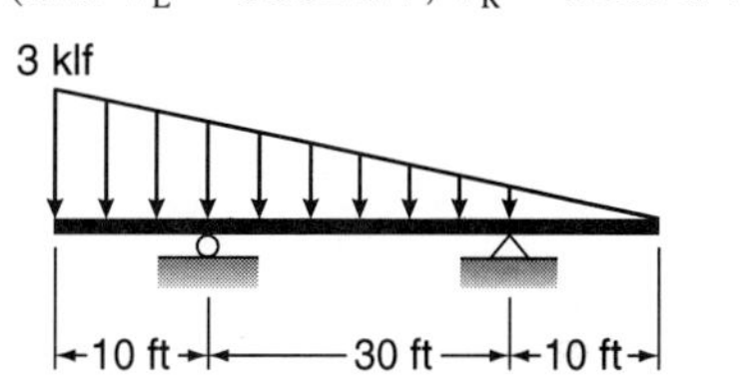

4.21

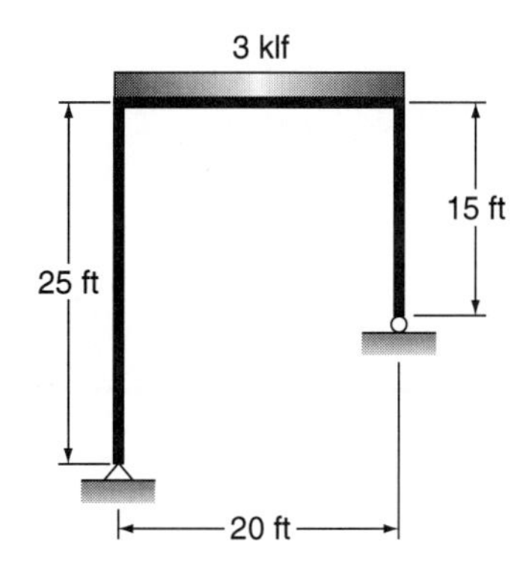

4.22 (*Ans.* $V_L = 0$, $H_L = 20$ k ←, $V_R = 40$ k ↑)

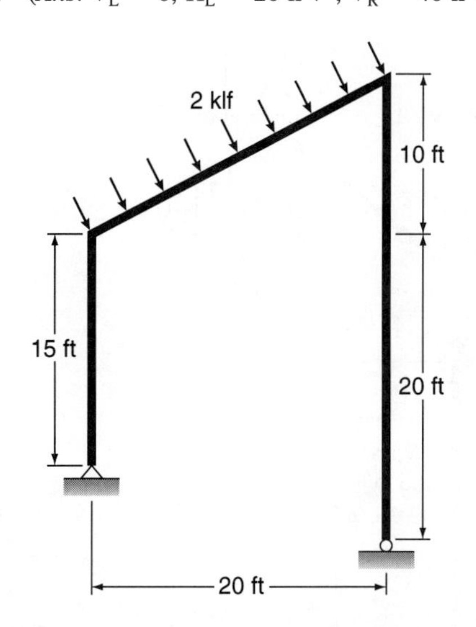

4.23

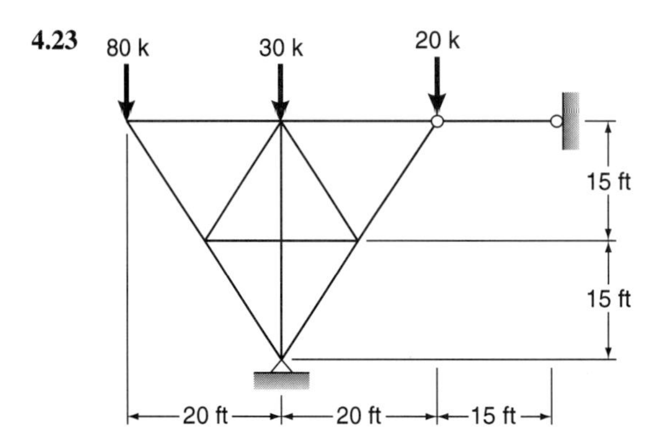

4.24 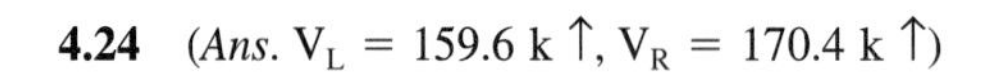
(*Ans.* V_L = 159.6 k ↑, V_R = 170.4 k ↑)

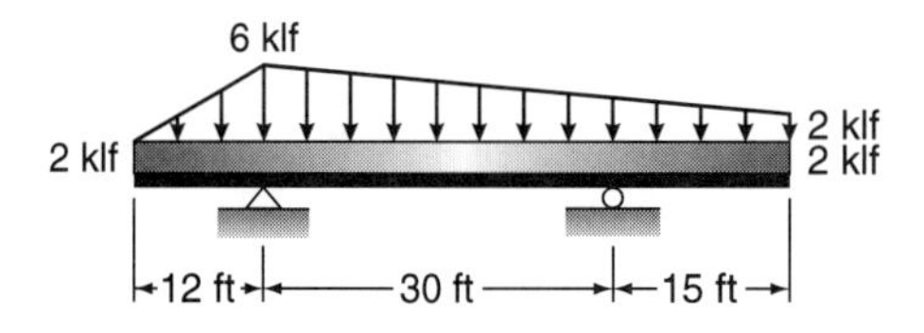

4.25

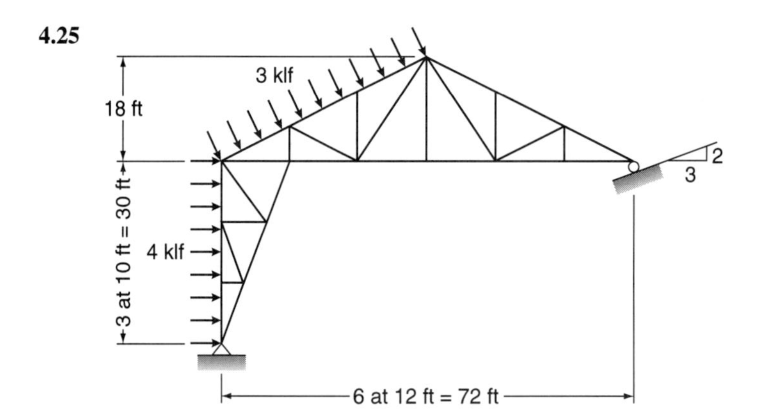

4.26 Consider only a 1 ft length of frame. (*Ans.* V_L = 66.96 k ↑, H_L = 22.30 k →, V_R = 93.04 k, H_R = 5.82 k →)

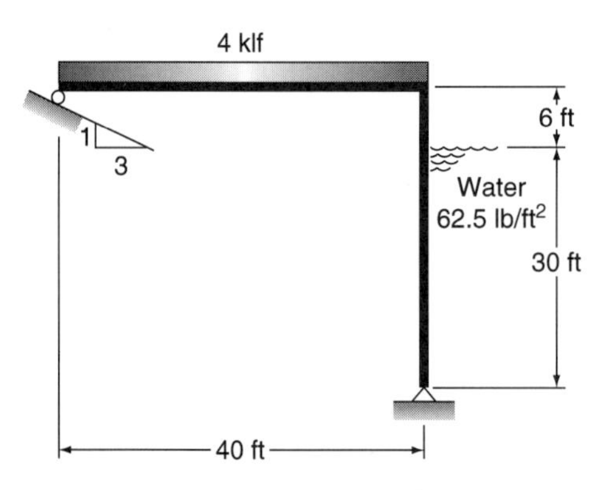

4.27

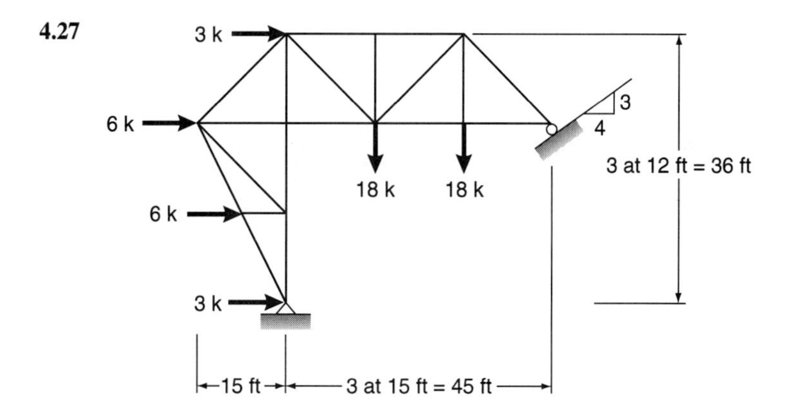

4.28 (*Ans.* $V_L = 75$ k ↑, $H_L = 37.5$ k ←, $V_R = 60$ k ↑, $H_R = 30$ k ←)

4.29

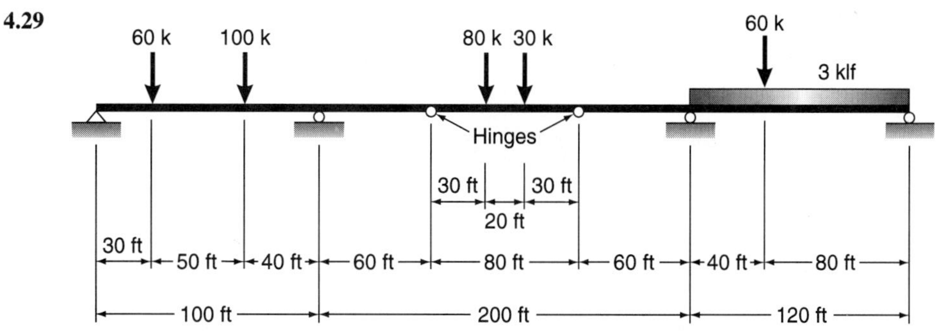

4.30 (*Ans.* $V_A = 60$ k ↓, $V_B = 280$ k, $V_C = 270$ k ↑, $V_D = 80$ k ↓)

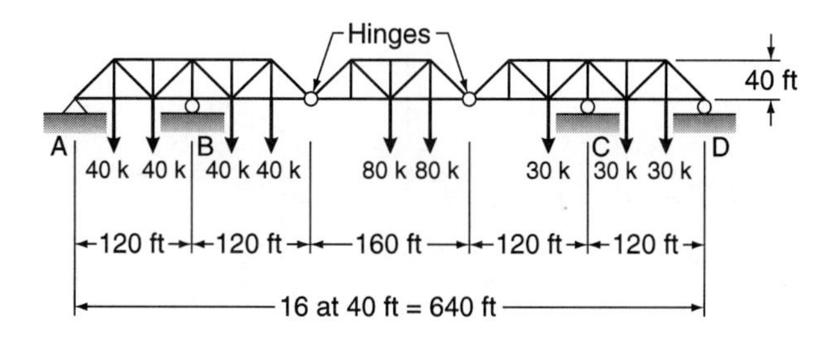

4.31

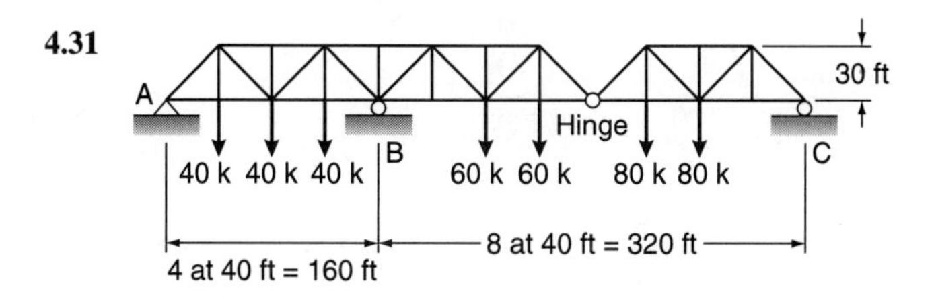

4.32 (*Ans.* $V_A = 198$ k ↑, $V_B = 753.2$ k, $V_C = 31.2$ k)

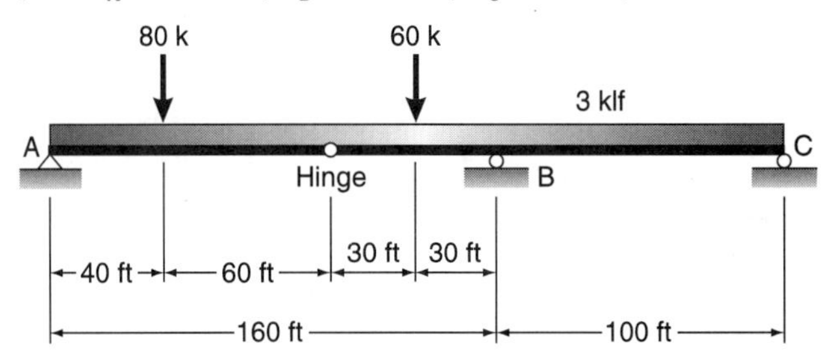

4.33

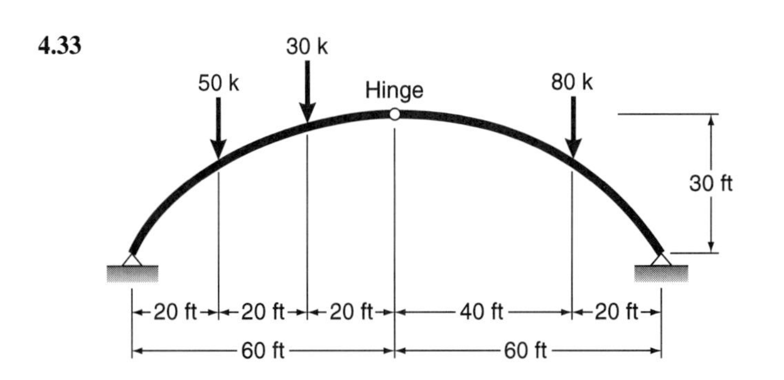

4.34 (*Ans.* $V_L = 94$ k ↑, $H_L = 108$ k →, $V_R = 66$ k ↑, $H_R = 108$ k ←)

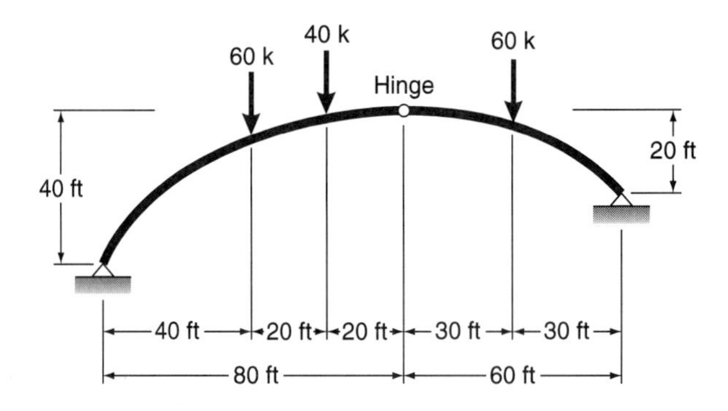

4.35

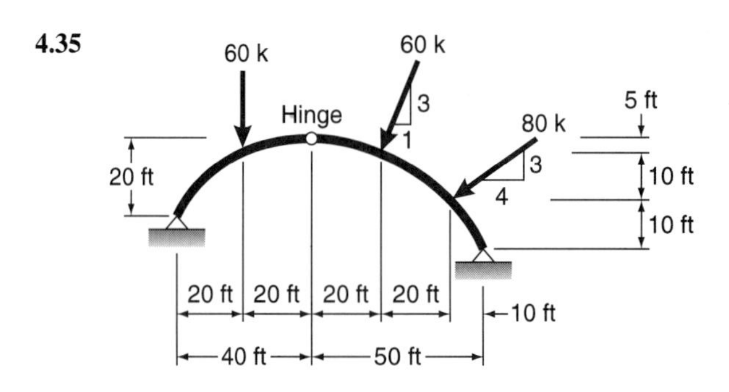

4.36 (*Ans.* $V_L = 73$ kN ↑, $H_L = 88.33$ kN →, $V_R = 137$ kN ↑, $H_R = 88.33$ kN ←)

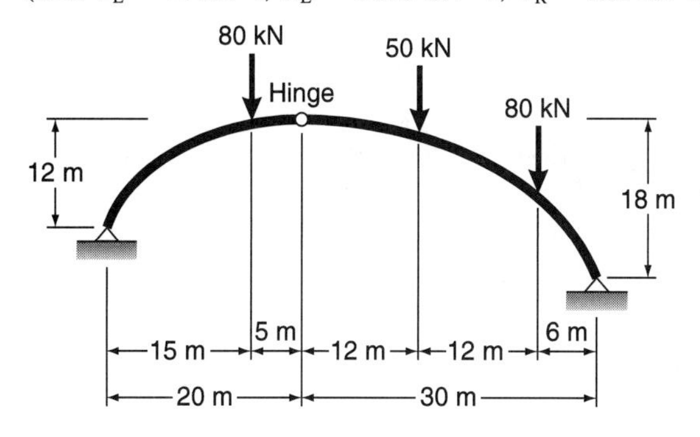

4.37

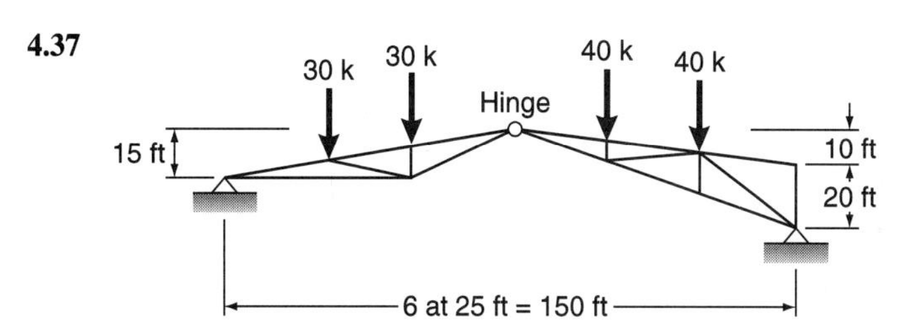

4.38 (*Ans.* $V_L = 5$ k ↑, $H_L = 90$ k →, $V_R = 5$ k ↑, $H_R = 90$ k →)

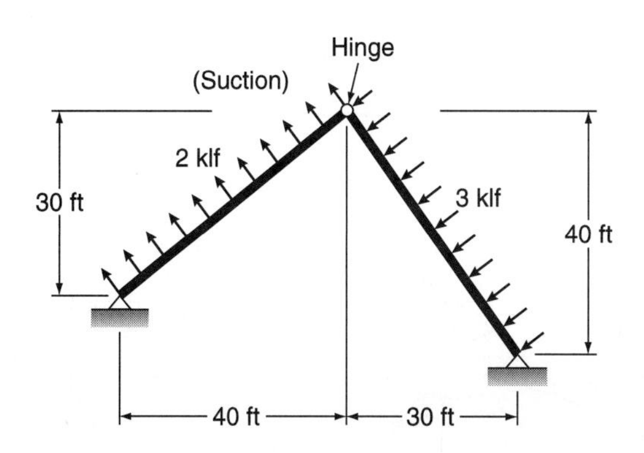

4.39 Repeat Problem 4.38 if the 2-klf load is increased to 3 klf and the 3-klf load is increased to 5 klf.

4.40 Repeat Problem 4.38 if the 3-klf load is removed. (*Ans.* $V_L = 40$ k ↓, $H_L = 30$ k →, $V_R = 40$ k ↓, $H_R = 30$ k →)

4.41

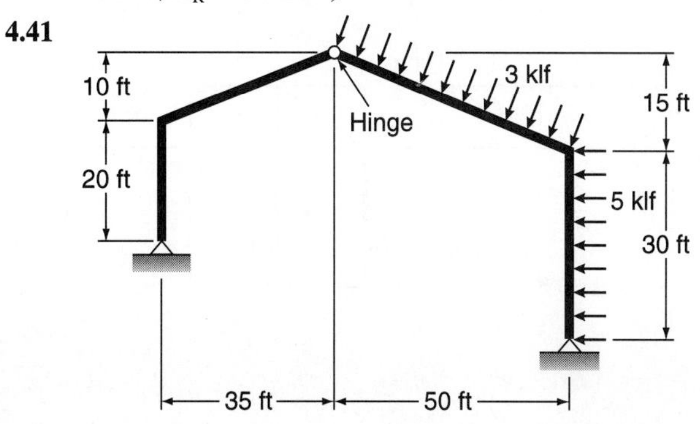

4.42 (*Ans.* $V_L = 64.5$ k ↓, $H_L = 101.9$ k ←, $V_R = 49.5$ k ↓, $H_R = 20.1$ k ←)

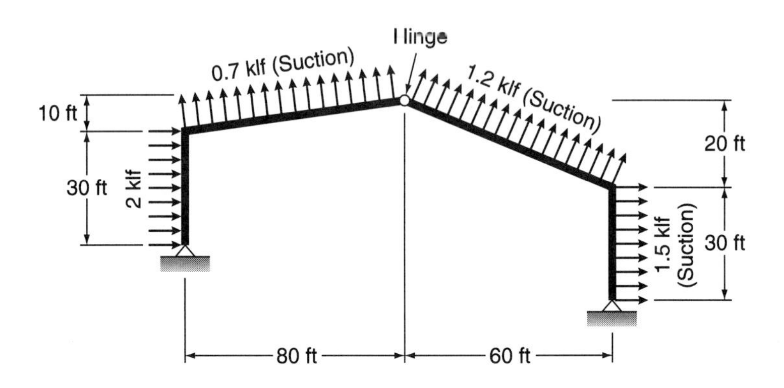

4.43

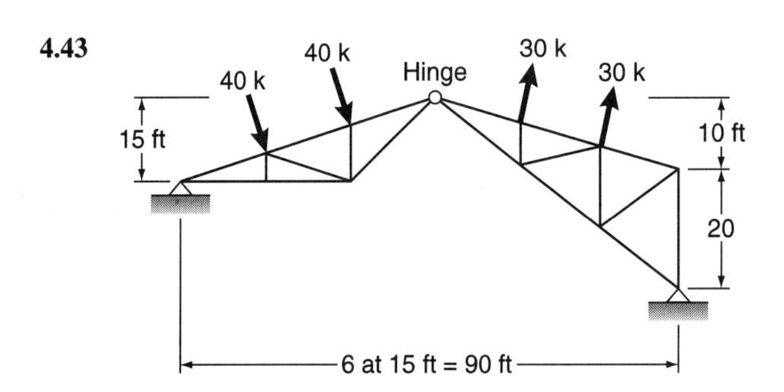

4.44 (*Ans.* $V_L = 96.3$ k ↑, $H_L = 97.2$ k →, $V_R = 93.7$ k ↑, $H_R = 97.2$ k ←)

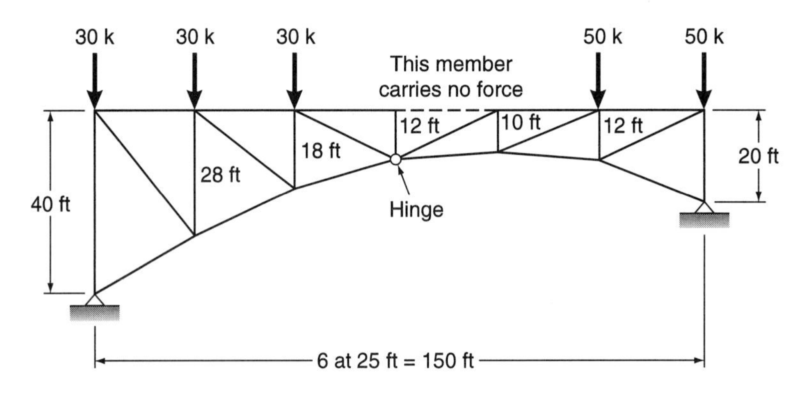

4.45

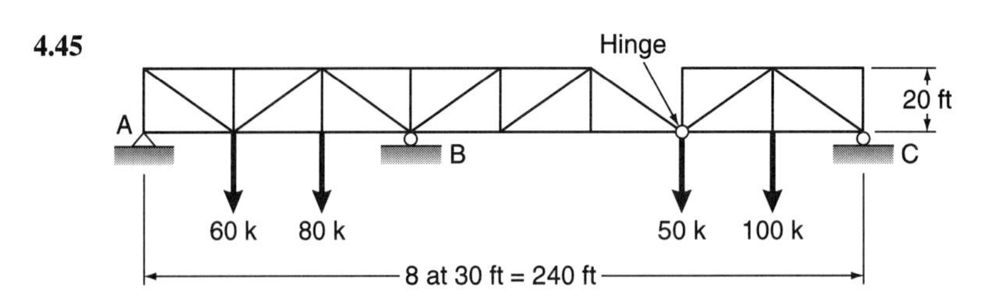

4.46 Determine the reaction components, cable sag at the 50-kN load, and the maximum tensile force in the cable. (*Ans.* V_L = 52.86 kN ↑, H_L = 142.8 kN ←, y = 5.70 m)

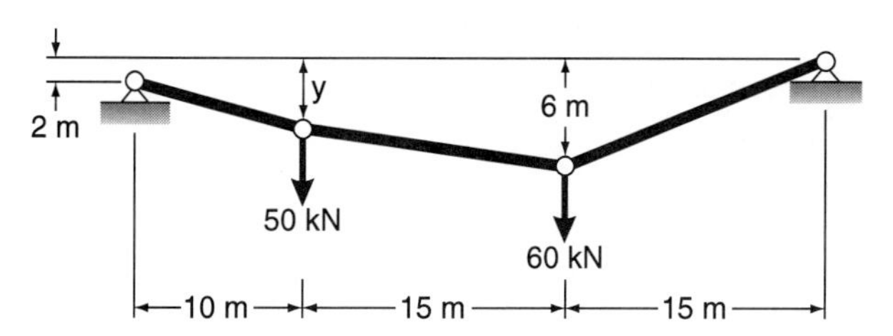

4.47 Determine the reaction components, cable sag at the 15- and 20-kip loads, and the maximum tensile force in the cable.

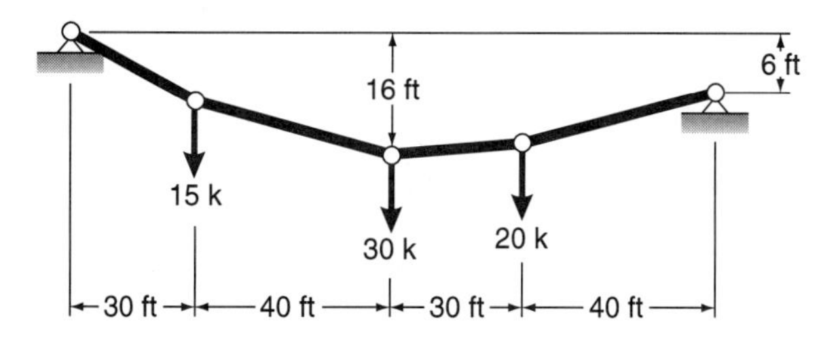

4.48 Determine the reaction components and the sag for the cable shown. The cable is 120 ft long.

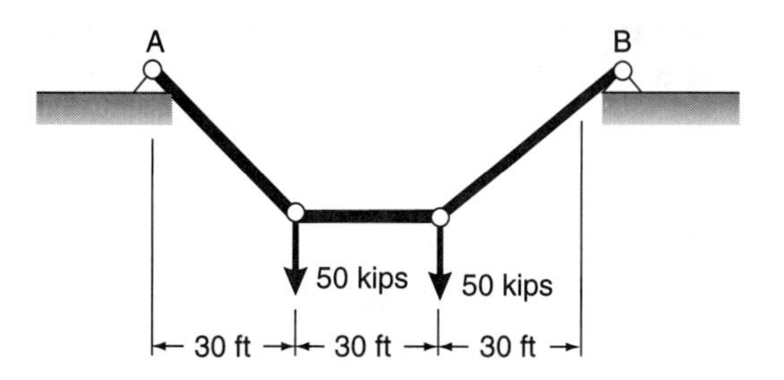

4.49 Repeat Problem 4.48 if the right-hand load is removed.

4.50 For Problem 4.49, prepare a graph of the force in the cable versus the length of the cable.

Chapter 5

Shearing Force and Bending Moment

5.1 INTRODUCTION

An important part of structural engineering, and indeed the understanding of structural behavior, is the understanding of the shearing forces and bending moments that exist within a structural system. Equations for shearing force and bending moment are needed to compute structural deflections. Very often the shearing force and bending moment are represented on diagrams to provide a visualization of structural response. These diagrams, from which the values of shearing force and bending moment at any point in a beam are immediately available, are very convenient in design since they visually provide the magnitude and location of maximum design forces. In this chapter we will examine methods to develop the equations for shearing force and bending moment in structural systems, as well as methods to construct shearing force and bending moment diagrams. It is doubtful that there is any other topic of which careful study will give more reward in structural engineering knowledge.

To examine the internal conditions in a structure, a free-body must be taken out and studied to see what forces must be present for the body to remain in a state of equilibrium. Shearing force and bending moments are two actions of the external loads on a structure that need to be understood to study properly the internal forces.

Shearing force is defined as the algebraic summation of the external forces that are perpendicular to the axis of the member to the left or to the right of a section. Shearing force is normally taken to be positive if the summation of the external forces to the left is up; the internal force on the left necessary for equilibrium is directed downward. Conversely, if the summation of external forces to the right is down, the internal force on the right is upward and is taken to be positive. This relationship is shown in Figure 5.1.

Bending moment is the algebraic sum of the moments caused by all of the external forces to the left or to the right of a particular section. The bending moment is computed about an axis through the centroid of the cross section. A positive sign indicates that the summation of external moments to the left is clockwise. That is, the internal bending moment on the left necessary for equilibrium is counterclockwise. Conversely, if the summation of external moments to the right is counterclockwise, then the internal bending

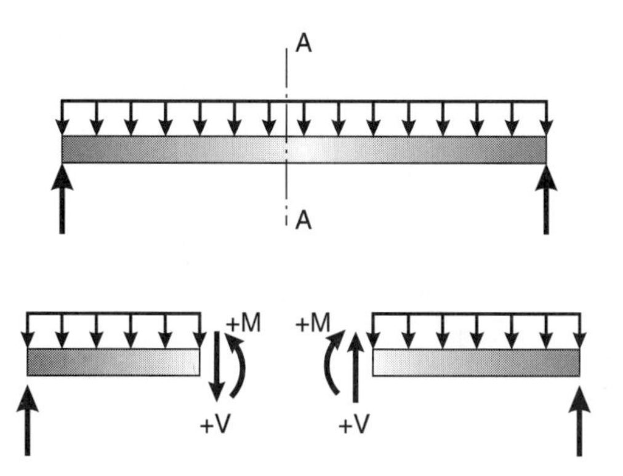

Figure 5.1 Positive shearing force and bending moment at section A–A in a beam

moment on the right necessary for equilibrium is clockwise and is considered positive. This relationship is also shown in Figure 5.1.

Throughout this textbook, the sign convention for internal forces as shown in Figure 5.1 will be referred to as the beam sign convention. Please note that while internal forces on opposite side of the cut section in this figure act in opposite directions, they are all positive. A study of Figure 5.1 shows that positive moment at a section causes tension in the bottom fibers and compression in the top fibers of the beam at the section.

The calculations for shearing force and bending moment at two sections in a simple beam are shown in Example 5.1. In each case, the summations are made using both the left and right sections to show that identical results are obtained.

The Tridge, a triple-span pedestrian bridge using glued laminated timber, Midland, Michigan (Courtesy of Unit Structures, Inc.)

EXAMPLE 5.1

Find the shearing force and the bending moment at sections A–A and B–B in the beam shown. We will use the positive sign convention that was shown in Figure 5.1: in this example, we will take forces acting upward and clockwise moments to be positive. Furthermore, we will write the summation of moments about the cut section.

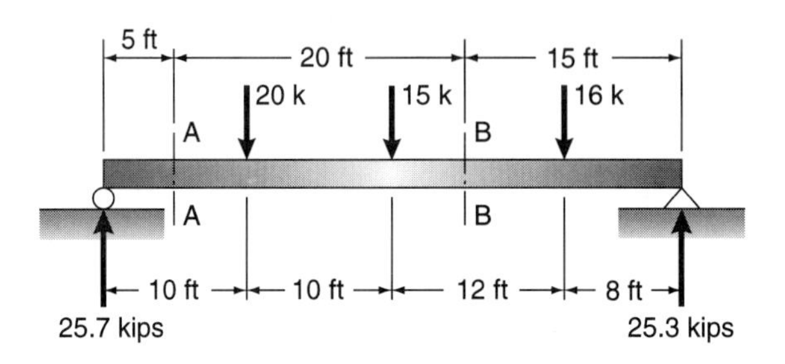

Solution. Free-body diagrams for section A–A:

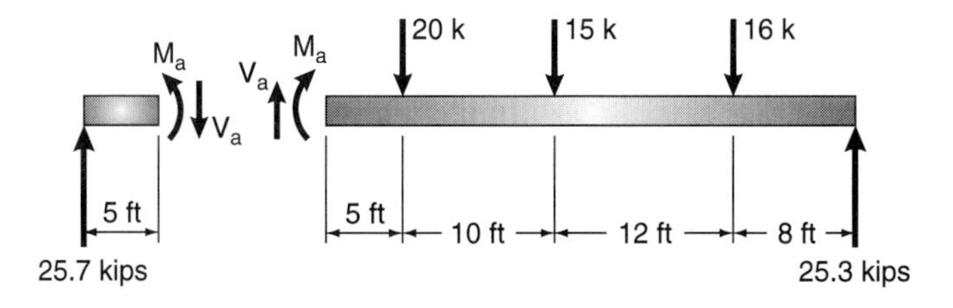

Shearing force at section A–A:

$$\sum F_{left} = 0 = 25.7 - V_a \qquad V_a = 25.7 \text{ kips}$$

$$\sum F_{right} = 0 = +V_a - 20 - 15 - 16 + 25.3 \qquad V_a = 25.7 \text{ kips}$$

Bending moment at section A–A:

$$\sum M_{left} = 0 = 25.7(5) - M_a \qquad M_a = 128.5 \text{ kip-ft}$$

$$\sum M_{right} = 0 = +M_a + 20(5) + 15(15) + 16(27) - 25.3(35) \qquad M_a = 128.5 \text{ kip-ft}$$

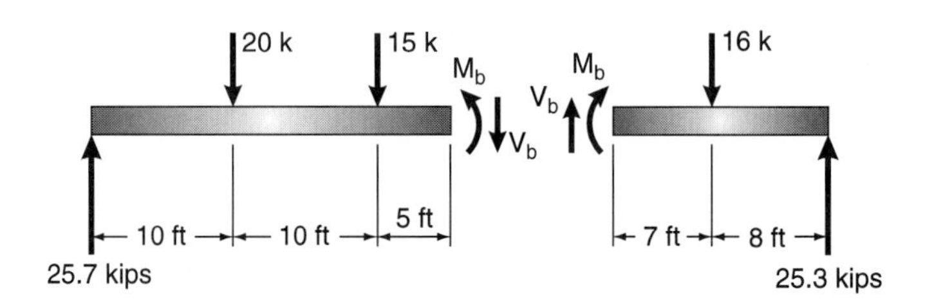

Free-body diagrams for section B–B:
Shearing force at section B–B:

$$\sum F_{left} = 0 = 25.7 - 20 - 15 - V_b \qquad V_b = -9.3 \text{ kips}$$

$$\sum F_{right} = 0 = +V_b - 16 + 25.3 \qquad V_b = -9.3 \text{ kips}$$

Bending moment at section B–B:

$$\sum M_{left} = 0 = 25.7(25) - 20(15) - 15(5) - M_b \qquad M_b = 267.5 \text{ kip-ft}$$

$$\sum M_{right} = 0 = M_b + 16(7) - 25.3(15) \qquad M_b = 267.5 \text{ kip-ft} \quad \blacksquare$$

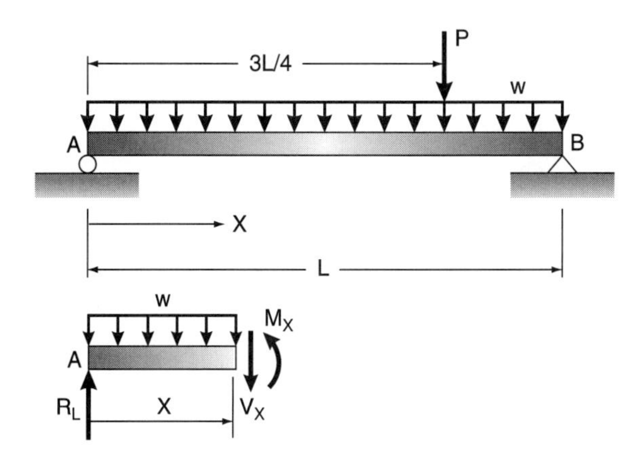

Figure 5.2

5.2 SHEARING FORCE AND BENDING MOMENT EQUATIONS

An important part of structural engineering is the ability to write, or derive, equations for the shearing force and bending moment that exist within a structural system. The ability to write these equations is important for understanding much of the information discussed in this book. Many of the methods used to compute structural deflections require an equation for the bending moment. We will learn how to write these equations for beams in this section. In the next section, we will learn about the relationships that exist between load, shearing force, and bending moment.

Let us consider the beam with the forces shown in Figure 5.2. Through the example that follows, we will develop equations for the shearing force and bending moment in the beam. Before doing so, let us explore the principles that are involved. First, we will be writing the equations in terms of position along the beam. We will denote this position by X, which is measured from some convenient reference point. Next, we will cut a free-body of the beam up to some arbitrary location of X as shown in Figure 5.2. We will then write the equations of equilibrium for the free-body and solve those equations for the shearing force and bending moment.

Every time the load along the length of the beam changes, the equations for shearing force and bending moment change. A particular set of equations developed for shearing force and bending moment, then, is valid for a particular range of the variable X. As such, we likely will need to perform the outlined steps several times along the beam or structural frame. Look at application of the method though an example.

EXAMPLE 5.2

Develop the equations for shearing force and bending moment for the beam shown in Figure 5.2. Use the left side of the beam as the reference point for measuring position along the beam.

Solution. First, we need to compute the reaction on the left side of the beam. By summing moments about B, we find that

$$\sum M_R = 0 = R_L L - wL\left(\frac{L}{2}\right) - P\left(\frac{L}{4}\right)$$

$$R_L = \frac{wL}{2} + \frac{P}{4}$$

Next, we will develop equations for shearing force and bending moment. Over the region $0 \leq X < 3L/4$ there are no changes in load so we can write one set of equations for this region. The free-body diagram for this section is shown on Figure 5.2. The equations of equilibrium for a section cut through this region are

$$\sum F_V = 0 = R_L - wX - V_x = \frac{wL}{2} + \frac{P}{4} - wX - V_X$$

$$\sum M_R = 0 = R_L X - wX\left(\frac{X}{2}\right) - M_x = \left(\frac{wL}{2} + \frac{P}{4}\right)X - \frac{wX^2}{2} - M_X$$

In these equations, the moments were summed about the right side of the free-body so that the equations did not include V_X, the unknown shearing force. By summing moments in this way, we did not need to solve simultaneous linear equations. These two equations can be solved for V_X and M_X. After doing so we obtain the equations

$$V_X = \frac{wL}{2} + \frac{P}{4} - wX$$

for $0 \leq X < 3L/4$

$$M_X = \left(\frac{wL}{2} + \frac{P}{4}\right)X - \frac{wX^2}{2}$$

These are the shearing force and bending moment equations for the region of the beam to the left of the concentrated load. At the concentrated load the equations for shearing force and bending moment change. We will cut a second free-body to the right of the concentrated load and write the equations of equilibrium. The free-body diagram that we will use is shown next.

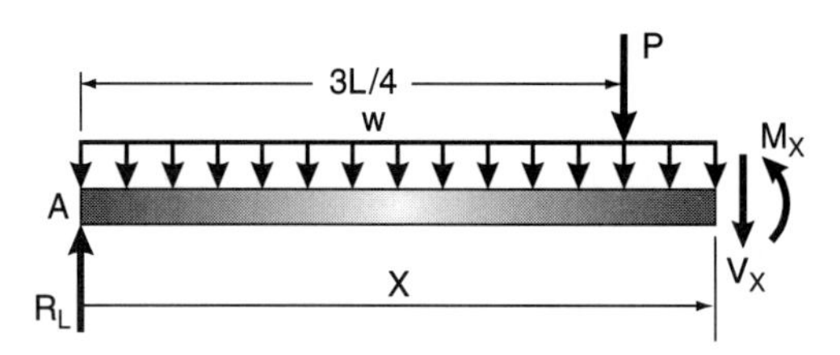

The location X will still be measured from the left end of the beam. The resulting equations for the region $3L/4 \leq X < L$ are

$$\sum F_V = 0 = R_L - wX - P - V_X = \frac{wL}{2} + \frac{P}{4} - wX - P - V_X$$

$$\sum M_R = 0 = R_L X - wX\left(\frac{X}{2}\right) - P\left(X - \frac{3L}{4}\right) - M_X$$
$$= \left(\frac{wL}{2} + \frac{P}{4}\right)X - \frac{wX^2}{2} - P\left(X - \frac{3L}{4}\right) - M_X$$

After solving these equations for the shearing force and bending moment we obtain

$$V_X = w\left(\frac{L}{2} - X\right) - \frac{3P}{4}$$

for $3L/4 \le X < L$

$$M_X = wX\left(\frac{L}{2} - \frac{X}{2}\right) + \frac{3P}{4}(L - X) \quad \blacksquare$$

Other beams are dealt with in exactly the same manner as the beam in the example. We cut a free-body in each region after a change in load, write the equations of equilibrium for that free-body, and solve the equations of equilibrium for the shearing force and bending moment.

Pedestrian bridge in Pullen Park, Raleigh, North Carolina (Courtesy of the American Institute of Timber Construction)

5.3 RELATIONS AMONG LOADS, SHEARING FORCES, AND BENDING MOMENTS

As you may have suspected from the previous section, there are significant mathematical relationships among the loads, shearing forces, and bending moments in a beam. These relationships are discussed in the following paragraphs, with reference to Figure 5.3 during the discussion. The beam is acted upon by a distributed load having a magnitude w, which does not have to be a uniformly distributed load.

To determine the relationships among shearing force, bending moment, and load, we will consider a segment of this beam having a length of ΔX. The forces that are acting

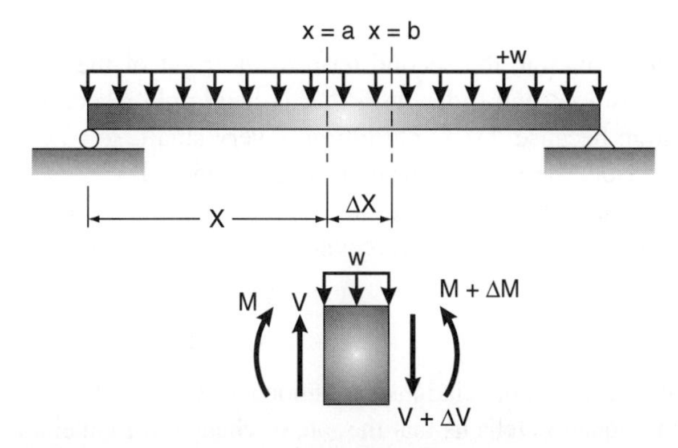

Figure 5.3 Internal forces acting on a section of a beam

on this segment are shown in Figure 5.3. This segment of the beam is loaded with a load of magnitude w. The only assumption regarding w is that ΔX is sufficiently small so that w is nearly uniform over the length ΔX. It does not have to be uniformly distributed. Over the length of this segment, the shearing force changes by an amount ΔV and the bending moment changes by an amount ΔM. The equations of equilibrium for the beam segment may be written as follows:

$$\sum F_v = 0 = V - w(\Delta X) - (V + \Delta V)$$
$$\sum M_c = 0 = M + V(\Delta X) - w\frac{(\Delta X)^2}{2} - (M + \Delta M) \qquad \textbf{Eq. 5.1}$$

The first of these equations can be solved for the change in shearing force, ΔV. By doing so we obtain the equation

$$\Delta V = -w(\Delta X) \qquad \textbf{Eq. 5.2}$$

After dividing both sides of this equation by ΔX and taking the limit as ΔX approaches zero, we obtain the relationship

$$\lim_{\Delta X \to 0} \frac{\Delta V}{\Delta X} = \frac{dV}{dX} = -w \qquad \textbf{Eq. 5.3}$$

This equation is the first of the relationships among load, shearing force, and bending moment. In words, it states that the rate of change of the shearing force at any location in the beam is equal to the negative of the applied distributed load at that same point. Please observe in Figure 5.3 that a load directed downward is a positive load.

From Equation 5.3 we can develop the next relationship between shearing force and load. If we integrate the distributed load over the length from $x = a$ to $x = b$ we obtain the equation

$$\Delta V = \int_a^b -w \, dX \qquad \textbf{Eq. 5.4}$$

This equation is the general expression for ΔV. In words, this integral equation tells us that the change in shearing force between any two points is equal to the area under the load diagram between those same two points.

Now let's return to the second of the two equations in Equation 5.1. By solving that equation for the change in moment, we obtain the equation

$$\Delta M = V(\Delta X) - w\frac{(\Delta X)^2}{2} \cong V(\Delta X) \qquad \textbf{Eq. 5.5}$$

Within this equation, the second term to the right of the equal sign deserves some discussion. It is a second-order term that is negligible in comparison to the other terms in the equation because Δx by definition is very small, so $(\Delta x)^2$ becomes extremely small. For this reason, the term is normally not considered.

Let us continue our discussion by dividing both sides of Equation 5.5 by ΔX and then taking the limit as ΔX approaches zero. Upon doing so we obtain the equation:

$$\lim_{\Delta X \to 0} \frac{\Delta M}{\Delta X} = \frac{dM}{dX} = V \qquad \textbf{Eq. 5.6}$$

This is the third of our relationships among load, shearing force, and bending moment. In words, this equation tells us that the rate of change of moment, the slope of the moment diagram, at any point along the beam is equal to the shearing force at that same point.

From Equation 5.5 we can obtain the next of the relationships between shearing force and bending moment. By integrating both sides of the equations over the segment from x = a to x = b we obtain the relationship

$$\Delta M = \int_a^b V\, dX \qquad \textbf{Eq. 5.7}$$

This is the last of the relationships for a distributed load. In words, this integral equation tells us that the change in bending moment between any two points along a beam is equal to the area under the shearing force diagram between those same two points.

EXAMPLE 5.3

Using the equations developed in Example 5.2 for the beam shown in Figure 5.2, verify that the relationships just developed are indeed true.

Solution. We will begin with the equation that describes the shearing force and bending moment in the region $0 \leq X < 3L/4$. In this region verify that the derivative of the shearing force and bending moment equations are indeed equal to the negative of the load and the shearing force, respectively.

$$V_X = \frac{wL}{2} + \frac{P}{4} - wX \qquad \frac{dV}{dX} = -w$$

$$M_X = \left(\frac{wL}{2} + \frac{P}{4}\right)X - \frac{wX^2}{2} \qquad \frac{dM}{dX} = \frac{wL}{2} + \frac{P}{4} - wX = V_X$$

The derivatives are what we expected them to be in each case. The derivative of the shearing force equation is the negative of the load that is acting on the beam. The derivative of the bending moment equation is the shearing force equation.

Now let's look at the region $3L/4 \leq X < L$. In this region we will first verify that the integral of load and shearing force represent the change in shearing force and bending moment over the same region. We will evaluate the relationships from just to the right of the concentrated load to the right end of the beam.

$$V_L - V_{\frac{3L}{4}} = -\left(\frac{3P}{4} + \frac{wL}{2}\right) + \left(\frac{3P}{4} + \frac{wL}{4}\right) = -\frac{wL}{4}$$

$$\int_{\frac{3L}{4}}^{L} w\, dx = -\frac{wL}{4}$$

The result is the same and as expected in both cases. Let us now verify that the change in bending moment over this region is equal to the integral of the shearing force over the region.

$$M_L - M_{\frac{3L}{4}} = 0 - \left(\frac{3wL^2}{32} + \frac{3PL}{16}\right) = -\left(\frac{3wL^2}{32} + \frac{3PL}{16}\right)$$

$$\int_{\frac{3L}{4}}^{L}\left(\frac{wL}{2} + \frac{3P}{4} - wX\right)dx = -\left(\frac{3wL^2}{32} + \frac{3PL}{16}\right)$$

Again, the result is as we expected it to be. The integral of shearing force over the segment is the change in bending moment. ■

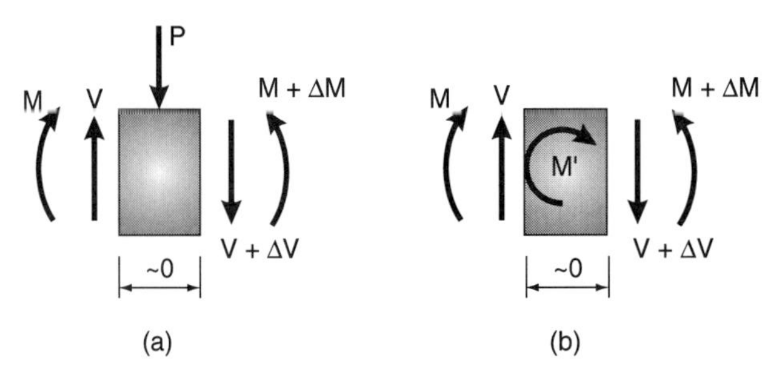

Figure 5.4 Forces acting on a section surrounding concentrated forces

The four relationships that we have developed are very useful for structural engineers. The procedure for drawing shear and moment diagrams described in Section 5.4 is based on these relationships, and is applicable to all structures regardless of loads or spans.

Before the process is described, we should talk about what occurs at the location of concentrated forces—of either a concentrated force or a concentrated moment. For this discussion, refer to Figure 5.4. Shown in this figure are a section of a beam with a concentrated force and a section of a beam with a concentrated moment. Concentrated forces are assumed to act at a point. That is, the length of beam on which the load acts is nearly equal to zero.

We will look first at Figure 5.4(a) to determine the change in shearing force and bending moment at the location of a concentrated force. By writing the equations of equilibrium for this segment, we obtain the relationships

$$\begin{aligned} &\sum F = 0 = V - P - (V + \Delta V) \\ &\therefore \Delta V = P \\ &\sum M = 0 = M - (M + \Delta M) \\ &\therefore \Delta M = 0 \end{aligned} \qquad \textbf{Eq. 5.8}$$

At the location of a concentrated force the change in shearing force is equal to the concentrated force and the change in bending moment is zero. Because concentrated forces are assumed to act at a point, the rate of change of the shearing force, and hence the slope of the shear force curve, at that point is infinite. At the point of application of the concentrated force, the shearing force curve is a vertical line.

Let us now investigate what happens at the location of a concentrated moment as shown in Figure 5.4(b). We will do this by writing the equilibrium equations and solving them for change in shearing force and change in bending moment. By doing so we find that

$$\begin{aligned} &\sum F = 0 = V - (V + \Delta V) \\ &\therefore \Delta V = 0 \\ &\sum M = 0 = M + M' - (M + \Delta M) \\ &\therefore \Delta M = M' \end{aligned} \qquad \textbf{Eq. 5.9}$$

At the location of a concentrated moment the change in shearing force is zero and the change in bending moment is equal to the applied moment. Again, because concentrated forces are assumed to act at a point, the rate of change of the bending moment, thus the slope of the bending moment curve, at that point is infinite. At the point of application of the concentrated moment, the bending moment curve is a vertical line.

5.4 SHEARING FORCE AND BENDING MOMENT DIAGRAMS

Shearing force and bending diagrams visually present the structural engineer with a wealth of information useful in design. Information about the overall behavior of a member, locations at which required structural cross-sections can change, and even places where a hinge can be used can be obtained from shearing force and bending moment diagrams. Shearing force and bending moment diagrams are quite simple to draw in most cases. They can be drawn by simply plotting the shearing force and bending moment equations that are developed using the methods discussed in Section 5.2, and shown in the Example 5.4. This can be a laborious process in many cases, however. As discussed in Section 5.5, the process can be simplified using the relationships between load, shearing force, and bending moment.

EXAMPLE 5.4

Draw the shearing force and bending moment diagrams for the beam shown. This is the same beam used in Example 5.2, except now values are assigned to each of the loads acting on the beam. Draw the shearing force and bending moment diagrams by plotting the equations.

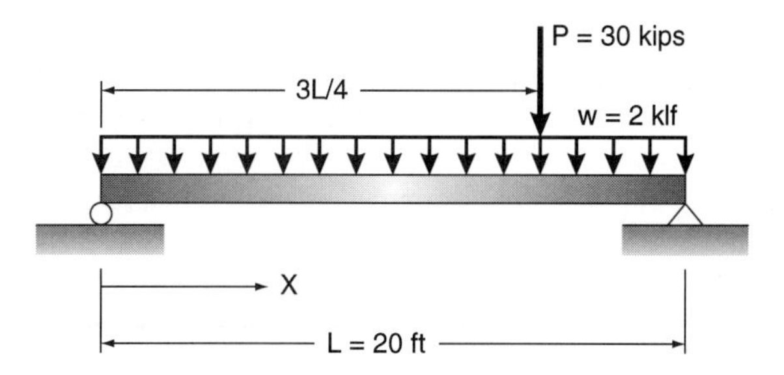

Solution. These diagrams were obtained by plotting the equations that were shown in Example 5.2. When looking at this solution, observe that the location of maximum moment is at the location of zero shearing force. This is consistent with the relationships developed in the previous section. We know from calculus that local maximum and minimum values of a function occur when the derivative of the function is equal to zero. In Section 5.3 we showed that shearing force is the first derivative of the bending moment. As such, when the shearing force is equal to zero, the bending moment is a local maximum. In some cases, such as at the end of a

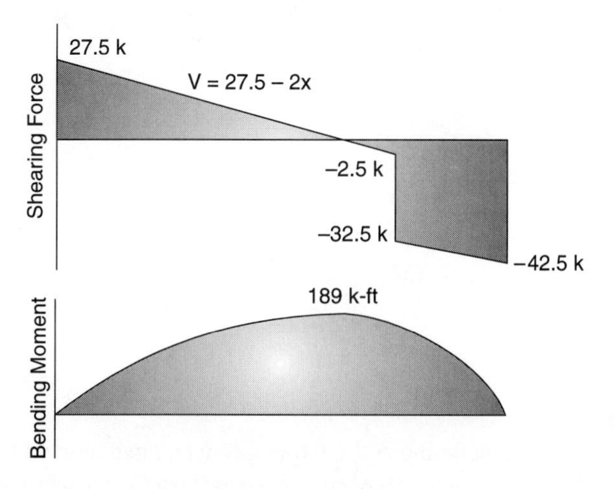

beam or at a point of discontinuity, the maximum moment can occur even though the shearing force is not equal to zero.

Before finishing with this example, let us look at one other aspect of the shearing force diagram. Look at the vertical jump in the shearing force at the location of the concentrated load. Observe that the change in the shearing force at that point is equal to the magnitude of the concentrated load. This occurs because the load is assumed to be acting at a point so the rate of change of that load is infinite as previously discussed. The infinite rate of change is indicated by the vertical line in the shearing force diagram at that point. ■

5.5 CONSTRUCTING SHEARING FORCE AND BENDING MOMENT DIAGRAMS

The usual method for obtaining shearing force and bending moment diagrams is to construct them using the relationships developed in Section 5.3. Using the relationships is much faster than plotting the equations, especially when there are multiple loads acting. When constructing the diagrams using these relationships, most engineers start at the left end of the structure and work their way to the right. The shearing force diagram is constructed first, followed by the bending moment diagram. The shapes of the diagrams depend upon the nature of the loads and can be determined from the relationships described in Section 5.3. The shearing force and bending moment diagrams for a simple beam are drawn in Example 5.5. When studying this and other examples, particular emphasis should be given to the shapes of the diagrams under uniform loads, between concentrated loads, and so on.

EXAMPLE 5.5

Draw the shearing force and bending moment diagrams for the beam with the loads acting as shown in the figure. Use the shearing force and bending moment relationships developed in Section 5.3 to construct the diagrams.

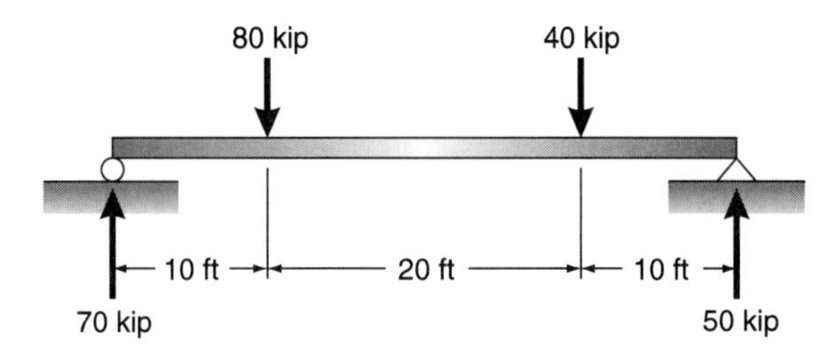

Solution. Shearing Force Diagram:

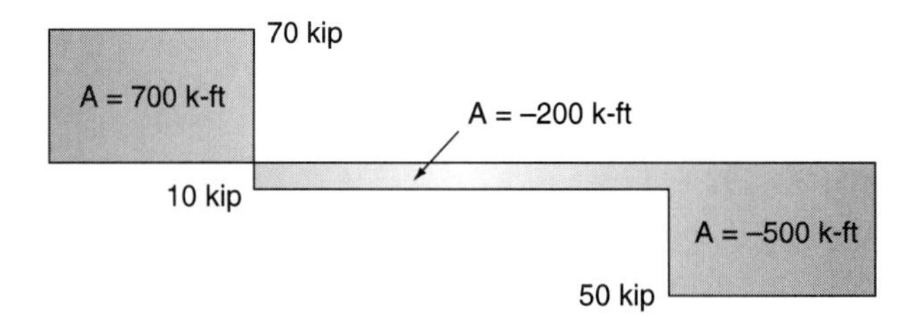

Beginning at the left is a concentrated upward force. Hence, the value of the shearing force at the left is 70-kips. There is no load acting between the left reaction and the 60-kip concentrated load so there is no change in shearing force in this region. At the location of the 60-kip force there is a negative change in shear of 60 kips. There is another negative change in shearing force at the 40-kip downward force. The right reaction is upward, causing a positive 50-kip change in shearing force. Bending Moment Diagram:

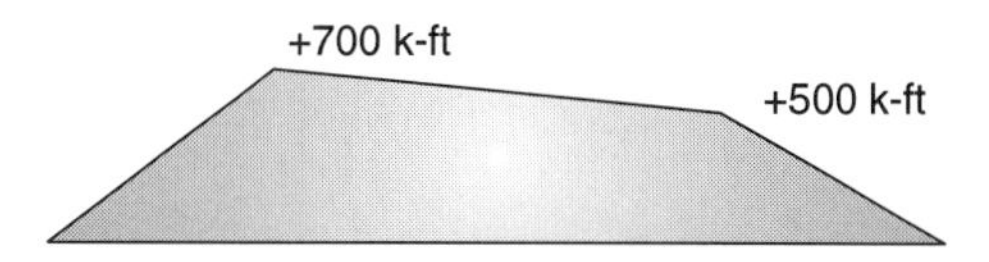

The bending moment at the left side of the beam is zero because there is a roller at that location. Between the left support and the 60-kip force the area under the shearing force is 700 k-ft, so there is a positive change in bending moment of 700 k-ft. In this region the shearing force is constant so the moment diagram in this region is a straight line. Between the two concentrated forces the area under the shearing force diagram is -200 k-ft, which indicates that the change in bending moment in this region is -200 k-ft. Again, the shearing force is constant so the bending moment diagram is a straight line. Finally, the area under the shearing force between the right concentrated force and the right support is -500 k-ft so the change in bending moment if -500 k-ft. At the right support the bending is zero, which is correct because of the pin support at that location. ■

The relationships among load, shearing force, and bending moment greatly simplify drawing the diagrams and determining maximum values. To determine the bending moment at a particular section, one needs only to compute the total area beneath the shearing force curve to the left or to the right of the section in question. When computing the total area, the algebraic signs of the various segments of the shear curve are taken into account. Shearing force and bending moment diagrams are self-checking. If they are initiated at one end of a structure, usually the left, and close to the proper value on the other end, the work probably is correct.

The authors usually find it convenient to compute the area of each part of a shearing force diagram and record that value on the part in question. This procedure is followed for the examples of this chapter where the values enclosed in the shear diagrams are areas. These recorded values simplify the construction of the bending moment diagrams.

Examples 5.6 through 5.8 illustrate the procedure for drawing shearing force and bending moment diagrams for ordinary beams. When shearing force and bending moment diagrams are drawn for inclined members, the components of loads and reactions perpendicular to the centroidal axes of the members are used and the diagrams are drawn parallel to the members.

EXAMPLE 5.6

Draw shear and moment diagrams for the beam shown in Figure 5.4. Also, determine the location of the maximum moment.

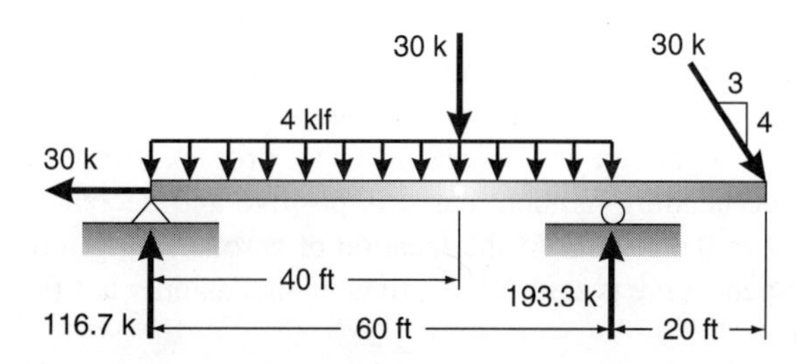

Solution. Note that acting on the beam is an axial tension force of 30 k for the entire length of the member. This force, however, does not affect the shearing force and bending moment in the beam. At the right side of this beam is a cantilevered segment. When constructing the shearing force and bending moment diagrams, use the relationships developed in Section 5.3.
Shearing Force Diagram:

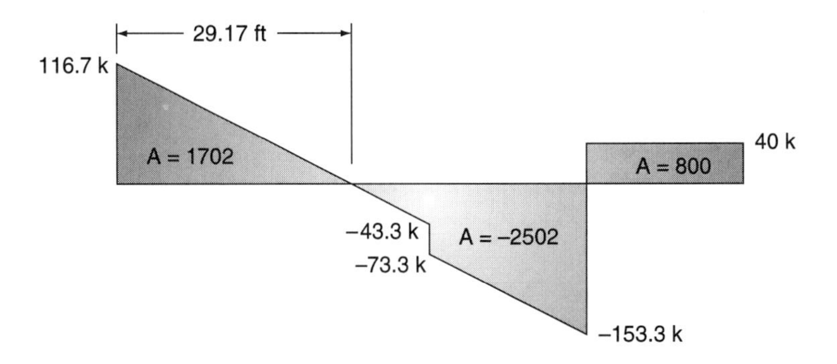

In this example, there is load acting between the left support and the concentrated load. The change in shearing force between these two points is the area under the load diagram between these two points, which is 160 kips. The area is positive so the change in shearing force is negative as indicated in Equation 5.4. The load is constant so the shearing force curve is a straight line. We continue constructing the rest of the shearing force diagram in the same manner.

A maximum moment may occur between the two supports at the location of zero shearing force. As such, we need to compute the location of the zero shearing force in that region. This can be accomplished using similar triangles.

$$\frac{40}{116.7 - (-43.3)} \approx \frac{x}{116.7}$$

$$\therefore x = 29.17\,\text{ft}$$

Bending Moment Diagram:

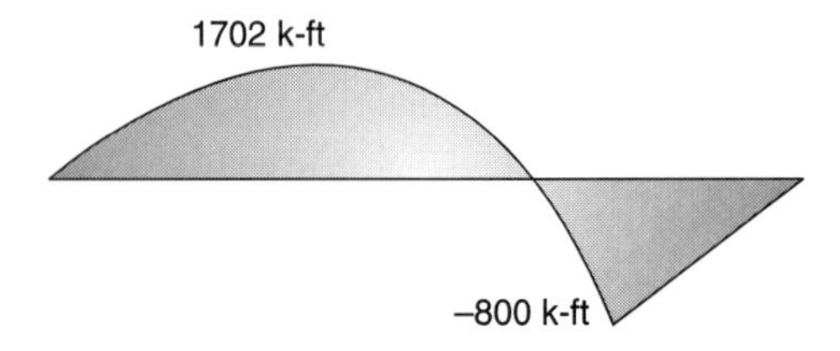

The bending moment at the left support is zero because there is a pin at that location and there are no concentrated moments acting. The area under the shearing force curve between the left support and the location of zero shearing force is 1702 k-ft so the change in bending moment is 1702 k-ft. The shearing force in this region is positive and decreasing as we progress from left to right. As such, the slope of the bending moment curve is positive and decreasing as we move from left to right in the region. At the location of zero shearing force, the slope of the bending moment curve is zero. We continue in this manner until we come to the right support.

From the location of zero shearing force to the right support, the shearing force is negative and its magnitude is increasing. The slope of the bending moment diagram, therefore, is negative and is increasing negatively.

Beyond the right support is the cantilevered portion of the beam. The area under the shearing force between the right support and the end of the cantilever is 800 k-ft so the change in bending moment along the cantilever is 800 k-ft. The shearing force is constant along the cantilever so the bending moment is a straight line. ■

Some structures have rigid arms (such as walls) fastened to them. If horizontal or inclined loads are applied to these arms, a twist or moment will suddenly be induced in the structure at the point of attachment. The fact that bending moment is computed about an axis through the centroid of the section becomes important because the lever arms of the forces applied must be measured relative to that centroid. Use the relationships developed in Equations 5.8 and 5.9 to draw the bending moment diagram at the point of resulting concentrated moment. From these relationships recall that at the location of a concentrated moment, the change in shearing force is zero and the change in bending moment is equal to the concentrated bending moment, taking into consideration the sign. Shearing force and bending moment diagrams are shown in Example 5.7 for a beam with a concentrated moment acting on it.

EXAMPLE 5.7

Draw shear and moment diagrams for the structure shown in the figure.

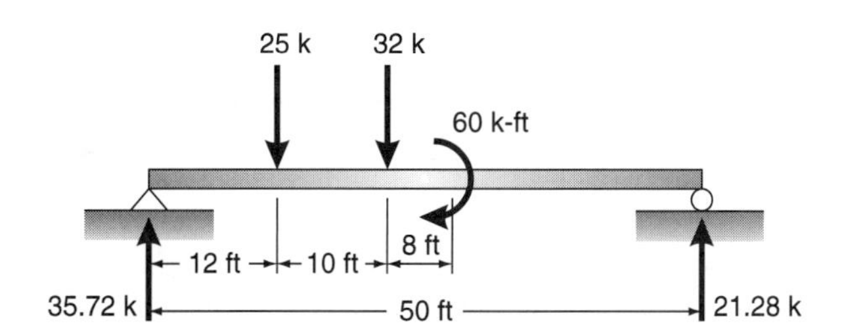

Solution

Shearing Force Diagram:

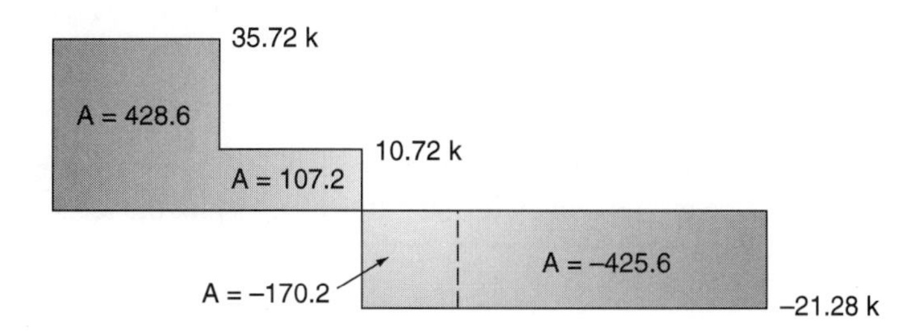

Bending Moment Diagram:

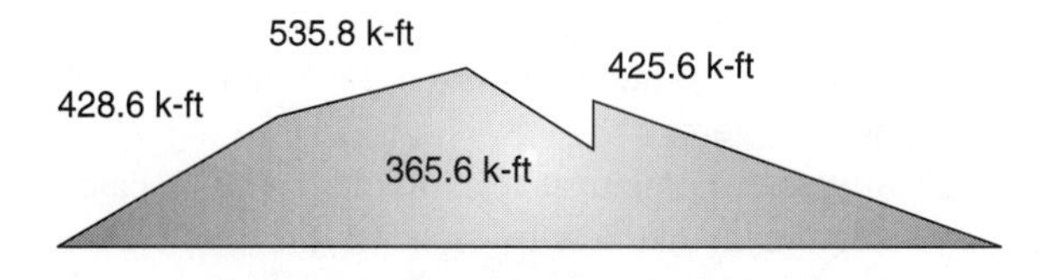

The bending moment diagram is drawn from left to right using the changes in bending moment, the area under the shearing force curve as we proceed. At the location of the concentrated moment, we observe a jump in bending moment diagram. The concentrated moment is clockwise. It therefore causes a positive change in bending moment at the point. ■

Shown in the next example is a beam with a linearly varying load acting on it. Drawing shearing force and bending moment diagrams for structures that are subjected to triangular loads may at first be difficult. Example 5.8 is presented to demonstrate how to cope with them.

EXAMPLE 5.8

Draw shearing force and bending moment diagrams for the beam shown in the figure.

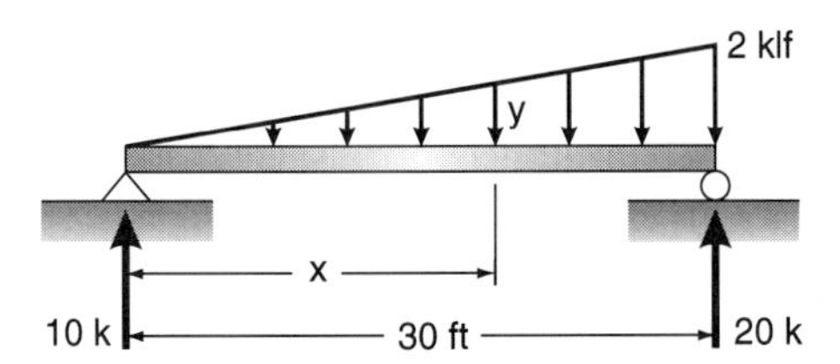

Solution. Calculate the reactions for the beam and sketch the shearing force diagram. Shearing Force Diagram:

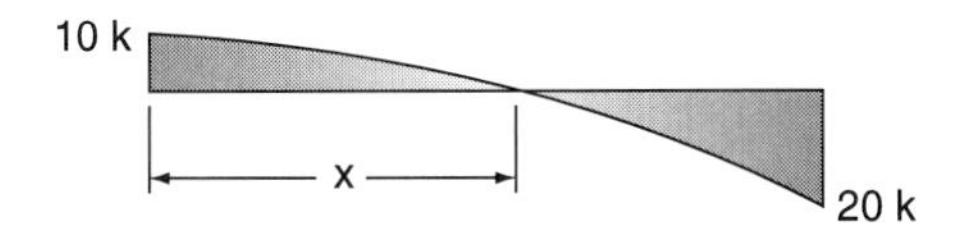

Note, however, that the point of zero shearing force is unknown and is shown in the figure as being located a distance x from the left support. The ordinate on the load diagram at this point is labeled y and can be expressed in terms of x by writing the following expression:

$$\frac{x}{30} \approx \frac{y}{2}$$

$$\therefore y = \frac{x}{15}$$

We can solve for x using the relationship between load and change in shearing force in Equation 5.2. Using the relationship, the location of x can be calculated as follows:

$$10 - (\Delta V) = 10 - \frac{1}{2}x\left(\frac{x}{15}\right) = 0$$

$$\therefore x = 17.32\,\text{ft}$$

Determine this value of x by writing an expression for bending moment in the beam and then determining the location on the beam at which the first derivative of that equation goes to zero.

Finally, the moment at a particular point is calculated from equilibrium by taking a section to the left or right of the point about which moment is to be determined. For this example, the moment at 17.32 ft is:

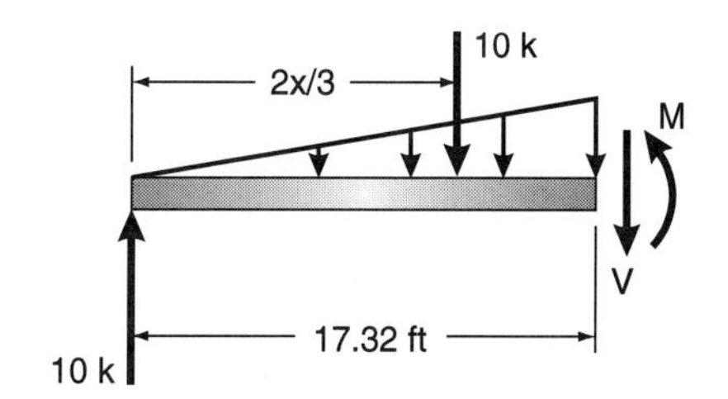

$$M = 10(17.32) - 10\left(\frac{17.32}{3}\right) = 115.5\,\text{k-ft}$$

The resulting bending moment diagram is then

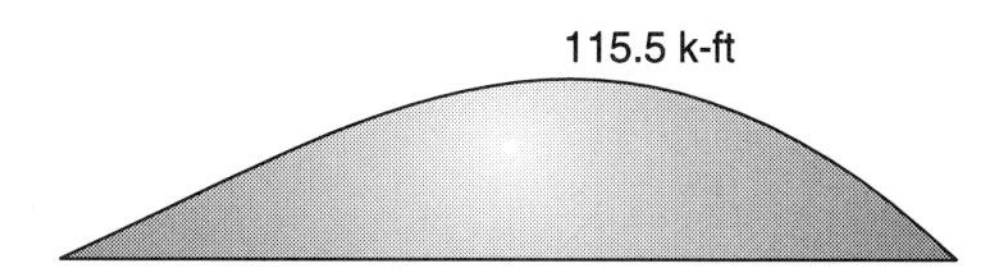

The reader should observe the shape of the shearing force and bending moment diagrams in this example. The load diagram is a first-order curve increasing from left to right. As such, the shearing force diagram is a second-order curve, a parabola, with increasing negative slope as we move from left to right. The bending moment curve is one order higher that the shearing force curve, so it is a third-order curve. ■

In the previous example, the student may wonder why, once the location of zero shearing force was determined, the authors did not simply use the area under the shear diagram from the left support to the point of zero shear. Such a procedure is correct, but the designer must be sure that he or she determines the area properly. When partial parabolas are involved, use of calculus may be necessary to determine the areas instead of standard parabolic formulas. Consequently, it may be simpler on some occasions simply to take moments of the loads and reactions about the points where moments are desired.

In Appendix E, the properties (centers of gravity and areas) of several geometric figures are presented. These values may be quite useful to the student preparing shearing force and bending moment diagrams for complicated loading situations.

EXAMPLE 5.9

Draw the shearing force and bending moment diagrams for the beam shown. Notice that this beam has a hinge located 8 ft to the right of the left support.

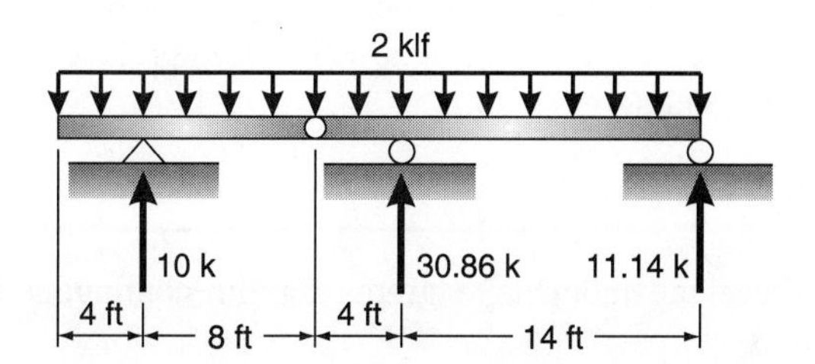

Solution

Shearing Force Diagram:

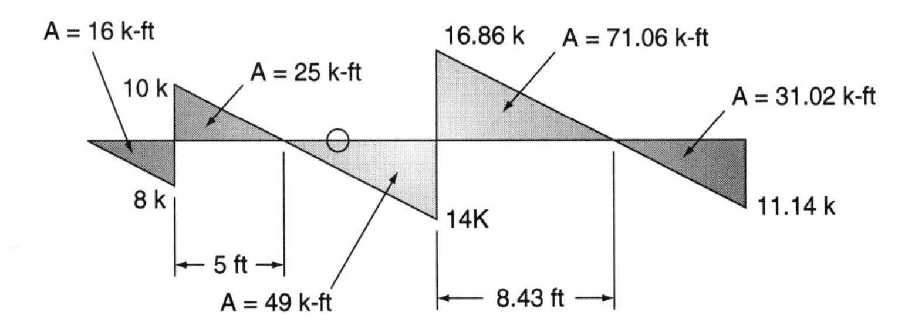

Bending Moment Diagram:

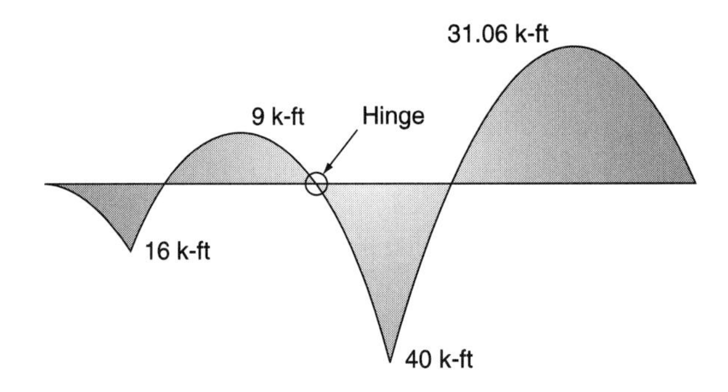

At the left side of the cantilever, both the shearing force and the bending moment are zero. Therefore, the shearing force and the bending moment diagrams both begin at zero at the left side of the beam. At the hinge, the bending moment is equal to zero. This is a necessary condition, since bending moment cannot be transferred across a hinge. Notice, though, that the shearing force at the hinge is not equal to zero, but in fact, is equal to six kips. Shearing force can be transferred across a hinge. Also observe that again the bending moment diagram has zero slope at the locations of zero shearing. ■

Now we will draw the shearing force and bending moment diagrams for a continuous beam. The reactions for the beam of Example 5.10 cannot be obtained using only the equations of static equilibrium. They have been computed by a method discussed in a later chapter. Once all of the forces acting on the structure are known, though, the process of drawing the shearing force and bending moment diagrams is the same, regardless of the number of supports in the structure. The same relationships exist among load, shearing force, and bending moment.

EXAMPLE 5.10

Draw the shear and moment diagrams for the continuous beam shown for which the reactions are given.

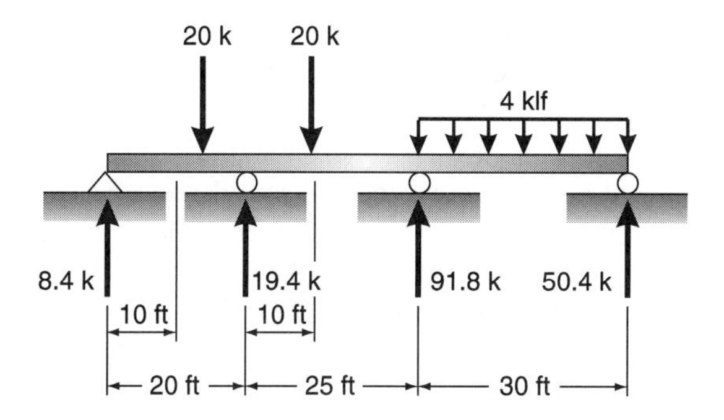

Solution
Shearing Force Diagram:

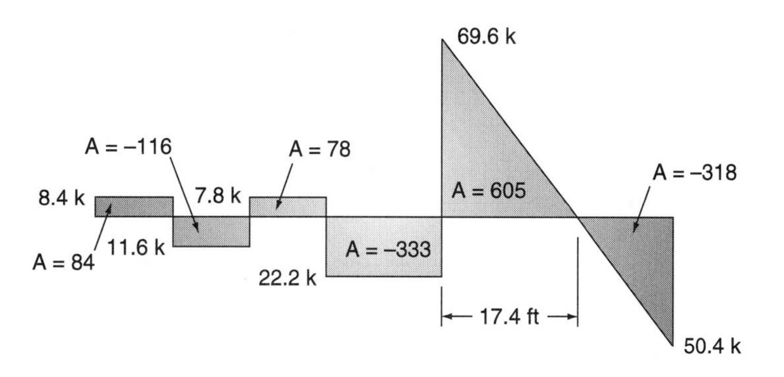

Bending Moment Diagram:

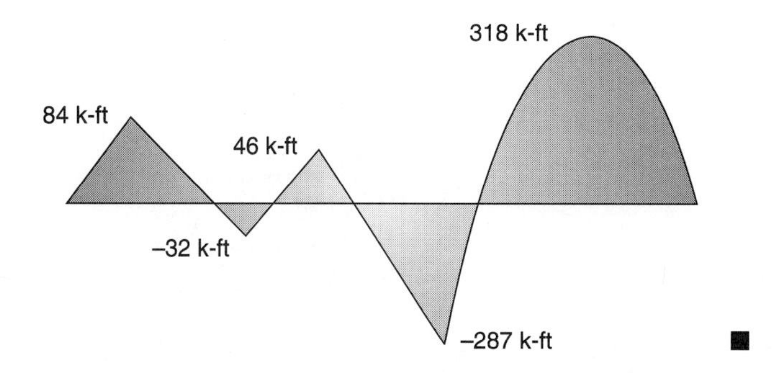

■

5.6 SHEAR AND MOMENT DIAGRAMS FOR STATICALLY DETERMINATE FRAMES

Shearing force and bending moment diagrams are as useful for rigid frames as they are for individual flexural members. The members of a rigid frame cannot rotate with respect to each other at their connections. Consequently, axial forces, shearing forces, and bending moments are transferred between the members at the joints. These values must be accounted for in the preparation of the shearing force and bending moment diagrams.

For a first example, the frame of Example 5.11 is considered. After the reactions are computed, free-body diagrams are prepared for each of the members. The forces involved at the bottom of the column are obviously the structural reactions. The forces at the top, which represent the effect of the rest of the frame on the top of the column, can easily be obtained from equations of static equilibrium. After the forces are known, the shearing force and bending moment diagrams are prepared.

Before beginning the example, the sign convention used for drawing the shearing force and bending moment diagrams must be considered. When drawing these diagrams for frames, some engineers assume they are standing underneath the frame and assume the sides of all members they see are analogous to the bottom side of horizontal beams. Thus, the right-hand side of the column to the left and the left-hand side of a column to the right are assumed to be the bottom side. This convention works well when the frames have one or two columns, but becomes very confusing, and inconsistent, when there are more than two columns.

To avoid the confusion that can result when a frame has more than two columns, and to be consistent for all frames, other engineers use the following convention when drawing shearing force and bending moment diagrams for frames:

- The lower side of sloping and horizontal members is considered to be the bottom side of a beam.
- The right side of vertical members (columns) is considered to be the bottom side of a beam.

We will use this convention when drawing shearing force and bending moment diagrams for all frames.

EXAMPLE 5.11

Draw shearing force and bending moment diagrams for the frame shown in the figure.

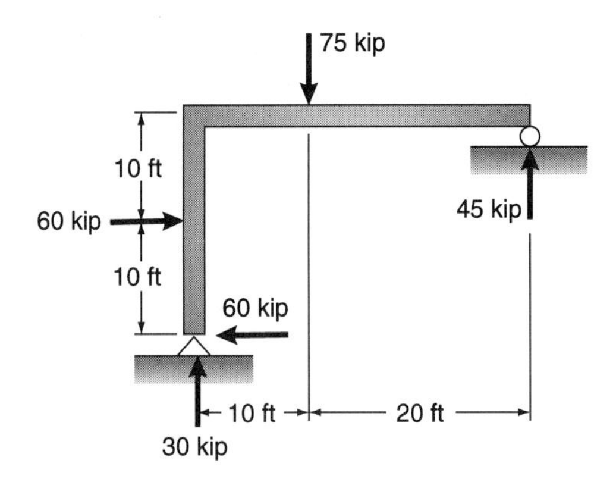

Solution

Free Body Diagrams:

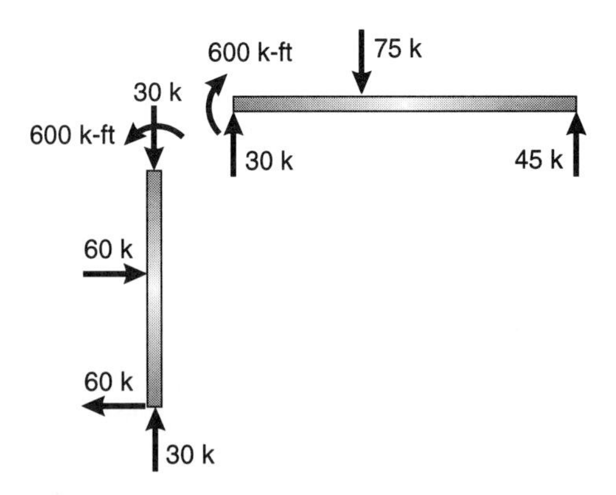

Shearing Force Diagrams:

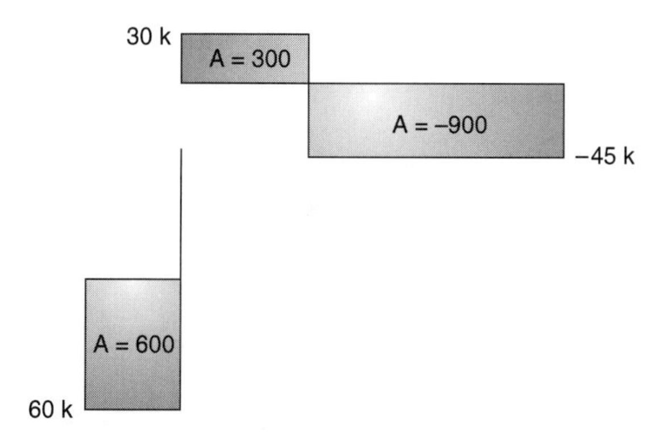

Bending Moment Diagrams:

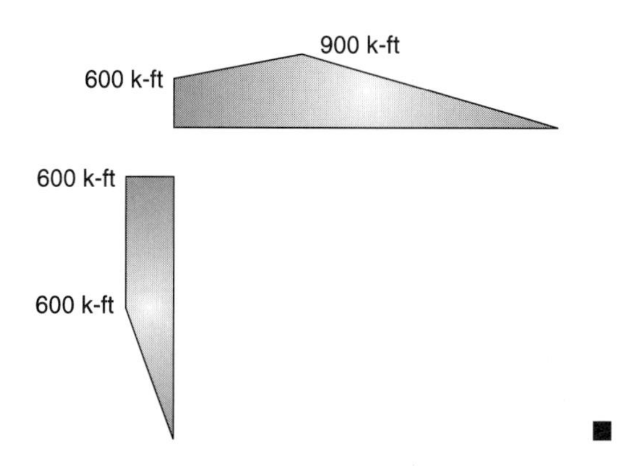

■

Shearing force and bending moment diagrams are drawn in Example 5.12 for a frame with two columns. Here again, the usual convention of drawing the diagrams is used.

EXAMPLE 5.12

Draw shearing force and bending moment diagrams for the frame shown in the figure.

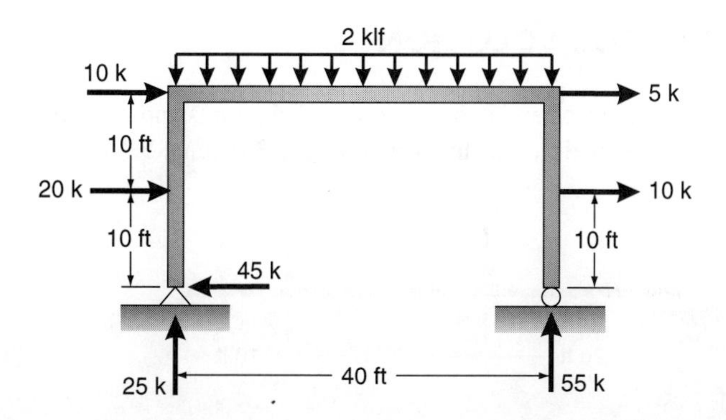

Solution. The free-body diagrams are not shown for the various members of this frame. The reader should now be capable of drawing the shearing force and bending moment diagrams directly from the external loads and reactions applied to the frame. Shearing Force Diagrams:

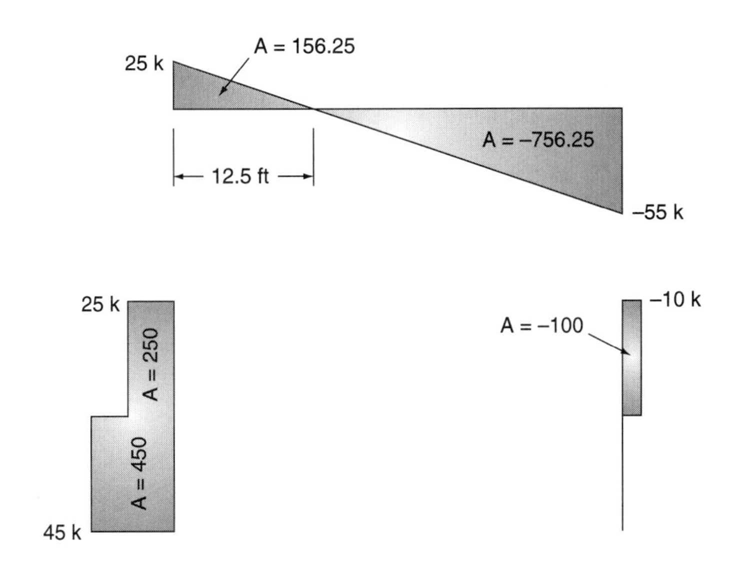

Bending Moment Diagrams:

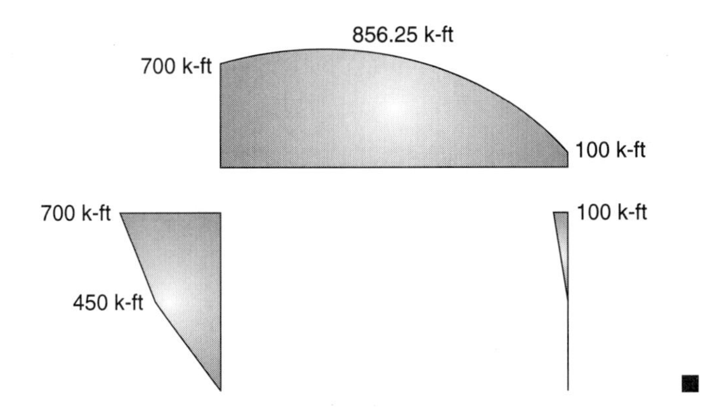

■

5.7 PROBLEMS FOR SOLUTION

Draw shearing force and bending moment diagrams and write the shearing force and bending moment equations for the structures in Problems 5.1 to 5.20.

5.1

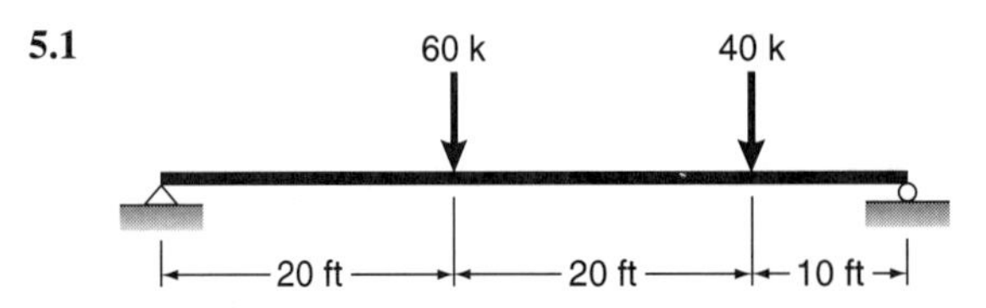

5.2 Repeat Problem 5.1 if the 40 k load is changed to 100 k. (*Ans.* max V = 104 k, max M = 1120 ft-k)

5.3

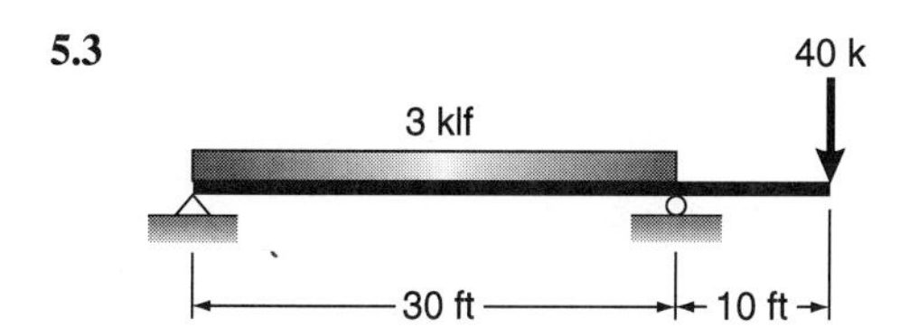

5.4 Repeat Problem 5.3 if the 3-klf load is changed to 1.5 klf. (*Ans.* max V = 40 k, max M = −400 ft-k)

5.5

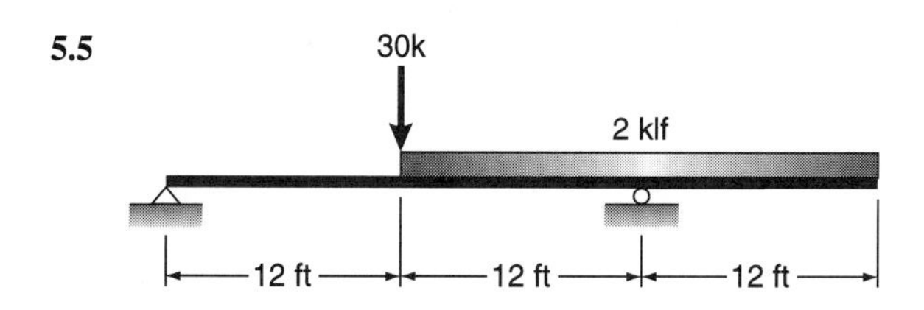

5.6 (*Ans.* max V = 170 kN, max M = 640 kN-m)

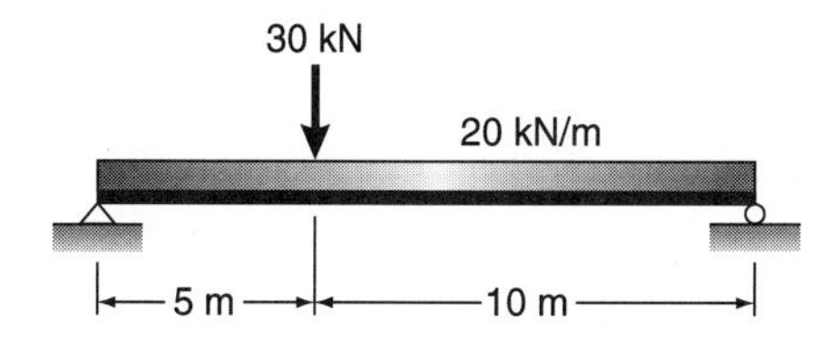

5.7

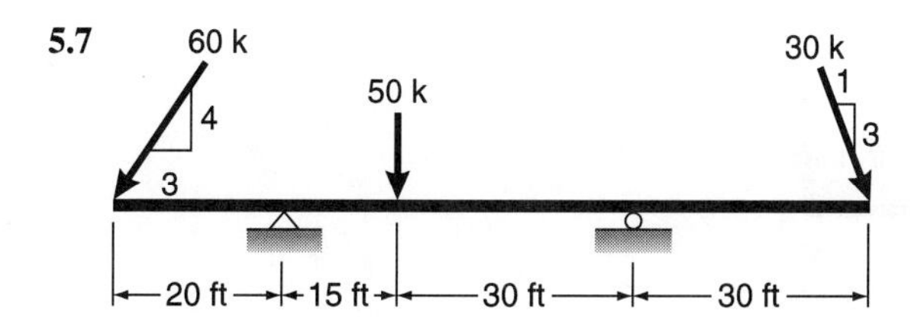

5.8 (*Ans.* max V = 100.5 k, max M = 968 ft-k)

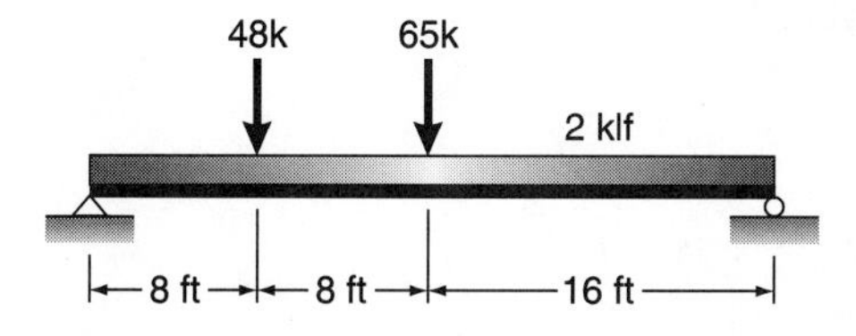

5.9

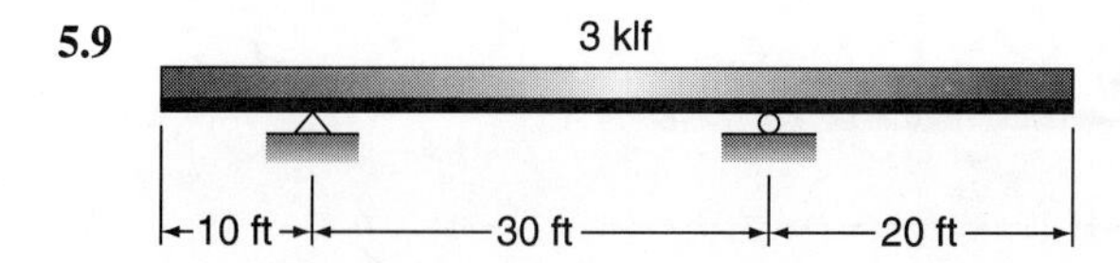

5.10 (*Ans*. max V = 160 kN, max M = −640 kN-m)

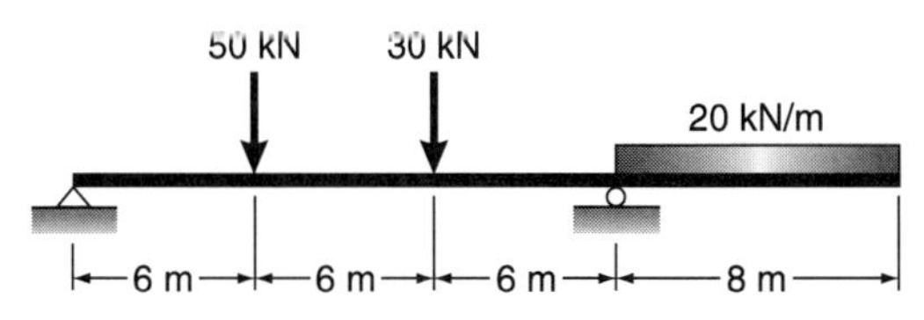

5.11

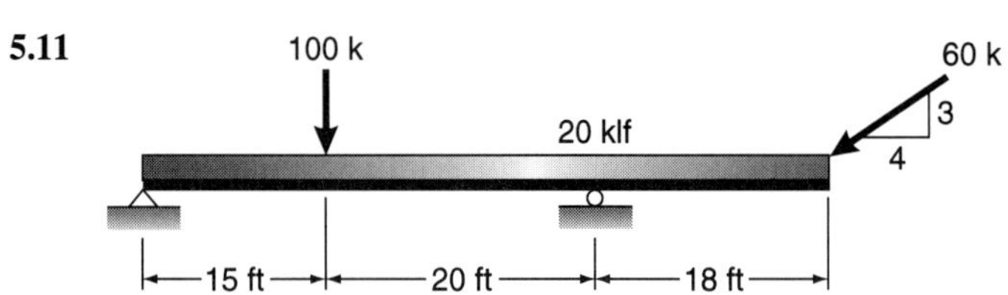

5.12 (*Ans*. max V = 54 k, max M = −1098 ft-k)

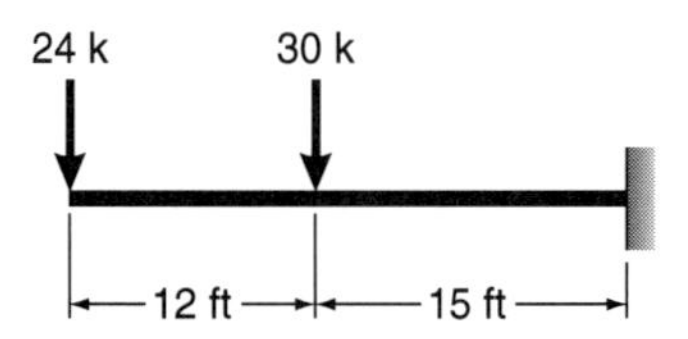

5.13

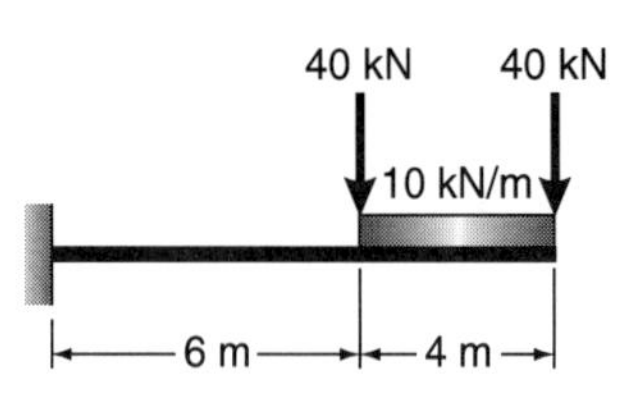

5.14 (*Ans*. max V = 120 kN, max M = −960 kN-m)

40 kN 40 kN

10 kN/m

6 m 4 m

5.15

4 klf

30 ft

5.16 (*Ans*. max V = 50 k, max M = −900 ft-k)

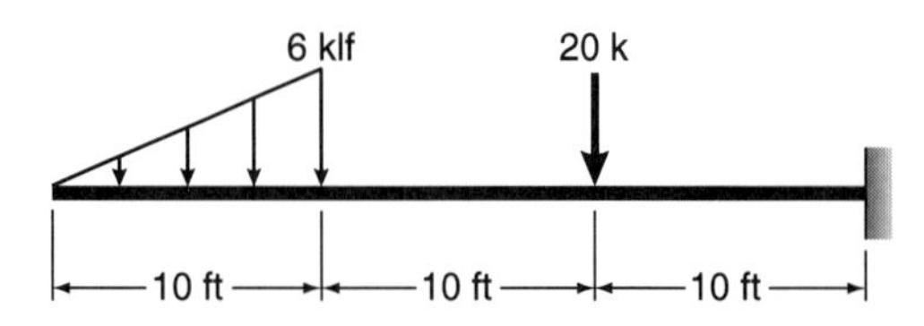

5.17

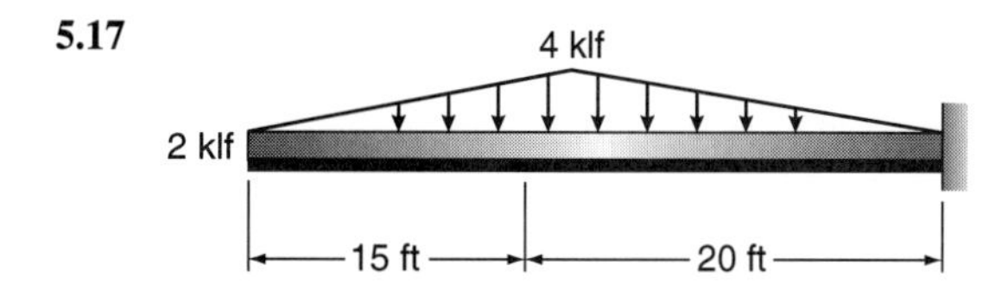

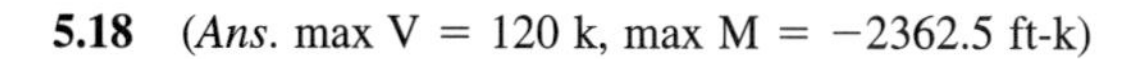

5.18 (*Ans.* max V = 120 k, max M = −2362.5 ft-k)

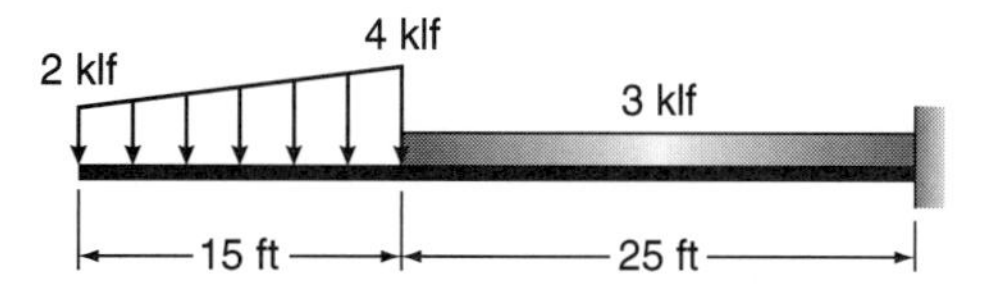

5.19

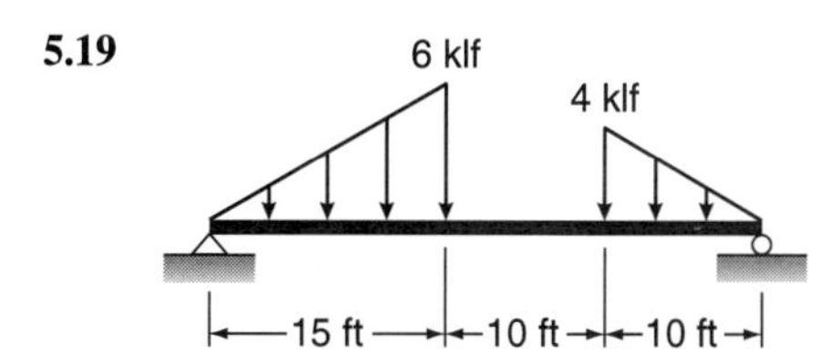

5.20 (*Ans.* max V = 57.68 k, max M = 278.6 ft-k)

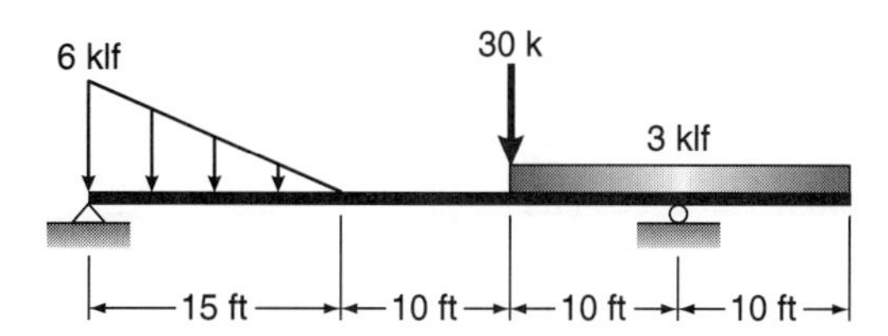

Draw shearing force and bending moment diagrams for the structures in Problems 5.21 to 5.36.

5.21

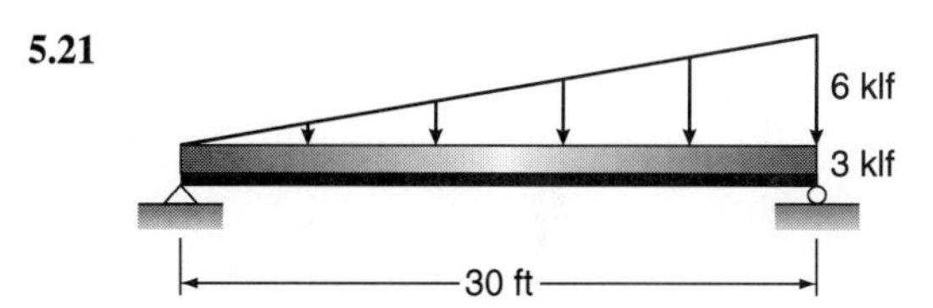

5.22 (*Ans.* max V = 79.64 k, max M = 644.4 ft-k)

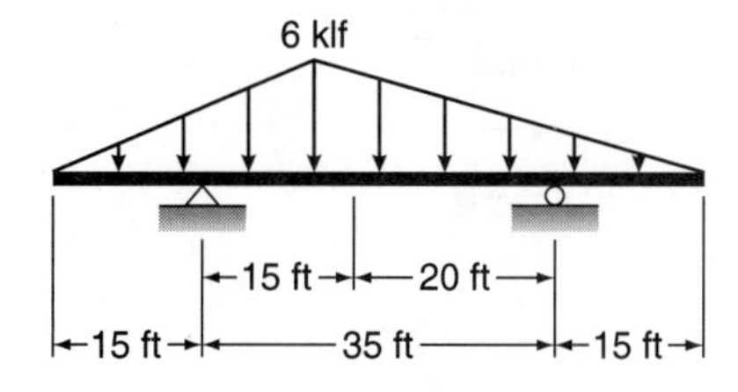

5.23

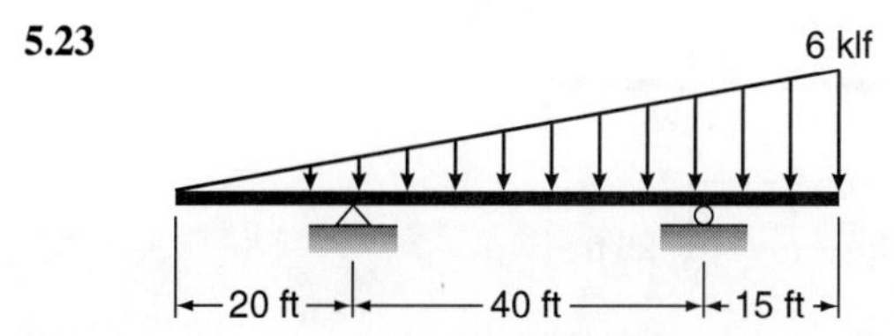

5.24 (*Ans.* max V = 317.9 kN, max M = 2526 kN-m)

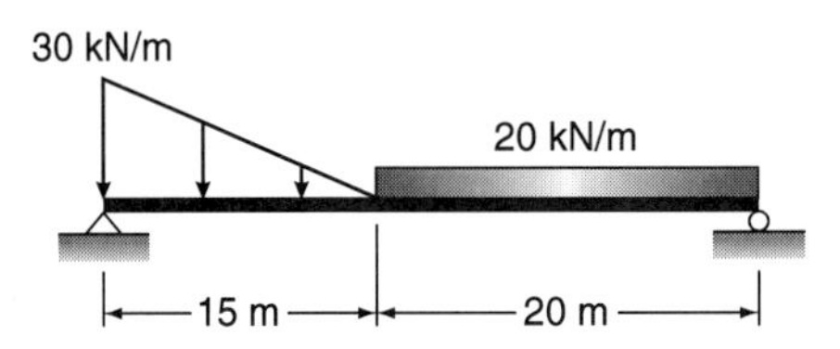

5.25

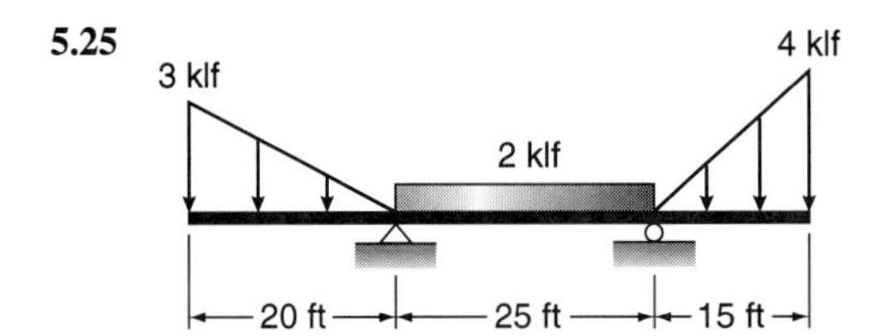

5.26 (*Ans.* max V = 115 k, max M = −1166.7 ft-k)

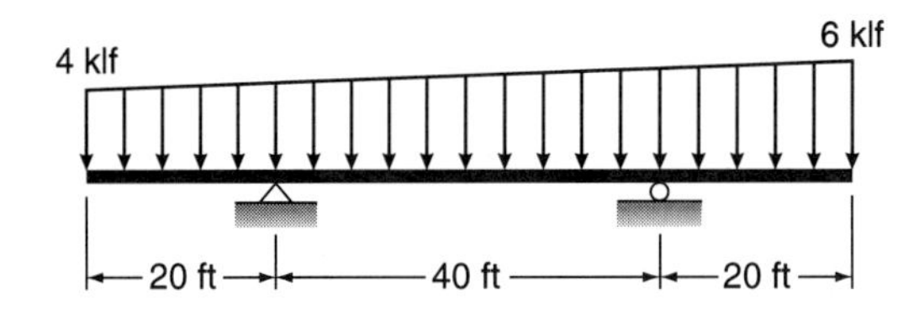

5.27

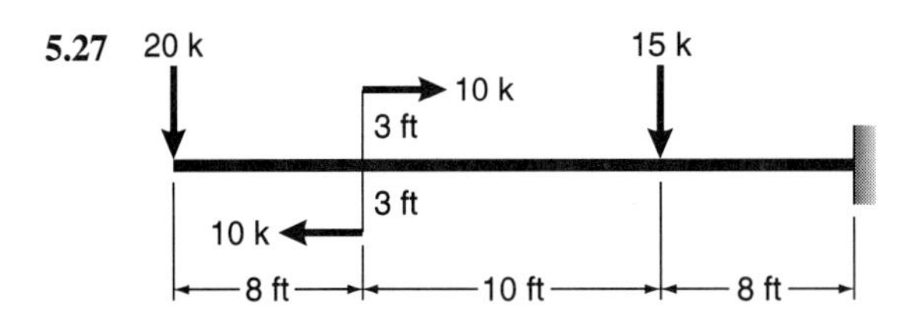

5.28 (*Ans.* max V = 22.4 k, max M = 83.7 ft-k)

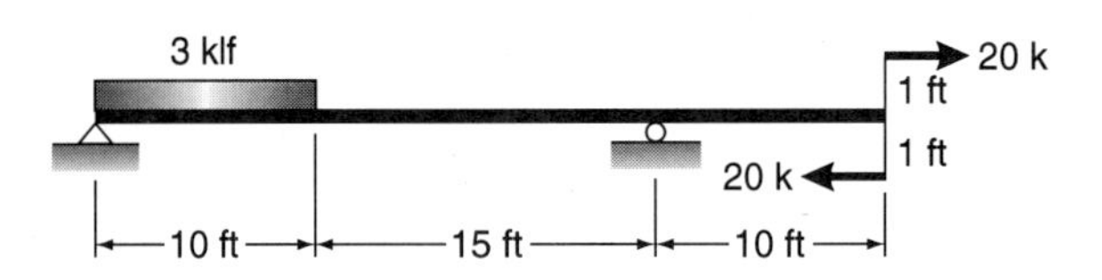

5.29

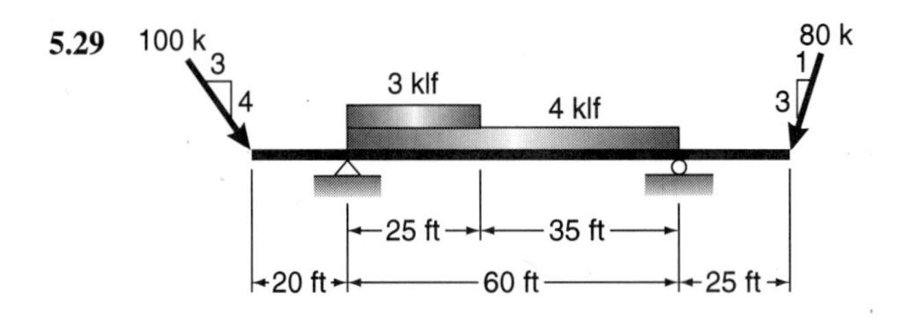

5.30 (*Ans.* max V = 100 k, max M = −2417.8 ft-k)

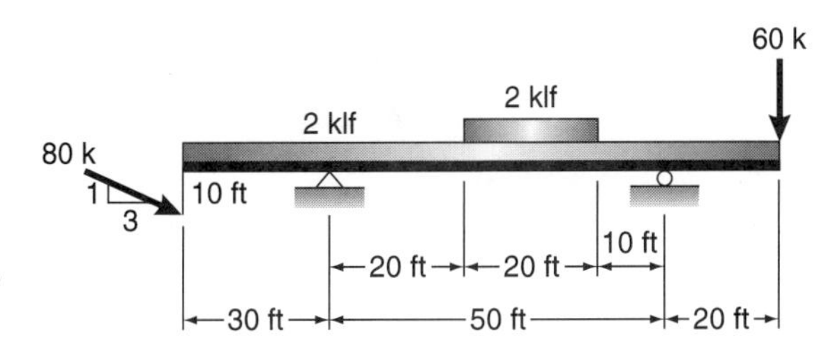

5.31

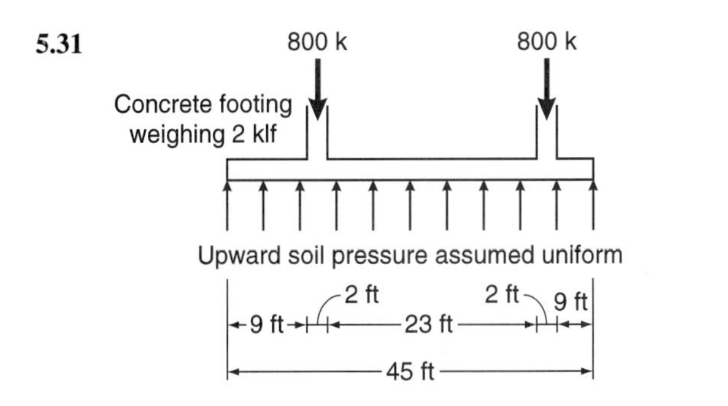

5.32 (*Ans.* max V = 89.66 k, max M = 2933 ft-k)

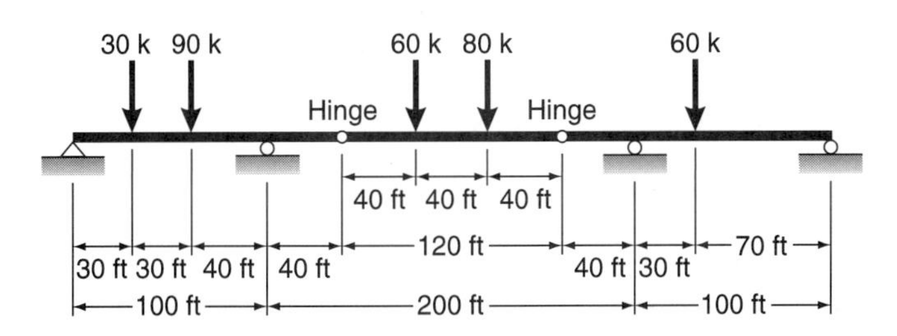

5.33

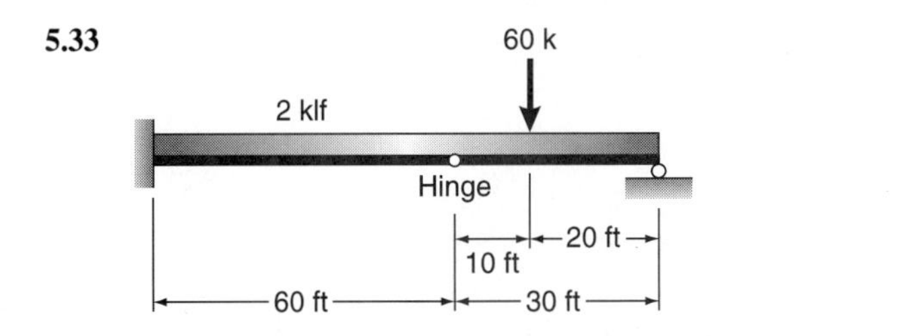

5.34 (*Ans.* max V = 80 k, max M = −2600 ft-k)

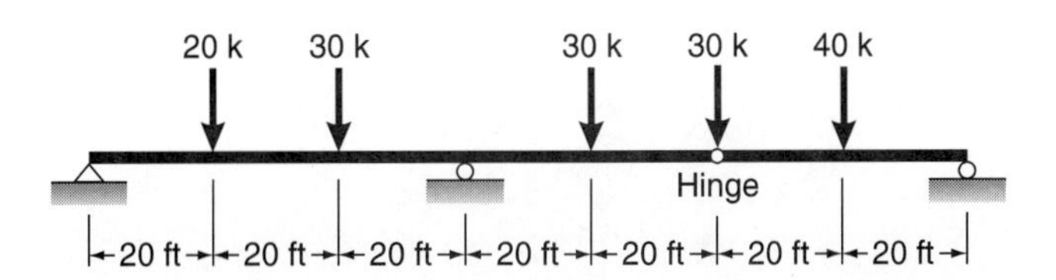

5.35 Given: Moment at interior support is −1274 ft-k

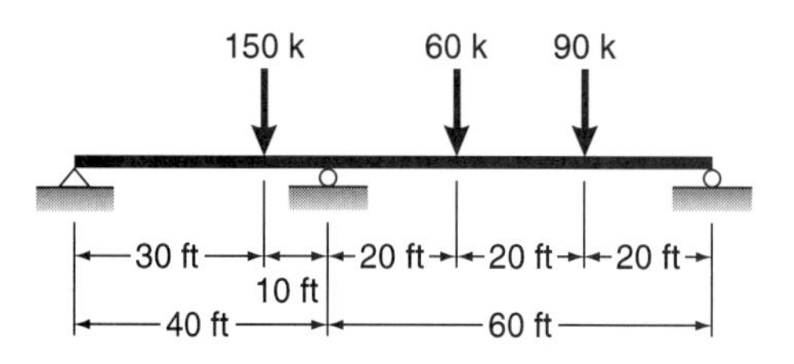

5.36 Given: Moment at fixed end is −147.7 ft-k; other reactions areas shown. (*Ans.* max V = 25.48 k, max M = −164.5 ft-k)

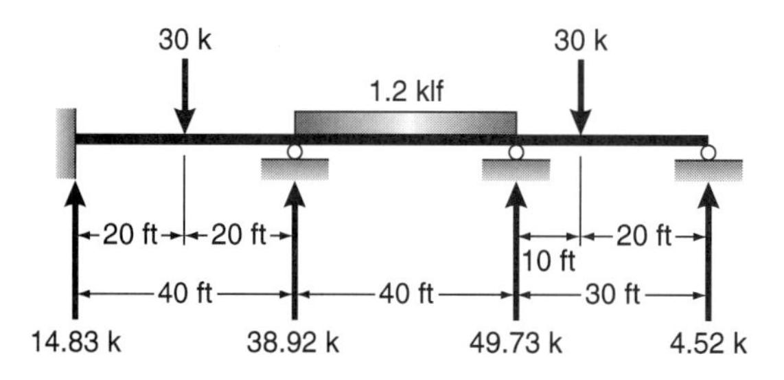

For Problems 5.37 to 5.39, draw the shear diagrams and load diagrams for the moment diagrams and dimensions given. Assume that upward forces are reactions.

5.37

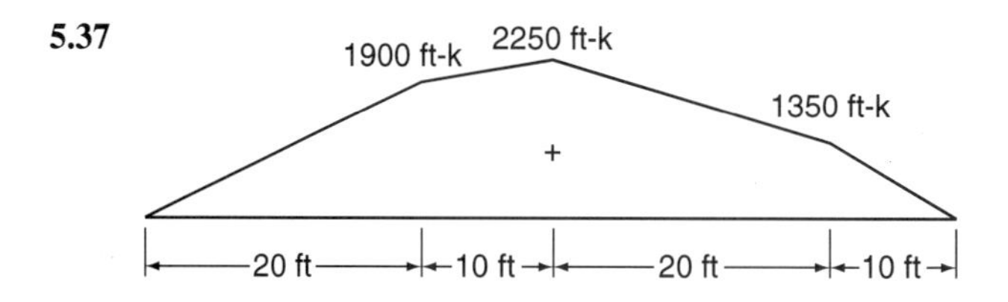

5.38 (*Ans.* Reactions and loads left to right: 96.5 kN, 80 kN, 40 kN, 50 kN, 73.5 kN)

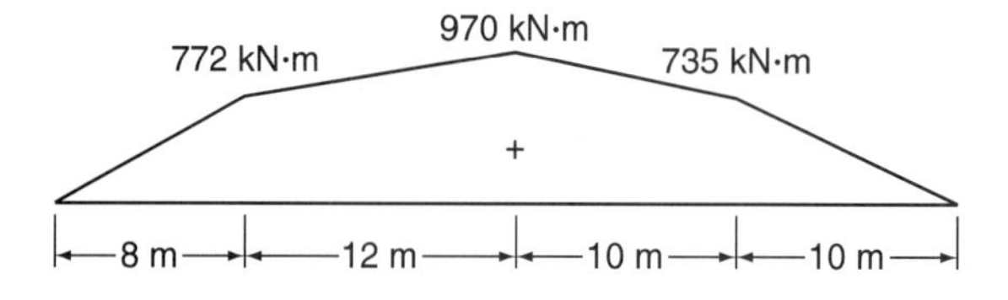

5.39

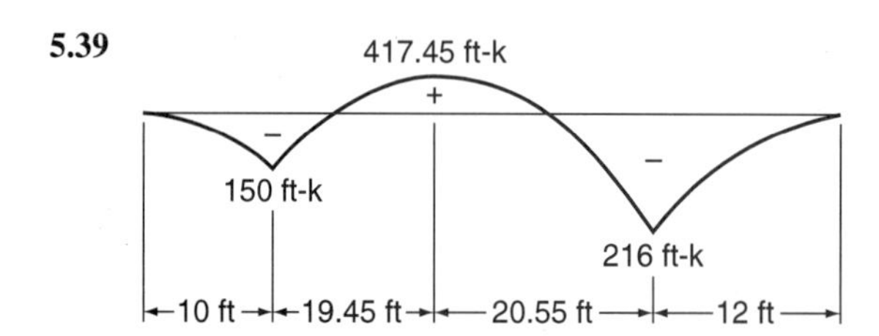

For Problems 5.40 to 5.49, draw shear and moment diagrams for the frames.

5.40 (*Ans*. max V = 58 k, max M = 841 ft-k)

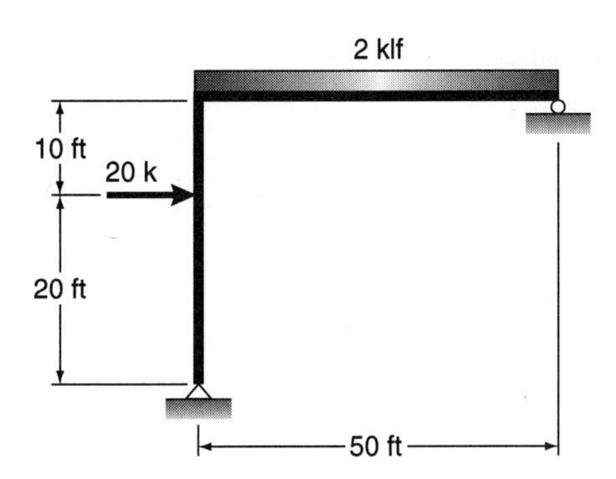

5.41

30 k
60 k
10 k
12 ft
20 k
12 ft
12 ft 12 ft 12 ft

5.42 Repeat Problem 5.41 if the roller and hinge supports are swapped. (*Ans*. max V = 46.67 k, max M = 520 ft-k)

5.43

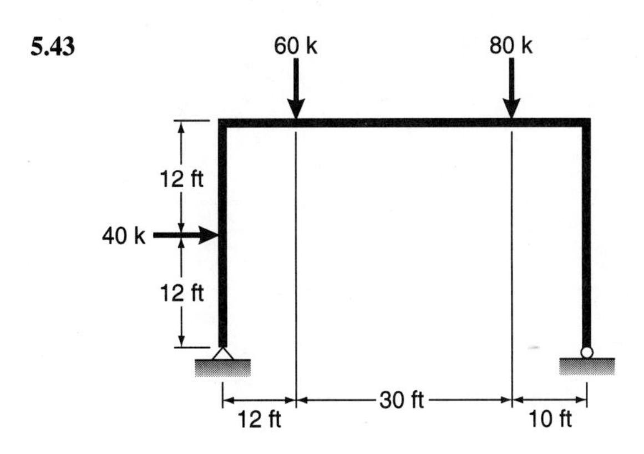

5.44 (*Ans*. max V = 113.33 k, max M = 1066.7 ft-k)

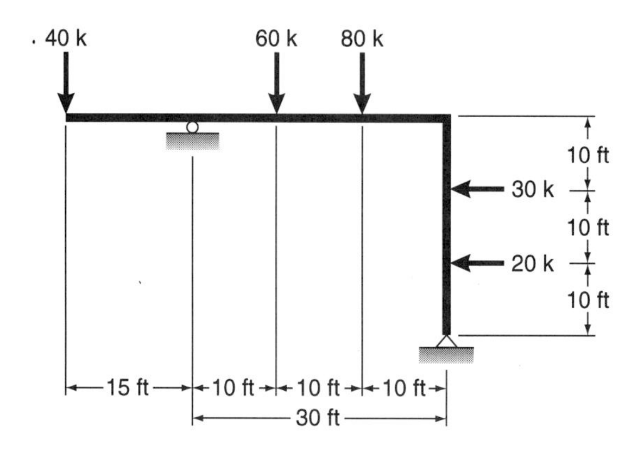

5.45

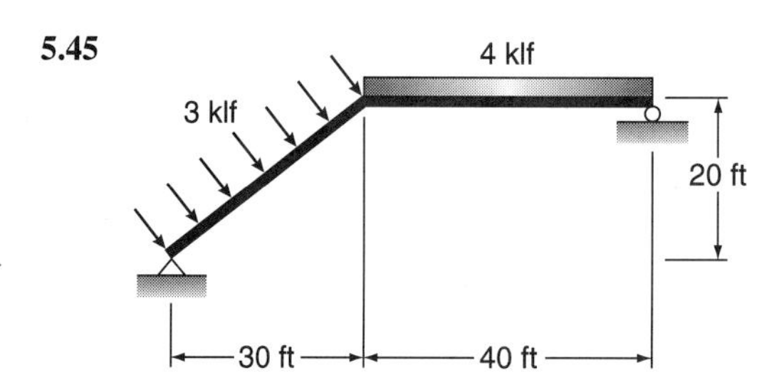

5.46 (*Ans*. max V = 90 k, max M = 1350 ft-k)

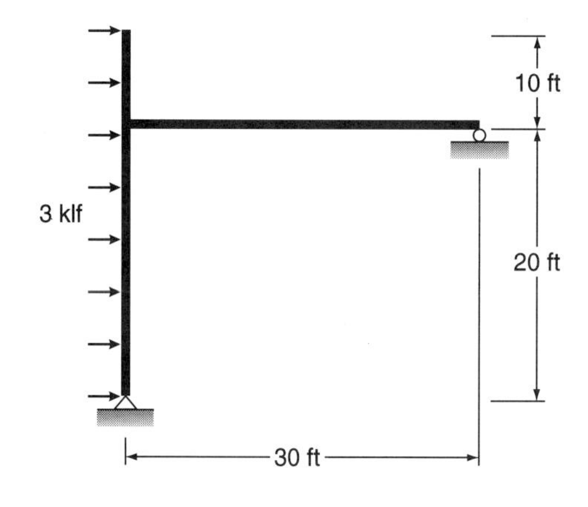

5.47

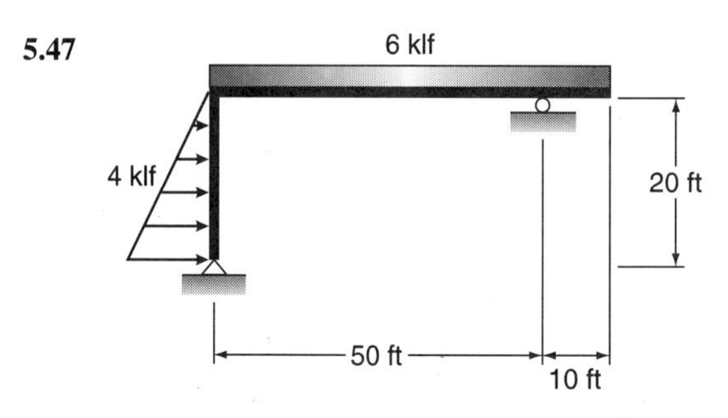

5.48 (*Ans.* max V = 88.85 k, max M = 2400 ft-k)

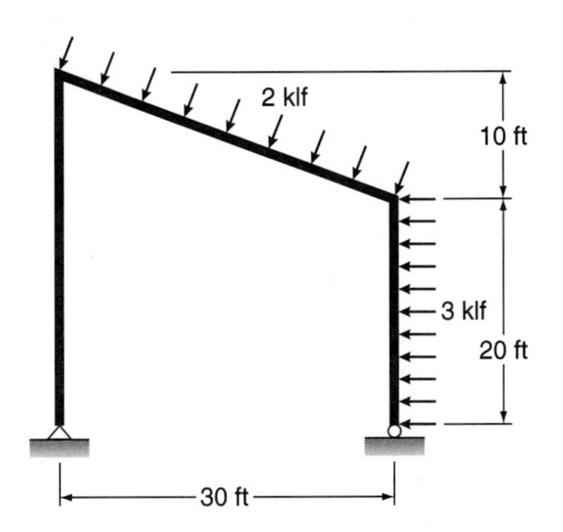

5.49

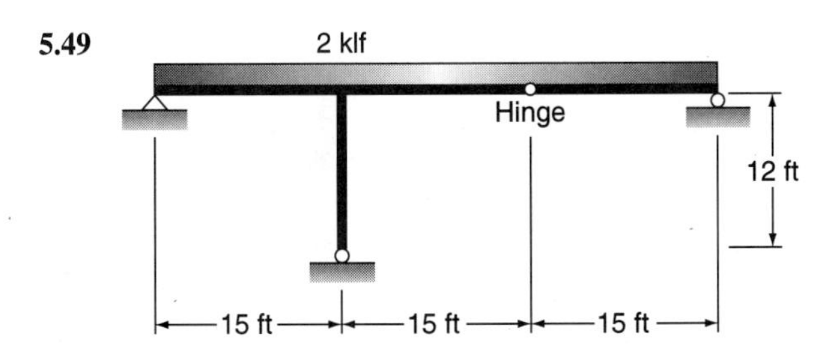

Chapter 6

Introduction to Plane Trusses

6.1 INTRODUCTION

Italian architect Andrea Palladio (1508–1580) is thought to have first used modern trusses, although his design basis is not known. He may have revived some old Roman designs and probably sized the truss component by some rules of thumb (perhaps old Roman rules). Palladio's extensive writing in architecture included detailed descriptions and drawings of wooden trusses quite similar to those used today. After his time, trusses were forgotten for 200 years, until they were reintroduced by Swiss designer Ulric Grubermann.

A truss is a structure formed by a group of members arranged in the shape of one or more triangles. Because the members are assumed to be connected with frictionless pins, the triangle is the only stable shape. Study of the truss of Figure 6.1(a) shows that it is impossible for the triangle to change shape under load—except through deformation of the members—unless one or more of the sides is bent or broken. Figures of four or more sides are not stable and may collapse under load, as seen in Figure 6.1(b) and 6.1(c). These structures may be deformed even if none of the members change length. We will see, however, that many stable trusses can include one or more shapes that are not triangles. Careful study reveals that trusses consist of separate groups of triangles that are connected together according to definite rules, forming non-triangular but stable figures in between.

Design engineers often are concerned with selecting between either a truss or a beam to span a given opening. Should no other factors be present, the decision probably would be based on consideration of economy. The smallest amount of material will nearly always be used if a truss is selected for spanning a certain opening; however, the cost of fabrication and erection of trusses probably will be appreciably higher than that required for beams. For shorter spans, the overall cost of beams (material cost plus fabrication and erection cost) will definitely be less, but as the spans become greater, the higher fabrication

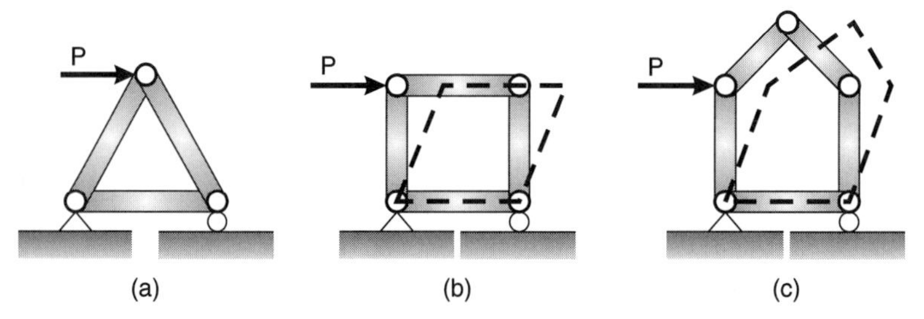

Figure 6.1 Stable truss versus unstable figures

and erection costs of trusses will be more than offset by the reduction in the total weight of material used. A further advantage of a truss is that for the same amount of material it can have greater stiffness than a beam with the same span.

A lower limit for the economical span of trusses is impossible to give. Trusses may be used for spans as small as 30 to 40 ft and as large as 300 to 400 ft. Beams may be economical for some spans much greater than the lower limits mentioned for trusses.

6.2 ASSUMPTIONS FOR TRUSS ANALYSIS

The following assumptions are made to simplify the analysis of trusses:

1. *Truss members are connected with frictionless pins.* In reality, pin connections are used for very few trusses erected today, and no pins are frictionless. A heavy bolted or welded joint is very different from a frictionless pin.
2. *Truss members are straight.* If they were not straight, the axial forces would cause them to have bending moments.
3. *The displacement of the truss is small.* The applied loads cause the members to change length, which then causes the truss to deform. The deformations of a truss are not of sufficient magnitude to cause appreciable changes in the overall shape and dimensions of the truss. Special consideration may have to be given to some very long and flexible trusses.
4. *Loads are applied only at the joints.* Members are arranged so that the loads and reactions are only applied at the truss joints.

Examination of roof and bridge trusses will prove this last statement to be generally true. Beams, columns, and bracing members frame directly into the truss joints of buildings with roof trusses. Roof loads are transferred to trusses by horizontal beams, called *purlins,* which span between the trusses. The roof is supported directly by the purlins. The roof may also be supported by rafters, or sub-purlins, which run parallel to trusses and are supported by the purlins. The purlins are placed at the truss joints unless the top-chord panel lengths become exceptionally long, in which case it is sometimes economical to place purlins between the joints, although some bending will develop in the top chords. Some types of roofing, such as corrugated steel and gypsum slabs, may be laid directly on the purlins. In this case, the purlins then have to be spaced at intermediate points along the top chord to provide a proper span for the supported roof. Similarly, the loads supported by a highway bridge are transferred to the trusses at the joints by floor beams and girders underneath the roadway.

The effect of the foregoing assumptions is to produce an ideal truss, whose members have only axial forces. As illustrated in Figure 6.2(a) and (b), a member with only axial force is subjected to axial tension or compression; there is no bending moment present as shown in Figure 6.2(c). Be aware, however, that even if all the assumptions about trusses were completely true, there still would be some bending in a member because of

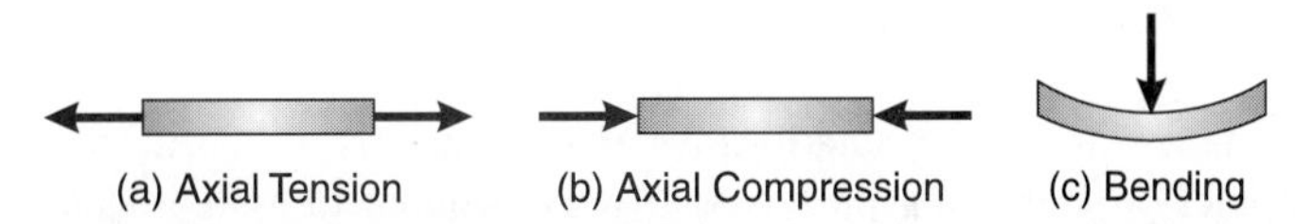

Figure 6.2 Forces acting on member

its own weight. The weight of the member is distributed along its length rather than being concentrated at the ends. Compared to the forces caused by the applied loads, the forces caused by self-weight are small and generally can be neglected when calculating the forces in the components.

Component forces obtained using some or all of these simplifying assumptions are very satisfactory in most cases, and are referred to as *primary forces*. Forces caused by conditions not considered in the primary force analysis are said to be *secondary forces*.

Truss for a conveyor at the West Virginia quarry of Pennsylvania Glass Sand Corporation (Courtesy Bethlehem Steel Corporation)

6.3 TRUSS NOTATION

A common system of denoting the members of a truss is shown in Figure 6.3. The joints are numbered from left to right. Those on the bottom are labeled L, for lower. The joints on the top are labeled U, for upper. Should there be, in more complicated trusses, joints between the lower and upper joints, they may be labeled M, for middle.

The various members of a truss often are referred to by the following names, as labeled in Figure 6.3.

1. *Chords* are those members forming the outline of the truss, such as members U_1U_2 and L_4L_5.
2. *Verticals* are named on the basis of their direction in the truss, such as members U_1L_1 and U_3L_3.
3. *Diagonals* also are named on the basis of their direction in the truss, such as members U_1L_2 and L_4U_5.
4. *End posts* are the members at the ends of the truss, such as members L_0U_1 and U_5L_6.
5. *Web members* include the verticals and diagonals of a truss. Most engineers consider the web members to include the end posts.

There are other notation systems frequently used for trusses. For instance, for computer-programming purposes, it is convenient to assign a number to each joint and each member of a truss. Such a system is illustrated in the next Chapter.

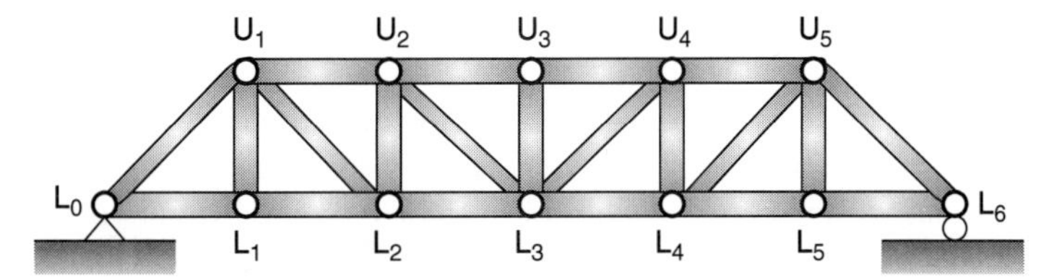

Figure 6.3 Common labeling joints of a truss

6.4 ROOF TRUSSES

The purposes of roof trusses are to support the roofs that keep out the elements (rain, snow, wind), the loads connected underneath (ducts, piping, ceiling), and their own weight.

Roof trusses can be flat or peaked. In the past, peaked roof trusses probably have been used more for short-span buildings and flatter trusses for longer spans. The trend today for both long and short spans, however, seems to be away from the peaked trusses and toward the flatter trusses. The change is predominantly due to the desired appearance of the building, and perhaps a more economical construction of roof decks. Figure 6.4 illustrates several types of roof trusses that have been used in the past.

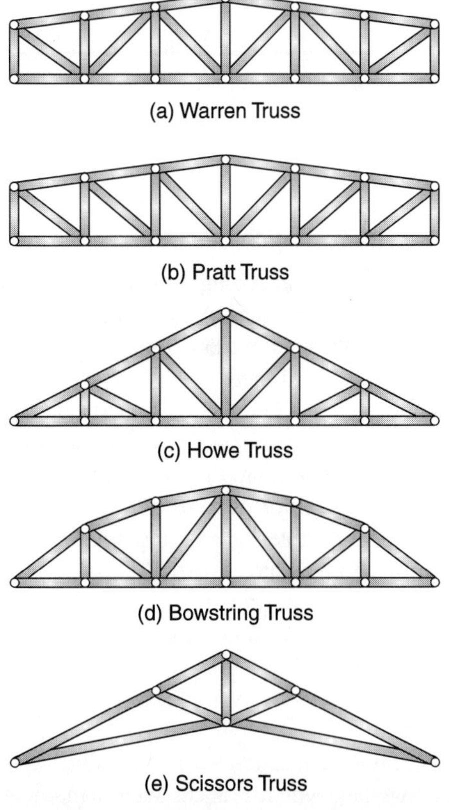

Figure 6.4 Several typical roof trusses

6.5 BRIDGE TRUSSES

As bridge spans become longer and loads heavier, trusses begin to be economically competitive with beams. Early bridge trusses were constructed with wood, and they had several disadvantages. First, they were subject to deterioration from wind and water. As a result covered bridges were introduced, and such structures often would last for quite a few decades. Nevertheless, wooden truss bridges, particularly railroad bridges, were subject to destruction by fire. In addition, there were some problems with the loosening over time of the fasteners under moving loads and as loads became heavier.

Willard Bridge over the Kansas River north of Willard, Kansas (Courtesy of the American Institute of Steel Construction, Inc.)

Because of the several preceding disadvantages, wooden truss bridges faded from use toward the end of the 19th century. Although there were some earlier iron truss bridges, structural steel bridges became predominant. Steel bridges did not require extensive protection from the elements and their joints had higher fatigue resistance.

Today existing steel truss bridges are being steadily replaced with steel, precast-concrete, or prestressed-concrete beam bridges. The age of steel truss bridges appears to be over, except for bridges with spans of more than several hundred feet, which represent a very small percentage of the total. Even for these longer spans, there is much competition from other types of structures such as cable-stayed bridges and prestressed-concrete box girder bridges.

Some highway bridges have trusses on the sides and overhead lateral bracing between the trusses. This type of bridge is called a *through bridge*. The floor system is supported by floor beams that run under the roadway and between the bottom-chord joints of the trusses. A through railroad bridge is shown in Figure 6.5(a)

In a *deck bridge,* the roadway is placed on top of the trusses or girders. Deck bridges have many advantages over through bridges: there is unlimited overhead horizontal and vertical clearance, future expansion is more feasible, and supporting trusses or girders can be placed close together, which reduces lateral moments in the floor system. Other advantages of the deck truss are simplified floor systems and possible reduction in the sizes of piers and abutments due to reductions in their heights. Finally, the very pleasing

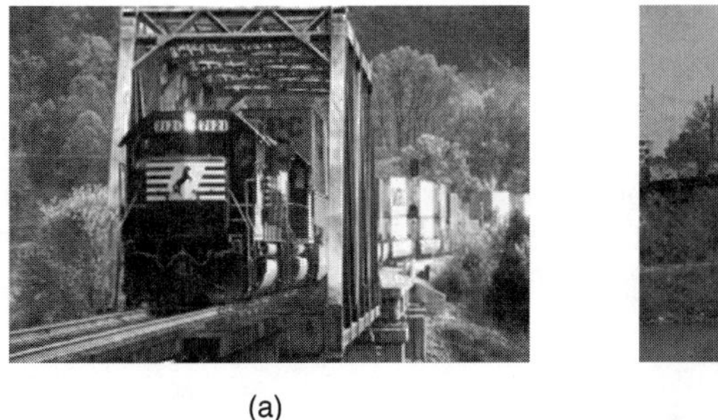

(a)

(b)

Figure 6.5 A through bridge (a) and deck bridge (b)

appearance of deck structures is another reason for their popularity. The only real disadvantage of a deck bridge is clearance beneath the bridge. The bridge may need to be set high to allow adequate clearance for ships and vehicles to pass underneath. A deck railroad bridge is shown in Figure 6.5(b). Deck bridges eliminate a sense of confinement exhibited by other types of bridges.

Sometimes short-span through-bridge trusses were so shallow that adequate depth was not available to provide overhead bracing and at the same time leave sufficient vertical clearance above the roadway for the traffic. As a result, the bracing was placed underneath the roadway. Through bridges without overhead bracing are referred to as *half-through* or *pony bridges*. One major problem with pony truss bridges is the difficulty of providing adequate lateral bracing for the top-chord compression members of the trusses. Pony trusses are not likely to be economical today because beams have captured the short-span bridge market.

Shown in Figure 6.6 are several types of bridge trusses. Parallel-chord trusses are shown in parts (a) through (d) of the figure. The Baltimore truss is said to be a subdivided truss because the unsupported lengths of some of the members have been

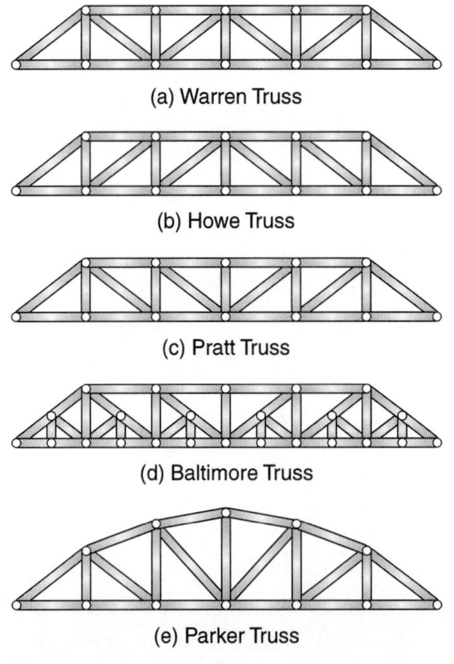

Figure 6.6 Some typical bridge trusses

reduced by short members called sub-diagonals and sub-verticals. The deeper a truss is made for the same-size chord members, the greater will be its resisting moment. If the depth of a truss is varied across its span in proportion to its bending moments, the result will be a lighter truss. The fabrication cost-per-pound of steel will be higher, though, than for a parallel-chord truss. As spans become longer, the reduced weight achieved by varying truss depths with bending moments will outweigh the extra fabrication costs, thus making economical the so-called curved-chord trusses. The Parker truss shown in Figure 6.6(e) is an example.

6.6 ARRANGEMENT OF TRUSS MEMBERS

The triangle has been shown to be the basic shape from which trusses are developed because it is the only stable shape. For the following discussion, remember that the members of trusses are assumed to be connected at their joints with frictionless pins.

Other shapes, such as those shown in Figure 6.7(a) and (b) are obviously unstable, and may possibly collapse under load. Structures such as these, however, can be made stable by one of the following methods:

- Add members so that the unstable shapes are altered to consist of triangles. The structures of Figure 6.7(a) and (b) are stabilized in this manner in (c) and (d), respectively.
- Use a member to tie the unstable structure to a stable support. Member AB performs this function in Figure 6.7(e).
- Make rigid some or all of the joints of an unstable structure, so they become moment resisting. A figure with moment-resisting joints, however, does not coincide with the definition of a truss; the members are no longer connected with frictionless pins.

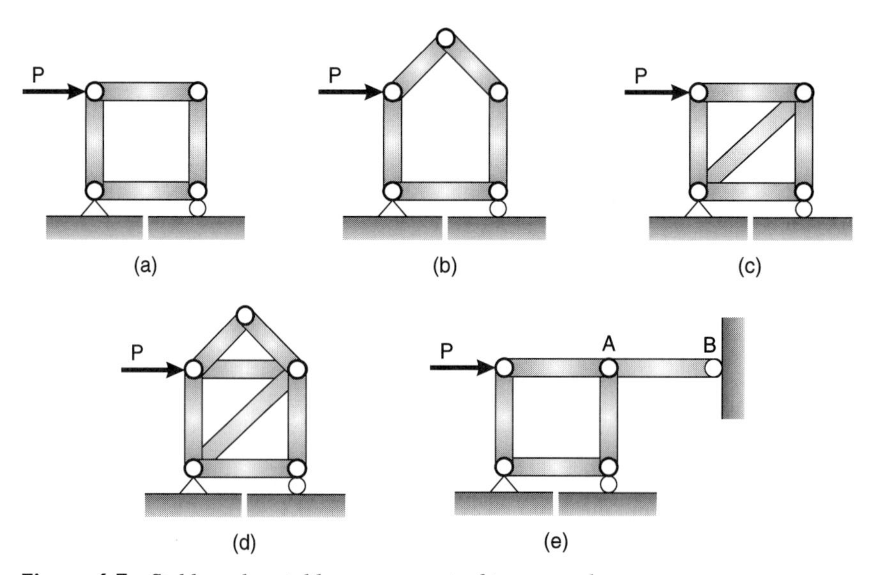

Figure 6.7 Stable and unstable arrangement of truss members

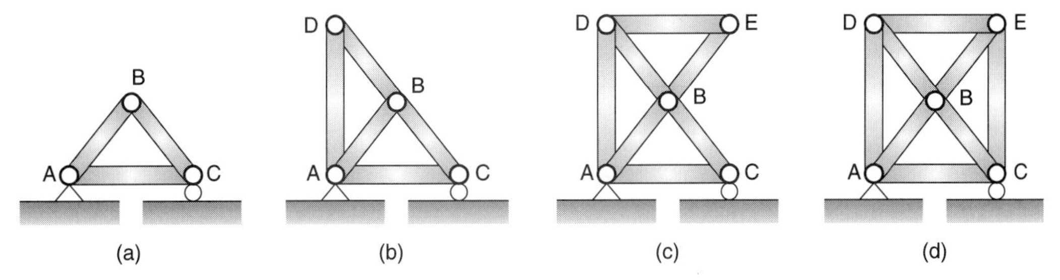

Figure 6.8 Expanding a simple truss

6.7 STATICAL DETERMINACY OF TRUSSES

The simplest form of truss, a single triangle, is illustrated in Figure 6.8(a). To determine the unknown forces and reaction components for this truss, it is possible to isolate the joints and write two equations of equilibrium for each. These equations of equilibrium, which involve the summation of vertical and horizontal forces, are:

$$\sum F_H = 0$$
$$\sum F_V = 0 \qquad \textbf{Eq. 6.1}$$

The single-triangle truss may be expanded into a two-triangle truss by the addition of two new members and one new joint. In Figure 6.8(b), triangle ABD is added by installing new members AD and BD, and the new joint D. A further expansion with a third triangle is made in Figure 6.8(c) by the addition of members BE, DE and joint E. For each of the new joints, D and E, a new pair of equations of equilibrium is available for calculating the two new-member forces. As long as this procedure of expanding the truss is followed, the truss will be statically determinate internally. Should new members be installed without adding new joints, such as member CE in Figure 6.8(d), the truss will become statically indeterminate because no new equations of equilibrium are available to find the new-member forces.

Brown's Bridge, Forsyth and Hall Counties, Gainesville, Georgia (Courtesy of the American Institute of Steel Construction)

Using this information, an expression can be written for the relationship that must exist between the number of joints, the number of members, and the number of reactions for a particular truss if it is to be statically determinate internally. The identification of externally determinate structures has previously been discussed in Chapter 4. In the following discussion, m is the number of members, j is the number of joints, and r is the number of reaction components.

If the number of available equations of static equilibrium, which is 2j, is sufficient to compute the unknown forces, the structure is statically determinate. As such, the following relation may be written:

$$2j = m + r \qquad \textbf{Eq. 6.2}$$

This equation is more commonly written as

$$m = 2j - r \qquad \textbf{Eq. 6.3}$$

For this equation to be applicable, the structure must be stable externally; otherwise the results are meaningless. Therefore, r is the least number of reaction components required for external stability. Should the structure have more external reaction components than necessary for stability, and thus be statically indeterminate externally, the value of r remains the least number of reaction components required to make it stable externally. This statement means that r is equal to 3 for the usual equations of static equilibrium, plus the number of any additional equations of condition that may be available.

It is possible to build trusses that have more members than can be analyzed using equations of static equilibrium. Such trusses are statically indeterminate internally, and m will exceed 2j − r because there are more members present than are necessary for stability. The extra components are said to be redundant members. If m is 3 greater than 2j − r, there are three redundant members, and the truss is internally statically indeterminate to the third degree. Should m be less than 2j − r, there are not enough members present for stability.

A brief glance at a truss usually will show if it is statically indeterminate. Trusses having members that cross over each other or members that serve as the sides for more than two triangles may quite possibly be indeterminate. The 2j − r expression should be used, however, if there is any doubt about the determinacy of a truss, because it is easy to be mistaken. The following example demonstrates the application of these concepts to a few common trusses.

EXAMPLE 6.1

Determine whether the trusses shown in the figures are statically determinate or indeterminate, and whether they are stable.

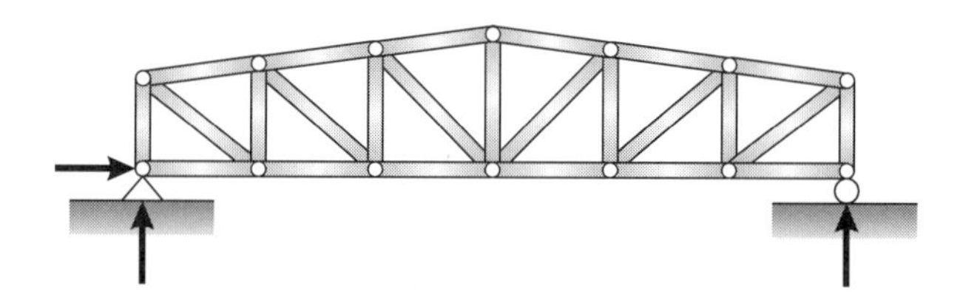

$m = 25$

$j = 14$

$r = 3$

This truss is stable and statically determinate internally and externally: m = 2j − r and r = 3.

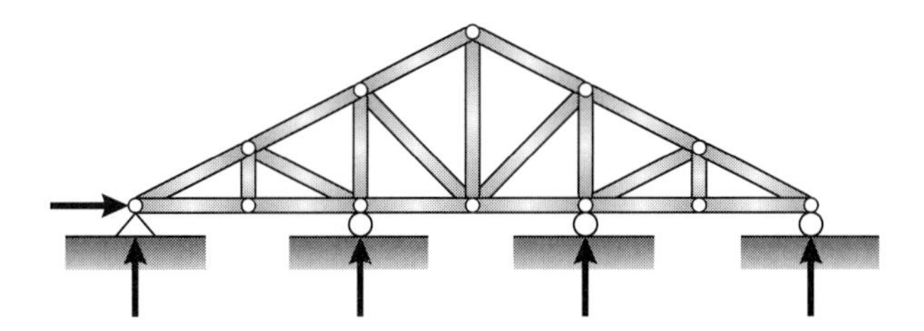

$$m = 21$$
$$j = 12$$
$$r = 5$$

This truss is stable because r > 3 and m > 2j − r. It is statically indeterminate externally because r > 3, but statically determinate internally because m = 2j − 3. In this last expression, 3 is used because that is the minimum number of reaction components needed for stability and statical determinacy.

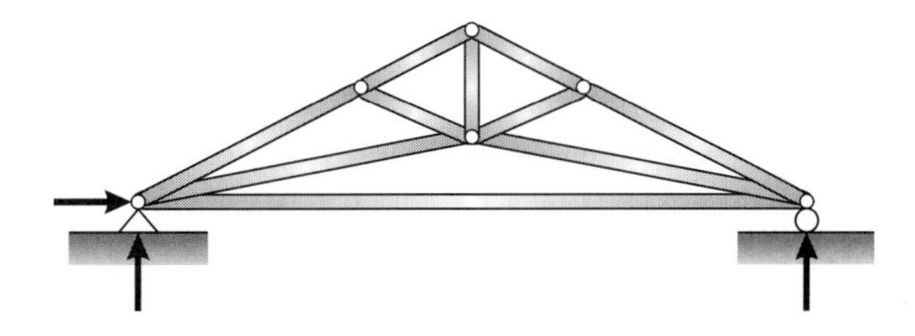

$$m = 10$$
$$j = 6$$
$$r = 3$$

This truss is stable and statically determinate externally because m = 3. It is statically indeterminate internally because m > 2j − 3. In fact, it is statically indeterminate internally to the first degree. ■

In Chapter 13, which pertains to the analysis of statically indeterminate structures, we will see that the values of the redundants may be obtained by applying certain simultaneous equations. The number of simultaneous equations used is equal to the total number of redundants, whether internal, external, or both. Therefore, it may be a little foolish to distinguish between internal and external determinacy. The separation is particularly questionable for some types of internally and externally redundant trusses where no solution of the reactions is possible independently of the member forces, and vice versa.

Nevertheless, if a truss is externally determinate and internally indeterminate, the reactions may be obtained using equations of static equilibrium. If the truss is externally indeterminate and internally determinate, the reactions are dependent on the internal-member forces and may not be determined by a method independent of those forces. If the truss is both externally and internally indeterminate, the solution of the member forces and reactions must be performed simultaneously. For any of these situations, it may be possible to obtain a few forces here and there by statics without going through the indeterminate procedure necessary for complete analysis. This subject is discussed in detail in later chapters.

6.8 METHODS OF ANALYSIS AND CONVENTIONS

An indispensable and essential tool in truss analysis is the ability to divide the truss into pieces, construct a free-body diagram for each piece, and then from these free-body diagrams determine the forces in the components. The free-body diagrams can be single joints in the truss or an entire section of the truss. As an example, consider again the truss that was shown in Figure 6.3. Let us cut a section around joint U_4 and through the truss just after joints U_1 and L_1 as shown in Figure 6.9(a). The resulting free-body diagrams are shown in Figure 6.9(b) and (c). When working with free-body diagrams such as that shown in Figure 6.9(b), we are said to be evaluating the member forces using the method of joints. On the other hand, if we evaluate the member forces using free-body diagrams such as that which is shown in Figure 6.9(c), we are said to be using the method of sections. The methods of joints will be considered in this chapter and the method of sections will be discussed in the next chapter.

Applying the equations of static equilibrium to isolated free-bodies enables us to determine the forces in the cut members. The free-bodies must be carefully selected so that the sections do not pass through too many members whose forces are unknown. When using the method of joints, there are only two relevant equations of equilibrium for each free-body: summation of forces vertically and summation of forces horizontally. There are three applicable equations of equilibrium for each free-body when using the method of sections: summation of forces vertically, summation of forces horizontally, and summation of moments. On any free-body we use, there cannot be more unknown forces than there are relevant equations of static equilibrium.

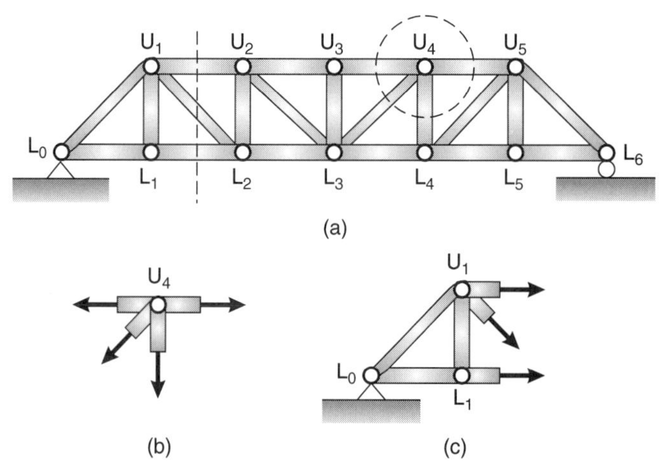

Figure 6.9 Free-body diagrams of a truss joint and a truss section

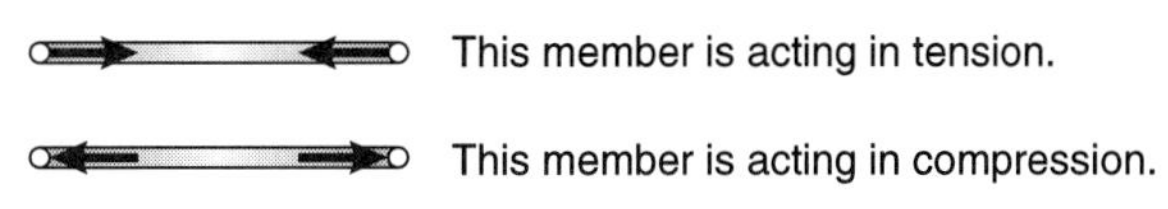

Figure 6.10 Arrows indicating a member's action on its joints

After you have analyzed a few trusses, you will have little difficulty in most cases in selecting satisfactory locations for the sections. You are not encouraged to remember specific sections for specific trusses, although you will probably fall into such a habit unconsciously as time passes. At this stage, you need to consider each case individually without reference to other similar trusses.

Good practice dictates that a sign convention be adopted when analyzing trusses, and that this convention be used consistently for all trusses. By doing so, many errors inherent with changing sign conventions are eliminated. We have adopted the sign convention that all unknown forces in a truss are assumed to be tensile forces. Further, tensile and compressive forces are indicated with a plus or minus sign (+ and −), respectively. This convention will be demonstrated through the example problems.

Arrows are often used to represent the character of forces. The arrows indicate what members are doing to resist the axial forces applied to them by the remainder of the truss. For example, if a particular member in a truss is acting in compression, it is pushing against the joints to which it is connected. Conversely, if a member is acting in tension, it is pulling on the joints to which it is connected. Arrows superimposed on the member indicate the action of the member on the joints as indicated in Figure 6.10.

After some practice analyzing trusses, the character of the forces in many of the members can be determined by examination. Try to picture whether a member is in tension or compression before making actual calculations: by doing so you will achieve a better understanding of the action of trusses under load. The next section demonstrates that it is possible to determine entirely by mathematical means the character as well as the numerical value of the forces.

6.9 METHOD OF JOINTS

An imaginary section may be passed around a joint in a truss, regardless of its location, completely isolating it from the remainder of the truss. The joint has become a free-body in equilibrium under the forces applied to it. The applicable equations of equilibrium, $\Sigma F_H = 0$ and $\Sigma F_V = 0$, may be applied to the joint to determine the unknown forces in members meeting there. It should be evident that no more than two unknowns can be determined at a joint with these two equations.

A person learning the method of joints may initially find it necessary to draw a free-body sketch for every joint in a truss he or she is analyzing. After you have computed the forces in two or three trusses, it will be necessary to draw the diagrams for only a few joints, because you will be able to easily visualize the free-bodies involved. Drawing large sketches helps visualization. The most important thing for the beginner to remember is that you are interested in only one joint at a time. Keep your attention away from the loads and forces at other joints. Your concern is only with the forces at the one joint on which you are working. The method of joints is illustrated through the examples that follow.

EXAMPLE 6.2

Using the method of joints, find all forces in the truss shown in the figure below

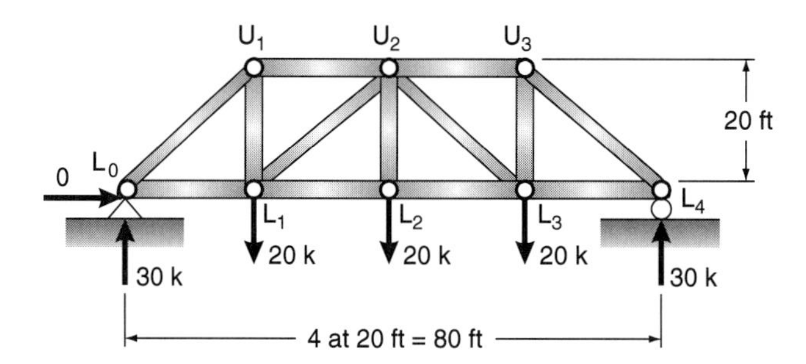

Solution. When applying the methods of joints, there can be only two unknown bar forces at each joint since there are only two equations of equilibrium for each joint—the summation of forces horizontally and summation of forces vertically. The free-body diagram used with each joint is shown along with the equations of equilibrium. Unknown bar forces will be assumed to be acting in tension.

We will begin with joint L_0.

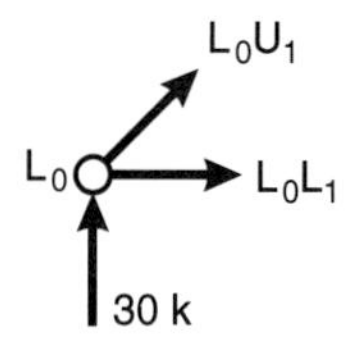

$$\left.\begin{aligned}\sum F_H &= L_0L_1 + L_0U_1\cos(45) = 0\\ \sum F_V &= 30 + L_0U_1\sin(45) = 0\end{aligned}\right\} \rightarrow \begin{aligned}L_0L_1 &= 30 \text{ kips}\\ L_0U_1 &= -42.43 \text{ kips}\end{aligned}$$

Next, we will move to joint U_1.

$$\left.\begin{aligned}\sum F_H &= -L_0U_1\cos(45) + U_1U_2 = 0\\ \sum F_V &= -L_0U_1\sin(45) - L_1U_1 = 0\end{aligned}\right\} \rightarrow \begin{aligned}L_1U_1 &= 30 \text{ kips}\\ U_1U_2 &= -30 \text{ kips}\end{aligned}$$

Now we will move to joint L_1; there are only two unknown forces acting on this joint:

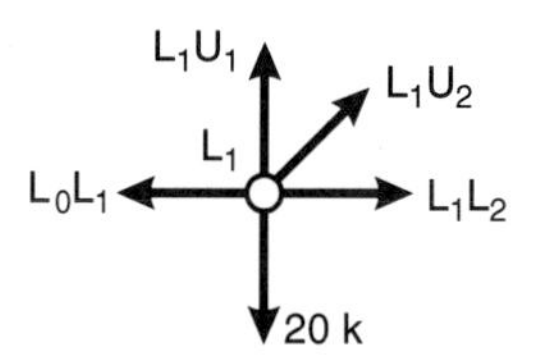

$$\left.\begin{aligned}\sum F_H &= -L_0L_1 + L_1L_2 + L_1U_2\cos(45) = 0\\ \sum F_V &= L_1U_1 - 20 + L_1U_2\sin(45) = 0\end{aligned}\right\} \rightarrow \begin{aligned}L_1U_1 &= 30 \text{ kips}\\ U_1U_2 &= -30 \text{ kips}\end{aligned}$$

Next we work with joint L_2. We cannot move to joint U_2 because there are three unknown bar forces acting at that joint:

L_2U_2
L_2
L_1L_2 L_2L_3
20 k

$$\left.\begin{aligned}\sum F_H &= -L_1L_2 + L_2L_3 = 0\\ \sum F_V &= L_2U_2 - 20 = 0\end{aligned}\right\} \rightarrow \begin{aligned}L_2L_3 &= 40 \text{ kips}\\ L_2U_2 &= 20 \text{ kips}\end{aligned}$$

We proceed with the remaining joints in this truss in the same manner. As we work through the joints, we are always moving to a joint that has no more than two unknown bar forces acting. The final bar forces are summarized in the following figure. A negative sign indicates that the bar is acting in axial compression.

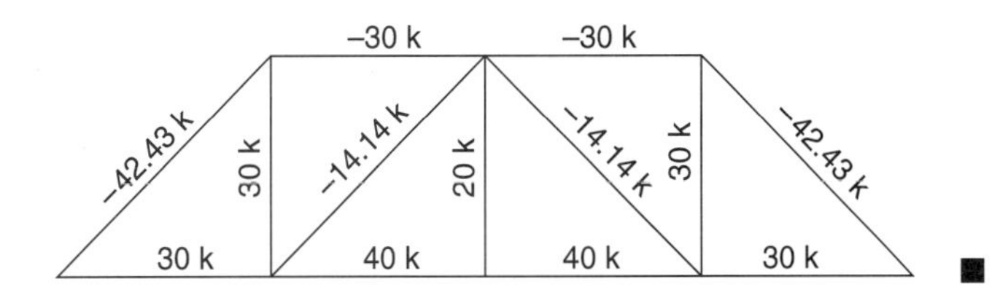

■

The method of joints may be used to compute the forces in all of the members of many trusses. The trusses of Examples 6.3 and 6.4 and all of the homework problems at the end of this chapter fall into this category. There are, however, a large number of trusses that must be analyzed using the method of joints as well as one of the methods discussed in Chapter 7. We prefer to calculate as many forces as possible in a truss by using the method of joints. At joints where there is a little difficulty, we take moments to obtain one or two forces (as discussed in Chapter 7), and then continue the calculations as far as possible by joints until we reach another point of difficulty where we again take moments, and so on.

EXAMPLE 6.3

Find all the forces in the truss shown in the following figure.

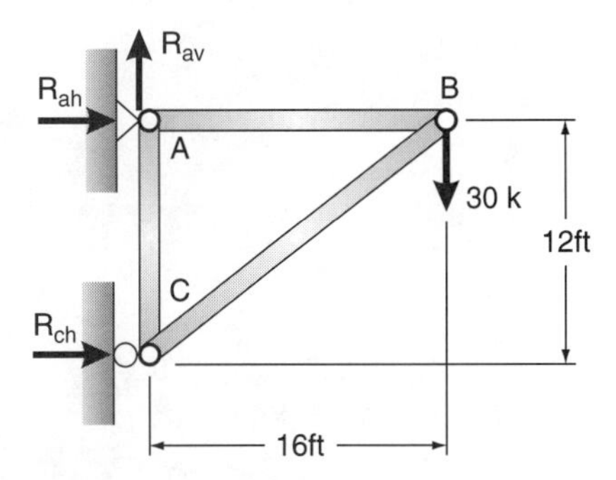

Solution. In this example, we will cast the solution as a set of simultaneous equations that can be represented in matrix form. When solving the example, we will determine the unknown bar forces and components of reaction. We will begin the solution by writing the equilibrium equations for each of the joints.

For joint A
$$\sum F_H = R_{ah} + F_{ab} = 0$$
$$\sum F_V = R_{av} - F_{ac} = 0$$

For joint B
$$\sum F_H = -F_{ab} - 0.8F_{cb} = 0$$
$$\sum F_V = -0.6F_{cb} - 30 = 0$$

For joint C
$$\sum F_H = R_{ch} + 0.8F_{cb} = 0$$
$$\sum F_V = F_{ac} + 0.6F_{cb} = 0$$

These equations can be represented in matrix form as

$$\begin{bmatrix} 1.0 & 0 & 0 & 1.0 & 0 & 0 \\ 0 & 1.0 & 0 & 0 & 0 & -1.0 \\ 0 & 0 & 0 & -1.0 & -0.8 & 0 \\ 0 & 0 & 0 & 0 & -0.6 & 0 \\ 0 & 0 & 1.0 & 0 & 0.8 & 0 \\ 0 & 0 & 0 & 0 & 0.6 & 1.0 \end{bmatrix} \begin{Bmatrix} R_{ah} \\ R_{av} \\ R_{ch} \\ F_{ab} \\ F_{cb} \\ F_{ac} \end{Bmatrix} = \begin{Bmatrix} 0 \\ 0 \\ 0 \\ 30 \\ 0 \\ 0 \end{Bmatrix}$$

This matrix equation, which represents six simultaneous linear equations, can be solved as

$$\begin{Bmatrix} R_{ah} \\ R_{av} \\ R_{ch} \\ F_{ab} \\ F_{cb} \\ F_{ac} \end{Bmatrix} = \begin{bmatrix} 1.0 & 0 & 0 & 1.0 & 0 & 0 \\ 0 & 1.0 & 0 & 0 & 0 & -1.0 \\ 0 & 0 & 0 & -1.0 & -0.8 & 0 \\ 0 & 0 & 0 & 0 & -0.6 & 0 \\ 0 & 0 & 1.0 & 0 & 0.8 & 0 \\ 0 & 0 & 0 & 0 & 0.6 & 1.0 \end{bmatrix}^{-1} \begin{Bmatrix} 0 \\ 0 \\ 0 \\ 30 \\ 0 \\ 0 \end{Bmatrix} = \begin{Bmatrix} -40.0 \\ 30.0 \\ 40.0 \\ 40.0 \\ -50.0 \\ 30.0 \end{Bmatrix}$$

These are the forces of reaction and the force in each of the bars in the truss. ■

Fairview-St. Mary's skyway system, Minneapolis, Minnesota (Courtesy of the American Institute of Steel Contruction)

EXAMPLE 6.4

Find all forces in the truss in the following figure. Use the method of joints for the solution.

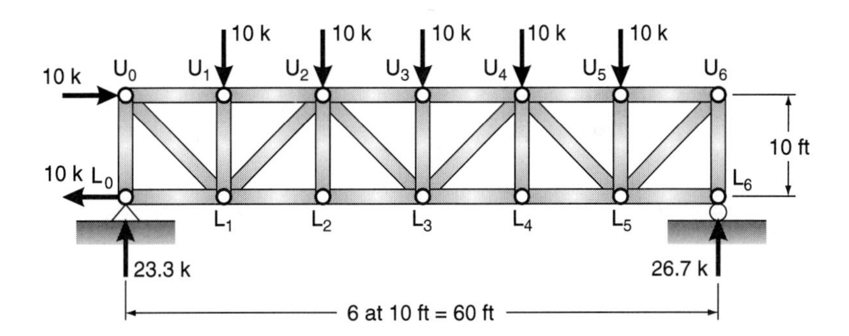

Solution. As in the previous examples, we will assume that all unknown bar forces are tensile forces. We will work our way through the joints one at a time looking for joints at which there are only two unknown forces, beginning with joint L_0.

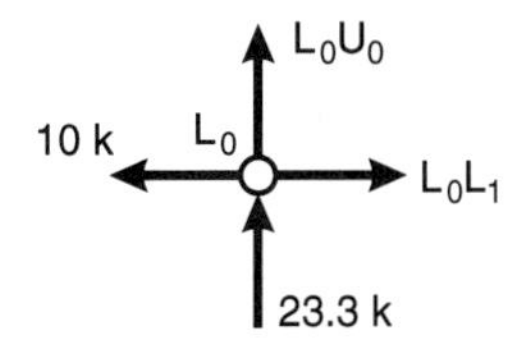

$$\left.\begin{array}{l}\sum F_H = L_0L_1 - 10 = 0 \\ \sum F_V = 23.3 + L_0U_0 = 0\end{array}\right\} \rightarrow \begin{array}{l}L_0L_1 = 10 \text{ kips} \\ L_0U_0 = -23.3 \text{ kips}\end{array}$$

We will next work with joint U_0.

$$\left.\begin{array}{l}\sum F_H = U_0U_1 + 10 + U_0L_1\cos(45) = 0 \\ \sum F_V = -L_0U_0 - U_0L_1\sin(45) = 0\end{array}\right\} \rightarrow \begin{array}{l}U_0U_1 = -33.3 \text{ kips} \\ U_0L_1 = 33.0 \text{ kips}\end{array}$$

We can now work with joint U_1.

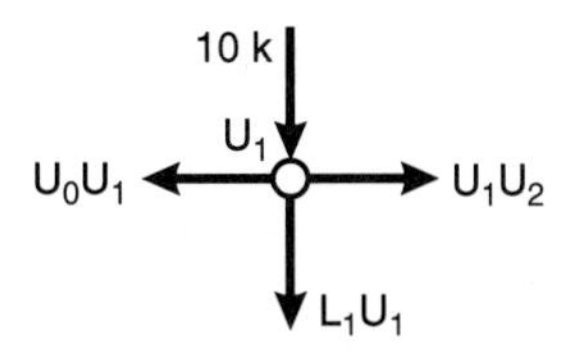

$$\left.\begin{array}{l}\sum F_H = -U_0U_1 + U_1U_2 = 0 \\ \sum F_V = -L_1U_1 - 10 = 0\end{array}\right\} \rightarrow \begin{array}{l}U_1U_2 = -33.3 \text{ kips} \\ L_1U_1 = -10.0 \text{ kips}\end{array}$$

Working with the remaining joints in a similar manner, we find that the forces in all of the bars in the truss are as shown in the following figure. A negative sign indicates that the bar is acting in axial compression.

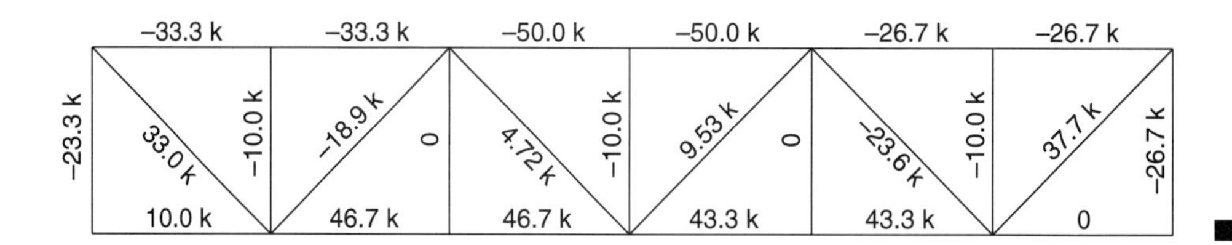

■

6.10 PROBLEMS FOR SOLUTION

For Problems 6.1 to 6.22 classify the structures as to their internal and external stability and determinacy. For statically indeterminate structures, include the degree of redundancy internally or externally. (The small circles on the trusses indicate the joints.)

6.1

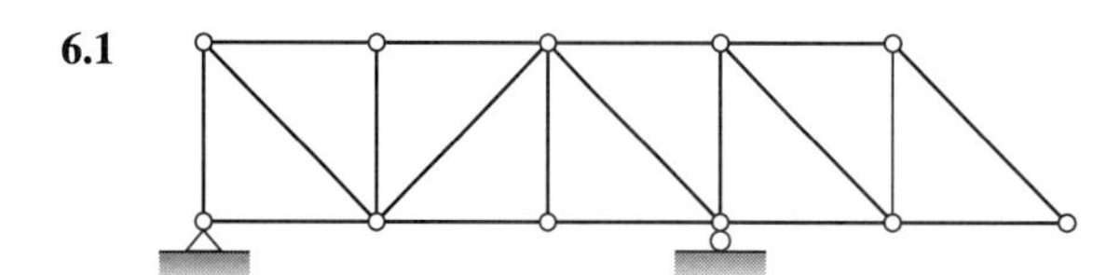

6.2 (*Ans.* Statically indeterminate internally to first degree)

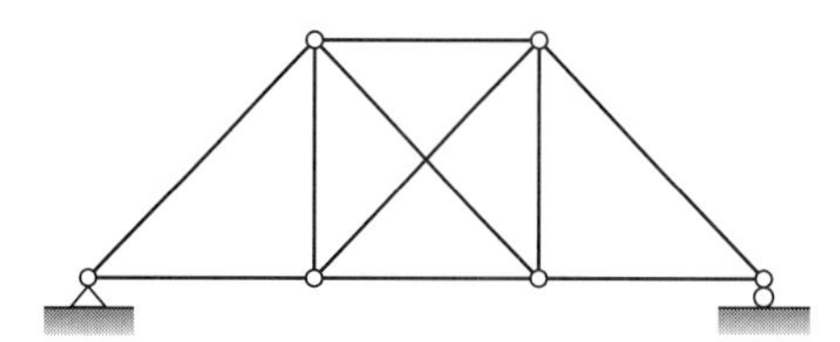

6.3

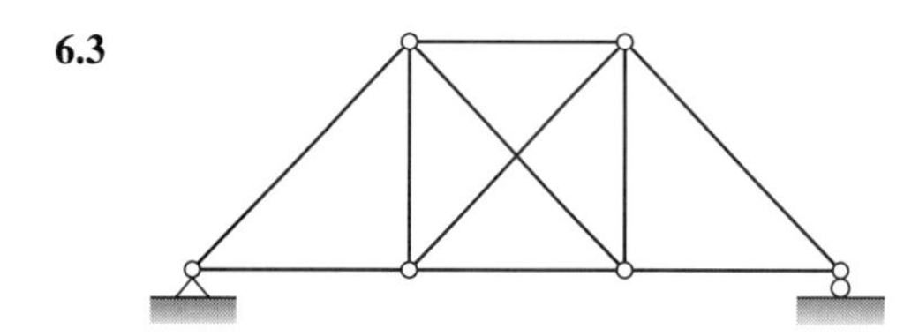

6.4 (*Ans.* Statically indeterminate internally to first degree)

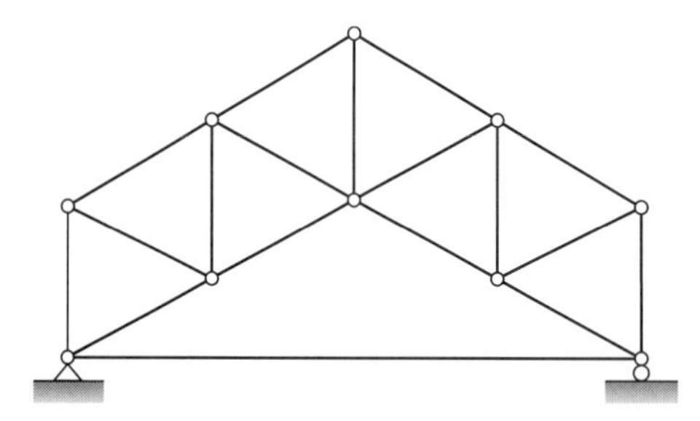

6.5

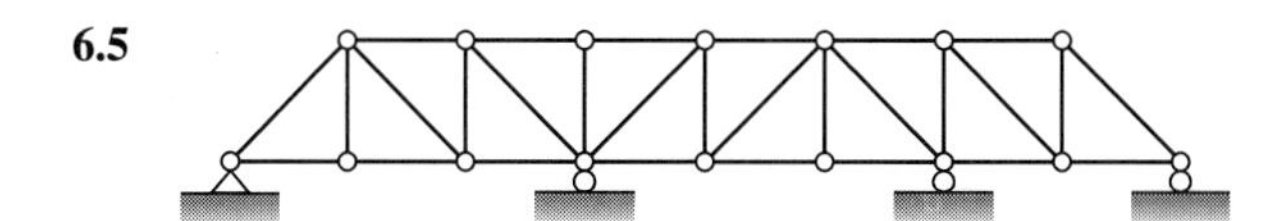

6.6 (*Ans*. Statically determinate externally and internally)

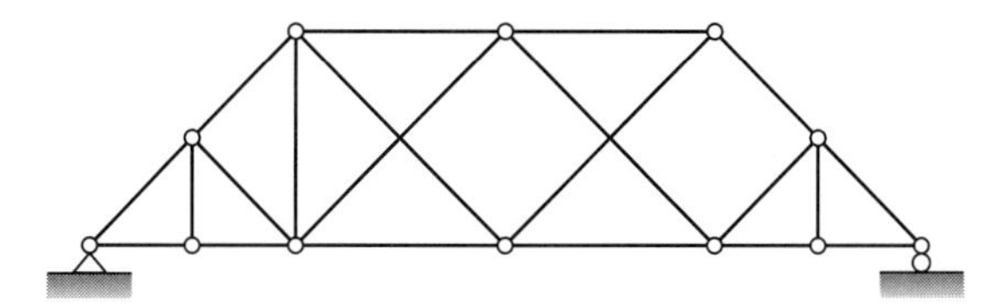

6.7

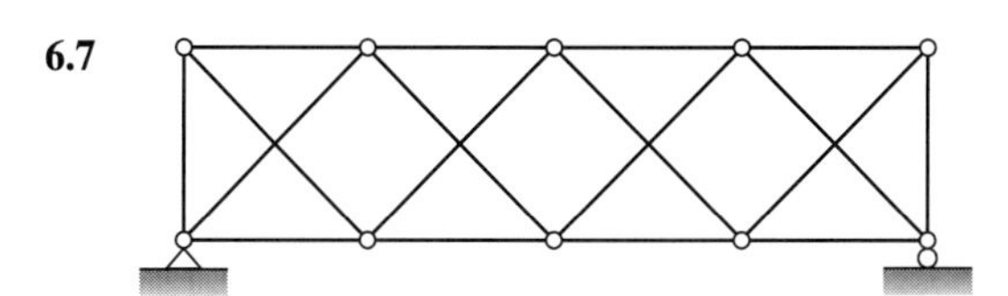

6.8 (*Ans*. Statically determinate externally and internally)

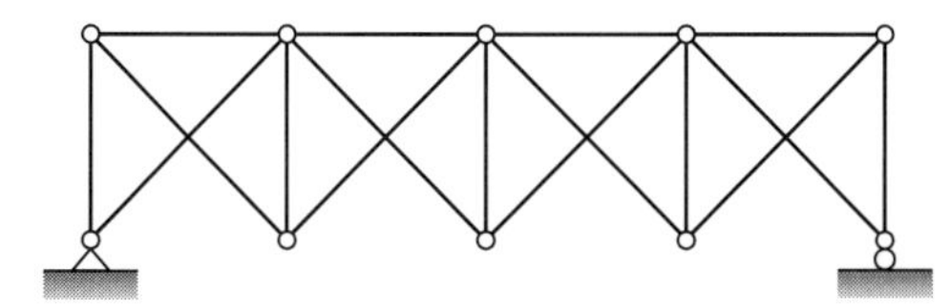

6.9

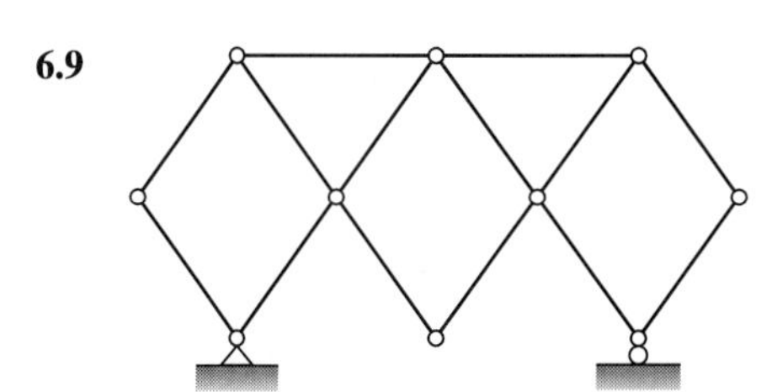

6.10 (*Ans*. Unstable)

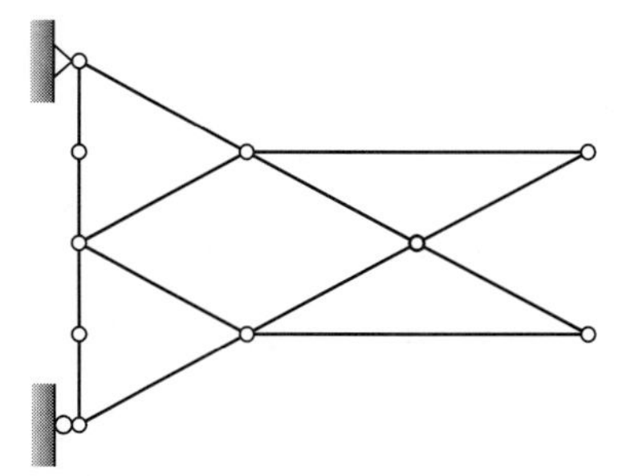

6.11

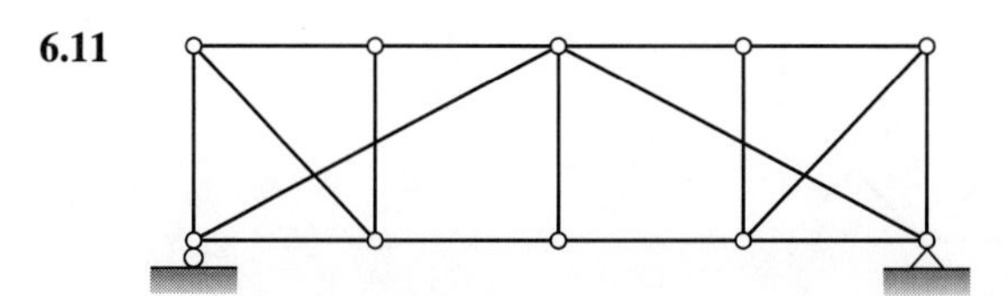

6.12 (*Ans*. Unstable externally)

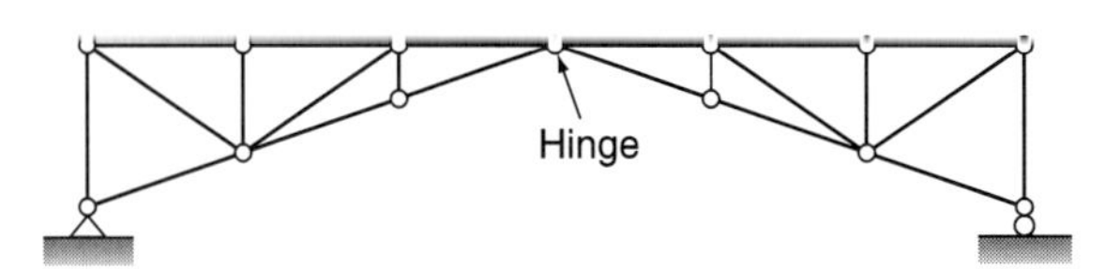

6.13

Hinge

6.14 (*Ans*. Statically determinate externally and internally)

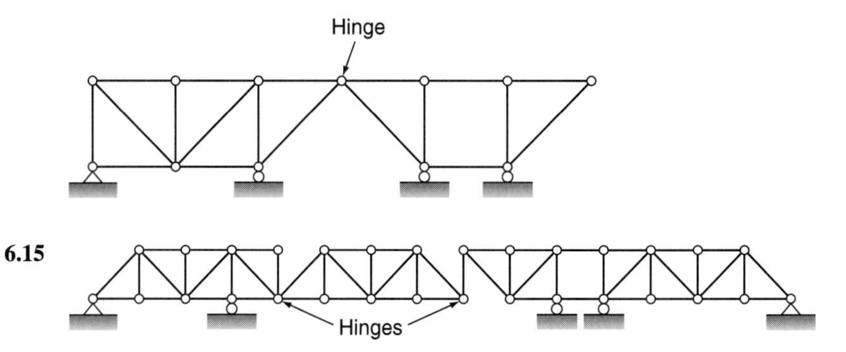

6.15

6.16 (*Ans*. Statically determinate externally and internally)

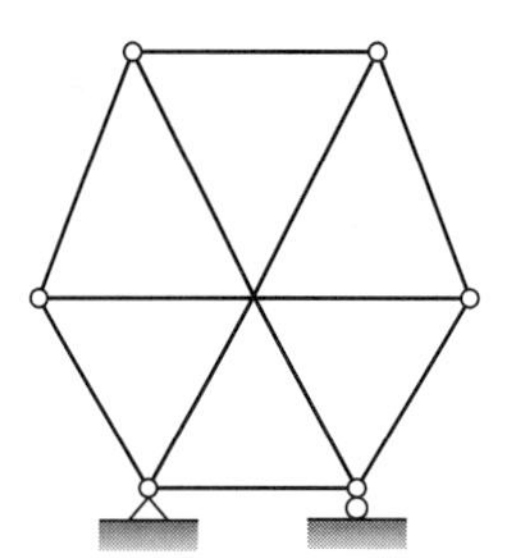

6.17

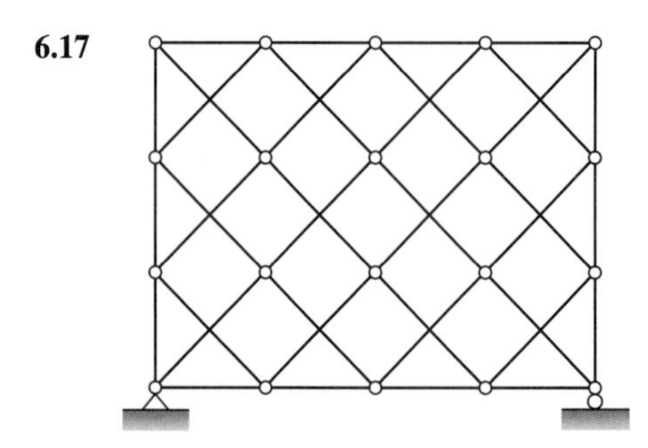

6.18 (*Ans*. Statically determinate externally and internally)

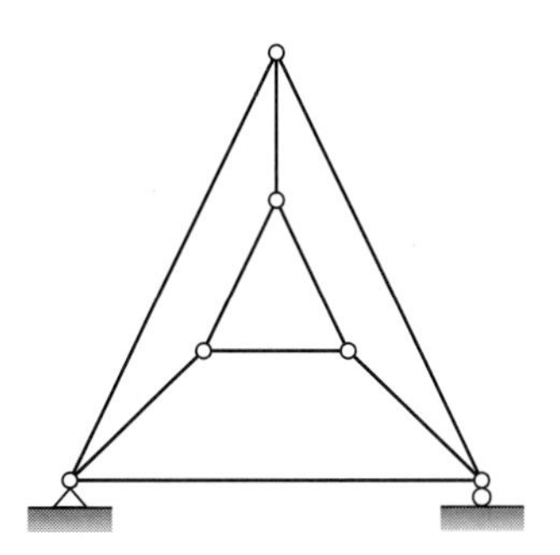

6.19

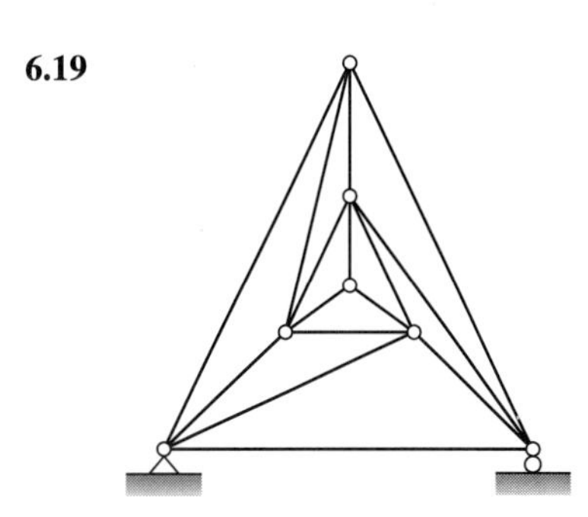

6.20 (*Ans*. Statically indeterminate internally to second degree)

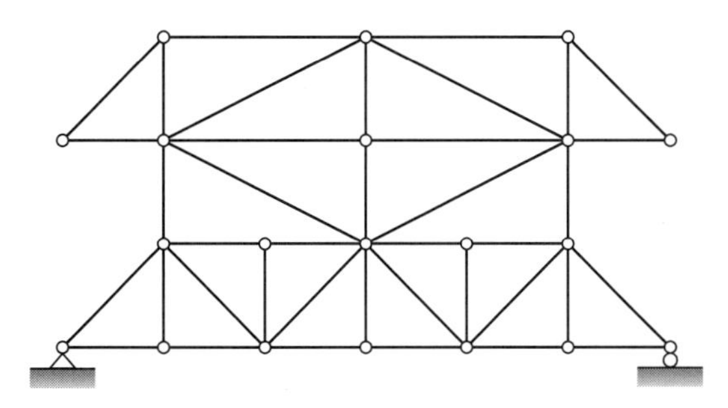

6.21 (*Ans*. Statically indeterminate internally to first degree)

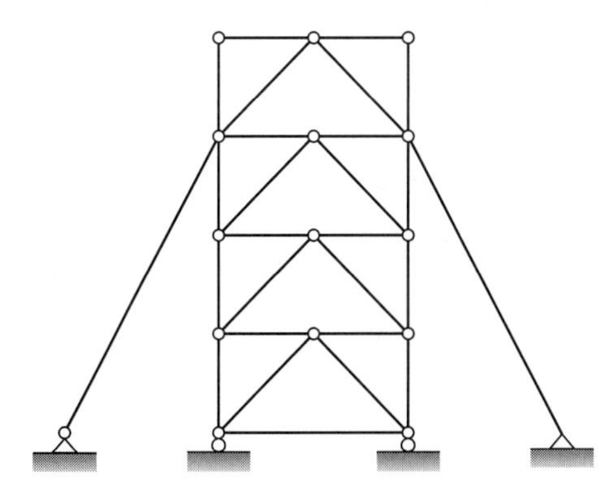

6.22

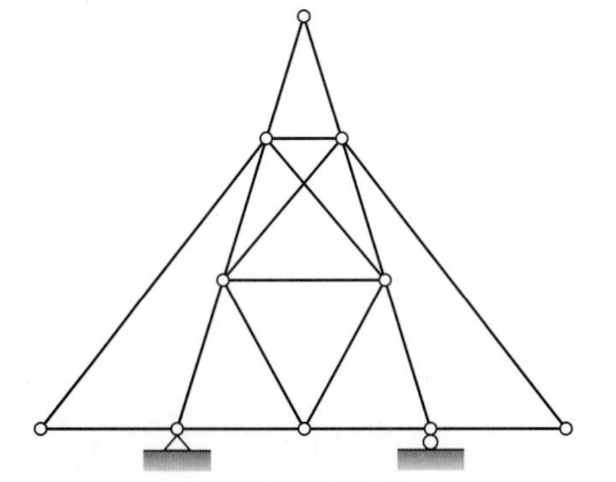

For Problems 6.23 to 6.37, compute the forces in all the members of the trusses using the method of joints.

6.23

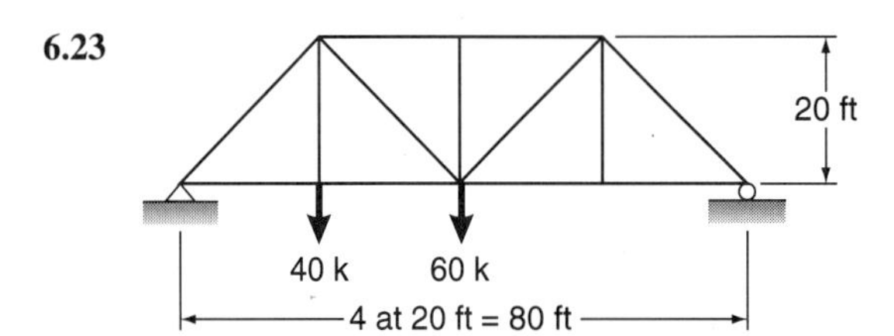

6.24 Rework Problem 6.23 if the truss depth is reduced to 15 ft and the loads doubled. (*Ans.* $L_0L_1 = +160$ k, $U_1U_2 = -213.3$ k, $L_2U_3 = +133.3$ k)

6.25

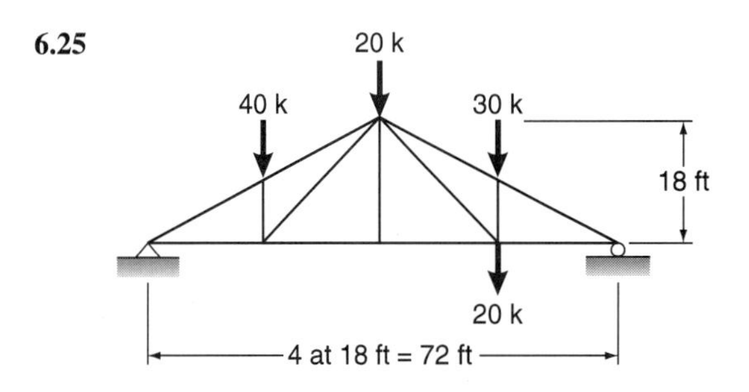

6.26 (*Ans.* $U_0U_1 = -110$ kN, $L_1L_2 = +140$ kN, $U_2L_3 = -62.5$ kN)

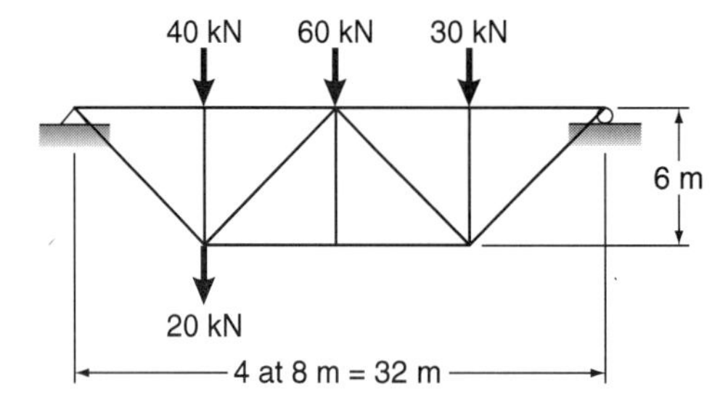

6.27

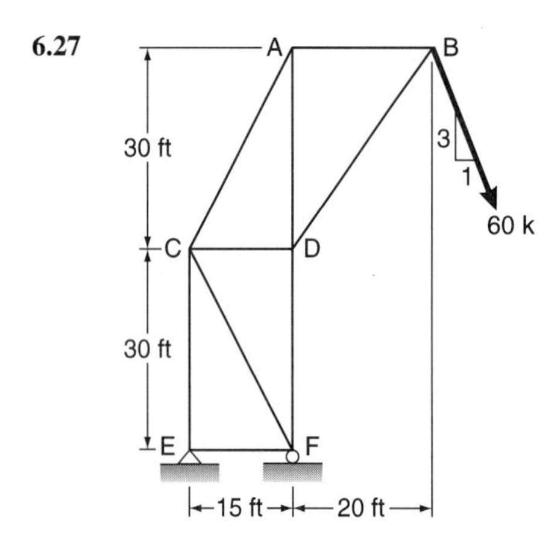

6.28 (*Ans.* $U_0L_0 = +55$ k, $U_0L_1 = +33.5$ k, $L_1L_2 = -40$ k)

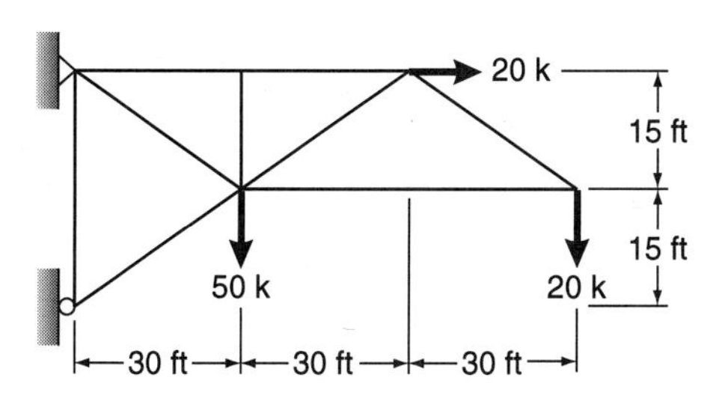

6.29

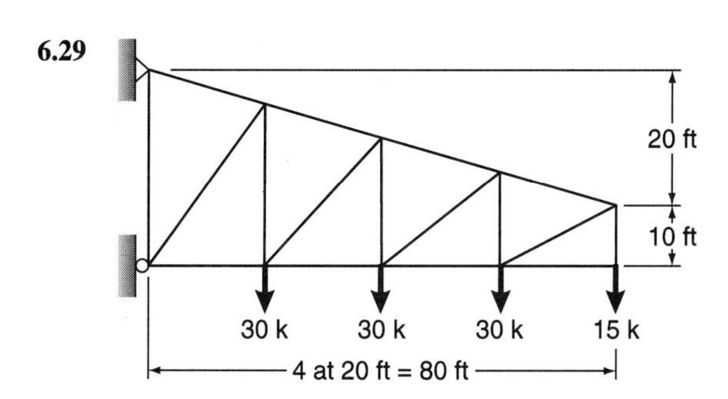

6.30 (*Ans.* $U_0L_0 = +10$ k, $L_0U_1 = -94.9$ k, $U_1L_2 = -126.5$ k)

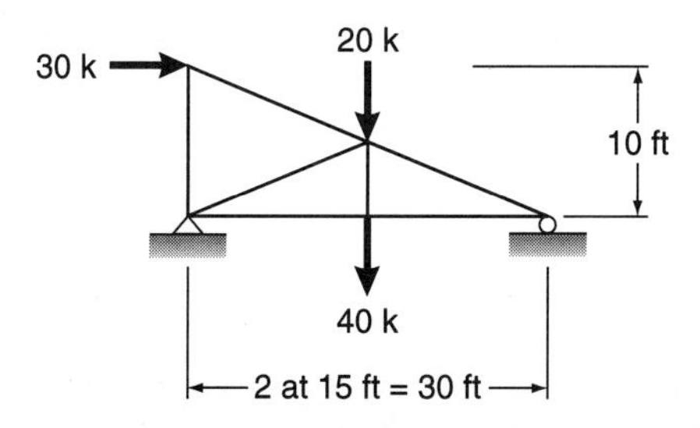

6.31 Repeat Problem 6.30 if the roller support (due to friction, corrosion, etc.) is assumed to supply one-third of the total horizontal force resistance needed, with the other two-thirds supplied by the pin support.

6.32 (*Ans.* $U_0U_1 = -28$ k, $U_1L_1 = -27.6$ k, $U_2L_2 = -10.8$ k)

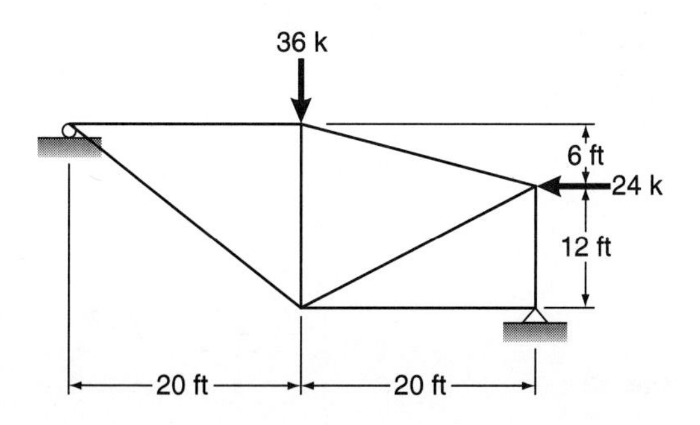

6.33

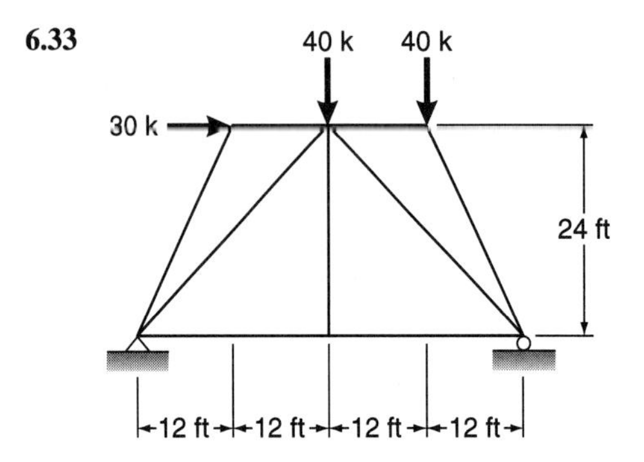

6.34 Rework Problem 6.33 if the supporting surface beneath the roller is changed as follows (*Ans.* $L_0U_2 = -21.2$ k, $U_2U_3 = -20$ k, $U_3L_4 = -44.7$ k)

6.35

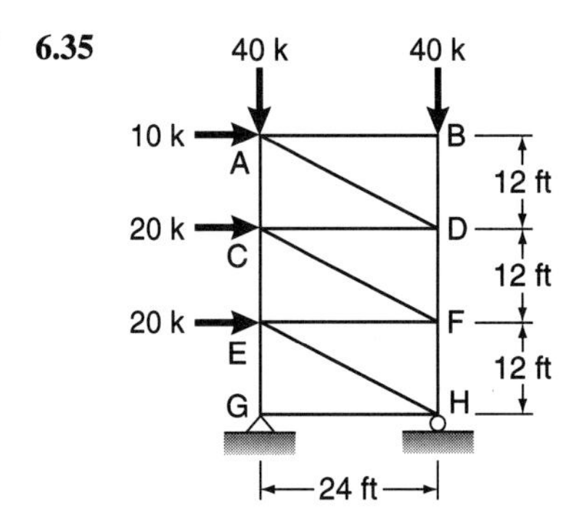

6.36 (*Ans.* $U_0U_1 = -86.7$ k, $U_1L_1 = -130$ k, $L_2U_3 = +150$ k)

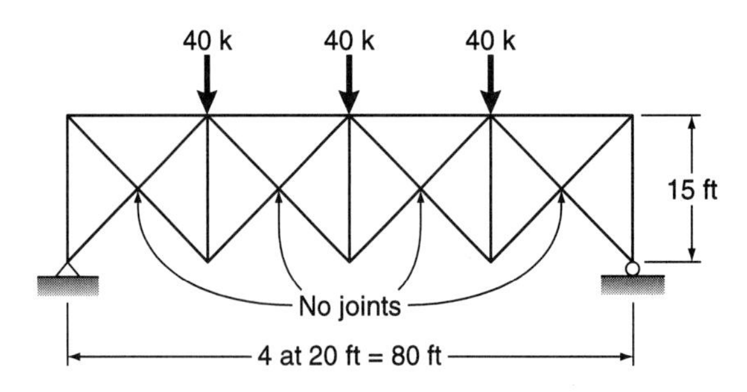

6.37

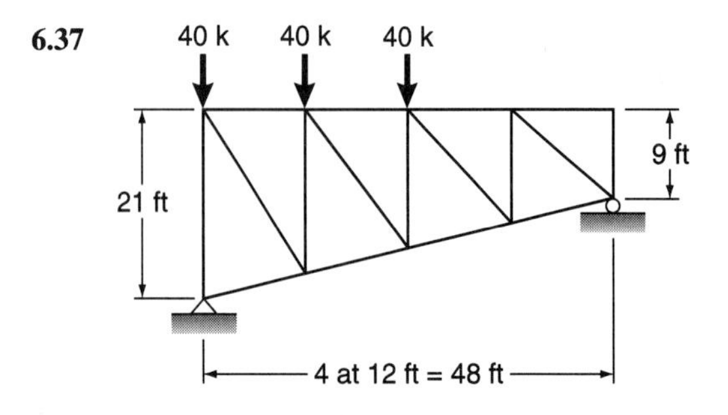

Chapter 7

Plane Trusses, Continued

7.1 ANALYSIS BY THE METHOD OF SECTIONS

Applying the equations of equilibrium to free-body diagrams of sections of a truss is the basis of force computation by the method of sections, just as it was when using the method of joints in the previous chapter. When using the method of sections to determine the force in a particular member, an imaginary plane is passed completely through the truss, which cuts it into two sections as shown in Figure 7.1(a). The resulting free-body diagrams are shown in Figure 7.1(b) and (c). The location at which the sections are cut is selected so there are at least as many equations of equilibrium as there are unknown forces.

The equations of static equilibrium are applied to each of the sections to determine the magnitude of the unknown forces. Particular attention should be paid to the point about which moments are summed when applying the equations. Often moments of the forces can be taken about a point so that only one unknown force appears in the equation. As such, the value of that force can be obtained. This objective usually can be attained by selecting a point along the line of action of one or more of the other members' forces and making it the point about which moments are summed. This point does not necessarily have to be on the cut section. Some familiar trusses have special locations for placing sections that greatly simplify the work involved. These cases will be discussed in the pages to follow.

The primary advantage of the method of sections is that the force in one member of a truss can be computed in most cases without having to compute the forces in other

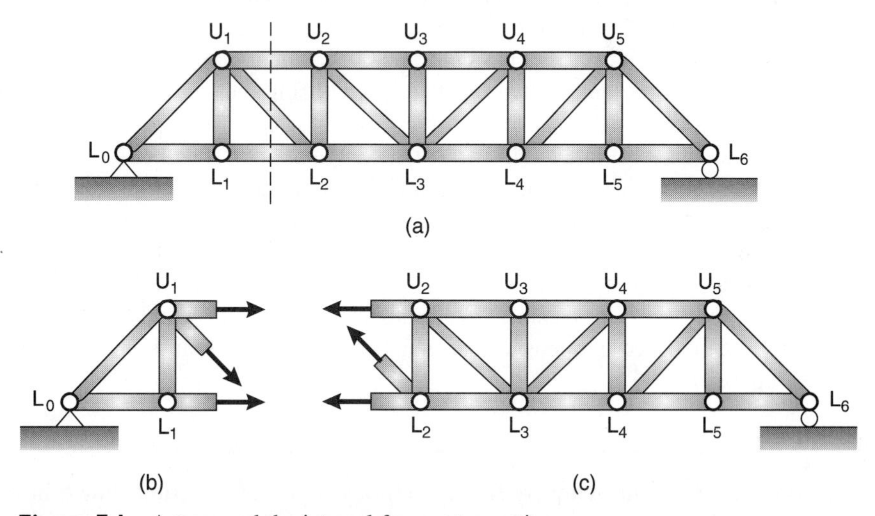

Figure 7.1 A truss and the internal forces at a section

members of the truss. If the method of joints were used, calculating the forces in other members, joint by joint from the end of the truss up to the member in question, would be necessary. Nevertheless, both methods are used in the analysis of trusses. In fact, both methods are often used at the same time. Depending upon the geometry of the truss, some member forces are more easily calculated with the method of sections than others that are more easily calculated using the method of joints.

Final truss slipped into place for Newport Bridge linking Jamestown and Newport, Rhode Island (Courtesy Bethlehem Steel Corporation)

7.2 APPLICATION OF THE METHOD OF SECTIONS

When using the method of sections, we need to establish a sign convention for the sense of the forces in the cut members just as we did when using the method of joints. As with the method of joints, assume that all unknown member forces are acting in tension. This sign convention is illustrated in Figure 7.1. Upon analysis, positive results indicate members that are in tension and negative results indicate members that are in compression. By using this sign convention, we will obtain consistent results with all of the methods of truss analysis: the errors that result from confusing the sense of member forces will be greatly reduced.

Examples 7.1 and 7.2 illustrate the computation of member forces using the method of sections. To demonstrate the principles, only the forces in selected members are computed. The forces in the other members could be computed by cutting additional sections or with the method of joints. When the equations of static equilibrium are written, note that the sense of the unknown forces is assumed to follow the sign convention previously discussed. Again, if the mathematical solution yields a negative number, the sense of the force is opposite that which was assumed, that is, the member is acting in compression.

EXAMPLE 7.1

Find the forces in members L_1L_2, L_1U_2, and U_1U_2 of the truss shown in the following figure using the method of sections.

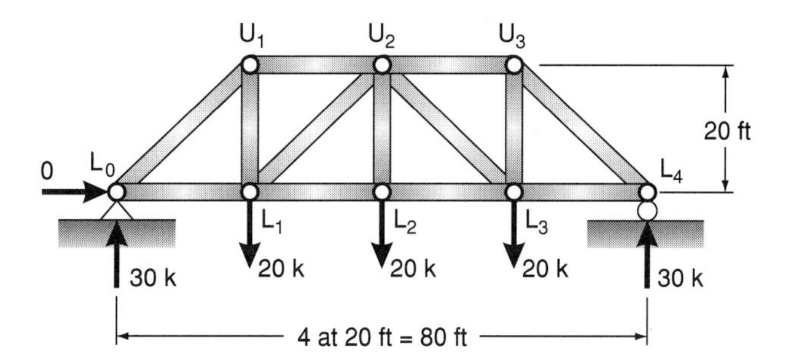

Solution. Cut a section through the second panel of the truss. The resulting free-body diagram is shown in the next figure.

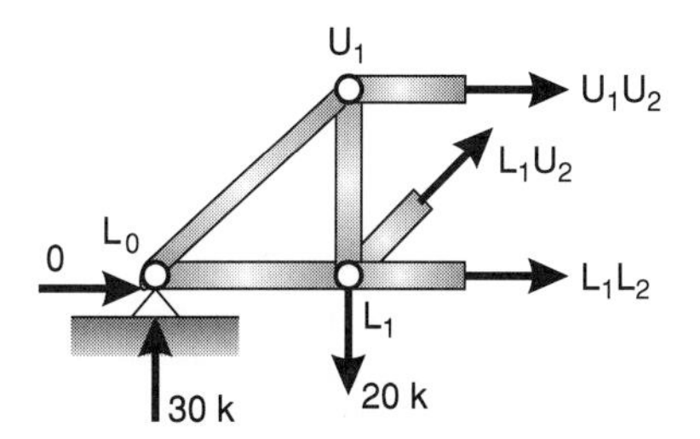

Using this free-body diagram calculate the unknown member forces. First, we will sum forces vertically to determine the force in L_1U_2.

$$\sum F_v = 0 = 30 - 20 + L_1U_2\cos(45)$$
$$\therefore\ L_1U_2 = -14.14\,\text{kips}$$

Next, we will sum moments about L_1 to determine the force in U_1U_2.

$$\sum M_{L_1} = 0 = 30(20) + U_1U_2(20)$$
$$\therefore\ U_1U_2 = -30\,\text{kips}$$

Lastly, we will sum forces horizontally to determine the force in L_1L_2.

$$\sum F_H = 0 = L_1L_2 + U_1U_2 + L_1U_2\cos(45) = L_1L_2 + (-30) + (-14.14)\cos(45)$$
$$\therefore\ L_1L_2 = 40\,\text{kips} \quad \blacksquare$$

EXAMPLE 7.2

Find the forces in members L_1L_2, U_1L_2, and U_1U_2 of the truss shown in the following figure using the method of sections.

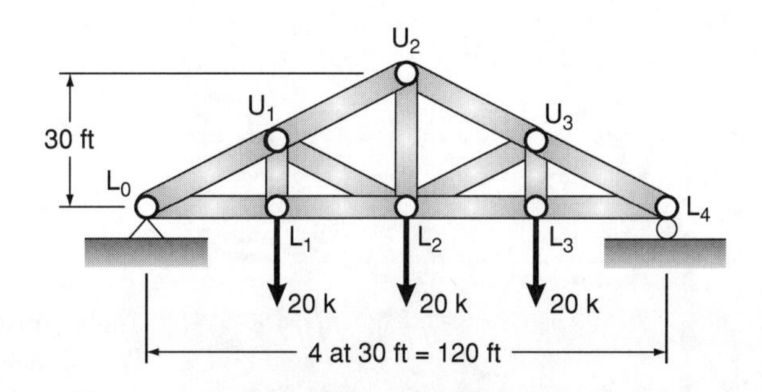

Solution. Cut a section through the second panel of the truss. The resulting free-body diagram is shown in the figure below.

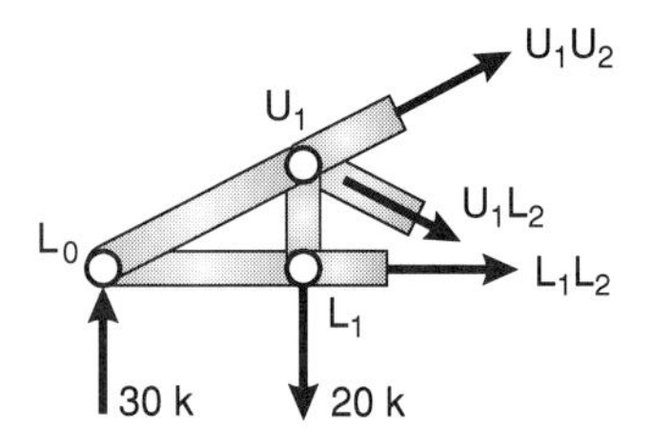

Using this free-body diagram calculate the unknown member forces. First, we will sum moments about U_1 to determine the force in L_1L_2.

$$\sum M_{U_1} = 0 = 30(30) - L_1L_2(15)$$

$$\therefore\ L_1L_2 = 60\,\text{kips}$$

Next, we will sum forces horizontally and vertically to obtain equations with which we can determine the forces in U_1U_2 and U_1L_2. These equations can be solved simultaneously to obtain the forces.

$$\left.\begin{aligned}\sum F_H &= L_1L_2 + U_1L_2\cos(26.56) + U_1U_2\cos(26.56)\\ \sum F_V &= 30 - 20 + U_1L_2\sin(26.56) + U_1U_2\sin(26.56)\end{aligned}\right\}\rightarrow\begin{aligned}U_1U_2 &= -44.72\,\text{kips}\\ U_1L_2 &= -22.36\,\text{kips}\end{aligned}$$

■

7.3 SHEARING FORCE AND BENDING MOMENT DIAGRAMS

In reality, a truss is nothing more than a beam, sometimes a tapered beam, without a solid web. Just as a beam carries bending moment and shearing forces, a truss carries bending moment and shearing forces. We have already seen these forces when we analyzed trusses using the method of sections. In a parallel chord truss, the bending moment is carried by the couple developed between the top and bottom chords. The shearing force is carried by the diagonal members.

Timber trusses for tannery, South Paris, Maine (Courtesy of the American Wood Preservers Institute)

As we did for beams in Chapter 5, we can construct shearing force and bending moment diagrams for trusses. Just as they did for beams, the shearing force and bending moment diagrams provide a visual understanding of the distribution of forces in the truss. The same rules we used to construct shearing force and bending moment diagrams for beams apply when constructing shearing force and bending moment diagrams for trusses. The specific rules that we should remember for constructing the diagrams are

- The change in shearing force between two points along the truss is equal to the change in load between those two points.
- The change in bending moment between two points along the truss is equal to area under the shearing force diagram between those two points.

When constructing shearing force and bending moment diagrams for trusses, we will use the same sign convention as we did for beams. This procedure is illustrated in Example 7.3.

EXAMPLE 7.3

Construct the shearing force and bending moment diagram for the truss shown in the figure.

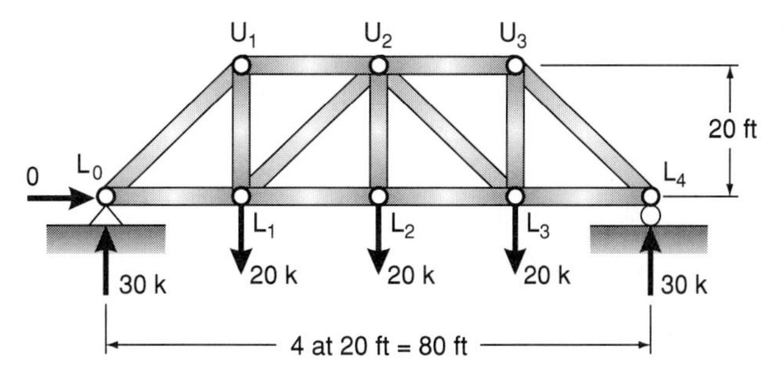

Solution. The shearing force diagram is

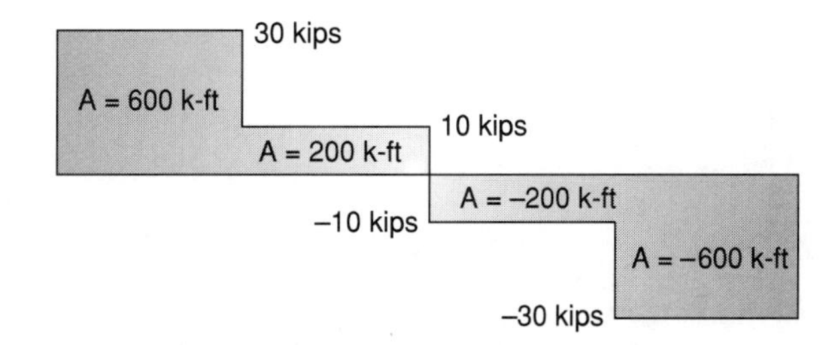

And the associated bending moment diagram is

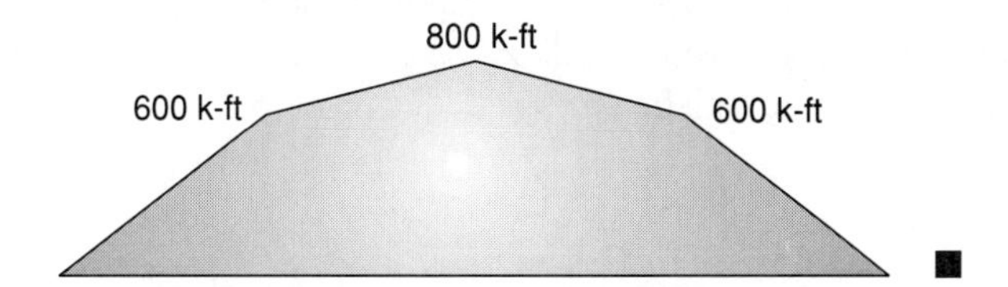

■

7.4 ZERO-FORCE MEMBERS

Frequently, some readily identifiable truss members have zero force in them if secondary forces due to member weights, load eccentricities, etc., are neglected. The ability to identify these members will often expedite truss analysis. Zero-force members usually can be

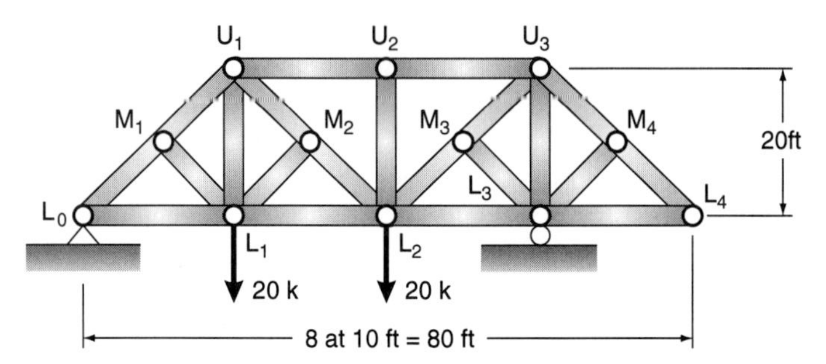

Figure 7.2 A truss with some zero-force members

identified by making brief examinations of the truss joints. The basic rule to identify zero-force members is:

> A member in a truss will have zero force if at a joint that member has a component of forces acting in a direction in which no other member at the joint has a component of force and in which there is no load acting.

This rule is best understood by applying it to specific cases. The three cases that follow are illustrations of the principle. In these illustrations, reference is made to the truss shown in Figure 7.2. The reader should note that these illustrations are not unique as they all stem from the same principle.

7.4.1 Case 1

If only one member at a joint has a component of force in a particular direction, and no external load applied to the joint has a component in the direction of the member, the force in the member must be zero. An examination of joint U_2 of the truss in Figure 7.2 reveals that member U_2L_2 has zero force because there is no applied load with a vertical component at the joint and no other member at the joint has a vertical component of force. If member U_2L_2 had a force in it, the sum of the vertical forces at U_2 could not be zero.

7.4.2 Case 2

Not only must the sum of the forces in the vertical and horizontal directions at a particular joint be equal to zero, but the sum of the forces in any direction at the joint must be equal to zero. Therefore, the force in member M_1L_1 of the truss in Figure 7.2 must be equal to zero. Member M_1L_1 has a component of force perpendicular to the line of action of members L_0M_1 and M_1U_1. Unless the force in member M_1L_1 was equal to zero, equilibrium could not be satisfied at M_1.

7.4.3 Case 3

Two members that are joined but not in line with each other will have zero force in them unless an external load is applied at the joint in the plane of the members. In the truss in Figure 7.2, members M_4L_4 and L_3L_4 are zero-force members. If the force in either or both members was non-zero, the equations of static equilibrium at L_4 could not be satisfied concurrently. That these members have zero force could also have been verified with Case 1 in Section 7.4.1.

7.5 STABILITY

The subject of truss stability was broached briefly in Chapter 6. As we develop complicated trusses such as those examined later in this chapter, the issue of stability and our ability to recognize unstable trusses becomes more important. For that reason, we will again discuss the stability of trusses.

The stability of a truss can always be determined through structural analysis. The members of a truss must be arranged to support the external loads. What will support the external loads satisfactorily is a rather difficult question to answer with only a glance at the truss under consideration, but an analysis of the structure will always provide the answer. If the structure is stable, the analysis will yield reasonable results and equilibrium will be satisfied at all of the joints in the truss. On the other hand, if a truss is unstable, equilibrium cannot be concurrently satisfied at all of the joints in the truss.

College fieldhouse, Largo, Maryland (Courtesy Bethlehem Steel Corporation)

Because analysis can be very time-consuming and frustrating if we finish the analysis only to find that the structure was unstable and could therefore not be used, a means to help identify unstable trusses before analyzing them is beneficial. Such means are identified in the following paragraphs. Included in the discussion are the zero-load test and evaluation of the determinant of the equations of equilibrium at the joints.

7.5.1 Less Than 2j − r Members

A truss that has less than 2j − r members is obviously unstable internally. Nevertheless, a truss can have as many or more than 2j − r members and still be unstable. The truss of Figure 7.3(a) satisfies the 2j − r relationship. It is statically determinate and stable. However, if the diagonal in the second panel is removed and added to the first panel as shown in Figure 7.3(b), the truss is unstable even though the number of members remains equal to 2j − r. The part of the truss to the left of the second panel can move with respect to the part of the truss to the right of the second panel because the second panel is a rectangle. As previously indicated, a rectangular shape is unstable unless restrained in some way.

Similarly, the addition of diagonals to the third and fourth panels, as in Figure 7.3(c), will not prevent the truss from being unstable. There are two more than 2j − r members,

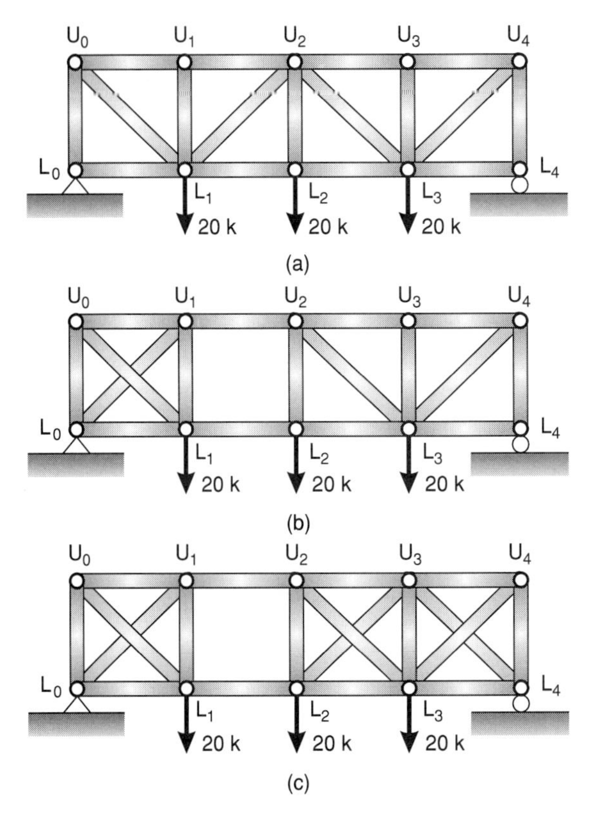

Figure 7.3 Stable (a) and unstable truss geometry (b and c)

and the truss is seemingly statically indeterminate to the second degree; but it is unstable because the second panel is unstable.

7.5.2 Trusses Consisting of Shapes That Are Not All Triangles

As you become more familiar with trusses, in most cases you will be able to tell with a brief glance if a truss is stable or unstable. For the present, though, it may be a good idea to study trusses in detail if you think there is a possibility of instability. When a truss has some non-triangular shapes in its geometry, we should be aware that instability is indeed a possibility. The trusses shown in Figure 7.3(b) and (c) fall into this category.

The fallacy of this statement about triangular shapes is that there is an endless number of perfectly stable trusses that can be assembled and do not consist entirely of triangles. As an example, consider the truss shown in Figure 7.4. The basic triangle $L_0U_1L_1$ has been extended by the addition of the joint M_1 and the members L_0M_1 and M_1L_1. A stable

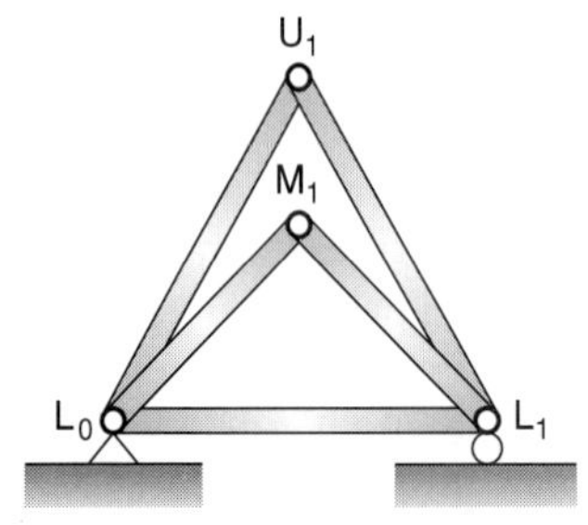

Figure 7.4 A stable truss not consisting entry of triangles

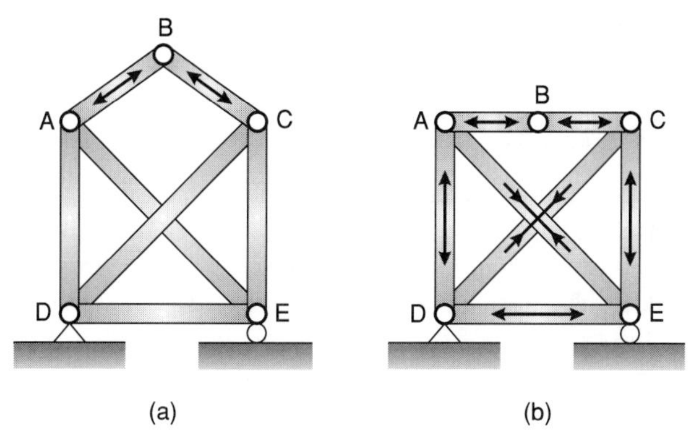

Figure 7.5

truss is maintained, even though the shape $L_0M_1L_1U_1$ is not a triangle. Joint M_1 cannot move without changing the length of one or more members. Therefore, the truss is stable.

Compound and complex trusses, which will be studied soon, often contain non-triangular shapes but are stable. The fundamental principle is that if a truss consists of shapes that are not triangular, it should be examined carefully to determine if any of the joints can displace without causing changes in length of one or more of the truss members.

7.5.3 Unstable Supports

A structure cannot be stable if its supports are unstable. To be stable, the truss must be supported by at least three nonparallel, non-concurrent forces. This subject was discussed in Chapter 4.

7.5.4 The Zero-Load Test

A statically determinate truss has only one possible set of forces for a given loading; it is said to have a unique solution. Therefore, if it is possible to show that more than one solution can be obtained for a structure for a given set of conditions, the structure is unstable. This concept leads to the idea of the *zero-load test* for inferring the stability of statically determinant trusses. Simply stated, the zero-load test is:

> If no external loads are applied to a truss, all of the members in the truss can logically be assumed to have zero force in them. Should an assumed non-zero force be assigned to one of the members of a truss that has no external loads and the forces are computed in the other members, the results must be incompatible if the truss is stable. If the calculated forces are compatible, the truss is unstable[1].

To illustrate this concept, consider the two statically determinant trusses shown in Figure 7.5. In both cases $m = 2j - r$. If we were to cause a compression force to occur in bar AB in the truss in Figure 7.5(a), achieving equilibrium at joint B would be impossible. We cannot achieve vertical equilibrium at the same time we achieve horizontal equilibrium. On the other hand, if we were to cause a compression force in bar AB of the truss in Figure 7.5(b), equilibrium could be achieved at all joints in the truss. The conclusion, then, is that the truss in Figure 7.5(a) is stable whereas that in Figure 7.5(b) is unstable.

[1]S. P. Timoshenko and D. H. Young, *Theory of Structures,* 2nd Ed., (New York: McGraw-Hill Book Company, 1965), p. 89.

This is a correct conclusion since we can tell by inspection that the first truss can carry a vertical load at joint B but the second truss cannot.

7.5.5 Stability and the Determinant

An analytical method to determine the stability of a truss is the determinant of the 2j equations of static equilibrium for the truss. If the determinant of the equations is nonzero, there exists a unique solution for the equations and the truss, therefore, is stable. On the other hand, if the determinant of the equations is equal to zero, there is not a unique solution for the equations of equilibrium. This indicates that the truss is unstable. This approach to evaluating stability is demonstrated in Example 7.4. The method has significant advantages because it is applicable to any truss, whether it is statically determinate or statically indeterminate. Further, it is a natural by-product of matrix structural analysis, which will be discussed beginning in Chapter 19.

EXAMPLE 7.4

Determine the stability of the truss shown in the figure by evaluating the determinant of the equation of equilibrium at the joints.

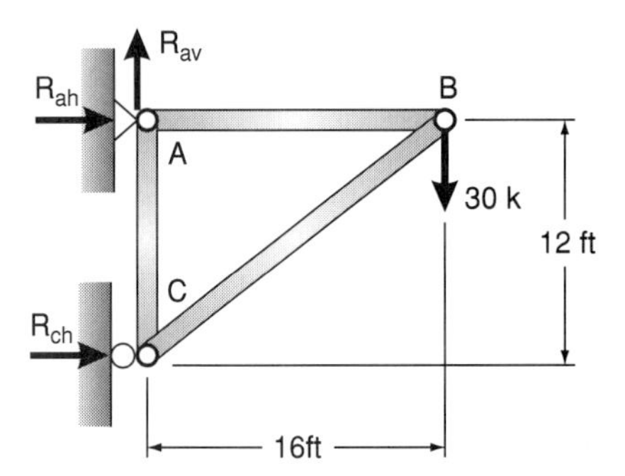

Solution. We studied this same truss in Example 6.3. The static equilibrium equations for each of the joints in the truss are

For joint A
$$\sum F_H = R_{ah} + F_{ab} = 0$$
$$\sum F_V = R_{av} - F_{ac} = 0$$

For joint B
$$\sum F_H = -F_{ab} - 0.8F_{cb} = 0$$
$$\sum F_V = -0.6F_{cb} - 30 = 0$$

For joint C
$$\sum F_H = R_{ch} + 0.8F_{cb} = 0$$
$$\sum F_V = F_{ac} + 0.6F_{cb} = 0$$

These equations can be represented in matrix form as:

$$\begin{bmatrix} 1.0 & 0 & 0 & 1.0 & 0 & 0 \\ 0 & 1.0 & 0 & 0 & 0 & -1.0 \\ 0 & 0 & 0 & -1.0 & -0.8 & 0 \\ 0 & 0 & 0 & 0 & -0.6 & 0 \\ 0 & 0 & 1.0 & 0 & 0.8 & 0 \\ 0 & 0 & 0 & 0 & 0.6 & 1.0 \end{bmatrix} \begin{Bmatrix} R_{ah} \\ R_{av} \\ R_{ch} \\ F_{ab} \\ F_{cb} \\ F_{ac} \end{Bmatrix} = \begin{Bmatrix} 0 \\ 0 \\ 0 \\ 30 \\ 0 \\ 0 \end{Bmatrix}$$

If the determinant of the coefficient matrix, the left matrix in the equation above, is equal to zero the truss is unstable. If the determinant of that matrix is non-zero the truss is stable.

$$\left\|\begin{matrix} 1.0 & 0 & 0 & 1.0 & 0 & 0 \\ 0 & 1.0 & 0 & 0 & 0 & -1.0 \\ 0 & 0 & 0 & -1.0 & -0.8 & 0 \\ 0 & 0 & 0 & 0 & -0.6 & 0 \\ 0 & 0 & 1.0 & 0 & 0.8 & 0 \\ 0 & 0 & 0 & 0 & 0.6 & 1.0 \end{matrix}\right\| = 0.6$$

The determinant is not equal to zero so we correctly conclude that the truss is stable. We can come to the same conclusion by applying the zero-load test. ■

7.6 SIMPLE, COMPOUND, AND COMPLEX TRUSSES

Often you will hear a truss be referred to as a simple truss or a compound truss. These are references to the geometric form of the truss, or the "building blocks" that form the truss. These are not references to the complexity of the analysis to determine member forces. For completeness, we will briefly discuss the categories of trusses from which these references arise. We will also discuss some issues regarding analysis of trusses in each category.

7.6.1 Simple Trusses

As we have seen, the first step in forming a truss is connecting three members at their ends to form a triangle. Subsequent segments are incorporated by adding two members and one joint; the new members meet at the new joint and each is pinned at its opposite ends into one of the existing joints. Trusses formed in this way are said to be *simple trusses*. The authors are not suggesting, however, that all trusses formed in this manner are "simple" to analyze.

7.6.2 Compound Trusses

A *compound truss* is a truss made by connecting two or more simple trusses. The simple trusses may be connected by three nonparallel non-concurrent links, by one joint and one link, by a connecting truss, by two or more joints, and so on. An almost unlimited number of trusses may be formed in this way. The Fink truss shown in Figure 7.6(a) consists of the two shaded simple trusses that are connected with a joint and a link. All of the methods for checking stability and for analysis can be used on compound trusses with equal success. The analysis is greatly simplified if the force in the link is first found using the method of sections and summing moments about the apex.

7.6.3 Complex Trusses

There are a few trusses that are statically determinate that do not meet the requirements necessary to fall within the classification of either simple or compound trusses. Such a truss is shown in Figure 7.6(b). These are referred to as *complex trusses*.

The members of simple and compound trusses usually are arranged so that sections may be passed through three members at a time; moments are taken about the intersection

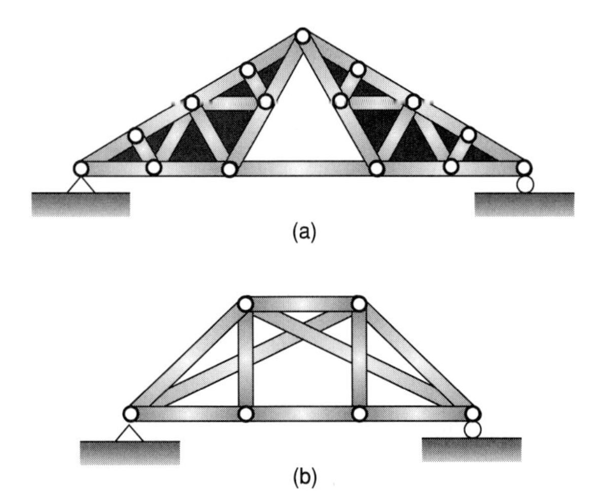

Figure 7.6 A compound truss (a) and complex truss (b)

of two of them and the force in the third member is found. Complex trusses may not be analyzed in this manner. Not only does the method of sections fail to simplify the analysis, the methods of joints is also of no avail. The difficulty lies in the fact that there are three members meeting at almost every joint. Consequently, there are too many unknowns at every location in the truss to pass a section and obtain the force in any member directly using the equations of static equilibrium.

One method of computing the forces in complex trusses is to write the equations of static equilibrium at each joint, which yields 2j equations. These equations may be solved simultaneously for the member forces and external reactions. It is often possible to calculate the external reactions initially, and their values may be used as a check against the results obtained from the solution of the simultaneous equations. This method will work for any statically determinate complex truss, but the solution of the equations is very tedious unless a digital computer is available. Note that the enclosed computer software can be used easily to analyze complex trusses, as well as other types of trusses. Should a truss be unstable, the computer will clearly indicate that fact and no analysis will be performed.

Generally, there is little need for complex trusses because it is possible to select simple or compound trusses that will serve the desired purpose equally well. Nevertheless, for a more comprehensive discussion of complex trusses refer to the method of substitute members described in *Theory of Structures* by S. P. Timoshenko and D. H. Young.[2]

7.7 EQUATIONS OF CONDITION

On some occasions, two or more separate structures are connected so that only one type of force can be transmitted through the connection. The three-hinged arch and cantilever types of structures described in Chapter 4 have been shown to fall into this class because they are connected with interior hinges unable to transmit rotation.

[2]S. P. Timoshenko and D. H. Young, *Theory of Structures*. 2nd Ed., (New York: McGraw-Hill, 1965), 92–103.

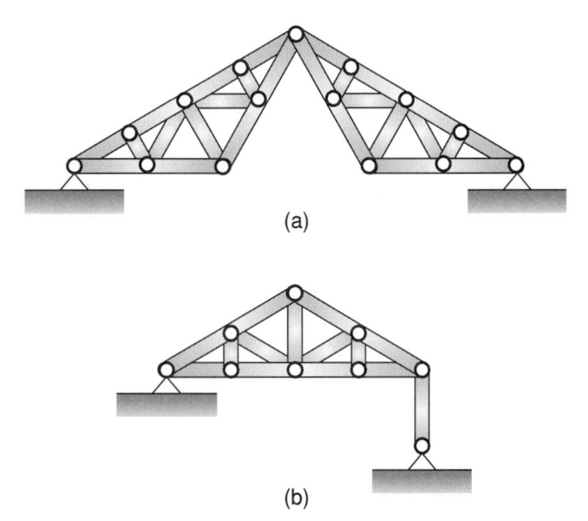

Figure 7.7 Trusses requiring equations of condition for analysis

Compound trusses are often formed in this manner. Equations of condition, also discussed in Chapter 4, provide additional equations that can be used to evaluate the unknown forces.

Perhaps the simplest way to produce a hinge in a truss is by omitting a chord member in one of the panels, as shown in Figure 7.7(a). One of the lower chord members was omitted from the Fink truss shown in Figure 7.6(a) to form this truss. It is obvious that the moment of all the external forces on the part of the structure to the left or the right of the pin connection at the top must be zero. The truss is statically determinate because there are three equations of static equilibrium and one condition equation available for calculating the four forces of reaction.

Figure 7.7(b) is another situation in which equations of condition are produced. In this truss, there are four forces of reaction, and the structure may appear to be statically indeterminate externally. However, the link that connects the truss to the right support provides an equation of condition. A link can only transmit force along its axis. As such, the resultant direction of the force caused by the right reaction is vertical. This equation of condition makes the structure statically determinate externally.

7.8 WHEN ASSUMPTIONS ARE NOT CORRECT

Every engineer should realize that often the assumptions regarding behavior of a structure (pinned joints, loads applied at joints only, and frictionless rollers, among others) may not be entirely valid. Consequently, we should consider what might happen to a structure if the assumptions made in analysis were appreciably in error. Perhaps because of friction, an expansion device or roller will resist a large proportion of any horizontal forces present. How would this affect the member forces of a particular truss? For this very reason the truss of Problem 6.31 was included at the end of Chapter 6, where it was assumed that one-third of the horizontal load was resisted by the roller.

The types of end supports used can have an appreciable effect on the magnitude of forces caused by lateral loads. Short roof trusses generally have no provisions made for temperature expansion and contraction and both ends of the trusses are bolted down to

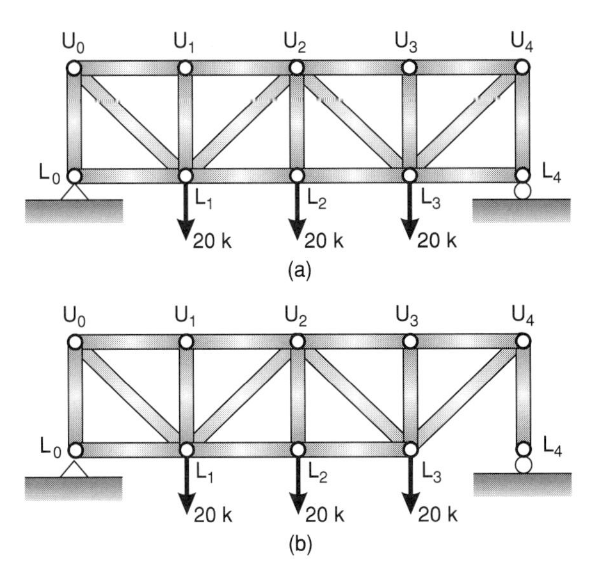

Figure 7.8

their supports. These trusses are actually statically indeterminate, but the usual practice is to assume that the horizontal loads split equally between the supports. For longer roof trusses, provisions for expansion and contraction are considered necessary. Usually the bolts at one end are set in a slotted hole to provide space for the anticipated length changes. A base plate is provided at the end on which the truss can slide.

Actually, it is impossible to provide a support that has no friction. From a practical standpoint, the maximum value of the horizontal reaction at the expansion end equals the vertical reaction times the coefficient of friction (one-third being a reasonable guess). Should corrosion occur, thus preventing movement (a rather likely prospect), an equal split of the lateral loads between the two ends of the truss may again be the best estimate.

An interesting case occurs when members have zero force from analysis. Consider for a moment the truss shown in Figure 7.8(a). From our previous discussion of zero-force members, and if we performed an analysis of the structure, we see that the force in member L_3L_4 is zero. This is true even under the influence of lateral loads. However, if we do not include the member, as shown in Figure 7.8(b), the truss is unstable. The member, therefore, is necessary. However, what is the force we use to design the member? We cannot find that answer from analysis and must resort to some other means. Normal practice would indicate that the member be designed for a force that is equal to some fraction of the shearing force in the panel.

7.9 COMPUTER ANALYSIS OF TRUSSES

Today, a large percentage of structural analysis is performed using computers. This is especially true for statically indeterminate structures of any significant size and complexity. For this reason, computer software for structural analysis has been included on the CD-ROM at the back of this book and use of the software is demonstrated through example and homework problems. The first of these problems is Example 7.5, which illustrates the analysis of a statically determinate plane truss. We will use SAP2000 in this first example. SABLE could also have been used.

Most structural analysis software uses matrix methods of analysis, which are discussed in the latter chapters of this book. These methods, which are based on displacements of the structure, require knowledge of the cross-sectional properties and material characteristics of the members. At this point in our study of structural analysis, we will be content with using the software as a "black-box" that can be used to obtain answers. Through study later in this course, we will develop an understanding of the theoretical basis of the software.

Nevertheless, before beginning the example, there are a number of concepts regarding computer analysis that must be explained. In the paragraphs to follow, these fundamental bits of information for analyzing a truss are discussed. Complete information about using the software is found by accessing the online help that is available after the software has been loaded onto your computer.

7.9.1 Coordinate System

When performing computer structural analysis, we must be concerned with two coordinate systems. The first of these is the global coordinate system, which is a right-hand Cartesian coordinate system in which we specify the geometry of the structure. We also use it to look at structural displacements and reaction forces. The origin of the global coordinate system is specified at some convenient location of the structure, often the lower left-hand corner. The global coordinate axes are normally aligned with the principle axes of the structure.

The second is the local coordinate system. It relates to the forces and deformations acting on, and in, individual members of the structure. The local coordinate system used depends on the type of member being used, and is established when the software is developed. The senses of the member forces obtained from the analysis are interpreted in the local coordinate system.

7.9.2 Joints and Member Connectivity

When beginning an analysis, we must define the geometry of the structure. This is accomplished by first specifying the location of the joints and then how the members are connected to the joints. The element used for the members must be appropriate for the analysis being conducted. When analyzing trusses we should select an element that can only carry axial forces since that is the only force theoretically resisted by a member in a truss. A graphical representation of the geometry, which is available in most software, is very useful in making sure that you have specified the geometry correctly.

7.9.3 Joint Restraints

A degree of freedom at a joint is a possible displacement of the joint in a particular direction. The structural supports are defined by restraining the appropriate degrees of freedom. Restraints specify that the displacement in the direction of that degree of freedom is zero. As such, there is a structural support in that direction. Also, degrees of freedom that are not considered in the analysis are restrained to zero displacement in order to remove them from the analysis. When analyzing trusses, all rotational degrees of freedom are restrained since all of the members in a truss are pin connected: one member cannot transfer moment to another member or to a joint.

7.9.4 Member Properties

The analyst may not feel all of these data are needed for a statically determinate truss. The computer programs, however, are general and applicable to both statically determinate and statically indeterminate structures. When analyzing trusses we must at least specify the cross-sectional area, the second moment of area, and modulus of elasticity for each member.

7.9.5 Loads Acting

We must specify all of the loads that are acting on the structure. Loads acting on the joints are specified in the global coordinate system; those acting directly on the members are specified in the local coordinate system. When analyzing trusses we only need to define joint loads since loads in trusses only act on the joints.

EXAMPLE 7.5

Using SAP2000, determine the forces in the members of the truss shown in the following figure. We analyzed this same truss in Examples 6.2 and 7.1. For the analysis, the units we will use are kips and inches. All member areas and moduli of elasticity are 4 in^2 and 29,000 ksi, respectively. The second moment of area can be taken as zero.

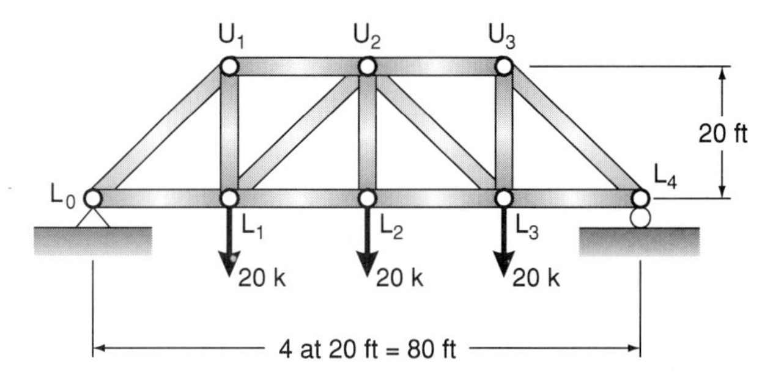

Solution. To begin the analysis we must associate a number with each of the members and nodes in the truss. The numbering scheme used in this analysis is shown in the next figure. These numbers are used to identify the joints and members when entering data into the computer and interpreting results of the analysis.

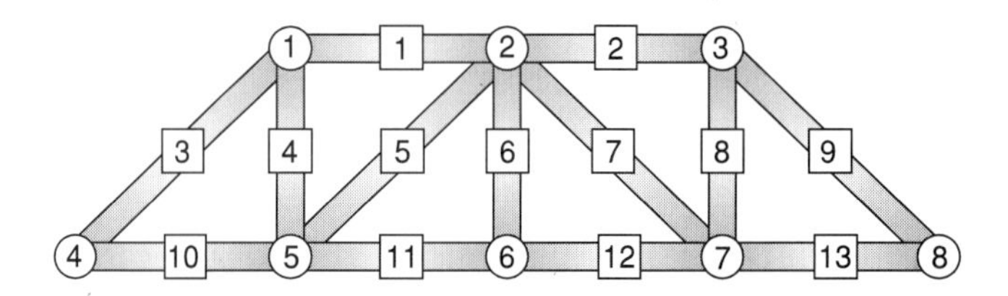

The complete data file for this example is not shown here but is included on the CD-ROM at the end of the book. As rotations are not considered in truss analysis, the rotational degree of freedom at each joint was restrained. Further, the X and Y translational degrees of freedom at joint 4 and the Y translational degree of freedom at joint 8 were restrained since these are the structural supports. A Pin–Pin member was selected as the member type because this member can only carry axial force, which is consistent with our assumption about the behavior of trusses. A Pin–Pin member is assumed to be pin-connected at both ends. Results obtained from the computer for the first four members are shown in the table below.

Beam	Case	End	Axial	Shear Y	Moment Z
1	1	i	30.00	0.0	0.0
		j	−30.00	0.0	0.0
2	1	i	30.00	0.0	0.0
		j	−30.00	0.0	0.0
3	1	i	42.43	0.0	0.0
		j	−42.43	0.0	0.0
4	1	i	−30.00	0.0	0.0
		j	30.00	0.0	0.0

Observe that the shearing force and bending moment at each end of every member are equal to zero. This is as expected and is consistent with our assumption about behavior. Also observe from the calculated results that the axial force at one end of each member is positive whereas the axial force the other end is negative. To determine whether a member is in axial tension or axial compression we must refer to the local coordinate system specified for the member. From the HELP information in SABLE, we can see that the local coordinate system is as shown in the next figure. These are the directions of positive forces at each end of a member.

From this sketch of the local coordinate system, we can see that a positive axial force at the left end of the member indicates axial compression whereas a positive axial force at the right end indicates axial tension. We will use the indicated results at the right end of the member, the j end, to interpret the magnitude and sense (tension or compression) of the axial forces in the members. Upon comparison with the results of Examples 6.2 and 7.1, we observe that the results are the same. ■

7.10 PROBLEMS FOR SOLUTION

For Problems 7.1 through 7.31, determine the forces in all the members of these trusses using method of sections or method of joints, as appropriate.

7.1

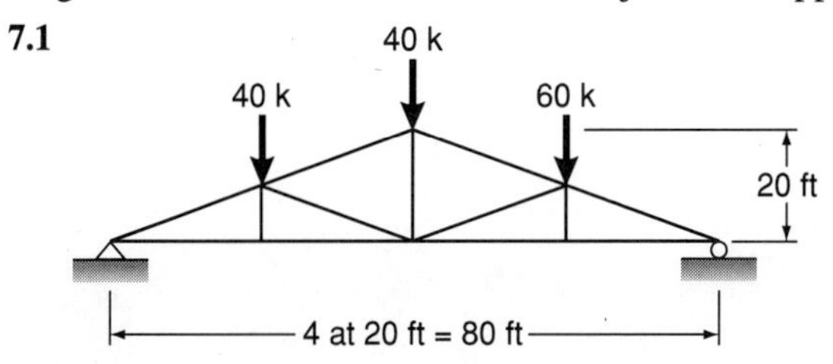

7.2 (*Ans.* $L_1L_2 = +300$ k, $U_3L_3 = +50$ k, $U_4U_5 = -158.1$ k)

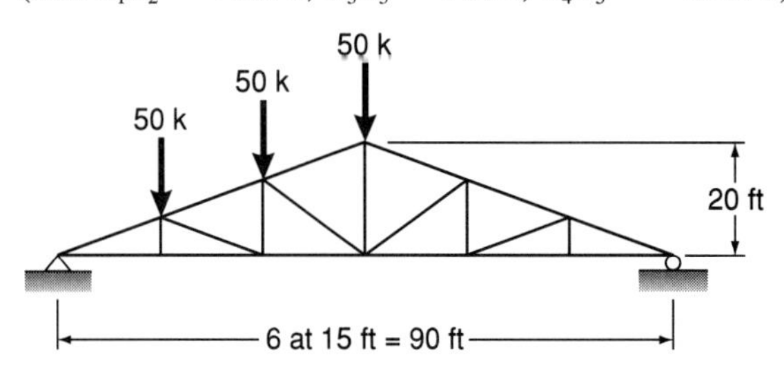

7.3 Rework Problem 7.1 if the panels are changed from 4 @ 20 ft to 4 @ 15 ft and the 40 k loads are doubled.

7.4 Rework Problem 7.1 if a uniform load of 2 klf is applied for the entire span in addition to the loads shown. This additional load is to be applied to the bottom-chord joints. (*Ans.* $L_0U_1 = -279.5$ k, $U_2L_2 = +130$ k, $L_2L_3 = +270$ k)

7.5

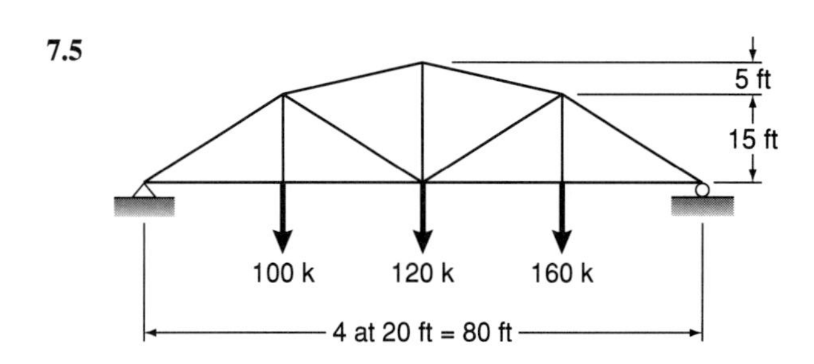

7.6 (*Ans.* $U_2L_2 = +109.3$ k, $L_3U_4 = +83.2$ k, $L_4L_5 = +213.3$ k)

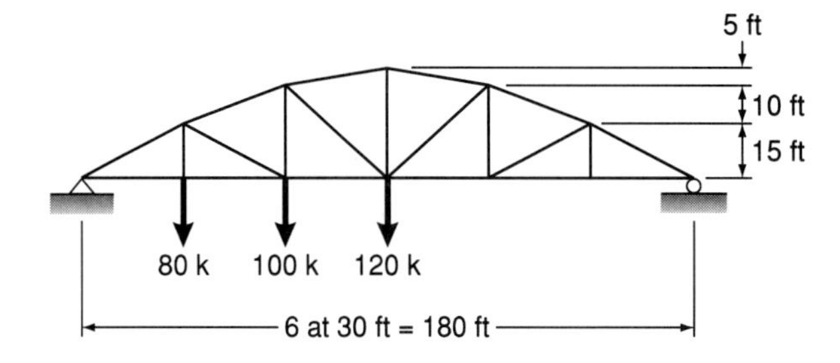

7.7

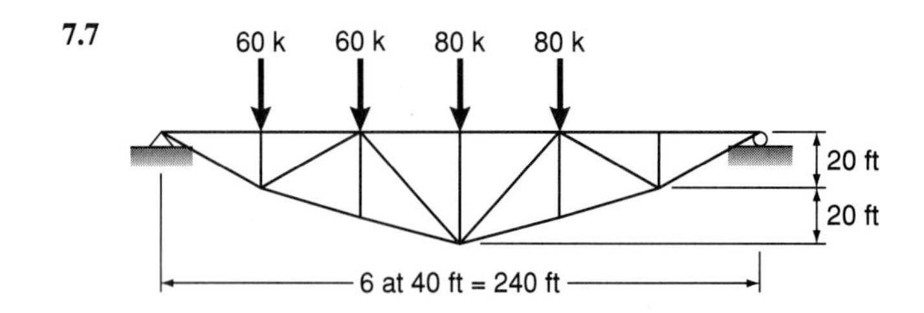

7.8 (*Ans.* AE = −42.4 k, BC = +10 k, DF = −14.1 k)

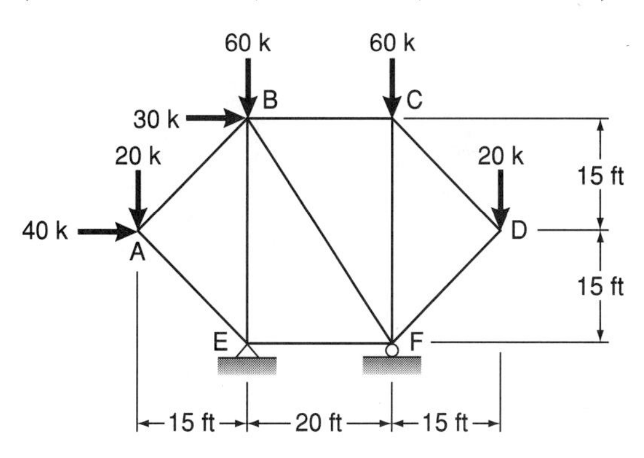

7.9

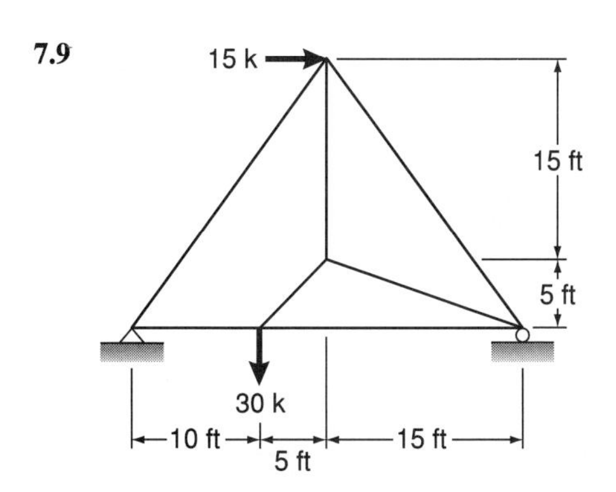

7.10 (*Ans.* AE = +20 k, AD = +58.3 k, DC = −51.0 k)

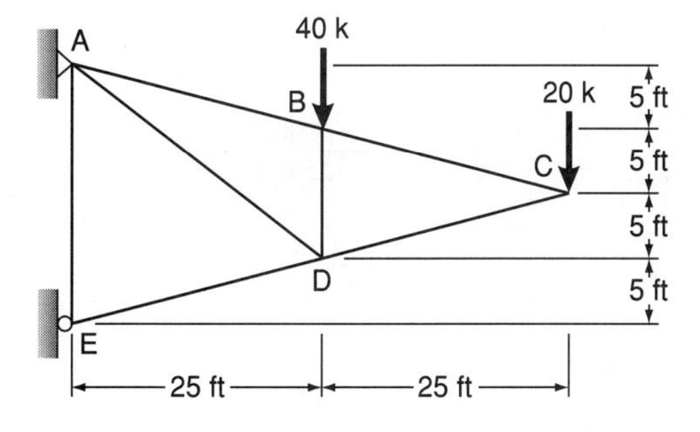

7.11

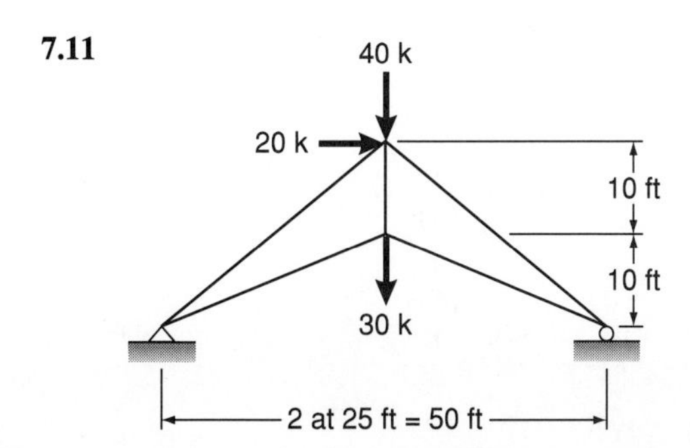

7.12 (*Ans.* $U_0L_1 = +561.4$ kN, $U_2L_2 = -466.7$ kN, $L_2L_3 = +421.2$ kN)

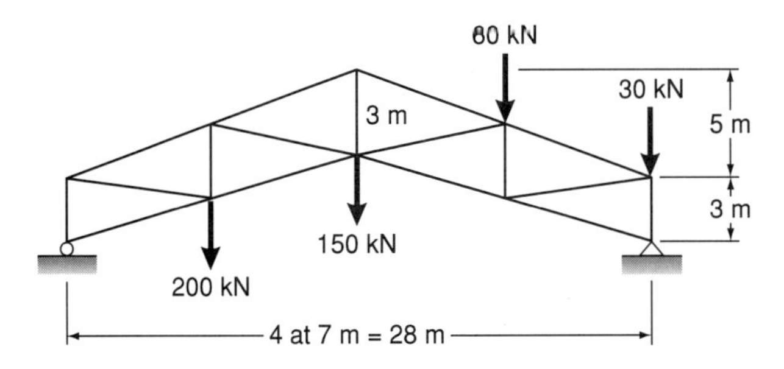

7.13

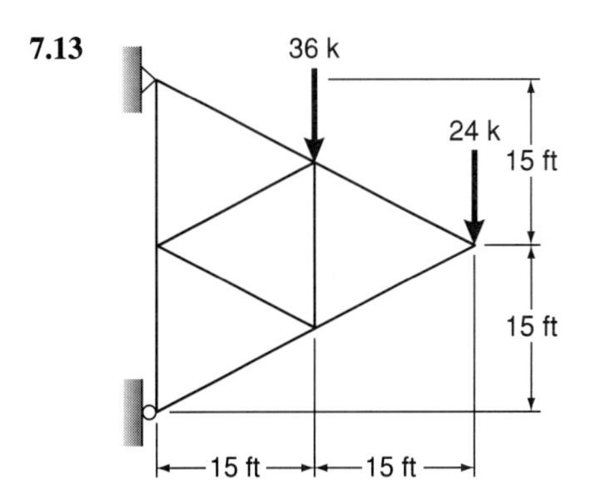

7.14 (*Ans.* $L_0U_1 = -164.4$ k, $U_2L_2 = +85$ k, $L_2L_4 = +146.3$ k)

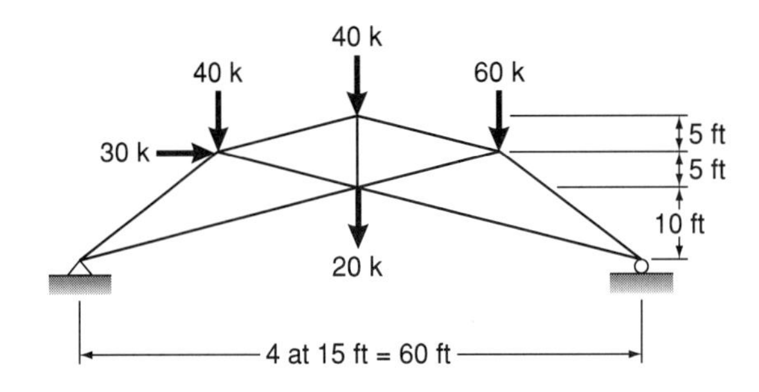

7.15

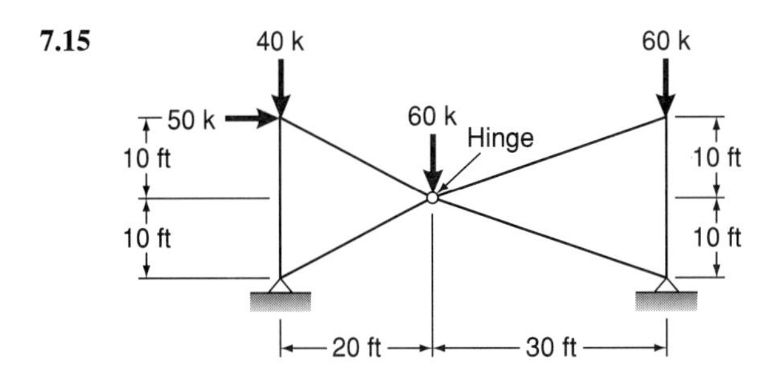

7.16 (*Ans.* AC = −254.9 k, BC = −152 k, DE = +36.5 k)

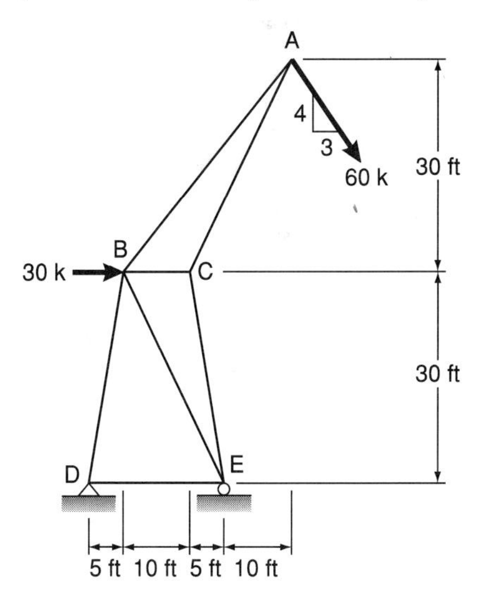

7.17

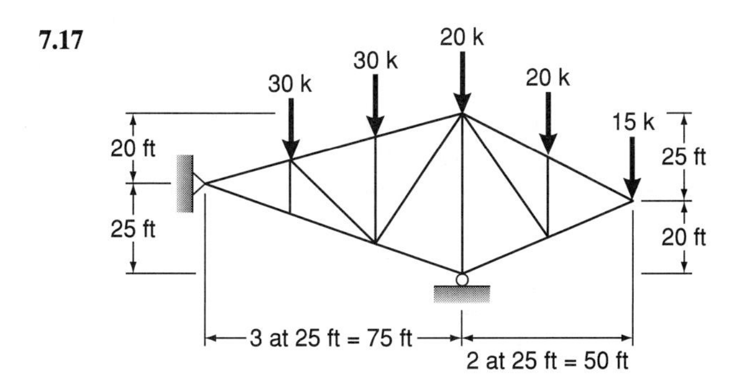

7.18 (*Ans.* L_0U_1 = −82.6 k, U_1L_1 = −9.2 k, U_1U_2 = −42.2 k)

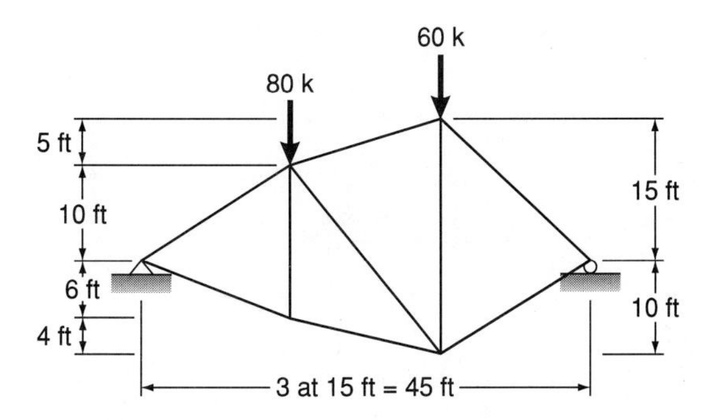

7.19

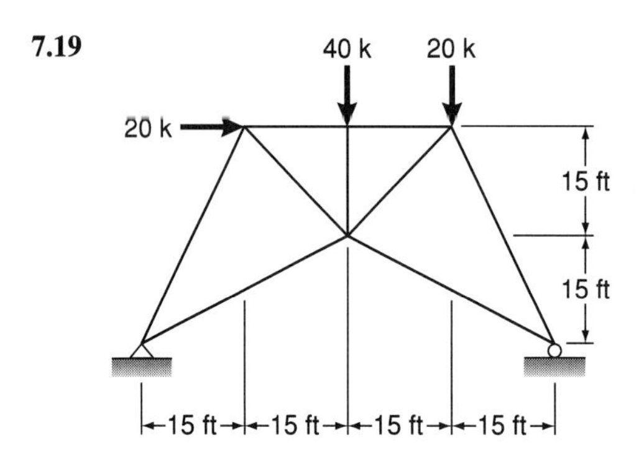

7.20 Repeat Problem 7.19 with the roller support inclined as shown. (*Ans.* $L_0U_1 = -20.5$ k, $U_2U_3 = -47.5$ k, $L_2L_4 = 0$)

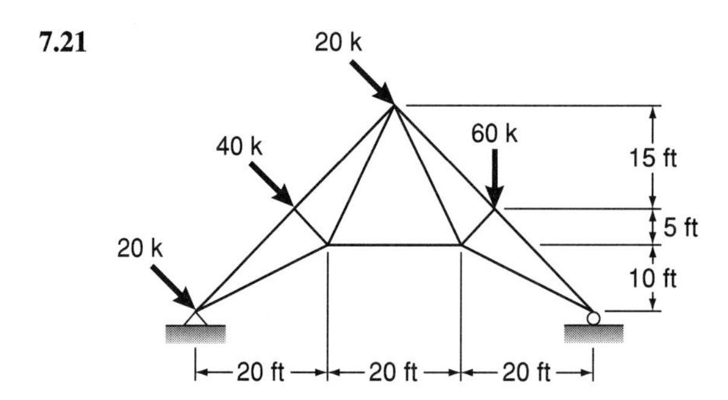

7.21

7.22 (*Ans.* AD $= -22.6$ k, BE $= -77.8$ k, CE $= -30$ k)

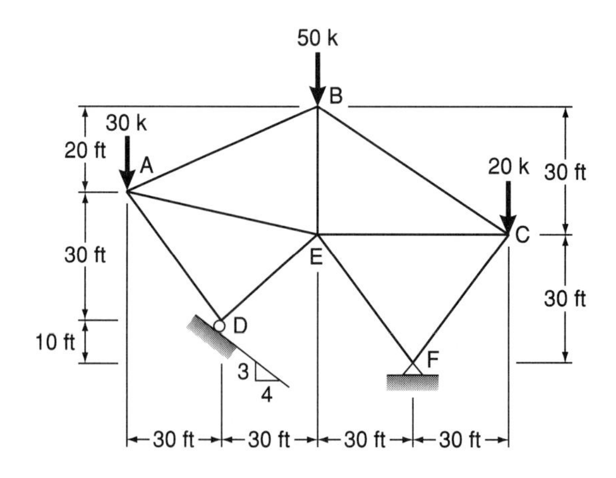

7.23

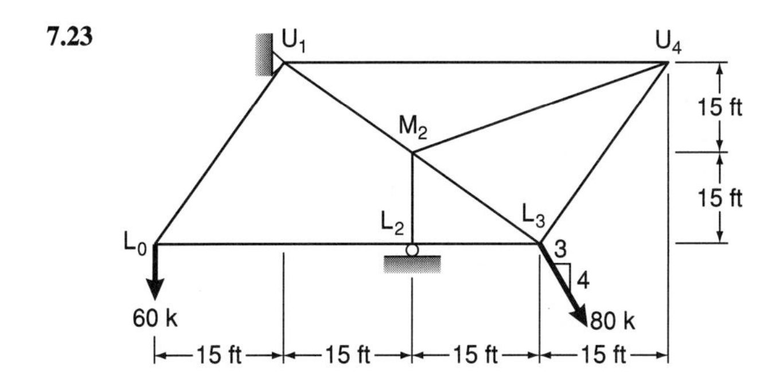

7.24 (*Ans.* $M_1L_2 = -42.4$ k, $U_2M_3 = +17.7$ k, $L_5L_6 = +87.5$ k)

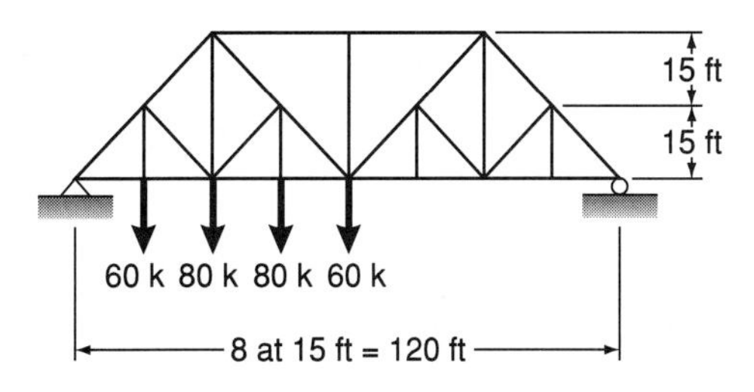

7.25

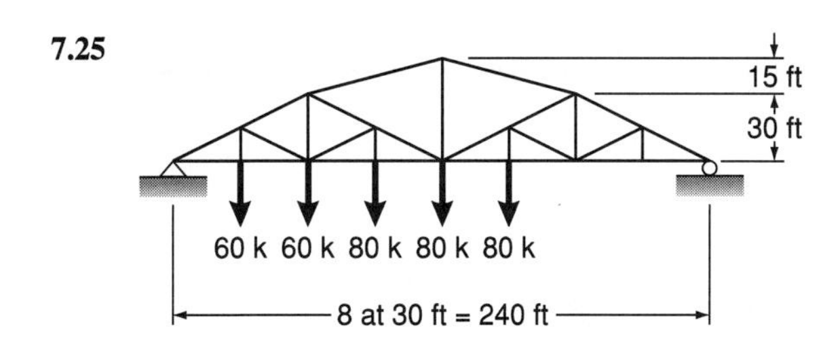

7.26 (*Ans.* $U_2M_2 = +86.2$ k, $L_3L_4 = +236.3$ k, $U_4M_5 = -34.9$ k)

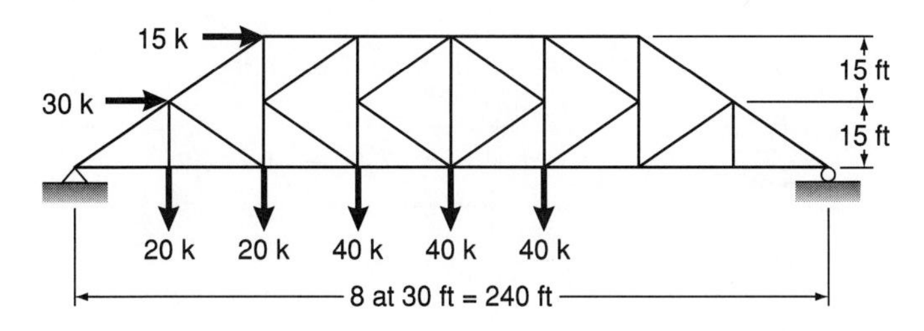

7.27

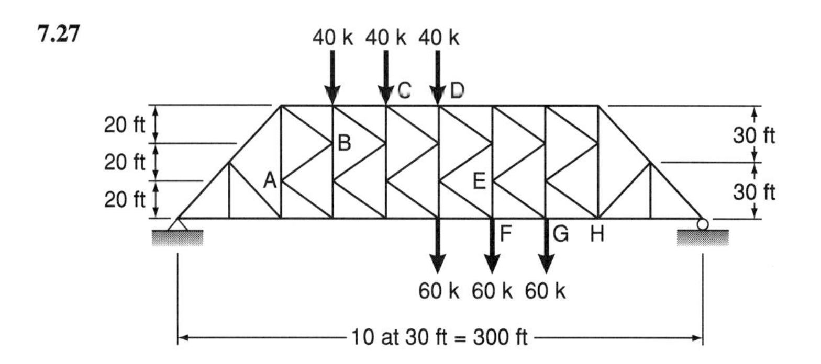

7.28 (*Ans.* $L_2U_3 = +10$ k, $U_3U_5 = -97.3$ k, $L_4L_{10} = +65$ k)

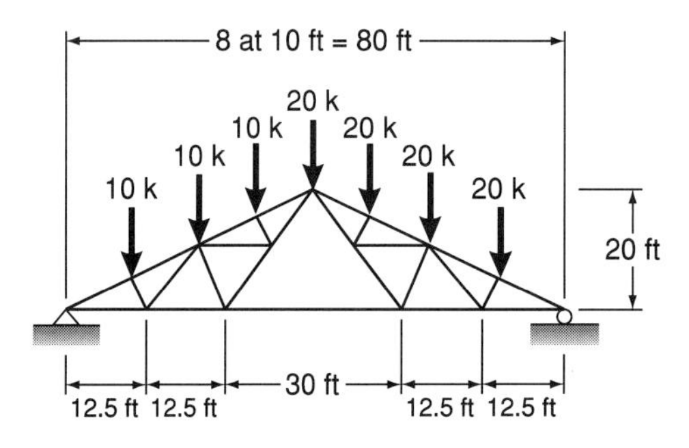

7.29

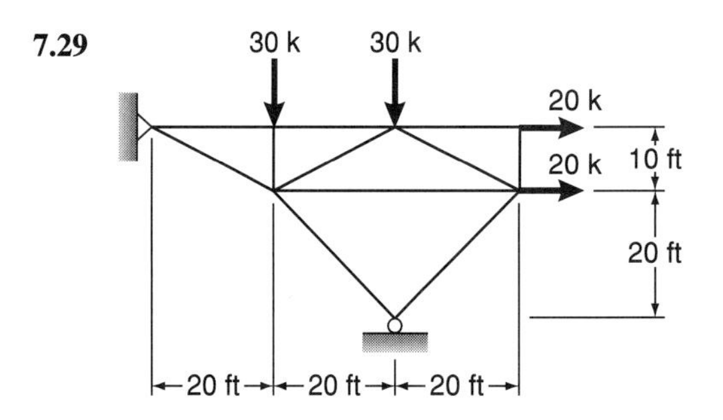

7.30 (*Ans.* AB = +19.6 k, DE = −18.7 k, CF = −6.7 k)

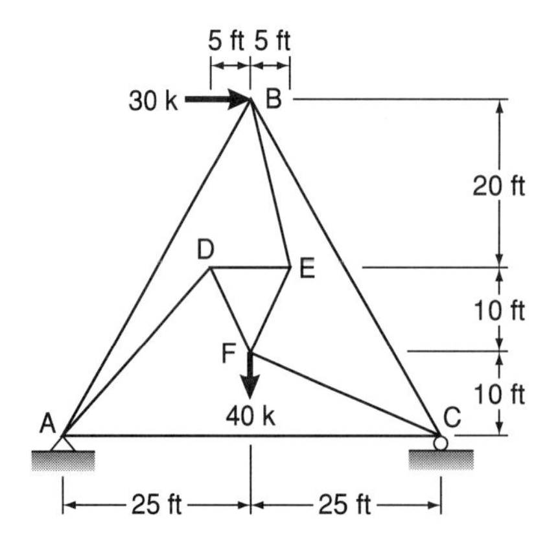

7.31

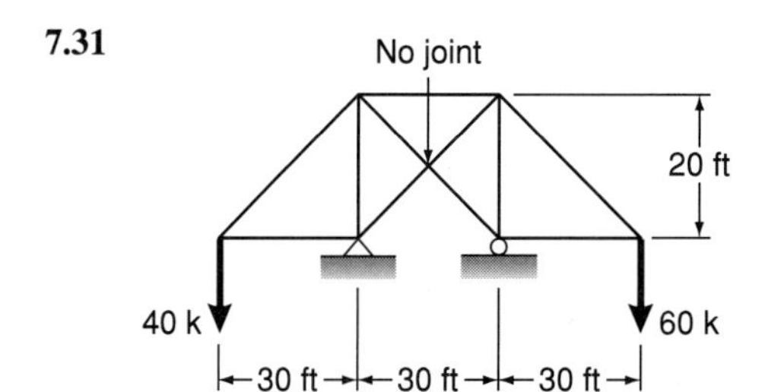

For Problems 7.32 through 7.35, determine directly the forces in each of the designated members using the method of sections.

7.32 Members U_2U_3, L_2L_3, and U_2L_3 of the truss of Problem 7.2. (*Ans.* −158.1 k, +225 k, −90.1 k)

7.33 Members L_2L_3, U_3U_4, and U_2L_3 of the truss of Problem 7.6.

7.34 Members U_2U_3, L_2L_3, and U_2L_3 of the truss of Problem 7.7. (*Ans.* −290 k, 348.2 k, and −67.6 k, respectively)

7.35 Members L_0L_1 and U_1U_2 of the truss of Problem 7.18.

For Problems 7.36 through 7.43, use the zero-load test to determine if the members have critical form.

7.36 (*Ans.* Unstable)

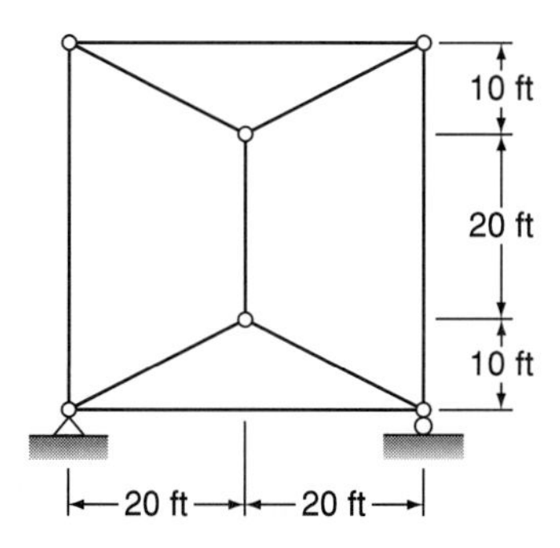

7.37

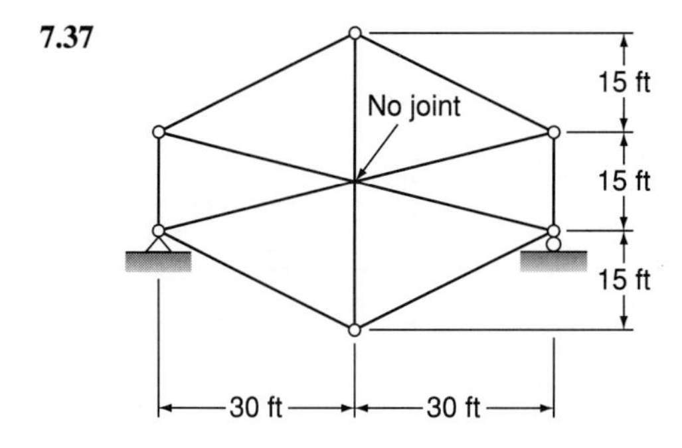

7.38 (*Ans.* Stable)

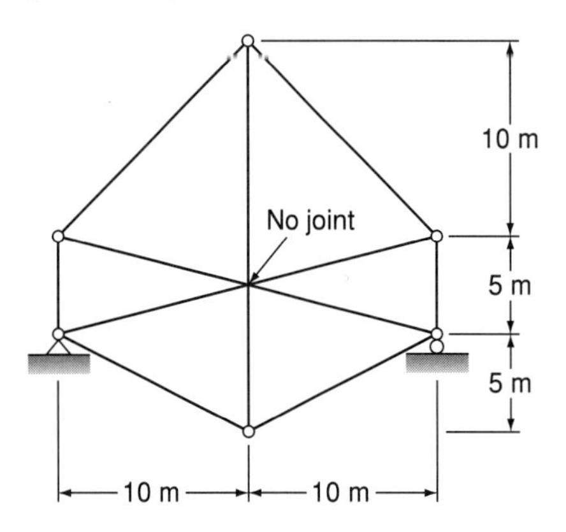

7.39

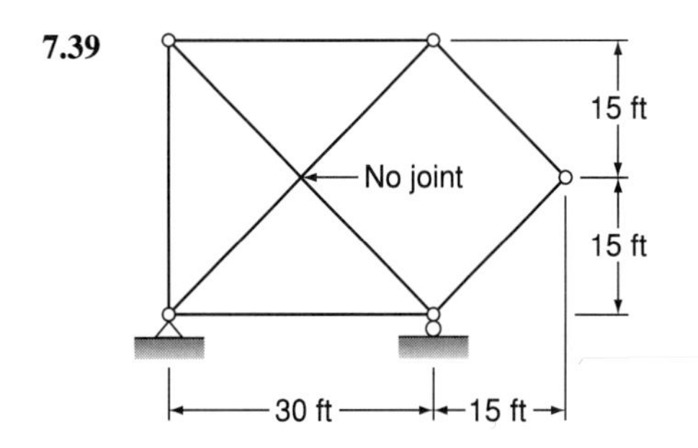

7.40 (*Ans.* Unstable)

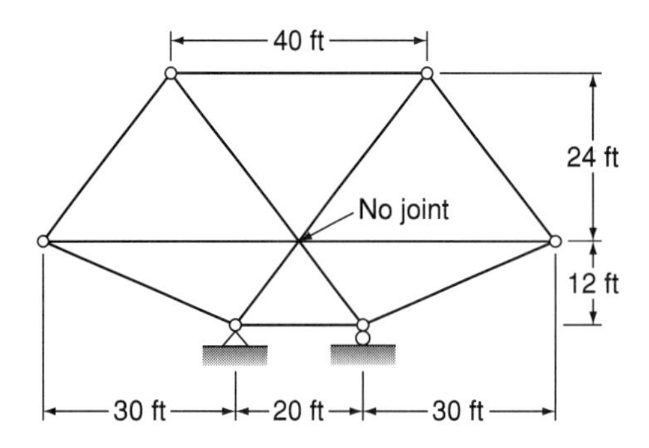

7.41

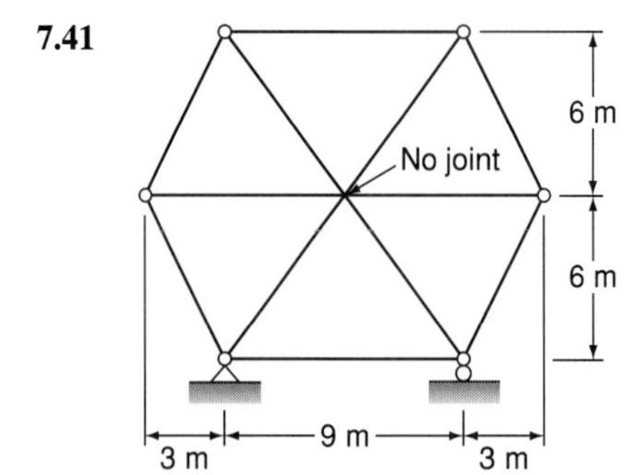

7.42 (*Ans.* Stable)

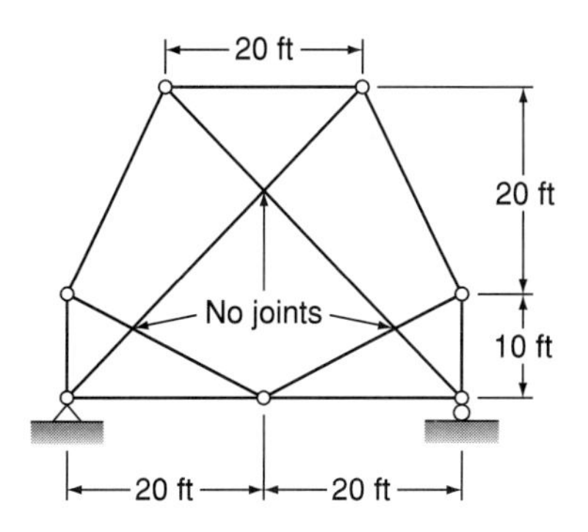

7.43

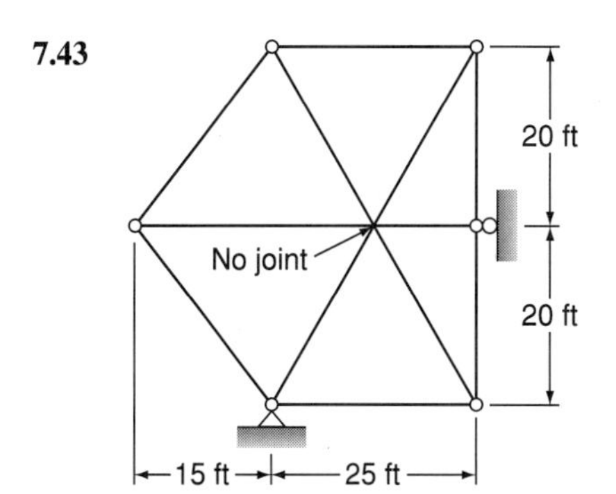

For Problems 7.44 through 7.45, determine the forces in all members.

7.44 (*Ans.* $U_1U_2 = +6.67$ k, $L_4L_5 = -113.1$ k, $L_8L_9 = +73.3$ k)

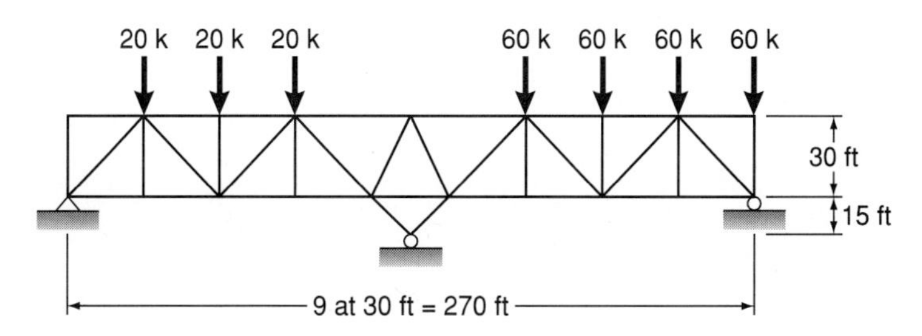

7.45

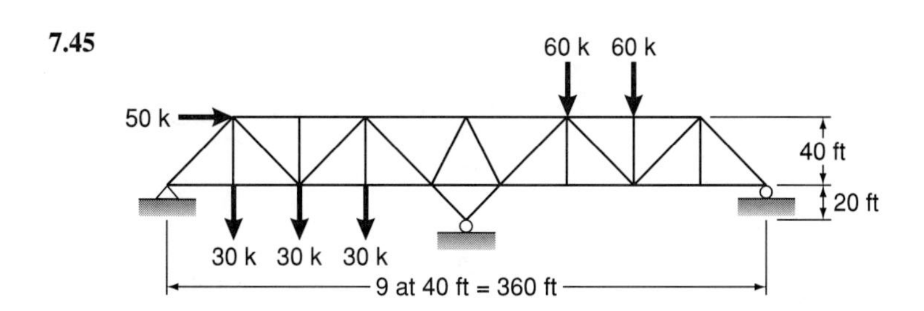

For Problems 7.46 through 7.51, analyze the trusses from previous Problems using the program SAP2000 or SABLE.

7.46 Problem 7.1 (*Ans.* $L_0L_1 = +130$ k, $U_2L_2 = +50$ k, $L_3L_4 = +150$ k)

7.47 Problem 7.8

7.48 Problem 7.13 (*Ans.* $U_0M_0 = +39$ k, $M_0L_1 = +20.1$ k, $L_1U_2 = -26.8$ k)

7.49 Problem 7.15

7.50 Problem 7.23 (*Ans.* $U_1U_4 = +3$ k, $M_2U_4 = -4.2$ k, $L_2L_3 = -45$ k)

7.51 Problem 7.30

Chapter 8

Three-Dimensional or Space Trusses

8.1 INTRODUCTION

A very brief and elementary discussion of small space trusses is presented in this chapter. This material hopefully will help the reader develop some understanding of the behavior of space structures and to recognize that the equations of statics are as applicable in three dimensions as they are in two dimensions.

For all but the very smallest space structures, analysis by the methods of joints and moments described herein is completely unwieldy. Consequently, a large percentage of colleges do not introduce the topic until students take a course in matrix analysis.

Nearly all structures are three- dimensional in nature. However, they usually can be broken down into separate systems, each lying in a single plane at right angles to the other. Because they are at right angles, the forces in one system have no affect on the forces in the other systems. The members joining two systems together serve as members of both systems, and their total force is obtained by combining the forces developed as a part of each of those systems. The end posts of bridge trusses having end portals (Figures 16.11 and 16.12) are one illustration. They serve as the end posts of the bridge trusses and as the columns of the portal.

Many towers, domes, and derricks are three-dimensional structures made up of members arranged so that it is impossible to divide them into different systems, each lying in a single plane, which then could be handled individually. Truss systems, for example, lie in planes that are not at right angles to one another. The forces in one truss framed into another at an angle other than 90° affect the forces in that second truss. For trusses of this type we must analyze the entire structure as a unit, rather than consider the systems in various planes individually. This chapter is devoted to these types of trusses.

Structural engineers are so accustomed to visualizing structures in one plane that when they encounter space trusses they may make frequent mistakes because their minds still are operating on a single-plane basis. If the layout of a space truss is not completely clear, the construction of a small model will probably clarify the situation. Even the simplest models of paper, cardboard, or wire are helpful.

8.2 BASIC PRINCIPLES

Prior to introducing a method of analyzing space trusses, a few of the basic principles pertaining to such structures need to be considered. Like two-dimensional trusses, Three-dimensional trusses are assumed to be composed of members subject to axial force only. In other words, the trusses are assumed to

- Have members that are straight between the joints;
- Have loads applied only at joints only; and
- Have members whose ends are free to rotate. (Note that for this situation to be true the members would have to be connected with universal joints, or at least with several frictionless pins).

Analyses based on these assumptions usually are quite satisfactory despite the welded and bolted connections used in actual practice.

A system of forces coming together at a single point can be combined into one resultant force. The forces so resolved do not necessarily have to be in the same plane. Similarly, a single force can be resolved into component forces in each of the three coordinate directions. The coordinate directions referred to here are the X, Y, and Z coordinates that compose the Cartesian coordinate system. Recall that the Cartesian coordinate system is a right-hand orthogonal coordinate system. That means that the coordinate axes are at right angles to each other. When a force is resolved into its coordinate components, as shown in Figure 8.1, the magnitude of the components is proportional to their length projections on the axes.

The values of the component forces can be computed algebraically from the following relationship:

> The force in a member is to the length of the member as the X, Y, or Z component of force is to the corresponding X, Y, or Z component of length.

This relationship can be expressed mathematically as:

$$\frac{F}{L} = \frac{F_x}{L_x} = \frac{F_y}{L_y} = \frac{F_z}{L_z}$$
$$L^2 = L_x^2 + L_y^2 + L_z^2 \qquad \textbf{Eq. 8.1}$$
$$F^2 = F_x^2 + F_y^2 + F_z^2$$

Space trusses may be either statically determinate or statically indeterminate; consideration is given here only to those that are statically determinate. The methods developed in later chapters for statically indeterminate structures apply equally to three-dimensional and two-dimensional trusses.

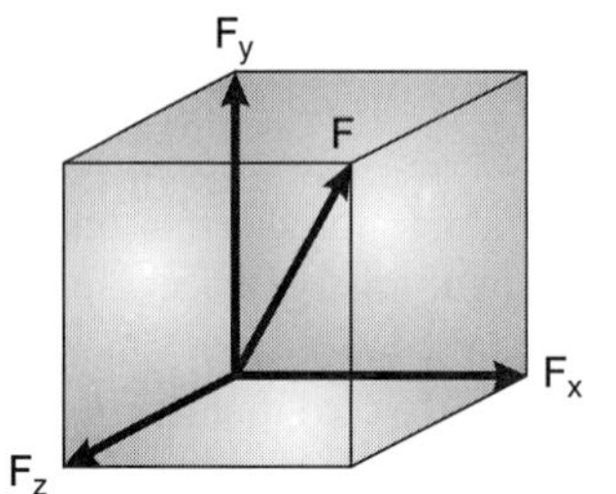

Figure 8.1 Resolution of a force into components

8.3 EQUATIONS OF STATIC EQUILIBRIUM

There are more equations of static equilibrium available for determining the reactions of three-dimensional structures because there are two more axes to take moments about and one new axis along which to sum forces. For equilibrium, the sum of the forces along each of the three reference axes must be equal to zero, as does the sum of the moments of all the forces about each of the axes. Six equations of equilibrium are available. These equations are

$$\sum F_x = 0 \quad \sum M_x = 0$$
$$\sum F_x = 0 \quad \sum M_x = 0 \qquad \textbf{Eq. 8.2}$$
$$\sum F_x = 0 \quad \sum M_x = 0$$

The six reaction components may be determined directly from these equations.

Denver, Colorado, Convention Center (Courtesy Bethlehem Steel Corporation)

Should a structure have more than six reaction components, it is statically indeterminate externally. If there are fewer than six components of reaction, the truss is unstable. If the number of components of reaction is equal to six, the truss is statically determinate externally. Many space trusses, however, have more than six reaction components and yet are statically determinate overall. Example 8.2 shows that the reactions for this type of structure may be determined by solving them concurrently with the member forces, using only equations of static equilibrium.

As with plane trusses, the basic geometric shape of the space truss is the triangle. A triangle can be extended into a space truss by adding three members and one joint. Each of the new members frames into one of the joints of the basic triangle; the other ends come together to form a new joint. The elementary space truss formed in this manner has six members and four joints. It is called a tetrahedron, a figure with four triangular surfaces. This fundamental space truss may be enlarged by the addition of three members and one joint. For each of the joints of a space frame three equations of static equilibrium ($\Sigma F_X = 0, \Sigma F_Y = 0$, and $\Sigma F_Z = 0$) are available for calculating the unknown forces. If we let j be the number of joints in the truss, m be the number of members, and r be the number of reaction components, we see that for a space truss to be statically determinate the following relation must be satisfied:

$$3j = m - r \qquad \textbf{Eq. 8.3}$$

Should there be joints in the truss where the members are all in one plane, only two equations are available at each, and it is necessary to subtract one from the left-hand side of the equation for each such joint. The omission of one member for each reaction component in excess of six will cause this equation to be satisfied, and the structure will be statically determinate internally. When this situation occurs, it is possible to compute the forces and reactions for the truss from three-dimensional static equilibrium, even though it is statically indeterminate externally.

8.4 STABILITY OF SPACE TRUSSES

The general rule for stability in regard to external forces is that the projection of the structure onto any of the three coordinate planes must itself be a stable truss. Therefore, as with two-dimensional structures, there must be at least three nonconcurrent components of reaction in any one plane. The results of reaction computations will be inconsistent for any other case.

In the preceding paragraphs, both external and internal stability and determinacy were treated as though they were completely independent subjects. The two have been separated here for clarity for the reader who has not previously encountered space trusses. Note however, from the example problems in the pages that follow that it is impossible in a majority of cases to consider the two separately. For instance, many trusses are statically indeterminate externally and statically determinate internally and can be completely analyzed using only the equations of static equilibrium. Few two-dimensional structures fall into this class.

In Section 8.3, it was stated that for a space truss to be statically determinate the following equation had to be satisfied:

$$3j = m - r \qquad \textbf{Eq. 8.4}$$

This equation is not sufficient to show whether a particular space truss is stable, however. Externally the reactions must be arranged to prevent movement of the structure: internally the members must be placed to prevent the joints from moving with respect to each other.

For external stability, the reactions must be placed so that they can resist translation along and rotation about each of the three coordinate axes. To achieve this goal there must be at least six nonparallel reactions and they must not intersect a common axis.

Internal stability can be achieved if the geometry of the truss is developed tetrahedron by tetrahedron, that is, by successively adding one joint and three members. In large space trusses it may be quite difficult to see if this condition has been met. An analysis of the truss, however, will provide a clear statement of stability. If we can obtain a *unique solution,* the truss is stable. If not, the truss is unstable. The zero-load test that was discussed in Section 7.5.4 also can be used to check stability.

8.5 SPECIAL THEOREMS APPLYING TO SPACE TRUSSES

From the principles of elementary statics, two theorems that are useful when analyzing space trusses may be developed. These are discussed in the following paragraphs.

1. The component of force in a member 90° to the direction of the member is equal to zero, because no matter how large the force may be it is equal to zero when

multiplied by the cosine of 90°. A force in one plane cannot have components in a plane normal to the original plane. Furthermore, a force in one plane cannot cause moment about any axis in its plane, because it will either intersect the axis or be parallel to it.

From the foregoing principle, if several members of a truss come together at a joint, and all but one lie in the same plane, the component of force in the member normal to the plane of the other members must be equal to the sum of the components of the external forces at the joint normal to the same plane. If no external forces are present, the member has a force of zero.

2. If there is a joint in a truss to which no external loads are applied, and all but two members framing into the joint have zero-force, these two members must have zero-force unless they form a straight line.

This latter theorem can be proven using the equations of static equilibrium.

8.6 TYPES OF SUPPORT

In Chapters 6 and 7, plane trusses were supported with rollers and hinges that provided one or two reaction forces. With three-dimensional trusses, the same types of supports are used, but the number of reaction forces can vary from one to three. These supports are described in the following paragraphs and are illustrated in Figure 8.2.

1. The *plane roller, steel ball,* and *flat plate* types of support provide resistance to movement perpendicular to the supporting surface. Thus, it has one component of reaction component, which can be toward or away from the supporting surface.
2. The *slotted roller* is free to move in one direction parallel to the supporting surface. Movement is prevented in the other direction parallel to the supporting surface as well as perpendicular to it, giving a total of two reaction components.
3. The *hinge* or *ball and socket joint* types of support provide resistance to movement in all three coordinate directions. Thus, they provide a total of three forces of reaction.
4. The *short link* provides resistance only in the direction of the link. As such, it provides only one force of reaction, and that force is parallel to the link.

This discussion indicates that it is possible to select a type of support having three reaction components or one that may have only one or two components. A little thought on the subject shows that the possibility of limiting the number of reaction components of a space truss is very advantageous. A truss that is statically indeterminate externally

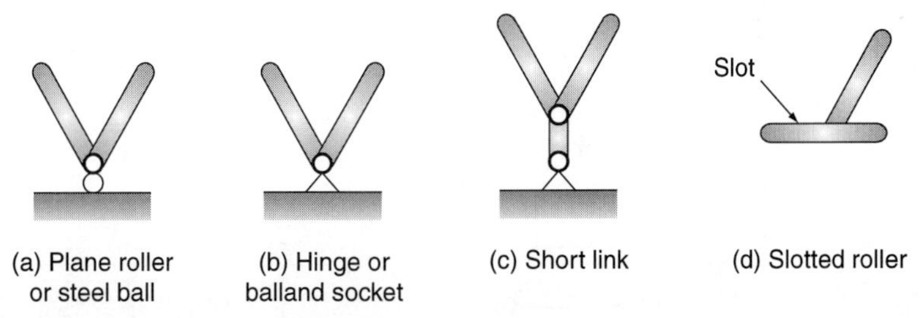

Figure 8.2 Types of support in a space truss

may have its total reaction components limited to six, making it statically determinate. The advantages of statically determinate and statically indeterminate structures are discussed in Chapter 12. For some structures, good design necessitates elimination of reaction forces in certain directions. The most obvious example occurs when a space truss is supported on walls where a reaction or thrust perpendicular to the wall is undesirable.

Broome County Veterans Memorial Arena, Binghamton, NY
(Courtesy Bethlehem Steel Corporation)

8.7 ILLUSTRATIVE EXAMPLES

Examples 8.1 and 8.2 illustrate the application of the foregoing principles to elementary space trusses. Example 8.1 considers a structure supported at three points with six reaction components, which can be computed directly. The second example presents a space truss supported at four points with seven reaction components, which cannot be solved directly. The directions in which reaction components are possible are indicated herein by dark heavy lines at the support points, as shown in the diagrams of the frames analyzed in Examples 8.1 to 8.3.

EXAMPLE 8.1

Determine the reactions and member forces in the space truss shown in the figure.

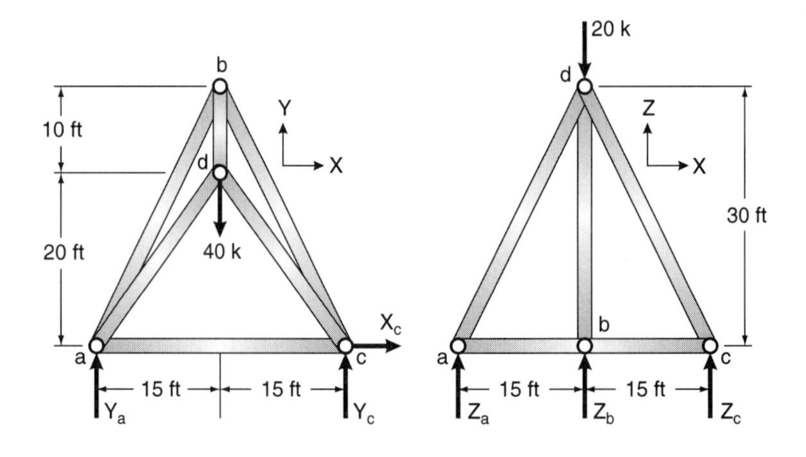

Solution. The truss is statically determinate and stable externally because there is a total of six reaction components with three nonconcurrent reaction forces in each plane. Internally it is statically determinate, as proven with the joint equation.

$$3j = m + r$$
$$s(4) = 6 + 6$$
$$12 = 12$$

The truss is statically determinate.

For a truss with three vertical reaction components, moments may be taken about an axis through any two of the components to find the third.

$$\sum M_x = 0 \quad \text{about line ac}$$
$$0 = 40(30) - 20(20) + 30Z_b$$
$$\therefore Z_b = -26.7\,k$$
$$\sum M_y = 0 \quad \text{about line of action of } Y_a$$
$$0 = 20(15) + 26.7(15) - 30Z_c$$
$$\therefore Z_c = 23.3\,k$$
$$\sum F_z = 0$$
$$0 = -20 - 26.7 + 23.3 + Z_a$$
$$\therefore Z_a = 23.4\,k$$

Similarly, where there are three unknown horizontal reaction components moments may be taken about a vertical axis passing through the point of intersection of two of the components.

$$\sum M_z = 0 \quad \text{about line of action of } Z_c$$
$$0 = -40(15) + 30Y_a$$
$$\therefore Y_a = 20.0\,k$$
$$\sum F_y = 0$$
$$0 = 20 - 40.0 + Y_c$$
$$\therefore Y_c = 20.0\,k$$
$$\sum F_x = 0$$
$$0 = 0 + X_c$$
$$\therefore X_c = 0.0$$

When the reactions have been found, the member forces can readily be computed using the method of joints. At joint a, member ad is the only member having a Z component of length; therefore, its component must be equal and opposite to Z_a, or 23.4-kip compression. The X and Y components of ad are proportional to its components of length in those directions. Setting up a table similar to the one shown simplifies the computation of components and resultant forces.

Considering joint a, the Y component of force in member ab can be determined by joints now that the Y component of ad is known:

$$\sum F_y = 0 \quad \text{at joint a}$$
$$0 = 20 - 15.6 - Y_{ab}$$
$$\therefore Y_{ab} = -4.4\,k$$

The other member forces are computed similarly using the method of joints. The results are shown in the table.

	Projection				Component Forces			
Bar	X	Y	Z	Length	X	Y	Z	Force
ab	15	30	0	33.5	−2.2	−4.4	0	−4.92
ac	30	0	0	30.0	13.9	0	0	13.9
ad	15	20	30	39.1	−11.7	−15.6	−23.4	−30.5
bc	15	30	0	33.5	−2.2	−4.5	0	−5.02
bd	0	10	30	31.6	0	8.9	26.7	28.1
cd	15	20	30	39.1	−11.7	−15.5	−23.3	−30.3

■

EXAMPLE 8.2

Find all reactions and member forces of the space truss shown in the figure.

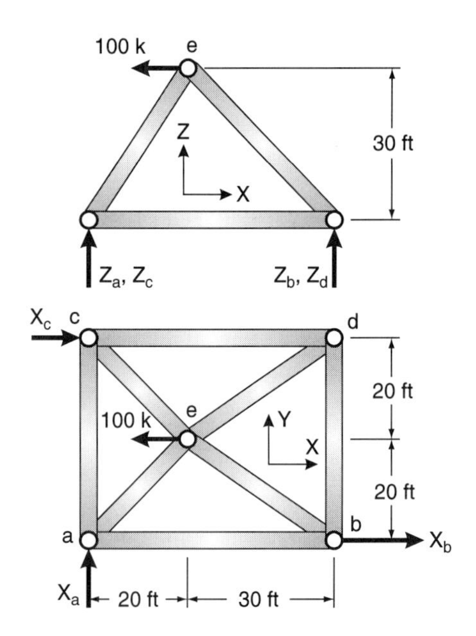

Solution. This truss appears to be statically indeterminate externally because there are seven definite forces of reaction and only six equations of static equilibrium. Internally, however, the truss is statically determinate as shown, and the analysis may be conducted using only equations of static equilibrium.

$$3j = m + r$$
$$3(5) = 8 + 7$$
$$15 = 15$$

Therefore the truss is statically determinate.

Although the truss is statically indeterminate externally, there are only three unknown reaction components in the XY plane, and these forces can be determined immediately. The other four components will be computed in conjunction with the member forces.

$$\sum M_z = 0 \quad \text{about line of action of } Z_a$$
$$0 = -100(20) + 40X_c$$
$$\therefore X_c = 50.0\,\text{k}$$
$$\sum F_x = 0$$
$$0 = 50 - 100.0 + X_b$$
$$\therefore X_b = 50.0\,\text{k}$$
$$\sum F_y = 0$$
$$0 = 0 + Y_a$$
$$\therefore Y_a = 0.0$$

If the value of one of the Z reaction components should be known, the values of the other three could be determined from static equilibrium. Assume Z_d has a value of S directed downward, and then compute the components of reaction in terms of S:

$$\sum M_y = 0 \quad \text{about line ac}$$
$$0 = -100(30) + 50S + 50Z_b$$
$$\therefore Z_b = 60 - S$$
$$\sum M_x = 0 \quad \text{about line ab}$$
$$0 = 40S - 40Z_c$$
$$\therefore Z_c = S$$
$$\sum F_z = 0$$
$$0 = S - S - (60 - S) + Z_a$$
$$\therefore Z_a = 60 - S$$
$$\text{Checking by } \sum M_x = 0 \quad \text{about line cd}$$
$$0 = (60 - S)(40) - 40Z_a$$
$$\therefore Z_a = 60 - S$$

The calculation of member forces may now be started using the reaction forces in terms of S. These computations are continued until the forces at both ends of one bar are determined in terms of S. The two values must be equal, and they are equated to calculate the correct value of S.

The Z component of force in member de is equal to S and is in tension, whereas the Z component of force in member be is equal to 60-S and is also in tension. The Y component of force in member de is equal to

$$Y_{be} = \frac{20}{30}(60 - S) = 40 - \frac{2}{3}S$$

By summing forces in the Y direction at joint d, member bd is found to be acting in compression and has a force of 2S/3. Similarly, by summing forces in the Y direction at joint b, member bd is found to have a compressive force of 40 − 2S/3. Because these two values are the force in member bd, they can be set equal to each other and the value of S can be computed.

$$\frac{2}{3}S = 40 - \frac{2}{3}S$$

$$\therefore S = 30\,\mathrm{k}$$

The numerical values of the Z-reaction components can now be found from S, and the forces in the truss can be determined using the method of joints. The use of a table to work with length and force components is again convenient. The results are shown in the following table.

	Projection				Component Forces			
Bar	X	Y	Z	Length	X	Y	Z	Force
ab	50	0	0	50.0	+20	0	0	+20.0
ae	20	20	30	41.2	−20	−20	−30	−41.2
ac	0	40	0	40.0	0	+20	0	+20.0
be	30	20	30	46.9	+30	+20	+30	+46.9
bd	0	40	0	40.0	0	−20	0	−20.0
de	30	20	30	46.9	+30	+20	+30	+46.9
cd	50	0	0	50.0	−30	0	0	−30.0
ce	20	20	30	41.2	−20	−20	−30	−41.2

■

Transmission towers for the country's first 345,000-volt transmission line, Chief Joseph-Snohomish Dam, Washington State (Courtesy Bethlehem Steel Corporation)

8.8 SOLUTION USING SIMULTANEOUS EQUATIONS

Three simultaneous equations:

$$\sum F_x = 0$$
$$\sum F_y = 0 \qquad \textbf{Eq. 8.5}$$
$$\sum F_z = 0$$

may be written for the forces meeting at each joint of a space truss. This results in 3j simultaneous equations. If the truss is statically determinate, the equations may be solved for both the member forces and the components of reaction components, which are the unknowns in the simultaneous equations. Despite the large number of equations that can result, this method of solution may be rather quick for small space trusses, because of the small number of unknowns that appear in each equation.

The preparation and solution of simultaneous equations for space trusses can be simplified by making use of tension coefficients.[1] The tension coefficient for a member is equal to the force in the member divided by its length. In each of the following expressions for force components, the value F/L is replaced by T, the tension coefficient.

$$F_x = \frac{L_x}{L}F = \frac{F}{L}L_x = TL_x$$
$$F_y = \frac{L_y}{L}F = \frac{F}{L}L_y = TL_y \qquad \textbf{Eq. 8.6}$$
$$F_z = \frac{L_z}{L}F = \frac{F}{L}L_z = TL_z$$

When tension coefficients are used, the resulting simultaneous equations are then in terms of the tension coefficient for each member. The equations are solved for the tension coefficients, which are then multiplied by the appropriate member length to obtain the final member forces. The analysis of a space truss through use of tension coefficients and simultaneous equations is demonstrated in the following example.

EXAMPLE 8.3

Using simultaneous equations, determine the forces of reaction and the member forces in the truss shown in the figure. This is the same truss analyzed in Example 8.1.

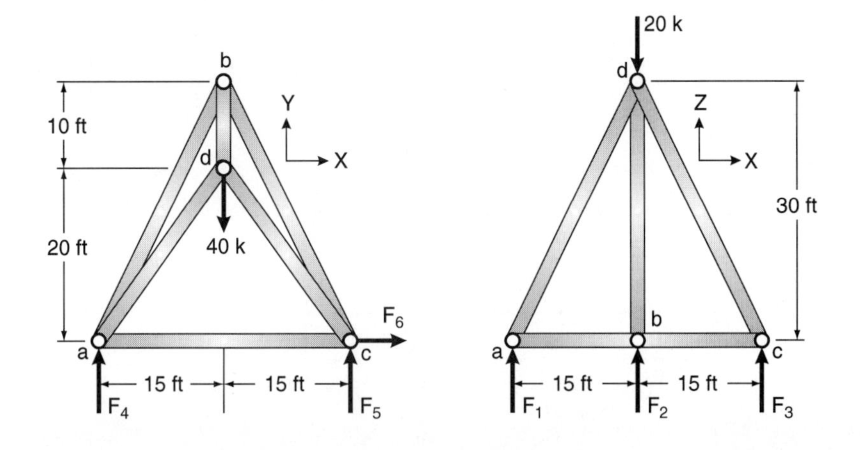

[1]R. V. Southwell, "Primary Stress Determination in Space Frames," *Engineering 109* (1920): 165.

Solution. We begin the analysis by writing the equations of equilibrium for each joint in the truss. Unknown member forces are assumed to be in tension and positive reactions are in the same direction as the coordinate axes.

$$\text{Joint a}\begin{cases} 30T_{ac} + 15T_{ab} + 15T_{ad} = 0 \\ F_4 + 30T_{ab} + 20T_{ad} = 0 \\ F_1 + 30T_{ad} = 0 \end{cases}$$

$$\text{Joint b}\begin{cases} -15T_{ab} + 15T_{bc} = 0 \\ -30T_{ab} - 30T_{bc} - 10T_{bd} = 0 \\ F_2 + 30T_{bd} = 0 \end{cases}$$

$$\text{Joint c}\begin{cases} F_6 - 30T_{ac} - 15T_{bc} - 15T_{cd} = 0 \\ F_5 + 30T_{bc} + 20T_{cd} = 0 \\ F_3 + 30T_{cd} = 0 \end{cases}$$

$$\text{Joint d}\begin{cases} -15T_{ad} + 15T_{cd} = 0 \\ -20T_{ad} + 10T_{bd} - 20T_{cd} = 40 \\ -30T_{ad} - 30T_{bd} - 30T_{cd} = 20 \end{cases}$$

These equations can be solved simultaneously to compute the unknown values. They can also be cast in matrix form for solution with a calculator or computer. The equations cast in matrix form are shown here. The dashed lines in the matrices are only to help keep track of the rows and columns.

$$\begin{bmatrix} 15 & 30 & 15 & 0 & 0 & 0 & 0 & 0 & 0 & 0 & 0 & 0 \\ 30 & 0 & 20 & 0 & 0 & 0 & 0 & 0 & 0 & 1 & 0 & 0 \\ 0 & 0 & 30 & 0 & 0 & 0 & 1 & 0 & 0 & 0 & 0 & 0 \\ -15 & 0 & 0 & 15 & 0 & 0 & 0 & 0 & 0 & 0 & 0 & 0 \\ -30 & 0 & 0 & -30 & -10 & 0 & 0 & 0 & 0 & 0 & 0 & 0 \\ 0 & 0 & 0 & 0 & 30 & 0 & 0 & 1 & 0 & 0 & 0 & 0 \\ 0 & -30 & 0 & -15 & 0 & -15 & 0 & 0 & 0 & 0 & 0 & 1 \\ 0 & 0 & 0 & 30 & 0 & 20 & 0 & 0 & 0 & 0 & 1 & 0 \\ 0 & 0 & 0 & 0 & 0 & 30 & 0 & 0 & 1 & 0 & 0 & 0 \\ 0 & 0 & -15 & 0 & 0 & 15 & 0 & 0 & 0 & 0 & 0 & 0 \\ 0 & 0 & -20 & 0 & 10 & -20 & 0 & 0 & 0 & 0 & 0 & 0 \\ 0 & 0 & -30 & 0 & -30 & -30 & 0 & 0 & 0 & 0 & 0 & 0 \end{bmatrix} \begin{Bmatrix} T_{ab} \\ T_{ac} \\ T_{ad} \\ T_{bc} \\ T_{bd} \\ T_{cd} \\ F_1 \\ F_2 \\ F_3 \\ F_4 \\ F_5 \\ F_6 \end{Bmatrix} = \begin{Bmatrix} 0 \\ 0 \\ 0 \\ 0 \\ 0 \\ 0 \\ 0 \\ 0 \\ 0 \\ 0 \\ 40 \\ 20 \end{Bmatrix}$$

Upon analysis, we find that the forces of reaction are as follows:

Reaction	Force	Reaction	Force
F_1	23.33 k	F_4	20 k
F_2	−26.67 k	F_5	20 k
F_3	23.33 k	F_6	0

The computed tension coefficients and the final member forces are:

Member	Tension Coefficient (T)	Member Length (L)	Final Force (T·L)
ab	−0.148	33.54 ft	−4.97
ac	0.463	30.00 ft	13.89
ad	−0.778	39.05 ft	−30.37
bc	−0.148	33.54 ft	−4.97
bd	0.889	31.62 ft	28.11
cd	−0.778	39.05 ft	−30.37

Observe that these are the same results obtained in Example 8.1. ■

8.9 SOLUTION USING COMPUTERS

From the preceding examples, we can see that if a space were of any significant size, analyzing that truss would be an extremely laborious process. Except for very simple trusses, most analysis of space trusses is performed with the use of a computer. Such an analysis proceeds similarly to the plane-truss analysis we saw in the previous chapter. In the following example, SAP2000 is used to analyze a space truss.

EXAMPLE 8.4

Analyze the space truss shown in the figure using SAP2000.

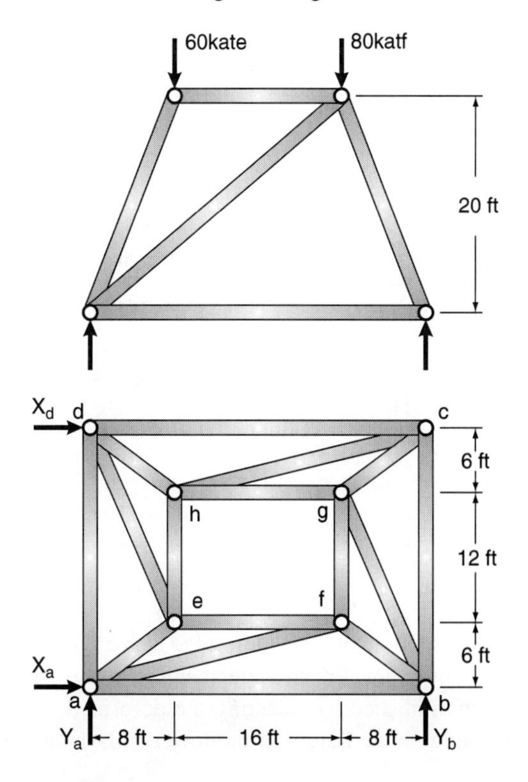

Solution. To analyze this truss using SAP2000, we must first establish the structural geometry. To do so, establish a grid that has four 8-ft spaces in the x directions and four 6-ft spaces in the y direction. There is one space in the z direction that is 20-ft high. We then establish the element connectivity. The joint and frame member geometry is shown in the figure. The complete input data file is contained in the example folder on the CD-ROM at the back of this book.

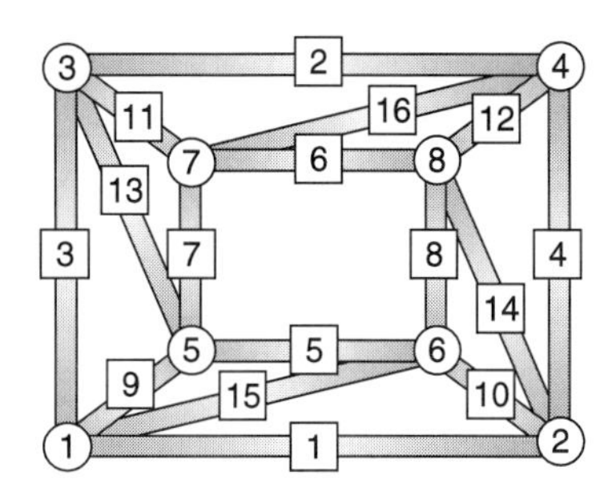

To analyze this truss, two modeling considerations need to be discussed. First, because truss members are assumed to only carry an axial force, each of the members in this structure needs to have the ends released against rotation in each direction. The members also need to have torsion released at the beginning of the member. These releases will result in a stable member that can only carry axial force.

Secondly, the three rotational degrees of freedom at each of the joints need to be restrained. Unless these are restrained, the computer will think that the structure is unstable, which it is not. The reasons for this will become clear when we begin studying matrix methods in Chapter 19. Restraining these degrees of freedom has no affect on the forces in the members since rotational forces cannot be transferred to the members because of the manner in which we have released the ends. Upon analysis, we find the forces in each of the members to be as shown in the table.

Member	Axial Force	Member	Axial Force
1	22.00	9	−50.31
2	8.00	10	−83.85
3	13.50	11	0.00
4	6.00	12	−22.36
5	−24.00	13	−21.05
6	0.00	14	28.07
7	0.00	15	−7.95
8	−24.00	16	0.00

You should display the deformed shape on the computer screen and then place it into motion to develop an understanding of how the structure deforms under load. This is accomplished by selecting the deformed shape from the display menu and clicking the animate button in the lower right-hand corner of the screen. ■

8.10 PROBLEMS FOR SOLUTION

For Problems 8.1 through 8.8, compute the reaction components and the member forces for the space trusses.

8.1

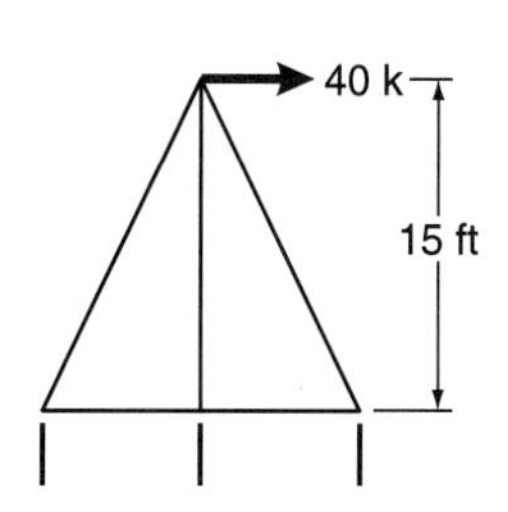

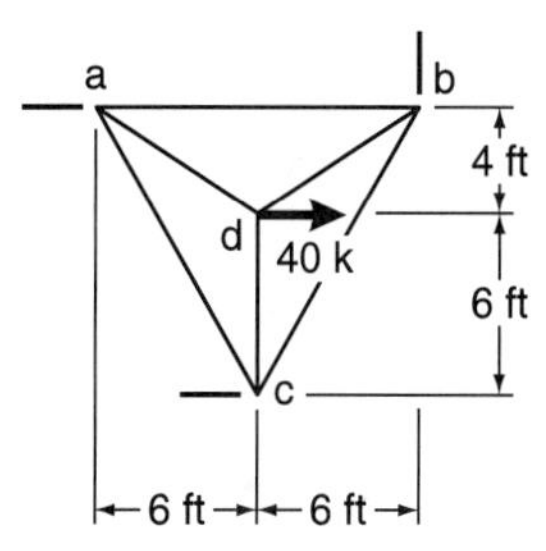

8.2 (*Ans*. ac = −43.16 k, bd = +50.31 k)

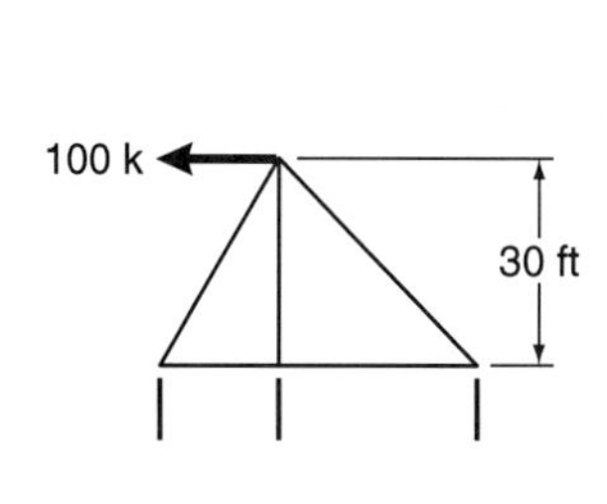

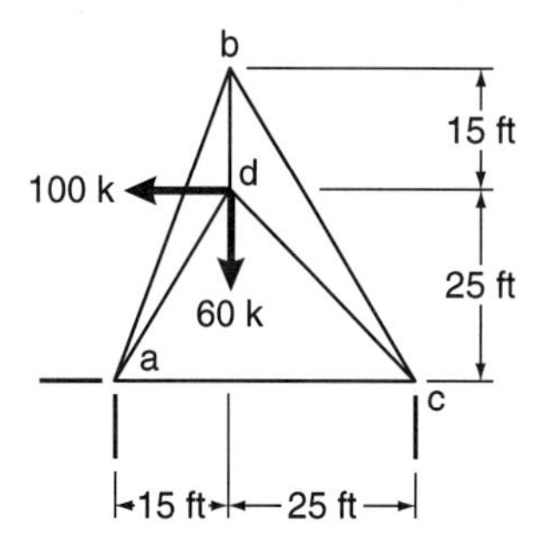

8.3

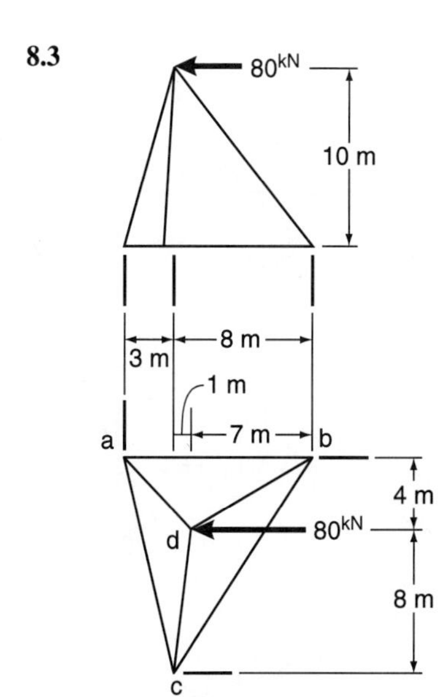

8.4 (*Ans.* ac = +6.67 k, ce = −15.64 k, ab = +20.00 k)

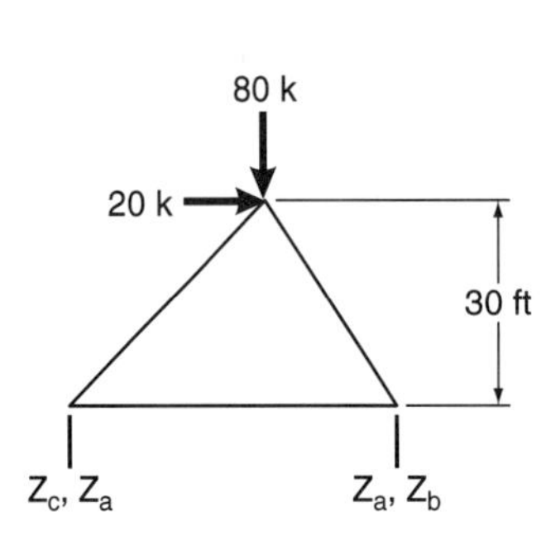

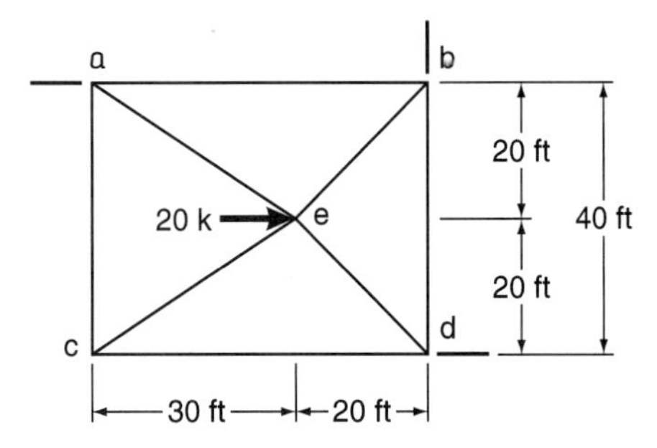

8.5

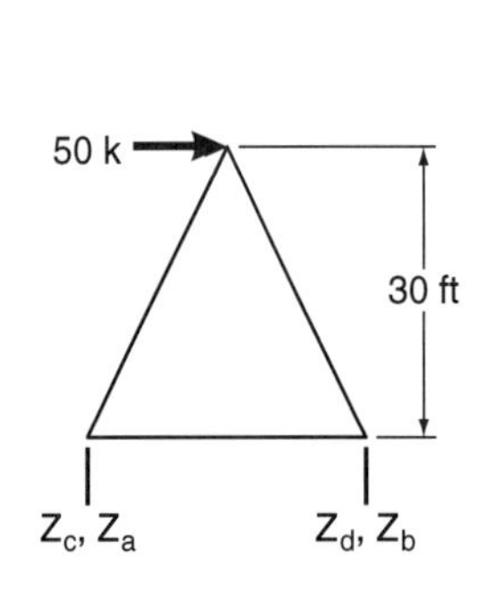

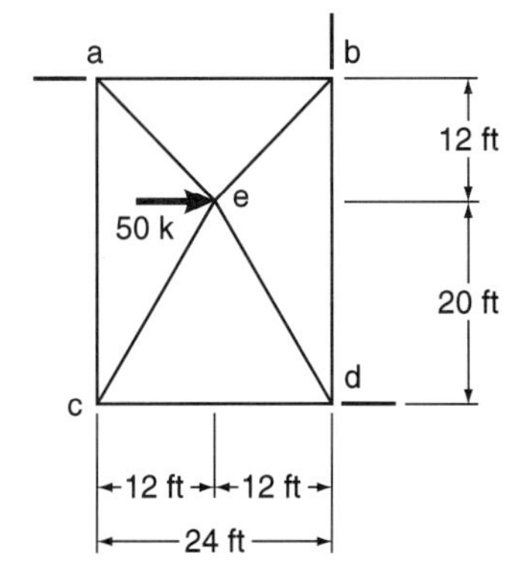

8.6 (*Ans.* cd = +12.5 k, gh = −50 k, bf = 0, ch = +63.7 k, Zc = 50 k, Yd = 12.5 k)

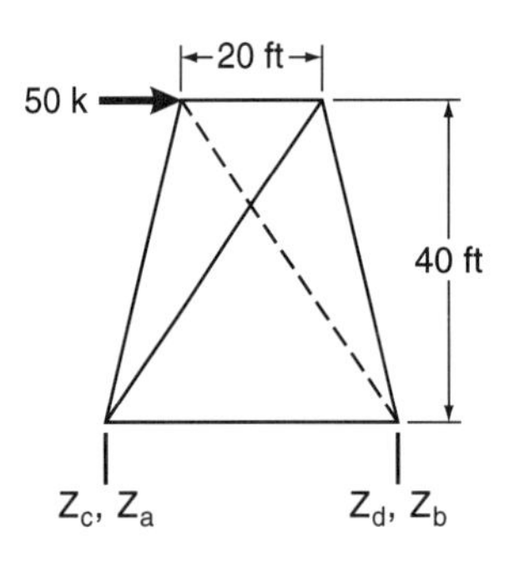

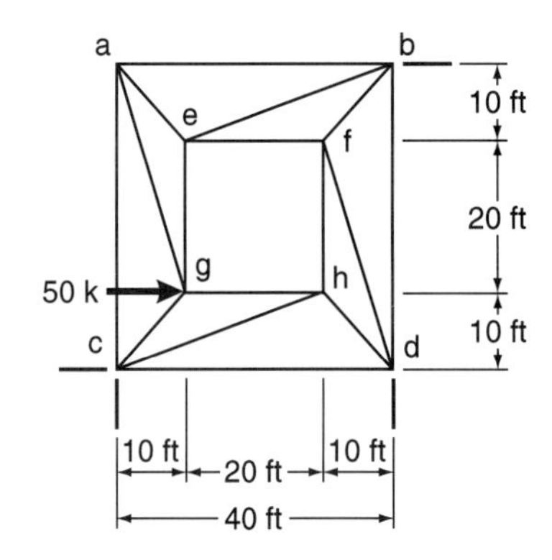

8.7

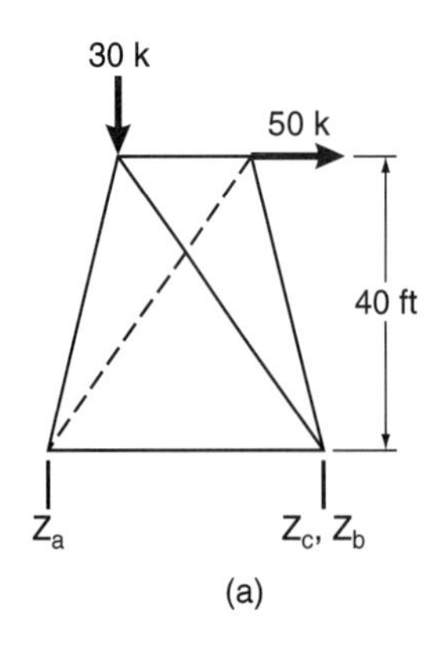

(a)

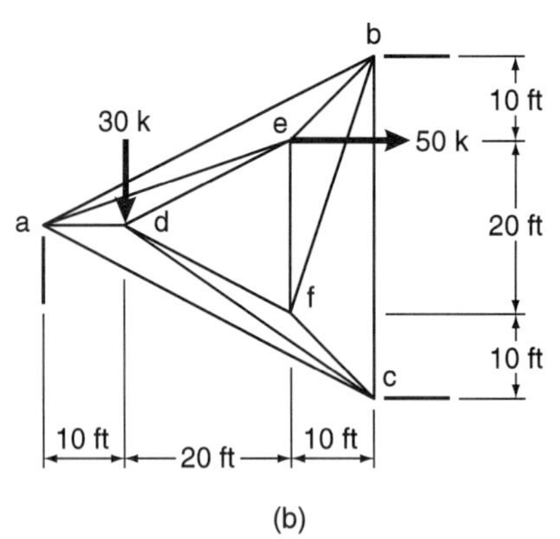

(b)

8.8 (*Ans.* de = −14.8 k, eh = +15.0 k, fj = −8.3 k hi = −32.7 k, Xb = 3.33 k, Ya = 6.67 k)

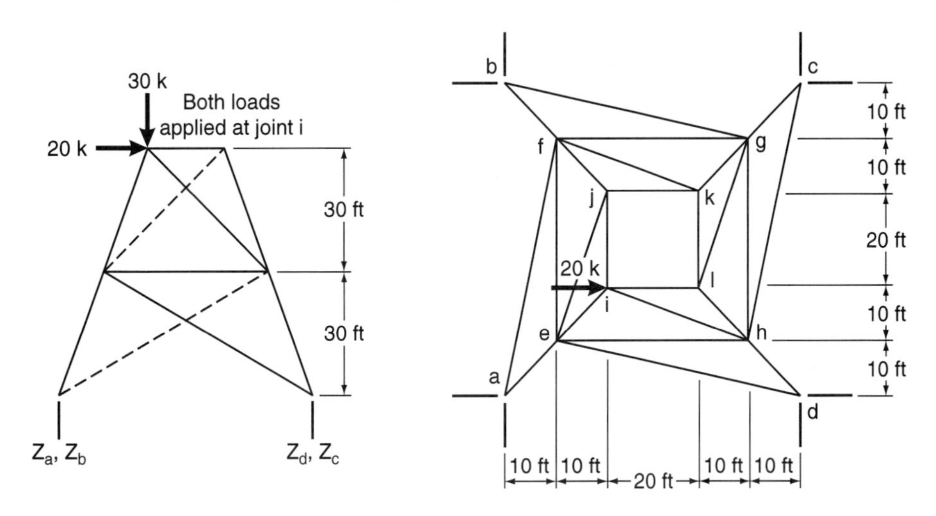

8.9 Solve Problem 8.1 using simultaneous equations for the entire truss.

8.10 Solve Problem 8.2 using simultaneous equations for the entire truss.

8.11 Solve Problem 8.5 using simultaneous equations for the entire truss.

Repeat the problems listed using SAP2000 on the enclosed computer diskette.

8.12 Problem 8.2 (*Ans.* ac = −43.2 k, bd = +50.3 k, cd = +89.8 k)

8.13 Problem 8.4

8.14 Problem 8.5

8.15 Problem 8.6

8.16 Problem 8.7

8.17 Problem 8.8 (*Ans.* de = −14.8 k, ej = +10.9 k, hi = −32.7 k)

Chapter 9

Influence Lines

9.1 INTRODUCTION

Structures supporting groups of loads fixed in one position have been discussed in the previous chapters. Regardless of whether beams, frames, or trusses were being considered and whether the functions sought were shears, reactions, or member forces, the loads were stationary. In practice, however, the engineer rarely deals with structures supporting only fixed loads. Nearly all structures are subject to loads moving back and forth across their spans. Perhaps bridges with their vehicular traffic are the most noticeable examples, but industrial buildings with traveling cranes, office buildings with furniture and human loads, and frames supporting conveyor belts are in the same category.

Each member of a structure must be designed for the most severe conditions that can possibly develop in that member. The designer places the live loads at the positions where they will produce these conditions. The critical positions for placing live loads will not be the same for every member. For example, the maximum force in one member of a bridge truss may occur when a line of trucks extends from end to end of the bridge. The maximum force in some other member, however, may occur when the trucks extend only from that member to one end of the bridge. The maximum forces in certain beams and columns of a building will occur when the live loads are concentrated in certain portions of the building. The maximum forces in other beams and columns will occur when the loads are placed elsewhere.

On some occasions, you can determine where to place the loads to give the most critical forces by inspection. On many other occasions, however, you will need to use certain criteria or diagrams to find the locations. The most useful of these devices is the influence line.

9.2 THE INFLUENCE LINE DEFINED

The influence line, which was first used by Professor E. Winkler of Berlin in 1867, shows graphically how the movement of a unit load across a structure influences some function of the structure[1]. The functions that may be represented include reactions, shears, moments, forces, and deflections.

[1]J. S. Kinney, *Indeterminate Structural Analysis* (Reading, Mass.: Addison-Wesley, 1957), Chapter 1.

Tennessee-Tombigbee Waterway Bridge in Mississippi (Courtesy of the Mississippi State Highway Department)

An influence line may be defined as a diagram whose ordinates show the magnitude and character of some function of a structure as a unit load moves across it. Each ordinate of the diagram gives the value of the function when the load is at that ordinate.

Influence lines are primarily used to determine where to place live loads to cause maximum forces. They may also be used to compute those forces. The procedure for drawing the diagrams is simply the plotting of values of the function under study as ordinates for various positions of the unit load along the span and then connecting those ordinates. You should mentally picture the load moving across the span and try to imagine what is happening to the function in question during the movement. The study of influence lines can immeasurably increase your knowledge of what happens to a structure under different loading conditions.

Study of the following sections should fix clearly in your mind what an influence line is. The actual mechanics of developing the diagrams are elementary, once the definition is completely understood. No new fundamentals are introduced here; rather, a method of recording information in a convenient and useful form is given.

9.3 INFLUENCE LINES FOR SIMPLE BEAM REACTIONS

Influence lines for the reactions of a simple beam are given in Figure 9.1. First consider the variation of the left-hand reaction, V_L, as a unit load moves from left to right across the beam. When the load is directly over the left support, $V_L = 1$. When the load is 2 ft to the right of the left support, $V_L = 18/20$, or 0.9. When the load is 4 ft to the right, $V_L = 16/20$, or 0.8, and so on.

Values of V_L are shown at 2-ft intervals as the unit load moves across the span. These values lie in a straight line because they change uniformly for equal intervals of the load. For every 2-ft interval, the ordinate changes 0.1. The values of V_R, the right-hand reaction, are plotted similarly for successive 2-ft intervals of the unit load. For each position of the unit load the sum of the ordinates of the two diagrams at any point equals (and for equilibrium certainly must equal) the unit load.

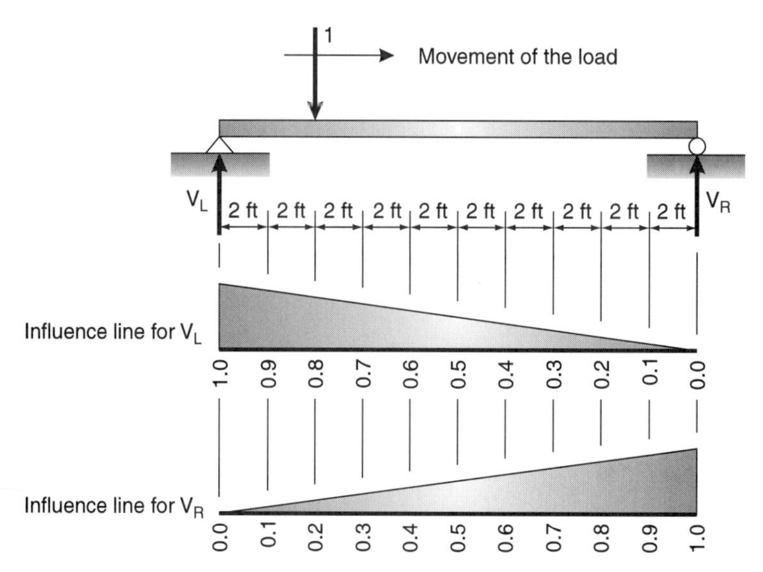

Figure 9.1 Influence line for reactions in a simple beam

9.4 INFLUENCE LINES FOR SIMPLE BEAM SHEARING FORCE

Influence lines are plotted in Figure 9.2 for the shearing force at two sections in a simple beam. The following sign convention for shearing force is used:

> Positive shearing force occurs when the sum of the transverse forces to the left of a section is up or when the sum of the forces to the right of the section is down.

This same sign convention was used for shearing force in Chapter 5. It is often referred to as the beam sign convention.

Precast-concrete Kalihiwai Bridge near Kilauea, Kauai, Hawaii (Courtesy of the Hawaii Department of Transportation)

Placing the unit load over the left support causes no shearing force at either of the two sections. Moving the unit load 2 ft to the right of the left support results in a left-hand reaction of 0.9. The sum of the forces to the left of section 1-1 is 0.1 downward; the shearing force is -0.1. When the load is 4 ft to the right of the left support and an

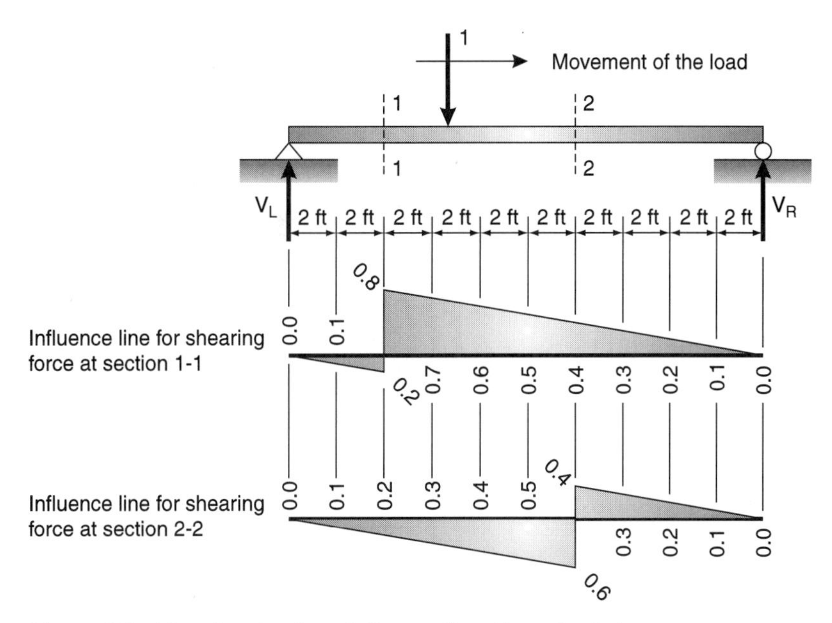

Figure 9.2 Two shearing force influence lines form simple beam

infinitesimal distance to the left of section 1-1, the shearing force to the left is -0.2. If the load is moved a very slight distance to the right of section 1-1, the sum of the forces to the left of the section becomes 0.8 upward; the shearing force is $+0.8$. Continuing to move the load across the span toward the right support results in the changing values of the shear at section 1-1. These values are plotted for 2-ft intervals of the unit load. The influence line for shear at section 2-2 is developed in the same manner.

Observe that the slope of the shearing force influence line to the left of the section in question is equal the slope of the influence line to the right of the section. In Figure 9.2, for instance, for the influence line at section 1-1, the slope to the left is $0.2/4 = 0.05$, while the slope to the right is $0.8/16 = 0.05$. This information is very useful in drawing other shear influence lines.

9.5 INFLUENCE LINES FOR SIMPLE BEAM MOMENTS

Influence lines for the bending moment are plotted in Figure 9.3 at the same sections of the beam used in Figure 9.1 for the shearing force influence lines. To review, a positive moment causes tension in the bottom fibers of a beam. It occurs at a particular section when the sum of the external moments of all the forces to the left is clockwise, or when the sum to the right is counterclockwise. Moments are taken at each of the sections for 2-ft intervals of the unit load.

The major difference between shearing force and bending moment diagrams as compared with influence lines should now be clear. A shearing force or bending moment diagram shows the variation of shearing force or bending moment across an entire structure for loads fixed in one position. An influence line for shearing force or bending moment, on the other hand, shows the variation of that function at one section in the structure caused by a unit load moving from one end of the structure to the other.

Influence lines for functions of statically determinate structures consist of a set of straight lines. An experienced analyst will be able to compute values of the function under

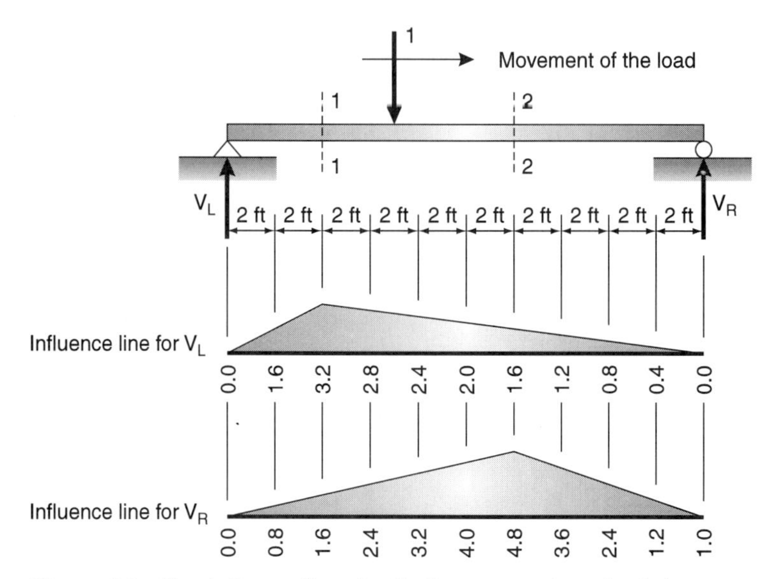

Figure 9.3 Two influence lines for binding moment in a simple beam

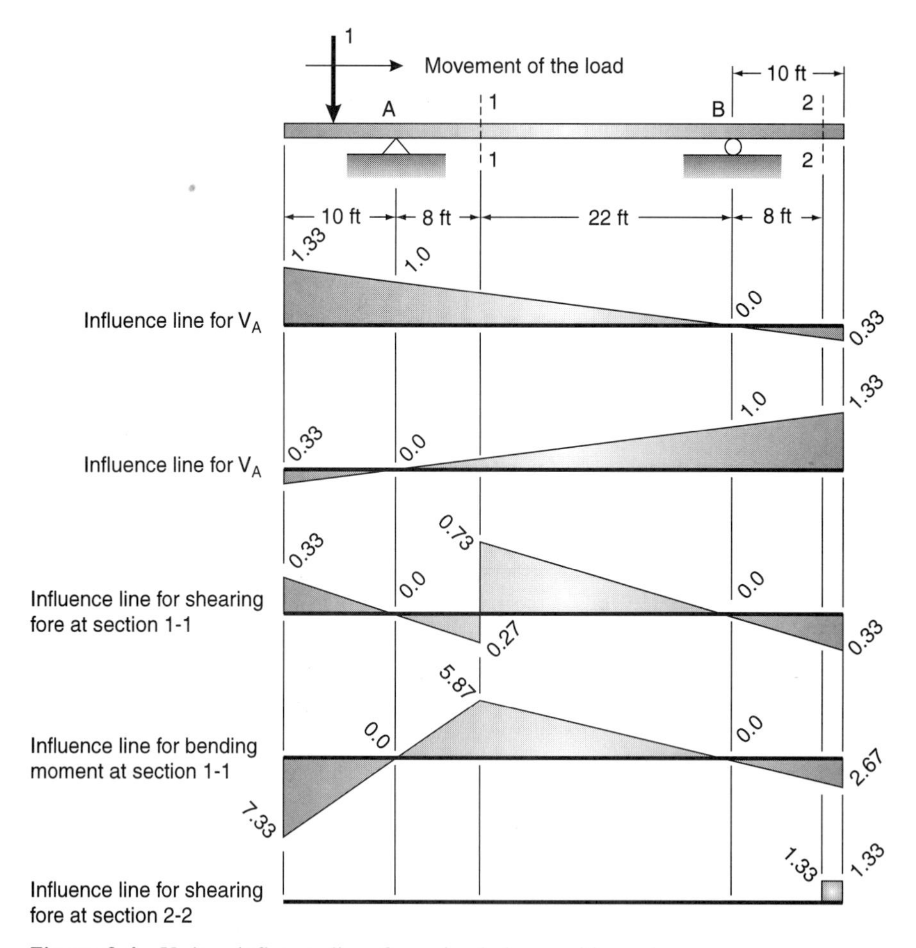

Figure 9.4 Various influence lines for a simple beam with overhangs

study at a few critical positions and connect the plotted values with straight lines. A student beginning his or her study, however, must be very careful to compute the value of the function for enough positions of the unit load. The shapes of influence lines for forces in truss members often are deceptive in their seeming simplicity. Plotting ordinates for several extra positions of the load is obviously better than failing to plot one essential value. Several influence lines for bending moment, shearing force, and reactions for an overhanging beam are plotted in Figure 9.4.

9.6 QUALITATIVE INFLUENCE LINES

The typical student initially has a great deal of difficulty drawing influence lines. Qualitative influence lines enable the correct shape of the desired figures to be obtained immediately and, hopefully, provide a better understanding of these useful diagrams.

The influence lines drawn in the previous sections for which numerical values were computed are referred to as *quantitative influence lines*. It is possible, however, to make rough sketches of these diagrams with sufficient accuracy for many practical purposes without computing any numerical values. These latter diagrams are referred to as *qualitative influence lines*.

A detailed discussion of the principle on which these sketches are made is presented in Chapter 14 together with a consideration of their usefulness.

It is important to explore deflection before persuing the concept of the qualitative influence line. Qualitative influence lines are based on a principle introduced by the German Professor Heinrich Müller-Breslau. This principle, derived in Section 14.2, is as follows:

> The deflected shape of a structure represents to some scale the influence line for a function such as reaction, shear, or moment if the function in question is allowed to act through a unit displacement.

In other words, the structure draws its own influence line when the proper displacement is applied.

As an example, consider the qualitative influence line for the left reaction of the beam of Figure 9.5(a). The constraint at the left support is removed and a displacement is introduced there in the direction of the reaction as shown in part (b) of the figure. When

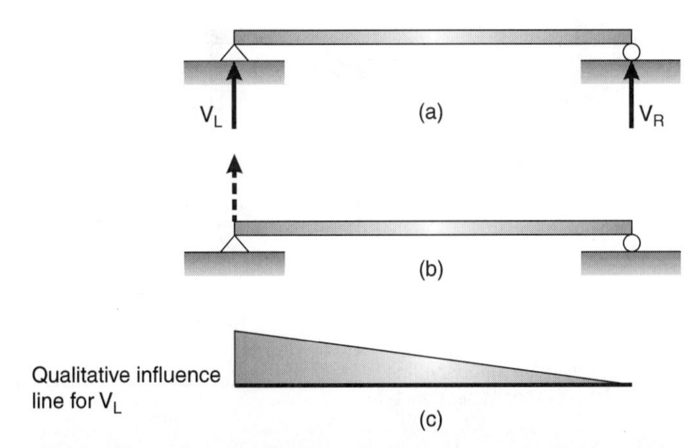

Figure 9.5 Qualitative influence line for reaction at left support

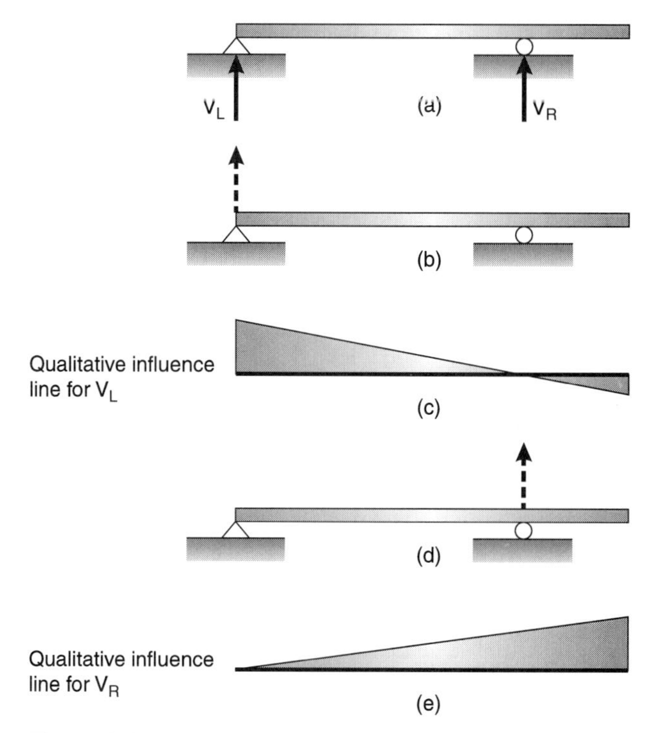

Figure 9.6 Qualitative influence lines for reactions in a beam with an overhang

the left end of the beam is pushed up, the area between the original and final position of the beam is the influence line for V_L to some scale. In a similar manner, the influence lines for the left and right reactions of the beam of Figure 9.6 are sketched.

As a third example, the influence line for the bending moment at section 1-1 in the beam of Figure 9.7 is considered. This diagram can be obtained by cutting the beam at the point in question and applying moments just to the left and just to the right of the cut section, as shown. In the figure the moment on each side of the section is positive

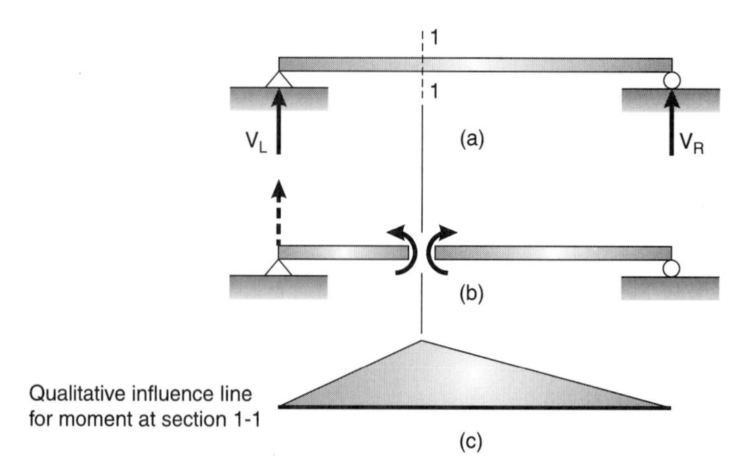

Figure 9.7 A qualitative influence line for moment in a simple beam

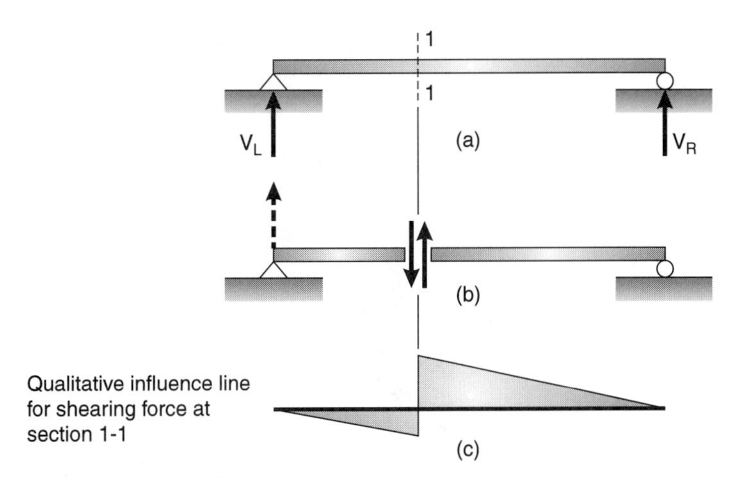

Figure 9.8 A qualitative influence line for shearing force in a simple beam

with respect to the segment of the beam on that side of the section. The resulting deflected shape of the beam is the qualitative influence line for the bending moment at section 1-1.

To draw a qualitative influence line for shear, the beam is cut at the point in question. A vertical force is applied to each side of the cut to provide positive shearing force, as shown in Figure 9.8 (b). To understand the direction used for these forces, note that they are applied to the left and to the right of the cut section to produce a positive shear for each segment. In other words, the force on the left segment is in the direction of a positive shear force applied from the right side and vice versa. Additional examples for qualitative influence lines are presented in Figure 9.9.

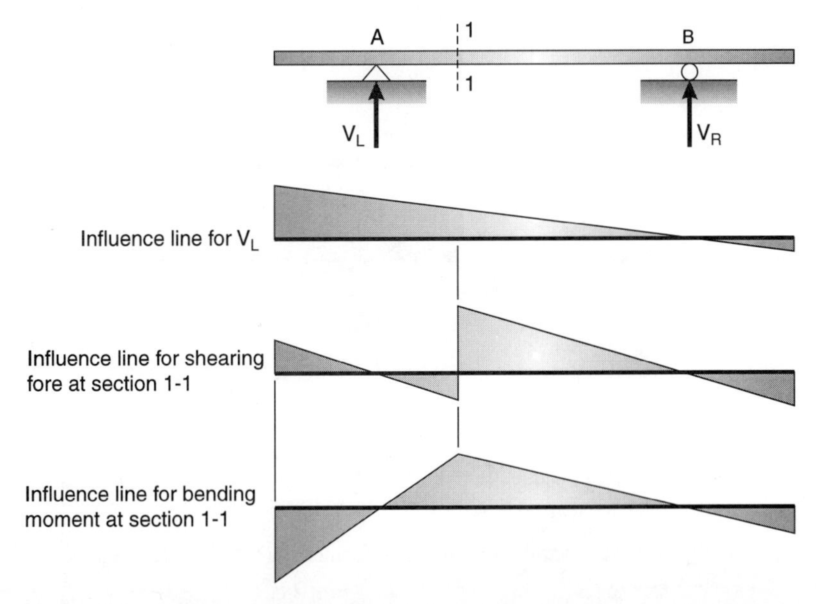

Figure 9.9 Various qualitative influence lines for a simple beam with overhangs

Overpass, Boise, Idaho (Courtesy of the American Concrete Institute)

Müller-Breslau's principle is useful for sketching influence lines for statically determinate structures, but its greatest value is for statically indeterminate structures. Though the diagrams are drawn exactly as before, note that in Figure 9.10 they

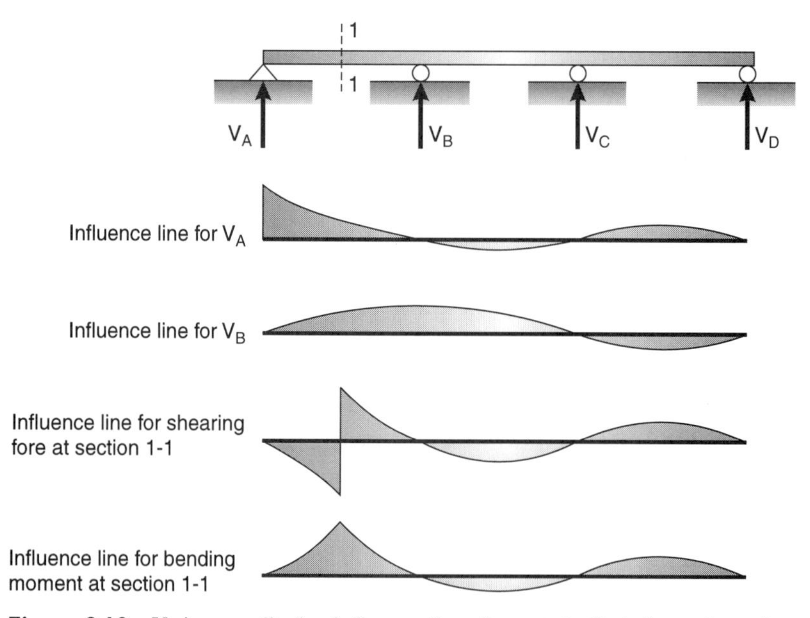

Figure 9.10 Various qualitative influence lines for a statically indeterminate 3 span beam

consist of curved lines instead of straight lines, as was the case for statically determinate structures.

9.7 USES OF INFLUENCE LINES; CONCENTRATED LOADS

Influence lines are the plotted functions of structural responses for various positions of a unit load. Having an influence line for a particular function of a structure makes immediately available the value of the function for a concentrated load at any position on the structure. The beam of Figure 9.1 and the influence line for the left reaction are used to illustrate this statement. A concentrated 1-kip load placed 4 ft to the right of the left support would cause V_L to equal 0.8 kip. Should a concentrated load of 175 kips be placed in the same position, V_L would be 175 times as great, or 140 kips.

The value of a function due to a series of concentrated loads is quickly obtained by multiplying each concentrated load by the corresponding ordinate of the influence line for that function. A 150-kip load placed 6 ft from the left support in Figure 9.1 and also a 200-kip placed 16 ft from the left support would cause V_L to equal (150)(0.7) + (200)(0.2), or 145 kips.

EXAMPLE 9.1

The influence lines for the left reaction and the centerline moment are shown for a simple beam in the figure. Determine the values of these functions for the several loads supported by the beam.

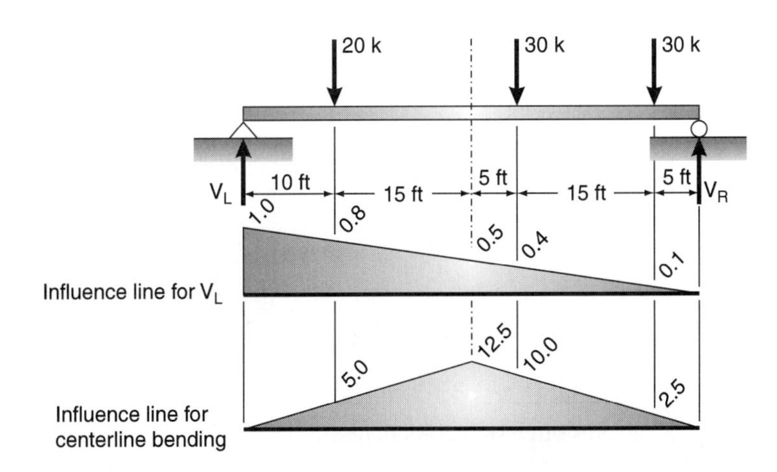

Solution. Compute the magnitude of the left reaction. That value is equal to the summation of each load times the ordinate of the influence diagram at the location of the load.

$$V_L = 20(0.8) + 30(0.4) + 30(0.1) = 31\,\text{kips}$$

Next, compute the magnitude of the moment at the centerline. It is computed in the same manner as the left reaction.

$$M_{CL} = 20(5.0) + 30(10.0) + 30(2.5) = 475\,\text{k-ft} \quad \blacksquare$$

9.8 USES OF INFLUENCE LINES; UNIFORM LOADS

The value of a certain function of a structure may be obtained from the influence line, when the structure is loaded with a uniform load, by multiplying the area of the influence line by the intensity of the uniform load. The following discussion proves this statement correct.

A uniform load of intensity w lb/ft is equivalent to a continuous series of smaller loads of (w)(0.1) lb on each 0.1 ft of the beam, or w dx lb on each length dx of the beam. Consider each length dx to be loaded with a concentrated load of magnitude w dx. The value of the function for one of these small loads is (w dx)(y) where y is the ordinate of the influence line at that point. The effect of all of these concentrated loads is equal to $\int wy\,dx$. This expression shows that the effect of a uniform is equal to the intensity of the uniform load, w, times the area of the influence line, $\int y\,dx$, along the section of the structure covered by the uniform load.

EXAMPLE 9.2

Assume the beam in Example 9.1 is loaded with a uniform load of 3 klf. Determine the magnitude of the left reaction and the centerline moment if the distributed load acts on the entire beam, and if it acts on only the left half of the beam.

Solution. Compute the values first for the load acting on the entire span of the beam.

$$V_L = (3)\left(\frac{1(50)}{2}\right) = 75\,\text{kips}$$

$$M_{CL} = (3)\left(\frac{12.5(50)}{2}\right) = 937.5\,\text{k-ft}$$

Now compute the same values for the uniform load when it acts only on the left half of the beam.

$$V_L = (3)\left(\frac{1 + 0.5}{2}(25)\right) = 56.25\,\text{kips}$$

$$M_{CL} = (3)\left(\frac{12.5(25)}{2}\right) = 468.75\,\text{k-ft} \quad \blacksquare$$

Structures often support both uniform and concentrated loads. The value of the function under study can be found by multiplying each concentrated load by its respective ordinate on the influence line. This result is added to the summation the magnitude of the uniform load multiplied by the area of the influence line opposite the section covered by the uniform load.

9.9 DETERMINING MAXIMUM LOADING EFFECTS

Beams must be designed to satisfactorily support the largest shearing forces and bending moments that can be caused by the loads to which they are subjected. As an example, consider the beam shown in Figure 9.11(a) and the influence line for bending moment at

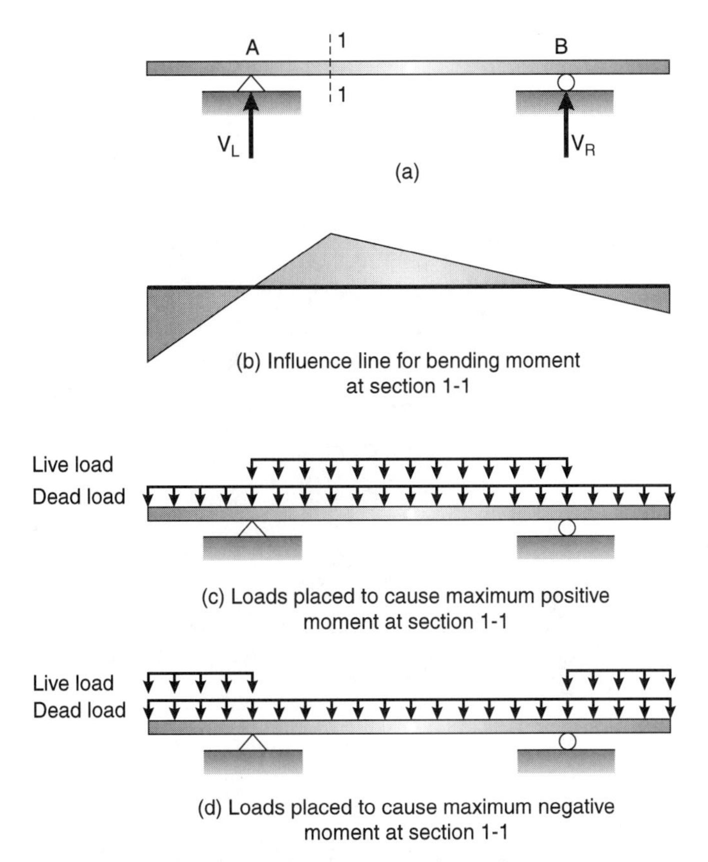

Figure 9.11 Placing loads to cause maximum forces

section 1-1 shown in Figure 9.11(b). This is the same beam shown in Figure 9.9. We now want to determine the maximum possible bending moments at section 1-1 for a uniform dead load and a uniform live load.

We will first determine the loading to cause maximum positive bending moment. The uniform dead load, which is the weight of the structure, will be applied from one end of the beam to the other as shown in Figure 9.11(c). The dead load is always acting on the entire structure. From the influence line we see that the unit load caused positive moment at section 1-1 only when it was located between the supports A and B. As such, the uniform live load is placed from A to B to determine the maximum positive bending moment, as shown in Figure 9.11 (c). If there had been a concentrated live load acting with the uniform live load, it would have been placed at section 1-1 since the unit load caused the greatest positive moment when it was located there. The bending moment caused by these loads can be calculated using the ordinates on the influence line or by using the equations of static equilibrium.

The maximum negative bending moment occurring at section 1-1 can be determined in a similar fashion. For this case, the loads would be placed as shown in Figure 9.11(d). When the unit load was placed on the cantilever portions of the beam, negative bending occurred at section 1-1. As such the distributed live load was placed at those locations. If there had been a live concentrated load, it would have been placed at the left or right end of the beam—whichever had the largest negative ordinate on the influence line.

Massachusetts Eye and Ear Infirmary in Boston. (Courtesy Bethlehem Steel Corporation.)

9.10 MAXIMUM LOADING EFFECTS USING CURVATURE

In the preceding section, an influence line was used to determine the critical positions for placing live loads to cause maximum bending moments. The same results can be obtained, and perhaps obtained more easily in many situations, by considering the deflected

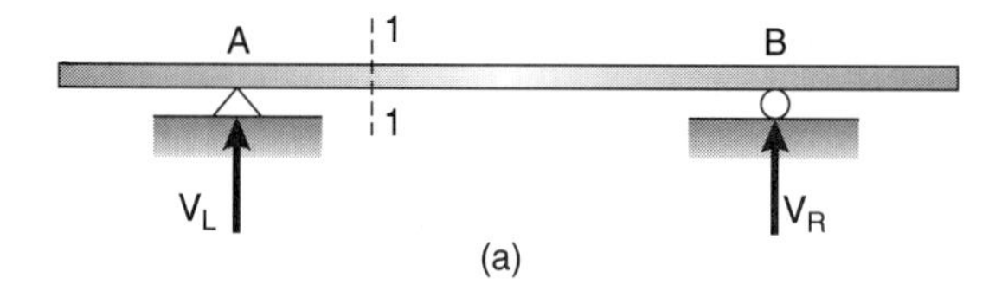

(a)

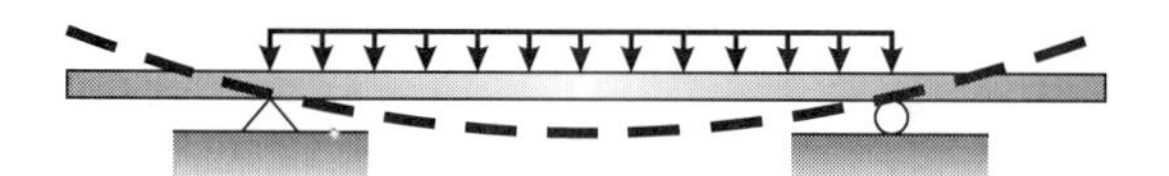

(b) Live loads placed to cause maximum positive moment at section 1-1

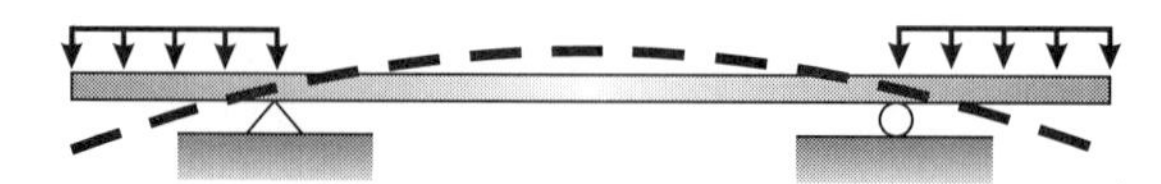

(c) Live loads placed to cause maximum negative moment at section 1-1

Figure 9.12 Obtaining maximum forces using curvature

shape or curvature of a member under load. If the live loads are placed so they cause the greatest curvature at a particular point, they will have bent the member to the greatest amount at that point. As such, the greatest bending moment will have been caused at that point.

For an illustration of this concept, let's determine the greatest positive bending moment at section 1-1 in the beam shown in Figure 9.12(a) due to the same loads considered in the last section. In Figure 9.12(b), the deflected shape of the beam is sketched, as it would be when a positive moment occurs at section 1-1. This deflected shape is shown by the dashed line. The dead load is placed all across the beam, while the live load is again placed from A to B; this location of the live load will magnify the deflected shape at section 1-1.

A similar situation is shown in Figure 9.12(c) to determine the maximum negative bending moment at section 1-1. The deflected shape of the beam, shown by the dashed line, is sketched consistent with a negative bending moment occurring at section 1-1. To magnify the negative or upward bending at section 1-1, the live load is placed on the cantilever portions of the beam, the parts outside the supports.

9.11 LIVE LOADS FOR HIGHWAY BRIDGES

Until about 1900, bridges in the United States were "proof loaded" before they were considered acceptable for use. Highway bridges were loaded with carts filled with stone or pig iron, and railway bridges were loaded with two locomotives in tandem. Such procedures were probably very useful in identifying poor designs and/or poor workmanship, but were no guarantee against overloads and fatigue stress situations.[2]

As discussed in *America's Highways 1776–1976,* during much of the 19th century highway bridges were designed to support live loads of approximately 80 to 100 psf applied to the bridge decks. These loads supposedly represented large, closely spaced crowds of people moving across the bridges. In 1875, the American Society of Civil Engineers (ASCE) recommended that highway bridges should be designed to support live loads varying from 40 to 100 psf—the smaller values were to be used for very long spans. The Office of Public Roads published a circular in 1913 recommending that highway bridges be designed for a live loading of: (a) a series of electric cars, or (b) a 15-ton road roller and a uniform live load on the rest of the bridge deck.

Although highway bridges must support several different types of vehicles, the heaviest possible loads are caused by a series of trucks. In 1931, the American Association of State Highway and Transportation Officials (AASHTO) Bridge Committee issued its first printed edition of the AASHTO Standard Specification for Highway Bridges. A very important part of these specifications was the use of the truck system of live loads. The truck loads were designated as H20, H15, and H10, representing two-axle design trucks of 20, 15, and 10 tons, respectively. Each lane of a bridge was to have an H-truck placed in it and was to be preceded and followed by a series of trucks weighing three quarters as much as the basic truck.[3]

Today the AASHTO specifies that highway bridges be designed for lines of motor trucks occupying 10-ft-wide lanes. Only one truck is placed in each span for each lane. The truckloads specified are designated with an H prefix (or M if SI units are used) followed by

[2]U.S. Department of Transportation, Federal Highway Administration, *America's Highways 1776–1976* (Superintendent of Documents, U.S. Government Printing Office, 1976), 429–432.

[3]*Ibid.*

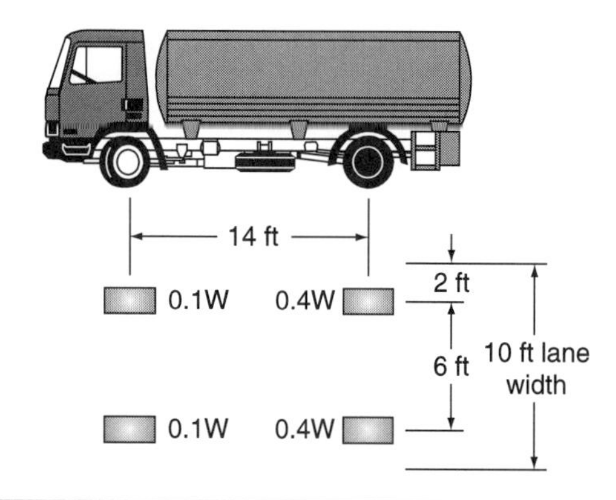

Truck	Weight (tons)	Front		Rear	
		Axle (kips)	Wheel (kips)	Axle (kips)	Wheel (kips)
H25-44	25	10,000	5,000	40,000	20,000
H20-44	20	8,000	4,000	32,000	16,000
H15-44	15	6,000	3,000	24,000	12,000
H10-44	10	4,000	2,000	16,000	8,000

Figure 9.13 AASHTO H-Truck Loads

a number indicating the total weight of the truck in tons (10^4 Newtons). The weight may be followed by another number indicating the year of the specifications. For example, an H20-44 loading indicates a 20-ton truck and the 1944 specification. A sketch of the truck and the pertinent dimensions is shown in Figure 9.13.

The selection of the particular truck loading to be used in design depends on the bridge location, anticipated traffic, and so on. These loadings may be broken down into three groups as follows.

9.11.1 Two-Axle Trucks: H10, H15, H20, and H25

The weight of an H truck is assumed to be distributed two-tenths to the front axle (for example, 4 tons, or 8 kips, for an H20 loading) and eight-tenths to the rear axle. The axles are spaced 14 ft on center, and the center-to-center lateral spacing of the wheels is 6 ft. Should a truck loading varying in weight from these be desired, one that has axle loads in direct proportion to the standard ones listed here may be used. A loading as small as the H10 may be used only for bridges supporting the lightest traffic.

9.11.2 Two-Axle Trucks Plus One-Axle Semi-Trailer: HS15-44, HS20-44, and HS25-44

For today's highway bridges carrying a great amount of truck traffic, the two-axle truck loading with a one-axle semi-trailer weighing 80% of the main truckload is commonly specified for design (Figure 9.14). The DOTs (Departments of Transportation) for many states today require that their bridges be designed for the HS25-44 trucks. This truck has 5 tons on the front axle, 20 tons on the rear axle, and 20 tons on the trailer axle. The distance from the rear truck axle to the semi-trailer axle is varied from 14 to 30 ft, depending on which spacing will cause the most critical conditions.

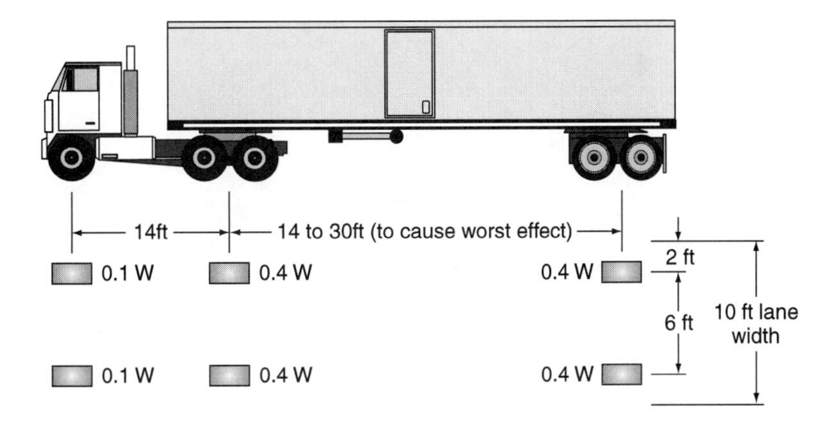

Truck	Weight (tons)	Front		Rear		Trailer	
		Axle (kips)	Wheel (kips)	Axle (kips)	Wheel (kips)	Axle (kips)	Wheel (kips)
HS25-44	25	10,000	5,000	40,000	20,000	40,000	20,000
HS20-44	20	8,000	4,000	32,000	16,000	32,000	16,000
HS15-44	15	6,000	3,000	24,000	12,000	24,000	12,000
HS10-44	10	4,000	2,000	16,000	8,000	16,000	8,000

Figure 9.14 AASHTO HS Truck Loads

9.11.3 Uniform Lane Loading

Computation of forces caused by a series of concentrated loads is a tedious job with a handheld calculator. This is true whether they represent two-axle trucks or two-axle trucks with semi-trailers. Therefore, a lane loading that will produce approximately the same forces frequently is used. The lane loading consists of a uniformly distributed load acting in combination with a single moving concentrated load. This load system represents a line of medium-weight traffic with a heavy truck somewhere in the line. The uniformly distributed load, per foot of traffic lane, is equal to 0.016 times the total weight of the truck to which the load is to be roughly equivalent. The concentrated moving load is equal to 0.45 times the truck weight for bending moment calculations and 0.65 times the truck weight for shearing force calculations. These values for an H20 loading would be as follows:

Lane Load	0.016(20 tons) = 640 lb/ft of lane	
Concentrated Load for Moment	0.45(20 tons) = 18 kips	**Eq. 9.1**
Concentrated Load for Shear	0.65(20 tons) = 26 kips	

For continuous spans, another concentrated load of equal weight is to be placed in one of the other spans at a location to cause maximum negative bending moment to occur. For positive moment, only one concentrated load is to be used per lane, with the uniform load placed on as many spans as necessary to produce the maximum positive bending moment.

The lane loading is more convenient, but it should not be used unless it produces bending moments and shearing forces that are equal to or greater than those produced by the corresponding H loading. Using information presented later in this chapter, in simple

John F. Fitzgerald Expressway, Mystic River Bridge, Boston, Massachusetts (Courtesy of the American Institute of Steel Construction, Inc.)

spans the equivalent lane loading for the HS20-44 truck will produce greater bending moments when the span length exceeds 145 ft and greater shearing forces when the span length exceeds 128 ft. Appendix A of the AASHTO specifications contains tables that give the maximum shears and moments in simple spans for the various H truck load or for their equivalent lane load, whichever controls.

The possibility of having a continuous series of heavily loaded trucks in every lane of a bridge that has more than two lanes does not seem as great for a bridge that has only two lanes. The AASHTO therefore permits the values caused by full loads in every lane to be reduced by a certain factor if the bridge has more than two lanes.

9.11.4 Interstate Highway System Loading

Another loading system can be used instead of the HS20-44 in the design of structures for the Interstate Highway System. This alternate system, which consists of a pair of 24-kip axle loads spaced 4 ft on center, is critical for short spans only. It is possible to show that this loading will produce maximum bending moments in simple spans ranging from 11.5 to 37 ft in length and maximum shearing forces in simple spans ranging from 6 to 22 ft in length. For other spans, the HS20-44 loading or its equivalent lane loading will be the more critical load.

9.12 LIVE LOADS FOR RAILWAY BRIDGES

Railway bridges are commonly analyzed for a series of loads devised by Theodore Cooper in 1894. His loads, which are referred to as E loads, represent two locomotives with their tenders followed by a line of freight cars as shown in Figure 9.15(a). A series of concentrated

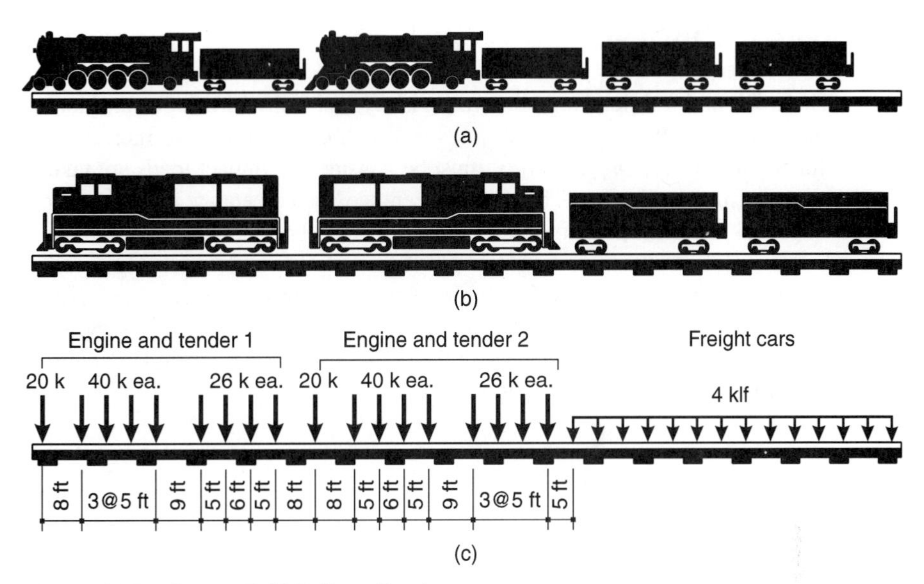

Figure 9.15 Cooper E-40 Railway Load

loads is used for the locomotives, and a uniform load represents the freight cars, as pictured in Figure 9.15(c). The E-40 train is assumed to have a 40-kip load on the driving axle of the engine. Since his system was introduced, the weights of trains have been increased considerably. Today bridges are designed based on E-80 load.[4]

If information is available for one E loading, the information for any other E loading can be obtained by direct proportion. The axle loads of an E-75 are 75/40 of those for an E-40; those for an E-60 are 60/72 of those for an E-72; and so on. As such, the axle loads for an E-80 load are twice those shown in Figure 9.15.

The American Railway Engineering Association also specifies an alternate loading. That loading is shown in Figure 9.16. This load or the E-80 load—whichever causes the greatest stress in the components—is to be used.

As we can see from the modern locomotives in Figure 9.15(b), Cooper's loads do not accurately picture today's trains. Nevertheless, they are still in general use despite the availability of several more modern and more realistic loads, such as Dr. D. B. Steinman's M-60 loading.[5]

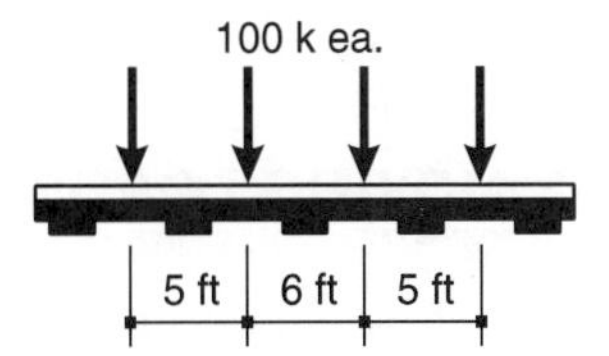

Figure 9.16 Alternate railroad loading

[4]American Railway Engineering Association, *AREA Manual,* 1996.

[5]"Locomotive Loadings for Railway Bridges," *Transactions of the American Society of Civil Engineers 86* (1923): 606–636.

9.13 IMPACT LOADING

The truck and trainloads applied to highway and railroad bridges are not applied gently and gradually; they are applied quickly, which causes forces to increase. As such, additional loads, called *impact loads,* must be considered. Impact loads are taken into account by increasing the live loads by some percentage, the percentage being obtained from purely empirical expressions. Numerous formulas have been presented for estimating impact. One example is the following AASHTO formula for highway bridges, in which I is the percent of impact and L is the length of the span, in feet, over which live load is placed to obtain a maximum stress. The AASHTO says that it is unnecessary to use an impact load greater than 30%, regardless of the value given by the formula. Note that as the span length increases the impact load decreases.

$$I = \frac{50}{L + 125} \qquad \textbf{Eq. 9.2}$$

Impact factors for railroad bridges are larger than those for highway bridges because of the much greater vibrations caused by the wheels of a train as compared to the relatively soft rubber-tired vehicles on a highway bridge. A person need only stand near a railroad bridge for a few seconds while a fast-moving and heavily loaded freight train passes over to see the difference. Tests have shown the impact on railroad bridges will often be as high as 100% or more. Not only does a train have a direct vertical impact, or bouncing up and down, but it also has a lurching or swaying back-and-forth type of motion. For vertical impact on beams, girders, and floor beams, the AREA provide the following impact factor:

$$I = 60 - \frac{L^2}{500} \qquad L < 100 \text{ ft}$$

$$I = \frac{1{,}800}{L - 40} + 10 \qquad L \geq 100 \text{ ft} \qquad \textbf{Eq. 9.3}$$

When designing trusses, the AREA requires the following factor for vertical impact:

$$I = \frac{4{,}000}{L + 25} + 15 \qquad \textbf{Eq. 9.4}$$

The AISC specification states that, unless otherwise specified, live loads shall be increased by certain percentages to account for impact. Some of these values are 100% for elevators, 33% for hangers supporting floors and balconies, and not less than 50% for supports of reciprocating machinery or power.

9.14 MAXIMUM VALUES FOR MOVING LOADS

In the pages of this chapter we have repeatedly indicated that to design a structure supporting moving loads, the engineer must determine where to place the loads to cause maximum forces at various points in the structure. If one can place the loads at the positions causing maximum forces to occur, one need not worry about any other positions the loads might take on the structure.

If a structure is loaded with a uniform live load and not more than one or two moving concentrated loads, the critical positions for the loads will be obvious from the influence lines. If, however, the structure is to support a series of concentrated loads of varying magnitudes, such as groups of truck or train wheels, the problem is not as simple. The influence line provides an indication of the approximate positions for placing the loads, because it is reasonable to assume that the heaviest loads should be grouped near the largest ordinates of the diagram.

Space is not taken herein to consider all of the possible situations that might be faced in structural analysis. We feel, however, that the determination of the absolute maximum bending moment caused in a beam by a series of concentrated loads is so frequently encountered by the engineer that discussion of this topic is warranted.

The absolute maximum bending moment in a simple beam usually is thought of as occurring at the beam centerline. Maximum bending moment does occur at the centerline if the beam is loaded with a uniform load or a single concentrated load located at the centerline. A beam, however, may be required to support a series of varying moving concentrated loads such as the wheels of a train, and the absolute maximum moment in all probability will occur at some position other than the centerline. To calculate the moment, it is necessary to find the point where it occurs and the position of the loads causing it. To assume that the largest moment is developed at the centerline of long-span beams is reasonable, but for short-span beams, this assumption may be considerably in error. It is therefore necessary to have a definite procedure for determining absolute maximum moment.

The bending moment diagram for a simple beam loaded with concentrated loads will consist of straight lines regardless of the position of the loads. Therefore, the absolute maximum moment occurring during the movement of these loads across the span will occur at one of the loads, usually at the load nearest the center-of-gravity of the group of loads. The beam shown in Figure 9.17 with the series of loads P_1, P_2, P_3, and so on is studied in the following paragraphs.

The load P_3 is assumed to be the one nearest the center of gravity of the loads on the span. It is located a distance L_1 from P_R, which is the resultant of all the loads on the span. The left reaction, R_L, is located a distance x from P_R. In the following paragraphs, maximum bending moment is assumed to occur at P_3, and a definite method is developed for placing this load to cause the maximum.

The bending moment at P_3 may be written as follows:

$$M = V_R(L - x - L_1) - P_2(L_2 - L_1) - P_1(L_3 - L_1) \qquad \textbf{Eq. 9.5}$$

After substituting the value of V_R, P_RX/L, we obtain the following equation:

$$M = \left(\frac{P_R x}{L}\right)(L - x - L_1) - P_2(L_2 - L_1) - P_1(L_3 - L_1) \qquad \textbf{Eq. 9.6}$$

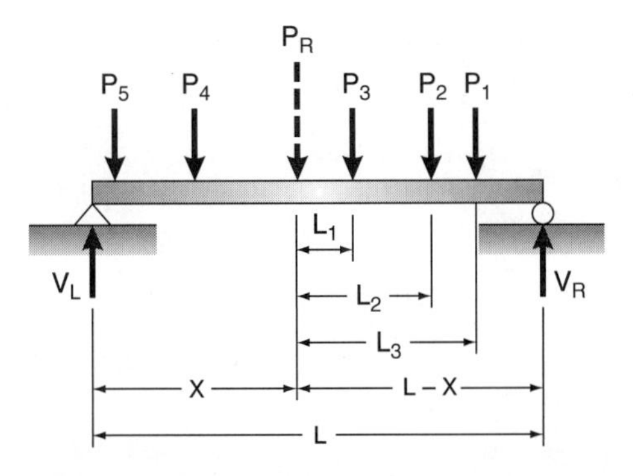

Figure 9.17 A simple beam with numerous applied loads

We want to find the value of X for which the moment at P_3 will be a maximum. Maximum bending moment at P_3, which occurs when the shearing force is zero, may be found by differentiating the bending moment equation with respect to X, setting the result equal to zero, and solving for X.

$$\frac{dM}{dx} = L - 2x - L_1 = 0$$

$$\therefore x = \frac{L_1}{2} - \frac{L}{2} \qquad \textbf{Eq. 9.7}$$

From the preceding derivation a general rule for absolute maximum moment may be stated:

> Maximum moment in a beam loaded with a moving series of concentrated loads usually will occur at the load nearest the center of gravity of the loads on the beam when the center of gravity is the same distance on one side of the centerline of the beam as the load nearest the center of gravity of the loads is on the other side.

Should the load nearest the center-of-gravity of the loads be relatively small, the absolute maximum moment may occur at some other load nearby. Occasionally two or three loads have to be considered to find the greatest value. Nevertheless, the problem is not a difficult one because another moment criteria not described herein—the average load to the left must be equal to the average load to the right—must be satisfied. There will be little trouble determining which of the nearby loads will govern. Actually, it can be shown that the absolute maximum moment occurs under the load that would be placed at the centerline of the beam to cause maximum moment there, when that wheel is placed as far on one side of the beam centerline as the center of gravity of all the loads is on the other.[6]

EXAMPLE 9.3

Determine the absolute maximum moment that can occur in the 50-ft simple beam shown in the following figure as the series of concentrated loads moves across the span.

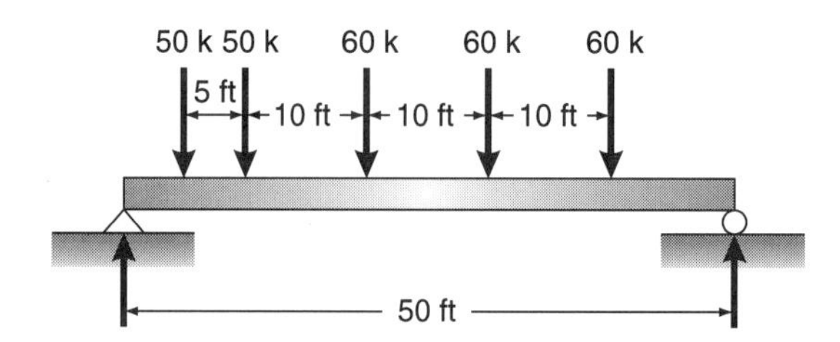

Solution. The center of gravity of the loads is determined from:

$$\frac{50(5) + 60(15) + 60(25) + 60(35)}{50 + 50 + 60 + 60 + 60} = 16.96 \text{ ft from left load}$$

Then the loads are placed as follows and the shearing force and bending moment diagrams are drawn.

[6] A. Jakkula, and H. K. Stephenson, *Fundamentals of Structural Analysis* (New York: Van Nostrand, 1953), 241–242.

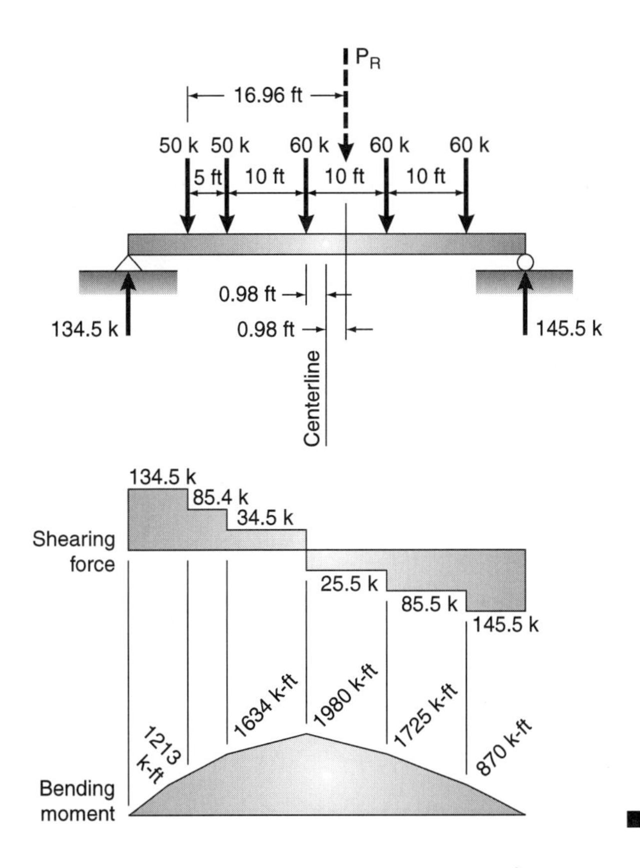

■

9.15 PROBLEMS FOR SOLUTION

For Problems 9.1 through 9.6, draw qualitative influence lines for all of the reactions and for shear and moment at section 1-1 for each of the beams.

9.1

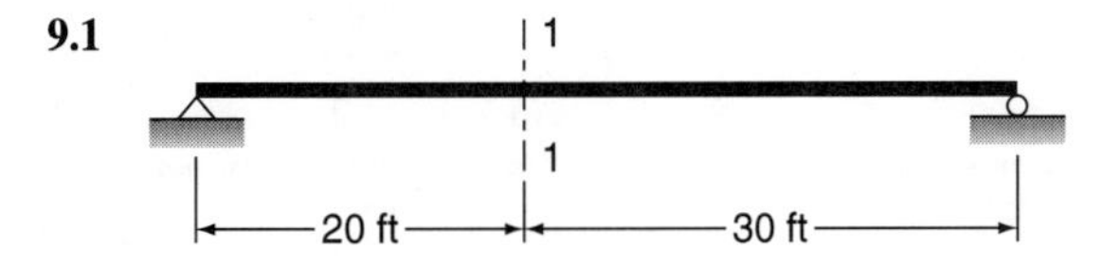

9.2

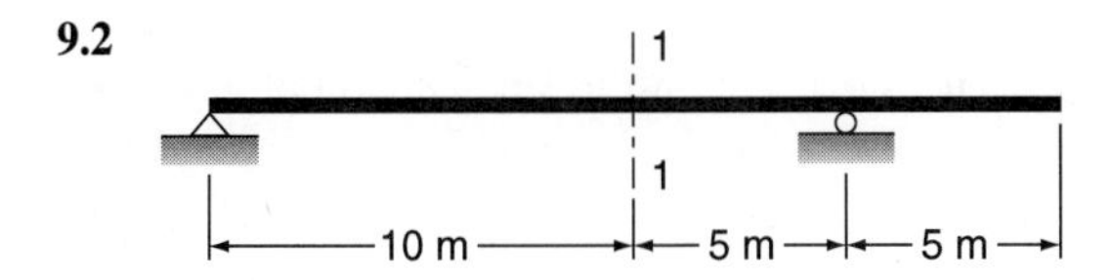

9.3

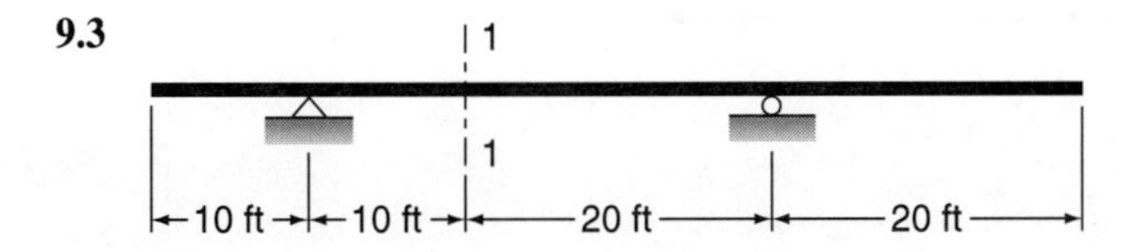

9.4

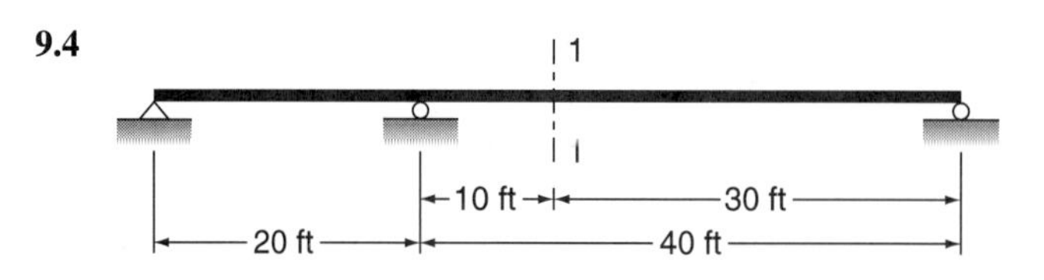

9.5

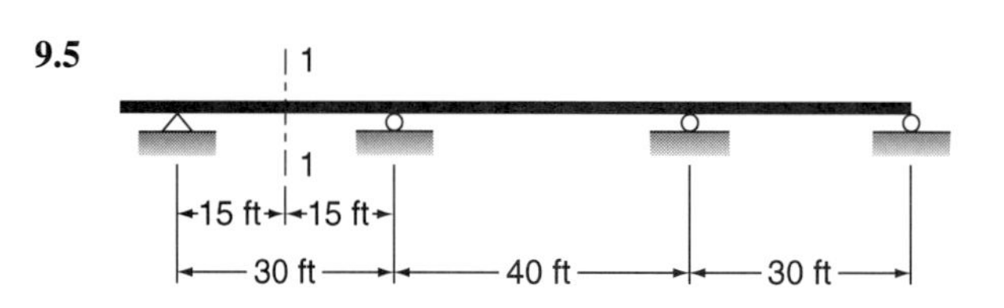

9.6

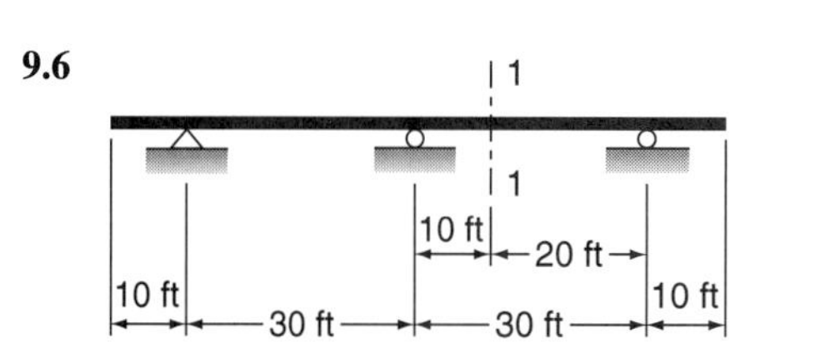

For Problems 9.7 through 9.18, draw quantitative influence lines for the situations listed.

9.7 Both reactions and also the shearing force and bending moment at section 1-1

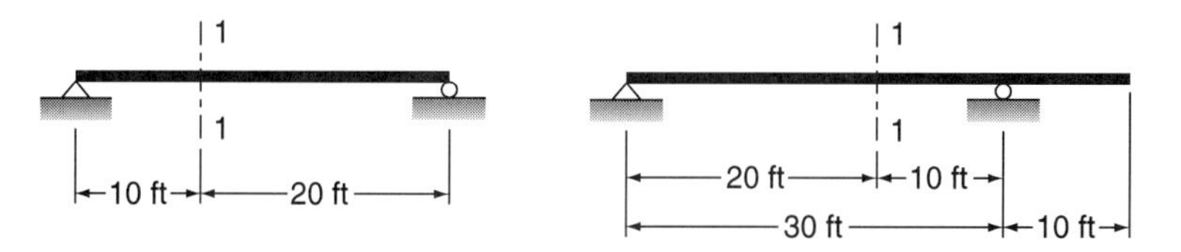

9.8 Both reactions and also the shearing force and bending moment at section 1-1. (*Ans.* Load at free end: $V_L = 0.33 \downarrow$, V<in><o>R<e> = 1.33 ↑, $V_{1\text{-}1} = -0.33$, $M_{1\text{-}1} = -6.67$)

9.9 Both reactions and also the shearing force and bending moment at section 1-1

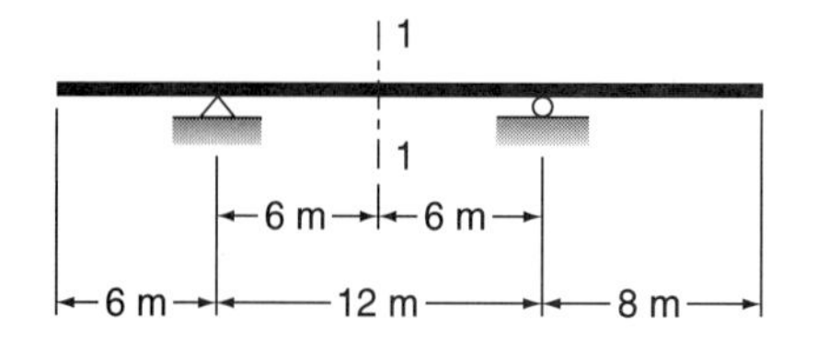

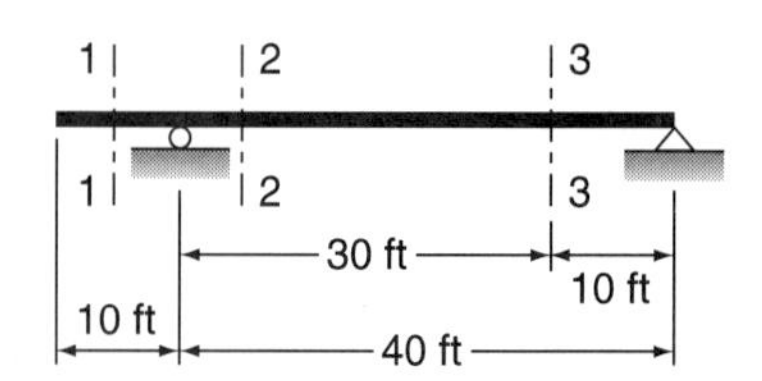

9.10 Both reactions and shear at sections 1-1, 2-2 (just to left and right of left support), and 3-3 (*Ans.* Load @ section 3-3: V<in><o>L<e> = +0.25, V<in><o>R<e> = +0.75, shear = 0 @ section 1-1)

9.11 Both reactions, shear at sections 1-1 and 2-2, and moment at section 2-2

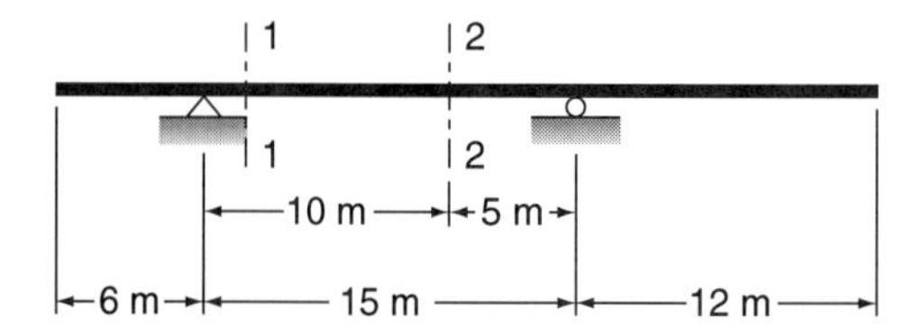

9.12 Vertical reaction and moment reaction at fixed-end, shear and moment at section 1-1 (*Ans.* Load @ free end: $V_L = +1.0$, $M_L = -15$, M @ section 1-1 $= -10$)

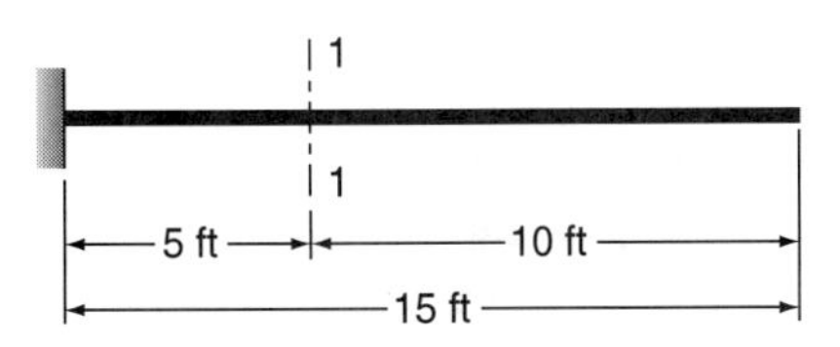

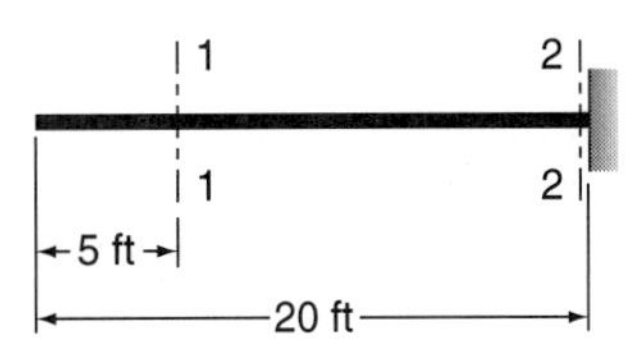

9.13 Shearing force and bending moment at sections 1-1 and 2-2

9.14 Both reactions as load moves from A to D (*Ans.* Load @ left end: $V_B = +1.50$, $V_E = -0.50$)

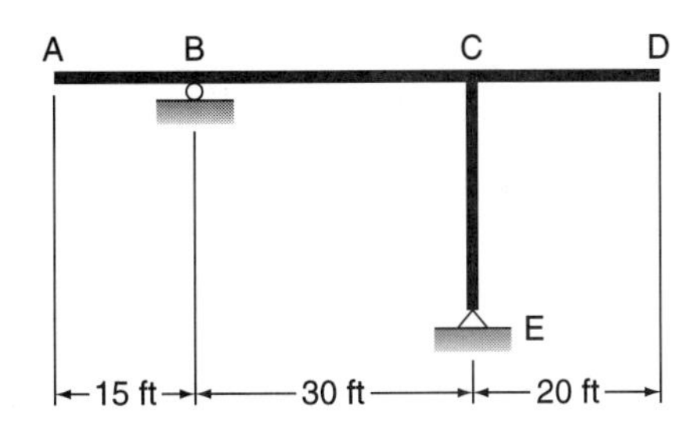

9.15 All reactions

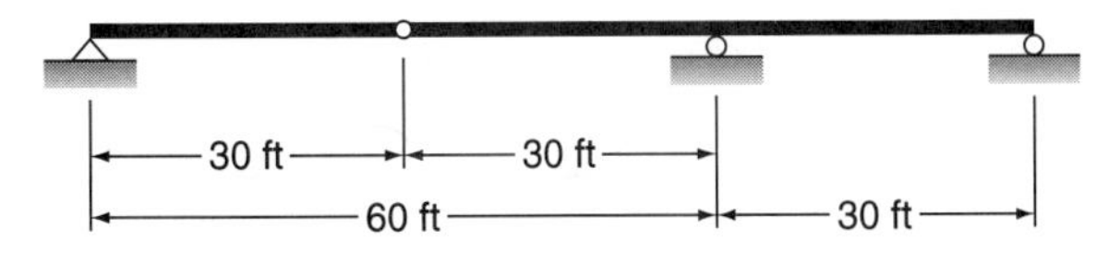

9.16 Vertical reactions at supports A and B (*Ans.* Load @ left end $= +1.40$, $V_B = -0.40$)

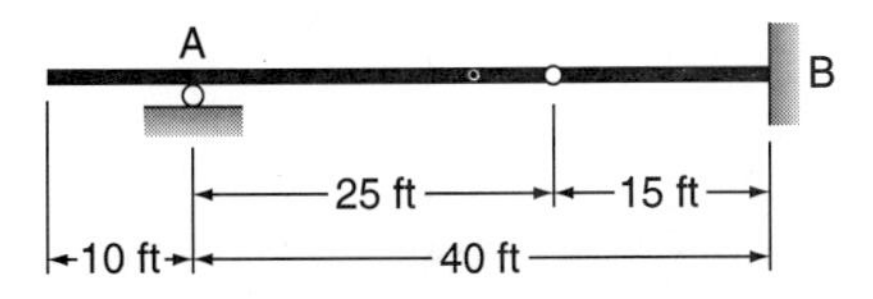

9.17 All vertical reactions, bending moments, and shearing forces at section 1-1

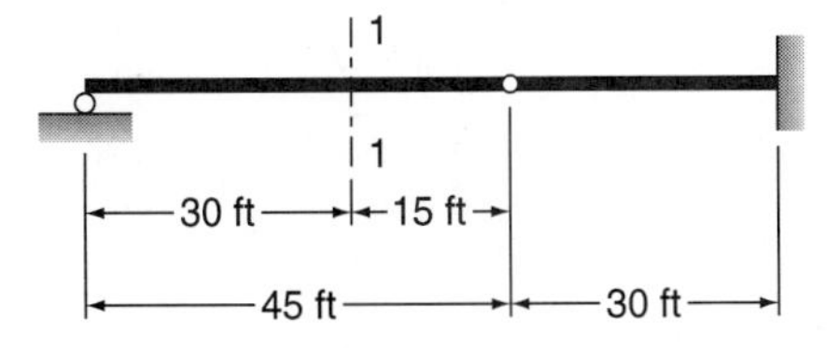

9.18 Reactions at supports A and B (*Ans.* Load @ left unsupported hinge: $V_A = -0.33$, $V_B = +1.33$)

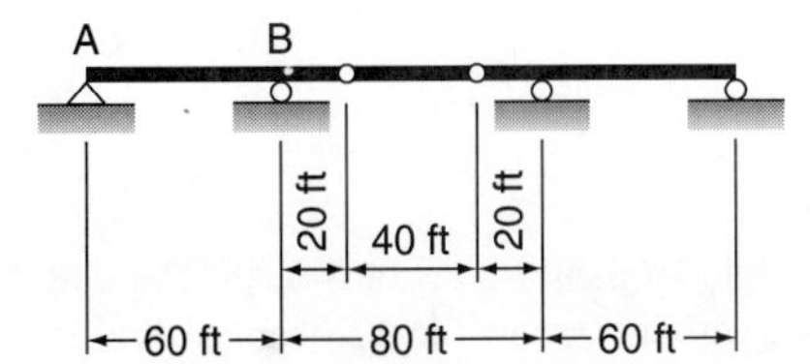

9.19 Draw influence lines for both reactions and for shear just to the left of the 16-kip load, and for moment at the 16-kip load. Determine the magnitude of each of these functions using the influence lines for the loads fixed in the positions shown in the accompanying illustration.

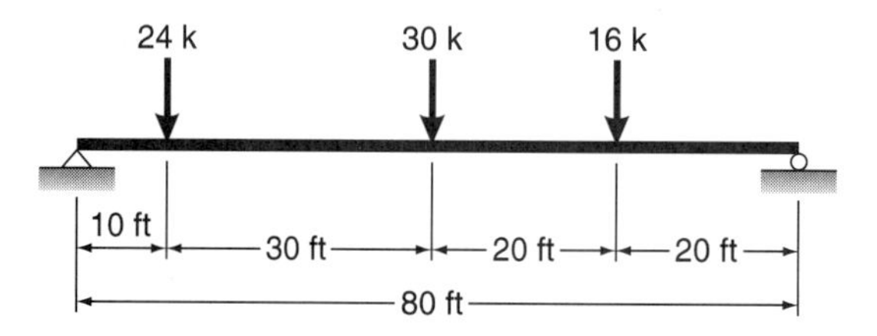

9.20 Draw influence lines for tension force in cable, for the shearing force in beam just to left of B, and for the bending moment at point C as a unit load moves from A to D. (*Ans.* Load @ B: Vertical component of cable tension = +1.0, M @ C = 0)

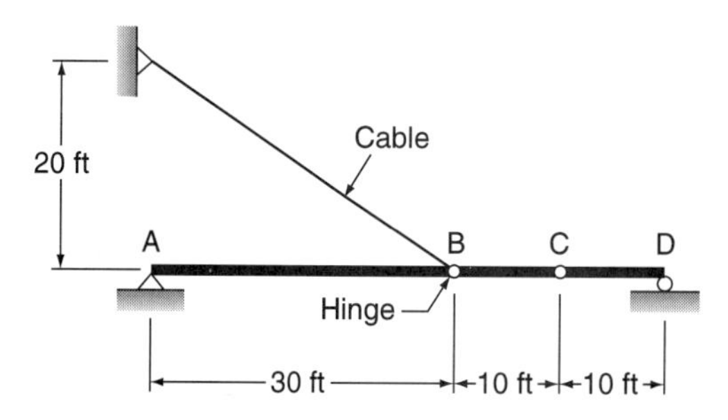

For Problems 9.21 through 9.27, using influence lines determine the quantities requested for a uniform dead load of 2 klf, a moving uniform live load of 3 klf, and a moving or floating concentrated live load of 20 kips. Assume impact = 25% in each live load case.

9.21 Maximum left reaction and maximum plus shear and moment at section 1-1

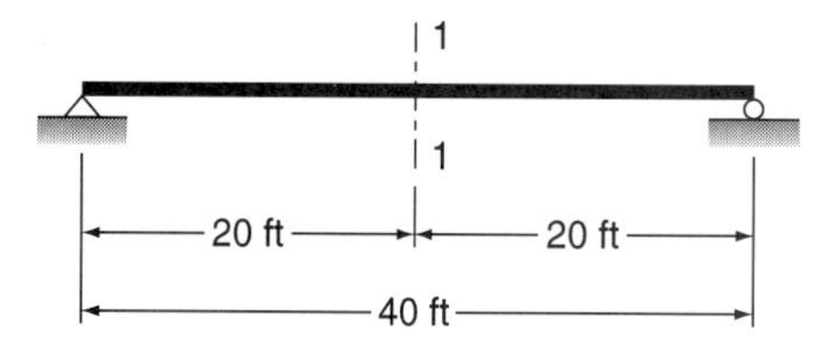

9.22 Maximum positive values of left reaction and shear and moment at section 1-1 (*Ans.* +149.37 k, −2.29 k, +1393.7 ft-k)

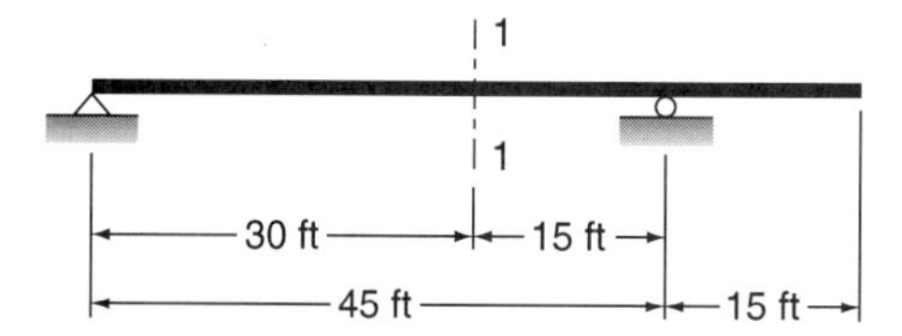

9.23 Maximum negative values of shear and moment at section 1-1, the shearing force just to left of support, and the bending moment at the support for the beam of Problem 9.13.

9.24 Maximum negative shear and moment at section 1-1 (*Ans.* −24.66 k, −180 ft-k)

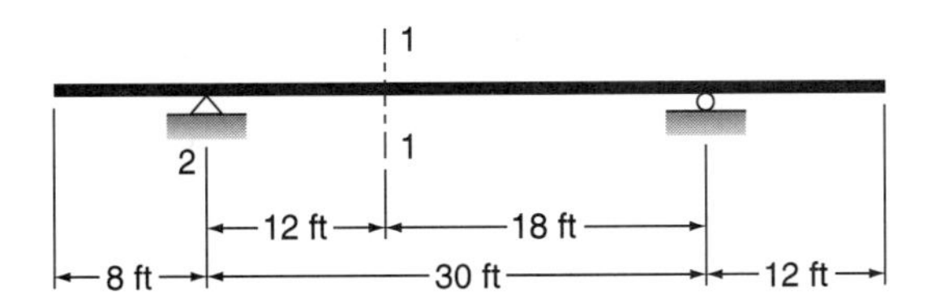

9.25 Maximum upward value of reactions at A and B and maximum negative shear and moment at section 1-1

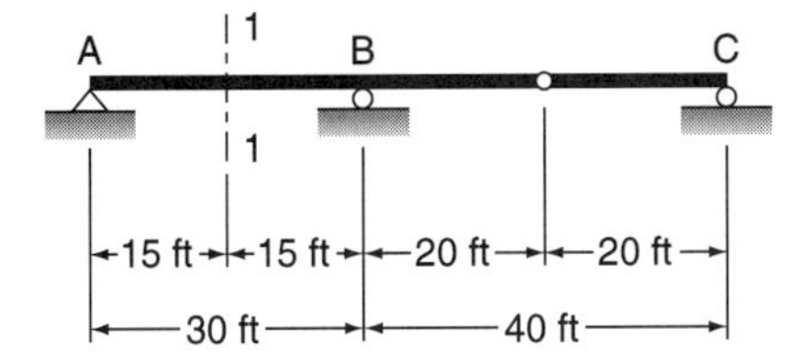

9.26 Maximum positive and negative values of shear and moment at section 3-3 in the beam of Problem 9.10 (*Ans.* Maximum positive values = −6.87 k, +1025 ft-k)

9.27 Maximum positive shear at unsupported hinge, maximum negative moment at support B, and maximum downward value of reaction at support A for the beam of Problem 9.25

9.28 Determine the absolute maximum shear and moment possible in a 30-ft simple beam due to the load system shown.

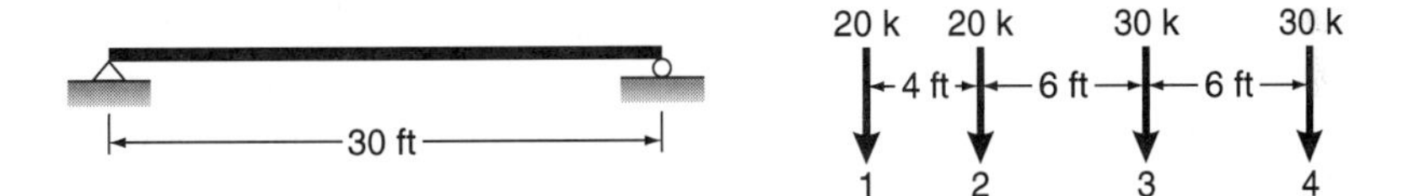

9.29 A simple beam of 20-m span supports a pair of 60-kN moving concentrated loads 4 m apart. Compute the maximum possible moment at the centerline of the beam and the absolute maximum moment in the beam. (*Ans.* 480 kN-m at the centerline, 486 kN-m absolute maximum)

9.30 What is the maximum possible moment that can occur in an 80-ft simple beam as the load system shown moves across the span?

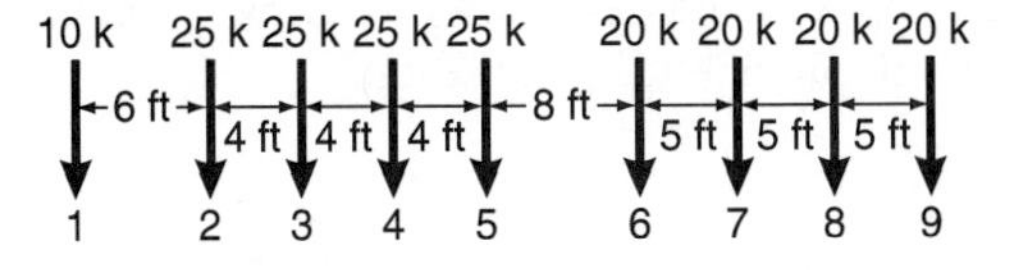

Chapter 10

Introduction to Calculating Deflections

10.1 INTRODUCTION

This chapter and the next are concerned with the elastic deformations of structures. Both the linear displacements of points (deflections) and the rotational displacements of lines (slopes) are considered. The word *elastic* is used to mean that:

1. Stresses are proportional to strains;
2. There is a linear variation of stress from the neutral axis of a beam to its extreme fibers; and
3. The members will return to their original geometry after loads are removed.

The deformations of structures are caused by bending moments, by axial forces, and by shearing forces. For beams and frames, the largest values are caused by bending moments, whereas for trusses the largest values are caused by axial forces. Deflections caused by shearing forces are neglected in this text, as they are quite small in almost all beam-like structures. Deflections caused by shearing forces, as a percentage of beam deflection, increase as the ratio of beam depth to span increases. For the usual depth/span ratio of 1/12 to 1/6, the percentages of shear deflections to bending deflections vary from about 1% to 8%. For a depth/span ratio of one-quarter, the percentages can be as high as 15% to 18%.[1]

In this chapter, displacements are computed using the moment-area method. Often, this method is referred to as geometric method because the deformations are obtained directly from the strains in the structure. In Chapter 11 displacements are determined with *energy methods,* which are based on the conservation of energy principle. Both the geometric and energy procedures will provide identical results.

10.2 SKETCHING DEFORMED SHAPES OF STRUCTURES

Before learning methods for calculating structural displacements, there is great benefit in learning to qualitatively sketch the expected deformed shape of structures. Understanding the displacement behavior of structural systems is a very important part of understanding how structures perform. A structural analyst should sketch the anticipated deformed shapes

[1]C. K. Wang, *Intermediate Structural Analysis* (New York: McGraw-Hill Book Company, 1983), 750.

of structures under load before making actual calculations. Such a practice provides an appreciation of the behavior of the structure and provides a qualitative check of the magnitudes and directions of the computed displacements.

To sketch the anticipated deformed shape of a structural system, there are only a few general rules. Some of these rules apply to the beams and columns and others apply to the joints between the components. By applying these simple rules, we can obtain reasonable qualitative indications of the deflection response of beams and frames. Only by applying the quantitative methods discussed in this and later chapters can we obtain the actual deflections.

10.2.1 Rules for Members

- A member deforms in the direction of the load applied to it.
- Deflections of loaded members are sketched first. Deflections of unloaded members are sketched after the deflections of the joints are sketched.
- Unless there is a hinge between a member and a joint, the end of the member and the joint displace in the same manner.
- Members with lower stiffness (EI/L) tend to deform more than do members with higher stiffness. That is to say, long slender members deform more than short stocky members do.
- When sketching the qualitative deformed shape, the beams and columns are assumed to remain the same length.

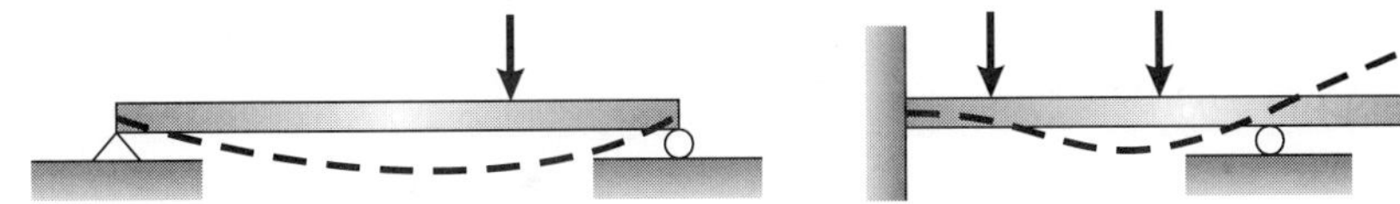

(a) Point of maximum deflection is some where to the left of this off-center load.

(b) Tangent at the fixed end is horizontal and the right end deflects upqard.

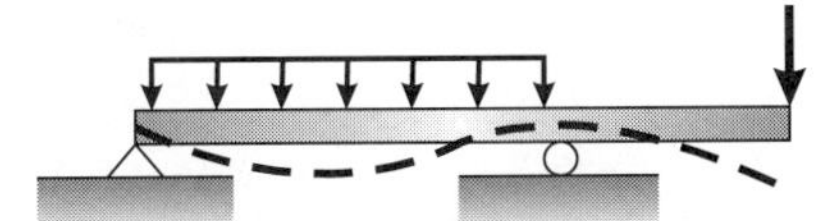

(c) Without calculations we do not know whether the deflection at the right end is up or down. The concentrated load tends to push the right end down while the uniform load tends to push it up.

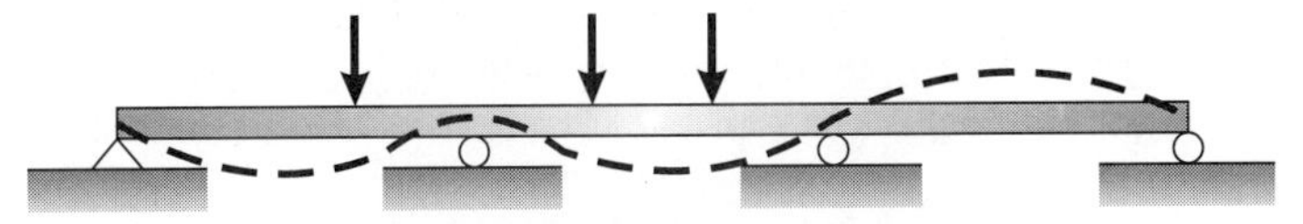

(d) Note the upward deflection in the third span.

(e) The frame sways to the right.

Figure 10.1 Qualitative deformed shapes of some structures under load

10.2.2 Rules for the Joints

- A joint in a structure is assumed to be rigid. A rigid joint can displace but it cannot deform—the joint does not change size or shape as it displaces. The relative orientation of the ends of the members connected to a joint is the same before and after displacement of the joint.
- A joint can only displace in accordance with the external supports acting on it. A joint at a fixed support can neither translate nor rotate. A joint at a pin support can rotate but it cannot translate. A joint at a roller can rotate, cannot translate perpendicular to the surface on which the rollers bears, and can translate parallel to the surface on which the roller bears.

In Figure 10.1, the approximate deflected shapes of several loaded structures are sketched by applying these rules. In each case, the member weight is neglected. You should note that the corners of the frame of part (e) of the figure are free to rotate, but the angles between the members meeting there are assumed to be constant. If the moment diagrams have previously been prepared, they can be helpful in making the sketches where we have both positive and negative moments. Several examples for sketching the qualitative deformed shapes of structures follows. In the examples, the thought process in preparing the sketches is discussed.

EXAMPLE 10.1

Consider the three-span continuous beam subjected to a concentrated force on one span and a distributed force on another span shown in the figure. Sketch the qualitative deflected shape for this beam.

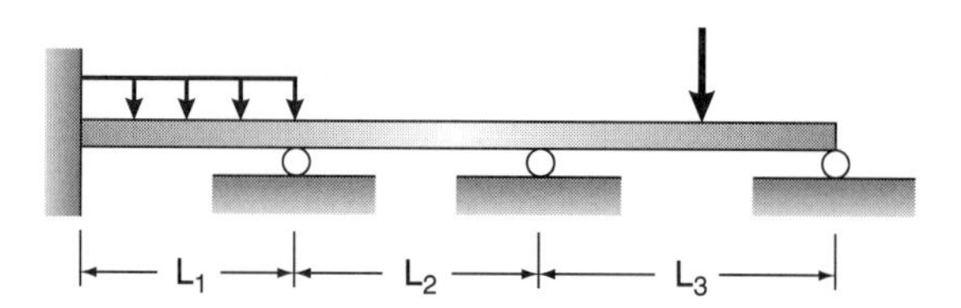

Solution. Sketch the left span first. The left side cannot rotate because it is connected to a fixed joint. The load is acting downward so the span will tend to deflect downward. The right end can rotate, but neither the left nor right ends can translate vertically. As the member deformation is sketched, sketch the displacement of the joints. Then sketch the deformation of the right span. Both ends of the right span can rotate but neither can translate. The result to this point is shown in the following figure.

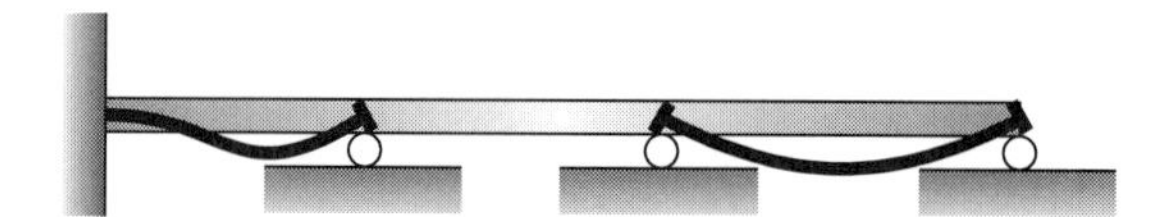

Now sketch the deformed shape of the middle span. Because the member has no external load acting on it, it deforms only in response to the displacement of the joints to which it is connected. The slope of the members connected at a particular joint must be the same.

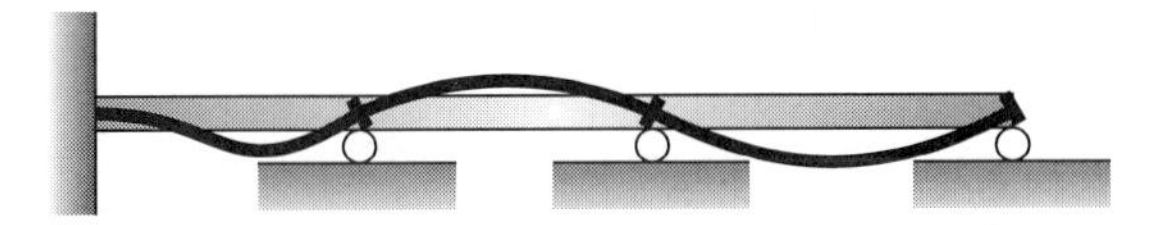

This is the qualitative deflected shape of the beam. ■

EXAMPLE 10.2

Sketch the qualitative deflected shape of the cantilevered beam subjected to a uniformly distributed that is shown in the figure.

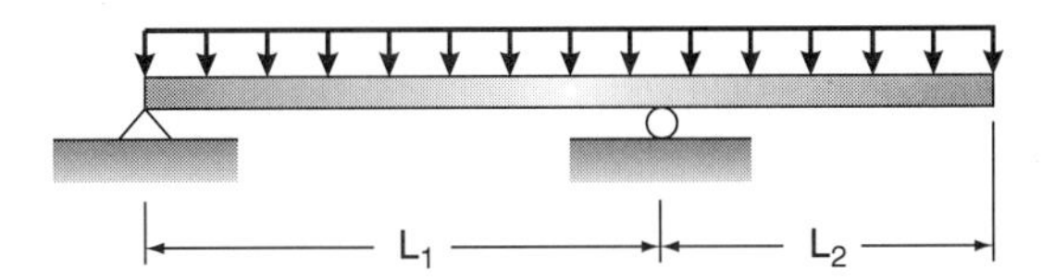

Solution. Sketching the deflected shape of this beam is a little more involved. The load on the left span tends to cause the joint at the right support to rotate counterclockwise while the load on the right span tends to cause that same joint to rotate clockwise. The longer span will tend to dominate the rotation; we can sketch the deformation accordingly.

Sketch the left span first and show the resulting rotation of the joints. The load is acting downward so the span will tend to deflect downward. Both ends can rotate, but neither end can translate vertically. As the member deformation is sketched, sketch the displacement of the joints.

Next, sketch the deformation of the right span, the cantilever span. There is not a support at the right side of the span so that end will displace in response to the applied load. Recall that the geometry of the joints does not change so the tangents of the deflected shapes of the two spans at the right support must be the same.

This is the qualitative deflected shape of the beam. Whether the right end moves upward or downward will depend on the magnitude of the loads and relative lengths of the two spans. ■

EXAMPLE 10.3

Sketch the qualitative deformed shape of the braced frame shown in the figure. Because this is a braced frame, the joints will not translate relative to one another.

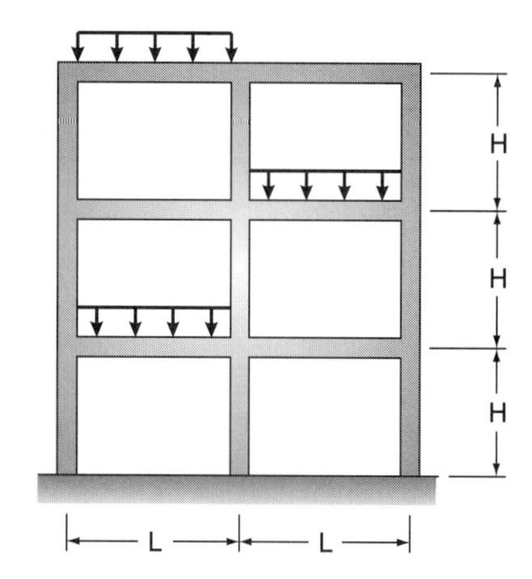

Solution. Sketch the deformed shape of the loaded members first. At the same time indicate the rotation of the joints. The loaded members will tend to deform in the direction of the applied loads.

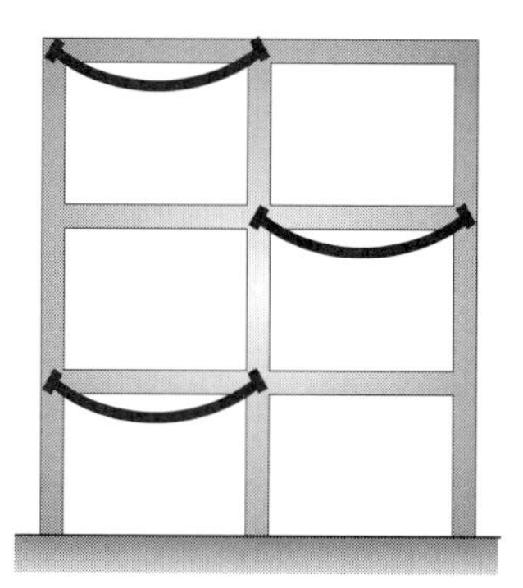

Now sketch the deformed shape of the other beams taking into account the displacements of the joints to which they are connected. Because the beams are fully connected at the joints, the tangents of the deformed shape of the beams connected at a joint must be the same.

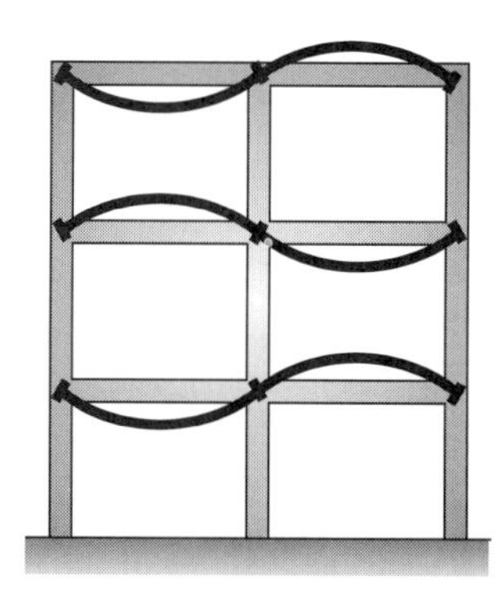

Now sketch the deformations of the columns. Recall that the joints do not deform so the right angles between the beams and columns must be maintained.

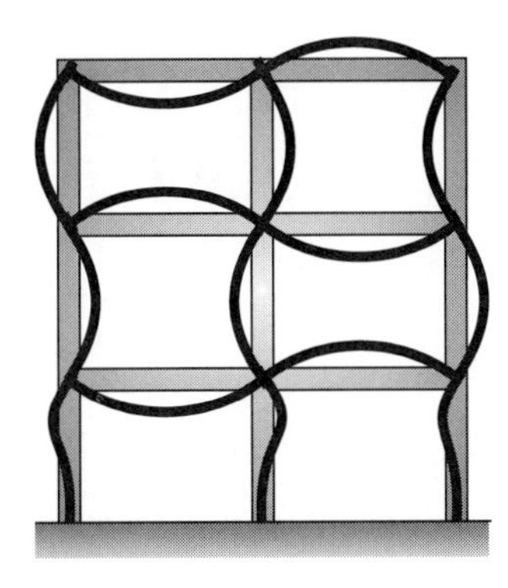

The result is the qualitative deformed shape of the braced frame. ■

EXAMPLE 10.4

Sketch the deformed shape of the unbraced frame shown in the following figure. Recall that an unbraced frame is a frame in which the joints can translate laterally.

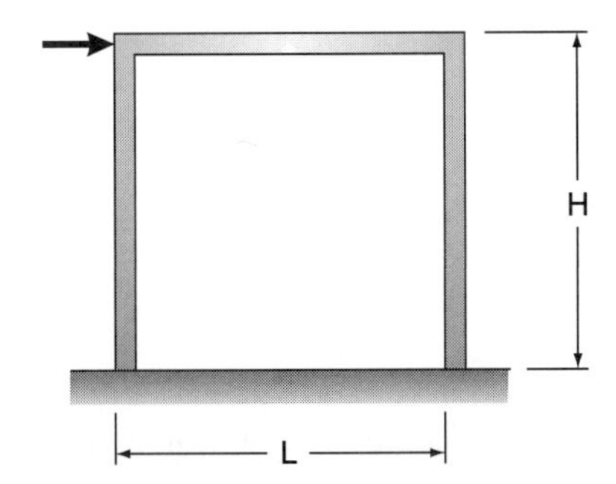

Solution. Sketch the deformed shape of the left column first. The load is applied on that side. The top of the column will move to the right and will tend to rotate clockwise. The beam will act to reduce that rotation. As the stiffness of the beam increases relative to the column, the rotation will decrease. Conversely, as the stiffness of the beam decreases relative to the column, the rotation will increase.

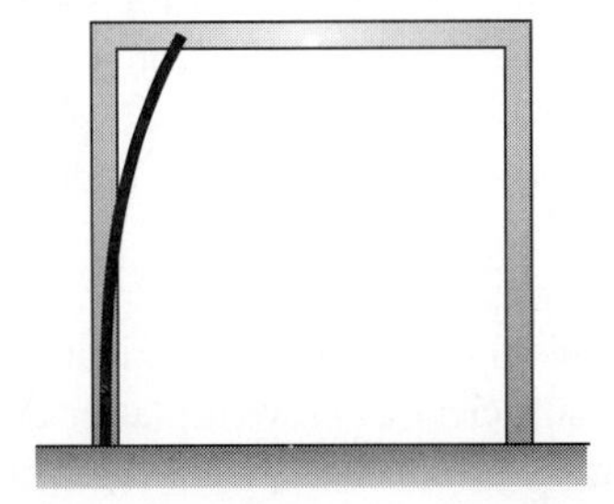

Now sketch the deformed shape of the right column. Because the beam is assumed not to shorten, the joint at the top tends to move to the right and rotate clockwise as it does so. The beam works to reduce that rotation.

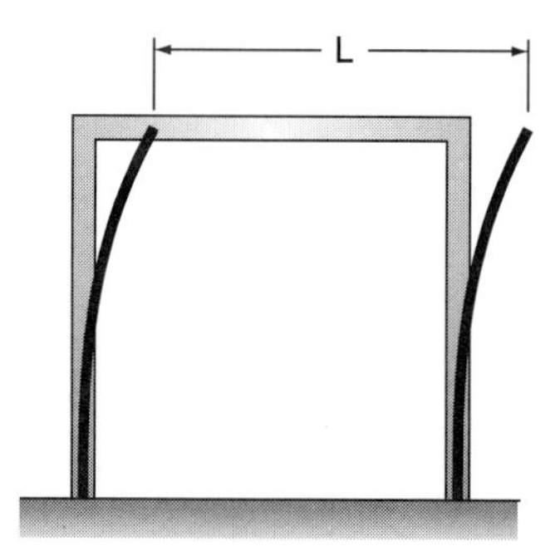

Lastly, sketch the deformed shape of the beam at the top. Because the joints do not deform, the beam and the columns must continue to form right angles at the joints.

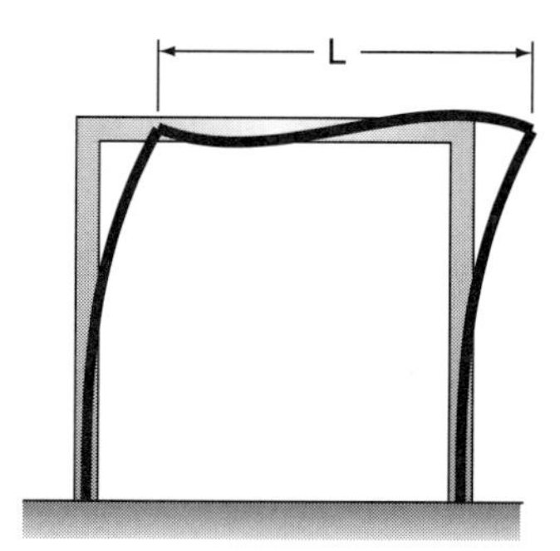

The resulting diagram is the qualitative deflected shape of this simple portal frame. This is the same frame that was shown in Figure 10.1(e). ■

10.3 REASONS FOR COMPUTING DEFLECTIONS

The members of all structures are made up of materials that deform when loaded. If deflections exceed allowable values, they may detract from the appearance of the structures, and the materials attached to the members may be damaged. For example, a floor beam that deflects too much may cause cracks in the ceiling below, or if it supports concrete or tile floors, it may cause cracks in the floors. In addition, the use of a floor supported by beams that "give" appreciably does not inspire confidence, although the beams may be perfectly safe. Excessive vibration may occur in a floor of this type, particularly if it supports machinery.

Standard American practice is to limit deflections caused by live load to $1/360$ of the span lengths. This figure probably originated for beams supporting plaster ceilings and was thought to be sufficient to prevent plaster cracks. Although a large part of the deflections in a building are due to dead load, most of which will have occurred before plaster is applied.

Houston Ship Channel Bridge, Houston, Texas (Courtesy of the Texas State Department of Highways and Public Transportation)

The $1/360$ deflection is only one of many maximum deflection values in use because of different loading situations, different designers, and different specifications. For situations in which precise and delicate machinery is supported, maximum deflections may be limited to $1/1500$ or $1/2000$ of the span lengths. The 1992 AASHTO specifications limit deflections in steel beams and girders due to live load and impact to $1/800$ of the span length. The value, which is applicable to both simple and continuous spans, is preferably reduced to $1/1000$ for bridges in urban areas that are used in part by pedestrians. Corresponding AASHTO values for cantilevered arms are $1/300$ and $1/375$.

Members subject to large downward deflections often are unsightly and may even cause users of the structure to be frightened. Such members may be *cambered* so their displacements do not appear to be so large. The members are constructed of such a shape that they will become straight under some loading condition (usually dead load). A simple beam would be constructed with a slight convex bend so that under gravity loads it would become straight as assumed in the calculations. Some designers take into account both dead and live loads when figuring the amount of camber.

Despite the importance of deflection calculations, rarely must structural deformations be computed—even for statically indeterminate structures—for modifying the original dimensions on which computations are based. The deformations of the materials used in ordinary work are quite small when compared to the overall dimensions of the structure. For example, the strain (ϵ) that occurs in a steel section that has a modulus of elasticity (E) of 29×10^6 psi when the stress (σ) is 20,000 psi is only

$$\varepsilon = \frac{\sigma}{E} = \frac{20 \times 10^3}{29 \times 10^6} = 0.000690 \qquad \textbf{Eq. 10.1}$$

This value is only 0.0690 percent of the member length.

Historically, several methods are available for determining deflections. A structural engineer should be familiar with several of these. For some structures one method may be easier to apply; for others another method is more satisfactory. In addition, the ability to evaluate a structural system by more than one method is important for checking results.

In this chapter and the next, the following methods of computing slopes and deflections are presented: (a) moment-area theorems, (b) virtual work, and (c) Castigliano's second theorem. These methods for computing deflections may be used for computing the reactions for statically indeterminate beams, frames, and trusses, as is shown in later chapters. Elastic weight and conjugate beam methods are included in Appendix F.

10.4 THE MOMENT-AREA THEOREMS

The moment-area method for calculating deflections was presented by Charles E. Greene of the University of Michigan about 1873. Under changing loads the neutral axis of a member changes in shape according to the positions and magnitudes of the loads. The elastic curve of a member is the shape the neutral axis takes under temporary loads. Professor Greene's theorems are based on the shape of the elastic curve of a member and the relationship between bending moment and the rate of change of slope at a point on the curve.

To develop the theorems, the simple beam of Figure 10.2 is considered. Under the influence loads of the beam deflects downward as indicated.

The segment of length dx, bounded on its ends by line a-a and line b-b, is shown in Figure 10.3. The size, degree of curvature, and distortion of the segment are tremendously exaggerated so that the slopes and deflections to be discussed can be seen easily. Line ac lies along the neutral axis of the beam and is unchanged in length. Line ce is drawn parallel to ab. Therefore, be is equal to ac and de represents the increase in length of be, the bottom fiber of the segment dx. Shown in Figure 10.1(b) is an enlarged view of triangle cde and the angle dθ. The angle dθ is the change in slope of the tangent to the elastic curve between the left and right ends of the section. Sufficient information is now available to determine dθ. In the derivation to follow, remember that the angle dθ being considered is very small. For a very small angle, the sine of the angle, the tangent of the angle, and the angle in radians are nearly identical, which permits their values to be used interchangeably. It is worthwhile to check a set of natural trigonometry tables to see the range of angles for which the functions almost coincide.

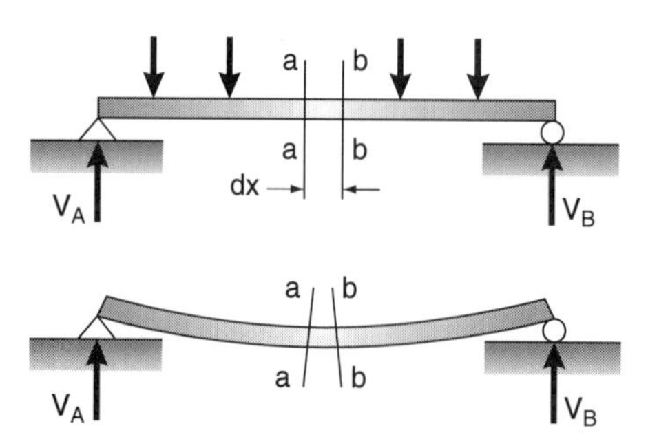

Figure 10.2

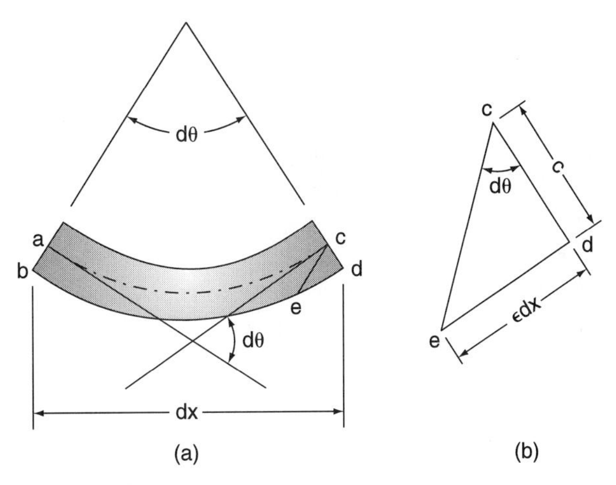

Figure 10.3

The bending moments developed by the external loads are positive and cause shortening of the upper beam fibers and lengthening of the lower fibers. The changes in fiber dimensions have caused the change in slope dθ. The modulus of elasticity is known, and the stress at any point can be determined by the elastic flexure formula. Therefore, the strain in any fiber can be found because it is equal to the stress divided by the modulus of elasticity. The value of dθ may be expressed as follows:

$$\tan(d\theta) = \frac{\text{strain}}{c} = \frac{ed}{cd} = d\theta$$

$$d\theta = \frac{\varepsilon\, dx}{c} \qquad \textbf{Eq. 10.2}$$

By substituting σ/E for ε, we find that:

$$d\theta = \frac{(\sigma/E)\,dx}{c} \qquad \textbf{Eq. 10.3}$$

However, stress can be calculated using the elastic flexure equation, so this equation can be expressed as:

$$d\theta = \frac{\left(\frac{Mc}{EI}\right)dx}{c} = \frac{M\,dx}{EI} \qquad \textbf{Eq. 10.4}$$

This equation represents the change in slope of the elastic curve between the two ends of a segment that has a length dx. The total change in slope from one point A in the beam to another point B can be expressed as the integral of dθ over the length AB, namely:

$$\theta_{AB} = \int_A^B \frac{M\,dx}{EI} \qquad \textbf{Eq. 10.5}$$

This equation represents the area of the M/EI diagram between the points A and B. The M/EI diagram is simply the bending moment diagram divided by EI. From this discussion the first moment-area theorem may be expressed as follows:

> The change in slope between the tangents to the elastic curve at two points on a member is equal to the area of the M/EI diagram between the two points.

Once we have a method by which changes in slopes between tangents to the elastic curve at various points may be determined, only a small extension is needed to develop a method for computing deflections between the tangents. In a segment of length dx, the neutral axis changes direction by an amount $d\theta$. The deflection $d\delta$, of one point on the beam with respect to the tangent at another point due to this angle change, then, is equal to:

$$d\delta = x\,d\theta \qquad \textbf{Eq. 10.6}$$

In this equation, x is the distance from the point at which deflection is desired to the point at which the tangent is computed. The value of $d\theta$ from Eq. 10.4 can be substituted into this expression to obtain:

$$d\delta = x\frac{M\,dx}{EI} = \frac{Mx\,dx}{EI} \qquad \textbf{Eq. 10.7}$$

To determine the total deflection from the tangent at one point, A, to the tangent at another point on the beam, B, Eq. 10.7 can be integrated over the distance AB, namely:

$$\delta_{AB} = \int_A^B \frac{Mx\,dx}{EI} \qquad \textbf{Eq. 10.8}$$

The preceding equation is a mathematical statement of the second moment-area theorem, which is:

> The deflection of a tangent to the elastic curve of a beam at one point with respect to a tangent at another point is equal to the first moment of the M/EI diagram between the two points about the point at which deflection is desired.

Stick-welding decking on a shopping center mall in Charlotte, North Carolina (Courtesy Lincoln Electric Company)

10.5 APPLICATION OF THE MOMENT-AREA THEOREMS

In the paragraphs that follow, you will see that the moment area method is most conveniently used for determining slopes and deflections for beams when the slope of the elastic curve at one or more points is known. Such beams include cantilevered beams, where the slope at the fixed end does not change. The method is applied quite easily to beams loaded with concentrated loads, because the moment diagrams consist of straight lines. These diagrams can be broken down into simple triangles and rectangles, which facilitates the mathematics. Beams supporting uniform loads or uniformly varying loads may be handled, but the mathematics is slightly more difficult.

Examples 10.5 to 10.9 illustrate the application of the moment-area theorems. On occasion, the mathematics may be simplified by drawing the moment diagram and making the calculations in terms of symbols. Such symbols would include P for a concentrated load, w for a uniform load, and L for span length. This concept is illustrated in Examples 10.5 and 10.8. The numerical values of each of the symbols are substituted in the final step to obtain the slope or deflection desired. To facilitate the solutions, properties of several areas frequently encountered are shown in Appendix E.

When solving problems, care must be taken to use consistent units in the calculations. Further, to prevent mistakes when applying the moment-area theorems, *always remember that the slopes and deflections obtained are with respect to tangents to the elastic curve at the points being considered.* The theorems do not directly give the slope or deflection at a point in the beam as compared to the horizontal (except in one or two special cases). Rather, the theorems give the change in slope of the elastic curve from one point to another, or the deflection of the tangent at one point with respect to the tangent at another point.

If a beam or frame has several loads applied, the M/EI diagram may be inconvenient to handle. This is especially true for beams and frames supporting with both concentrated loads and uniform loads. When distributed load and concentrated loads are applied concurrently, the M/EI diagram becomes rather complex, which causes difficulty in computing the needed properties of the areas involved. The calculations may be simplified by drawing a separate M/EI diagram for each of the loads and determining the slopes and deflections caused by each load separately. The final result for a particular point can be found by adding the values obtained for the individual loads. The principle of superposition applies to the moment-area method, and to any of the other methods for computing deflections.

EXAMPLE 10.5

Determine the slope and deflection of the right end of the cantilevered beam shown in the figure.

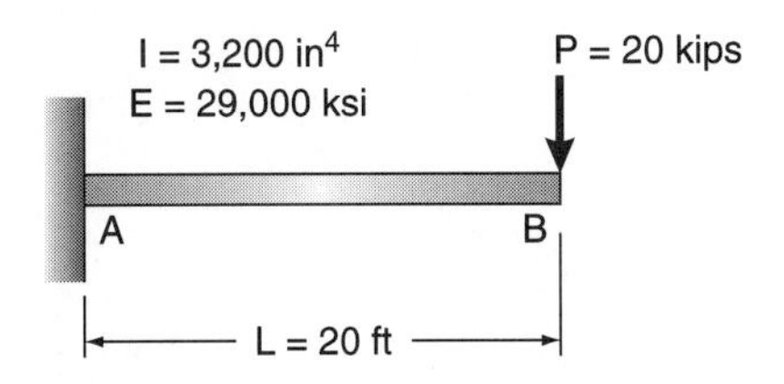

Solution. The tangent to the elastic curve at the fixed end is horizontal. Therefore, the changes in slope and deflection of the tangent at the free end, with respect to a tangent at the fixed end, are the slope and deflection of that point. The bending moment diagram, and the associated M/EI diagram, for the beam are shown in the next figure.

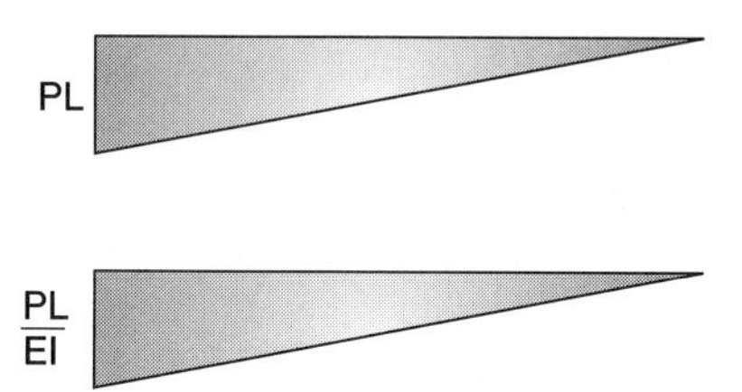

The slope at B is equal the area under the M/EI diagram between A to B

$$\theta_B = \left(\frac{1}{2}\right)L\left(\frac{PL}{EI}\right) = \frac{PL^2}{2EI} = \frac{(20)(1000)[20(12)]^2}{2(3{,}200)(29 \times 10^6)}$$

$$\theta_B = 0.00621 \text{ rad} = 0.36°$$

The deflection at B is equal to the first moment of the M/EI diagram between A to B about B, namely:

$$\delta_B = \left(\frac{1}{2}\right)L\left(\frac{PL}{EI}\right)\left(\frac{2L}{3}\right) = \frac{PL^3}{3EI} = \frac{(20)(1000)[20(12)]^3}{3(3{,}200)(29 \times 10^6)}$$

$$\delta_B = 0.99 \text{ inches} \quad ■$$

EXAMPLE 10.6

Determine the slope and deflection of the beam at point B, which is 10 ft from the left end of the beam as shown in figure.

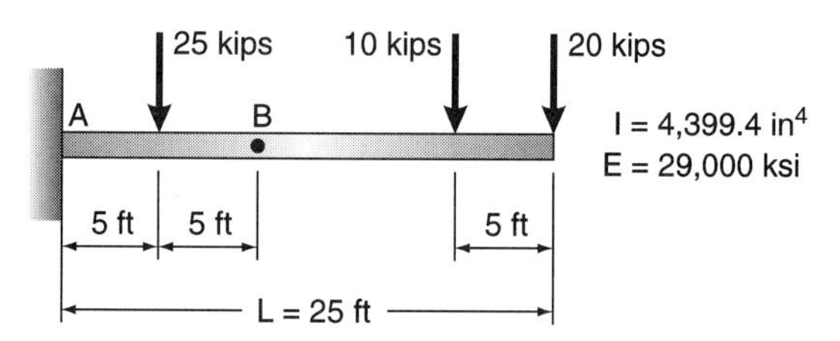

Solution. The left end of the beam is again fixed. As such the slope at B is equal to the area of the M/EI diagram from A to B. The M/EI diagram is broken down into convenient triangles, as shown, for making the calculations.

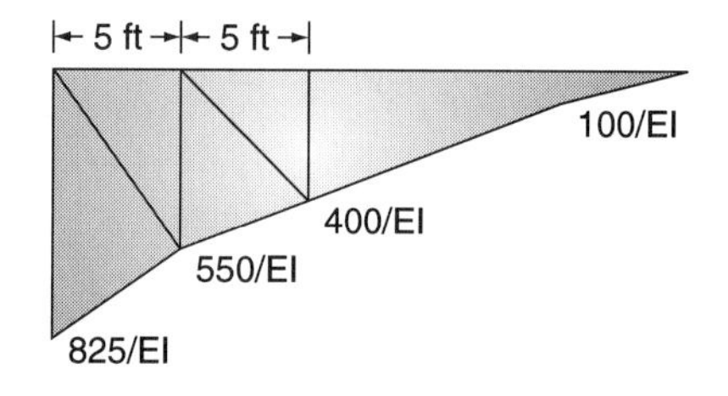

$$\theta_B = \frac{\frac{1}{2}(825)(5) + \frac{1}{2}(550)(5) + \frac{1}{2}(550)(5) + \frac{1}{2}(400)(5)}{EI} = \frac{5812.5 \text{ k-ft}^2}{EI}$$

$$\theta_B = \frac{5812.5(12)^2(1000)}{(29 \times 10^6)(4399.4)} = 0.00656 \text{ rad} = 0.38°$$

The deflection at B is equal to the first moment of the M/EI diagram between A to B about B.

$$\delta_B = \frac{\frac{1}{2}(825)(5)(8.33) + \frac{1}{2}(550)(5)(6.67) + \frac{1}{2}(550)(5)(3.33) + \frac{1}{2}(400)(5)(1.67)}{EI} = \frac{32{,}600\,\text{k-ft}^3}{EI}$$

$$\delta_B = \frac{32{,}600(12)^3(1000)}{(29 \times 10^6)(4399.4)} = 0.442\,\text{inches} \quad ■$$

EXAMPLE 10.7

Determine the slope and deflection at the free end of the cantilevered beam shown in the figure.

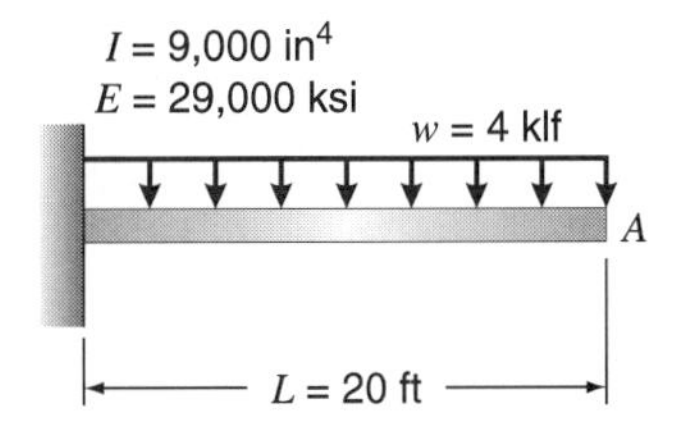

Solution. The M/EI diagram used for the solution is shown here.

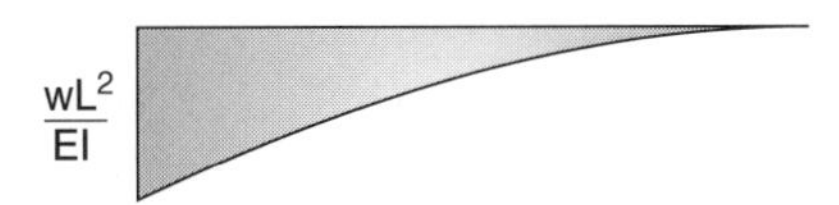

Carefully note the units used to solve the slope and deflection equations developed. Inches and pounds are used throughout the equations. The value of w, then, is 4000/12 lb/in. and not just 4000 lb/ft. Be careful with the units for the distributed load, since the slope or deflection can easily be miscalculated by a multiple of 12.

As before, the slope at the left side is horizontal because that end is fixed. The slope of the elastic curve at A, then, is equal to the area of the M/EI diagram between the left end and A. The properties of the area under the M/EI diagram can be found in Appendix E.

$$\theta_A = \left(\frac{1}{3}\right)L\left(\frac{wL^2}{2EI}\right) = \frac{wL^3}{6EI}$$

$$\theta_A = \frac{\left(\frac{4000}{12}\right)[20(12)]^3}{6(9000)(29 \times 10^6)} = 0.00294\text{ rad} = 0.17°$$

The deflection at A is equal to the first moment of the M/EI diagram between the left end and A about A.

$$\delta_A = \left(\frac{1}{3}\right)L\left(\frac{wL^2}{2EI}\right)\left(\frac{3L}{4}\right) = \frac{wL^4}{8EI}$$

$$\delta_A = \frac{\left(\frac{4000}{12}\right)[20(12)]^4}{8(9000)(29 \times 10^6)} = 0.530\,\text{inches} \quad ■$$

EXAMPLE 10.8

Compute the slope and deflection at the free end of the cantilevered beam shown in the figure. The moment of inertia of the beam has been increased near the support where the bending moment is greatest.

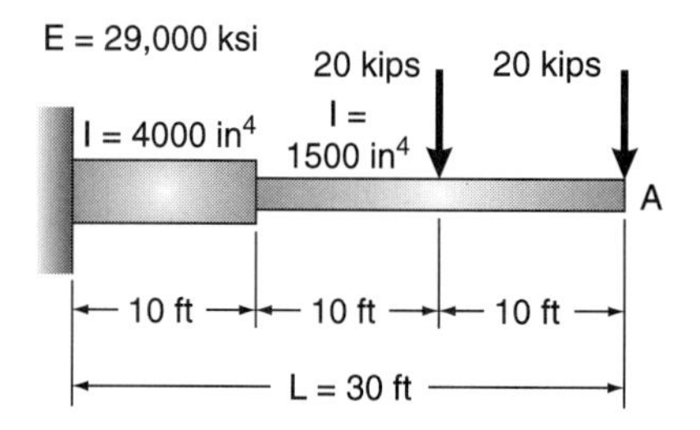

Solution. The moment diagram and the corresponding M/EI diagram are shown below. The M/EI diagram is drawn by keeping the constant E as a symbol and dividing the ordinates by the proper moments of inertia. The resulting figure is conveniently divided into triangles and the computations made as before.

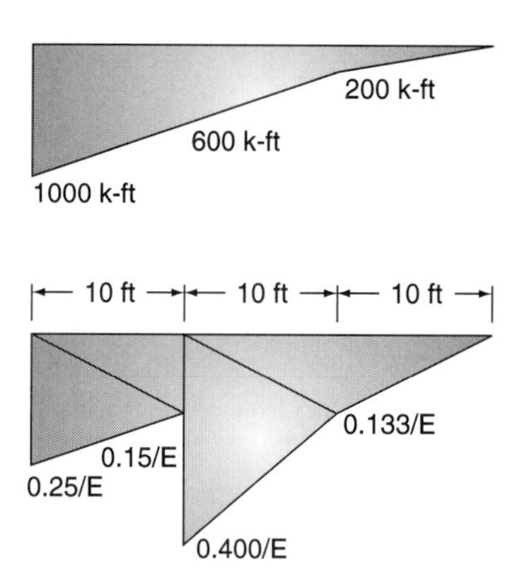

The slope at the left end of the beam is horizontal because that end is a fixed support. The slope of the elastic curve at A, then, is equal to the area of the M/EI diagram between the left end and A.

$$\theta_B = \frac{\frac{1}{2}(10)(0.25) + \frac{1}{2}(10)(0.15) + \frac{1}{2}(10)(0.40) + \frac{1}{2}(20)(0.133)}{E} = \frac{5.333\,\text{k-ft}^2}{E}$$

$$\theta_B = \frac{5.333(12)^2(1000)}{29 \times 10^6} = 0.0265\,\text{rad} = 1.52°$$

Again, the deflection at B is equal to the first moment of the M/EI diagram between about B.

$$\delta_B = \frac{\frac{1}{2}(10)(0.25)(26.67) + \frac{1}{2}(10)(0.15)(23.33) + \frac{1}{2}(10)(0.40)(16.67) + \frac{1}{2}(20)(0.133)(10)}{E}$$

$$\delta_B = \frac{97.47\,\text{k-ft}^3}{E} = \frac{97.47(12)^3(1000)}{29 \times 10^6} = 5.81\,\text{inches} \quad ■$$

EXAMPLE 10.9

Compute the deflection at the centerline of the uniformly loaded simply supported beam.

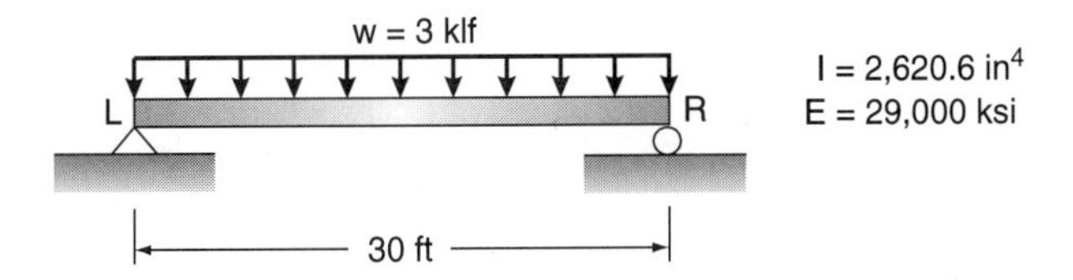

Solution. The M/EI diagram used in the solution is shown here.

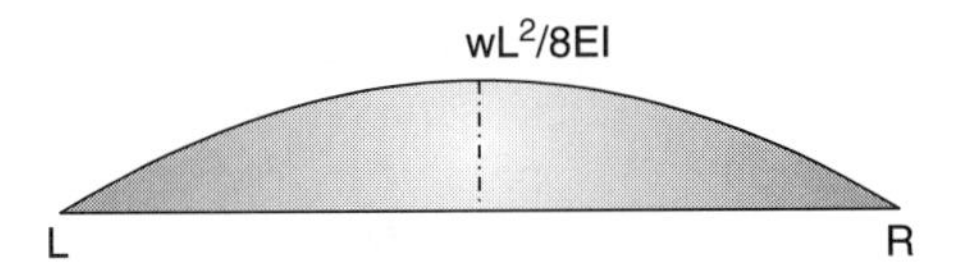

The tangents to the elastic curve at each end of the beam are inclined. Computing the deflection between a tangent at the centerline and one of the end tangents is a simple matter, but the result is not the actual deflection at the centerline of the beam. To obtain the correct deflection, a somewhat roundabout procedure is used. That procedure is described and demonstrated in the following paragraphs.

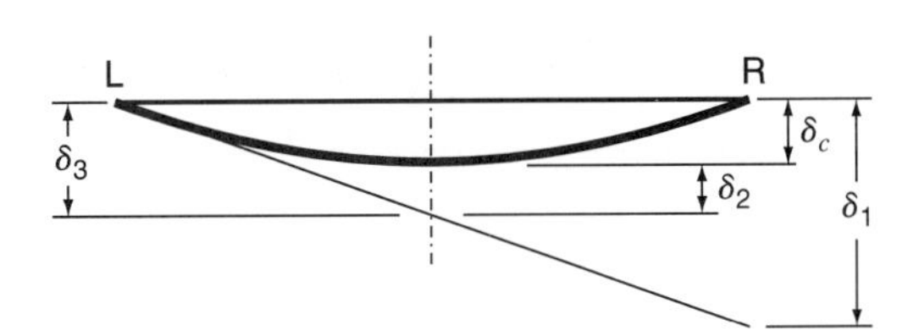

First, the deflection of the tangent at the right end from the tangent at the left end, δ_1, is found:

$$\delta_1 = \left(\frac{2}{3}\right)L\left(\frac{wL^2}{8EI}\right)\left(\frac{L}{2}\right) = \frac{wL^4}{24EI}$$

Next, the deflection of a tangent at the centerline from a tangent at L, δ_2, is found:

$$\delta_2 = \left(\frac{2}{3}\right)\left(\frac{L}{2}\right)\left(\frac{wL^2}{8EI}\right)\left(\frac{3}{8}\right)\left(\frac{L}{2}\right) = \frac{wL^4}{128EI}$$

Then, by proportions, the distance from the original chord between L and R and the tangent at L, δ_3, can be computed:

$$\delta_3 = \frac{1}{2}\delta_1 = \left(\frac{1}{2}\right)\left(\frac{wL^4}{24EI}\right) = \frac{wL^4}{48EI}$$

The difference between δ_3 and δ_2 is the centerline deflection.

$$\delta_C = \delta_3 - \delta_2 = \frac{wL^4}{48EI} - \frac{wL^4}{128EI} = \frac{5wL^4}{384EI}$$

$$\delta_C = \frac{5\left(\frac{3000}{12}\right)[30(12)]^4}{384(2620.6)(29 \times 10^6)} = 0.719\text{ inches} \quad \blacksquare$$

10.6 MAXWELL'S LAW OF RECIPROCAL DEFLECTIONS

The deflections of two points in a beam have a surprising relationship to each other. This relationship was first published by James Clerk Maxwell in 1864. Maxwell's law may be stated as follows:

> The deflection at one point A in a structure due to a load applied at another point B is exactly the same as the deflection at B if the same load is applied at A.

The rule is perfectly general and applies to any type of structure, whether it is a truss, beam, or frame, which is made up of elastic materials following Hooke's law. The displacements may be caused by flexure, by shear, or by torsion. When preparing influence lines for continuous structures, when analyzing statically indeterminate structures and with model-analysis problems, this useful tool is frequently applied.

The law is not only applicable to the deflections in all of these types of structures but is also applicable to rotations. For instance, a unit couple at A will produce a rotation at B equal to the rotation caused at A if the same couple is applied at B.

Example 10.10 demonstrates that the law is correct for a simple cantilevered beam in which the deflections at two points are determined using the moment-area method.

EXAMPLE 10.10

Demonstrate Maxwell's law of reciprocal deflections by comparing the deflection at point A caused by a 10 load applied first at B with the deflection at B caused by a 10-kip load at A. The cantilever beam used in this example and the first loading condition are shown in the figure.

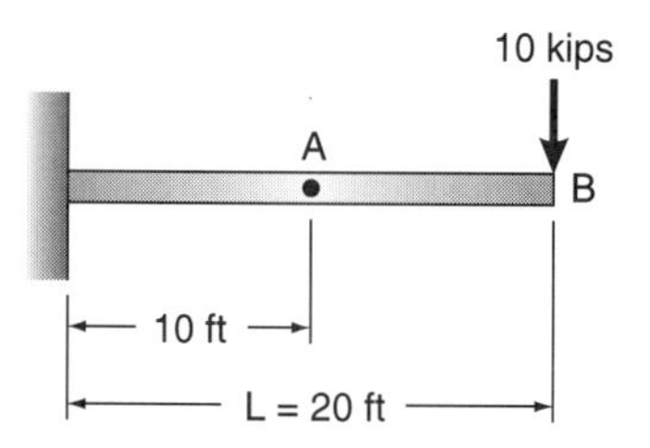

Solution. The M/EI diagram for this beam with a 10-kip load applied at B is shown next. The M/EI diagram has been divided into simple triangles to facilitate computation.

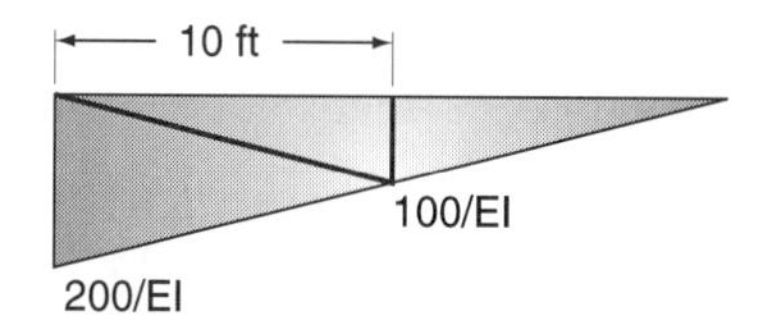

Because the left side is a fixed support, the deflection at A is equal to the first moment of the area under the M/EI diagram between the left support and A about A.

$$\delta_A = \frac{200(10)(6.667)}{2EI} + \frac{100(10)(3.333)}{2EI} = \frac{8{,}332}{EI}$$

When the 10-kip load is applied at A, the loaded beam and the corresponding M/EI diagram are as shown.

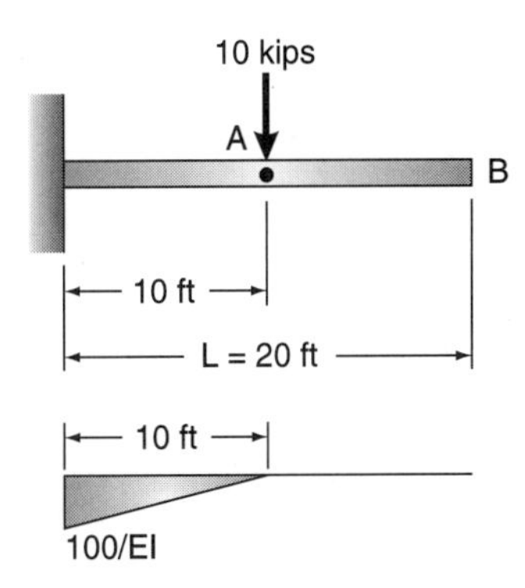

The deflection caused at B by this load is the first moment of the M/EI diagram about B.

$$\delta_B = \frac{100(10)(16.667)}{2EI} = \frac{8{,}335}{EI}$$

These two computed deflection are the same, which demonstrates Maxwell's law of reciprocal deflections: *The deflection at A caused by a load at B is equal to the deflection at B caused by the same load at A.* ■

10.7 PROBLEMS FOR SOLUTION

For problems 10.1 through 10.11, qualitatively sketch the deformed shape of the structures.

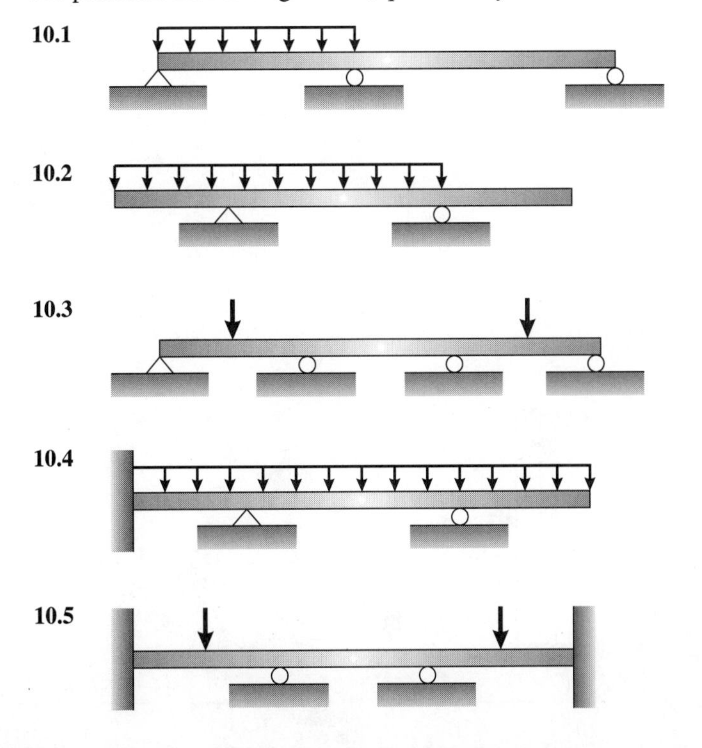

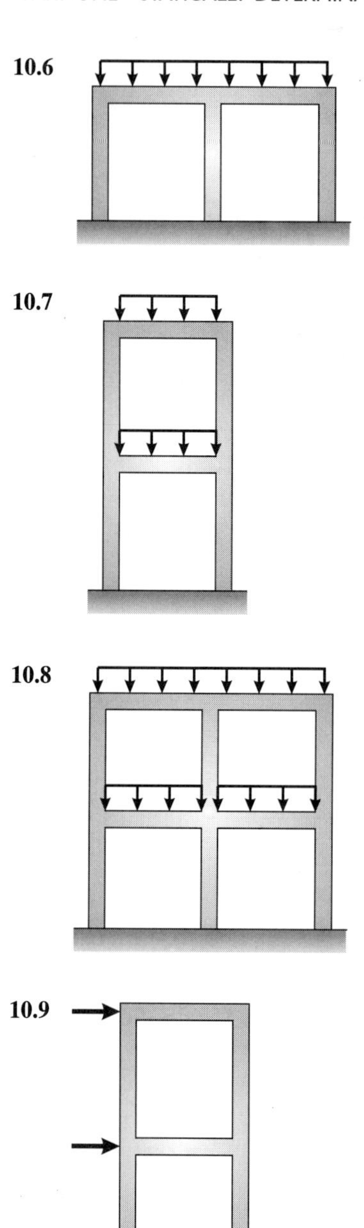
10.6
10.7
10.8
10.9

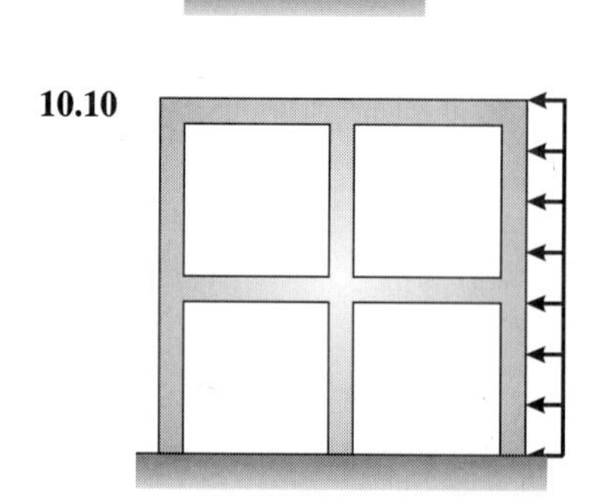
10.10

10.11

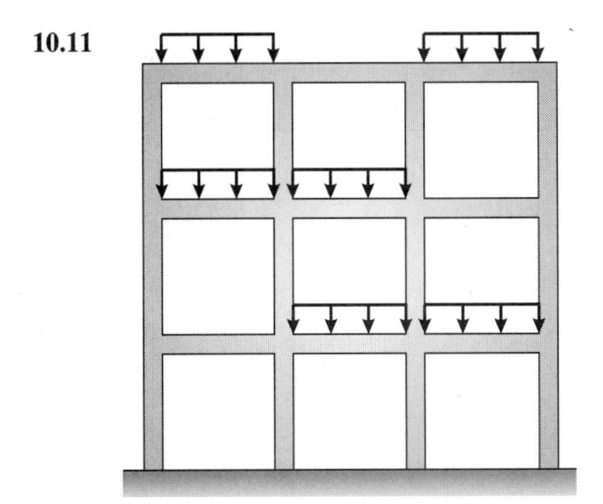

Using the moment-area method, determine the quantities asked for each of Problems 10.12 through 10.27.

10.12 θ_A, Δ_A: E = 29 × 10^6 psi. I = 1000 in^4

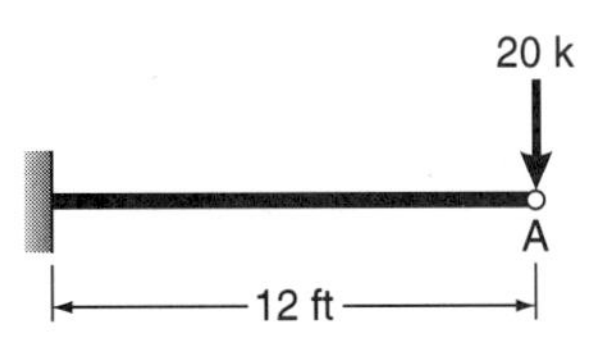

10.13 θ_A, θ_B, Δ_A, Δ_B: E = 29 × 10^6 psi. I = 1500 in^4 (*Ans.* $\theta_A = \theta_B$ = 0.00477 rad, Δ_A = 0.801 in. ↓, Δ_B = 0.458 in. ↓)

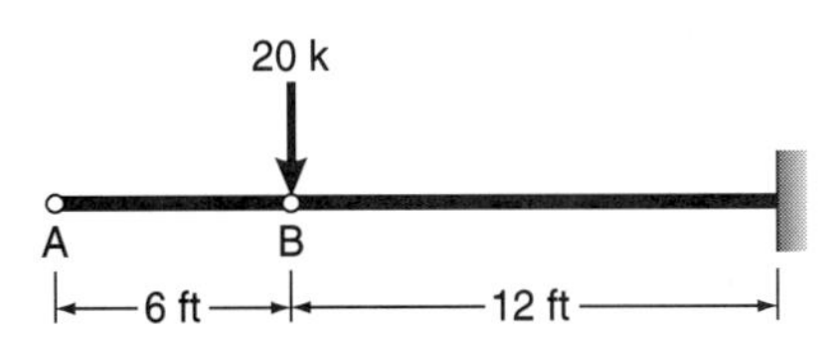

10.14 θ_A, θ_B, Δ_A, Δ_B: E = 29 × 10^6 psi. I = 4100 in^4

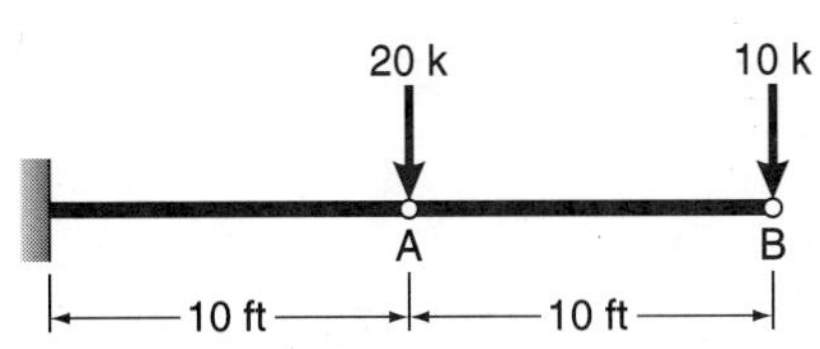

10.15 θ_A, Δ_A: E = 29 × 10^6 psi. I = 843 in^4 (*Ans.* θ_A = 0.00295 rad, Δ_A = 0.265 in. ↓)

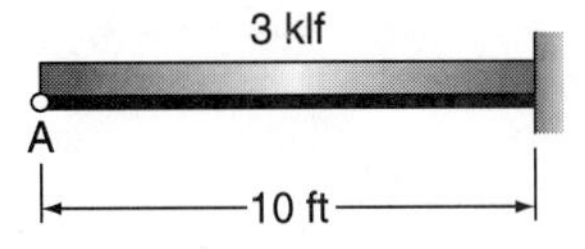

10.16 θ_A, θ_B, Δ_A, Δ_B: E = 29 × 10^6 psi. I = 1140 in^4

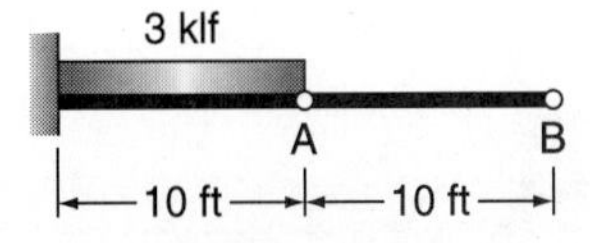

10.17 θ_A, θ_B, Δ_A, Δ_B: $E = 29 \times 10^6$ psi. $I = 4000$ in^4 (*Ans.* $\theta_A = 0.00430$ rad, $\theta_B = 0.00502$ rad, $\Delta_A = 0.300$ in. ↓, $\Delta_B = 0.760$ in. ↓)

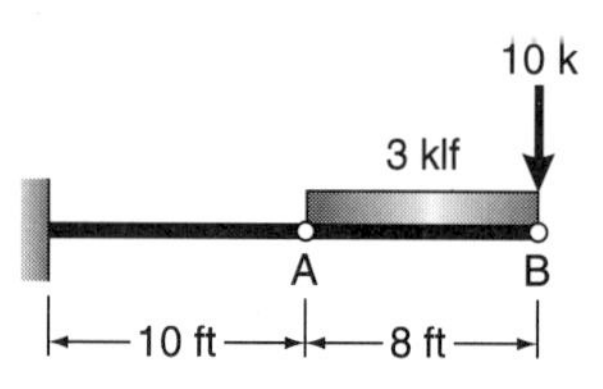

10.18 θ_A, θ_B, Δ_A, Δ_B: $E = 29 \times 10^6$ psi. $I = 5200$ in^4

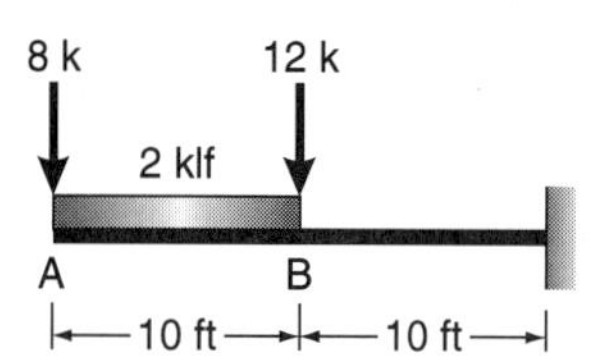

10.19 θ_A, Δ_A, $E = 29 \times 10^6$ psi. $I = 3000$ in^4 (*Ans.* $\theta_A = 0.00297$ rad, $\Delta_A = 0.231$ in. ↓)

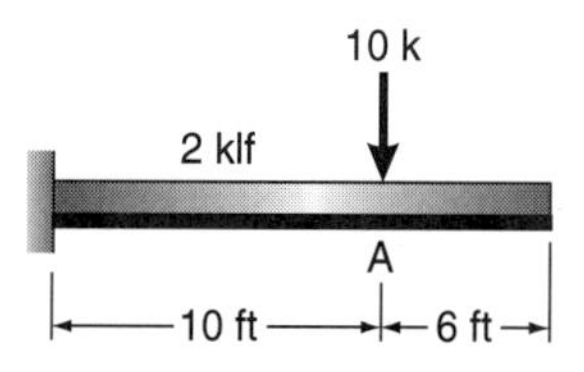

10.20 θ_A, θ_B, θ_C, Δ_A, Δ_B, Δ_C: $E = 29 \times 10^6$ psi. $I = 1330$ in^4

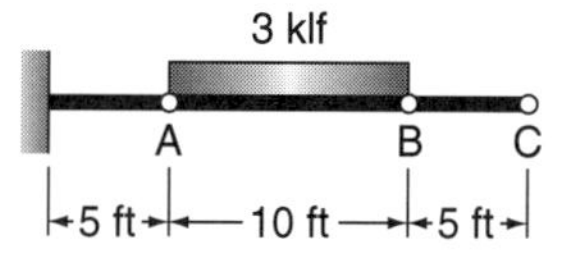

10.21 θ_A, Δ_A: $E = 5 \times 10^6$ psi. $I = 6000$ in^4 (*Ans.* 0.00104 rad, 0.119 in. ↓)

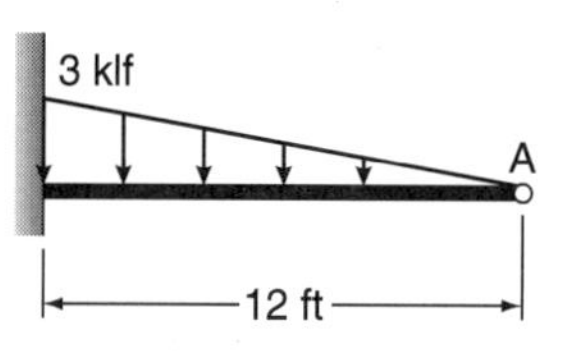

10.22 θ_A, θ_B, Δ_A, Δ_B: $E = 29 \times 10^6$ psi

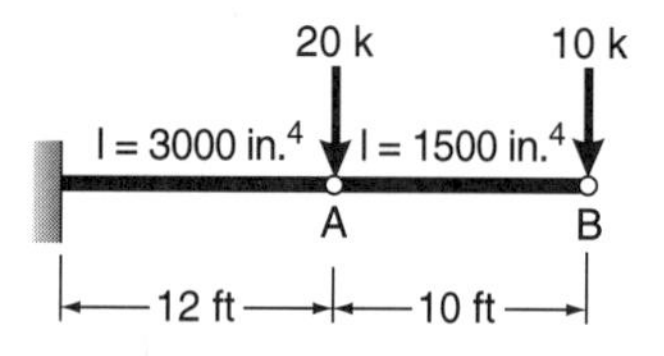

10.23 θ_A, θ_B, Δ_A, Δ_B: $E = 29 \times 10^6$ psi. (*Ans.* 0.00426 rad, 0.00331 rad, 0.724 in. ↓, 0.241 in. ↓)

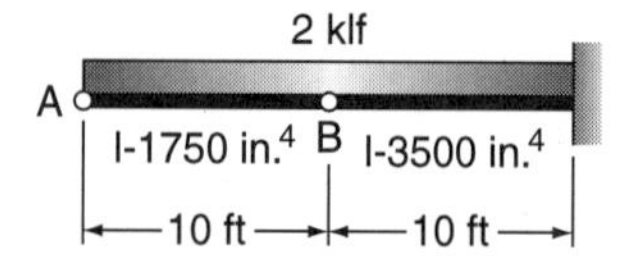

10.24 θ_A, Δ_A: E = 200 000 MPa, I = 3.0 × 10^8 mm^4

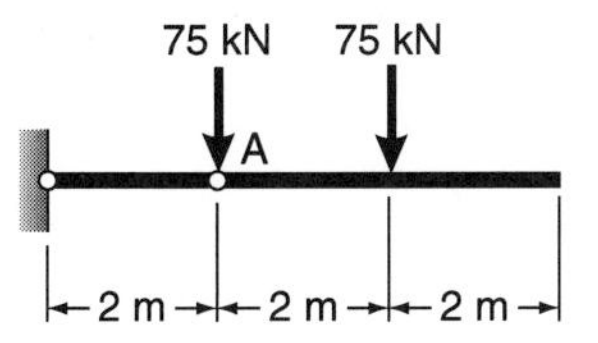

10.25 θ_A, Δ_A: E = 200 000 MPa (*Ans.* θ_A = 0.008 67 rad, Δ_A = 48 mm)

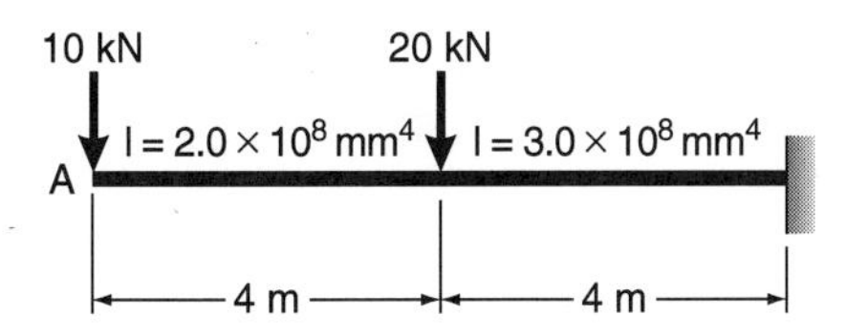

10.26 θ_A, Δ_A: E = 29 × 10^6 psi. I = 1500 in^4

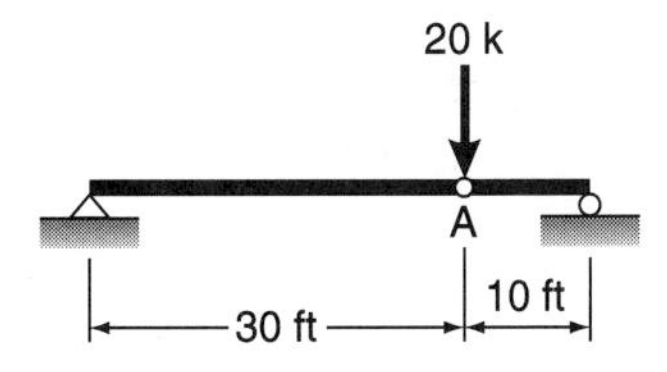

10.27 Δ_A: E = 29 × 10^6 psi. I = 3200 in^4 (*Ans.* 0.631 in. ↓)

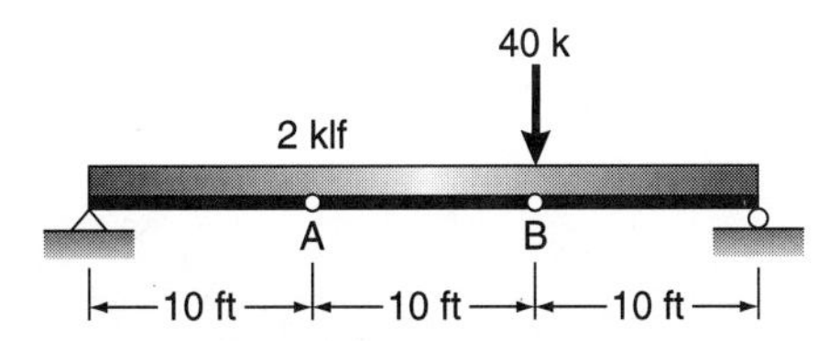

For Problems 10.28 through 10.32 compute the fixed-end moments for the beams: E and I constant except as shown.

10.28

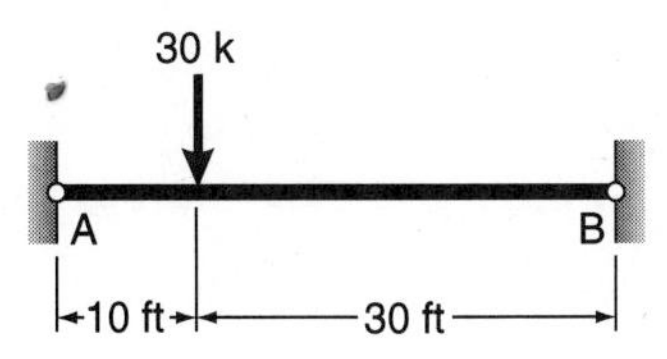

10.29 (*Ans.* M_A = −66 ft-k, M_B = −30 ft-k)

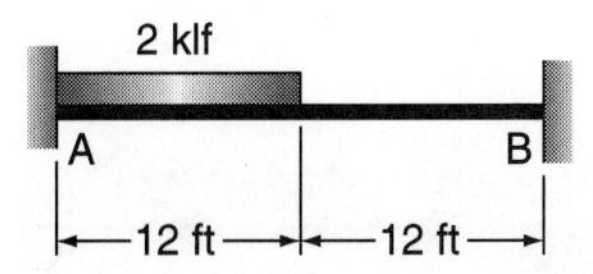

10.30

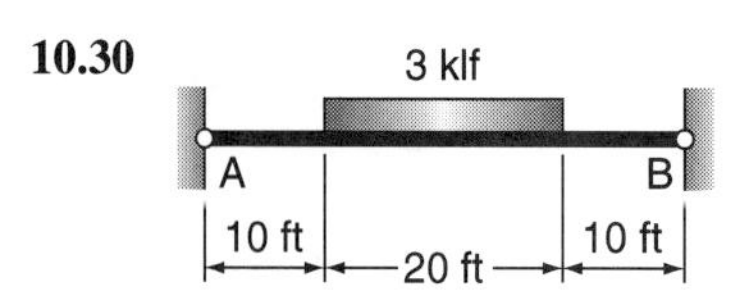

10.31 (*Ans.* $M_A = M_B = -233$ ft-k)

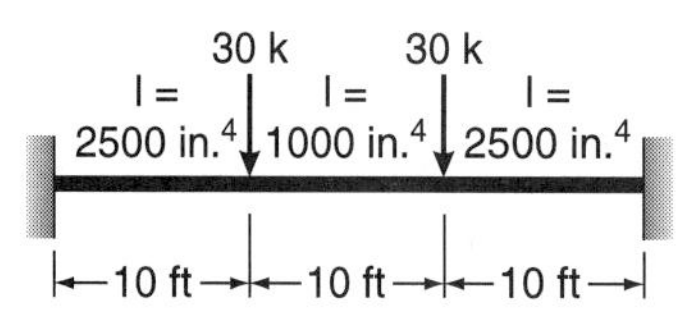

10.32

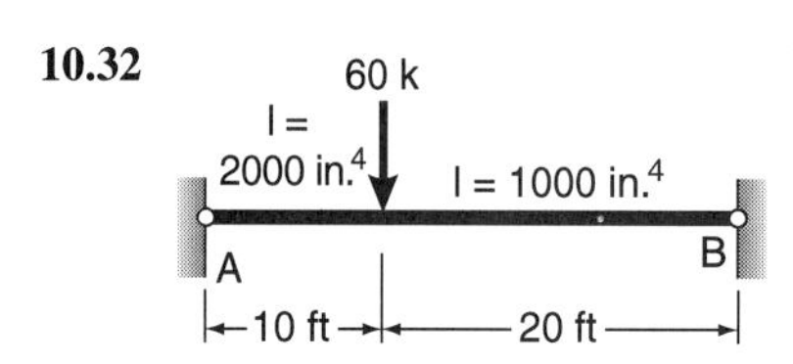

Chapter 11

Deflection and Angle Changes—Energy Methods

11.1 INTRODUCTION TO ENERGY METHODS

In Chapter 10 deflections and angular changes were computed using a geometric method, the moment-area method. This procedure is satisfactory for many structures, including some that are rather complicated. However, it is not easily applicable to all types of structures. For instance, it cannot be used at all for trusses.

In this chapter, we will calculate deflections and angular changes using the principle of conservation of energy. Two energy methods, virtual work and Castigliano's theorem, will be introduced. For some complicated structures, these methods may be more desirable than the moment-area method because of the simplicity with which the expressions may be set up for making the solutions. In addition, energy methods are applicable to more types of structures.

11.2 CONSERVATION OF ENERGY PRINCIPLE

Before the conservation of energy principle is introduced, a few comments are presented concerning the concept of work. For this discussion, *work* is defined as the product of a force and its displacement in the direction in which the force is acting. Should the force be constant and act only in the direction of the displacement, the work done by the force is equal to the force times the total displacement as follows:

$$W = F\Delta \qquad \textbf{Eq. 11.1}$$

On many occasions, the force changes in magnitude, as does the deformation. For such a situation, it is necessary to sum up the small increments of work for which the force can be assumed to be constant, that is

$$W = \int F\, d\Delta \qquad \textbf{Eq. 11.2}$$

The *conservation of energy principle* is the basis of all energy methods. When a set of external loads is applied to a deformable structure, the points at which the loads are applied move. The result is that the members or elements making up the structure become deformed. According to the conservation of energy principle, the work done by the external loads, W, is equal to the work done by the internal forces acting on the elements of the structure, U. Thus:

$$W = U \qquad \textbf{Eq. 11.3}$$

As the deformation of a structure takes place, the internal work—commonly referred to as strain energy—is stored within the structure as potential energy. If the elastic limit of the material is not exceeded, the elastic strain energy will be sufficient to return the structure to its original undeformed state when the loads are removed. Should a structure be subjected to more than one load, the total energy stored in the structure will equal the sum of the energies stored in the structure as caused by each of the loads.

The conservation of energy principle is applicable only when static loads are applied to elastic systems. If the loads are not applied gradually, acceleration may occur and some of the external work will be transferred into kinetic energy. If inelastic strains are present, some of the energy will be lost in the form of heat.

11.3 VIRTUAL WORK OR COMPLEMENTARY VIRTUAL WORK

With the virtual work principle, a system of forces in equilibrium is related to a compatible system of displacements in a structure. The word *virtual* means "in essence but not in fact." Thus, the virtual quantities discussed in this chapter do not really exist. When we speak of a virtual displacement we are speaking of a fictitious displacement imposed on a structure. *The work performed by a set of real forces during a virtual displacement is called virtual work.*

We will see that for each virtual work theorem there is a corresponding theorem that is based on complementary virtual work. As a result the method is sometimes referred to as complementary virtual work.

Virtual work is based on the principle of virtual velocities, which was introduced by Johann Bernoulli of Switzerland in 1717. The virtual work theorem can be stated as follows:

> If a displacement is applied to a deformable body that is in equilibrium under a known load or loads, the external work performed by the existing load or loads due to this new displacement is equal to the internal work performed by the stresses existing in the body that were caused by the original load or loads.

The virtual work or complementary work method also is often referred to as the *dummy-load* or *unit-load method*. The name dummy-load or unit-load method is used because a fictitious or dummy load, usually of unit magnitude, is used in the solution.

Virtual work is based on the law of conservation of energy. To make use of this law in the derivation to follow, it is necessary to make the following assumptions:

1. The external and internal forces are in equilibrium.
2. The elastic limit of the material is not exceeded.
3. There is no movement of the supports.

Should a load F_1 be gradually applied to the bar shown in Figure 11.1(a) so that the load and the bar's deformation (or increase in length) increases from zero, the force

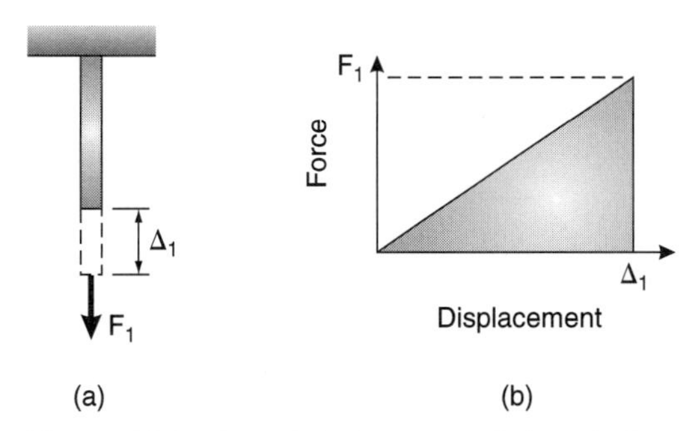

Figure 11.1 Force-Displacement relationship for an elastic bar Force-Displacement curve

displacement curve is shown in Figure 11.1(b). The bar will elongate an amount Δ_1 under the influence of the load F_1. The external work done by the load is equal to the area under the force-displacement curve, the shaded area in Figure 11.1(b), which is equal to:

$$W = \tfrac{1}{2}F_1\Delta_1 \qquad \textbf{Eq. 11.4}$$

Should another force F_2, in addition to the force F_1, be gradually applied to the bar as shown in Figure 11.2(a), the deformation will increase by an amount Δ_2. The resulting force-displacement curve is shown in Figure 11.2(b). Additional external work will be performed. The total external work is again equal to the area under the force-displacement curve, which is

$$W = \tfrac{1}{2}F_1\Delta_1 + F_1\Delta_2 + \tfrac{1}{2}F_2\Delta_2 \qquad \textbf{Eq. 11.5}$$

The first term on the right-hand side of Equation 11.5 is equal to the work done by F_1 as it is gradually applied. This is the area I in Figure 11.2(b). The third term on the right-hand side of the equation is the work done by F_2 as it is gradually applied. This is the area III in Figure 11.2(b). Because F_1 is present when F_2 is applied, it moves through the displacement Δ_2, and the work it does is the second term or the right-hand side of Equation 11.5. This work is the Area II in Figure 11.2(b), and is referred to as virtual work if the force F_2 is a virtual (not real) force.

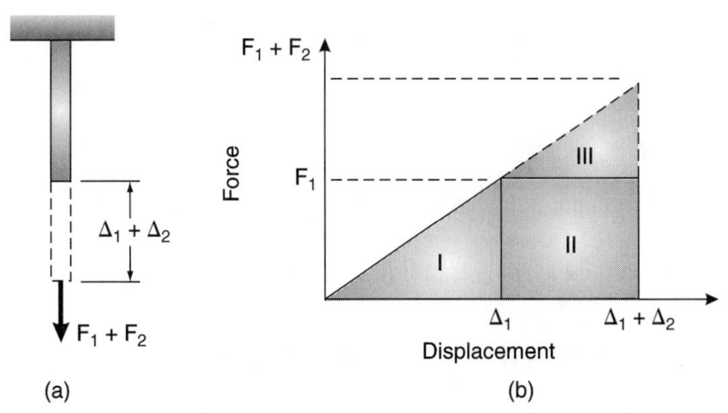

Figure 11.2 Force-Displacement relationship for an elastic bar with multiple loads

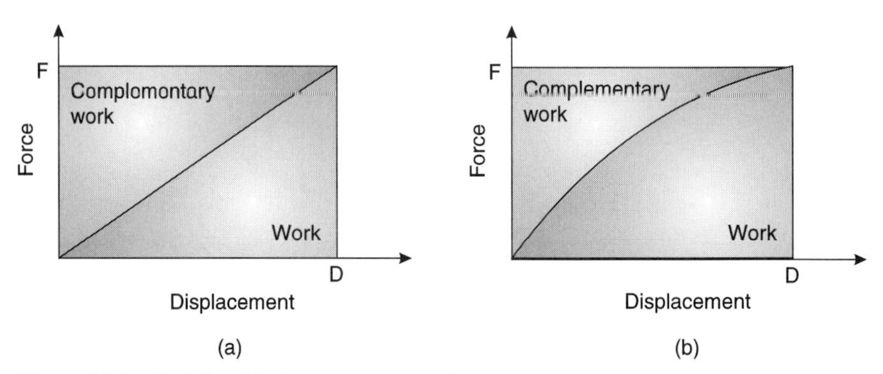

Figure 11.3 Relationship between work and complementary work

Another type of work is complementary work. It is equal to the area above the force-displacement curve as shown in Figure 11.3(a). In this textbook, complementary work is denoted with the symbol $\overline{W}$. Complementary work does not have a physical meaning as does the external work, but we can see that for a gradually applied load with linear system response

$$\overline{W} = W = \tfrac{1}{2}F\Delta \qquad \textbf{Eq. 11.6}$$

For the system and forces shown in Figure 11.2(a), the complementary work is

$$\overline{W} = \tfrac{1}{2}F_1\Delta_1 + F_2\Delta_1 + \tfrac{1}{2}F_2\Delta_2 \qquad \textbf{Eq. 11.7}$$

The second term on the right-hand side of this equation is referred to as complementary virtual work if the force F_2 is a virtual force.

Thus, $\overline{W}$ is the complement of the work W because it completes the rectangle. As long the structure behaves linearly and elastically, work and complementary work are equal. Should the system response not be linear, however, the load-displacement curve will not be a straight line as shown in Figure 11.3(b). As can be seen in that figure, work and complementary work are not equal.

From the principle of conservation of energy previously discussed, the complementary virtual work is equal to the complementary virtual strain energy, $\overline{U}_v$, in the structure, namely

$$\overline{W}_v = \overline{U}_v = F_2\Delta_1 \qquad \textbf{Eq. 11.8}$$

From this expression, we observe that if we can compute the virtual stain energy we can compute the structural displacement Δ_1 as

$$\Delta_1 = \frac{\overline{U}_v}{F_2} \qquad \textbf{Eq. 11.9}$$

The displacement at the point of application is Δ_1 and is in the same direction as the virtual force F_2. In the following sections of this chapter, we will use this principle to compute displacements in various types of structures.

11.4 TRUSS DEFLECTIONS BY VIRTUAL WORK

The truss shown in Figure 11.4 will be considered for this discussion. Loads P_1 to P_3 are applied to the truss as shown and cause forces in the truss members. Each member of the truss shortens or lengthens depending on the character of the force acting

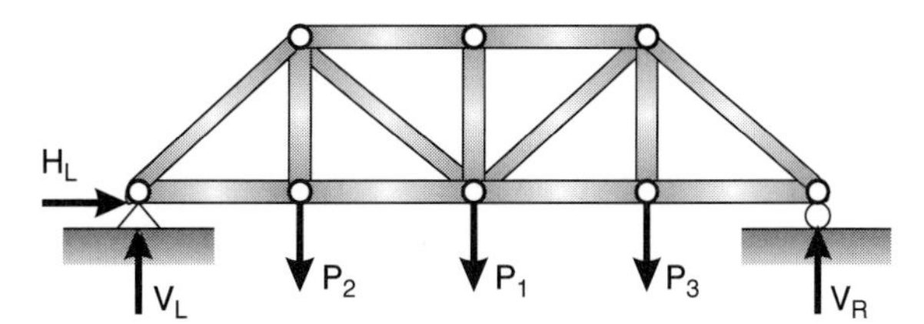

Figure 11.4 A typical truss for which deflections are computed

in it. These internal deformations cause external deflections, and each external load moves through a short distance. In this section, the principle of complementary virtual work is used to determine truss deflections, with a virtual force having unit magnitude.

To apply this method we must be able to write an expression for the strain energy in the structure. To do so, we need an expression for the deformation of a bar in a truss. That expression, from previous knowledge of mechanics, is

$$\delta = \frac{fL}{AE} \qquad \textbf{Eq. 11.10}$$

where δ is the deformation of the bar, f is the force in the bar, A is the area of the bar and E is the modulus of elasticity of the material. If the force f is a virtual force, the deformation δ is a virtual deformation. The strain energy (U) in a bar, then, is equal to

$$U = \frac{1}{2}f\left(\frac{fL}{AE}\right) = \left(\frac{1}{2}\right)\frac{f^2L}{AE} \qquad \textbf{Eq. 11.11}$$

For a truss composed of many bars, the strain energy in the structure is equal to the sum of the strain energies in each of the bars, namely

$$U = \frac{1}{2}\sum_{i=1}^{n}\left(\frac{f^2L}{AE}\right)_i \qquad \textbf{Eq. 11.12}$$

Now we are ready to begin computing displacements in trusses. To determine displacements using the principle of complementary virtual work, two loading systems are considered. The first system consists of the structure subjected to the force or forces for which the deflections are to be calculated. The second system consists of the structure subjected only to the virtual force acting at the point and in the direction in which displacement is desired. We will consider a virtual force having a magnitude of unity. It is placed on the structure at the location and in the direction of which we want to compute displacement.

The complementary virtual strain energy in a bar is equal to the deformation of the bar caused by the real forces times the force in the bar caused by the virtual force. The complementary virtual strain energy in a truss composed of several bars, then, is

$$\overline{U}_v = \sum_{i=1}^{n}\left[\left(\frac{fL}{AE}\right)f_v\right]_i \qquad \textbf{Eq. 11.13}$$

The first term inside the summation is the deformation of a bar caused by the applied loads. The second term is the force in the bar caused by the virtual force applied to the structure. Using Equation 11.9, the displacement of the truss at and in the same direction as the virtual force F_v is

$$\Delta = \frac{\sum_{i=1}^{n}\left[\left(\frac{fL}{AE}\right)f_v\right]_i}{F_v} \qquad \textbf{Eq. 11.14}$$

The advantage of using a virtual force with a unit magnitude should be obvious from this equation—we will be dividing by unity. Let us now apply this principle to the computation of displacements in trusses.

11.5 APPLICATION OF VIRTUAL WORK TO TRUSSES

Examples 11.1 and 11.2 illustrate the application of virtual work to trusses. In each case the forces due to the external loads are computed initially. Next, the external loads are removed and a unit load is placed at the point and in the direction in which deflection is desired (not necessarily horizontal or vertical). The forces due to the unit load are determined, and, finally, the value of fL/AE for each of the members is found. To simplify the analysis, a table is used. The modulus of elasticity is carried through as a constant until the summation is made for all of the members, at which time its numerical value is used. Should there be members of different moduli of elasticity, the actual or relative values of the moduli must be used in the individual operations. A positive resulting value indicates a deflection in the direction of the unit load; a negative value indicates the deflection is in the opposite direction.

Aluminum trusses, Reynolds Hangar, Byrd field, Richmond, Virginia (Courtesy of Reynolds Metals Company)

EXAMPLE 11.1

Determine the horizontal and vertical components of deflection at joint L_4 in the truss shown in the figure. The number next to each bar is the area of the bar in square inches, $E = 29 \times 10^6$ psi.

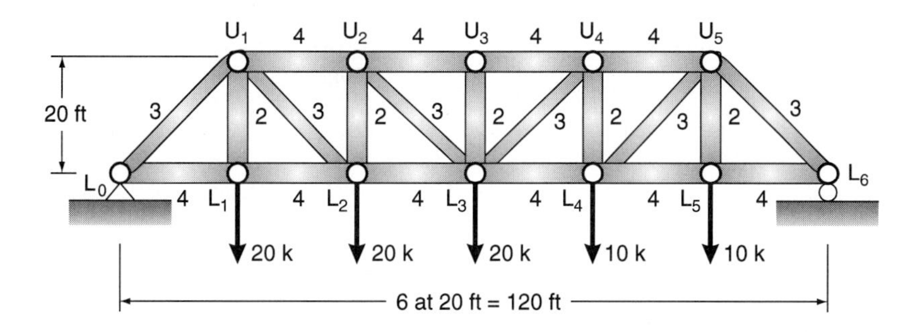

Solution. First determine the forces in the bars caused by the applied loads. This can be accomplished using any of the methods discussed in Chapter 6. The result is

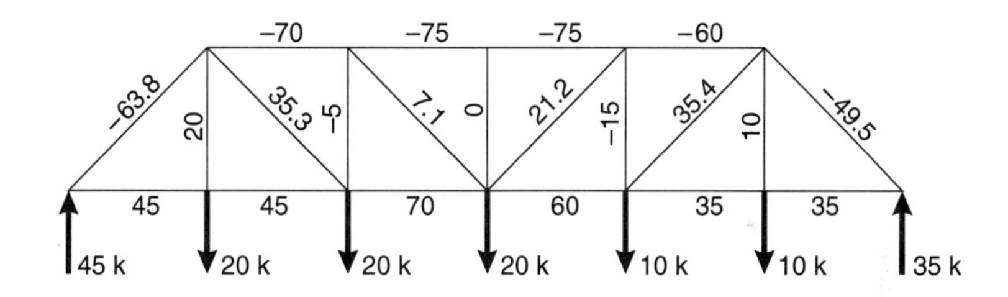

Determine the force in the bars caused by the vertical virtual unit force at L_4:

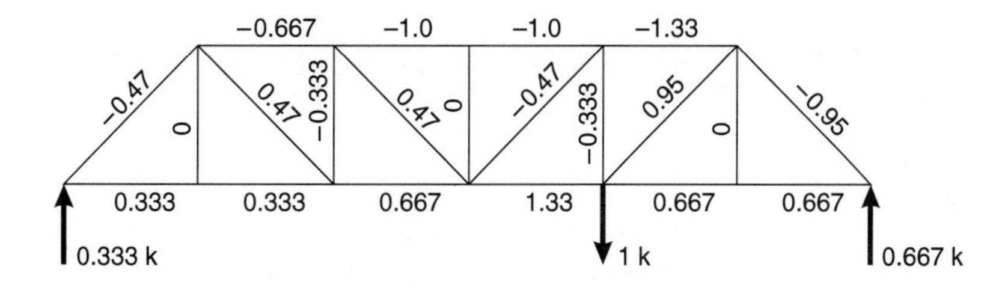

Determine the force in the bars caused by the horizontal virtual unit force at L_4:

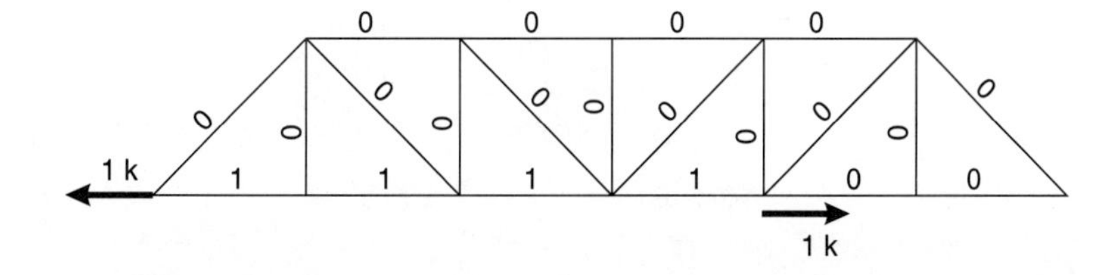

Next, determine the virtual strain energy in the truss in each case.

					Vertical Displacement		Horizontal Displacement	
Member	L (in)	A (in^2)	L/A	f (kips)	f_v (kips)	$fL(f_v)/A$	f_h (kips)	$fL(f_h)/A$
L_0L_1	240	4	60.0	45	0.333	900	1.0	2700
L_1L_2	240	4	60.0	45	0.333	900	1.0	2700
L_2L_3	240	4	60.0	70	0.667	2800	1.0	4200
L_3L_4	240	4	60.0	60	1.33	4800	1.0	3600
L_4L_5	240	4	60.0	35	0.667	1400	0	0
L_5L_5	240	4	60.0	35	0.667	1400	0	0
L_0U_1	340	3	113.3	−63.8	−0.47	3400	0	0
U_1U_2	240	4	60.0	−70	−0.667	2800	0	0
U_2U_3	240	4	60.0	−75	−1.0	4500	0	0
U_3U_4	240	4	60.0	−75	−1.0	4500	0	0
U_4U_5	240	4	60.0	−60	−1.33	4800	0	0
U_5L_6	340	3	113.3	−49.5	−0.95	5340	0	0
U_1L_1	240	2	120.0	20	0	0	0	0
U_1L_2	340	3	113.3	35.3	0.47	1880	0	0
U_2L_2	240	2	120.0	−5	−0.33	200	0	0
U_2L_3	340	3	113.3	7.1	0.47	378	0	0
U_3L_3	240	2	120.0	0	0	0	0	0
L_3U_4	340	3	113.3	21.2	−0.47	−1130	0	0
U_4L_4	240	2	120.0	−15	0.333	−600	0	0
L_4U_5	340	3	113.3	35.4	0.95	3810	0	0
U_5L_5	240	2	120.0	10	0	0	0	0
						42,078		13,200

$$(\Delta_{L_4})_V = \frac{42{,}078}{29{,}000} = 1.45 \text{ in} \downarrow$$

$$(\Delta_{L_4})_H = \frac{13{,}200}{29{,}000} = 0.455 \text{ in} \rightarrow$$

Both displacements are positive so they are in the direction of the applied virtual forces. ■

EXAMPLE 11.2

Determine the vertical component of deflection of joint L_4 in the following figure using the method of virtual work. The number next to each bar is the area of the bar in square inches; $E = 29 \times 10^6$ psi.

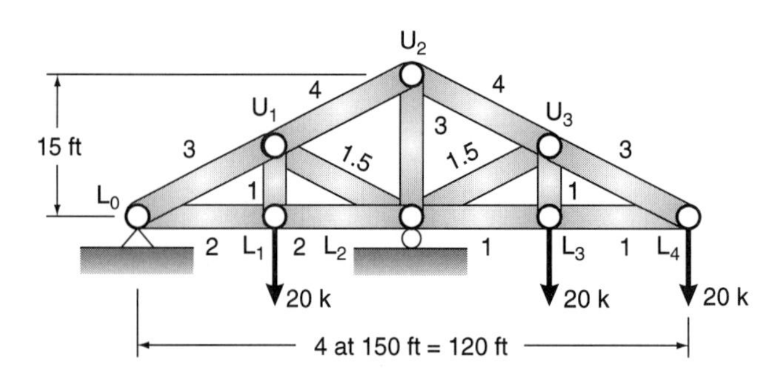

Solution. First, compute the forces in the bars due to the external loads:

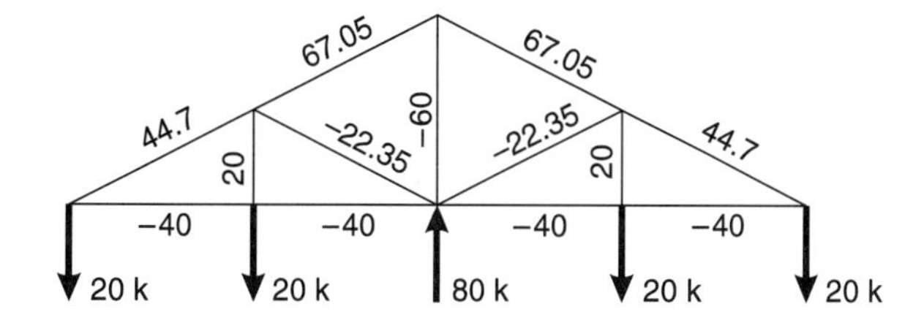

Next compute the forces in the bars caused by a vertical virtual unit force at L_4:

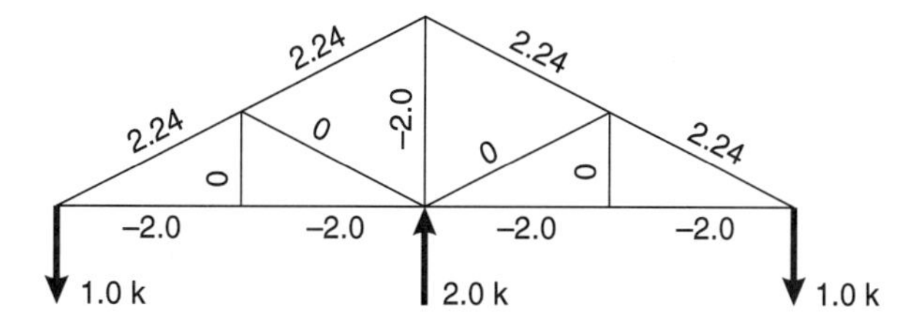

Next, determine the virtual strain energy in the truss in each case:

Member	L (in)	A (in^2)	L/A	f (kips)	f_v (kips)	fLf_v/A
L_0L_1	180	2.0	90.0	−40.0	−2.0	7,200
L_1L_2	180	2.0	90.0	−40	−2.0	7,200
L_2L_3	180	1.0	180.0	−40	−2.0	14,400
L_3L_4	180	1.0	180.0	−40	−2.0	14,400
L_0U_1	202	3.0	67.3	44.7	2.24	6,730
U_1U_2	202	4.0	50.5	67.05	2.24	7,575
U_2U_3	202	4.0	50.5	67.05	2.24	7,575
U_3U_4	202	3.0	67.3	44.7	2.24	6,730
U_1L_1	90	1.0	90.0	20.0	0	0
U_1L_2	202	1.5	134.5	−22.35	0	0
U_2L_2	180	3.0	60.0	−60.0	−2.0	7,200
L_2U_3	202	1.5	134.5	−22.35	0	0
U_3L_3	90	1.0	90.0	20.0	0	0
						79,010

Now compute the vertical deflection at L_4.

$$(\Delta_{L4})_V = \frac{79{,}010}{29{,}000} = 2.72 \text{ in} \downarrow$$

The value is positive, so the resulting displacement is downward. The resulting displacement is in the same direction as the applied virtual force. ■

11.6 DEFLECTIONS OF BEAMS AND FRAMES BY VIRTUAL WORK

The law of conservation of energy may be used to develop an expression for the deflection at any point in a beam or frame. Each fiber in the beams or columns of a structure can be thought of as a very small member similar to the members of the trusses considered in

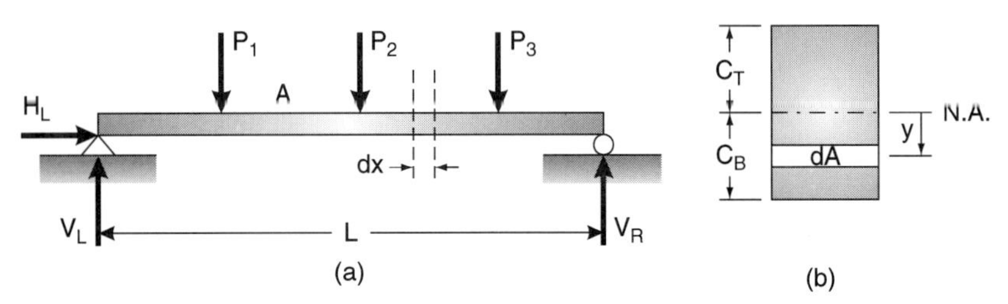

Figure 11.5 System for development of virtual work for beam

the preceding sections. Each member has a length dx and an area of dA. The summation of the virtual strain energy in each of these small members must equal the virtual work performed by the external loads.

For the following discussion, the beam of Figure 11.5(a) is considered. Shown in Figure 11.5(b) is the beam cross section. We want to determine the vertical deflection, Δ, at point A in the beam caused by the external loads P_1 to P_3. To do so, we will place a unit virtual force at A and compute the virtual work done by that load and the virtual strain energy caused by that load. Recall from the previous discussion of trusses that the virtual strain energy in a particular member is equal to the force in the member caused by the virtual unit force times the deformation of the member caused by the external loads. The total virtual strain energy in the truss is equal to the summation of the virtual strain energy in each member.

The following symbols are used in writing an expression for the virtual strain energy in the beam:

M—the moment at any section in the beam caused by the external loads, and

m—the moment at any section in the beam caused by the unit virtual force.

We will begin the discussion by computing the virtual force in dA caused by the applied unit virtual force. The stress in a differential area of the beam cross section caused by applied virtual force can be found from the flexure formula as follows:

$$\sigma_v = \frac{my}{I}$$

$$f = \sigma_v \, dA = \frac{my}{I} dA \qquad \textbf{Eq. 11.15}$$

Continuing, the deformation of the differential area dA caused by the applied loads is determined as follows:

$$\sigma = \frac{My}{I}$$

$$\delta = \varepsilon \, dx = \frac{\sigma}{E} dx = \frac{1}{E}\left(\frac{My}{I}\right) dx = \frac{My}{EI} dx \qquad \textbf{Eq. 11.16}$$

The virtual strain energy in the differential, then, is equal to

$$u_v = f_v \delta = \left(\frac{my}{I}\right)\left(\frac{My}{EI}\right) dA \, dx = \frac{mMy^2}{EI^2} dA \, dx \qquad \textbf{Eq. 11.17}$$

This is the incremental virtual strain energy. The virtual strain energy in the beam is equal to the incremental strain integrated over volume, namely

$$U_v = \int\int_V \frac{mMy^2}{EI^2} dA\, dx \qquad \textbf{Eq. 11.18}$$

However, from previous knowledge of structural mechanics, we know that

$$I = \int_A y^2 dA \qquad \textbf{Eq. 11.19}$$

If we substitute Equation 11.19 into Equation 11.18, we find that:

$$U_v = \int_L \frac{mM}{EI} dx$$

The virtual work performed by a unit virtual load at A as it moves through the displacement Δ, the displacement at A caused by the external loads, is:

$$W_v = 1(\Delta) \qquad \textbf{Eq. 11.20}$$

By equating the virtual work and the virtual strain energy, an expression for the deflection at point A in the beam, the point at which the unit virtual force was applied, is obtained. That expression is

$$W_v = U_v$$

$$1(\Delta) = \int_L \frac{Mm}{EI} dx \qquad \textbf{Eq. 11.21}$$

$$\Delta = \int_L \frac{Mm}{EI} dx$$

These principles will be applied to beams and frames to compute displacements, deflections and rotations, in beams and frames in the following section.

11.7 EXAMPLE PROBLEMS FOR BEAMS AND FRAMES

Examples 11.3 to 11.7 illustrate the application of virtual work to beams and frames. To apply the method, a unit virtual force is placed at the point and in the direction in which deflection is desired. Expressions are written for M and m throughout the structure and the results are integrated over the length of the structure. To apply these principles, the ability to write equations for the bending moment in the structure caused by the virtual and applied forces is necessary. The principles for writing moment equations were developed in Chapter 5 and should be referred to, if necessary.

EXAMPLE 11.3

Determine the deflection at the free end of the beam shown in the figure using virtual work. For this problem, $I = 5{,}000\ in^4$ and $E = 29 \times 10^3$ ksi.

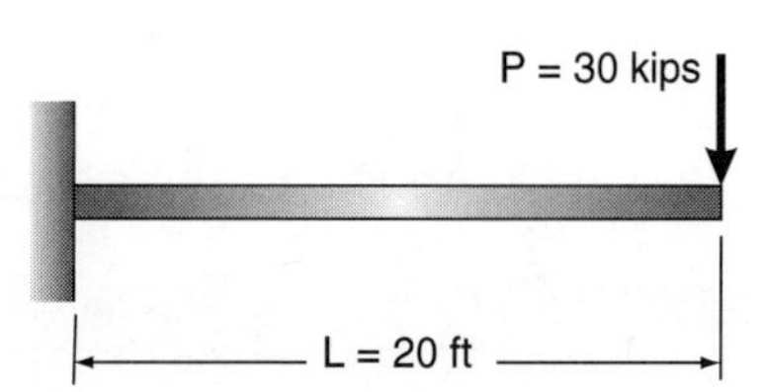

Solution. An expression is written for M_x, the moment caused by the 30-kip load at some location x along the beam. The free-body diagram we will use is shown here. The distance x is measured from the free end.

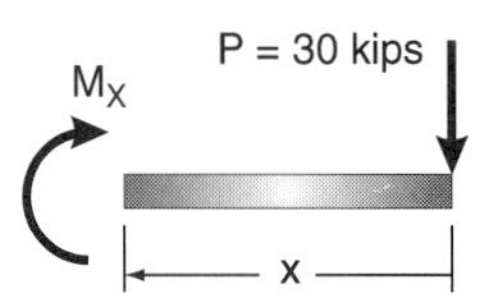

The resulting expression for M is

$$M_x = -Px$$

A unit virtual force is placed at the free end as shown in the next figure. Using the free-body diagram, an expression is written for the moment m_x at some location x along the beam caused by this unit virtual force.

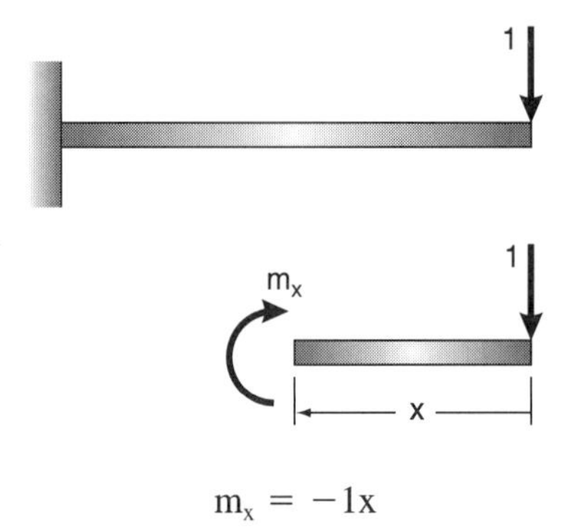

$$m_x = -1x$$

Note that the origin of x may be selected at any point as long as the same point is used for writing M and m for that portion of the beam.

Next we can compute the deflection at the free end of the cantilever.

$$(1)\ \Delta = \int_0^L \frac{M_x m_x}{EI} dx = \int_0^L \frac{(-Px)(-1x)}{EI} dx = \int_0^L \frac{Px^2}{EI} dx$$

$$\Delta = \frac{P}{EI}\left[\frac{x^3}{3}\right]_0^L = \frac{PL^3}{3EI} = \frac{30[20(12)]^3}{3(29{,}000)(5{,}000)} = 0.955 \text{ in.}$$

The positive sign indicates that the deflection is downward because the virtual force is applied downward. ■

EXAMPLE 11.4

Determine the deflection at the free end of the beam shown in the figure using virtual work. For this problem, $I = 5{,}000 \text{ in}^4$ and $E = 29 \times 10^3$ ksi.

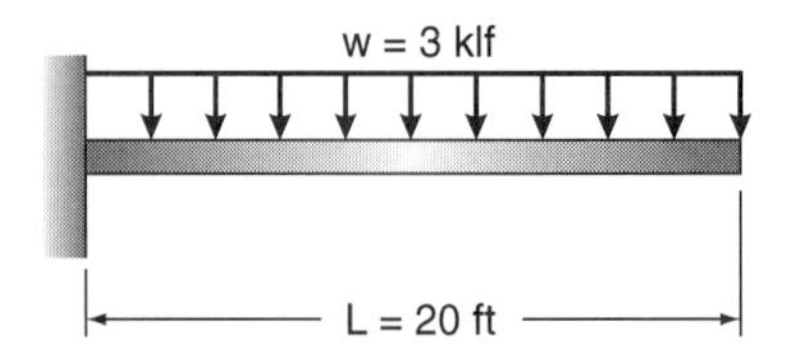

Solution. As in the previous example, an expression is written for M_x, the moment caused by the uniformly distributed load at some location x along the beam. The free-body diagram we will use is shown here. The distance x is measured from the free end.

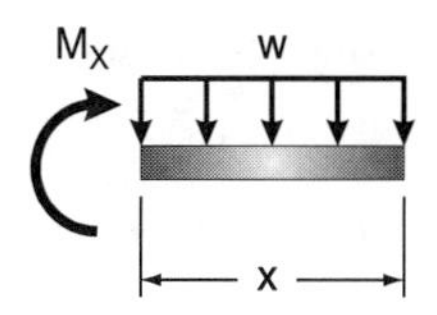

$$M_x = -wx\left(\frac{x}{2}\right) = \frac{-wx^2}{2}$$

Again, a unit virtual force is placed at the free end as shown in the next figure. Using the free-body diagram following after the figure, an expression is written for the moment m_x at some location x along the beam caused by this unit virtual force.

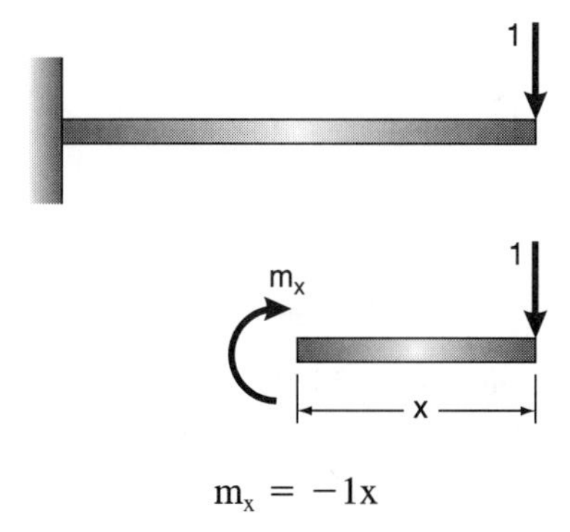

$$m_x = -1x$$

Next we can compute the deflection at the free end of the cantilever.

$$(1)\ \Delta = \int_0^L \frac{M_x m_x}{EI} dx = \int_0^L \frac{\left(\frac{-wx^2}{2}\right)(-1x)}{EI} dx = \int_0^L \frac{wx^3}{2EI} dx$$

$$\Delta = \frac{w}{2EI}\left[\frac{x^4}{4}\right]_0^L = \frac{wL^4}{8EI} = \frac{\left(\frac{3}{12}\right)[20(12)]^4}{8(29{,}000)(5{,}000)} = 0.714 \text{ in.}$$

The positive sign indicates that the deflection is downward because the virtual force is applied downward. ■

EXAMPLE 11.5

Using virtual work, determine the deflection in the beam shown here at the point of application of the 30 kip load. For this beam, $I = 1{,}000 \text{ in}^4$ and $E = 29 \times 10^3$ ksi.

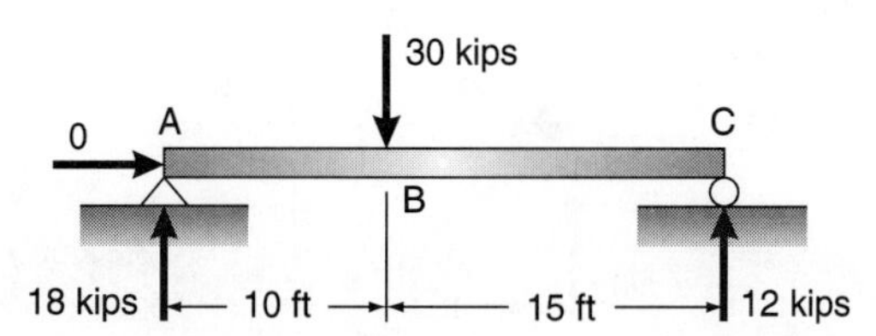

Solution. Because the concentrated load acting on the beam causes a discontinuity in the bending moment equations, one expression for the bending moment from A to B and another for the bending moment from B to C must be written. The same is true for m.

The bending moments caused by the applied load are

$$M_{AB} = 18x$$
$$M_{BC} = 18x - 30(x - 10) = -12x + 300$$

The virtual force that we will use to compute the deflection at the 30 kip load is shown as

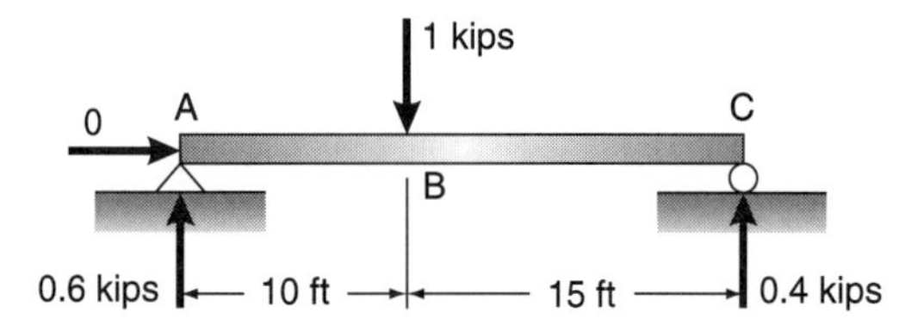

The equations for the bending moment in the beam caused by this virtual force are

$$m_{AB} = 0.6x$$
$$m_{BC} = 0.6x - 1(x - 10) = -0.4x + 10$$

The deflection at the 30 kip load can now be computed. That deflection is

$$1\Delta_B = \frac{1}{EI}\int_0^{10} M_{AB}m_{AB}\,dx + \frac{1}{EI}\int_{10}^{25} M_{BC}m_{BC}\,dx$$
$$= \frac{3{,}600}{EI} + \frac{25{,}000 - 19{,}600}{EI}$$
$$= \frac{9{,}000\text{ k-ft}^3}{EI} = \frac{9{,}000(1{,}728)}{(29{,}000)(1{,}000)}\text{ in}$$
$$\Delta = 0.536\text{ in.} \downarrow$$

The deflection is downward because the resulting sign is positive and the unit virtual force is applied in a downward direction, the implicit positive direction. ■

EXAMPLE 11.6

Determine the deflection at the free end of the cantilever on the beam shown in the figure using virtual work. For this problem, $I = 1{,}000\text{ in}^4$ and $E = 29 \times 10^3$ ksi.

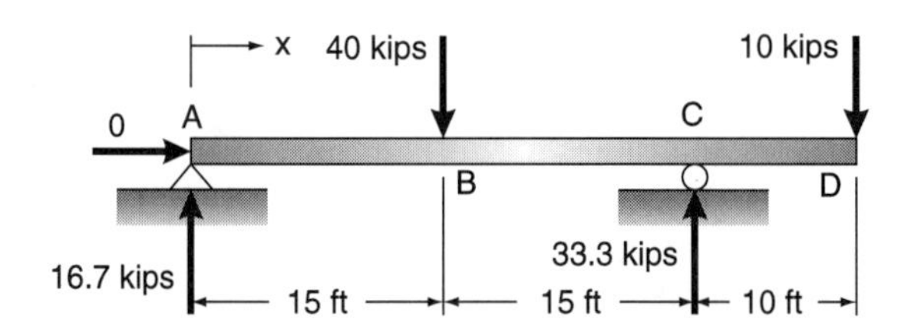

Solution. In solving this example problem, we will introduce the Heaviside step function, $\Phi(x)$. This function evaluates to zero when x is less than zero. Otherwise it evaluates to unity. For example, consider the following function evaluated at x = 5, x = 10, and x = 15:

$$f(x) = 5 + \Phi(x - 10)(10x)$$
$$f(5) = 5 + 0(10)5 = 5$$
$$f(10) = 5 + 1(10)(10) = 105$$
$$f(15) = 5 + 1(10)(15) = 155$$

By using this function we can write a single moment equation that applies over the entire beam.

We will write the equation for bending moment from the left support to the right support; the left support is used for the origin of x. The applicable free-body diagram is shown in the figure. We cut the section after the last change in force acting on the beam—either applied load or force of reaction.

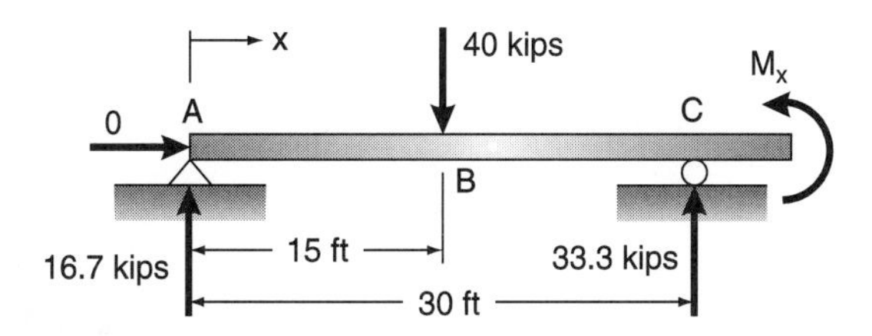

Using this free-body diagram, and incorporating the Heaviside step function, the equation for the bending moment caused by the external loads is

$$M_x = 16.7x - \Phi(x - 15)[40(x - 15)] + \Phi(x - 30)[33.3(x - 30)]$$

Because we want to compute the deflection at the free end of the cantilever portion of the beam, we will place a unit virtual force at that location as shown.

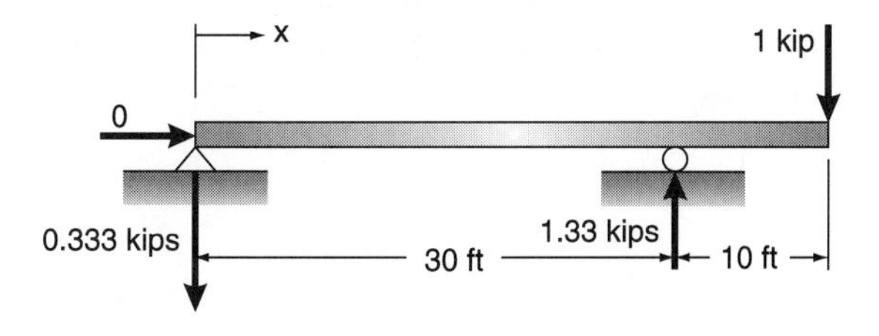

The free-body diagram that we will use to determine the equation for virtual bending moment is shown next. Again, we cut the section after the last change in load.

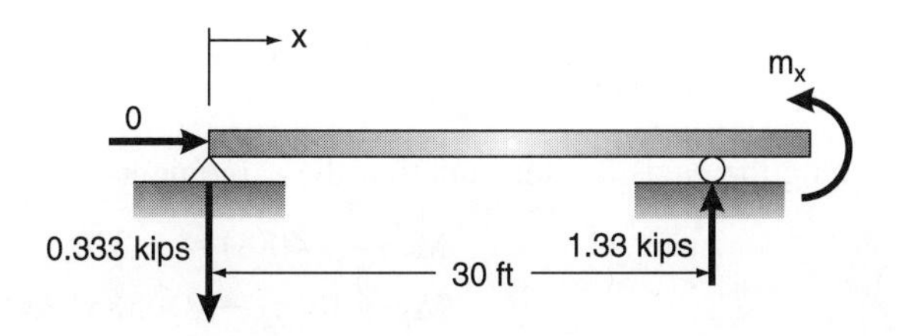

The virtual bending moment equation, with the heaviside step function, is

$$m_x = -0.333x + \Phi(x - 30)[1.33(x - 30)]$$

The deflection at the free-end of the cantilever is then computed as

$$\Delta = \frac{1}{EI}\int_0^{40} M_x m_x \, dx = \frac{-9309(1728)}{29{,}000(1{,}000)} = -0.555 \text{ in.}$$

The negative sign indicates that the computed deflection is in the opposite direction of the applied unit virtual force. The deflection, therefore, is 0.555 inches upward. ■

EXAMPLE 11.7

Determine the horizontal displacement at point D in the frame shown in the next figure. For this problem, E = 29,000 ksi. The values of the second moments of area for the different members that compose the frame are also shown.

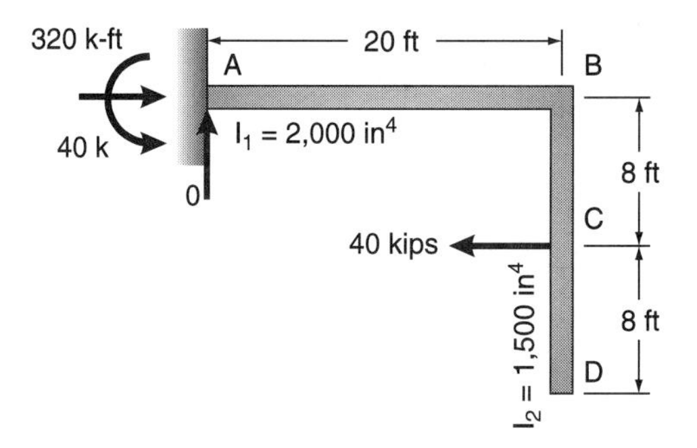

Solution. One equation for the bending moment caused by the external load will be written for the beam and another will be written for the column. The free-body diagrams that we will use to write these equations are shown. The directions in which position along the members will be taken are indicated on the free-body diagrams.

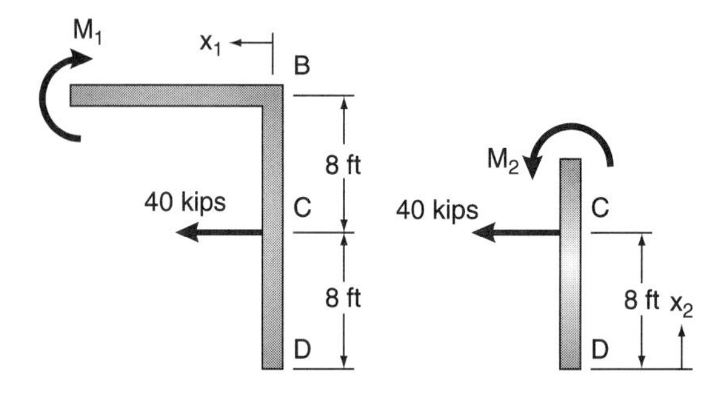

Using the eaviside step function these moment equations are

$$M_1 = -40(8) = -320$$
$$M_2 = \Phi(x_2 - 8)[40(x_2 - 8)]$$

Even though there is no moment in the column between the bottom and the 40-kip load, we needed to write the bending moment equation from the bottom because there will be virtual moment in the column starting from the bottom.

The virtual force that we will use in the solution is shown in the next figure. The force is placed in the horizontal direction at D because we want to compute the horizontal displacement at D.

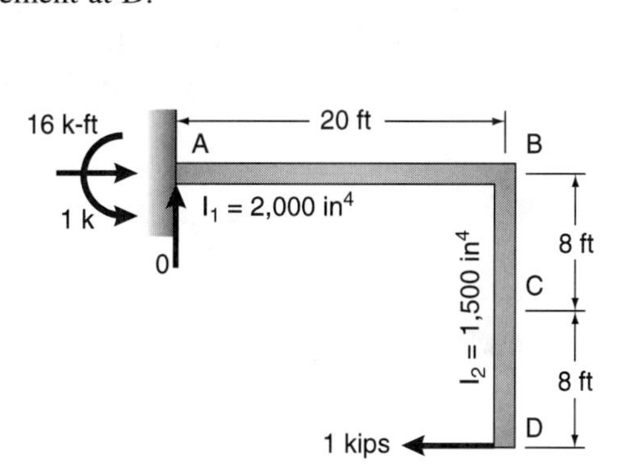

As with the bending moments caused by the external load, there will be two virtual bending moment equation—one for the beam and one for the column. The free-body diagrams that we will use to write the equations are shown here.

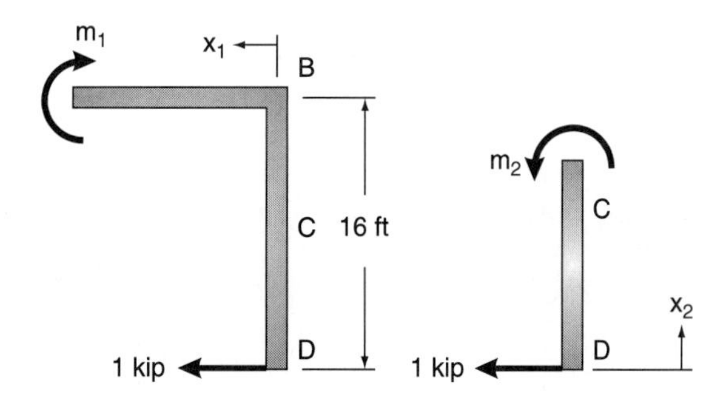

Using these free-body diagrams, the virtual bending moment equations are

$$m_1 = -1(16) = -16$$
$$m_2 = 1x_2$$

The horizontal deflection at D can then be found from

$$\Delta = \frac{1{,}728}{29{,}000(2{,}000)}\int_0^{20} -320(-16)\,dx_1 + \frac{1{,}728}{29{,}000(1{,}500)}\int_0^{16} \Phi(x_2 - 8)[40(x_2 - 8)][1(x_2)]\,dx_2$$

$$\Delta = 3.729\,\text{in}$$

The positive sign on the computed deflection indicates that it is to the left since that is the direction of the applied virtual force. ■

I-40, I-240 interchange, Oklahoma City, Oklahoma (Courtesy of the State of Oklahoma Department of Transportation)

11.8 ROTATIONS OR ANGLE CHANGES BY VIRTUAL WORK

In the previous section, we used virtual work to determine displacements in beams and frames. Virtual work may also be used to determine the rotations or slopes at various points in a structure. To compute the deflection of a particular location in the structure, we applied a unit virtual force at the location. Similarly, to compute a rotation, we apply a virtual unit moment at the location where the rotation is needed. This concept is shown in Figure 11.6 for finding the slope at point A in the beam.

The virtual bending moment m is the bending moment in the beam or structure caused by this unit virtual moment. Similarly to the manner in which we computed displacements, when a unit virtual moment is applied to the beam, the rotation is computed from

$$\theta = \int_L \frac{Mm}{EI} dx$$

A positive angle θ indicates that the rotation is in the same direction as the applied unit virtual moment. A negative sign indicates that the rotation is in the opposite direction. This principle is illustrated in the following two examples.

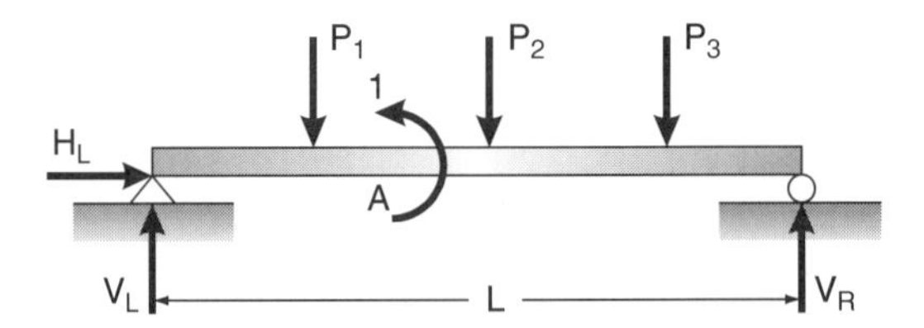

Figure 11.6 Use of virtual work to compute rotation

EXAMPLE 11.8

Find the slope at the free end of the cantilever beam shown in the figure.

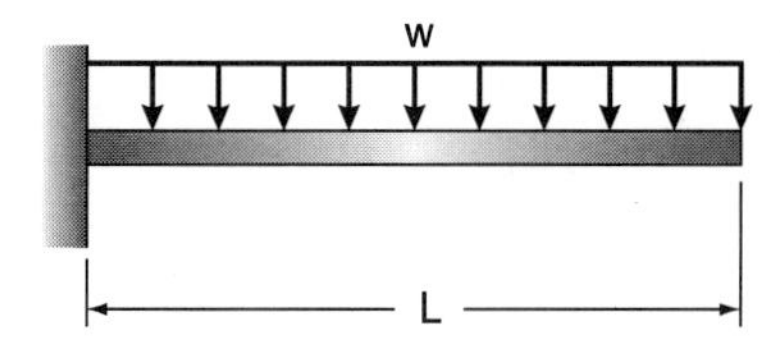

Solution. We must first determine the equations for bending moment in the beam caused by the applied distributed load. The free-body diagram that we will use is shown here.

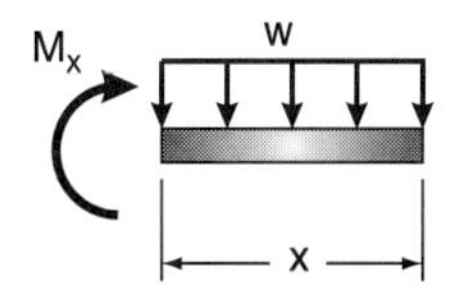

The bending moment equation for the applied load is

$$M_x = -wx\left(\frac{x}{2}\right) = \frac{-wx^2}{2}$$

Observe that this is the same free-body diagram and bending moment equation that we obtained in Example 11.4.

The virtual force system that we will use is shown in the following figure. We want to know the rotation at the end of the cantilever so we apply a unit virtual rotation at that point.

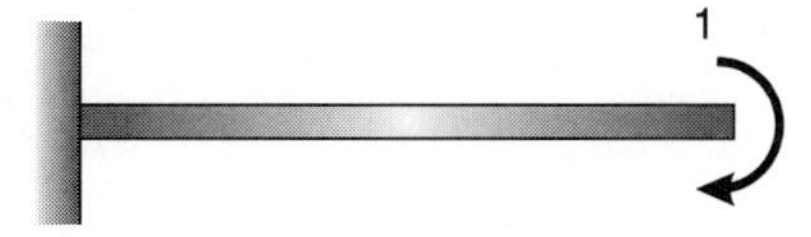

The free-body diagram that we use to determine the bending moment in the beam caused by this virtual moment is next.

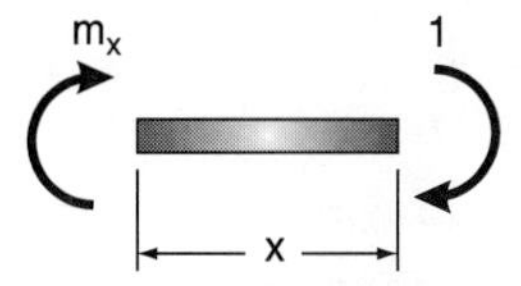

The equation for virtual bending moment is

$$m_x = -1$$

Throughout the entire beam the virtual bending moment is constant.

We can then compute the rotation from

$$\theta = \frac{1}{EI}\int_0^L Mm\,dx = \frac{1}{EI}\int_0^L\left(\frac{-wx^2}{2}\right)(-1)\,dx$$

$$\theta = \frac{wL^3}{6EI}\text{radians}$$

The positive sign indicates that the rotation is in the same direction as the applied virtual moment. The free end of the beam rotates in a clockwise direction. ■

EXAMPLE 11.9

Find the slope at the 30-kip load in the beam shown in the figure below. EI in this beam is constant.

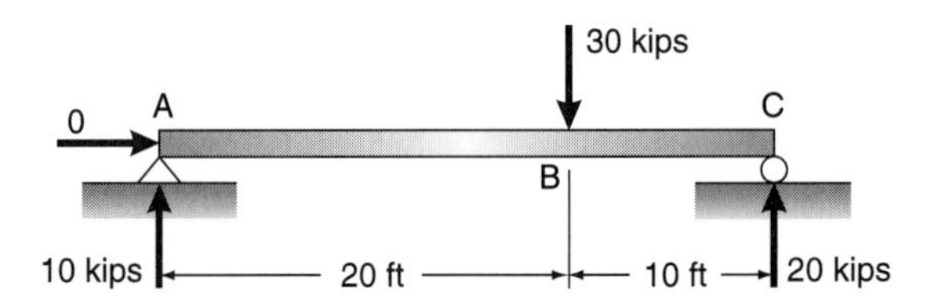

Solution. First we will write the equation for the bending moment in the beam caused by the applied load using the free-body shown next. For convenience in the solution, we will use the Heaviside step function.

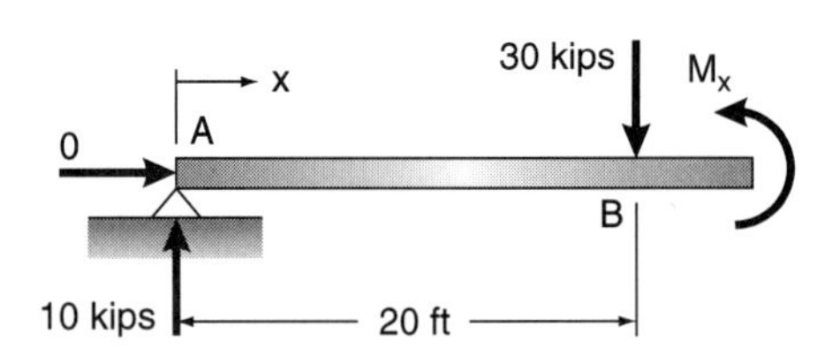

$$M'_x = 10x - \Phi(x - 20)[30(x - 20)]$$

To compute the slope at the location of the 30-kip load using virtual work we must apply a unit virtual force at that location:

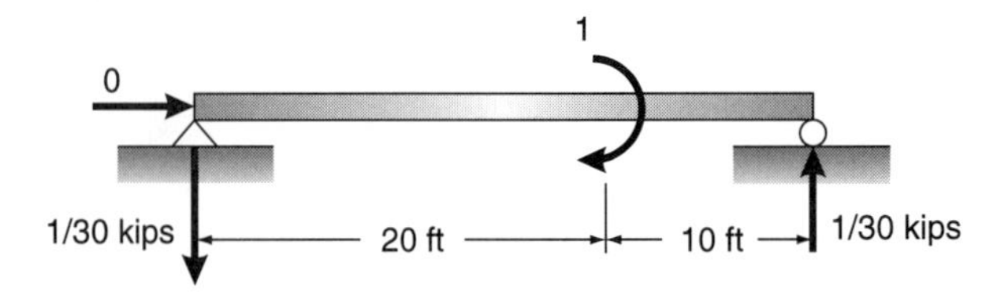

Using the following free-body diagram, we can determine the virtual bending moment in the beam caused by the unit virtual moment:

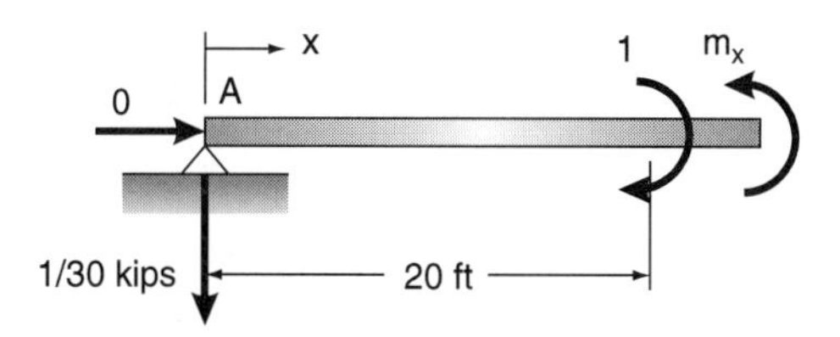

$$m_x = \frac{-1}{30}x + \Phi(x - 20)(1)$$

We can now compute the rotation at the point of application of the 30-kip load from

$$\theta = \frac{1}{EI}\int_0^L [10x - \Phi(x - 20)(30)(x - 20)]\left[\frac{-1}{30}x + \Phi(x - 20)(1)\right]dx$$

$$\theta = \frac{-666.7\ \text{ft}^2\text{-k}}{EI}$$

The negative sign indicates that the actual rotation is counter-clockwise, opposite the direction of the virtual unit moment. ■

11.9 INTRODUCTION TO CASTIGLIANO'S THEOREMS

Alberto Castigliano, an Italian railway engineer, published an original and elaborate book in 1879 on the study of statically indeterminate structures.[1] In this book were included the two theorems that are known today as Castigliano's first and second theorems. The first theorem, commonly known as the method of least work, is discussed in Chapter 13. It has played an important role historically in the development of the analysis of statically indeterminate structures.

Castigliano's second theorem, which provides an important method for computing deflections, is considered in this section and the next. Its application involves equating deflection to the first partial derivative of the total strain energy of the structure with respect to a load at the point where deflection is desired. In detail, the theorem can be stated as follows:

> For any linear elastic structure subject to a given set of loads, constant temperature, and unyielding supports, the first partial derivative of the strain energy with respect to a particular force will equal the displacement of that force in the direction of its application.

In presenting the virtual-work method for beams and frames in Section 11.6, the following equation was derived:

$$U = \int \frac{Mm}{EI}dx \qquad \textbf{Eq. 11.22}$$

[1]A. Castigliano, *Théorie de l' équilibre des systèmes élastique et ses applications,* Turin, 1879. (There was an English translation by E. S. Andrews entitled *Elastic Stresses in Structures,* published by Scott, Greenwood & Son in London in 1919. This translation was then reprinted in the United States by Dover Publications, Inc., New York, and entitled *The Theory of Equilibrium of Elastic Systems and Its Application.*)

This is an expression for the virtual strain energy caused by a virtual unit load placed at the point and in the direction of a desired deflection when the external loads have been replaced on the beam.

The internal work of the actual stresses caused by gradually applied external loads can be determined in a similar manner. The work in each fiber equals the average stress, (My/2I) da (as the external loads vary from zero to full value), times the total strain in the fiber, (My/EI) dx. Integrating the product of these two expressions for the cross section of the member and throughout the structure gives the total internal work or strain energy stored for the entire structure.

$$U = \int_L \frac{M^2}{2EI} dx \qquad \textbf{Eq. 11.23}$$

In a similar manner, the strain energy in the members of a truss caused by a set of gradually applied loads can be shown to equal

$$U = \sum \frac{f^2 L}{2AE} \qquad \textbf{Eq. 11.24}$$

These expressions will be used frequently in the following section, in which a detailed discussion of Castigliano's second theorem is presented.

11.10 CASTIGLIANO'S SECOND THEOREM

As a general rule the other methods of deflection computation (such as moment–area or virtual-work) are a little easier to use and more popular than Castigliano's second theorem. For certain structures this method is very useful, and from the standpoint of the student its study serves as an excellent background for the extremely important fist theorem.

The derivation of the second theorem (presented in this section) is very similar to that given by Kinney.[2] Shown in Figure 11.7 is a beam that has been subjected to the gradually applied loads P_1 and P_2. These loads cause the deflections Δ_1 and Δ_2. We want to determine the magnitude of the deflection Δ_1 at load P_1.

The external work performed during the application of the loads must equal the average load times the deflection. It also must equal the strain energy of the beam.

$$W = \frac{P_1 \delta_1}{2} + \frac{P_2 \delta_2}{2} \qquad \textbf{Eq. 11.25}$$

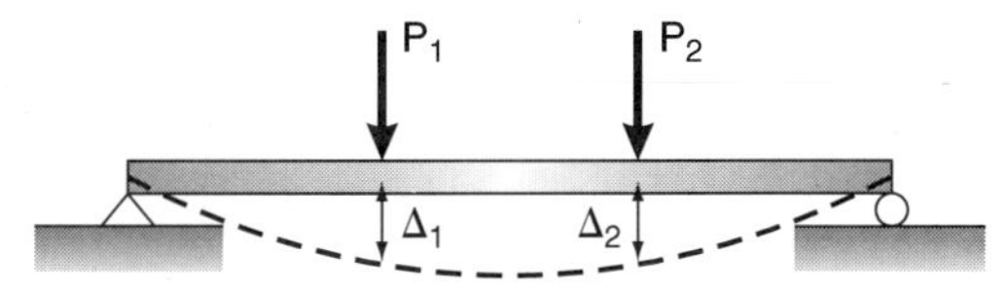

Figure 11.7 Beam for development of Castigliano's second theorem

[2] J. S. Kinney, *Indeterminate Structural Analysis* (Reading, Mass.: Addison-Wesley, 1957), 84–86.

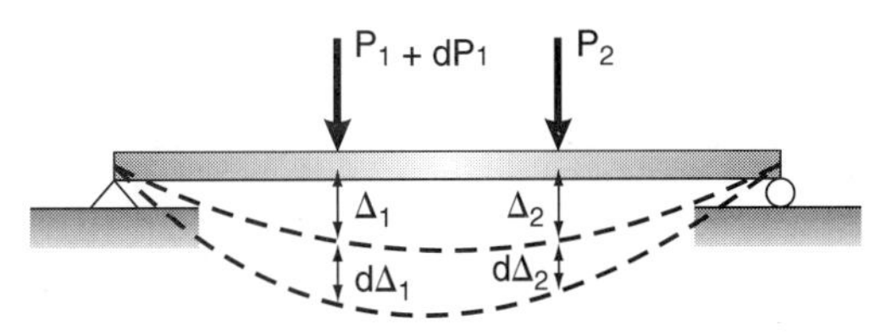

Figure 11.8 Addition deflection caused by dP_1

Should the load P_1 be increased by the small amount dP_1, the beam will deflect an additional amount. Figure 11.8 illustrates the additional deflections, $d\Delta_1$ and $d\Delta_2$, at each of the loads.

The additional work performed, or strain energy stored, during the application of dP_1 is as follows:

$$dW = \left(P_1 + \frac{dP_1}{2}\right) d\Delta_1 + P_2 d\Delta_2 \qquad \textbf{Eq. 11.26}$$

This equation represents the variation in work caused by a small variation in load. Performing the indicated multiplication and neglecting the product of the differentials, which are negligibly small, we find that:

$$dW = P_1 d\Delta_1 + P_2 d\Delta_2 \qquad \textbf{Eq. 11.27}$$

Now, instead of adding dP_1 after the loads P_1 and P_2, let's place all three loads on the beam at the same time. In doing so, the work done by the loads can be represented by

$$W = \left(\frac{P_1 + dP_1}{2}\right)(\Delta_1 + d\Delta_1) + \left(\frac{P_2}{2}\right)(\Delta_2 + d\Delta_2) \qquad \textbf{Eq. 11.28}$$

Again, if we perform the indicated multiplication and neglect the product of the differentials, we obtain the equation:

$$W = \frac{P_1\Delta_1}{2} + \frac{P_1 d\Delta_1}{2} + \frac{dP_1\Delta_1}{2} + \frac{P_2\Delta_2}{2} + \frac{P_2 d\Delta_2}{2} \qquad \textbf{Eq. 11.29}$$

The change in the work done in this case can be obtained by subtracting Equation 11.25 from Equation 11.29. In doing so we find that:

$$dW = \frac{P_1\Delta_1}{2} + \frac{P_1 d\Delta_1}{2} + \frac{dP_1\Delta_1}{2} + \frac{P_2\Delta_2}{2} + \frac{P_2 d\Delta_2}{2} - \frac{P_1\Delta_1}{2} - \frac{P_2\Delta_2}{2}$$

$$dW = \frac{P_1 d\Delta_1}{2} + \frac{dP_1\Delta_1}{2} + \frac{P_2 d\Delta_2}{2} \qquad \textbf{Eq. 11.30}$$

From Equation 11.27, we find that the $P_2 d\Delta_2$ is

$$P_2 d\Delta_2 = dW - P_1 d\Delta_1$$

By substituting this result into Equation 11.30, and solving for the deflection Δ_1, the deflection we wanted to compute, we find that

$$\Delta_1 = \frac{dW}{dP_1} \qquad \textbf{Eq. 11.31}$$

Since the work done by the external loads is equal to the strain energy in the structure caused by the loads, and since more than one load is usually applied to a structure, this deflection can be written as a partial derivative in terms of strain energy as follows:

$$\Delta = \frac{\partial W}{\partial P} = \frac{\partial U}{\partial P} \qquad \textbf{Eq. 11.32}$$

This is a mathematical statement of Castigliano's second theorem. By substituting Equation 11.23 into Equation 11.32, the deflection at a point in a beam or frame using Castigliano's second theorem can be written as

$$\Delta = \frac{\partial}{\partial P} \int_L \frac{M^2}{2EI} dx \qquad \textbf{Eq. 11.33}$$

When applying the method represented by this equation, M would be squared and integrated; then the first partial derivative would be taken. If M is rather complicated, as it frequently is, the process becomes very tedious. For this reason usually it is simpler to differentiate under the integral sign with the following results:

$$\Delta = \int_L \frac{M}{EI}\left(\frac{\partial M}{\partial P}\right) dx \qquad \textbf{Eq. 11.34}$$

For a truss, the corresponding relationships are:

$$\Delta = \frac{\partial}{\partial P} \sum \frac{f^2 L}{2AE}$$
$$\Delta = \sum \frac{fL}{AE}\left(\frac{\partial f}{\partial P}\right) \qquad \textbf{Eq. 11.35}$$

Examples 11.10 through 11.12 illustrate the application of Castigliano's second theorem. In applying the theorem, the load at the point where the deflection is desired is referred to as P. After the operations required in the equation are completed, the numerical value of P is replaced in the expression. Should there be no load at the point or in the direction in which deflection is desired, an imaginary force P will be placed there in the direction desired. After the operation is completed, the correct value of P (zero) will be substituted in the expression. This principle is demonstrated in Example 11.11.

If slope or rotation is desired in a structure, the partial derivative is taken with respect to an assumed moment P acting at the point where rotation is desired. A positive sign on the answer indicates that the rotation is in the assumed direction of the moment P.

Burro Creek Bridge, Wickieup, Mohave County, Arizona. (Courtesy of the American Institute of Steel Construction, Inc.)

EXAMPLE 11.10

Determine the vertical deflection at the free end of the cantilever beam shown in the figure. For this problem; E = 29,000 ksi and I = 1,200 in^4

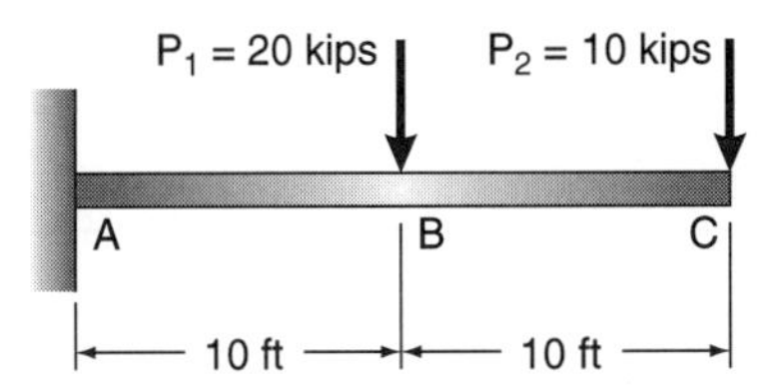

Solution. The free-body diagrams that we will use in the solution for this example are

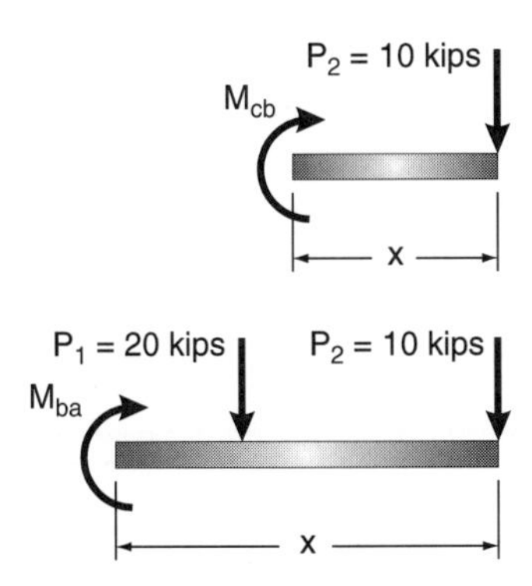

The moment equations obtained using these free-body diagrams are

$$M_{cb} = -P_2x$$
$$M_{ba} = -P_2x - P_1(x - 10)$$

We can use the load P_2 to determine the deflection at the free end of the beam because P_2 is acting at that location. By taking the partial derivative of the moment equations with respect to P_1 we obtain

$$\frac{\partial M_{cb}}{\partial P_2} = \frac{\partial}{\partial P_2}(-P_2x) = -x$$
$$\frac{\partial M_{ba}}{\partial P_2} = \frac{\partial}{\partial P_2}[-P_2x - P_1(x - 10)] = -x$$

The displacement at the free end of the beam can then be found from

$$\Delta = \frac{1}{EI}\int_0^{10} M_{cb}\left(\frac{\partial M_{cb}}{\partial P_2}\right)dx + \frac{1}{EI}\int_{10}^{20} M_{ba}\left(\frac{\partial M_{ba}}{\partial P_2}\right)dx$$
$$\Delta = \frac{1}{EI}\left[\frac{8{,}000}{3}P_2 + \frac{2{,}500}{3}P_1\right]$$

After making the appropriate substitutions for the known values, we find that

$$\Delta - \frac{1{,}728}{29{,}000(1{,}200)}\left[\left(\frac{8{,}000}{3}\right)10 + \left(\frac{2{,}500}{3}\right)20\right]$$

$$\Delta = 2.152 \text{ in.}$$

The positive sign indicates that the deflection is downward because the load P_2 that we used is acting downward. ■

EXAMPLE 11.11

Determine the vertical deflection at the free end of the cantilever beam shown in the figure. For this problem, E = 29,000 ksi and I = 4,000 in^4

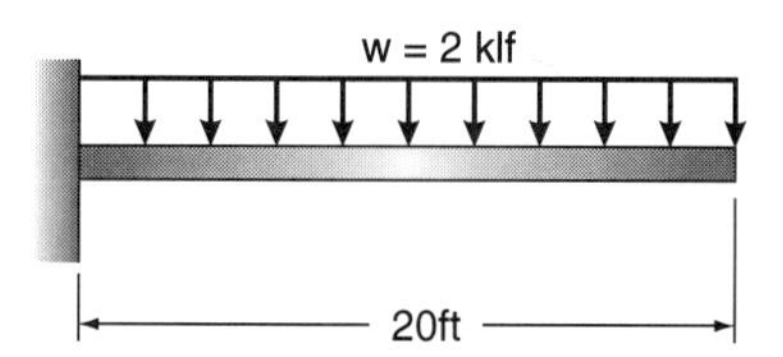

Solution. To determine the deflection at the free end of the beam, we must place a load of magnitude zero at that location as shown in the figure. The free-body diagram that we will use to write the moment equation is also shown.

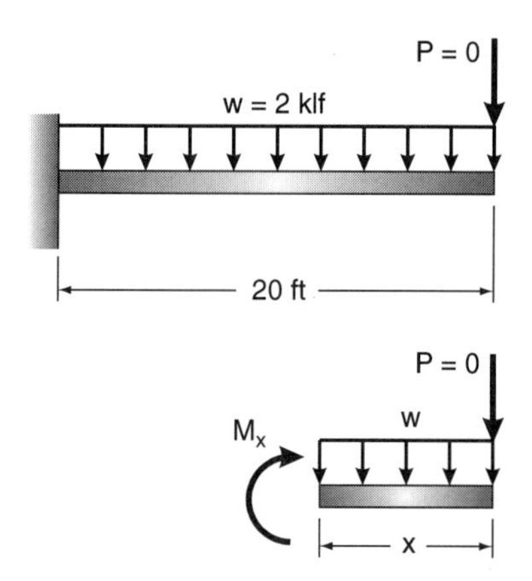

Using this free-body diagram, the equation for the bending moment in the beam is

$$M_x = -Px - 2x\left(\frac{x}{2}\right) = -Px - x^2$$

The partial derivative of this equation with respect to P is

$$\frac{\partial M_x}{\partial P} = \frac{\partial}{\partial P}(-Px - x^2) = -x$$

The deflection at the end of the cantilever can then be found from

$$\Delta = \frac{1}{EI}\int_0^{20} (-Px - x^2)(-x)\,dx$$

$$\Delta = \frac{1}{EI}\left[40{,}000 + \frac{8{,}000P}{3}\right]$$

However, because P is equal to zero, this equation reduces to

$$\Delta = \frac{1{,}728}{29{,}000(4{,}000)}\left[40{,}000 + \frac{8{,}000(0)}{3}\right] = 0.596 \text{ in.}$$

The positive sign indicates that the deflection is downward because the load P that we used was acting downward. ■

EXAMPLE 11.12

Determine the vertical deflection at the 30-kip load in the beam shown in the figure. Use the Heaviside step function in the solution. For this problem, E = 29,000 ksi and $I = 2{,}250 \text{ in}^4$.

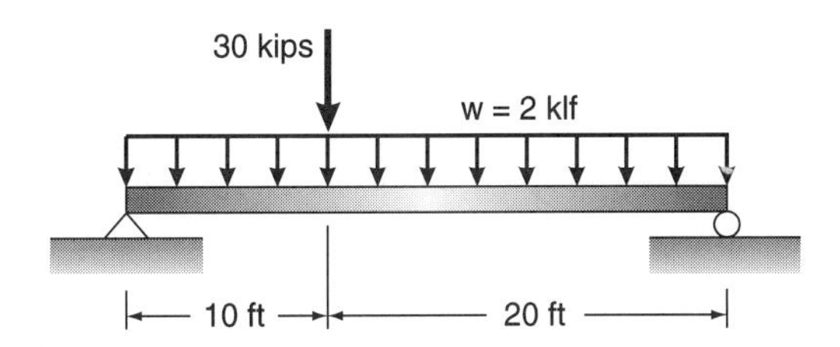

Solution. The free-body diagram that we will use in the solution is shown in the following figure. Because we want to compute the vertical deflection at the 30-kip load, we will represent that load as P when we write the bending moment equation.

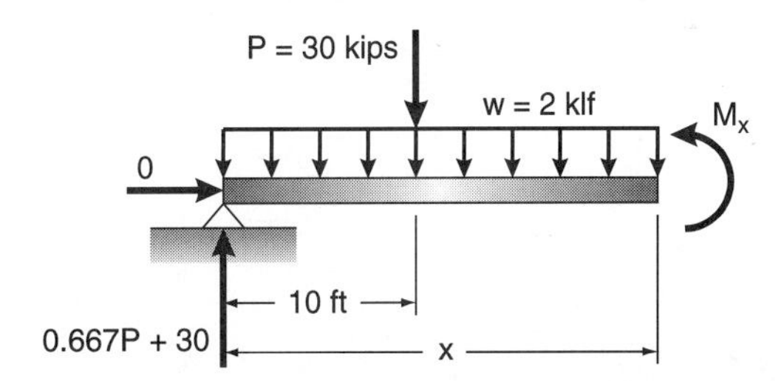

The bending moment equation obtained from this free-body diagram is

$$M_x = \left(\frac{2}{3}P + 30\right)x - wx\left(\frac{x}{2}\right) - \Phi(x - 10)[P(x - 10)]$$

$$M_x = \left(\frac{2}{3}P + 30\right)x - \frac{wx^2}{2} - \Phi(x - 10)[P(x - 10)]$$

The deflection at and in the direction of the 30-kip load can be obtained from

$$\Delta = \frac{1}{EI}\int_0^{30} M_x\left(\frac{\partial M_x}{\partial P}\right)dx = \frac{1}{EI}\left[\frac{4{,}000}{9}P + \frac{55{,}000}{3}\right]$$

$$\Delta = \left(\frac{1{,}728}{29{,}000(2{,}250)}\right)\left(\frac{4{,}000}{9}(30) + \frac{55{,}000}{3}\right) = 0.839 \text{ in.}$$

The positive sign indicates that the deflection is downward because the 30-kip that we used for the partial derivatives is acting downward. ■

11.11 PROBLEMS FOR SOLUTION

For Problems 11.1 through 11.6 use the virtual work method to determine the deflection of each of the joints marked on the trusses shown in the accompanying illustrations. The circled figures are areas in square inches and $E = 29 \times 10^6$ psi unless otherwise indicated.

11.1 U_2 vertical, U_2 horizontal

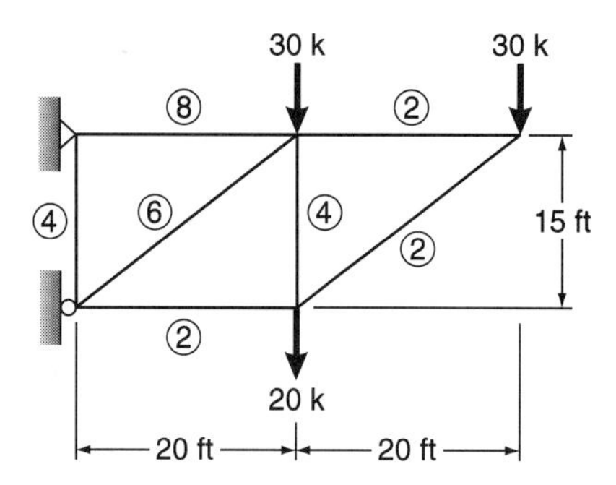

11.2 L_2 vertical, L_4 horizontal (*Ans.* $L_2 = 1.64$ in. ↓, $L_4 = 0.552$ in. →)

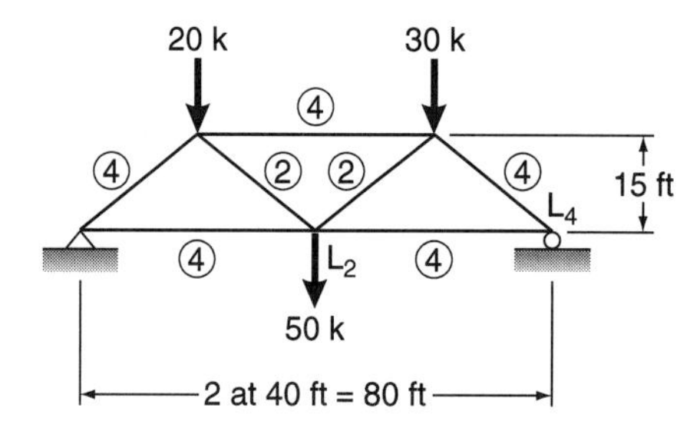

11.3 U_2 vertical, U_1 horizontal

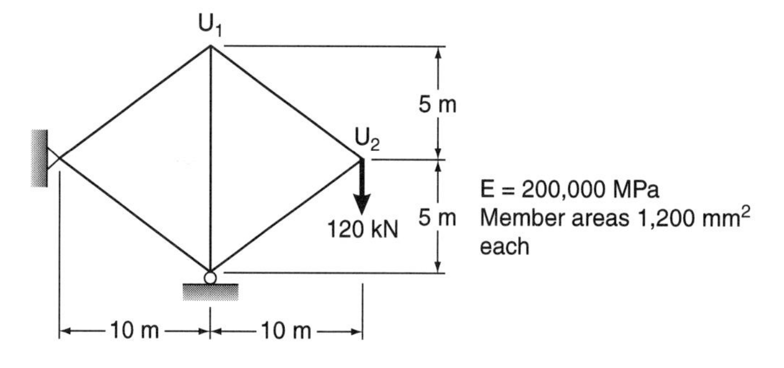

11.4 L_3 vertical (*Ans.* $L_3 = 5.54$ in. ↓)

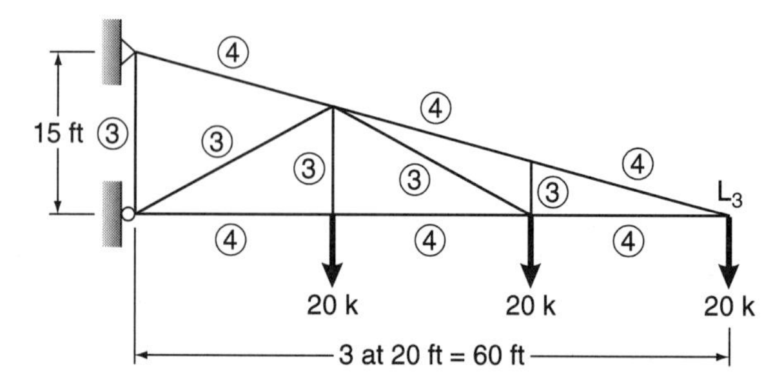

11.5 U_1 vertical, L_1 horizontal

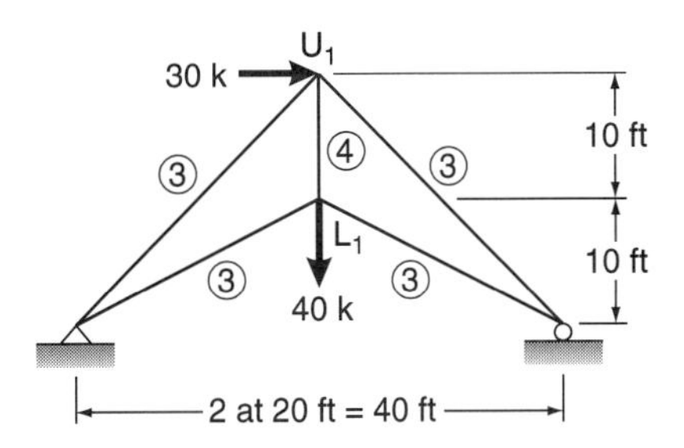

11.6 L_1 vertical, L_1 horizontal (*Ans.* $L_1 = 0.143$ in. ↓, 0.130 in. →)

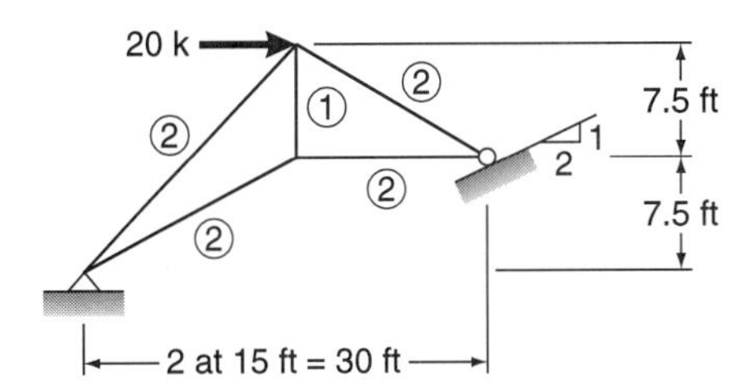

Use the virtual-work method for solving Problems 11.7 through 11.30. $E = 29 \times 10^6$ psi for all problems unless otherwise indicated.

11.7 Determine the slope and deflection at points A and B of the structure shown in the accompanying illustration. $I = 2000\ \text{in}^4$

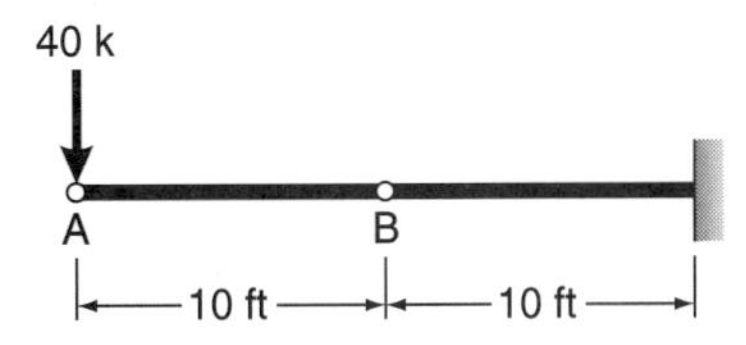

11.8 Determine the slope and deflection underneath each of the concentrated loads shown in the accompanying illustration. $I = 3500\ \text{in}^4$ (*Ans.* $\theta_{20} = 0.00372$ rad, $\theta_{30} = 0.00231$ rad, $\Delta_{20} = 0.468$ in. ↓, $\Delta_{30} = 0.078$ in. ↓)

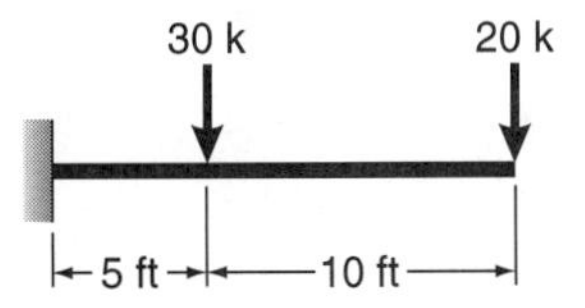

11.9 Find the slope and deflection at the free end of the beam shown in the accompanying illustration. $I = 2.35 \times 10^9\ \text{mm}^4$, $E = 200\ 000$ MPa.

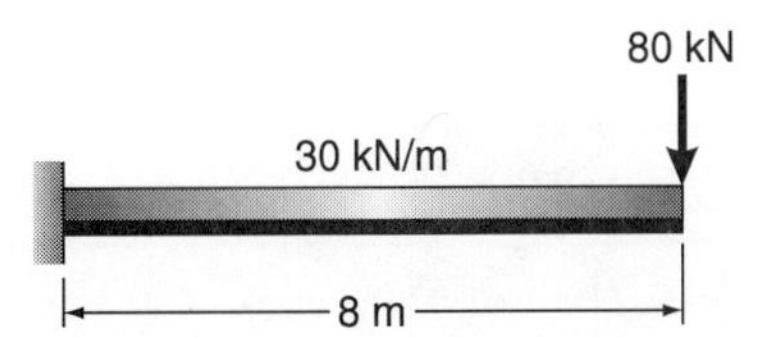

11.10 Find the deflections at points A and B on the beam shown in the accompanying illustration. I = 3500 in^4 (*Ans.* Δ_A = 1.61 in. ↓, Δ_B = 0.553 in. ↓)

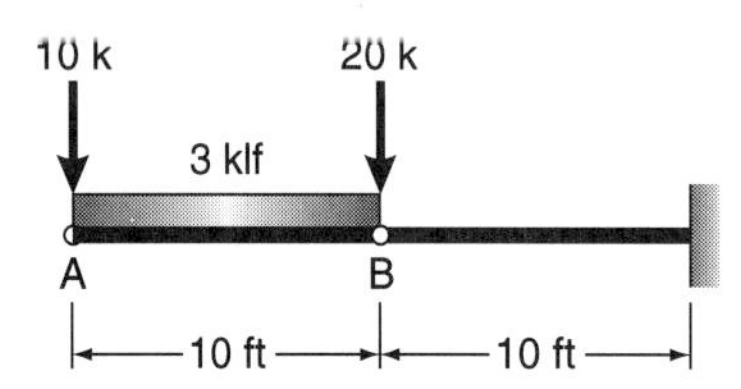

11.11 Rework Problem 10.26.

11.12 Determine the slope and deflection 10 ft from the right end of the beam of Problem 10.26.

11.13 Find the slope and deflection at the 20-kip load in the beam of Problem 10.27.

11.14 Calculate the slope and deflection at a point 3 m from the left support of the beam shown in the accompanying illustration. I = 2.5 × 10^8 mm^4, E = 200 000 MPa. (*Ans.* θ = 0.018 9 rad, Δ = 84.7 mm ↓)

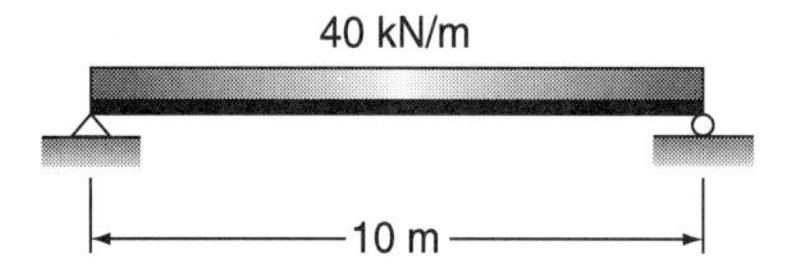

11.15 Rework Problem 10.28.

11.16 Rework Problem 10.30.

11.17 Rework Problem 10.22.

11.18 Determine the deflection at the centerline and the slope at the right end of the beam shown in the accompanying illustration. I = 2100 in^4 (*Ans.* Δ = 0.419 in. ↓, θ = 0.00244 rad)

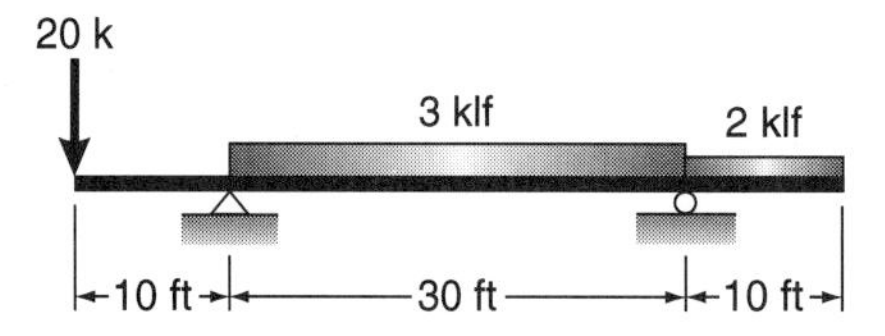

11.19 Calculate the deflection at the 60-kip load in the structure shown in the accompanying illustration.

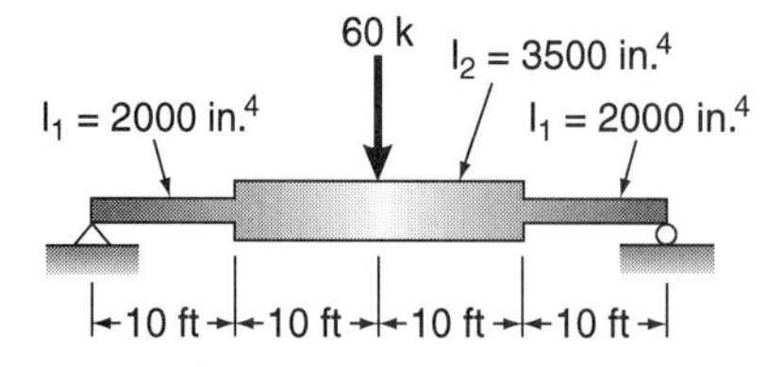

11.20 Calculate the slope and deflection at the 60-kip load on the beam shown in the accompanying illustration. (*Ans.* θ_{60} = 0.00069 rad, Δ_{60} = 0.0274 in. ↓)

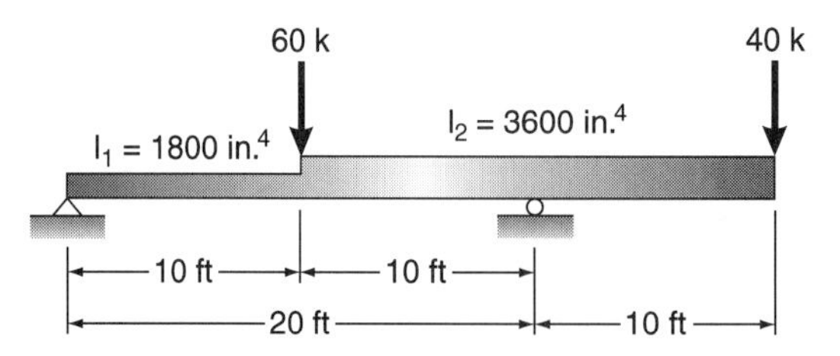

11.21 Calculate the slope and deflection at the 60-kN load on the structure shown in the accompanying illustration. $I = 1.46 \times 10^9$ mm^4. E = 200 000 MPa

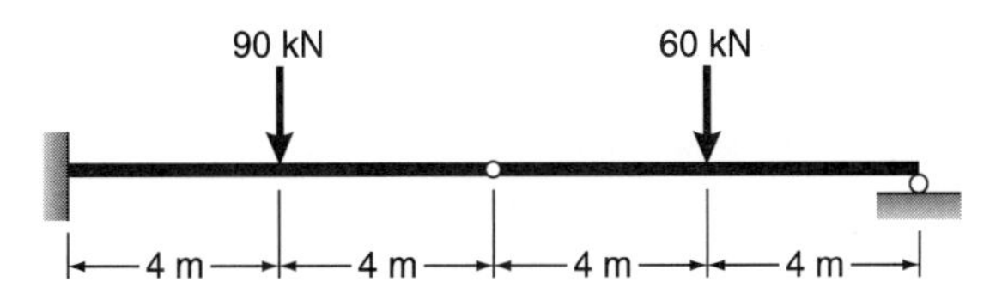

11.22 Find the slope and deflection 20 ft from the left end of the simple beam shown in the accompanying illustration. $I = 2250$ in^4 (*Ans.* $\theta_{20} = 0.00221$ rad, $\Delta_{20} = 0.459$ in. ↓)

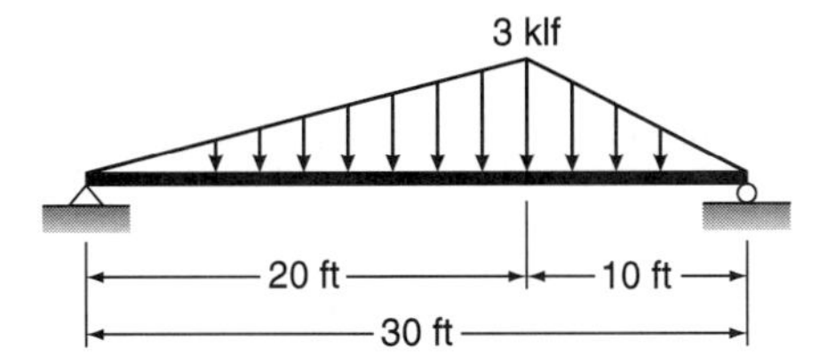

11.23 Calculate the slope and deflection at the free end of the beam shown in the accompanying illustration. $I = 3200$ in^4

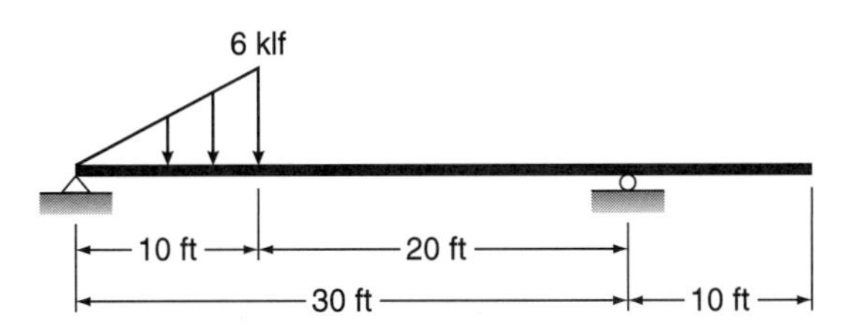

11.24 Find the vertical deflection at point A and the horizontal deflection at point B on the frame shown in the accompanying illustration. $I = 4000$ in^4 (*Ans.* $\Delta_A = 0.92$ in. ↓, $\Delta_B = 0.298$ in. ←)

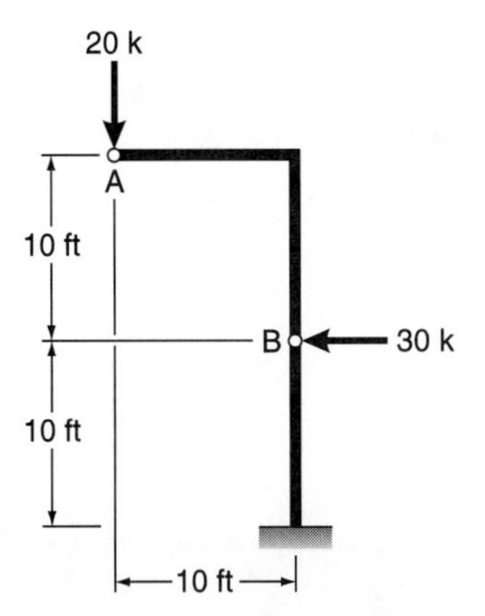

11.25 Determine the horizontal deflection at the 20-kip load and the vertical deflection at the 30-kip load on the frame shown in the accompanying illustration.

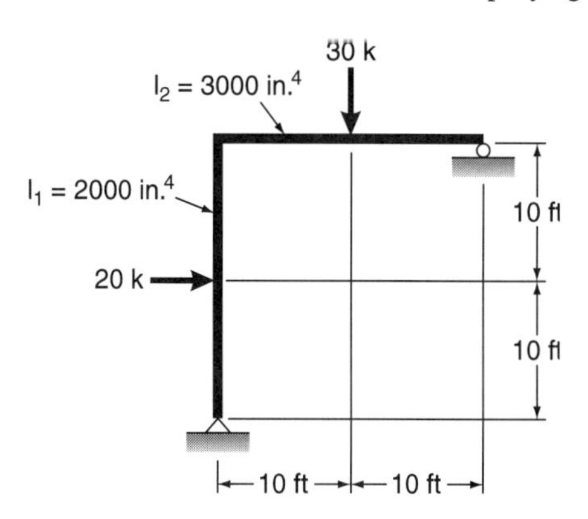

11.26 Find the vertical and horizontal components of deflection at the 30-kip load in the frame shown in the accompanying illustration. $I = 1500$ in^4 (*Ans.* $\Delta_H = 4.37$ in. $\rightarrow$, $\Delta_V = 2.38$ in. $\downarrow$)

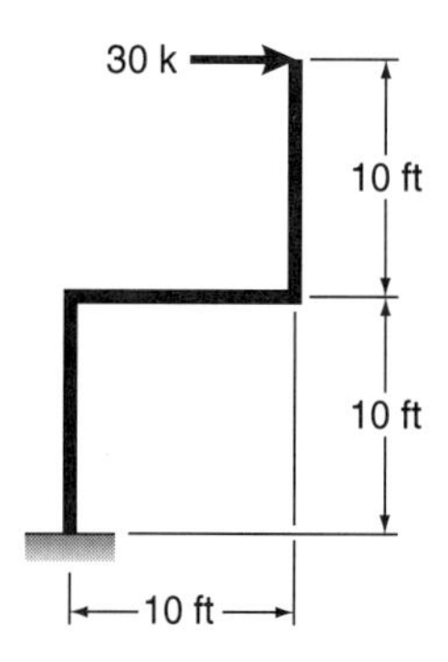

11.27 Determine the vertical and horizontal components of deflection at the free end of the frame shown in the accompanying illustration. $I = 2.60 \times 10^9$ mm^4. $E = 200\ 000$ MPa

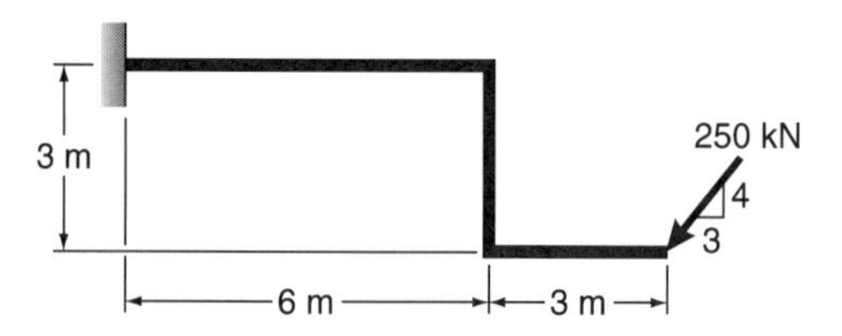

11.28 Find the horizontal deflection at point A and the vertical deflection at point B on the frame shown in the accompanying illustration. $I = 1200$ in^4 (*Ans.* $\Delta_B = 2.98$ in. $\leftarrow$, $\Delta_n = 1.99$ in. $\uparrow$)

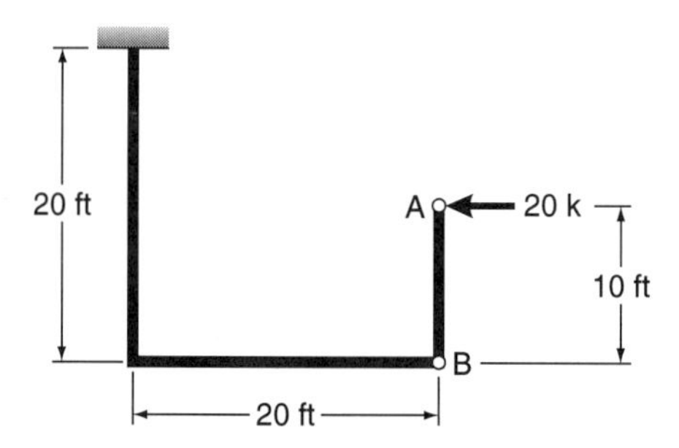

11.29 Calculate the horizontal deflection at the roller support and the vertical deflection at the 30-kip load for the frame shown in the accompanying illustration. I = 1500 in^4

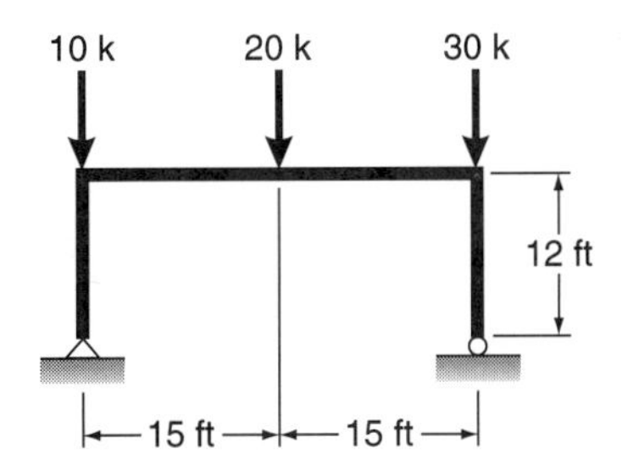

11.30 Find the horizontal deflections at points A and B on the frame shown in the accompanying illustration. I = 3000 in^4 (*Ans.* $\Delta_A = 0.463$ in. →, $\Delta_B = 0.430$ in. →)

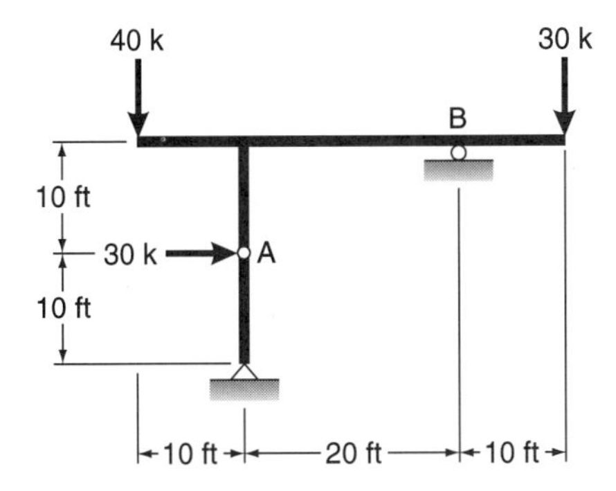

For Problems 11.31 through 11.46, use Castigliano's second theorem to determine the slopes and deflections indicated. E = 29 × 10^6 psi for the problems having customary units and 200,000 MPa for those with SI units. Similarly circled values are cross-sectional areas in square inches or in square millimeters.

11.31 Deflection at each load; I = 3250 in^4

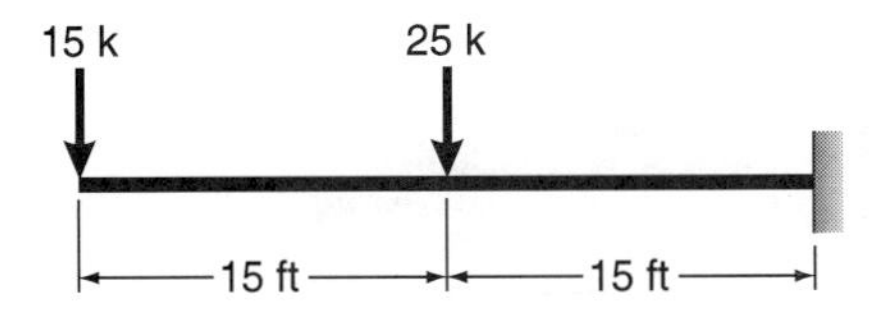

11.32 Slope and deflection at free end; I = 1750 in^4 (*Ans.* $\theta = 0.00638$ rad, $\Delta = 1.53$ in. ↓)

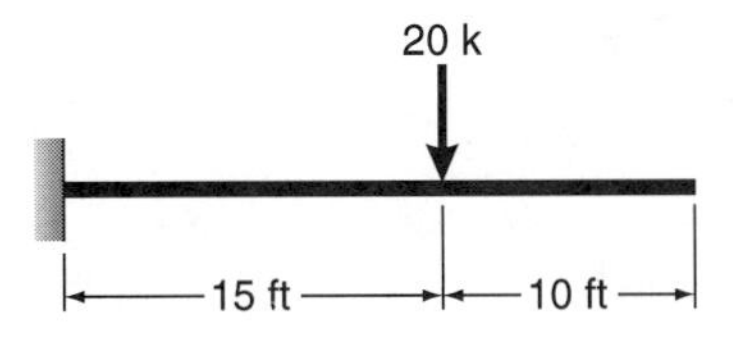

11.33 Slope and deflection at 30 k load; I = 5500 in^4

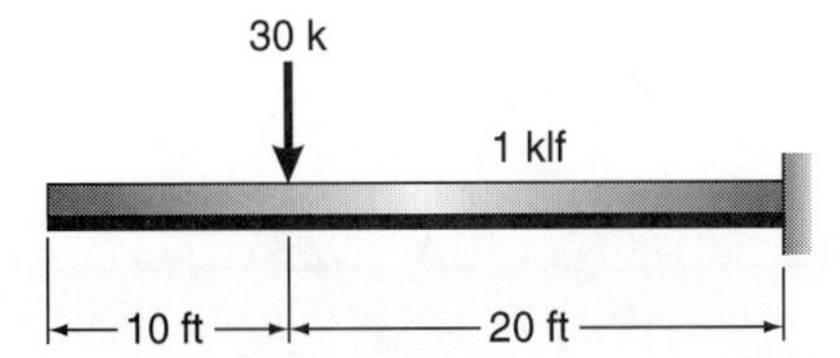

11.34 Deflection at each end of uniform load; $I = 1250 \text{ in}^4 >$ (*Ans.* $\Delta_{20\text{ft}} = 2.44$ in. ↓, $\Delta_{10\text{ft}} = 0.83$ in. ↓)

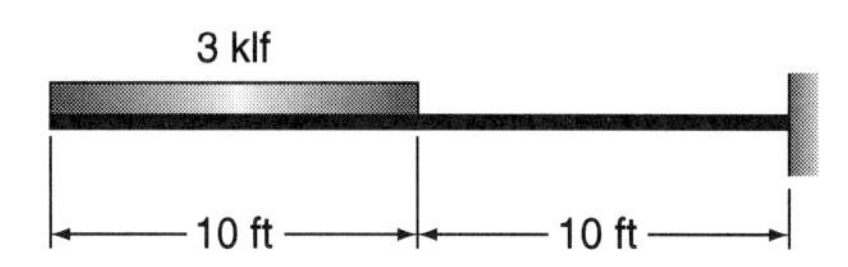

11.35 Deflection at each load; $I = 4.0 \times 10^8 \text{ mm}^4$

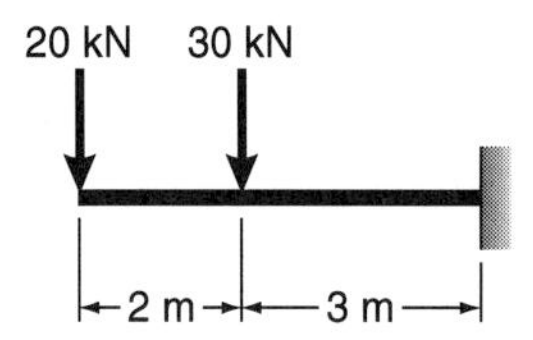

11.36 Slope and deflection at 100-kN load. $I = 1.798 \times 10^8 \text{ mm}^4$ (*Ans.* $\theta = -0.003\ 47$ rad, $\Delta = 26.12$ mm ↓)

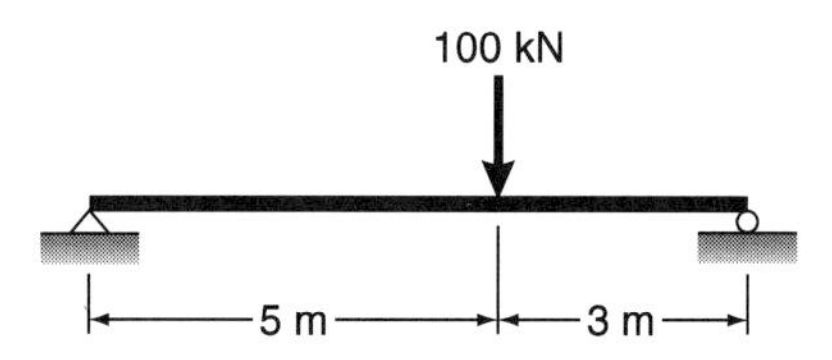

11.37 Rework Problem 10.24.

11.38 Deflection at each load; $I = 1750 \text{ in}^4$ (*Ans.* $\Delta_{20\text{k}} = 2.16$ in. ↓, $\Delta_{40\text{k}} = 1.65$ in. ↓)

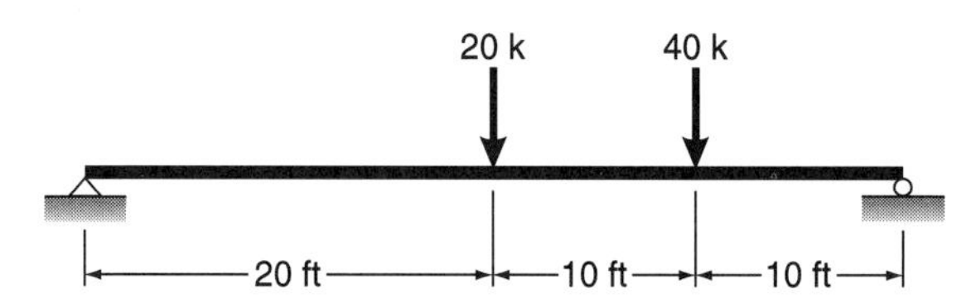

11.39 Slope and deflection at the 60 k load: $I = 2700 \text{ in}^4$

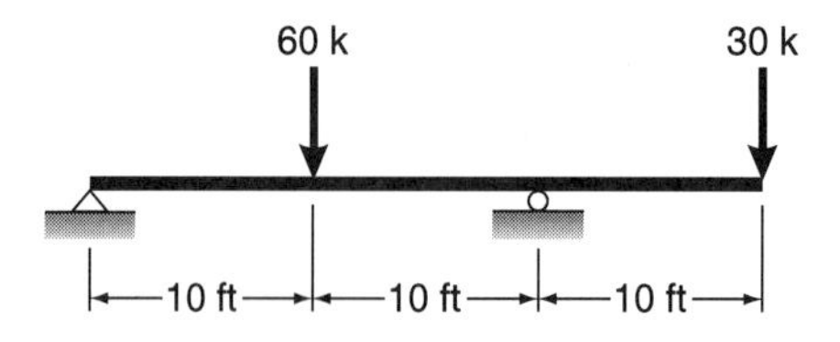

11.40 Rework Problem 10.25.

11.41 Rework Problem 11.26.

11.42 Rework Problem 11.23 (*Ans.* $\theta = -0.000772$ rad, $\Delta = 0.0868$ in. ↑)

11.43 Rework Problem 11.27.

11.44 Vertical deflection at joint L_1 (*Ans.* $\Delta = 0.652$ in. $\downarrow$)

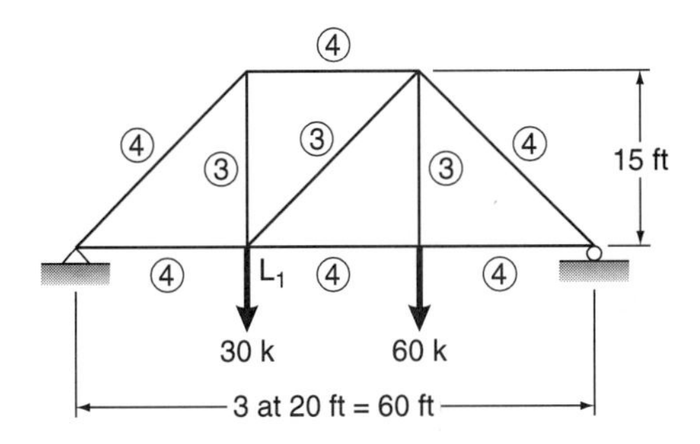

11.45 Vertical and horizontal deflection at L_0

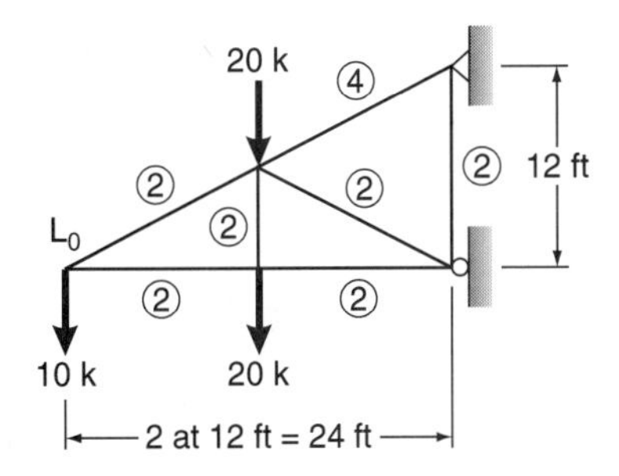

11.46 Vertical deflection at L_1 and horizontal deflection at L_2 (*Ans.* $\Delta_V = 45.2$ mm $\downarrow$, $\Delta_H = 21$ mm $\rightarrow$)

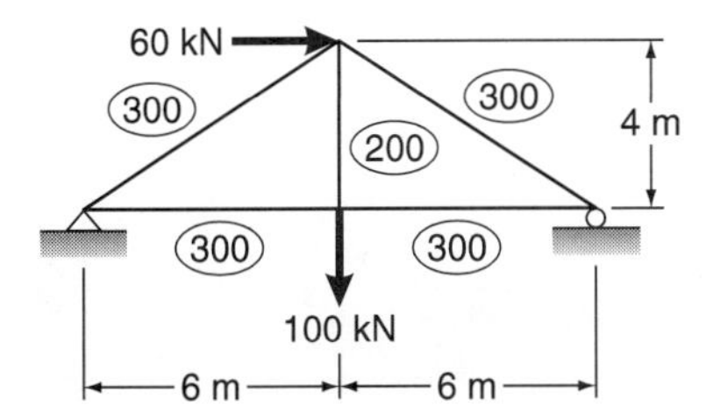

PART TWO

STATICALLY INDETERMINATE STRUCTURES

CLASSICAL METHODS

Chapter 12

Introduction to Statically Indeterminate Structures

12.1 INTRODUCTION

When a structure has too many external reactions and/or internal forces to be determined using only equations of static equilibrium (including any equations of condition), it is statically indeterminate. A load placed on one part of a statically indeterminate or continuous structure will cause shearing forces, bending moments, and deflections in other parts of the structure. In other words, loads applied to a column affect the beams, slabs, and other columns and vice versa. This is often true, but not necessarily true, with statically determinate structures.

Up to this point, the text has been so completely devoted to statically determinate structures that the reader may have been falsely led to believe that statically determinate beams and trusses are the rule in modern structures. In truth, it is difficult to find an ideal, simply supported beam. Probably the best place to look for one would be in a textbook on structures, for bolted or welded beam-to-column connections do not produce ideal simple supports with zero moments.

The same holds for statically determinate trusses. Bolted or welded joints are not frictionless pins, as previously assumed. The other assumptions made about trusses in the earlier chapters of this book are not altogether true either, and thus in a strict sense all trusses are statically indeterminate because they have some bending and secondary forces.

Almost all reinforced-concrete structures are statically indeterminate. The concrete for a large part of a concrete floor, including the support beams, and girders, and perhaps parts of the columns, may be placed at the same time. The reinforcing bars extend from member to member, as from one span of a beam into the next. When there are construction joints, the reinforcing bars are left protruding from the older concrete so they may be lapped or spliced to the bars in the newer concrete. In addition, the old concrete is cleaned and perhaps roughened so that the newer concrete will bond to it as well as possible. The result of all these facts is that reinforced-concrete structures are generally monolithic or continuous and thus are statically indeterminate.

About the only way a statically determinate reinforced-concrete structure could be built would be with individual sections precast at a concrete plant and assembled on the job site. Even these structures could have some continuity in their joints.

Until the early part of the 20th century, statically indeterminate structures were avoided as much as possible by most American engineers. However, three great developments completely changed the picture. These developments were monolithic reinforced-concrete structures, arc welding of steel structures, and modern methods of analysis.

12.2 CONTINUOUS STRUCTURES

As the spans of simple structures become longer, the bending moments increase rapidly. If the weight of a structure per unit length remained constant, regardless of the span, the dead-load moment would vary in proportion to the square of the span length ($M = wl^2/8$). This proportion, however, is not correct, because the weight of structures must increase with longer spans to be strong enough to resist the increased bending moments. Therefore, the dead-load moment increases at a greater rate than does the square of the span.

For economy, it pays in long spans to introduce types of structures that have smaller moments than the tremendous ones that occur in long-span, simply supported structures. One type of structure that considerably reduced bending moments, cantilever-type construction, was introduced in Chapter 4. Two other moment-reducing structures are discussed in the following paragraphs.

In some locations, a beam with fixed ends, rather than one with simple supports, may be possible. A comparison of the moments developed in a uniformly loaded simple beam with those in a uniformly loaded, fixed-ended beam is made in Figure 12.1.

The maximum bending moment in the fixed-ended beam is only two thirds of that in the simply supported beam. Usually it is difficult to fix the ends, particularly in the case of bridges. For this reason, *flanking spans* are often used, as illustrated in Figure 12.2. These spans will partially restrain the interior supports, thus tending to reduce the moment in the center of the span. This figure compares the bending moments that occur in three uniformly loaded simple beams (spans of 60 ft, 100 ft, and 60 ft) with the moments of a uniformly loaded beam continuous over the same three spans.

The maximum bending moment in the continuous beam is approximately 40% less than that for the simple beams. Unfortunately, there will not be a corresponding 40% reduction in total cost. The cost-reduction probably is only a small percentage of the total cost of the structure because such items as foundations, connections, and floor systems are not reduced a great deal by the moment reduction. Varying the lengths of the flanking spans will change the magnitude of the largest moment occurring in the continuous member. For a constant uniform load over the 3 spans, the smallest moment will occur when the flanking spans are from 0.3 to 0.4 times as long as the center span.

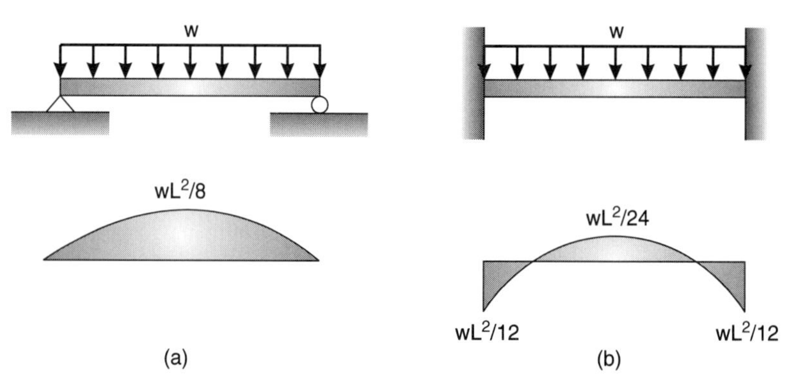

Figure 12.1 A simple beam (a) and a continuous beam (b)

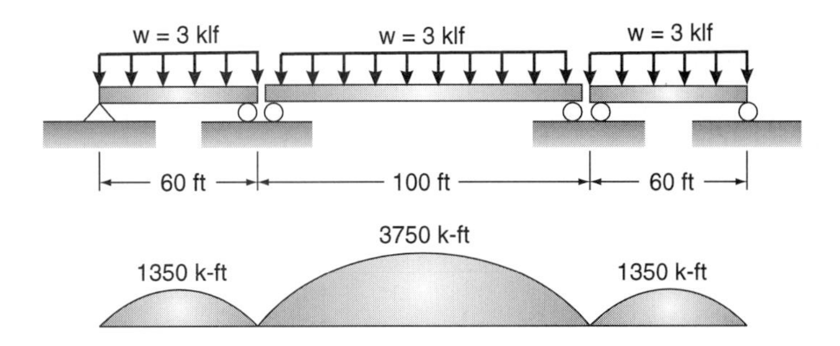

(a) Moment diagrams if 3 separate spans are used

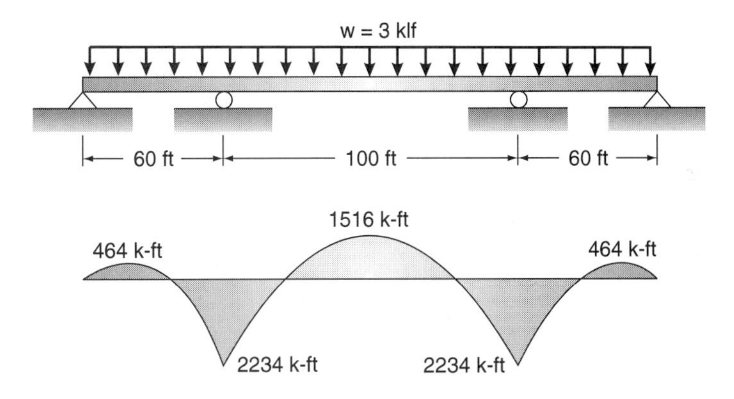

(b) Moment diagrams if one continuous span is used

Figure 12.2 Comparison of bending moments in 3 simple beams versus one continuous beam

In the foregoing discussion, the moments developed in beams were reduced appreciably by continuity. This reduction occurs when beams are rigidly connected to each other or where beams and columns are rigidly connected. There is a continuity of action in resisting a load applied to any part of a continuous structure because the load is resisted by the combined efforts of all the members of the frame.

Colorado River arch bridge, Utah Route 95 (Courtesy of the Utah Department of Transportation)

12.3 ADVANTAGES OF STATICALLY INDETERMINATE STRUCTURES

When comparing statically indeterminate structures with statically determinate structures, the first consideration for most engineers would likely be cost. However, making a general economic statement favoring one type of structure over another is impossible without reservation. Each structure presents a different and unique situation, and all factors must be considered—economic or otherwise. In general, statically indeterminate structures have the following advantages.

12.3.1 Savings in Materials

The smaller bending moments developed often enable the engineer to select smaller members for the structural components. The material saving could possibly be as high as 10 to 20% for highway bridges. Because of the large number of force reversals that occur in railroad bridges, the cost saving is closer to 10%.

A structural member of a given size can support more load if it is part of a continuous structure than if it is simply supported. The continuity permits the use of smaller members for the same loads and spans or increased spacing of supports for the same size members. The possibility of fewer columns in buildings, or fewer piers in bridges, may permit a reduction in overall costs.

Continuous concrete or steel structures are cheaper without the joints, pins, and so on required to make them statically determinate, as was frequently the practice in past years. Monolithic reinforced-concrete structures are erected so that they are naturally continuous and statically indeterminate. Installing the hinges and other devices necessary to make them statically determinate would be a difficult construction problem, and very expensive. Furthermore, if a building frame consisted of columns and simply supported beams, objectionable diagonal bracing between the joints would be necessary to provide a stable frame with sufficient rigidity.

12.3.2 Larger Safety Factors

Statically indeterminate structures often have higher safety factors than statically determinate structures. When parts of statically indeterminate steel or reinforced-concrete structures are overstressed they will often have the ability to redistribute portions of those stresses to less-stressed areas. Statically determinate structures generally do not have this ability.[1] If the bending moments in a component of a statically determinate structure reach the ultimate moment capacity of that component, the structure would fail. This is not the case for statically indeterminate structures since load may be redistributed to other parts of the structure.

As can be clearly shown, a statically indeterminate beam or frame normally will not collapse when the ultimate moment capacity is reached at just one location. Instead, there is a redistribution of the moments in the structure. This behavior is quite similar to the case in which three men are walking along with a log on their shoulders and one of the men gets tired and lowers his shoulder just a little. The result is redistribution of load to the other men with corresponding changes in the shearing forces and bending moments along the log.

[1] J. C. McCormac, *Structural Steel Design LRFD Method,* 2nd ed. (New York: HarperCollins, 1995), 225–235.

12.3.3 Greater Rigidity and Smaller Deflections

Statically indeterminate structures are more rigid and have smaller deflections than statically determinate structures. Because of their continuity, they are stiffer and have greater stability against all types of loads (horizontal, vertical, moving, etc.).

12.3.4 More Attractive Structures

It is difficult to imagine statically determinate structures having the gracefulness and beauty of many statically indeterminate arches and rigid frames being erected today.

Broadway Street Bridge showing cantilever erection, Kansas City, Missouri (Courtesy of USX Corporation)

12.3.5 Adaptation to Cantilever Erection

The cantilever method of erecting bridges is of particular value where conditions underneath (probably marine traffic or deep water) hinder the erection of falsework. Continuous statically indeterminate bridges and cantilever-type bridges are conveniently erected by the cantilever method.

12.4 DISADVANTAGES OF STATICALLY INDETERMINATE STRUCTURES

A comparison of statically determinate and statically indeterminate structures shows the latter have several disadvantages that make their use undesirable on many occasions. These disadvantages are discussed in the following paragraphs.

12.4.1 Support Settlement

Statically indeterminate structures are not desirable where foundation conditions are poor, because seemingly minor support settlements or rotations may cause major changes in the bending moments, shearing forces, reaction forces, and member forces. Where statically indeterminate bridges are used despite the presence of poor foundation conditions, it is occasionally necessary to physically measure the dead-load reactions. The supports of the bridge are jacked up or down until the calculated reaction forces are obtained. The support is then built to that elevation.

12.4.2 Development of Other Stresses

Support settlement is not the only condition that causes stress variations in statically indeterminate structures. Variation in the relative positions of members caused by temperature changes, poor fabrication, or internal deformation of members in the structure under load may cause significant force changes throughout the structure.

12.4.3 Difficulty of Analysis and Design

The forces in statically indeterminate structures depend not only on their dimensions but also on their elastic and cross-sectional properties (moduli of elasticity, moments of inertia, and areas). This situation presents a design difficulty: the forces cannot be determined until the member sizes are known, and the member sizes cannot be determined until their forces are known. The problem is handled by assuming member sizes and computing the forces, designing the members for these forces and computing the forces for the new sizes, and so on, until the final design is obtained. Design by this method—*the method of successive approximations*—takes more time than the design of a comparable statically determinate structure, but the extra cost is only a small part of the total cost of the structure. Such a design is best done by interaction between the designer and the computer. Interactive computing is now used extensively in the aircraft and automobile industries.

12.4.4 Force Reversals

Generally, more force reversals occur in statically indeterminate structures than in statically determinate structures. Additional material may be required at certain sections to resist the different force conditions and to prevent fatigue failures.

12.5 METHODS OF ANALYZING STATICALLY INDETERMINATE STRUCTURES

Statically indeterminate structures contain more unknown forces than there are equations of static equilibrium. As such, they cannot be analyzed using only the equations of static equilibrium; additional equations are needed. Forces beyond those needed to maintain a stable structure are redundant forces. The redundant forces can be forces of reaction or forces in member that compose the structure. There are two general approaches used to find the magnitude of these redundant forces: *force methods* and *displacement methods*. The bases of these methods are discussed in this section.

12.5.1 Force Methods

With force methods, equations of condition involving displacement at each of the redundant forces in the structure are introduced to provide the additional equations necessary for solution. Equations for displacement at and in the direction of the redundant forces are written in terms of the redundant forces; one equation is written for the displacement condition at each redundant force. The resulting equations are solved for the redundant forces, which must be sufficiently large to satisfy the boundary conditions. As we will soon see, the boundary conditions do not necessarily have to be zero displacement. Force methods are also called *flexibility methods* or the *compatibility methods*.

I-180 viaduct, Cheyenne, Wyoming (Courtesy of the Wyoming State Highway Department)

In 1864, James Clerk Maxwell published the first consistent force method for analyzing statically indeterminate structures. His method was based on a consideration of deflections, but the presentation (which included the reciprocal deflection theorem) was rather brief and attracted little attention. Ten years later Otto Mohr independently extended the theory to almost its present stage of development. Analysis of redundant structures with the use of deflection computations is often referred to as the *Maxwell–Mohr method* or the *method of consistent distortions*.[2,3]

The force methods of structural analysis are somewhat useful for analyzing beams, frames, and trusses that are statically indeterminate to the first or second degree. They also are convenient for some single-story frames with unusual dimensions. For structures that are highly statically indeterminate, such as multistory buildings and large complex trusses, other methods are more appropriate and useful. These methods, which include moment distribution and the matrix methods, are more satisfactory and are introduced in later chapters. As such, the force methods have been almost completely superseded by the methods of analysis described in Part III of this text. Nevertheless, study of the force methods will provide an understanding of the behavior of statically indeterminate structures that might not otherwise be obtained.

12.5.2 Displacement or Stiffness Methods

In the displacement methods of analysis, the displacement of the joints (rotations and translations) necessary to describe fully the deformed shape of the structure are used in

[2]J. I. Parcel and R. B. B. Moorman, *Analysis of Statically Indeterminate Structures* (New York: Wiley, 1955), 48.

[3]J. S. Kinney, *Indeterminate Structural Analysis* (Reading, Mass.: Addison-Wesley, 1957), 12–13.

the equations instead of the redundant forces used in the force methods. When the simultaneous equations that result are solved, these displacements are determined and then substituted into the original equations to determine the various internal forces. The most commonly used displacement method is the matrix method discussed in Chapters 19 through 21.

12.6 LOOKING AHEAD

In Chapters 13–15, classical methods of analyzing statically indeterminate structures are presented. These methods include virtual work, Castigliano's theorems, and slope deflection. The first two methods are force methods, whereas the latter method is a displacement method of analysis. These methods are primarily of historical interest and are almost never used in practice. However, they do form the basis for modern methods of analysis.

The authors have placed in Part Three (Chapters 16–21) methods for analyzing statically indeterminate structures that are frequently used today in the structural engineering profession. First, several approximate methods of analysis are introduced, and then moment distribution and the matrix methods are discussed at length.

Chapter 13

Energy Methods for Statically Indeterminate Structures

13.1 BEAMS AND FRAMES WITH ONE REDUNDANT

The propped cantilever beam in Figure 13.1(a) is statically indeterminate to the first degree. Beams of this type are not seen very often in practice, but they are an excellent example for learning to analyze statically indeterminate structures. The beam supports a uniformly distributed load, and is supported by forces of reaction at points A and B. Removal of support B would leave a statically determinate beam as shown in Figure 13.1(b). The beam would be a simple cantilever beam, which proves that the structure is statically indeterminate to the first degree. As such, the reaction at B is redundant; it is not needed for stable equilibrium.

To calculate the reactions, we need another equation in addition to the three equations of static equilibrium. The vertical deflection at B can be used as the additional equation, the equation of condition, needed to calculate reactions. The deflection at B, δ_B, caused by all of the external forces acting on the beam, including the forces of reaction, is equal to zero.

We can use virtual work to calculate the deflection at B. The magnitude of V_B is unknown, but the magnitude necessary to cause the displacement at B to be equal to zero can be determined. Saying this in another way, we will evaluate V_B so that the boundary condition of zero displacement at B is satisfied. During the discussion that follows, we will refer to Figure 13.2.

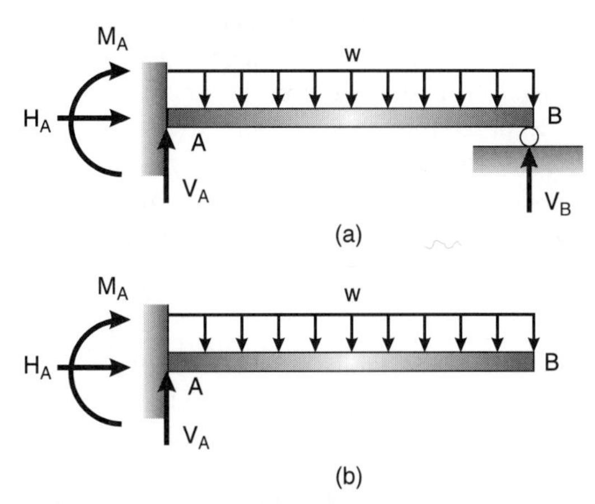

Figure 13.1 A statically indeterminate propped cantilever and a statically determinate cantilever

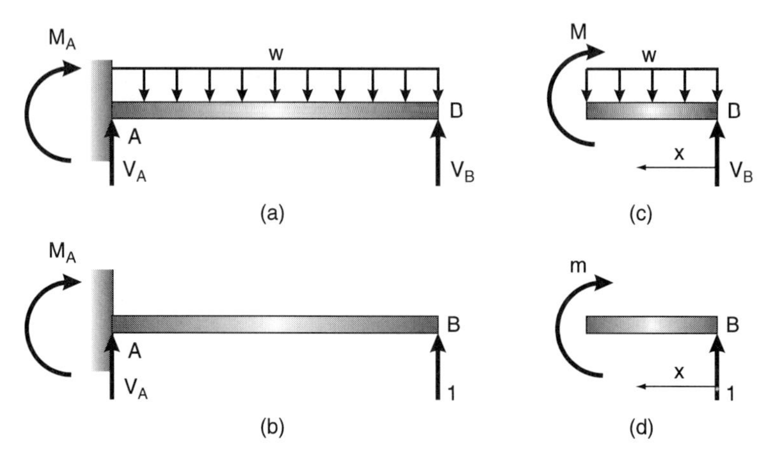

Figure 13.2 Illustration for development of virtual work

We remove the redundant support and replace it with an unknown force, as shown in Figure 13.2(a). Next we apply a unit virtual force to the beam, with all other forces removed, as shown in Figure 13.2(b). We then write moment equations for the real and virtual forces using the free-body diagrams shown in Figure 13.2(c) and Figure 13.2(d), respectively. If x is measured from the right, the resulting moment equations are

$$M = V_B x - \frac{wx^2}{2} \qquad \textbf{Eq. 13.1}$$

$$m = x$$

Observe in these equations that the direction of the real and virtual moments is the same, and that x is measured from the same location in the real and virtual systems. Observe also that the first term on the right-hand side of the top equation is equal to the force we are trying to evaluate, V_B, times the virtual moment. Using these two moment equations, the deflection at B can be written as

$$\delta_B = 0 = \frac{1}{EI}\int_0^L \left(V_B x - \frac{wx^2}{2}\right) x \, dx$$

$$= \frac{V_B}{EI}\int_0^L x^2 \, dx + \frac{1}{EI}\int_0^L -\frac{wx^2}{2} x \, dx \qquad \textbf{Eq. 13.2}$$

Consider the last of these two equations. The first term on the right-hand side is V_B times the integral of the virtual moment squared. The second term on the right-hand side is the integral of the virtual moment times the moment caused by all forces acting on the beam except the force we are trying to evaluate, V_B. We will refer to the forces in the beam excluding the force at the redundant reaction as the primary forces. We can substitute the virtual moment and moment caused by the primary forces from Equation 13.1 into the bottom equation in Equation 13.2. When doing so, we will define the moment in the beam caused by the primary forces to be M_0. The result is

$$0 = \frac{V_B}{EI}\int_0^L m^2 \, dx + \frac{1}{EI}\int_0^L M_0 m \, dx \qquad \textbf{Eq. 13.3}$$

This equation can be rearranged to the form

$$V_B = -\frac{\dfrac{1}{EI}\displaystyle\int_0^L M_0 m \, dx}{\dfrac{1}{EI}\displaystyle\int_0^L m^2 \, dx} \qquad \textbf{Eq. 13.4}$$

Physically, the denominator of this equation is the deflection in the beam caused by the unit virtual force. This deflection will be identified with the notation δ_v. The student should note that displacements caused by unit loads often are called *flexibility coefficients*. The numerator of the equation is the deflection caused by the primary forces and will be identified with the notation as δ_0. Using this notation, Equation 13.4 could have been written as

$$V_B = -\frac{\delta_0}{\delta_v} \qquad \textbf{Eq. 13.5}$$

If the resulting sign on V_B is positive, the reaction acts in the same direction as the virtual force. If it is negative, the reaction acts in the direction opposite that of the virtual force.

Examples 13.1 to 13.3 illustrate the force method of computing the reactions for statically indeterminate beams having one redundant reaction component. Example 13.4 shows that the method may be extended to include statically indeterminate frames as well. After the value of the redundant reaction in each problem is found, the other reactions are determined using the equations of static equilibrium.

EXAMPLE 13.1

Determine the reactions and draw shear and moment diagrams for the propped cantilever beam shown in the figure.

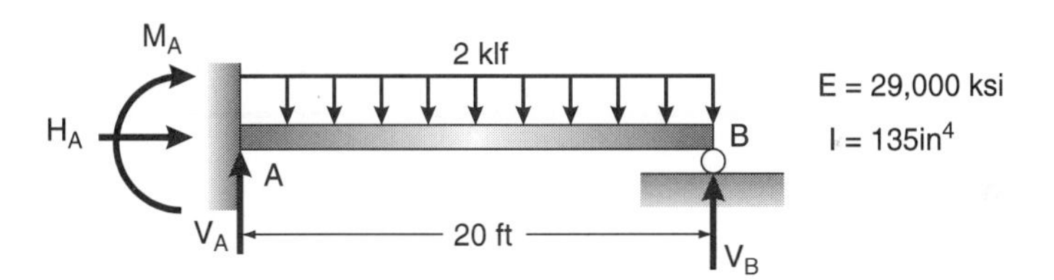

Solution. This beam is the same as we used in development of the theory. We will again use V_B as the redundant reaction. The primary and virtual force systems that we will use in the analysis are shown next. In this example we will use units of inches and kips.

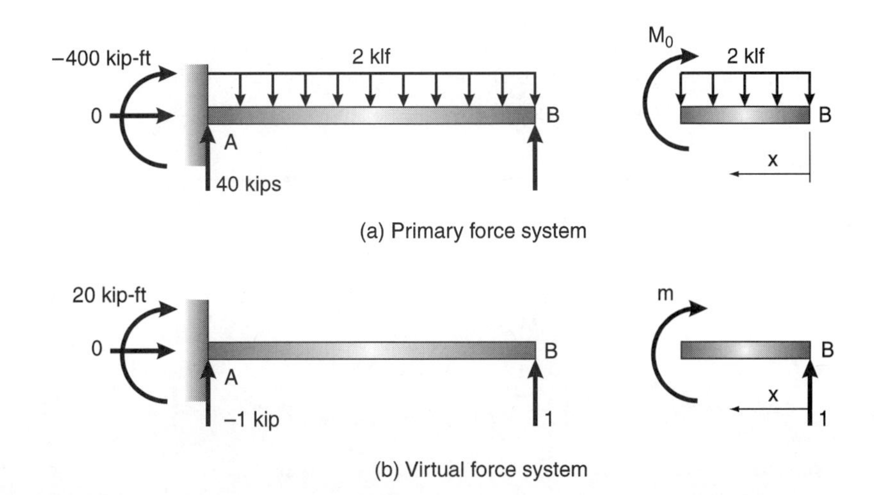

(a) Primary force system

(b) Virtual force system

First, we must determine the primary and virtual moment equations. For this beam, these equations are

$$M_0 = -\frac{\frac{2}{12}x^2}{2} = -\frac{x^2}{12}$$

$$m = x$$

Next we will compute the primary and virtual deflections, which are

$$\delta_0 = \frac{1}{EI}\int_0^L M_0 m\,dx = \frac{1}{(29{,}000)(135)}\int_0^{240}\left(-\frac{x^2}{12}\right)x\,dx = -17.66\,\text{in.}$$

$$\delta_v = \frac{1}{EI}\int_0^L m^2\,dx = \frac{1}{(29{,}000)(135)}\int_0^{240} x^2\,dx = 1.177\,\text{in.}$$

We can now compute the force V_B, which is

$$V_B = -\frac{\delta_0}{\delta_v} = -\frac{-17.66}{1.177} = 15\,\text{kips}$$

The sign on V_B is positive so it is acting upward; it is acting in the same direction as the virtual force.

Now that the force at the redundant reaction has been computed, we can compute the remaining reactions using the equations of static equilibrium. We will first sum forces horizontally to determine the horizontal reaction at A:

$$\sum F_H = H_A = 0$$
$$\therefore H_A = 0$$

Next we will sum forces vertically to determine the vertical reaction at A:

$$\sum F_V = V_A - wL + V_B = V_A - 2(20) + 15 = 0$$
$$\therefore V_A = 25\,\text{kips}$$

Lastly, we will sum moments clockwise about the left support to determine the magnitude of the reaction moment:

$$\sum M = M_A + wL\left(\frac{L}{2}\right) - V_B(20) = M_A + 2(20)\left(\frac{20}{2}\right) - 15(20) = 0$$
$$\therefore M_A = -100\,\text{k-ft}$$

The negative sign on the computed reaction moment indicates that it is acting opposite the direction indicated on the drawing. ■

EXAMPLE 13.2

Determine the reactions for the two-span continuous beam shown in the figure.

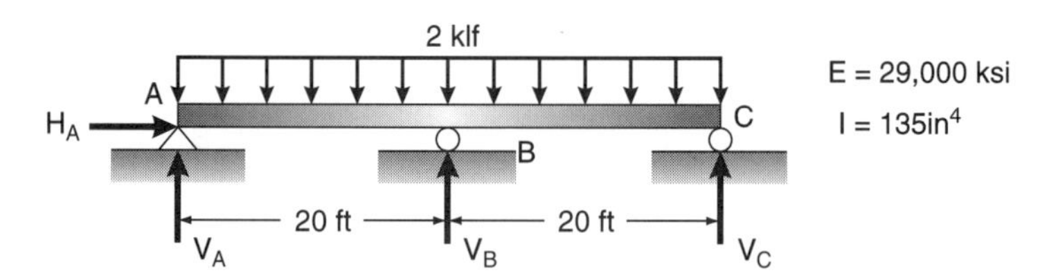

Solution. This beam is statically indeterminate to the first degree; there is one more unknown force of reaction than there are equations of equilibrium. To compute the reactions, we will use the displacement at B as the needed equation of condition. The primary force system and the free-body diagram used to calculate the primary moment, is then

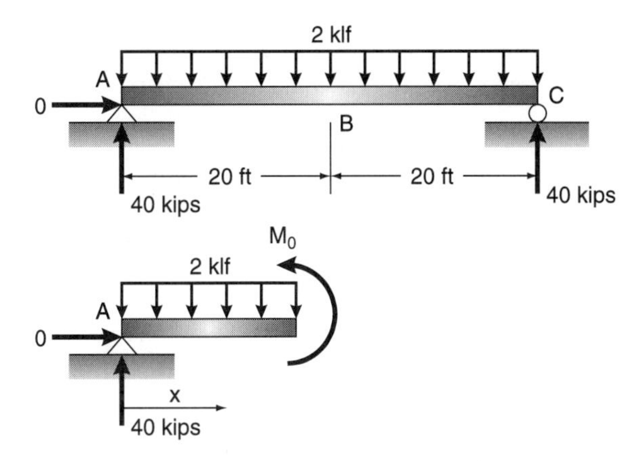

Using the free-body diagram, we can determine the equation for moment caused by the primary forces. That moment equation is

$$M_0 = 40x - \frac{\frac{2}{12}x^2}{2} = 40x - \frac{x^2}{12}$$

The units used in this moment equation are inches and kips.

The virtual force system we will use is shown in the following figure as are the free-body diagrams we will need to determine the virtual moment in the beam. Two equations will be needed because the moment equation changes at the location of the virtual force.

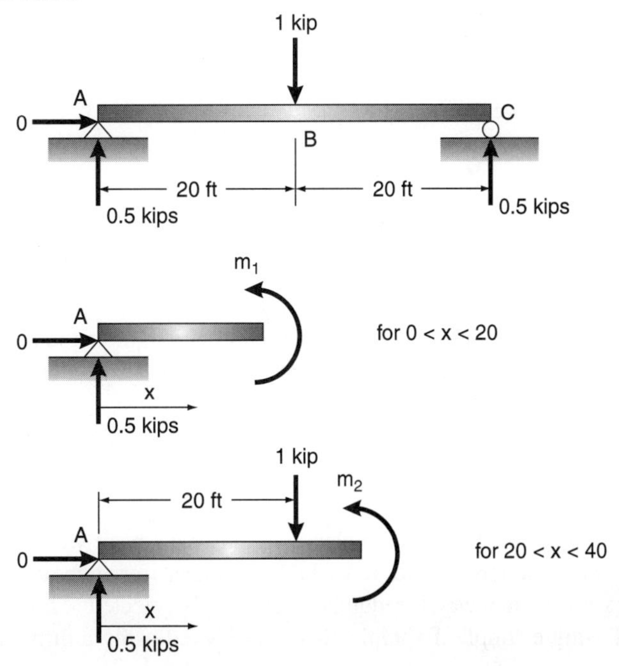

The virtual moment equations obtained using these free-body diagrams, using inch and kip units, are

$$m_1 = 0.5x \qquad \text{for } 0 < x \le 240$$
$$m_2 = 0.5x - 1(x - 240) = 240 - 0.5x \qquad \text{for } 240 < x \le 480$$

Using these primary and virtual force moment equations, we can compute the primary and virtual deflections of this beam. The primary deflection is

$$\delta_0 = \frac{1}{EI}\int_0^{240} M_0 m_1\,dx + \frac{1}{EI}\int_{240}^{480} M_0 m_2\,dx = \frac{1}{29{,}000(135)}\int_0^{240}\left(40x - \frac{x^2}{12}\right)(0.5x)\,dx$$
$$+ \frac{1}{29{,}000(135)}\int_{240}^{480}\left(40x - \frac{x^2}{12}\right)(240 - 0.5x)\,dx \qquad \delta_0 = 29.425\text{ in}$$

$$\delta_v = \frac{1}{EI}\int_0^{240} m_1^2\,dx + \frac{1}{EI}\int_{240}^{480} m_2^2\,dx = \frac{1}{29{,}000(135)}\int_0^{240}(0.5x)^2\,dx$$
$$+ \frac{1}{29{,}000(135)}\int_{240}^{480}(240 - 0.5x)^2\,dx \qquad \delta_v = 0.589\text{ in}$$

Using these computed deflections, we can compute the vertical reaction at B, which is

$$V_B = -\frac{\delta_0}{\delta_v} = -\frac{29.425}{0.589} = -50\text{ kips}$$

The negative sign indicates that the reaction at B is upward, opposite the direction of the virtual force that was applied at B.

We can now compute the magnitude of the remaining forces of reaction using equations of static equilibrium. First, we will compute the force H_A:

$$\sum F_H = H_A = 0$$
$$\therefore H_A = 0$$

Next, we can compute the vertical reaction at C by summing moments clockwise about A:

$$\sum M = wL\left(\frac{L}{2}\right) - V_B(20) - V_C(40) = \frac{2(40)^2}{2} - 50(20) - 40V_C$$
$$\therefore V_C = 15\text{ kips}$$

In this equation, L represents the total length of the beam, which is 480 in. Finally we can determine the vertical reaction at A by summing forces vertically:

$$\sum F_V = V_A - wL + V_B + V_C = V_A - 2(40) + 50 + 15 = 0$$
$$\therefore V_A = 15\text{ kips} \quad \blacksquare$$

In the preceding examples, you may have been wondering why we selected the redundant reaction we did and which redundant force should be selected. The truth is, it does not matter which redundant reaction is selected. This point is demonstrated in the following example. Example 13.1 is reworked using a different redundant reaction.

EXAMPLE 13.3

Solve the problem in Example 13.1 again, but this time use the moment A as the redundant reaction.

Solution. The primary and virtual force systems that we will use in the analysis are illustrated in the figure.

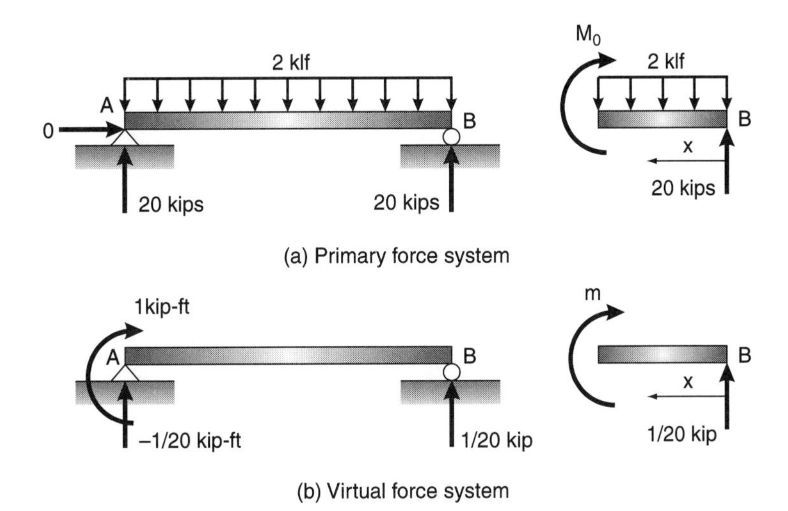

Observe now that instead of placing a unit force at B we have now placed a unit moment at A in the virtual system. Observe also that in the primary force system there is no moment at A and that there is a force at B. Using the indicated free-body diagrams, and working in inch and kip units, we can find the primary and virtual moments to be

$$M_0 = 20x - \frac{\frac{2}{12}x^2}{2} = 20x - \frac{x^2}{12}$$

$$m = \left(\frac{1}{20(12)}\right)x = \frac{x}{240}$$

Next we will compute the primary and virtual deflections, which are

$$\delta_0 = \frac{1}{EI}\int_0^L M_0 m\,dx = \frac{1}{(29{,}000)(135)}\int_0^{240}\left(20x - \frac{x^2}{12}\right)\left(\frac{x}{240}\right)dx = 0.025\,\text{rad}$$

$$\delta_v = \frac{1}{EI}\int_0^L m^2\,dx = \frac{1}{(29{,}000)(135)}\int_0^{240}\left(\frac{x}{240}\right)^2 dx = 2.043 \times 10^{-5}\,\text{rad}$$

We can now compute the force M_A, which is

$$M_A = -\frac{\delta_0}{\delta_v} = -\frac{0.025}{2.043 \times 10^{-5}} = -1{,}200\,\text{kip-in} = -100\,\text{kip-ft}$$

The sign on M_A is negative so it is acting counter-clockwise, in the direction opposite that of the virtual force. This is the same magnitude for M_A that we obtained in Example 13.1. The other forces of reaction can be evaluated using the equations of static equilibrium and the results would be the same as we previously obtained. ■

The principles discussed can be applied to a frame as well as they can to a beam. Example 13.4 demonstrates application of the principles to a simple frame.

EXAMPLE 13.4

Compute the reactions for the frame structure shown in the figure.

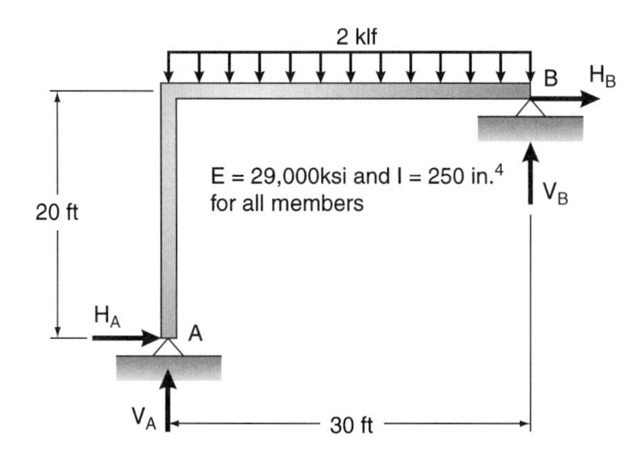

Solution. The horizontal reaction at B will be selected as the redundant reaction. The primary force system, then, is as shown in the next figure. One moment equation will be written for the column and another moment equation will be written for the horizontal member. Using the moment conventions indicated on the illustration, these moment equations are

$$M_1 = 0$$

$$M_2 = 30x_2 - \frac{\frac{2}{12}x_2^2}{2} = 30x_2 - \frac{x_2^2}{12}$$

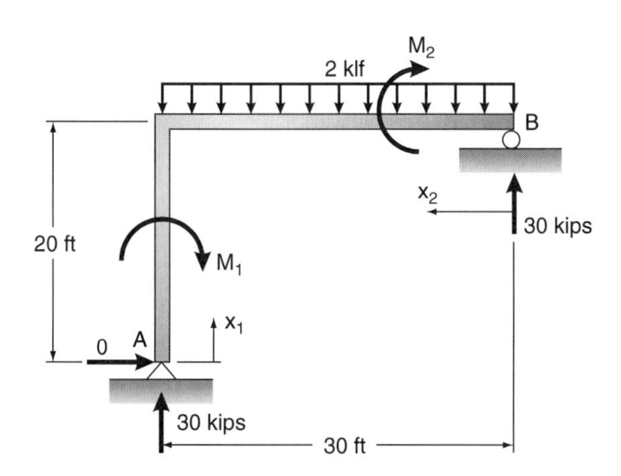

A horizontal virtual load of unit magnitude acting to the right is applied at B. The virtual force system that results is

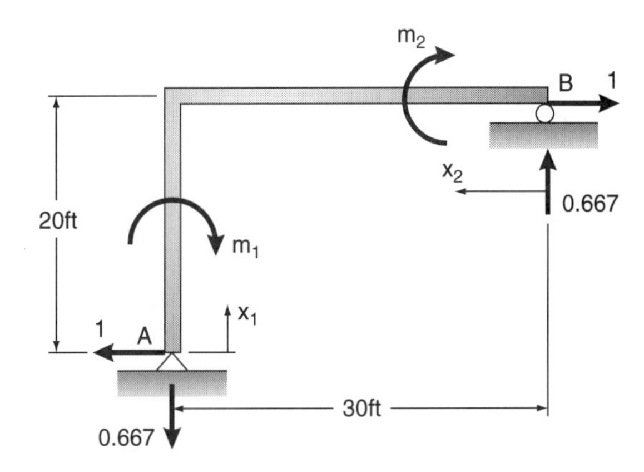

Using this virtual-force system and the indicated conventions for the virtual moments, the virtual-moment equations are

$$m_1 = -0.667x_1$$
$$m_2 = 0.667x_2$$

The primary and virtual deflections can now be computed using these moment equations. Those displacements are

$$\delta_0 = \frac{1}{EI}\int_0^{240} M_1 m_1\, dx_1 + \frac{1}{EI}\int_0^{360} M_2 m_2\, dx_2 = \frac{1}{29{,}000(250)}\int_0^{240} (0)(-0.667x_1)dx_1$$
$$+ \frac{1}{29{,}000(250)}\int_0^{360}\left(30x_2 - \frac{x_2^2}{12}\right)(0.667x_2)dx_2 \qquad \delta_0 = 10.52\,\text{in}$$

The virtual displacement is found to be

$$\delta_v = \frac{1}{EI}\int_0^{240} m_1 m_1\, dx_1 + \frac{1}{EI}\int_0^{360} m_2 m_2\, dx_2$$
$$= \frac{1}{29{,}000(250)}\int_0^{240} (-0.667x_1)^2 dx_1 + \frac{1}{29{,}000(250)}\int_0^{360} (0.667x_2)^2 dx_2$$
$$\delta_v = 1.213\,\text{in.}$$

The force of reaction H_B can now be computed.

$$H_B = -\frac{\delta_0}{\delta_v} = -\frac{10.52}{1.213} = -8.674\,\text{kips}$$

The negative indicates that the force is acting to the left, opposite the direction of the virtual force. The remaining reactions can be calculated using the equations of static equilibrium. First, we will compute H_A by summing forces horizontally:

$$\sum F_H = H_A + H_B = H_A + (-8.674) = 0$$
$$\therefore\ H_A = 8.674\,\text{kips}$$

Next, the vertical reaction at B can be computed by summing moments clockwise about A.

$$\sum M = \frac{w(30)^2}{2} - V_B(30) + H_B(20) = \frac{2(30)^2}{2} - V_B(30) + (-8.674)(20) = 0$$

$$\therefore\ V_B = 24.22\,\text{kips}$$

Lastly, we can sum forces vertically to solve for V_A.

$$\sum F_v = V_A - w(30) + V_B = V_A - 2(30) + 24.217 = 0$$

$$\therefore\ V_A = 35.78\,\text{kips}\quad \blacksquare$$

Raritan River Bridge, New Jersey (Courtesy of Steinman, Boynton, Gronquist & Birdsall Consulting Engineers)

13.2 BEAMS AND FRAMES WITH TWO OR MORE REDUNDANTS

The force method of analyzing beams and frames with one redundant may be extended to beams and frames having two or more redundants. The three-span continuous beam shown in Figure 13.3, which has two redundant reactions, is considered for discussion.

Because this beam is statically indeterminate to the second degree, two equations of condition are necessary to solve for the reactions. We will be working with the primary force system and two virtual force systems; there will be one virtual force system for each of the redundant reactions. Let us select the reaction at B and the reaction at C as the forces we will evaluate using the equations of condition. The primary

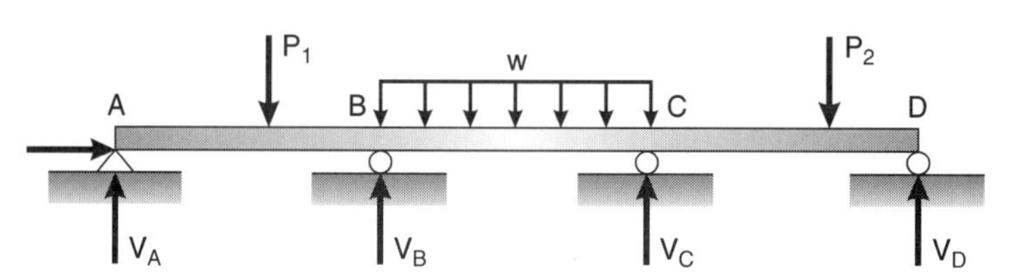

Figure 13.3 A continuous beam with two redundant reactions

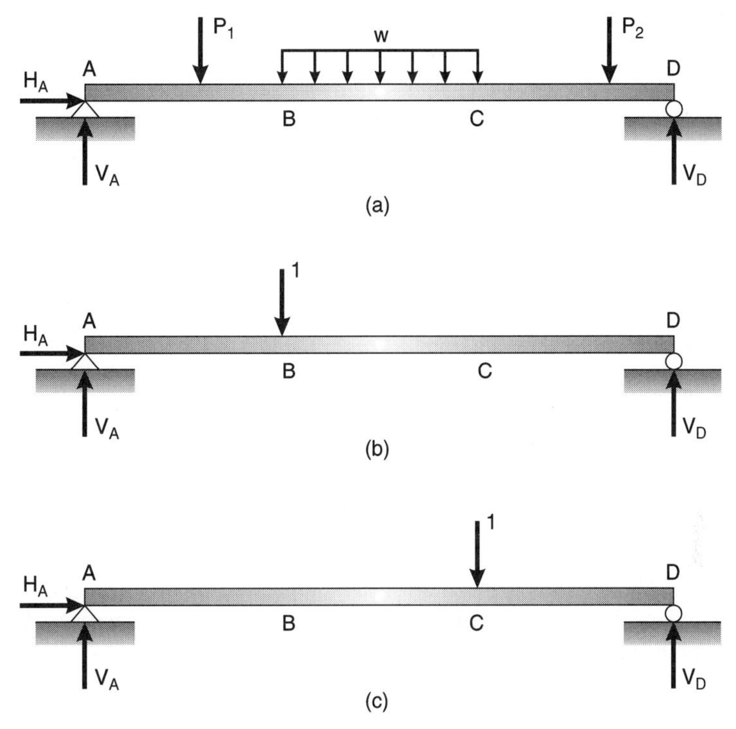

Figure 13.4 Primary and virtual force system for a continuous beam with two redundant reactions

force system, then, is shown in Figure 13.4(a). The virtual force systems are shown in Figure 13.4(b) for the redundant reaction at B and in Figure 13.4(c) for the redundant reaction at C.

The equations of condition are written by considering the displacement of the beam at each of the redundant reactions. The displacement at B is the result of the displacement at B caused by the primary forces plus the displacement at B caused by the reaction at B *plus* the displacement at B caused by the reaction at C. A similar statement can be made about the displacement at C. From the boundary conditions, the displacement at both B and C must be equal to zero. These relationships can be represented mathematically as:

$$\delta_B = \frac{1}{EI}\int_L M_0 m_b \, dx + \frac{V_B}{EI}\int_L m_b m_b \, dx + \frac{V_C}{EI}\int_L m_c m_b \, dx = 0$$
$$\delta_C = \frac{1}{EI}\int_L M_0 m_c \, dx + \frac{V_B}{EI}\int_L m_b m_c \, dx + \frac{V_C}{EI}\int_L m_c m_c \, dx = 0 \qquad \textbf{Eq. 13.6}$$

These equations are analogous to Equation 13.2. Consider for a moment the first of these two equations. The first term on the right-hand side is the displacement at B caused by the primary forces. The second term is the deflection at B caused by the reaction at B; the third term is the deflection at B caused by the reaction at C. Similar statements can be made about the second equation. In terms of displacements, these equations could have been represented as

$$\delta_{0_B} + V_B\delta_{bb} + V_C\delta_{bc} = 0$$
$$\delta_{0_C} + V_B\delta_{cb} + V_C\delta_{cc} = 0 \qquad \textbf{Eq. 13.7}$$

Both equations contain the two unknowns, V_B and V_C, whose values may be obtained by solving the equations simultaneously. In these equations, small subscripts are used to represent the virtual displacements. The first subscript indicates the point at which deflection is being computed and the second subscript indicates the location of the virtual force. For example, δ_{bc} is the displacement at B caused by a unit virtual force at C. This deflection is determined from

$$\delta_{bc} = \frac{1}{EI}\int_L m_b m_c \, dx \qquad \textbf{Eq. 13.8}$$

These principles can be expanded to systems involving n redundant reactions. The generalization of the equations for multiple redundant reactions is

$$\begin{aligned} \delta_{0_1} + V_1\delta_{1,1} + V_2\delta_{1,2} + \cdots + V_n\delta_{1,n} &= 0 \\ \delta_{0_2} + V_1\delta_{2,1} + V_2\delta_{2,2} + \cdots + V_n\delta_{2,n} &= 0 \\ \vdots \\ \delta_{0_n} + V_1\delta_{n,1} + V_2\delta_{n,2} + \cdots + V_n\delta_{n,n} &= 0 \end{aligned} \qquad \textbf{Eq. 13.9}$$

These equations can be solved simultaneously for the unknown reaction forces.

Example 13.5 demonstrates the application of the principles for a beam with two redundant reactions. Because of the number of equations involved, the solution is cast in terms of matrices.

EXAMPLE 13.5

Determine the forces of reaction in the three-span continuous beam shown in the figure.

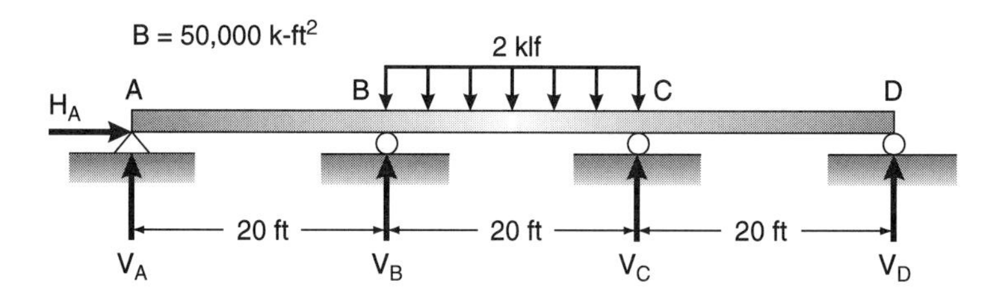

Solution. We will use the reactions at B and C as the redundant reactions. The primary force system then becomes

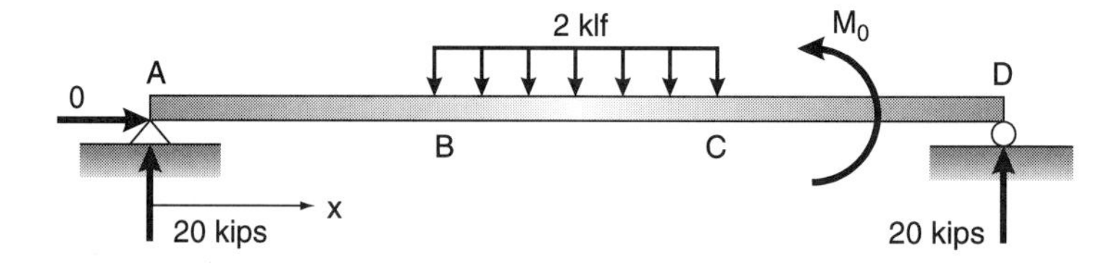

Using the indicated convention for positive internal bending moment, the equations for primary moment in the beam are

$$M_0 = 20x \qquad \text{for } 0 \le x < 20$$

$$M_0 = 20x - \frac{2(x-20)^2}{2} = 20x - (x-20)^2 \qquad \text{for } 20 \le x < 40$$

$$M_0 = 20x - 20(2)(x-30) = 1{,}200 - 20x \qquad \text{for } 40 \le x < 60$$

The virtual-force system and the corresponding moment equations for a unit virtual force applied at B are

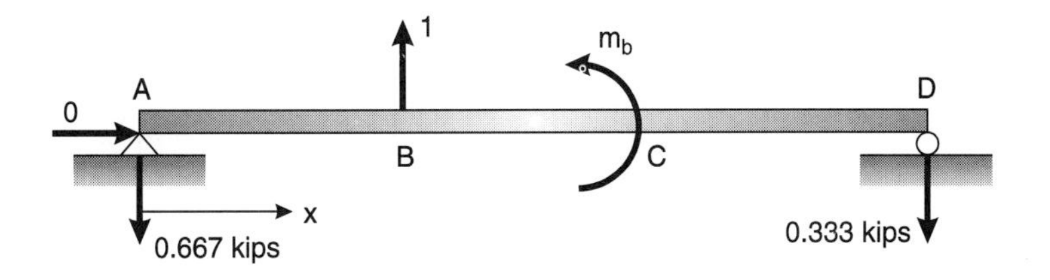

$$m_b = -0.667x \qquad \text{for } 0 \leq x < 20$$
$$m_b = -0.667x + (x - 20) = 0.333x - 20 \qquad \text{for } 20 \leq x < 60$$

The virtual force system and the corresponding moment equations for a unit virtual force applied at C are

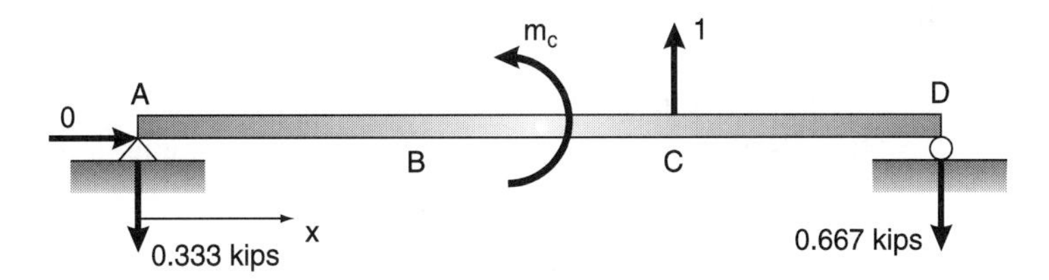

$$m_c = -0.333x \qquad \text{for } 0 \leq x < 40$$
$$m_c = -0.333x + (x - 40) = 0.667x - 40 \qquad \text{for } 40 \leq x < 60$$

Note that in both of the virtual force systems the direction in which the unit virtual force acts is the same as the assumed direction of the reaction force.

Now that the moment equations are known, we can compute the necessary terms in the equations of condition, which are

$$\delta_{0_B} + V_B\delta_{bb} + V_C\delta_{bc} = 0$$
$$\delta_{0_C} + V_B\delta_{cb} + V_C\delta_{cc} = 0$$

The displacement terms in these equations are

$$\delta_{0_B} = \frac{1}{EI}\left[\int_0^{20}(20x)\left(-\frac{2x}{3}\right)dx + \int_{20}^{40}(20x - (x-20)^2)\left(\frac{x}{3} - 20\right)dx + \int_{40}^{60}(1200 - 20x)\left(\frac{x}{3} - 20\right)dx\right] = -2.933 \text{ in.}$$

$$\delta_{0_C} = \frac{1}{EI}\left[\int_0^{20}(20x)\left(-\frac{x}{3}\right)dx + \int_{20}^{40}(20x - (x-20)^2)\left(\frac{x}{3}\right)dx + \int_{40}^{60}(1200 - 20x)\left(\frac{2x}{3} - 40\right)dx\right] = -2.933 \text{ in.}$$

$$\delta_{bb} = \frac{1}{EI}\left[\int_0^{20}\left(-\frac{2x}{3}\right)^2 dx + \int_{20}^{60}\left(\frac{x}{3} - 20\right)^2 dx\right] = 0.071 \text{ in.}$$

$$\delta_{bc} = \frac{1}{EI}\left[\int_0^{20}\left(-\frac{2x}{3}\right)\left(-\frac{x}{3}\right)dx + \int_{20}^{40}\left(\frac{x}{3} - 20\right)\left(-\frac{x}{3}\right)dx + \int_{40}^{60}\left(\frac{x}{3} - 20\right)\left(\frac{2x}{3} - 40\right)dx\right] = 0.062 \text{ in.}$$

$$\delta_{cb} = \frac{1}{EI}\left[\int_0^{20}\left(-\frac{x}{3}\right)\left(-\frac{2x}{3}\right)dx + \int_{20}^{40}\left(-\frac{x}{3}\right)\left(\frac{x}{3} - 20\right)dx + \int_{40}^{60}\left(\frac{2x}{3} - 40\right)\left(\frac{x}{3} - 20\right)dx\right] = 0.062 \text{ in.}$$

$$\delta_{cc} = \frac{1}{EI}\left[\int_0^{40}\left(-\frac{x}{3}\right)^2 dx + \int_{40}^{60}\left(\frac{2x}{3} - 40\right)^2 dx\right] = 0.071 \text{ in.}$$

Given these computed values, the equations of condition become

$$-2.933 + 0.071V_B + 0.062V_C = 0$$
$$-2.933 + 0.062V_B + 0.071V_C = 0$$

Next we can write the applicable equations of static equilibrium. The equations of static equilibrium and the equations of condition can be solved simultaneously to obtain all of the reactions. The equations of static equilibrium are

$$\sum F_V = V_A + V_B + V_C + V_D - 2(20) = 0$$
$$\sum F_H = H_A = 0$$
$$\sum M_D = 60V_A + 40V_B + 20V_C - 2(20)(30) = 0$$

The equations of static equilibrium and the equations of condition can be represented in matrix form as

$$\begin{bmatrix} 1 & 1 & 1 & 1 & 0 \\ 60 & 40 & 20 & 0 & 0 \\ 0 & 0.071 & 0.062 & 0 & 0 \\ 0 & 0.062 & 0.071 & 0 & 0 \\ 0 & 0 & 0 & 0 & 1 \end{bmatrix} \begin{Bmatrix} V_A \\ V_B \\ V_C \\ V_D \\ H_A \end{Bmatrix} = \begin{Bmatrix} 40 \\ 1{,}200 \\ 2.933 \\ 2.933 \\ 0 \end{Bmatrix}$$

This matrix can be solved for the unknown reactions as:

$$\begin{Bmatrix} V_A \\ V_B \\ V_C \\ V_D \\ H_A \end{Bmatrix} = \begin{bmatrix} 1 & 1 & 1 & 1 & 0 \\ 60 & 40 & 20 & 0 & 0 \\ 0 & 0.071 & 0.062 & 0 & 0 \\ 0 & 0.062 & 0.071 & 0 & 0 \\ 0 & 0 & 0 & 0 & 1 \end{bmatrix}^{-1} \begin{Bmatrix} 40 \\ 1{,}200 \\ 2.933 \\ 2.933 \\ 0 \end{Bmatrix} = \begin{Bmatrix} -2.053 \\ 22.053 \\ 22.053 \\ -2.053 \\ 0 \end{Bmatrix}$$ ■

13.3 SUPPORT SETTLEMENT

Continuous beams with unyielding supports have been considered in the preceding sections. Should the supports settle or deflect from their theoretical positions, major changes may occur in the reactions, shearing forces, bending moments, and stresses in the structure. These changes are easily understood by considering three men walking with a log on their shoulders (a statically indeterminate situation). If one of the men lowers his shoulder slightly, he will not have to support as much of the total weight as before. He has, in effect, backed out from under the log and thrown more of its weight to the other men. The settlement of a support in a statically indeterminate continuous beam has the same effect. Regardless of the factors that cause the displacements at the supports (weak foundations, temperature changes, poor erection or fabrication, and so on), analysis may be made with the deflection expressions previously developed for continuous beams.

I-81 river relief route interchange, Harrisburg, Pennsylvania
(Courtesy of Gannett Fleming)

We can adapt the principles previously studied to include the effects of support settlement. Let us consider again the propped-cantilever beam in Figure 13.1(a). An expression for deflection at point B was developed in Section 13.2. That expression was

$$\delta_B = 0 = \frac{V_B}{EI}\int_0^L x^2\,dx + \frac{1}{EI}\int_0^L -\frac{wx^2}{2}x\,dx \qquad \textbf{Eq. 13.10}$$

This expression was developed on the assumption that support B did not displace—its displacement was zero. If the support were to settle, or displace, some amount Δ, Equation 13.6 would be modified as follows:

$$\delta_B = \Delta = \frac{V_B}{EI}\int_0^L x^2\,dx + \frac{1}{EI}\int_0^L -\frac{wx^2}{2}x\,dx \qquad \textbf{Eq. 13.11}$$

If we make the same substitutions for primary and virtual moments as before, this equation becomes

$$\Delta = \frac{V_B}{EI}\int_0^L m^2\,dx + \frac{1}{EI}\int_0^L M_0 m\,dx \qquad \textbf{Eq. 13.12}$$

This equation can be rearranged to the form

$$V_B = \frac{\Delta - \dfrac{1}{EI}\displaystyle\int_0^L M_0 m\,dx}{\dfrac{1}{EI}\displaystyle\int_0^L m^2\,dx} \qquad \textbf{Eq. 13.13}$$

Recognize that, as before, the denominator in this equation is the deflection caused by the virtual unit force and that the numerator is the support settlement minus the deflection caused by the primary forces. This equation can be represented as

$$V_B = \frac{\Delta - \delta_0}{\delta_v} \qquad \textbf{Eq. 13.14}$$

When applying this equation, be very aware of units and the direction of the settlement. Positive settlement is in the same direction as the applied virtual force; negative settlement is in the direction opposite the virtual force. Compatible units should always be used in the analysis to obtain the correct forces. Both of these issues are very important and are common sources of error in problem solution.

Example 13.6 illustrates the analysis of the propped-cantilever beam of Example 13.1 with settlement at the right support. The student should note that a seemingly small displacement has significantly changed the bending moment in the beam.

EXAMPLE 13.6

Determine the reactions for the beam in Example 13.1, which is reproduced in the following figure if support B settles 1 in.

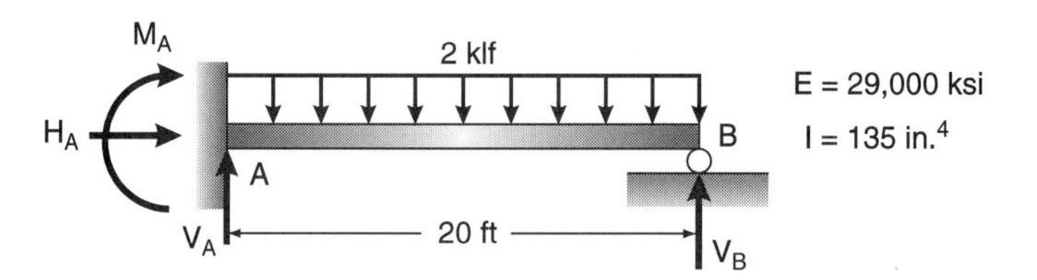

Solution. Because the settlement occurs at B, we will use V_B as the redundant force. This example, then, proceeds in exactly the same manner as Example 13.1 through the calculation of δ_0 and δ_v. From that example these values are

$$\delta_0 = -17.66 \text{ in.}$$
$$\delta_v = 1.177 \text{ in.}$$

We are now ready to compute V_B. The support has settled 1 in., which implies that it has moved downward. Normally if the support had moved upward the support displacement would have been referred to as heave instead of settlement. Downward support displacement is opposite the direction of the virtual force used, so it has a negative sign associated with it. That is to say

$$\Delta = -1 \text{ in.}$$

Using the values of calculation of δ_0 and δ_v, and this settlement, V_B is calculated from

$$V_B = \frac{\Delta - \delta_0}{\delta_v} = \frac{(-1) - (-17.66)}{1.177} = 14.15 \text{ kips}$$

Now that the force at the redundant reaction has been computed, including the effects of support settlement, we can compute the remaining reactions using the equations of static equilibrium. We will first sum forces horizontally to determine the horizontal reaction at A:

$$\sum F_H = H_A = 0$$
$$\therefore H_A = 0$$

Next we will sum forces vertically to determine the vertical reaction at A:

$$\sum F_V = V_A - wL + V_B = V_A - 2(20) + 14.15 = 0$$

$$\therefore V_A = 25.85 \text{ kips}$$

As a final step, we will sum moments clockwise about the left support to determine the magnitude of the reaction moment:

$$\sum M = M_A + wL\left(\frac{L}{2}\right) - V_B(20) = M_A + 2(20)\left(\frac{20}{2}\right) - 14.15(20) = 0$$

$$\therefore M_A = -117.0 \text{ kip-ft}$$

The negative sign on the computed reaction moment indicates that it is acting opposite the direction indicated on the drawing.

Observe that a settlement of only $1/240^{th}$ of the span length changed the bending moment at the left support by 17%. Support settlement can have a significant influence on the forces in a statically indeterminate structural system. ■

When several or all of the supports are displaced, the analysis may be conducted based on relative settlement values. For example, if all of the supports of the beam of Figure 13.5(a) were to settle 1.5 in., the stress conditions would be unchanged. If the supports settle different amounts but remain in a straight line, as illustrated in Figure 13.5(b), the situation theoretically is the same as before settlement. The beam has undergone rigid-body displacement.

When inconsistent settlements occur and the supports no longer lie on a straight line, the stress conditions change because the beam is distorted. The situation may be handled by drawing a line through the displaced positions of two of the supports, usually the end ones. This is the dashed line in Figure 13.5(c). The distances of the other supports from this line are determined and used in the calculations, as illustrated in Figure 13.5(c). In this illustration, support B has displaced relative to the other supports; the relative support displacement is used in analysis.

The same principles can be applied to structural systems with more than one redundant force of reaction. Consider the three-span continuous beam shown in Figure 13.6(a).

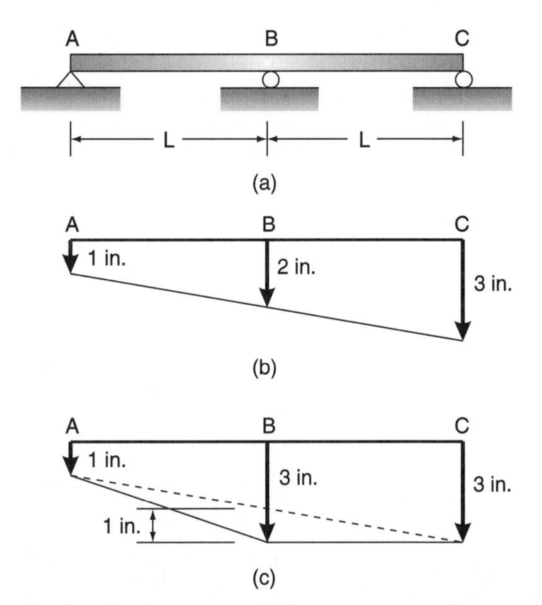

Figure 13.5 Example of relative support settlement

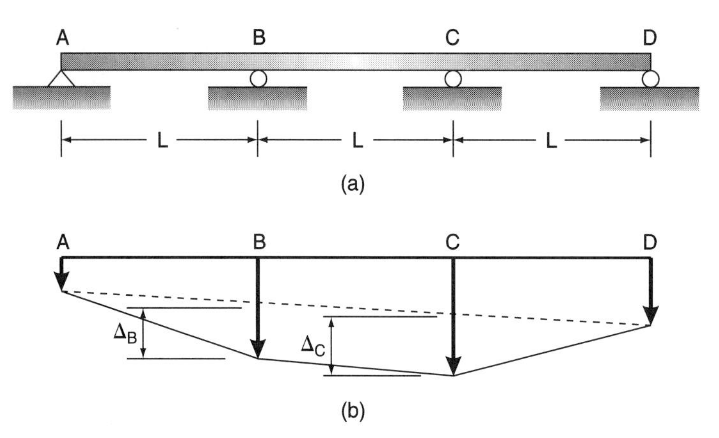

Figure 13.6 Differential settlement in a cloudly redundant beam

Assumed that the supports settle as shown in Figure 13.6(b). The supports at B and C settle relative to the other two supports in the structure. These supports should be taken as the redundant supports because of the differential settlement. When doing so the equations of condition are written as:

$$\begin{aligned} \delta_{0_B} + \Delta_B + V_B\delta_{bb} + V_C\delta_{bc} &= 0 \\ \delta_{0_C} + \Delta_C + V_B\delta_{cb} + V_C\delta_{cc} &= 0 \end{aligned} \qquad \textbf{Eq. 13.15}$$

If the relative settlements had been determined differently, different redundants would have been used when developing the equations of condition, but the process would have been the same.

13.4 ANALYSIS OF EXTERNALLY REDUNDANT TRUSSES

Trusses may be statically indeterminate because of redundant reactions, redundant members, or a combination of both. Externally redundant trusses will be considered initially, and they will be analyzed using deflection computations in a manner closely related to the procedure used for statically indeterminate beams.

The two-span truss shown in Figure 13.7 is considered for the following discussion. This truss is statically indeterminate to the first degree externally. One reaction component, for example, V_B, is selected at the redundant and is evaluated through use of an equation of condition, just as we did with statically indeterminate beams. The same equation of condition that we used with beams is also applicable to trusses. That equation was

$$\delta_0 + V_B\delta_v = 0$$

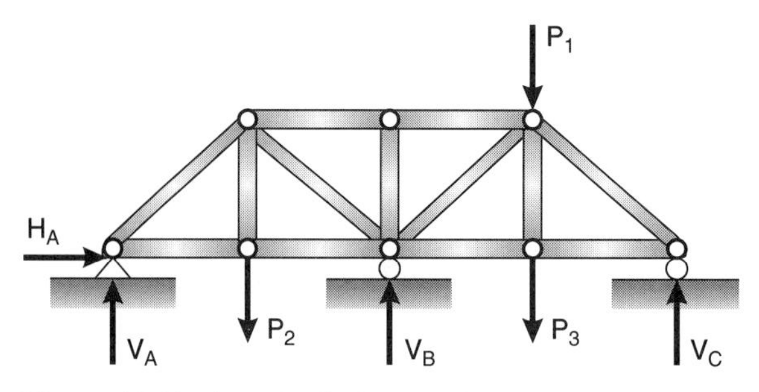

Figure 13.7 A statically indeterminate truss

In this equation, δ_0 is the deflection at the redundant reaction caused by all of the forces acting on the truss except the redundant reaction. These forces were referred to as the primary forces. The deflection is calculated using virtual work with a unit virtual force placed at the redundant reaction. The term δ_v is the deflection at the redundant reaction caused by the virtual force at that point. Recall from Chapter 11 that these two deflections can be calculated from

$$\delta_0 = \sum_{i=1}^{n}\left(\frac{f'f_vL}{AE}\right)_i \qquad \delta_v = \sum_{i=1}^{n}\left(\frac{f_v^2L}{AE}\right)_i \qquad \textbf{Eq. 13.16}$$

The resulting expression for the redundant reaction force is then

$$V_B = -\frac{\sum_{i=1}^{n}\left(\frac{f'f_vL}{AE}\right)_i}{\sum_{i=1}^{n}\left(\frac{f_v^2L}{AE}\right)_i} \qquad \textbf{Eq. 13.17}$$

In these last two equations, the term f' represents the force in a bar of the truss caused by the primary forces and the term f_v represents the force in the bar caused by the virtual force.

After the magnitude of the redundant reaction has been computed, the other reaction forces can be evaluated using the equations of static equilibrium. After all of the reactions are known, the final forces in each bar can be computed. Instead of analyzing the entire truss, the actual bar forces can be computed using the already known primary force, the virtual forces in the bar, and the computed force of reaction. The actual force in a particular bar n of the truss is

$$f_n = f'_n + V_B(f_v)_n \qquad \textbf{Eq. 13.18}$$

Example 13.7 illustrates the complete analysis of a two-span truss by the method just described. After the redundant reaction is found, the other reactions and the final member forces may be determined using the equations of static equilibrium.

EXAMPLE 13.7

Compute the reactions and member forces for the two-span continuous truss shown in the figure. The numbers by the bars are member areas in square inches.

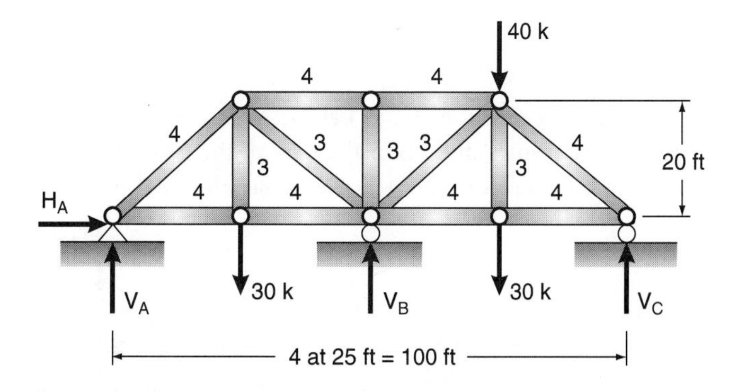

Solution. Select the center support at the redundant support. The primary forces are then:

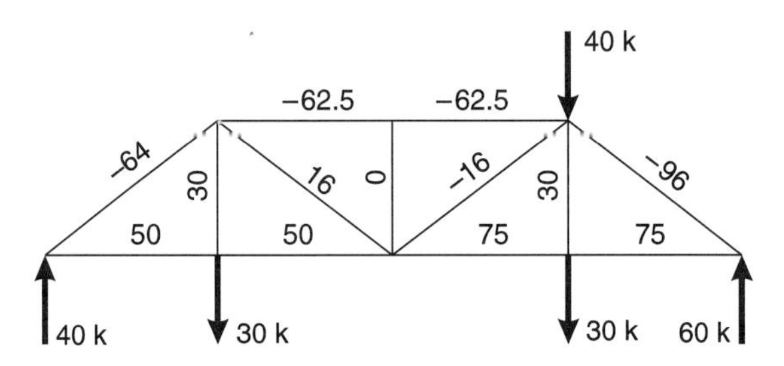

Using a unit virtual force applied at the redundant support, the virtual force system is

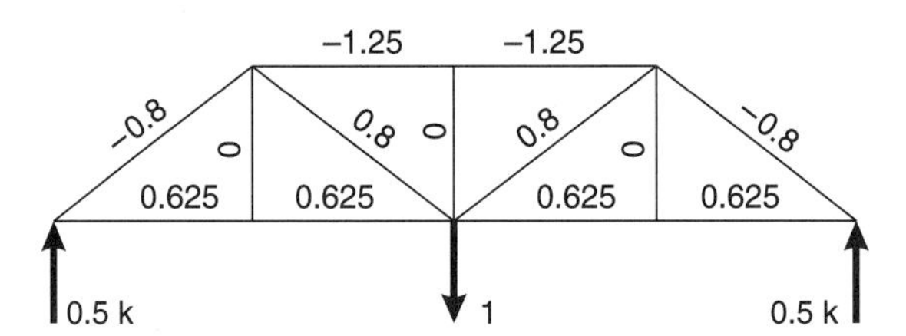

The computations necessary to determine V_B are shown in the table.

Member	L (in)	A in^2	$\frac{L}{A}$	f′ (kips)	f_v(kips)	$\frac{f'f_vL}{A}$	$\frac{f_v^2L}{A}$	$f' + V_Bf_v$
	300	4	75	+50	+0.625	+2340	+29.2	+15.0
	300	4	75	+50	+0.625	+2340	+29.2	+15.0
	300	4	75	+75	+0.625	+3510	+29.2	+40.0
	300	4	75	+75	+0.625	+3510	+29.2	+40.0
	384	4	96	−64	−0.800	+4920	+61.4	−19.2
	300	4	75	−62.5	−1.25	+5850	+117.0	+7.5
	300	4	75	−62.5	−1.25	+5850	+117.0	+7.5
	384	4	96	−96	−0.800	+7370	+61.4	−51.2
	240	3	80	+30	0.00	0	0	+30.0
	384	3	128	+16	+0.800	+1640	+82.0	−28.8
	240	3	80	0	0.00	0	0	0
	384	3	128	−16	+0.800	−1640	+82.0	−60.8
	240	3	80	+30	0	0	0	+30.0
					Σ =	35,690	637.6	

$$V_B = -\frac{\dfrac{35{,}690}{E}}{\dfrac{637.6}{E}} = -56.0 \text{ kips}$$

The remaining reaction forces can be evaluated using the equations or static equilibrium. The forces in each of the bars of the truss are computed in the last column of the table. Shown in the next figure are the final forces acting on the truss.

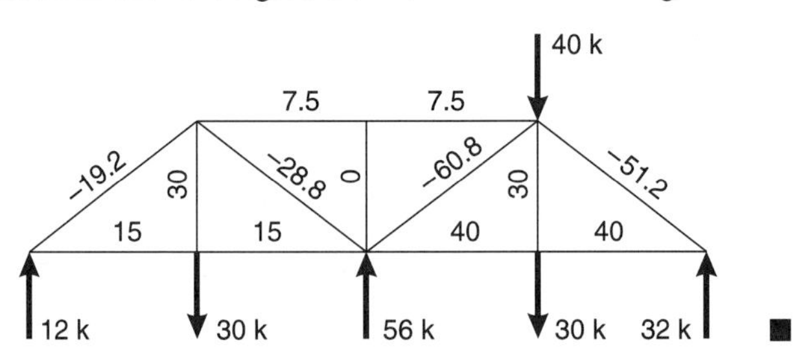

■

Rio Grande Gorge Bridge, Taos County, New Mexico (Courtesy of the American Institute of Steel Construction, Inc.)

It should be evident that the deflection procedure may be used to analyze trusses that have two or more redundant reactions. The truss shown in Figure 13.8 is continuous over three spans. The reactions at the interior supports, V_B and V_C can be considered as the redundant reactions. The following equations of condition, previously written for a three-span continuous beam, are applicable to the truss:

$$\begin{aligned} \delta_{0_B} + V_B\delta_{bb} + V_C\delta_{bc} &= 0 \\ \delta_{0_C} + V_B\delta_{cb} + V_C\delta_{cc} &= 0 \end{aligned} \qquad \textbf{Eq. 13.19}$$

If the forces due to a unit load at B are referred to as the f_{vB} forces and those due to a unit load at C are called the f_{vB} forces, the deflection terms in these equations are

$$\begin{aligned} \delta_{0B} &= \sum_{i=1}^{n}\left(\frac{f' f_{vb} L}{AE}\right)_i & \delta_{0C} &= \sum_{i=1}^{n}\left(\frac{f' f_{vc} L}{AE}\right)_i \\ \delta_{bb} &= \sum_{i=1}^{n}\left(\frac{f_{vb}^2 L}{AE}\right)_i & \delta_{cb} &= \sum_{i=1}^{n}\left(\frac{f_{vc} f_{vb} L}{AE}\right) \\ \delta_{bc} &= \sum_{i=1}^{n}\left(\frac{f_{vb} f_{vc} L}{AE}\right) & \delta_{cc} &= \sum_{i=1}^{n}\left(\frac{f_{vc}^2 L}{AE}\right) \end{aligned} \qquad \textbf{Eq. 13.20}$$

Observe that a unit load at B will cause a deflection at C that is equal to the deflection at B caused by a unit load at C. This is another illustration of Maxwell's law of reciprocal deflections. If the terms in Equation 13.20 are substituted into Equation 13.19,

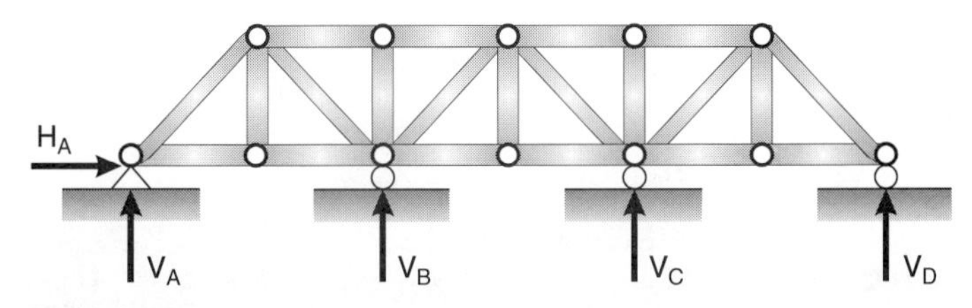

Figure 13.8

the equations of condition become

$$\sum_{i=1}^{n}\left(\frac{f'f_{vb}L}{AE}\right)_i + V_B\sum_{i=1}^{n}\left(\frac{f_{vb}^2L}{AE}\right) + V_C\sum_{i=1}^{n}\left(\frac{f_{vb}f_{vc}L}{AE}\right) = 0$$

$$\sum_{i=1}^{n}\left(\frac{f'f_{vc}L}{AE}\right) + V_B\sum_{i=1}^{n}\left(\frac{f_{vc}f_{vb}L}{AE}\right) + V_C\sum_{i=1}^{n}\left(\frac{f_{vc}^2L}{AE}\right) = 0 \qquad \textbf{Eq. 13.21}$$

Solving these equations simultaneously will yield the values of the redundants. Should support settlement occur, the deflections would have to be worked out numerically in the same units given for the settlements.

Tennessee River Bridge, Stevenson, Alabama (Courtesy of USX Corporation)

13.5 ANALYSIS OF INTERNALLY REDUNDANT TRUSSES

The truss shown in Figure 13.9 has one more member than is necessary for stability and is therefore statically indeterminate internally to the first degree, as can be proved by applying the equation

$$m = 2j - 3 \qquad \textbf{Eq. 13.22}$$

Internally redundant trusses may be analyzed in a manner closely related to the one used for externally redundant trusses. One member is taken to be the redundant and the force in that member is determined from an equation of condition. The truss, if the redundant

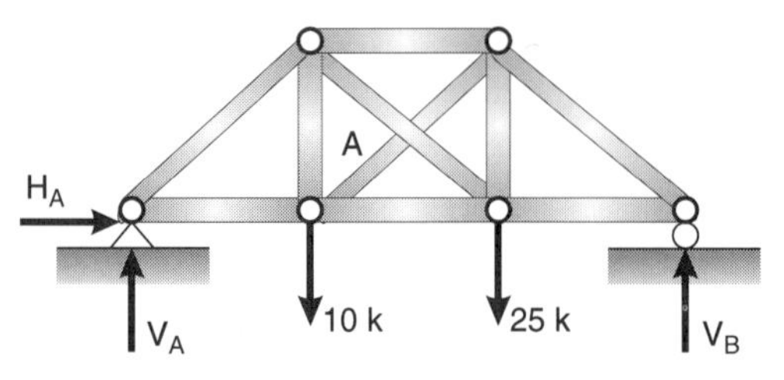

Figure 13.9 An internally redundant truss

member were missing, must form a statically determinate and stable truss. For our discussion we will take the member labeled A in Figure 13.9 to be the redundant member.

The equation of condition that we will use is the deformation of the redundant member. The equation of condition is then

$$\sum_{i=1}^{n}\left(\frac{ff_vL}{AE}\right)_i = \delta_A = \frac{f_AL}{AE} \qquad \textbf{Eq. 13.23}$$

In this equation f are the forces in each of the bars caused by the forces acting on the truss. The forces f_v are the virtual forces in the truss when the virtual force in A is equal to unity. The force f_A is the force in bar A, the redundant bar. If we were to let the primary forces in the truss be the forces in the members if the force in the redundant bar were equal to zero, Equation 13.23 could be expressed as

$$\sum_{i=1}^{n}\left[\frac{(f' + f_af_v)f_vL}{AE}\right]_i = \frac{f_AL}{AE} \qquad \textbf{Eq. 13.24}$$

In this equation, the forces f′ are the forces in the bars if the force in the redundant bar is equal to zero. We can separate the redundant bar from the summation. When doing so we obtain the equation

$$\sum_{i=1}^{n}\left[\frac{(f' + f_Af_v)f_vL}{AE}\right]_{i\neq A} + \frac{f_AL}{AE} = \frac{f_AL}{AE} \qquad \textbf{Eq. 13.25}$$

Observe that all bars except bar A are included in the summation. The second term in the equation was obtained because in bar A the force f′ is equal to zero and the force f_v is equal to unity. Upon reducing and rearranging Equation 13.25, we obtain the expression

$$\sum_{i=1}^{n}\left(\frac{(f'f_v)L}{AE}\right)_i + f_A\sum_{i=1}^{n}\left(\frac{f_v^2L}{AE}\right)_i = 0 \qquad \textbf{Eq. 13.26}$$

This last equation can then be solved for f_A, which is

$$f_A = \frac{\sum_{i=1}^{n}\left(\frac{(f'f_v)L}{AE}\right)_i}{\sum_{i=1}^{n}\left(\frac{f_v^2L}{AE}\right)_i} \qquad \textbf{Eq. 13.27}$$

The application of this method for analyzing internally redundant trusses is illustrated in Example 13.8. After the force in the redundant member is found, the force in the other members can be found. These forces, as before, are $f_i = f_i' + f_A(f_v)_i$

EXAMPLE 13.8

Determine the forces in the members of the internally redundant truss shown in the figure. The number by each bar is the area of that bar.

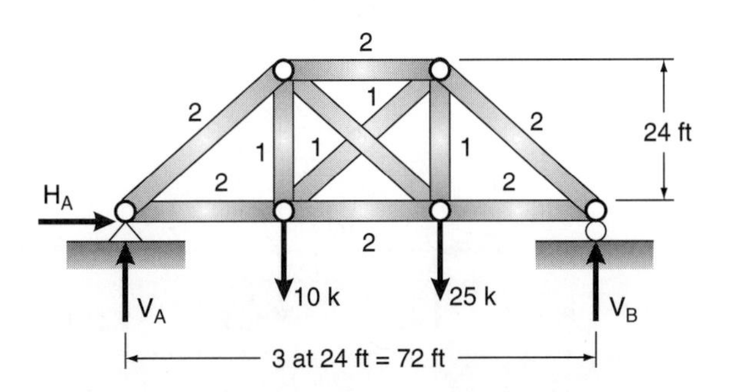

Solution. Assume L_1U_2 to be the redundant member. The primary forces then become

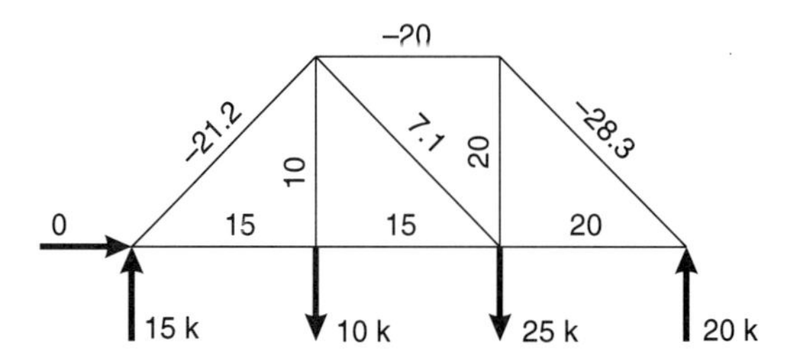

Applying a positive unit virtual force at L_1U_2, the virtual forces in the truss become

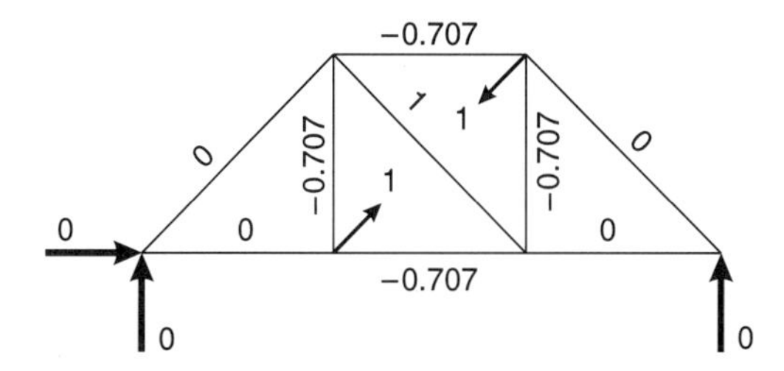

After performing the calculations in the following table and computing f_A, we can determine the following final forces.

Member	L (in)	A in^2	$\frac{L}{A}$	f′ (kips)	f_v(kips)	$\frac{f'f_vL}{A}$	$\frac{f_v^2L}{A}$	$f' + f_Af_v$
	288	2	144	+15.0	0	0	0	+15.0
	288	2	144	+15.0	−0.707	−1530	+72.0	+13.47
	288	2	144	+20.0	0	0	0	+20.0
	408	2	204	−21.2	0	0	0	−21.0
	288	2	144	−20.0	−0.707	+2040	+72.0	−21.53
	408	2	204	−28.3	0	0	0	−28.3
	288	1	288	+10.0	−0.707	−2040	+144.0	+8.47
	408	1	408	+7.1	+1.0	+2900	+408.0	+9.26
	408	1	408	0	+1.0	0	+408.0	+2.16
	288	1	288	+20.0	−0.707	−4070	+144.0	+18.47
					Σ =	−2,700	1,248	

$$f_A = -\frac{\frac{-2{,}700}{E}}{\frac{1{,}248}{E}} = +2.16$$

The final forces in the truss are

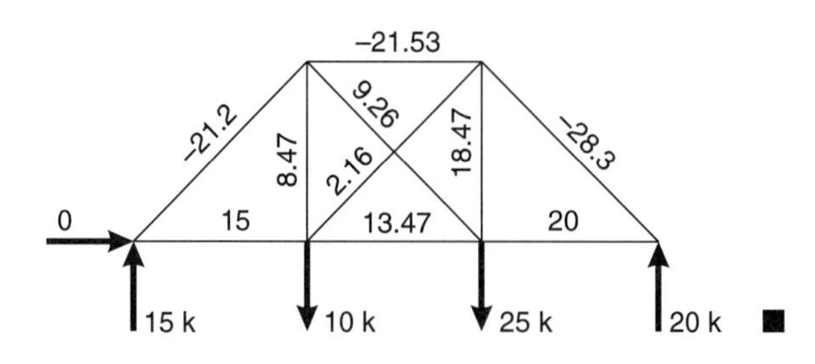

For trusses that have more than one redundant internally, simultaneous equations are necessary in the solution. Two members, with forces of f_A and f_B, are assumed to be the redundant members. Following the principles developed, and recalling previous solutions for systems with multiple redundants, the equations of condition used in the solution are

$$\sum_{i=1}^{n}\left(\frac{f'f_{va}L}{AE}\right)_i + f_A\sum_{i=1}^{n}\left(\frac{f_{va}^2L}{AE}\right) + f_B\sum_{i=1}^{n}\left(\frac{f_{va}f_{vb}L}{AE}\right) = 0$$
$$\sum_{i=1}^{n}\left(\frac{f'f_{vb}L}{AE}\right) + f_A\sum_{i=1}^{n}\left(\frac{f_{vb}f_{va}L}{AE}\right) + f_B\sum_{i=1}^{n}\left(\frac{f_{vb}^2L}{AE}\right) = 0 \qquad \textbf{Eq. 13.28}$$

13.6 ANALYSIS OF TRUSSES REDUNDANT INTERNALLY AND EXTERNALLY

Deflection equations have been written so frequently in the past few sections that the reader probably is able to set up his or her own equations for types of statically indeterminate beams and trusses not previously encountered. Nevertheless, one more group of equations is developed here; the ones necessary for the analysis of a truss that is statically indeterminate internally and externally. For the following discussion the truss shown in Figure 13.10, which has two redundant members and one redundant reaction component, will be considered.

The diagonals labeled D and E and the interior reaction V_B can be removed from the truss, which leaves a statically determinate structure. Combining the principles we learned in Sections 13.4 and 13.5, the following equations of condition can be written:

$$\delta_B + V_B\delta_{bb} + f_D\delta_{bd} + f_E\delta_{be} = 0$$
$$\delta_D + V_B\delta_{db} + f_D\delta_{dd} + f_E\delta_{de} = 0 \qquad \textbf{Eq. 13.29}$$
$$\delta_E + V_B\delta_{eb} + f_D\delta_{ed} + f_E\delta_{ee} = 0$$

Using principles previously discussed and used, the deflection terms in these equations are

$$\delta_B = \sum_{i=1}^{n}\frac{f'f_bL}{AE} \quad \delta_{bb} = \sum_{i=1}^{n}\frac{f_b^2L}{AE} \quad \delta_B = \sum_{i=1}^{n}\frac{f_bf_dL}{AE} \quad \delta_B = \sum_{i=1}^{n}\frac{f_bf_eL}{AE}$$
$$\delta_D = \sum_{i=1}^{n}\frac{f'f_dL}{AE} \quad \delta_B = \sum_{i=1}^{n}\frac{f_df_bL}{AE} \quad \delta_{dd} = \sum_{i=1}^{n}\frac{f_d^2L}{AE} \quad \delta_B = \sum_{i=1}^{n}\frac{f_df_eL}{AE} \qquad \textbf{Eq. 13.30}$$
$$\delta_E = \sum_{i=1}^{n}\frac{f'f_eL}{AE} \quad \delta_B = \sum_{i=1}^{n}\frac{f_ef_bL}{AE} \quad \delta_B = \sum_{i=1}^{n}\frac{f_ef_dL}{AE} \quad \delta_{ee} = \sum_{i=1}^{n}\frac{f_e^2L}{AE}$$

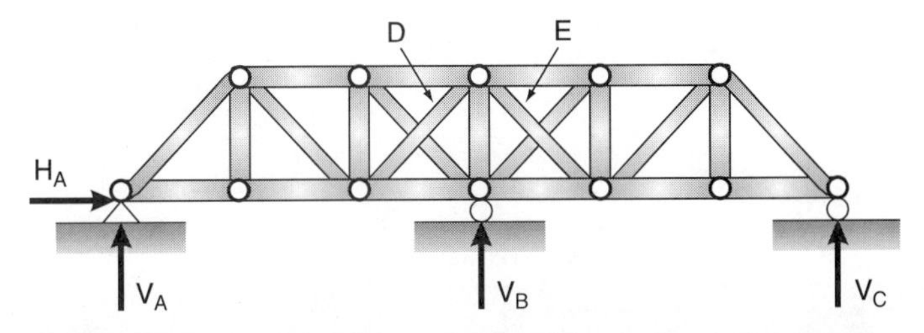

Figure 13.10 An internally and externally statically indeterminate truss

Computation of these sets of deflections permits the calculation of the numerical values of the redundant forces. Nothing new is involved in the solution of this type of problem, and space is not taken for the lengthy calculations necessary for an illustrative example.

Delaware River Turnpike Bridge (Courtesy of USX Corporation)

13.7 TEMPERATURE CHANGES, SHRINKAGE, FABRICATION ERRORS, AND SO ON

Structures are subject to deformations due not only to external loads but also to temperature changes, support settlements, inaccuracies in fabrication dimensions, shrinkage in reinforced concrete members caused by drying and plastic flow, etc. Such deformations in statically indeterminate structures can cause the development of large additional forces in the members. As an example, assume that the top-chord members of the truss in Figure 13.11 are exposed to the sun much more than are the other members. Consequently, on a hot sunny day, they may have much higher temperatures than the other members, and the member forces may undergo some appreciable changes.

Problems such as these may be handled exactly as were the previous problems in this chapter. The changes in the member length due to temperature are computed. These values correspond to the f′L/AE values, each equal the temperature change multiplied by the coefficient of expansion of the material and the member length. By analogy, these are the primary forces times the virtual forces times the member stiffness. A virtual unit force

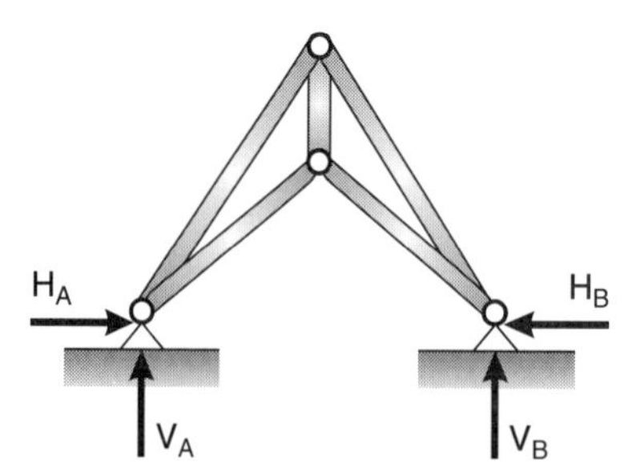

Figure 13.11 Truss used for discussion of effects of temperature changes

is placed at the redundant support in the direction of the redundant, and the f_v forces are computed. The solution then proceeds as before and the usual deflection expression is written. Such a problem is illustrated in Example 13.9.

EXAMPLE 13.9

The top-chord members of the statically indeterminate truss shown in the figure increase in temperature by 60°F. If $E = 29 \times 10^6$ psi and the coefficient of linear temperature expansion is 0.0000065/°F, determine the forces induced in each of the truss members. The number by each member is the area of that member in square inches.

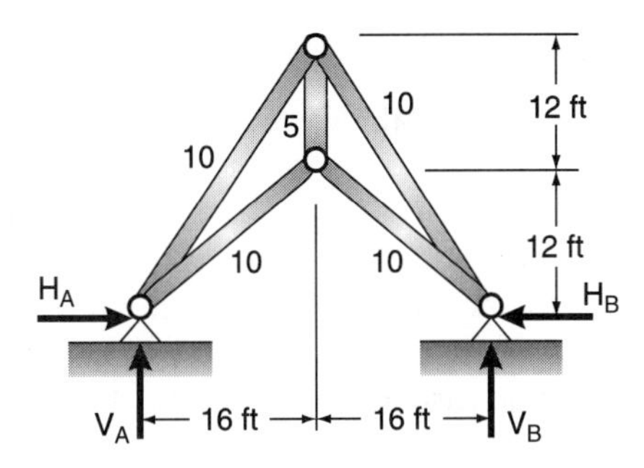

Solution. The temperature change is the only load acting on the truss. Assume that H_B is the redundant reaction and compute the virtual forces, which are

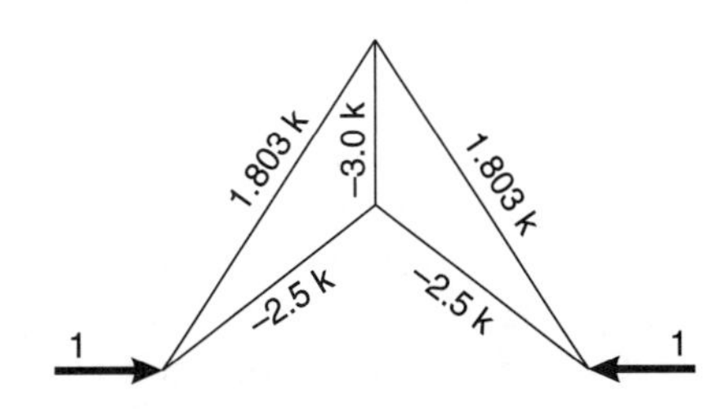

Member	L (in.)	A in²	$\frac{L}{A}$	f_v	$\frac{f_v^2 L}{A}$	$\Delta L = \Delta T \mu L$	$f_v \Delta L$
L_0L_1	240	10	24	−2.5	+150.0		
L_1L_2	240	10	24	−2.5	+150.0		
L_0U_1	346	10	34.6	+1.803	+112.48	0.1349	0.2433
U_1L_2	346	10	34.6	+1.803	+112.48	0.1349	0.2433
U_0L_1	144	5	28.8	−3.0	+259.2		
				Σ =	784.16		0.4866

$$H_B = -\frac{\delta_B}{\delta_{bb}} = \frac{0.4866}{\dfrac{784.16}{29{,}000}} = -18.02 \text{ kips}$$

The final forces in the truss members due to the temperature change are as follows:

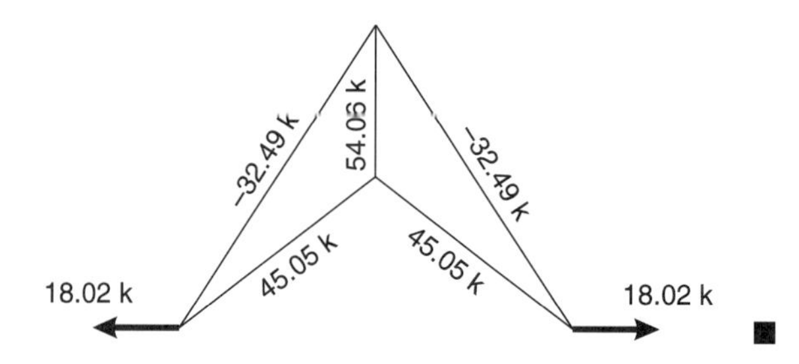

We can see from the last example that changes in temperature can cause very significant forces to occur in a structure.

13.8 CASTIGLIANO'S SECOND THEOREM

Castigliano developed two theorems that are useful in structural analysis. In Chapter 11, we saw that the first partial derivative of strain energy with respect to a load P (real or virtual) is equal to the deflection at and in the direction of P. This is a statement of Castigliano's second theorem. We used it to compute deflections in statically determinate structures. This theorem also is applicable to the analysis of statically indeterminate structures.

For this discussion the continuous beam of Figure 13.12 and the vertical reaction at support B, V_B, are considered. If the first partial derivative of the strain energy in this beam, which is equal to the complementary strain energy in a linearly elastic structure, is taken with respect to the reaction V_B, the deflection at B will be obtained. In this problem, because B is a structural support, that deflection is zero.

$$\frac{\partial U}{\partial V_B} = \delta_B = 0 \qquad \textbf{Eq. 13.31}$$

Equations of this type can be written for each point of constraint in a statically indeterminate structure. A structure will deform in a manner consistent with its physical limitations and so that the internal work of deformation will be at a minimum. These equations are the equations of condition needed to analyze the structure.

To analyze an internally redundant statically indeterminate structure with Castigliano's second theorem, certain members are assumed to be the redundant members. There must be a sufficient number of other members to form a statically determinate and stable base structure. The forces in the redundant members are evaluated using equations of condition for the deformation of these members in a manner similar to that used for virtual work in Chapter 11. The forces in the structure, f, are determined from the external loads and the assumed forces of X_1, X_2, X_3 and so on in the redundant members. The total strain energy can be written in terms of the forces f, and the result can be differentiated

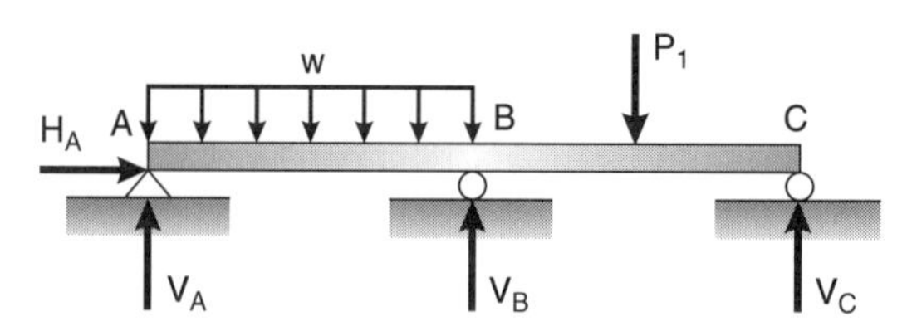

Figure 13.12

with respect to each of the redundants. The derivatives are set equal to zero and the forces in the redundant members are determined. The forces in the other members can then be determined.

Examples 13.10 to 13.15 illustrate the analysis of statically indeterminate structures using Castigliano's second theorem. Although consistent distortion methods are the most general methods for analyzing various types of statically indeterminate structures, they are not frequently used today because other methods previously discussed are more easily applied.

EXAMPLE 13.10

Determine the reaction at support C in the beam shown in the following figure using Castigliano's second theorem.

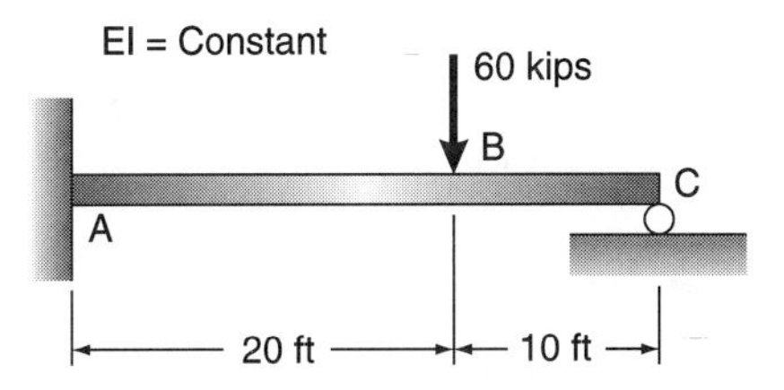

Solution. This beam is statically indeterminate to the first degree. We will take the reaction at C, V_c, to be the redundant reaction. Given this assumption, the moment equations we will use in the solution are:

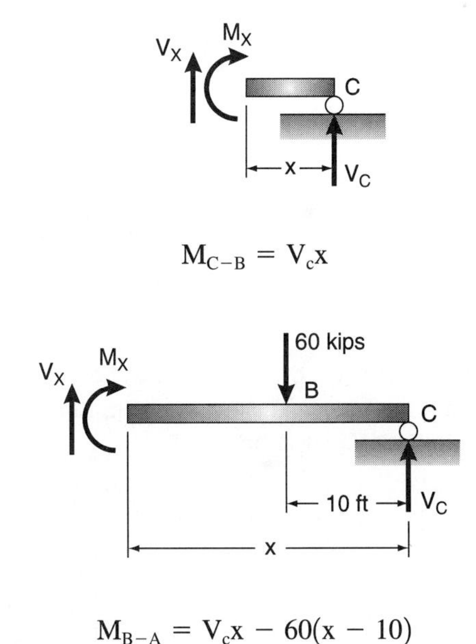

$$M_{C-B} = V_c x$$

$$M_{B-A} = V_c x - 60(x - 10)$$

After determining these moment equations, we can solve for the redundant reaction. Recall that:

$$\frac{\partial U}{\partial F} = \int_0^L \frac{1}{EI} M \frac{\partial M}{\partial F} dx$$

Section	M	$m = \frac{\partial M}{\partial V_c}$	$\frac{1}{EI}\int_L M\left(\frac{\partial M}{\partial V_c}\right)dx$
C to B	$V_c(x)$	x	$\frac{1}{EI}\int_0^{10} Mm\,dx$ $\frac{1{,}000}{3}V_c$
B to A	$V_c(x) - 60(x - 10)$	x	$\frac{1}{EI}\int_{10}^{30} Mm\,dx$ $\frac{26{,}000}{3}V_c - 280{,}000$
			$9{,}000V_c - 280{,}000 = 0$ $V_c = 31.1 \uparrow$

Observe in the solution that $\partial M/\partial V_c$ is identical to the virtual moment m if a virtual unit force is used in the solution to solve for the same redundant reaction. ■

EXAMPLE 13.11

Determine the force in member CD of the truss shown in the figure. The cross-sectional areas of the members are indicated on the diagram and E is constant.

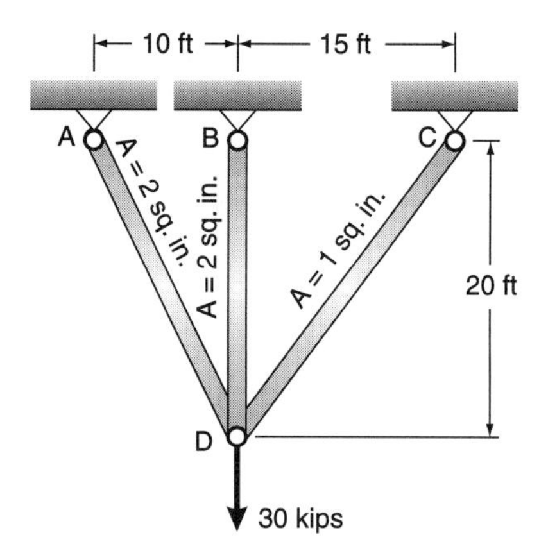

Solution. Member CD is assumed to be the redundant member and is assumed to have a force of T in it. The forces, then, in all of the bars in the truss are as shown on the following illustration.

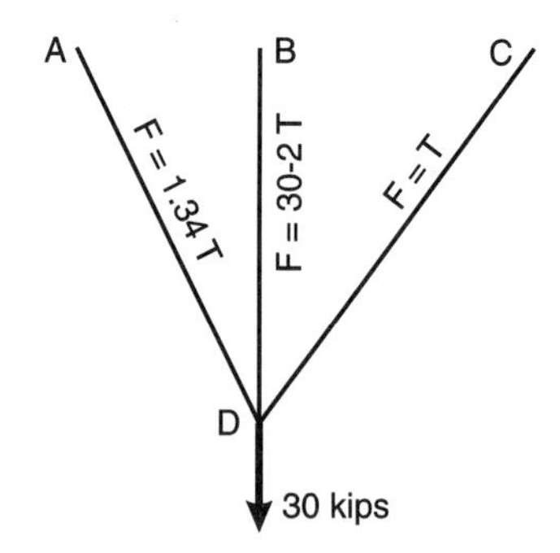

Using the next equation, the results shown in the following table are obtained.

$$\frac{\partial W}{\partial T} = \frac{\partial}{\partial T}\sum \frac{F^2L}{2AE} = \sum \frac{FL}{AE}\frac{\partial F}{\partial T}$$

Member	L/A	F	$\frac{\partial F}{\partial T}$	$\frac{FL}{A}\left[\frac{\partial F}{\partial T}\right]$
AD	134	1.34T	1.34	240T
BD	240	30 − 2T	−2.0	480T − 7200
CD	300	T	1.0	300T
				$\Sigma = 1020T - 7200 = 0$
				$T = 7.60$ k

After the force T is known, the force in all of the other members can be easily computed. ■

It should be obvious from the preceding examples that the amount of work involved in the analysis of statically indeterminate trusses by either Castigliano's second theorem or by consistent distortions is about equal. For statically indeterminate beams and frames, other methods such as the moment distribution method presented in Chapters 17 and 18 are much simpler to apply.

Castigliano's second theorem, however, is particularly useful for analyzing composite structures, such as that considered in Example 13.12. In these types of structures both bending and truss action take place. The reader will be convinced of the advantage of Castigliano's second theorem for the analysis of composite structures if he or she attempts to solve the following two problems by consistent distortions.

EXAMPLE 13.12

Using Castigliano's second theorem, calculate the force in the steel rod of the composite structure shown in the following figure.

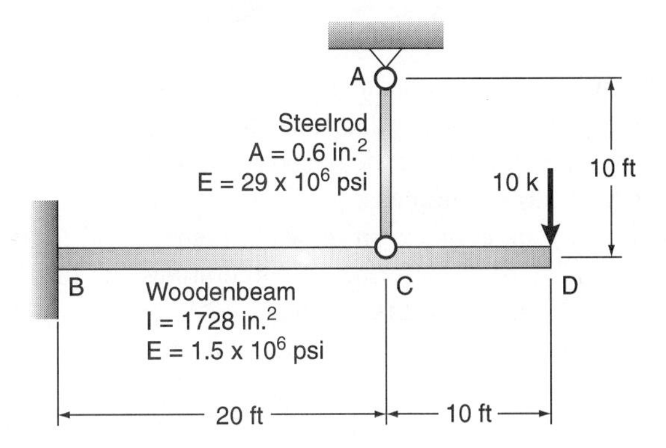

Solution. To evaluate this structure, we will assume that the rod is the redundant member and the force in it is equal to T. When solving this problem we must consider axial strain energy as well as flexural strain energy. The rod has only an axial force in it whereas the beam contains only flexural strain energy. The strain energy in the rod and the beam can be computed from

$$U_{rod} = \frac{T^2L}{2AE_{rod}}$$

$$U_{beam} = \int_0^{360} \frac{[-10x + T\Phi(x - 120)(x - 120)]^2}{2E_{beam}I} dx$$

The partial derivative of the total strain energy with respect to the force in the rod, T, must be equal to zero. As such, we can solve for T from

$$0 = \frac{\partial U_{rod}}{\partial T} + \frac{\partial U_{beam}}{\partial T} = 1.785T - 31.11$$

$$\therefore T = 17.43 \text{ kips} \quad \blacksquare$$

13.9 CASTIGLIANO'S FIRST THEOREM: THE METHOD OF LEAST WORK

As we discussed briefly in Chapter 11, Castigliano developed two theorems. We have already discussed in detail his second theorem for evaluating statically determinate and statically indeterminate structural systems. As you recall from that discussion, Castigliano's second theorem is based on complementary strain energy, which, conveniently for a linearly elastic structure, is equal to strain energy. Castigliano's first theorem, which is really the converse of the second theorem, states that the partial derivative of strain energy with respect to a displacement yields the magnitude of the force at that point moving through that displacement.

Castigliano's first theorem, which is commonly known as the *method of least work,* has played an important role in the development of structural analysis through the years. Nevertheless, it is only occasionally used today. Conceptually, the columns and girders meeting at a joint in a building will all deflect the same amount—the smallest possible value. Each member does no more work than is necessary, and the total work performed by all of the members and loads is the least possible. The method of least work has the disadvantage that it is not applicable in its usual form to forces caused by displacements due to temperature changes, support settlements, and fabrication errors.

The method of least work states that from all the admissible configurations of an elastic structure that satisfy the conditions of equilibrium, the configuration with the minimum potential energy is the stable equilibrium configuration. An admissible configuration in general is one that does not violate the boundary conditions, the equilibrium, and the continuity of the system. An important phrase in the definition of least work is *admissible system configuration.* Let us look at the beam shown in Figure 13.13 to better understand this concept.

Possible admissible deformed configurations for this beam are shown on the left side of Figure 13.14. Notice that although we know from our experience that the probable configuration is the upper configuration, the other configuration is possible because it

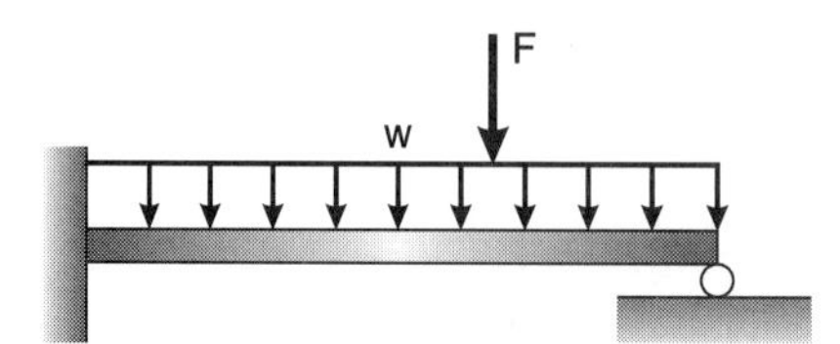

Figure 13.13

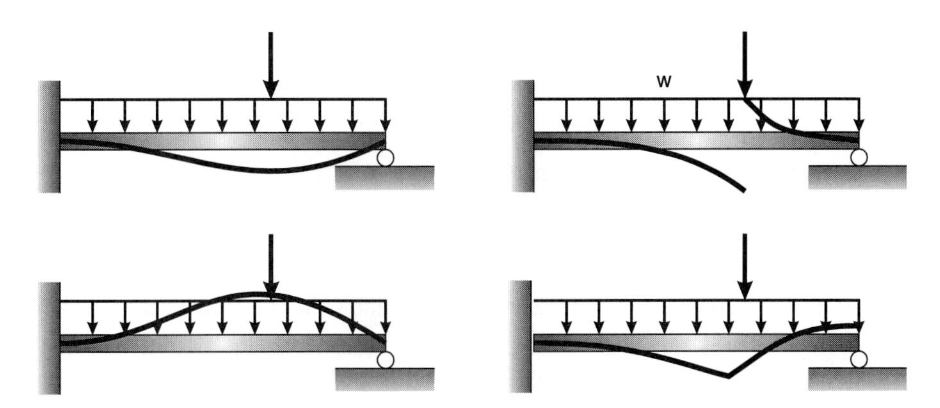

Figure 13.14 Admissible and inadmissible deflected shapes for a dropped cantilever beam

has not violated continuity or the boundary conditions. The configurations shown on the right side of Figure 13.14 are inadmissible because continuity of the beam or the boundary conditions have been violated.[1]

The potential energy in a structure is

$$\Pi = U + \Omega \qquad \textbf{Eq. 13.32}$$

The first term on the right side of the equation, U, is the strain energy stored in the structure as a result of the applied loads. The second term, Ω, is the potential energy of the applied loads. From this equation, we observe that if we can minimize the term Ω, the entire expression takes on a minimum value because, as we have already seen, the strain energy is a function of the distortion caused by the applied loads.

The work done by the applied loads, which is a loss of potential to do work, is equal to the sum of all the loads times the distances through which they move. Substituting this into Equation 15.2, it becomes

$$\Pi = U - \sum_{i=1}^{n} P_i D_i \qquad \textbf{Eq. 13.33}$$

The minus sign in Equation 13.33 exists because a force loses potential energy as it does work: its potential to do work decreases.

The potential energy in the system is minimum for the configuration that satisfies the equation

$$d\Pi = 0 = \frac{\partial \Pi}{\partial D_1} dD_1 + \frac{\partial \Pi}{\partial D_2} dD_2 + \cdots \qquad \textbf{Eq. 13.34}$$

This equation is the complete derivative of the potential energy with respect to the displacements in the structure. For a nontrivial solution to exist, at least one of the displacement derivatives, dD_i, must be nonzero. In fact, they may all be nonzero. As such, the only way for this expression always to be true is for all of the coefficients, $\partial\Pi/\partial D_i$, modifying the displacement derivatives to be zero. This leads to as many simultaneous linear equations as there are degrees of freedom in the problem being solved.

[1]Robert D. Cook and Warren C. Young, *Advanced Mechanics of Materials* (New York: Macmillan Publishing Company, 1985), 231–235.

The coefficients are

$$\frac{\partial \Pi}{\partial D_i} - \frac{\partial}{\partial D_i}[U - \sum P_i D_i] - 0 - \frac{\partial U}{\partial D_i} - P_i \qquad \textbf{Eq. 13.35}$$

From this equation, we see that

$$P_i = \frac{\partial U}{\partial D_i} \qquad \textbf{Eq. 13.36}$$

This last equation is a statement of Castigliano's first theorem. That theorem is:

> The derivative of strain energy with respect to a displacement is equal to the force at and in the direction of that displacement.

Castigliano's first theorem is not as easy to use as the second theorem, or, for that matter, as easy as other energy methods such as virtual work, because strain energy is more easily formulated in terms of load than in terms of displacement. However, for some structural responses, and for the development of system matrices when using matrix methods (which we will study beginning in Chapter 19), Castigliano's first theorem is the easiest method to use. This method is particularly useful for the evaluation of structures with nonlinear response.

13.10 ANALYSIS USING COMPUTERS

As should be quite apparent to you by now, the analysis of statically indeterminate structural systems is very tedious, especially when structures of any significant size are involved. In many cases, hand solutions are not feasible even if computational software is used. Computer programs are usually employed in engineering offices when analyzing large structures. Analysis of statically indeterminate structures using a computer is demonstrated in the following examples. In Examples 13.13 and 13.14, the computer program SAP2000 is used to analyze two statically indeterminate structures.

EXAMPLE 13.13

Using SAP2000, determine the total tension in the cable and the shearing forces and bending moments in the beam of the structure shown in the figure. For this structure E is equal to 29,000 ksi.

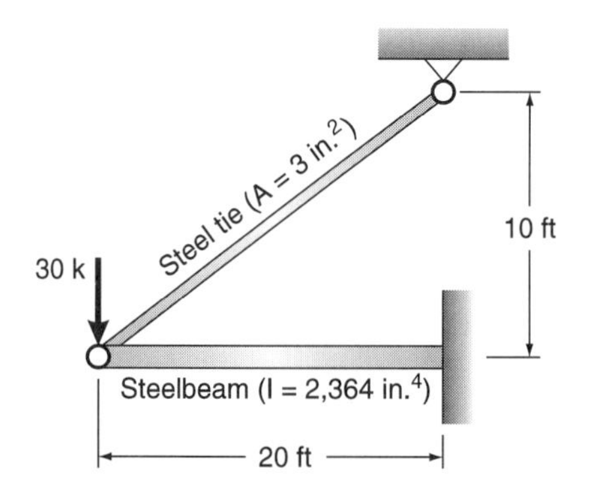

Solution. The computer model that we will use in the solution is shown in the figure. The steel tie was treated as a truss member—the flexural degrees of freedom at each end of the member were released, as was the torsional degree of freedom at the beginning of the member. Be sure that the self-weight multiplier is set to zero before performing the analysis.

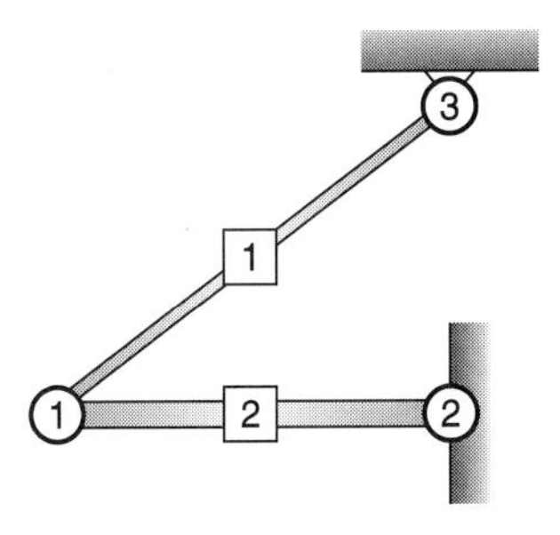

Upon analysis, we find the forces in the member to be as listed in the table.

		Beginning		End	
Member	P	M	V	M	V
1	43.42	0	0	0	0
2	−34.73	0	3.95	−947.8	3.95

■

EXAMPLE 13.14

Determine the forces in the truss shown in the next figure using the computer program SAP2000. This is the same truss that was analyzed if Example 13.8. The number by each member is the area of that member. The complete input data for this example is contained in the *EXAMPLES* directory on the CD-ROM at the back of the book.

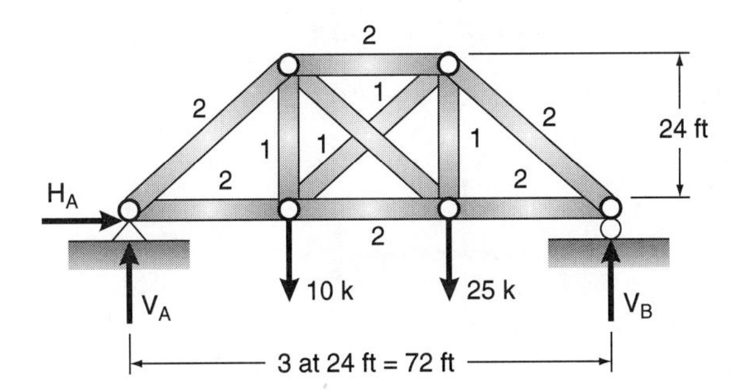

This truss was modeled in the same manner as the trusses in the previous chapters. The flexural degrees of freedom at the ends of each member were released as were the torsional degrees of freedom at the beginning of each member. The self-weight factor was set equal to zero so that self-weight did not influence the results. This is consistent with the analysis performed in Example 13.8. The resulting structural model is shown.

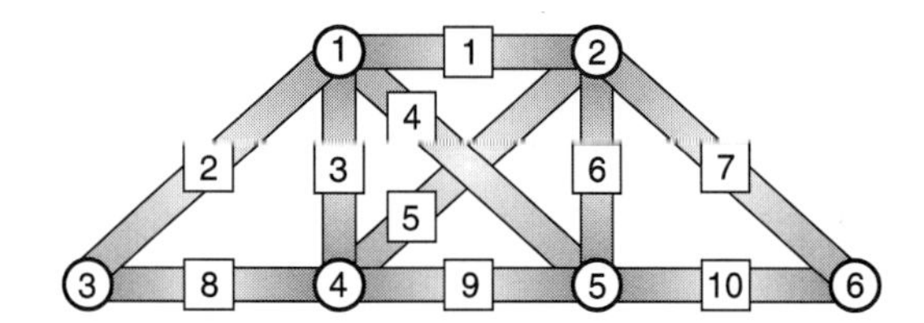

The bar forces that resulted from the analysis are shown in the table.

		Beginning		End	
Member	P	M	V	M	V
1	−21.54	0	0	0	0
2	−21.21	0	0	0	0
3	8.46	0	0	0	0
4	9.25	0	0	0	0
5	2.18	0	0	0	0
6	18.46	0	0	0	0
7	−28.28	0	0	0	0
8	15.00	0	0	0	0
9	13.46	0	0	0	0
10	20.00	0	0	0	0

■

13.11 PROBLEMS FOR SOLUTION

For Problems 13.1 to 13.23 compute the reactions and also draw the shearing force and bending moment diagrams for the continuous beams or frames. For each beam or frame, E and I are constant unless noted otherwise. The method of virtual work is to be used.

13.1

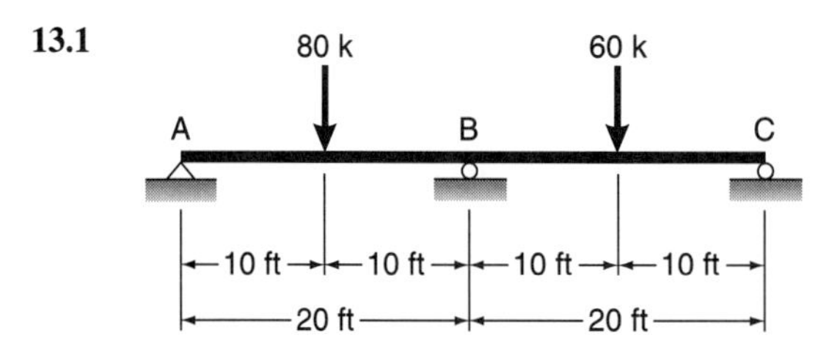

13.2 (*Ans.* V_A = 95 k, M_A= −400 k-ft)

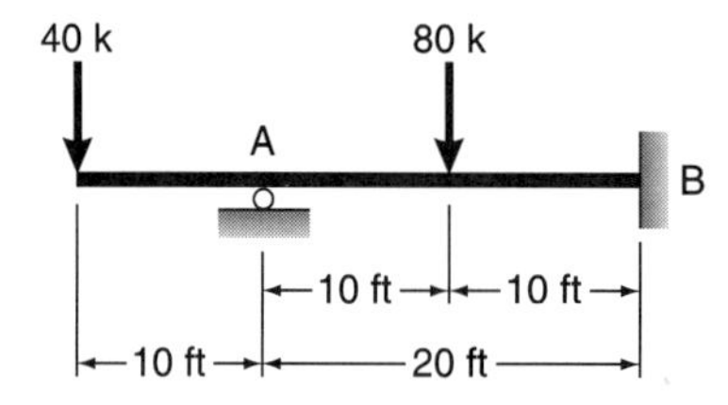

13.3

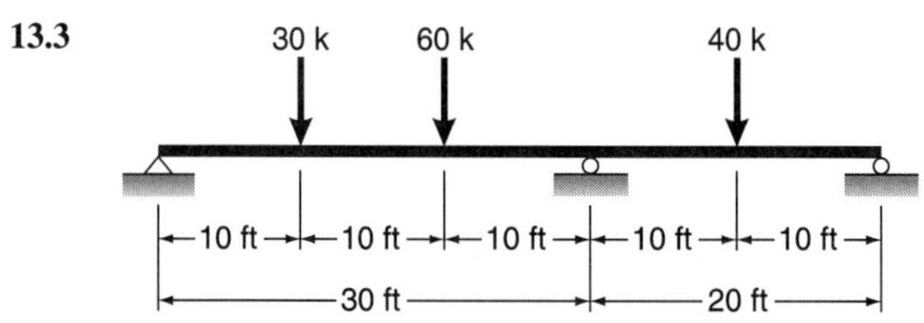

13.4 
(*Ans.* $V_A = 45.00$ k, $M_A = -270$ ft-k)

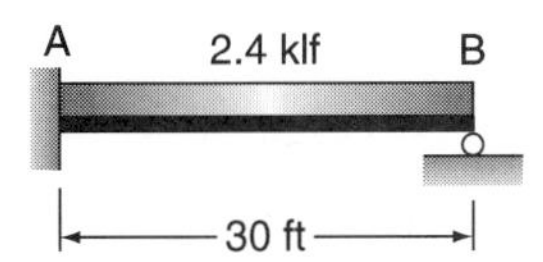

13.5

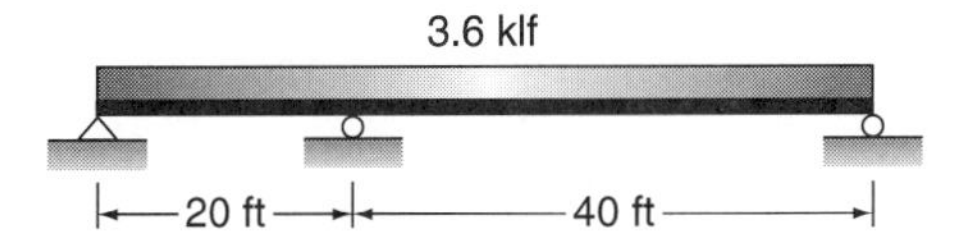

13.6 (*Ans.* $M_B = 204.2$ ft-k, $V_C = 29.19$ k)

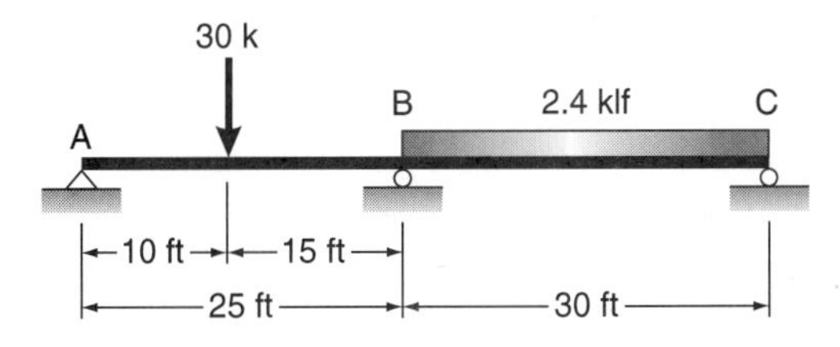

13.7

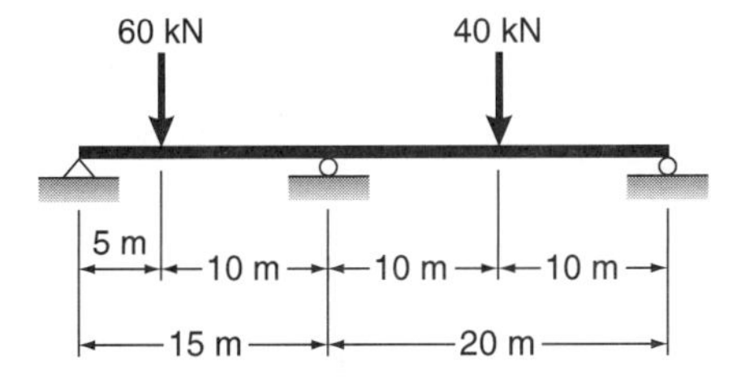

13.8 (*Ans.* $V_A = 9.86$ k ↑, $M_B = -154.2$ k, $V_C = 83.92$ k ↑, $M_C = -276.2$ ft-k)

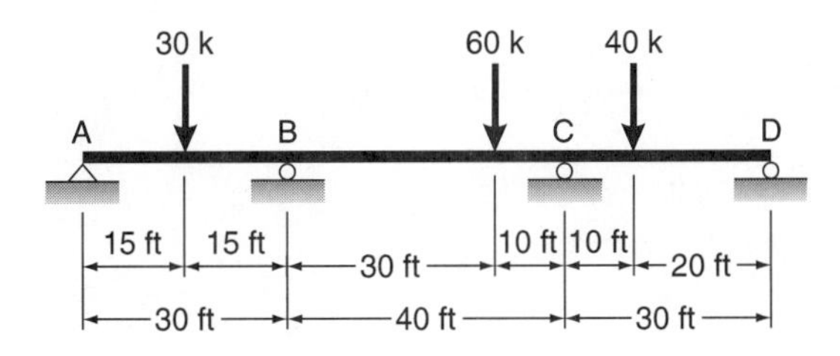

13.9

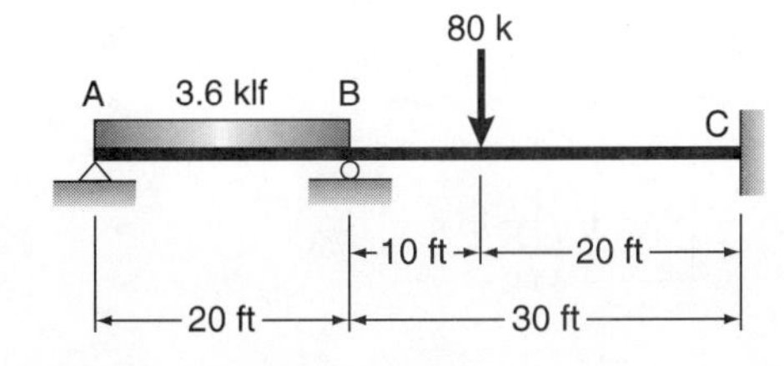

13.10 (*Ans.* $V_A = 6.00$ k ↑, $V_B = 91.60$ k, $M_B = -480$ ft-k)

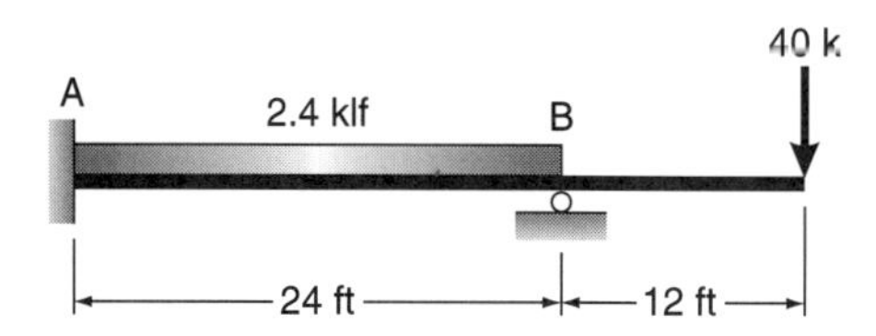

13.11

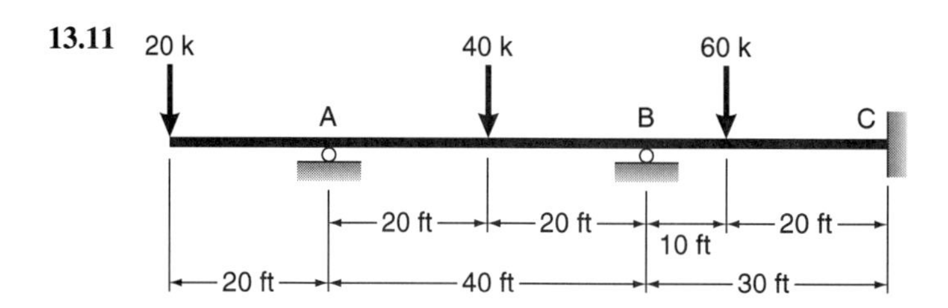

13.12 The beam of Problem 13.1 assuming support B settles 2.50 in. $E = 29 \times 10^6$ psi. $I = 1200$ in^4 (*Ans.* $V_B = 58.51$ k ↑, $M_B = +115$ ft-k)

13.13 The beam of Problem 13.1 assuming the following support settlements: A = 4.00 in., B = 2.00 in., and C = 3.50 in. $E = 29 \times 10^6$ psi. $I = 1200$ in^4.

13.14 The beam of Problem 13.8 assuming the following support settlements: A = 1.00 in., B = 3.00 in., C = 1.50 in., and D = 2.00 in. $E = 29 \times 10^6$ psi. $I = 3200$ in^4. (*Ans.* $M_B = 156$ ft-k, $M_C = -508.8$ ft-k)

13.15

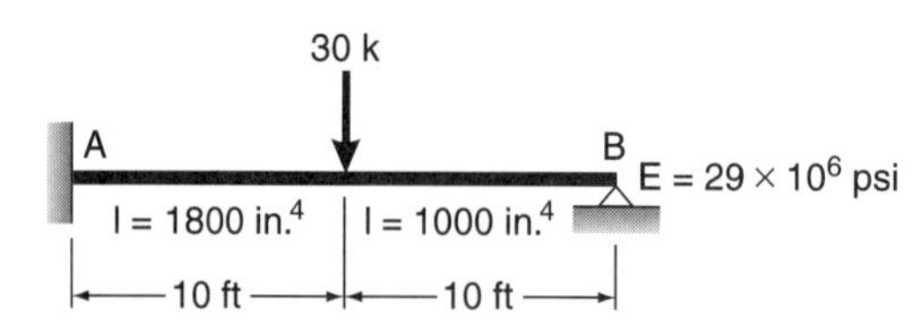

13.16 (*Ans.* $M_A = -129.6$ ft-k, $V_B = 8.52$ k ↑)

30 k

A B E = 29 × 10⁶ psi

I = 1800 in.⁴ I = 1000 in.⁴

10 ft 10 ft

13.17 $E = 29 \times 10^6$ psi

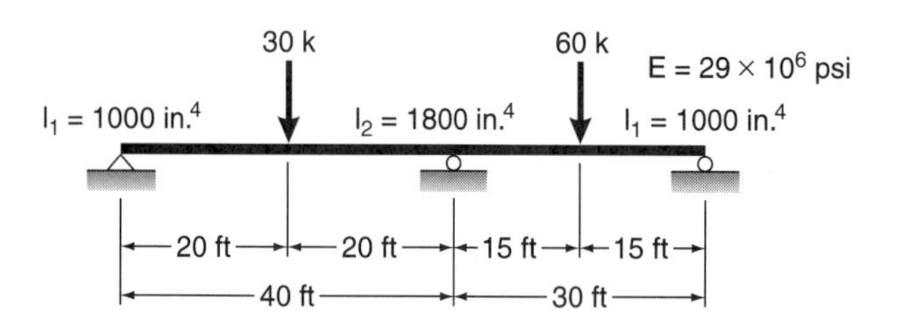

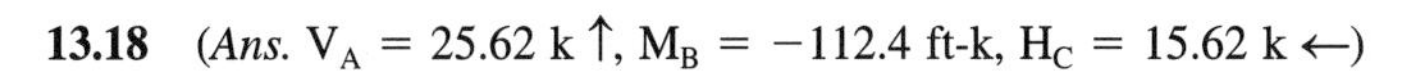
13.18 (*Ans.* $V_A = 25.62$ k ↑, $M_B = -112.4$ ft-k, $H_C = 15.62$ k ←)

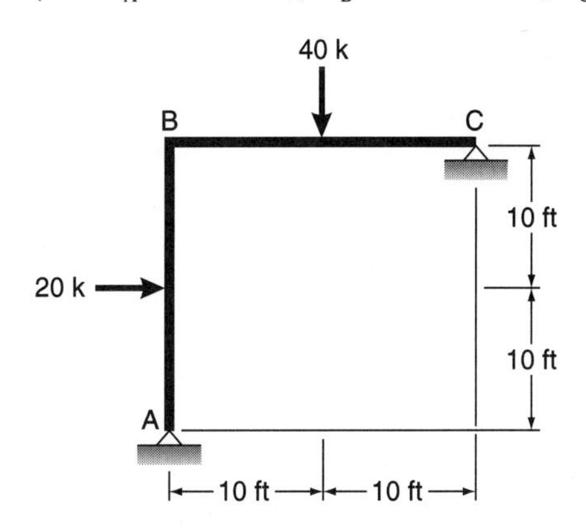

13.19

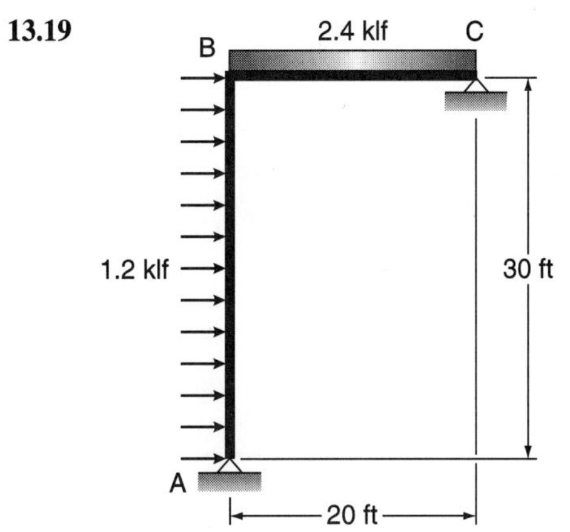

13.20 (*Ans.* $V_A = 3$ k ↓, $V_D = 27$ k ↑, $M_B = 123.4$ ft-k)

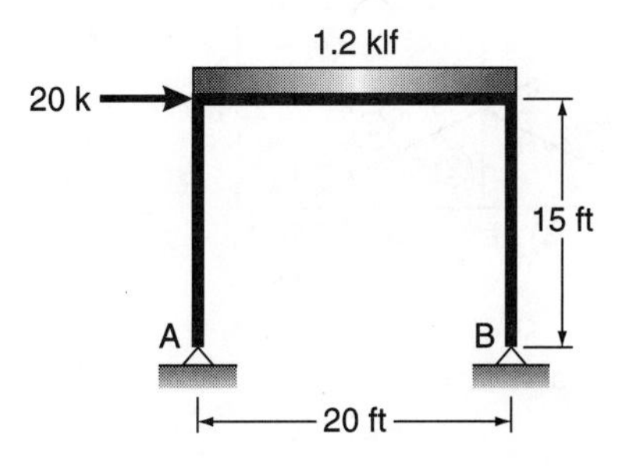

13.21

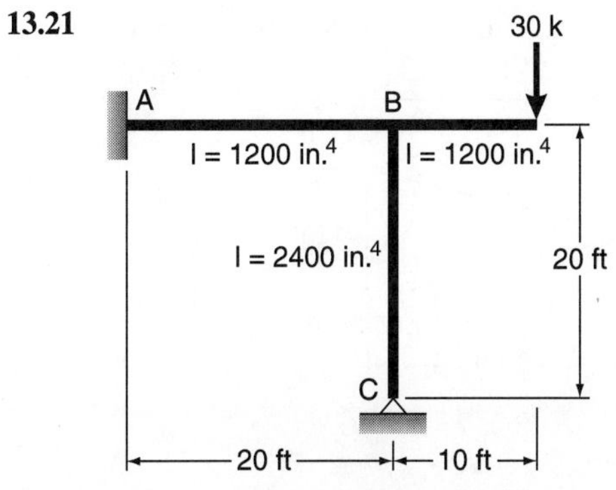

13.22 Repeat Problem 13.21 if the column is fixed at its base, (*Ans.* $M_A = 50$ ft-k, $V_C = 37.50$ k ↑)

13.23

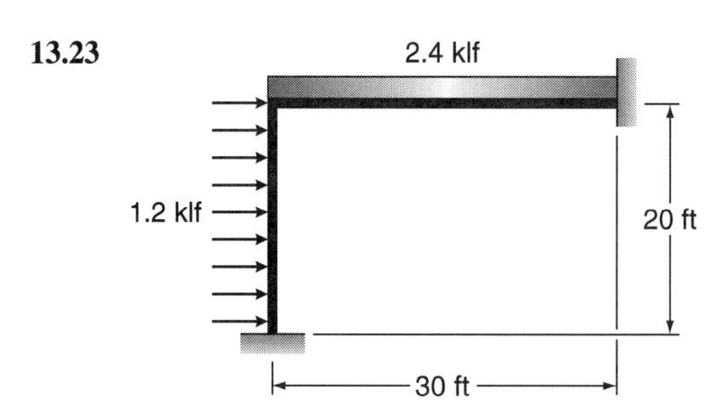

For Problems 13.24 to 13.39, determine the reactions and member forces for the trusses. Circled figures are member areas, in square inches unless shown otherwise. E is constant.

13.24 (*Ans.* $V_B = 19.8$ k ↑, $L_0L_1 = +15.1$ k, $U_1L_2 = -21.35$ k)

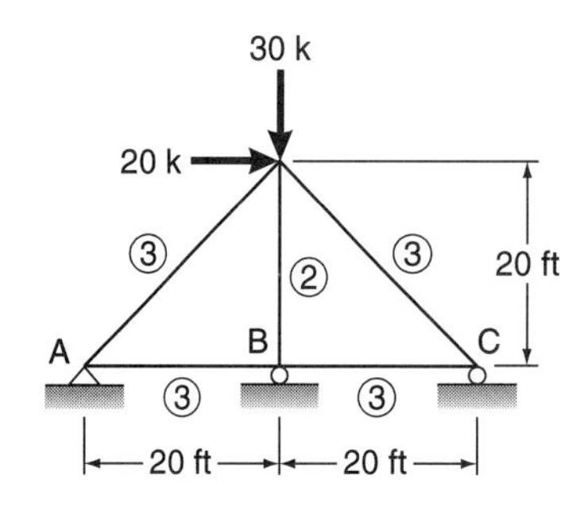

13.25

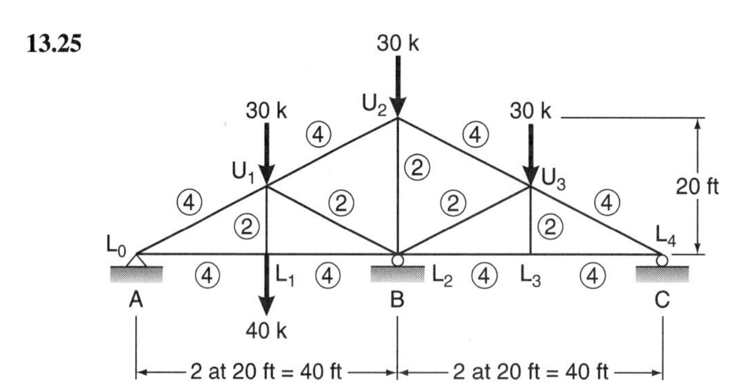

13.26 All areas are equal (*Ans.* $H_B = 23.2$ k, $U_0L_1 = -32.55$ k, $U_1L_1 = +4.2$ k)

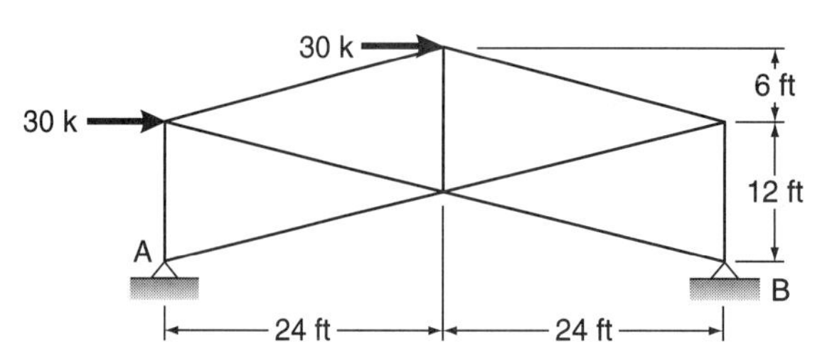

13.27 All areas are equal

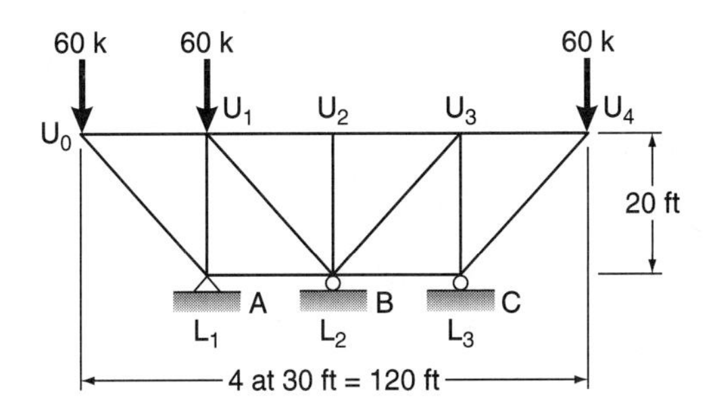

13.28 (*Ans.* V_B = 80.67 kN, L_1L_2 = +42.86 k, U_1L_1 = −80.67 k)

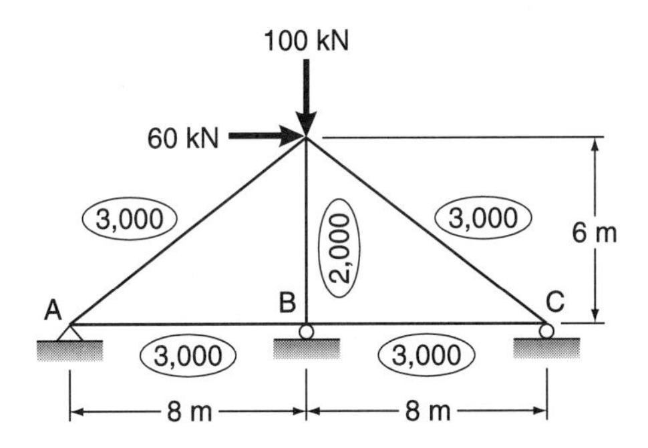

13.29

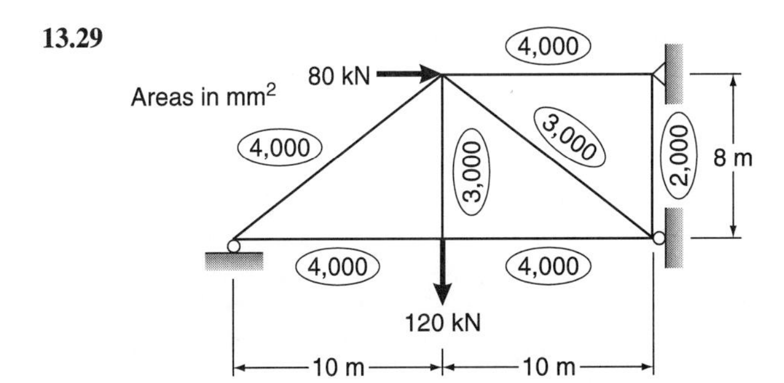

13.30 All areas are equal (*Ans.* L_1L_2 = +47.5 k, U_2U_3 = +41.5 k, U_2L_2 = +60 k)

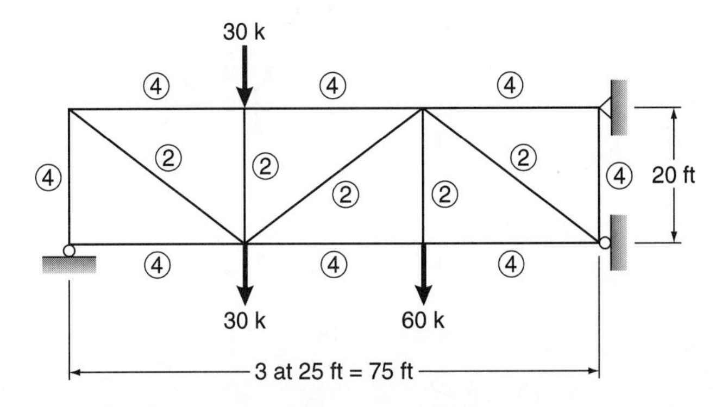

13.31

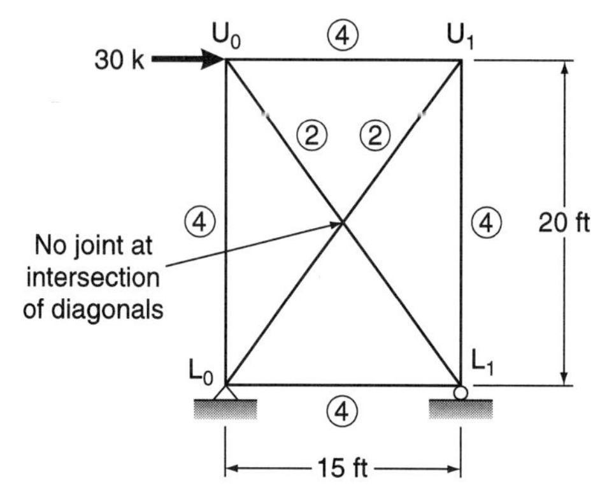

13.32 (*Ans.* $U_0L_1 = +8.88$ k, $L_1U_2 = +7.15$ k)

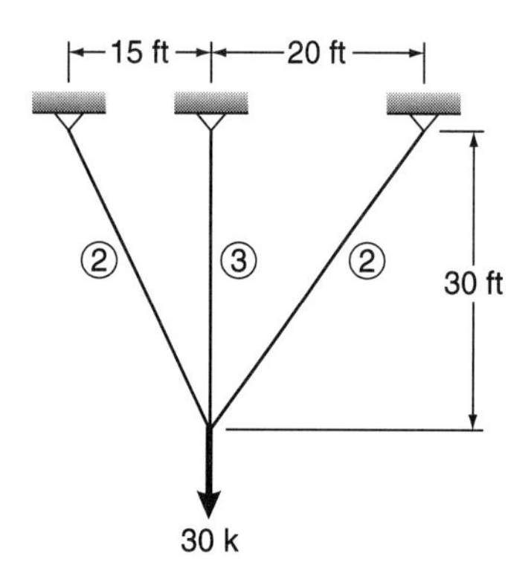

13.33 All areas are equal

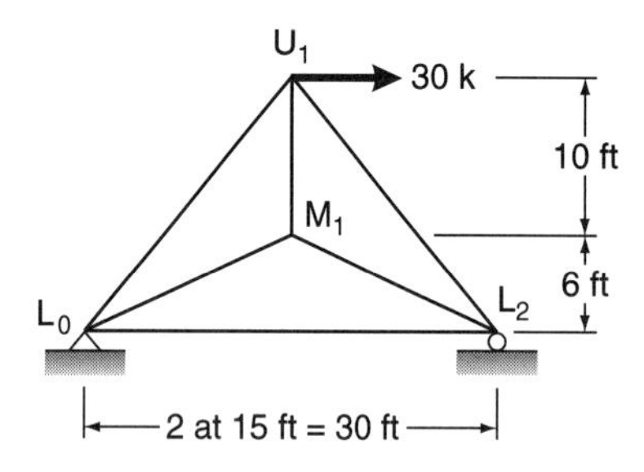

13.34 All areas are equal (*Ans.* $L_1U_2 = -180.3$ kN, $U_0L_0 = -40.2$ kN, $U_1L_1 = -140.2$ kN)

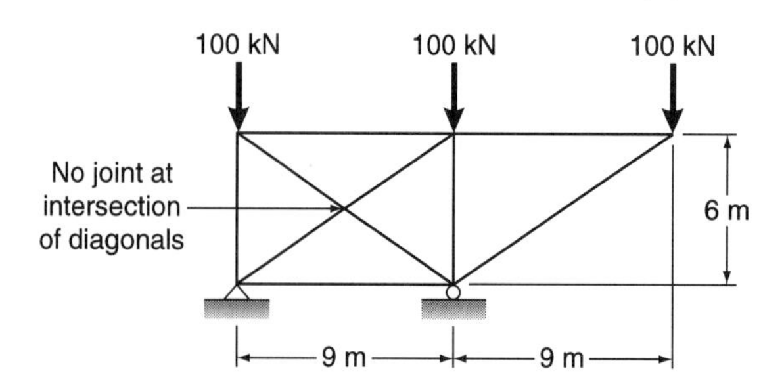

13.35

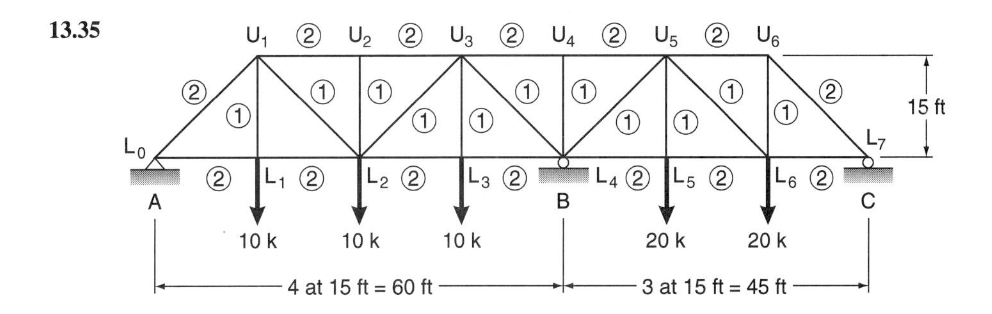

13.36 (*Ans.* $L_3L_4 = -192$ k, $U_3U_4 = +192$ k)

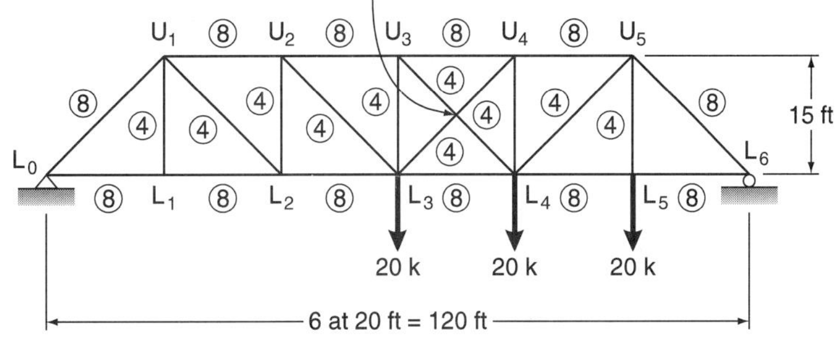

13.37

No joint at intersection of diagonals

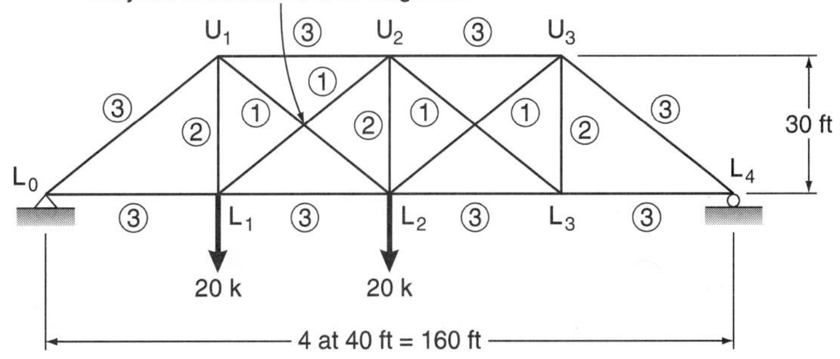

13.38 (*Ans.* $V_A = 1.1$ k ↓, $U_1U_2 = +2.2$ k, $U_2L_3 = +22.0$ k)

No joint at intersection of diagonals

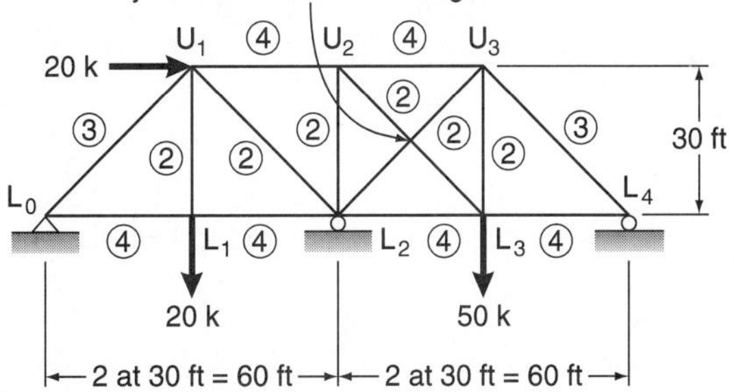

13.39 Determine the forces in all the members of the truss shown in the accompanying illustration if the top-chord members, U_0U_1 and U_1U_2, have an increase in temperature of 75°F and there is no change in temperature in other members. The coefficient of linear expansion, ε, is 0.0000065.

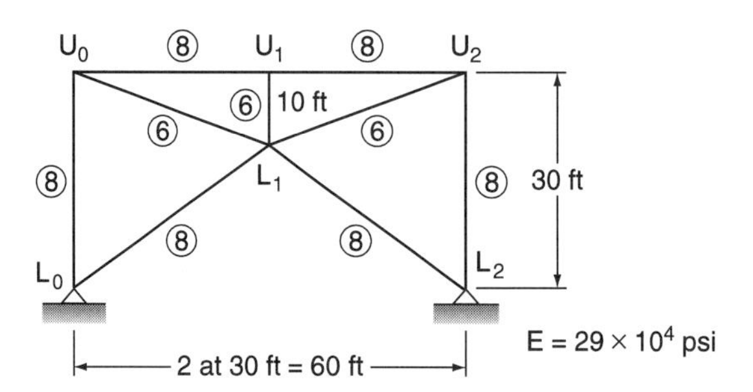

For Problems 13.40 through 13.56 analyze the structures using Castigliano's theorems. The values of E and I are constant unless otherwise indicated. The circled values are areas, in square inches.

13.40

30 k 60 k
A B C
15 ft 15 ft 10 ft 10 ft
30 ft 20 ft

13.41 (*Ans.* $V_A = 18.25$ k ↑, $V_C = 18.12$ k ↑, $M_B = -236$ ft-k

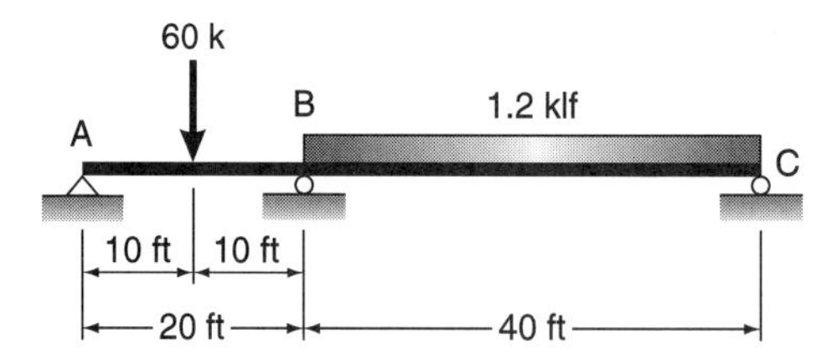

13.42

30 k
A 2.4 klf B
C
10 ft 20 ft
20 ft 30 ft

13.43 Problem 13.32 (*Ans.* Member forces left to right +8.88 k, 16.10 k, 7.15 k)

13.44

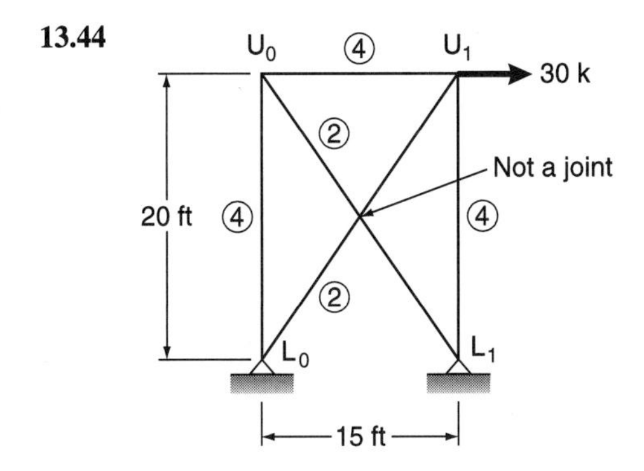

13.45 (*Ans.* $L_0L_1 = -209.0$ kN, $U_0L_1 = +161.3$ kN)

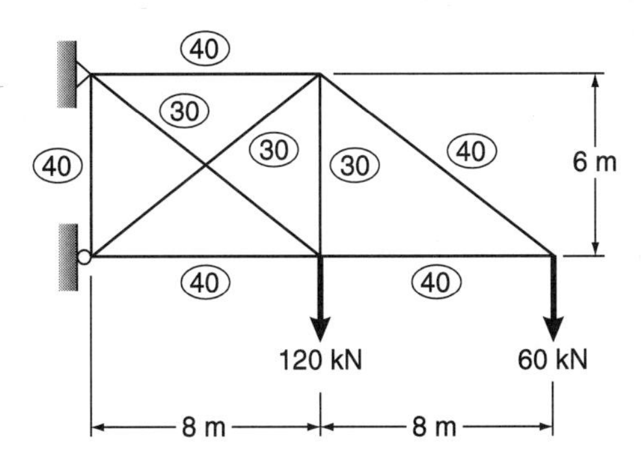

13.46 Problem 13.33

13.47 Find the force in tie (*Ans.* Cable tension +43.4 k)

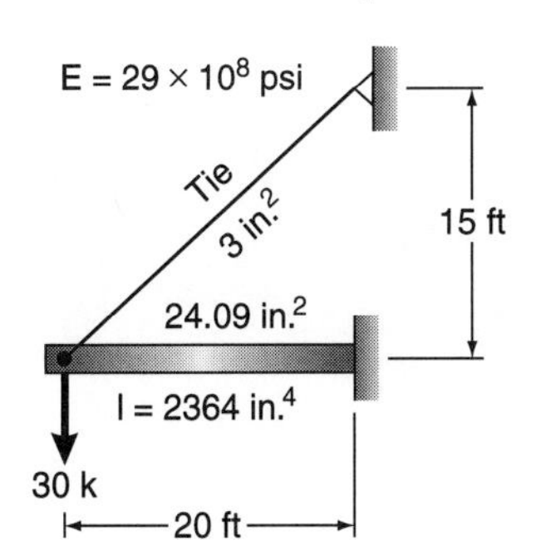

13.48

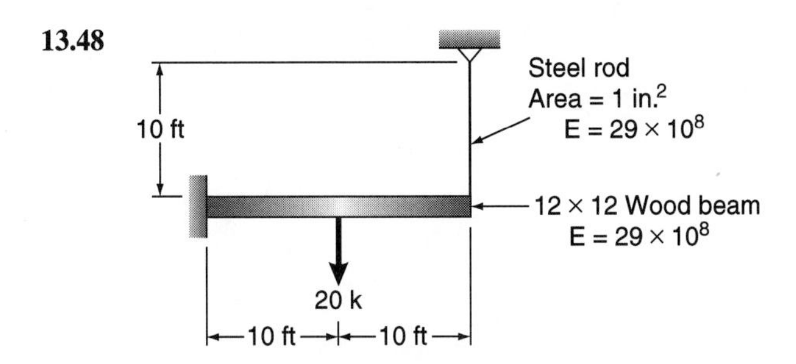

13.49 (*Ans.* $L_0U_1 = -39.2$ k, $L_1L_2 = -4.68$ k, cable tension $= +6.48$ k)

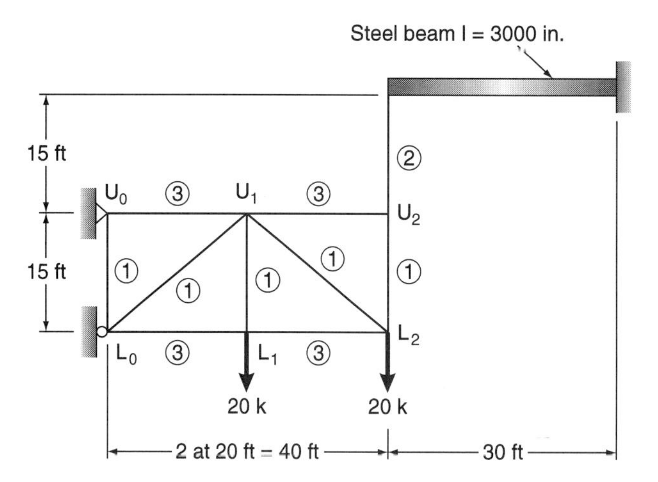

13.50 Problem 13.24

13.51 Problem 13.30 (*Ans.* $L_1L_2 = +47.5$ k, $U_2U_3 = +41.5$ k, $U_2L_2 = +60$ k)

13.52 Problem 13.34

13.53 All member areas = 10 in.^{2} (*Ans.* $L_1L_2 = -18.1$ k, $U_1U_2 = +31.9$ k, $U_1L_2 = -16.8$ k)

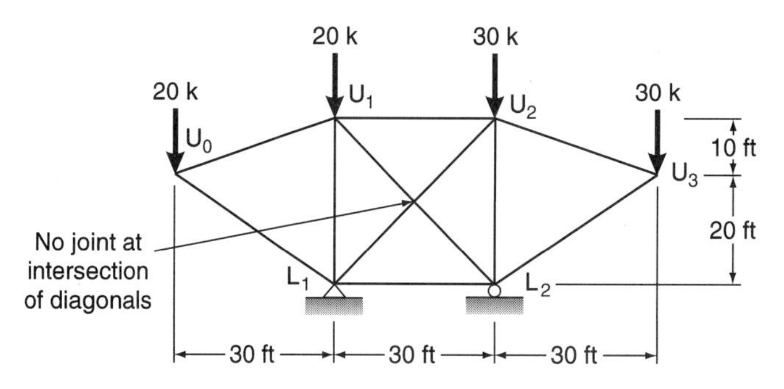

13.54

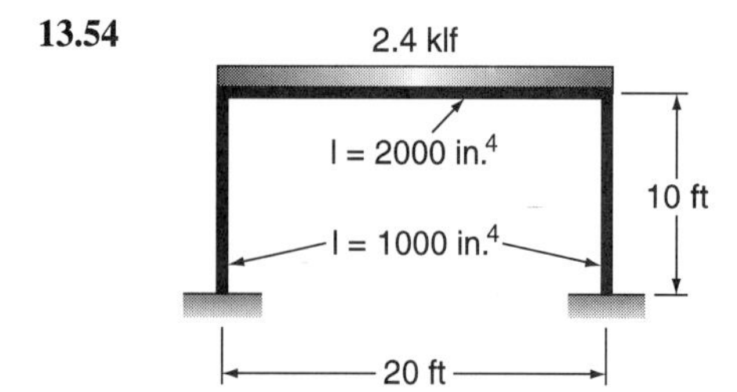

13.55 (*Ans.* $H_A = 18.16$ k →, $M_B = 363.2$ ft-k, $M_C = 18.4$ ft-k)

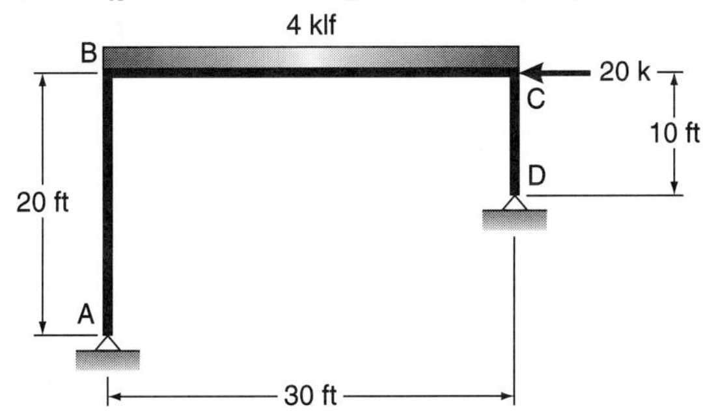

13.56

For Problems 13.57 through 13.65, use the computer program SAP2000 or SABLE and repeat the following problems.

13.57 Problem 13.1 ($V_B = 96.27$ k, $M_B = -262.6$ ft-k)

13.58 Problem 13.2

13.59 Problem 13.24 ($U_1L_1 = -19.8$ k, $L_1L_2 = +15.1$ k)

13.60 Problem 13.38

13.61 Problem 13.44

13.62 Problem 13.47 (*Ans.* Cable tension $= +43.4$ k)

13.63 Problem 13.38

13.64 Problem 13.39 (*Ans.* $U_1U_2 = 0$, $U_1L_2 = +6.49$ k)

13.65 Problem 13.56

Chapter 14

Influence Lines for Statically Indeterminate Structures

14.1 INFLUENCE LINES FOR STATICALLY INDETERMINATE BEAMS

The uses of influence lines for statically indeterminate structures are the same as those for statically determinate structures. They enable the design engineer to locate the critical positions for live loads and to compute forces for various positions of the loads. Influence lines for statically indeterminate structures are not as simple to draw as they are for statically determinate structures. For the latter case, it is possible to compute the ordinates for a few controlling points and connect those values with a set of straight lines. Unfortunately, influence lines for continuous structures require the computation of ordinates at a large number of points because the diagrams are either curved or made up of a series of chords. The chord-shaped diagram occurs where loads can only be transferred to the structure at intervals, as at the panel points of a truss or at joists framing into a girder.

The problem of preparing the diagrams is not as difficult as the preceding paragraph indicates because a large percentage of the work may be eliminated by applying Maxwell's law of reciprocal deflections. The preparation of an influence line for the interior reaction of the two-span beam shown in Figure 14.1 is considered in the following discussion.

The procedure for calculating V_B has been to write a compatibility of displacement equation of the form

$$V_b\delta_{bb} - \delta_B = 0 \qquad \textbf{Eq. 14.1}$$

Recall from Chapter 13 that δ_B is the deflection at B caused by the primary forces and δ_{bb} is the deflection at B caused by a virtual unit force at B. The same procedure may be used in drawing an influence line for V_B. A unit load is placed at some point x along the beam. From Maxwell' law of reciprocal deflections in Section 10.6, we know that this

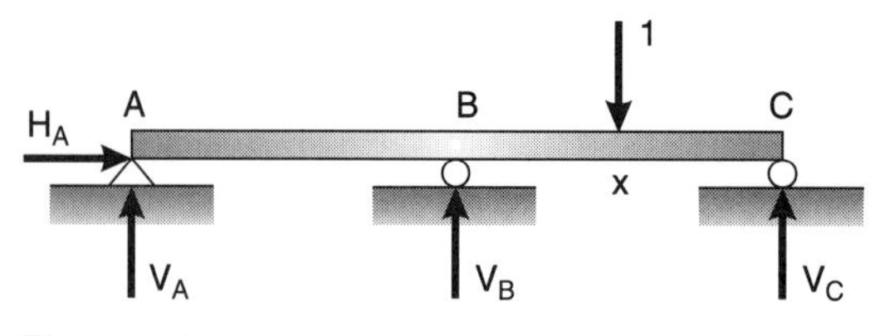

Figure 14.1

load causes a deflection δ_B that is equal to δ_{bx}. As such, the following relationship can be written

$$V_B = -\frac{\delta_B}{\delta_{bb}} = -\frac{\delta_{bx}}{\delta_{bb}} \qquad \textbf{Eq. 14.2}$$

At first glance it appears that the unit load will have to be placed at numerous points on the beam and the value of δ_{bx} laboriously computed for each location. A study of the deflections caused by a unit load at point x, however, will show that these computations are unnecessary. By Maxwell's law the deflection, δ_{bx} at B due to a unit load at x is identical to the deflection at x caused by a unit load at B, δ_{xb}. The expression for V_B thus becomes

$$V_B = -\frac{\delta_{xb}}{\delta_{bb}} \qquad \textbf{Eq. 14.3}$$

By now, you should realize that the unit load need only be placed at B, and the deflections at various points across the beam computed. Dividing each of these values by δ_{bb} yields the ordinates for the influence line. Essentially, we are removing the support and placing a unit load at that location, computing the deflected shape of the beam, and then scaling that deflection so the maximum value is unity. Another way of expressing this principle is as follows:

> If a unit deflection is caused at a support for which the influence line is desired, the beam will draw its own influence line because the deflection at any point along the beam is the ordinate of the influence line at that point for the reaction in question.

Maxwell's presentation of his theorem in 1864 was so brief that its value was not fully appreciated until 1886 when Heinrich Müller-Breslau more clearly showed its true worth as described in the preceding paragraph.[1] Müller-Breslau's principle may be stated in detail as follows:

> The deflected shape of a structure represents to some scale the influence line for a function such as stress, shear, moment, or reaction component if the function is allowed to act through a unit displacement.

This principle is applicable to statically determinate and indeterminate beams, frames, and trusses. Its correctness is proven in the next section of this chapter.

The influence line for the reaction at the interior support of two-span beam is presented in Example 14.1. Influence lines also are shown for the end reactions, the values for ordinates having been obtained by statics from those computed for the interior reaction.

EXAMPLE 14.1

Draw influence lines for reactions at each support of the structure shown in the figure.

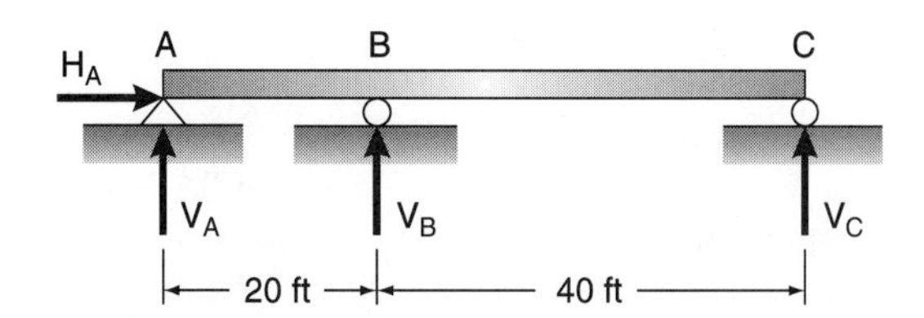

[1]J. S. Kinney, *Indeterminate Structural Analysis* (Reading, Mass.: Addison-Wesley, 1957), 14.

Solution. Remove V_B, place a unit load at B, and compute the deflections caused by that load at 10-ft intervals along the beam as indicated on the next figure.

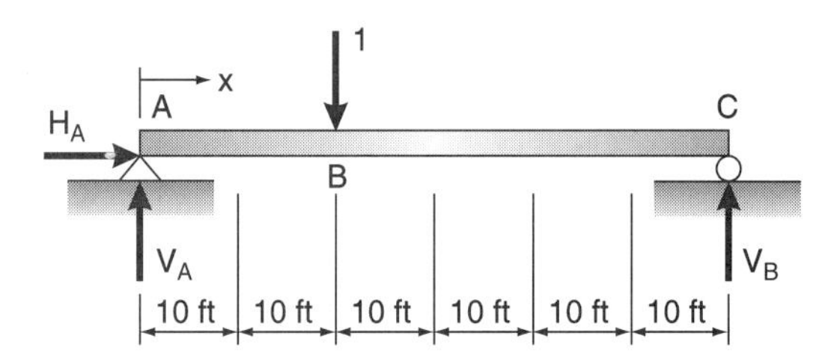

The bending moment in the beam at any location caused by a unit force placed at some distance L_1 from the left side can be computed from

$$R_A = 1\left(\frac{L - L_1}{L}\right)$$

$$M_x = R_A x - \Phi(x - L_1)(1)(x - L_1) = \left(\frac{L - L_1}{L}\right)x - \Phi(x - L_1)(x - L_1)$$

Using this equation and Müller-Breslau's principle, the magnitude of the reaction at B for a unit load placed at A can be computed from:

$$V_B = \frac{\int_0^{60}\left[\left(\frac{60 - a}{L}\right)x - \Phi(x - a)(x - a)\right]\left[\left(\frac{60 - 20}{L}\right)x - \Phi(x - 20)(x - 20)\right]dx}{\int_0^{60}\left[\left(\frac{60 - 20}{L}\right)x - \Phi(x - 20)(x - 20)\right]^2 dx}$$

Using this last equation, we then find that the ordinates of the influence line at 10-ft intervals along the beam are as shown in the table.

Position	V_B Influence Value	Position	V_B Influence Value
0	0.000	40 ft	0.875
10 ft	0.594	50 ft	0.484
20 ft	1.000	60 ft	0.000
30 ft	1.078		

The resulting influence line for the reaction at B is

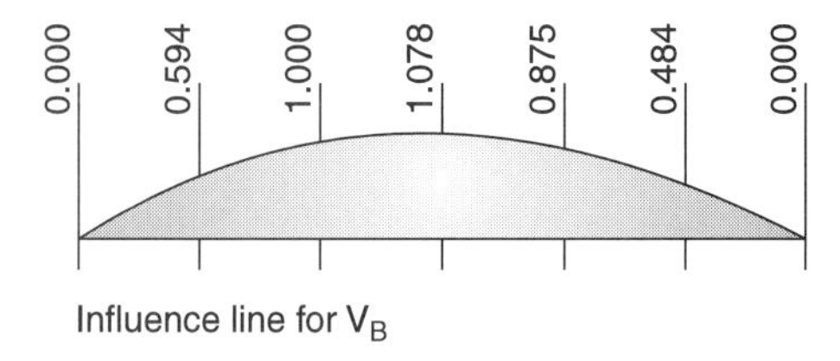

Influence line for V_B

Having determined the values of V_B for various positions of the unit load, the values of V_A for each position of load can be determined from static equilibrium. The equation for the reaction at A and the resulting influence line are

$$V_{A \text{ at } X=a} = \frac{(L - a) - (V_{B \text{ at } X=a})(L - 20)}{L}$$

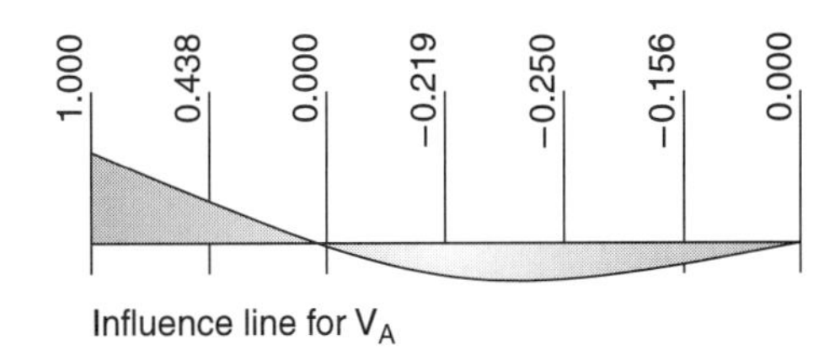

Influence line for V_A

The influence line for the reaction at C can be computed in a similar manner. The resulting equation for the reaction and the influence line are

$$V_{C \text{ at } X=a} = \frac{a - (20)(V_{B \text{ at } X=a})}{L}$$

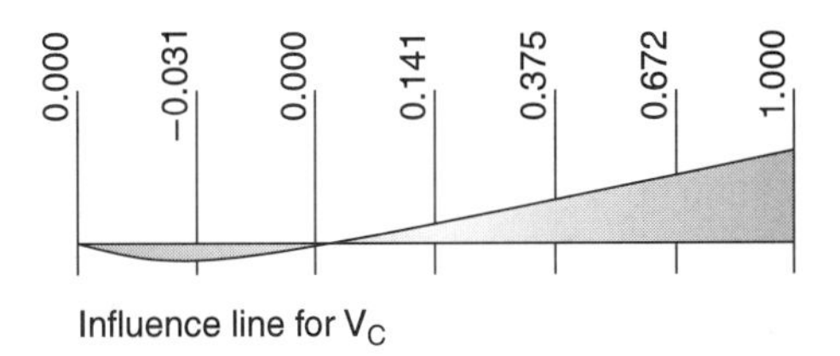

Influence line for V_C

Next, we will investigate the method for drawing the influence lines for beams that are continuous over three spans, which have two redundants. For this discussion, the beam shown in Figure 14.2 is considered, and the reactions V_B and V_C are assumed to be the redundants.

Again, we will remove the redundant forces and compute the deflections at various locations along the beam for a unit load at B and a unit load at C. A unit load at any point x causes a deflection at B of δ_{bx}. From Maxwell's law, that deflection is equal to the deflection at x due to a unit load at B, which is δ_{xb}. Similarly, δ_{cx} is equal to δ_{xc}. After computing δ_{xb} and δ_{xc} at the several locations along the beam, their values at each section may be substituted into the following simultaneous equations, whose solution will yield the values of V_B and V_C.

$$\delta_{xb} + V_B\delta_{bb} + V_C\delta_{bc} = 0$$
$$\delta_{xc} + V_B\delta_{cb} + V_C\delta_{cc} = 0 \qquad \textbf{Eq. 14.4}$$

The simultaneous equations are solved quickly, although a large number of ordinates are being computed, because the only variables in the equations are δ_{xb} and δ_{xc}. After the influence lines are prepared for the redundant reactions of a beam, the ordinates for any other function (moment, shear, and so on) can be determined from considerations of static equilibrium. Illustrated in Example 14.2 are the calculations

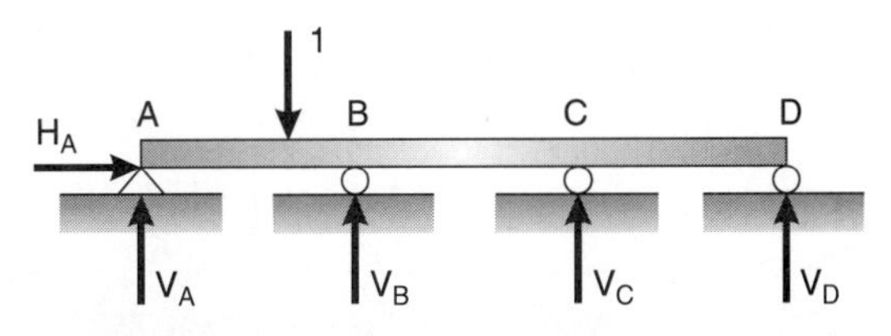

Figure 14.2

necessary for preparing influence lines for several functions of a three-span continuous beam.

EXAMPLE 14.2

Draw influence lines for V_B, V_C, V_D, $M_{x=70}$, and the shearing force at x = 60 ft for the beam shown in the figure.

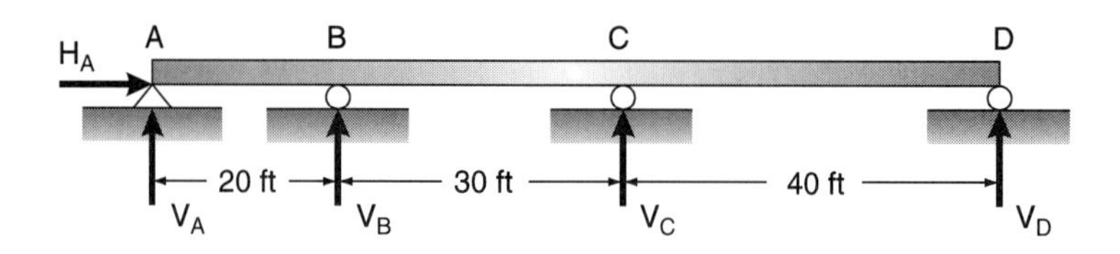

Solution. To obtain the needed influence lines, we will remove reactions V_B and V_C, place a unit load at various locations along the beam, and compute the necessary deflections. The locations at which we will compute deflections are shown here.

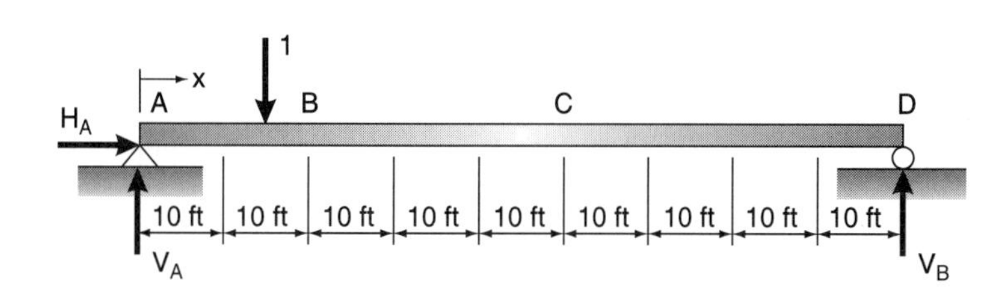

From Example 14.1, we know that the moment in a beam at some location x caused by a unit load place at L_1 is

$$M_x = \left(\frac{L - L_1}{L}\right)x - \Phi(x - L_1)(x - L_1)$$

Using this equation, the necessary deflections, as indicated in Equation 14.4, can be computed. These deflections are

$$EI\delta_{bb} = \int_0^{90} \left[\left(\frac{90 - 20}{90}\right)x - \Phi(x - 20)(x - 20)\right]^2 dx = 7.259 \times 10^3$$

$$EI\delta_{cc} = \int_0^{90} \left[\left(\frac{90 - 50}{90}\right)x - \Phi(x - 50)(x - 50)\right]^2 dx = 1.481 \times 10^4$$

$$EI\delta_{bc} = \int_0^{90} \left[\left(\frac{90 - 20}{90}\right)x - \Phi(x - 20)(x - 20)\right]\left[\left(\frac{90 - 50}{90}\right)x - \Phi(x - 50)(x - 50)\right]dx$$

$$= 9.037 \times 10^3$$

$$EI\delta_{cb} = \int_0^{90} \left[\left(\frac{90-50}{90}\right)x - \Phi(x-50)(x-50)\right]\left[\left(\frac{90-20}{90}\right)x - \Phi(x-20)(x-20)\right]dx$$

$$= 9.037 \times 10^3$$

$$EI\delta_{xb} = \int_0^{90} \left[\left(\frac{90-a}{90}\right)x - \Phi(x-a)(x-a)\right]\left[\left(\frac{90-20}{90}\right)x - \Phi(x-20)(x-20)\right]dx$$

$$EI\delta_{xc} = \int_0^{90} \left[\left(\frac{90-a}{90}\right)x - \Phi(x-a)(x-a)\right]\left[\left(\frac{90-50}{90}\right)x - \Phi(x-50)(x-50)\right]dx$$

Using these deflections, we can solve the following simultaneous equations for V_B and V_C:

$$\delta_{xb} - V_B\delta_{bb} - V_C\delta_{bc} = 0$$
$$\delta_{xc} - V_B\delta_{cb} - V_C\delta_{cc} = 0$$

The negative sign occurs in the equations because the unit load was assumed to be acting downward while the reactions at B and C are assumed to be acting upward. The calculated values of V_B and V_C for the various positions of the unit load are summarized in the following table.

Unit Load at	V_B Influence Value	V_C Influence Value	Unit Load at	V_B Influence Value	V_C Influence Value
0	0.000	0.000	50 ft	0.000	1.000
10 ft	0.645	−0.073	60 ft	−0.234	1.024
20 ft	1.000	0.000	70 ft	−0.267	0.813
30 ft	0.870	0.309	80 ft	−0.167	0.446
40 ft	0.443	0.710	90 ft	0.000	0.000

The influence lines for these two reaction forces are

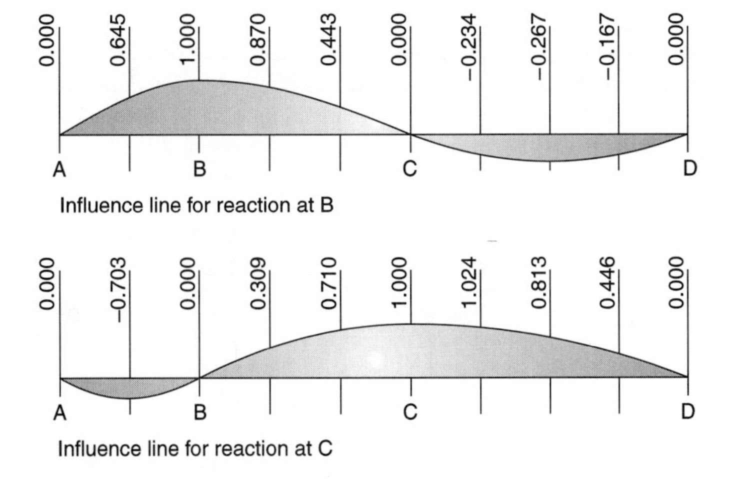

Influence line for reaction at B

Influence line for reaction at C

Using the ordinates on these influence lines, the influence for the other actions can be computed using equations of static equilibrium. The resulting influence lines are shown in the following figures.

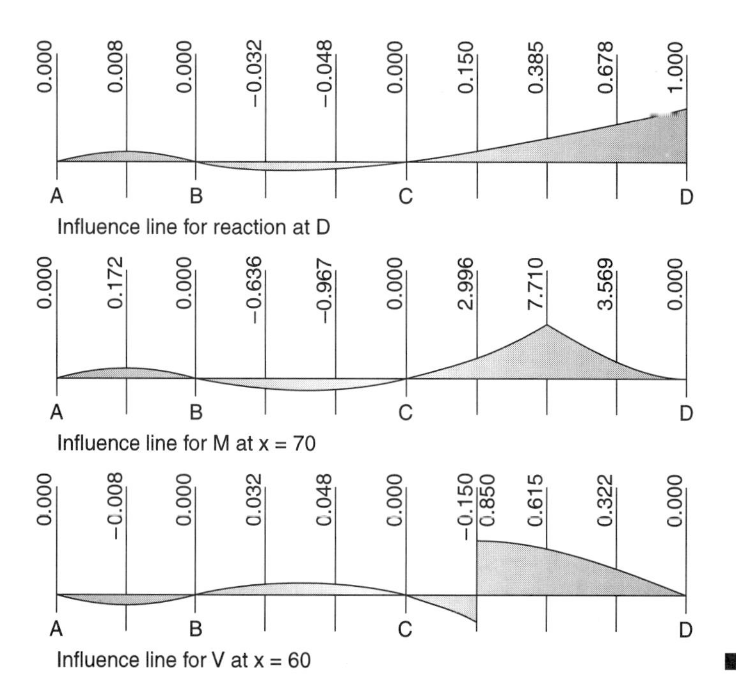

Influence lines for the reactions, shearing forces, and bending moments for frames can be prepared in the same manner as they were for the statically indeterminate beams just considered. Space is not taken here to present such calculations, however. The preparation of qualitative influence lines is discussed in the next section of this chapter. From this discussion, you will see how to place live loads on building frames to cause maximum values of a particular response.

14.2 QUALITATIVE INFLUENCE LINES

Müller-Breslau's principle is based upon Castigliano's theorem of least work, which was presented in Chapter 13. This theorem can be expressed as: *when a displacement is induced into a structure, the total virtual work done by all the active forces is equal to zero.*

Müller-Breslau's principle can be proven by considering the beam shown in Figure 14.3(a). This beam is subjected to a moving unit load. To determine the magnitude of

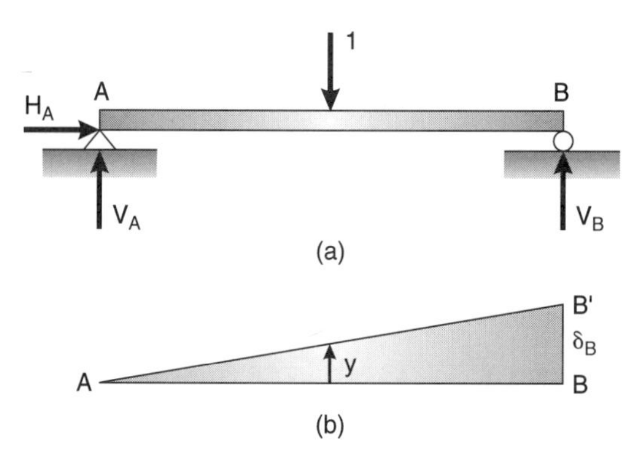

Figure 14.3 Illustration for development of qualitative influence lines

the reaction V_B, we can move the support and reaction at B through a small virtual displacement δ_B, as shown in Figure 14.3(b). The beam's position is now represented by line AB′ and the unit load has been moved through the distance y.

The virtual work equation for the active forces on the beam

$$(V_B)(\delta_B) = 1.0(y)$$

$$V_B = \frac{y}{\delta_B} \quad \textbf{Eq. 14.5}$$

Should the magnitude of δ_b be unity, V_B will be equal to y.

$$V_B = \frac{y}{1.0} = y \quad \textbf{Eq. 14.6}$$

You can now see that y is the ordinate of the deflected shape beam at the location of the unit load. It is also the magnitude of the right-hand reaction, V_B, due to that moving unit load. Therefore, the deflected beam position AB′ represents the influence line for V_B. Similar proofs can be developed for the influence lines for other functions of a structure, such as shearing forces and bending moments. This is consistent with what was stated in the Section 14.1.

Müller-Breslau's principle is of great importance. The shape of the usual influence line needed for continuous structures is so simple to obtain from his principle that in many situations it is unnecessary to perform the computations needed to obtain the numerical values of the ordinates. Using this principle, you can sketch influence lines with sufficient accuracy to locate the critical positions for live load for various functions of the structure. This capability is of particular importance for building frames.

If the influence line is desired for the left reaction of the continuous beam shown in Figure 14.4(a), its general shape can be determined by letting the reaction move upward through a unit displacement. If the left end of the beam displaced upward in this manner, the beam would take the shape shown in Figure 14.4(b). This deformed shape can be easily sketched, remembering the other supports are unyielding. From Müller-Breslau's principle, this deformed shape is representative of the influence line. Influence lines obtained by sketching the deformed shape are said to be *qualitative influence lines,* whereas influence lines obtained by calculating the values of the ordinates are said to be *quantitative influence lines*. The influence line for V_C in Figure 14.4(c) is another example of a qualitative influence for a reaction force.

The influence line in Figure 14.4(d) is for positive bending moment at point 1 near the center of the first span. The beam is assumed to have a pin or hinge inserted at 1 and a couple applied adjacent to each side of the pin that will cause compression on the top fibers (positive moment). The applied couple on each side of the pin causes the left span to take the shape indicated: the deflected shape of the remainder of the beam may be sketched approximately. The result is the qualitative influence line for positive moment in span 1. A similar procedure is used to draw the qualitative influence line for negative moment at point 2, which is located on the third span. The difference is that the moment couple applied at the assumed pin will tend to cause compression on the bottom fibers, which corresponds with negative moment.

Finally, qualitative influence lines are drawn for positive shearing force at points 1 and 2. At point 1, the beam is cut, and the cut ends are displaced in a direction consistent with positive shearing force on the respective ends, as shown in Figure 14.4(f). The beam will take that shape when this occurs. The same procedure is used to draw the qualitative influence line for positive shearing force at point 2 on the beam.

From these diagrams, considerable information is available concerning critical conditions for live loads acting on the beam. To obtain the maximum positive value of V_A caused

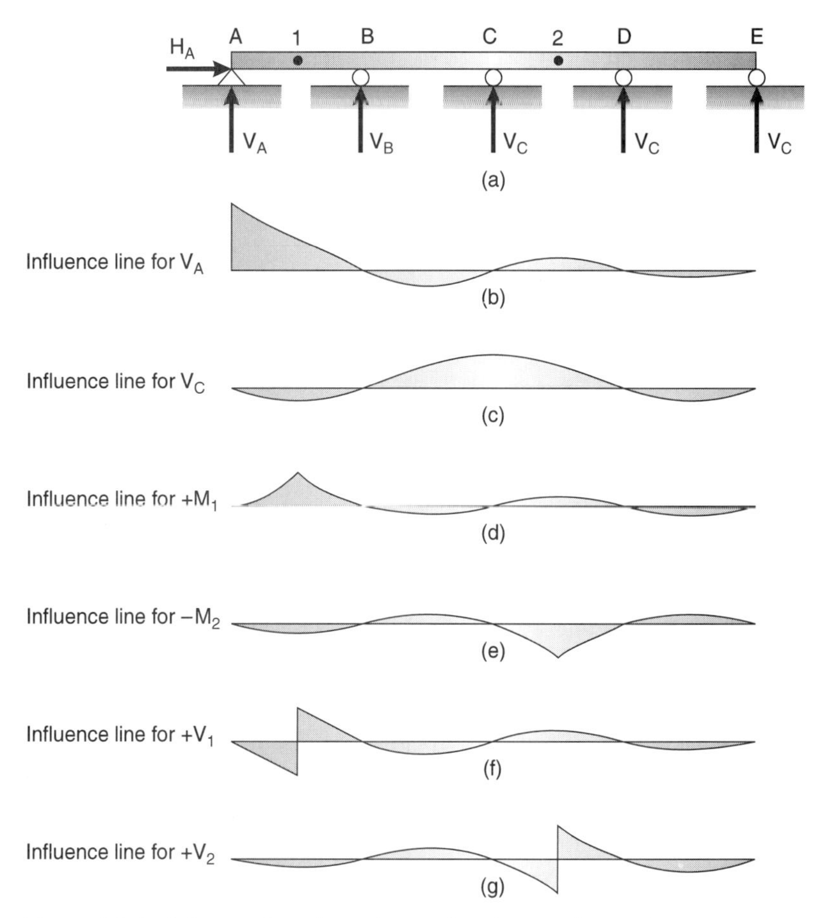

Figure 14.4 Influence lines for various actions in a continuous beam

by a uniform live load, the load should be placed on spans 1 and 3. The influence line has positive ordinates on these spans. To obtain the maximum negative moment in span 1, spans 2 and 4 should be loaded. Other maximum conditions can be obtained in similar fashion.

Qualitative influence lines are particularly valuable for determining critical positions for loads in buildings, as illustrated by the moment influence line for the building frame in Figure 14.5(a). When drawing the diagrams for an entire frame, the joints are assumed

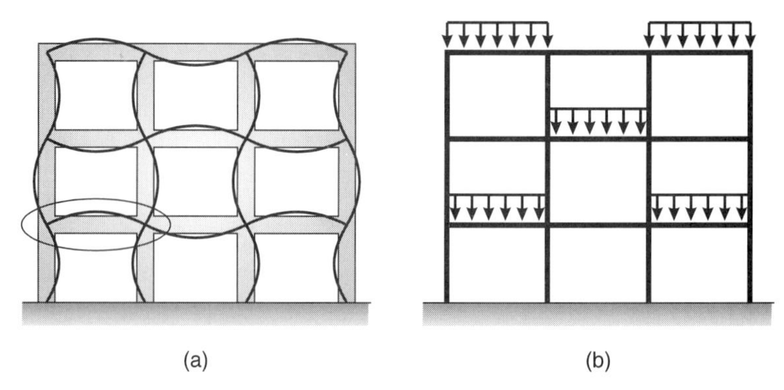

Figure 14.5 Qualitative influence for positive moment in a member and position for loads in a building frame

to be free to rotate. The members at each joint are assumed to be rigidly connected to each other so that the angles between them do not change during rotation. The particular moment influence line shown in Figure 14.5(a) is for positive moment at the center of the circled beam.

The spans that should be loaded to cause maximum positive moment are obvious from the diagram in Figure 14.5(b). It should be noted that loads on a beam more than approximately three spans away have little effect on the function under consideration. This fact was seen in the influence lines constructed in Example 14.2; the ordinates even two spans away were quite small.

Be warned that qualitative influence lines should be drawn only for functions near the center of spans or at the supports. Influence lines for sections near quarter points of members should not be sketched without a good deal of study. Near the quarter point of a span is a point called the *fixed point,* at which the influence line changes in type. The subject of fixed points is discussed at some length in the book *Continuous Frames of Reinforced Concrete* by H. Cross and N. D. Morgan.[2]

14.3 INFLUENCE LINES FOR STATICALLY INDETERMINATE TRUSSES

When analyzing statically indeterminate trusses, influence lines are as necessary to determine the critical positions for live loads as they were for statically determinate trusses. The discussion of construction of these diagrams for a statically indeterminate truss is quite similar to the one presented for statically indeterminate beams in Sections 14.1 and 14.2. To prepare the influence line for a reaction of a continuous truss, the support is removed and a unit load is placed at the support point. For this position of the unit load, the deflection at each of the truss joints is determined. For example, consider the preparation of an influence line for the interior reaction of the truss shown in Figure 14.6. The value of the reaction when the unit load is at joint x may be expressed as

$$V_B = -\frac{\delta_{xb}}{\delta_{bb}} = -\frac{\sum \dfrac{f'_x f_v L}{AE}}{\sum \dfrac{f_v^2 L}{AE}} \qquad \textbf{Eq. 14.7}$$

After the influence line for V_B has been plotted, the influence line for another reaction may be prepared by repeating the process of removing it as the redundant, introducing

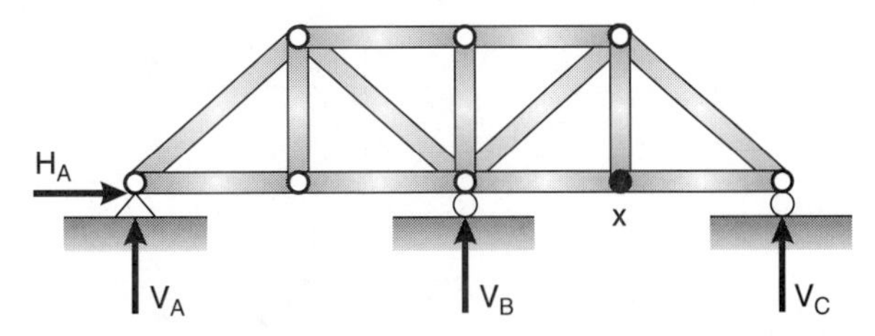

Figure 14.6 A statically indeterminate truss used to domenstrate qualitative influence lines

[2]H. Cross and N. D. Morgan, *Continuous Frames of Reinforced Concrete* (New York: Wiley, 1932).

a unit load there, and computing the necessary deflections. A simpler procedure is to compute the other reactions, or any other functions for which influence lines are desired, using the equations of static equilibrium after the diagram for V_B is prepared. This method is used in Example 14.3 for a two-span truss for which influence lines are desired for both the reactions and the forces in several members.

EXAMPLE 14.3

Draw influence lines for the three vertical reactions and for force in member U_1U_2 in the truss shown in the figure. The load moves along the bottom chord. The numbers beside each member are the areas of the members in square inches.

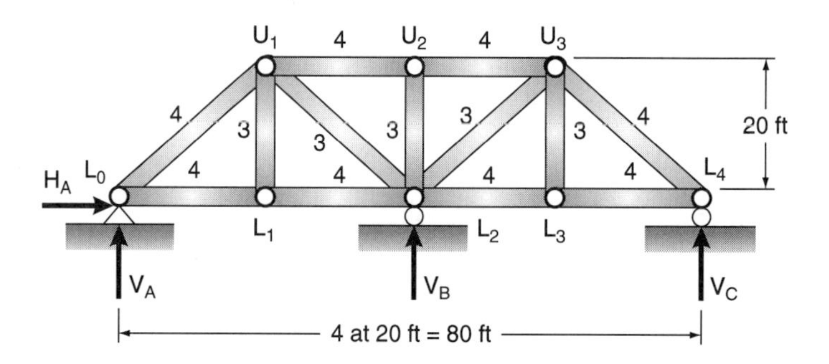

Solution. Determine the influence lines in a manner similar to the way we determined the influence lines for statically indeterminate beams. Remove the interior support, apply a unit load at that location, and then compute the deflection at each of the joints along the bottom chord. To compute the deflections we need to know the forces in the members caused by a unit load at each joint on the bottom chord. Those forces are shown in the following figures.

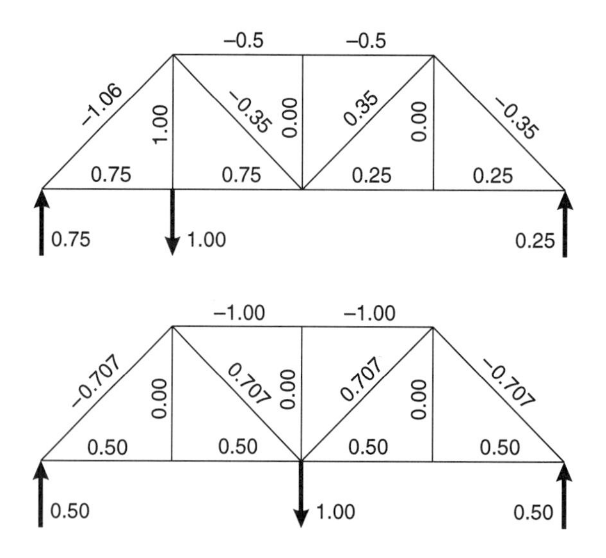

Note that the deflection at L_1 caused by a unit load at L_2 is the same as the deflection L_3 because the structure is symmetric.

Member	L/A	f_b	f_v	$\frac{f_b^2 L}{A}$	$\frac{f_b f_v L}{A}$
L_0L_1	60	0.50	0.75	15	22.5
L_1L_2	60	0.50	0.75	15	22.5
L_2L_3	60	0.50	0.25	15	7.5
L_3L_4	60	0.50	0.25	15	7.5
L_0U_1	85	−0.707	−1.06	42.5	63.6
U_1U_2	60	−1.00	−0.50	60	30.0
U_2U_3	60	−1.00	−0.50	60	30.0
U_3L_4	85	−0.707	−0.35	42.5	21.0
U_1L_1	80	0.0	1.00	0.0	0.0
U_1L_2	113	0.707	−0.35	56.5	−28.0
U_2L_2	80	0.0	0.0	0.0	0.0
L_2U_3	113	0.707	0.35	56.5	28.0
U_3L_3	80	0.0	0.0	0.0	0.0
				$\Sigma = 378$	$\Sigma = 204.6$

Using the data in this table, we can calculate the ordinates of the influence line for the reaction at B:

$$IL_{L1,L3} = \frac{204.6}{378} = 0.542$$

$$IL_{L1} = \frac{204.6}{204.6} = 1.000$$

The ordinate of the influence line at the two supports is zero because the reaction force at B is zero when the load is over supports A and C. Further, the negative sign was dropped because the assumed direction of the reaction at B is opposite the direction of the applied unit load. The resulting influence line for the reaction at B is.

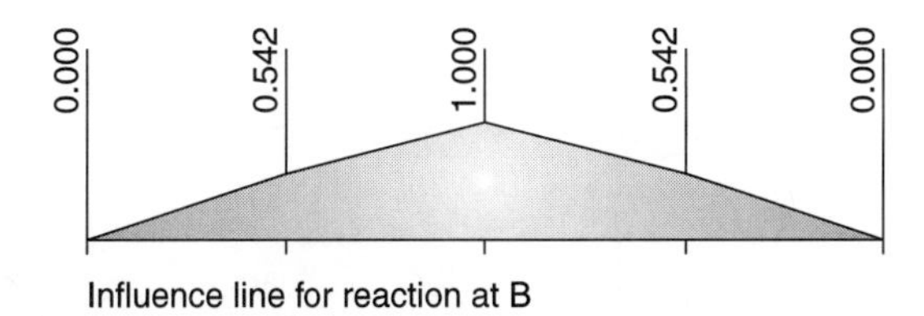

Influence line for reaction at B

After the influence line for the reaction at B is known, the influence lines for the other reactions can be computed from the equations of equilibrium. These influence lines are shown here.

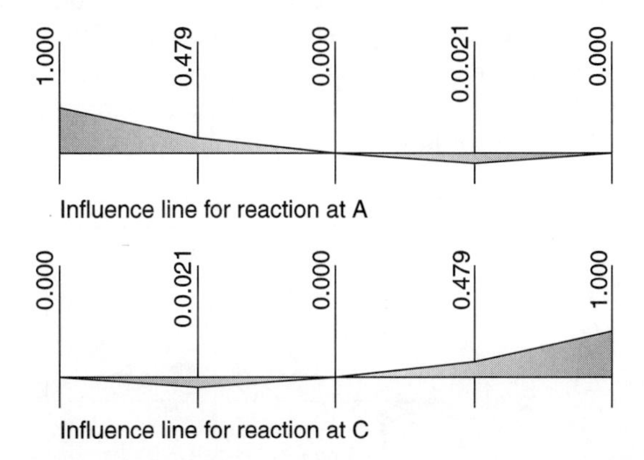

Influence line for reaction at A

Influence line for reaction at C

Again, after the influence lines for each of the reactions is known, the influence line for the force in member U_1U_2 can be computed from equations of static equilibrium. That influence line follows:

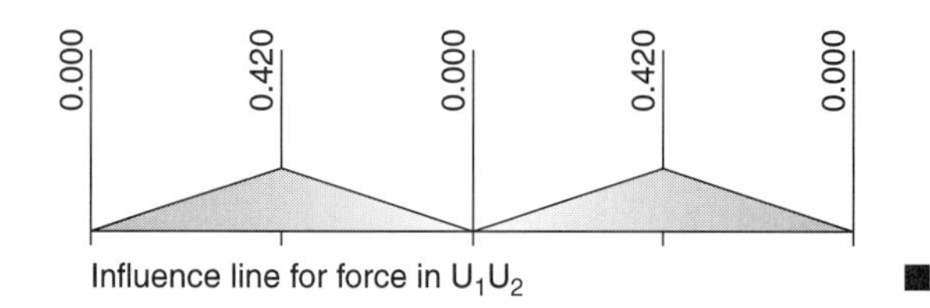

Influence line for force in U_1U_2 ■

Example 14.4 shows that influence lines for members of an internally redundant truss may be prepared by an almost identical procedure. The member assumed to be the redundant has a unit force induced in it thereby causing deflections at each of the joints, which need to be calculated. The ordinates of the diagram for the member are obtained by dividing each of these deflections by the deflection at the member. All other influence lines are prepared from considerations of static equilibrium.

EXAMPLE 14.4

Prepare influence lines for force in members L_1U_2, U_1U_2, and U_2L_2 of the truss shown in the figure. The load is assumed to move along the bottom chord of the truss. We analyzed this same truss in Example 13.8. The number beside each member is the area of that member in square inches.

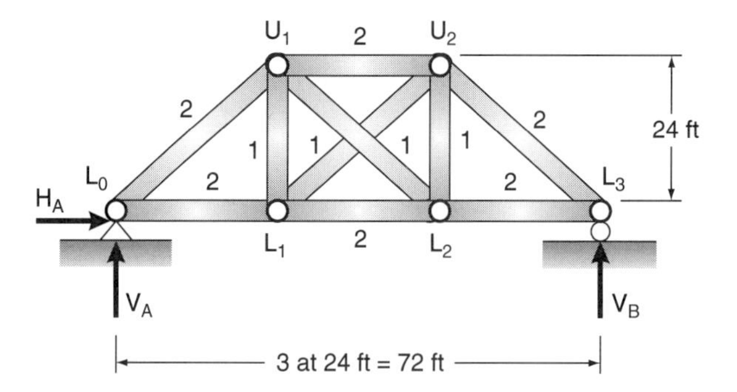

Solution. To compute the desired influence lines, we will treat member L_1U_2 as the redundant member and compute the deflections along the bottom chord caused by a unit force in this member. To accomplish this task, we need the forces in the truss, without the redundant member, caused by a unit force at L_1 and at L_2. We also need the forces in the truss caused by a unit force in the redundant member. Those forces are shown in the figures that follow.

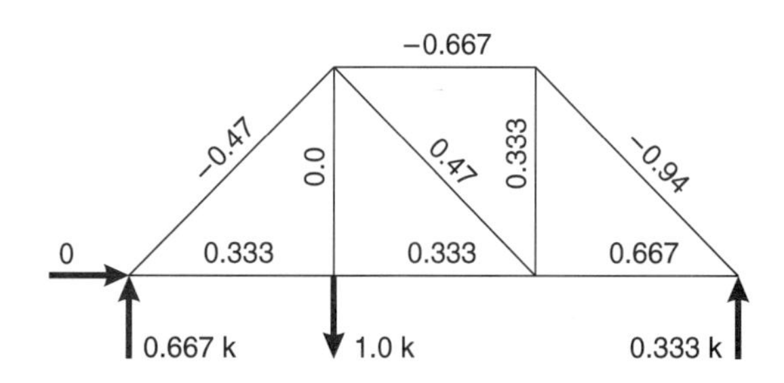

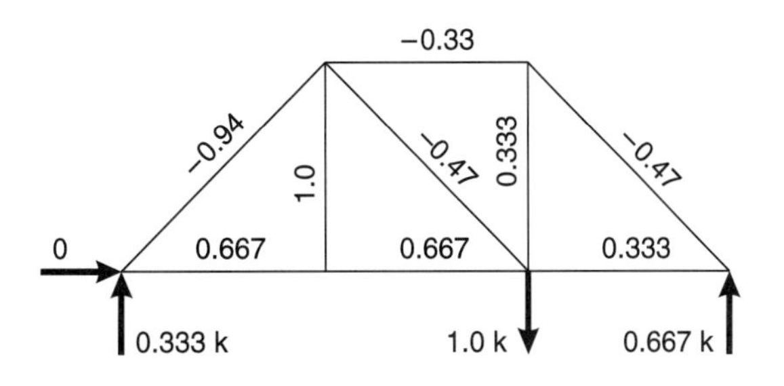

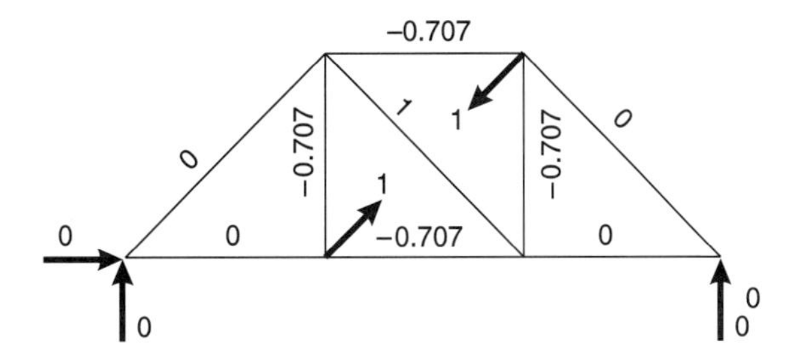

Next we will compute the deflections along the bottom chord and then calculate the ordinates of the influence lines.

Member	L/A	f_{L1}	f_{L2}	f_v	$\frac{f_{L1}f_vL}{A}$	$\frac{f_{L2}f_vL}{A}$	$\frac{f_v^2L}{A}$
L_0L_1	144	0.67	0.33	0.0	0	0	0
L_1L_2	144	0.67	0.33	−0.707	−68	−34	72
L_2L_3	144	0.33	0.67	0.0	0	0	0
L_0U_1	204	−0.94	−0.47	0.0	0	0	0
U_1U_2	144	−0.33	−0.67	−0.707	34	68	72
U_2L_3	204	−0.47	−0.94	0.0	0	0	0
U_1L_1	288	1.00	0.0	−0.707	−204	0	144
U_1L_2	408	−0.47	0.47	1.00	−192	192	408
L_1U_2	408	0.00	0.0	1.00	0	0	408
U_2L_2	288	0.33	0.67	−0.707	−34	−136	144
					$\Sigma = -498$	$\Sigma = 90$	$\Sigma = 1248$

Using this data, we can calculate the ordinates of the influence line for L_1U_2 as follows:

$$IL_{L1} = -\frac{498}{1248} = 0.398$$

$$IL_{L2} = -\frac{90}{1248} = -0.073$$

The ordinate of the influence line at the two supports is zero because the force in L_1U_2 is zero when the load is over the supports. The resulting influence line for L_1U_2 is

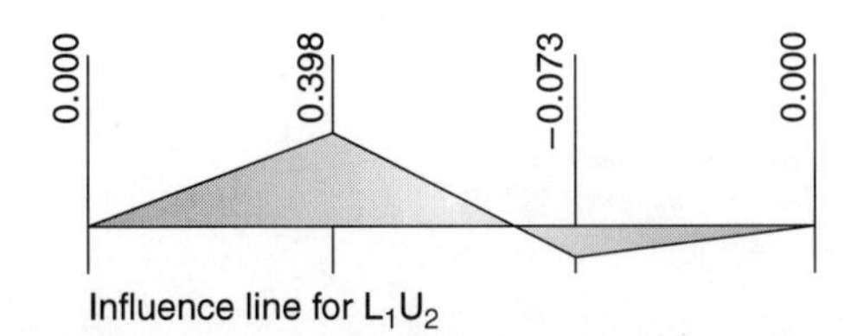

Influence line for L_1U_2

Again, once the forces in L_1U_2 are known as a function of the position of the load, the forces in other members of the truss can be determined and the influence lines for those members plotted. The influence lines thus constructed for the other two members are

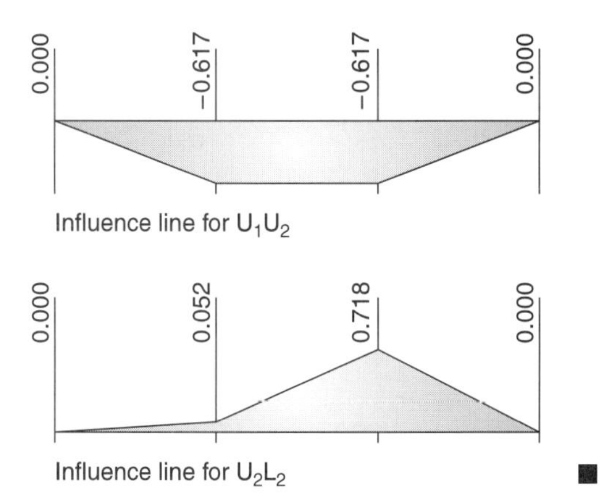

14.4 INFLUENCE LINES USING SABLE

As we have seen in this chapter, the hand calculations necessary for preparing quantitative influence lines for statically indeterminate beams, frames, and trusses can be extremely tedious. For applications such as this, computers can be very helpful by saving time and avoiding math mistakes. By the way, the preparation of qualitative influence lines will provide rather good checks on the computer-generated diagrams.

Using the *behavior analysis* feature, SABLE, a unit load can be placed at different locations on a structure and the analysis automatically performed. As the unit load is moved, the magnitude of a particular force, as a function of the position of the load, can be determined. The resulting solutions can be easily plotted to obtain quantitative influence lines.

14.5 PROBLEMS FOR SOLUTION

Draw quantitative lines for the situations listed in Problems 14.1 to 14.6.

14.1 Reactions for all supports of the beam shown. Place unit load at 10-ft intervals.

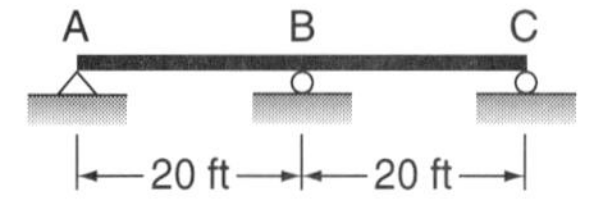

14.2 The left vertical reaction and the moment reaction at the fixed end of the beam in the accompanying illustration. Place unit load at 5-ft intervals. (*Ans.* Load at c_L: $V_A = 0.687$ ↑, $M_A = -3.75$)

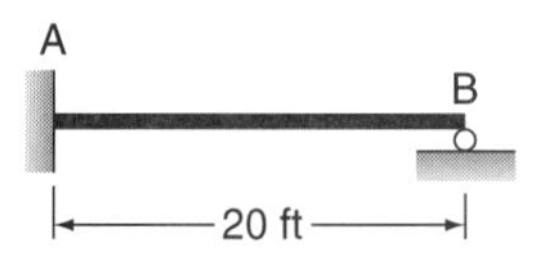

14.3 Shear immediately to the left of support B and moment at support B for the beam shown. Place unit load at 10-ft intervals.

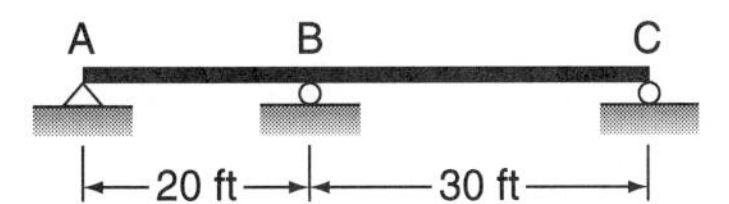

14.4 Shear and moment at a point 20 ft to the left of the fixed-end support C of the beam shown. Place unit load at 10-ft intervals. (*Ans.* Load at left end: V = −0.32, M = +3.2; load halfway from B to C: V = +0.24, M = −2.4)

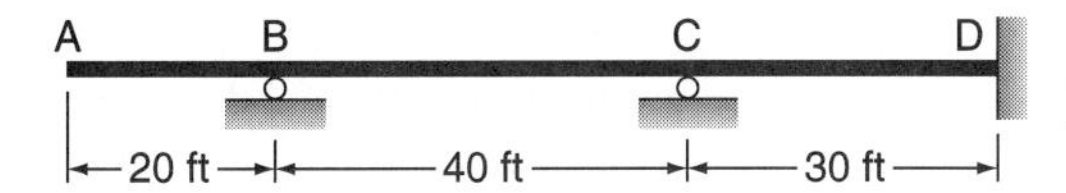

14.5 Vertical reactions and moments at A and B. Place unit load at 10-ft intervals

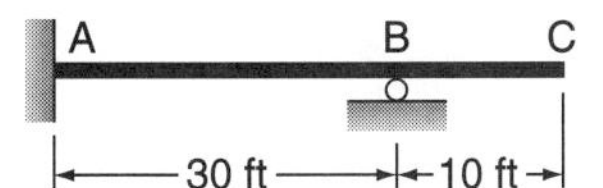

14.6 Vertical reactions at A and B, moment at A and x, and shear just to left of x. Place unit load at 10-ft intervals. (*Ans.* Load at x: V_A = 0.722 ↑, V_B = 0.278 ↑, M_A = −8.88, M_x = 5.56)

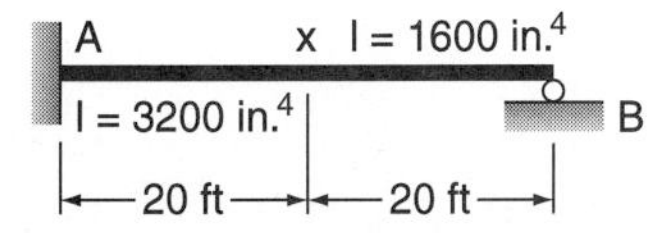

Using Müller-Breslau's principle, sketch influence lines qualitatively for the functions indicated in the structures of Problems 14.7 through 14.14.

14.7 With reference to the accompanying illustration: (a) reactions at A and C, (b) positive moment at x and y, and (c) positive shear at x

14.8 With reference to the accompanying illustration: (a) reaction at A, (b) positive and negative moments at x, and (c) negative moment at B

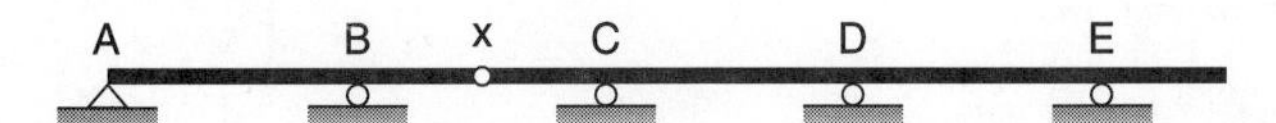

14.9 With reference to the accompanying illustration: (a) vertical reactions at A and C, (b) negative moment at A and positive moment at x, and (c) shear just to left of C

14.10 With reference to the accompanying illustration: (a) vertical reactions at A and C, (b) negative moments at x and C, and (c) shear just to right of B

14.11 With reference to the accompanying illustration: (a) reaction at A, (b) positive moment at x, (c) positive shear at y, and (d) positive moment at z, assuming right side of column is bottom side

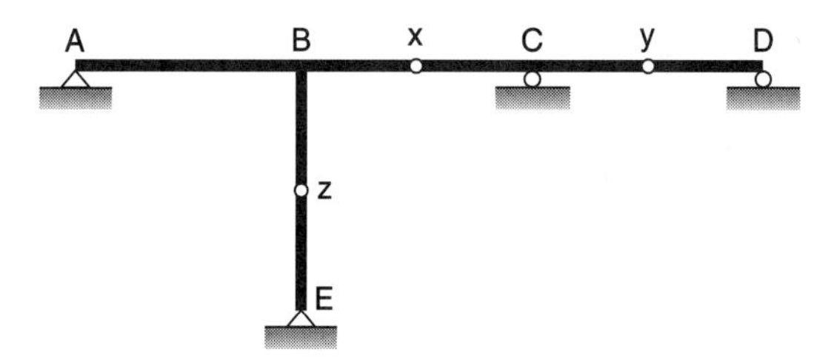

14.12 With reference to the accompanying illustration: (a) positive moment at x, (b) positive shear at x, and (c) negative moment just to the right of y

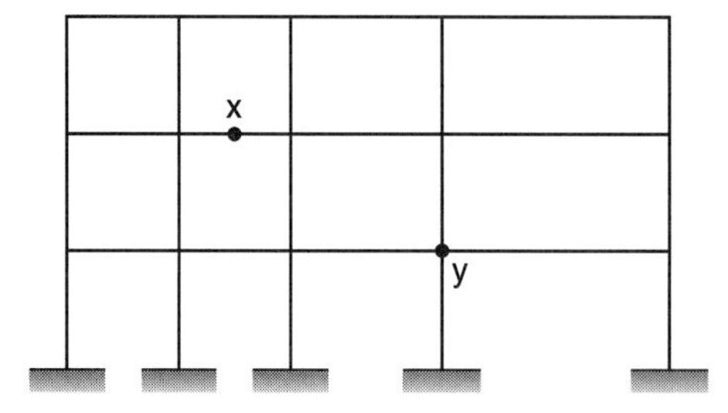

14.13 With reference to the accompanying illustration: (a) positive moment and shear at x, and (b) negative moment just to the right of y

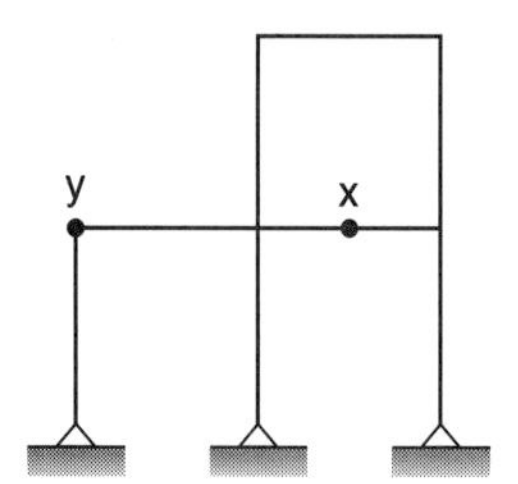

14.14 With reference to the accompanying illustration: (a) positive and negative moment at x and (b) positive shear just to right of y

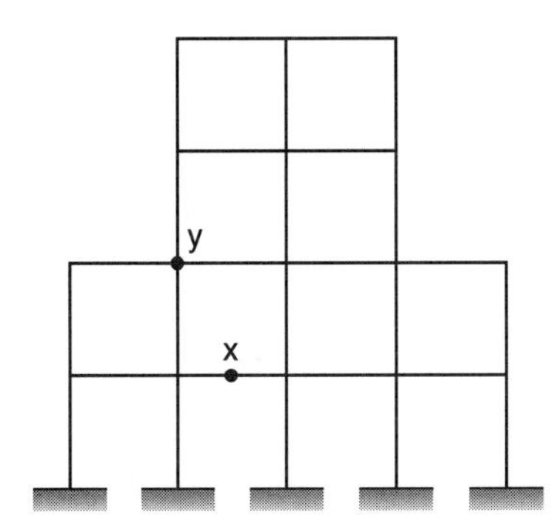

Draw quantitative influence lines for the situations listed in Problems 14.15 through 14.20.

14.15 Reactions at all supports for the truss of Problem 13.35

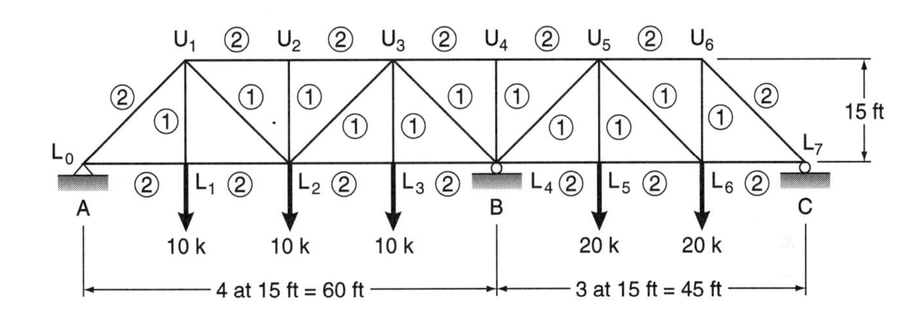

14.16 Forces in members U_1L_2, U_3U_4, and L_4L_5 of the truss of Problem 14.15 (*Ans.* Load at L_2: $U_1L_2 = +0.634$, $U_3U_4 = +0.208$, $L_4L_5 = -0.140$)

14.17 Reactions for all supports for the truss. Assume the loads are moving across the top of the structure.

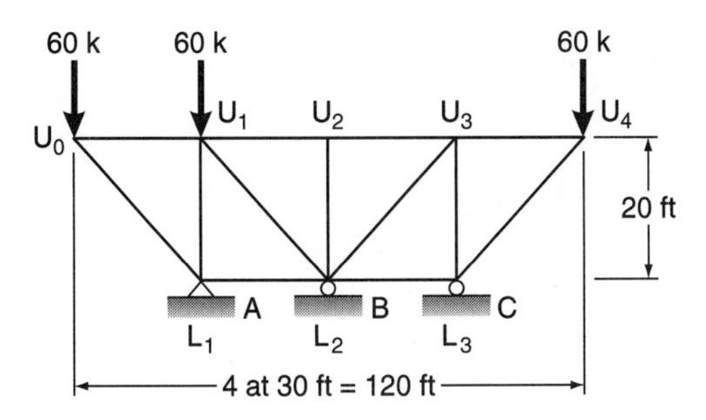

14.18 Forces in members U_3L_4, L_3L_4, U_1U_2, and U_2L_2 of the truss (*Ans.* Load at L_3: $U_3L_4 = -0.417$, $L_3L_4 = +1.67$; Load at L_2: $U_1U_2 = -1.78$, $U_2L_2 = +0.33$)

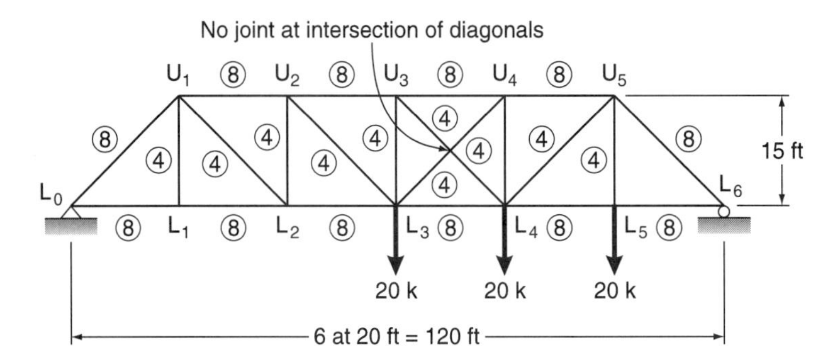

14.19 Forces in members L_1U_2 and U_2L_3 of the truss

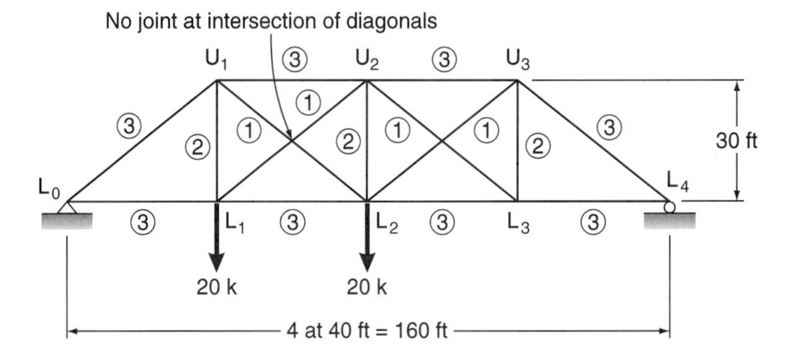

14.20 Force in member L_2U_3 and the center reaction of the truss (*Ans.* Load at L_1: $V_B = +0.504 \uparrow$, $L_2U_3 = -0.0078$)

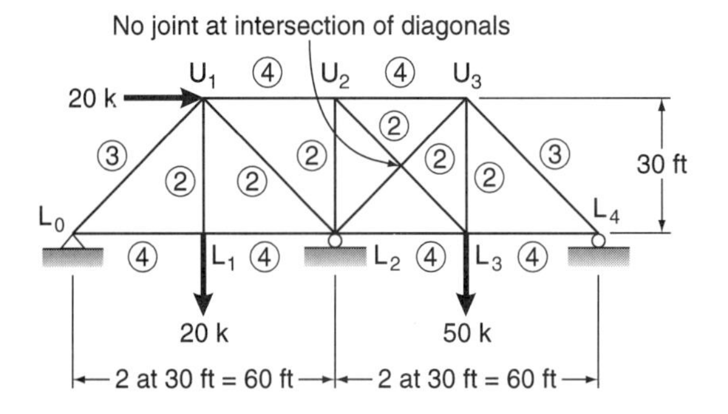

Chapter 15

Slope Deflection: A Displacement Method of Analysis

15.1 INTRODUCTION

George A. Maney introduced slope deflection in a 1915 University of Minnesota engineering publication.[1] His work was an extension of earlier studies of secondary stresses by Manderla[2] and Mohr.[3] For nearly 15 years, until the introduction of moment distribution, slope deflection was the popular "exact" method used for the analysis of continuous beams and frames in the United States.

Slope deflection is a method that takes into account the flexural deformations of beams and frames (rotations, settlements, etc.), but that neglects shear and axial deformations. Although this classical method is generally considered obsolete, its study can be useful for several reasons. These include:

1. Slope deflection is convenient for hand analysis of some small structures.
2. Knowledge of the method provides an excellent background for understanding the moment distribution method discussed in Chapters 17 and 18.
3. It is a special case of the displacement, or stiffness, method of analysis previously defined in Section 12.5 and provides a very effective introduction for the matrix formulation of structures described in Chapters 19 to 21.
4. The slopes and deflections determined by slope deflection enable the analyst to easily sketch the deformed shape of a particular structure. The result is that he or she has a better "feel" for the behavior of structures.

[1]G. A. Maney, *Studies in Engineering,* No. 1 (Minneapolis: University of Minnesota, 1915).

[2]H. Manderla, "Die Berechnung der Sekundarspannungen," *Allg. Bautz* 45 (1880): 34.

[3]O. Mohr, "Die Berechnung der Fachwerke mit starren knotenverbingungen," *Zivilinginieur,* 1892.

Hospital, Stamford, Connecticut (Courtesy of the American Concrete Institute)

15.2 DERIVATION OF SLOPE-DEFLECTION EQUATIONS

The name *slope deflection* comes from the fact that the moments at the ends of the members in statically indeterminate structures are expressed in terms of the rotations (or slopes) and deflections of the joints. For developing the equations, members are assumed to have constant cross section between supports. Although it is possible to derive expressions for members of varying section, the results are so complex they are of little practical value. It is further assumed that the joints in a structure may rotate or deflect, but the angles between the members meeting at a joint remain unchanged.

Span AB of the continuous beam of Figure 15.1(a) is considered for the following discussion. If the span is completely fixed at each end, the slope of the elastic curve of the beam at the ends is zero. External loads produce fixed-end moments, and these moments cause the span to take the shape shown in Figure 15.1(b). Joints A and B, however, are not actually fixed. Under load, they will rotate slightly to positions such as those shown in Figure 15.1(c). In addition to the rotation of the joints, one or both of the supports may possibly settle. Settlement of the joints will cause a chord rotation of the member as shown in part (d), where support B is assumed to have settled an amount Δ.

From Figure 15.1 we see that the values of the final end moments at A and B (M_{AB} and M_{BA}) are equal to the sum of the moments caused by the following:

1. The fixed-end moments (FEM_{AB} and FEM_{BA}) which can be determined with the moment-area theorems as illustrated in Example 10.4. A detailed list of fixed-end moment expressions is presented in Appendix D.
2. The moments caused by the rotations of joints A and B (θ_A and θ_B).
3. The moments caused by chord rotation ($\psi = \Delta/L$) if one or both of the joints settles or deflects.

When a joint in a structure rotates, the slopes of the tangents to the elastic curves of the members connected to that joint change. For a particular beam, the change in slope is equal to the end shear of the beam when it is loaded with the M/EI diagram. The beam is assumed to have the end moments FEM_{AB} and FEM_{BA}, shown in Figure 15.2.

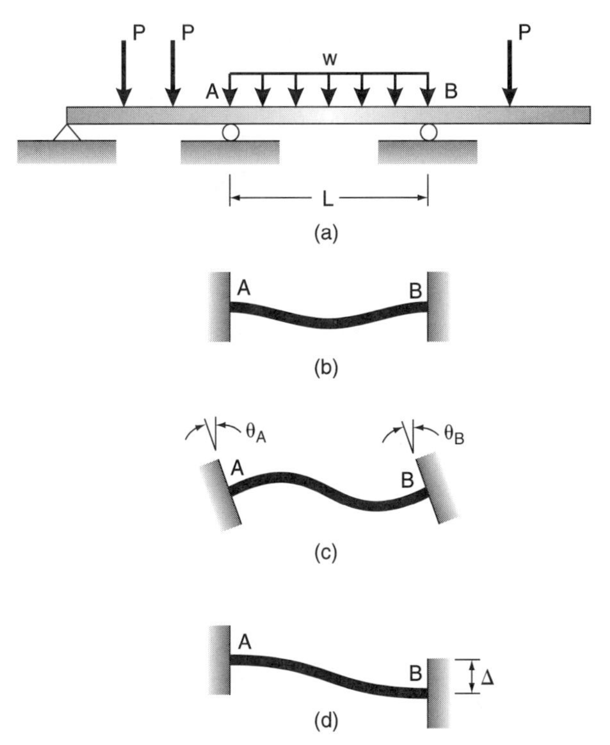

Figure 15.1 Illustration for development of equations for slope-deflection method

The end reactions or end slopes are as follows:

$$\theta_A = \frac{\frac{1}{2}\left(\frac{M_{AB}}{EI}\right)(L)\left(\frac{2L}{3}\right) - \frac{1}{2}\left(\frac{M_{BA}}{EI}\right)(L)\left(\frac{L}{3}\right)}{L} = \frac{L}{6EI}(2M_{AB} - M_{BA})$$

$$\theta_B = \frac{\frac{1}{2}\left(\frac{M_{BA}}{EI}\right)(L)\left(\frac{2L}{3}\right) - \frac{1}{2}\left(\frac{M_{AB}}{EI}\right)(L)\left(\frac{L}{3}\right)}{L} = \frac{L}{6EI}(2M_{BA} - M_{AB})$$

Eq. 15.1

If one of the supports of the beam settled or deflected an amount Δ, the angles θ_A and θ_B caused by joint rotation would be changed by Δ/L(or ψ), as was illustrated in Figure 15.1(d). When the chord rotation is added to the expressions in Equation 15.1, the following equations are obtained for the slopes of the tangents to the elastic curves at the

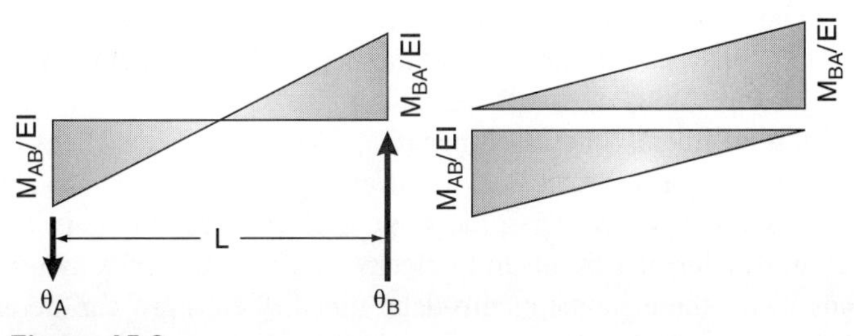

Figure 15.2

ends of the beams:

$$\theta_A = \frac{L}{6EI}(2M_{AB} - M_{BA}) + \psi$$
$$\theta_B = \frac{L}{6EI}(2M_{BA} - M_{AB}) + \psi \qquad \textbf{Eq. 15.2}$$

These equations can be solved simultaneously for M_{AB} and M_{BA}. The result are the following end moments in the beam due to the joint rotation and deflection:

$$M_{AB} = 2EK(2\theta_A + \theta_B - 3\psi)$$
$$M_{AB} = 2EK(\theta_A + 2\theta_B - 3\psi) \qquad \textbf{Eq. 15.3}$$

In these equations, the term I/L has been replaced with K. This term is called the *stiffness factor.* (We will see it again when we talk about moment distribution in Chapter 17.)

The final end moments are equal to the moments caused by rotation and deflection of the joints plus the fixed-end moments acting on the ends of the beam. The final slope-deflection equations are as follows:

$$M_{AB} = 2EK(2\theta_A + \theta_B - 3\psi) + FEM_{AB}$$
$$M_{AB} = 2EK(\theta_A + 2\theta_B - 3\psi) + FEM_{BA} \qquad \textbf{Eq. 15.4}$$

With these equations, we can express the end moments in a structure in terms of joint rotations and settlements. Using the methods previously studied, we have to write one equation for each redundant in the structure. The number of unknowns in these equations is equal to the number of redundants. The effort to solve these equations is considerable for highly redundant structures. The slope-deflection method appreciably reduces the amount of work involved in analyzing multiredundant structures because the unknown moments are expressed in terms of only a few unknown joint rotations and settlements. Even for multistory frames, the number of unknown slopes and chord rotations that appear in any one equation is rarely more than five or six, whereas the degree of indeterminacy of the structure is many times that figure.

15.3 APPLICATION OF SLOPE DEFLECTION TO CONTINUOUS BEAMS

Examples 15.1 to 15.4 illustrate the analysis of statically indeterminate beams using the slope-deflection equations. Let us define a procedure for applying the slope-deflection equations to the analysis of statically indeterminate beams. Before doing so, however, we need to establish a sign convention for the different moments, slopes, and displacements. The sign convention is the major difficulty experienced in applying the slope-deflection equations. It is essential to understand these signs before attempting to apply the equations.

For previous work in this book a plus sign for the moment has indicated tension in the bottom fibers, whereas a negative sign has indicated tension in the top fibers. This sign convention was necessary for drawing moment diagrams and has been referred to as beam notation. When using the slope-deflection equations it is simpler to use a convention in which signs are given to clockwise and counterclockwise moments at the member ends. Once these moments are determined, their signs can be easily converted to beam notation for drawing shearing force and bending moment diagrams.

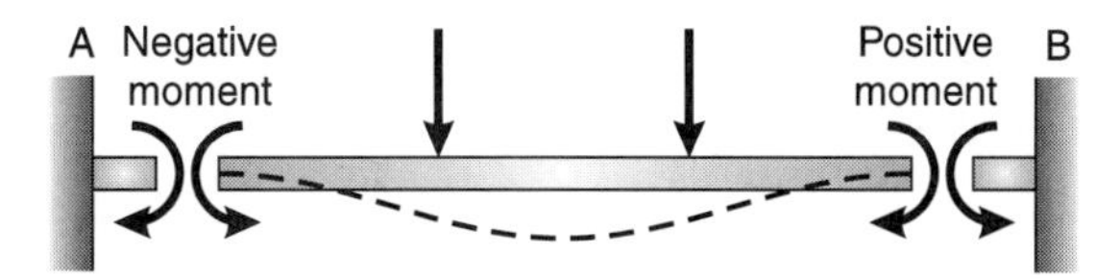

Figure 15.3 Sign convention for the slope-deflection method

The following convention is used in this chapter for slope deflection, and in Chapters 17 and 18 for moment distribution:

- Should a member cause a moment that tends to rotate a joint clockwise, the moment acting of the joint is considered negative.
- Should a member cause a moment that tends to rotate a joint counterclockwise, the moment acting on the joint is considered positive.

In other words, a clockwise resisting moment on the member is considered positive, and a counterclockwise resisting moment is considered negative. This sign convention is illustrated in Figure 15.3. The moment at the left end, M_{AB}, tends to rotate the joint clockwise; it is considered a negative moment. Observe that the resisting moment on the end of the member is counterclockwise. At the right end of the beam, the loads have caused a moment that tends to rotate the joint counterclockwise. As such, M_{BA} is considered positive.

Using this sign convention, the procedure that we will use when evaluating statically indeterminate structures using the slope-deflection equation is as follows:

1. For each member in the beam, compute the fixed-end moments.
2. Compute the chord rotation, ψ, for each member in the beam. Chord rotation is considered positive when the chord of a beam is rotated clockwise by the settlements; the sign is the same no matter which end is being considered since the entire beam rotates in that direction.
3. Determine the boundary conditions for the beam. The boundary conditions are points on the beam at which we know the moment or rotation at the end of a member. For example, at fixed ends the rotation is zero while at pin supports the moment is zero.
4. Taking into consideration the boundary conditions on the beam, write the two moment equations for each span. That is, equations are written for M_{AB} and M_{BA} on span AB; for M_{BC} and M_{CB} on span BC, and so on. These moment equations are written in terms of the unknown values of θ at the supports.
5. Write the compatibility equations at the supports. The summation of moments at a support must equal zero. For example, assuming two members are connected at support B, $M_{AB} + M_{BA} = 0$.
6. Simultaneously solve the equations that result for the unknown rotations and then compute the moments at the ends of the members.
7. Compute the shearing forces at the ends of the members and the support reactions.

When span lengths, moduli of elasticity, and moments of inertia are constant for the spans of a continuous beam, the term 2EK is constant for all members and may be canceled from the equations. Should the values of K vary from span to span, as they often do, it is convenient to express them in terms of relative values, as is done in Example 15.6.

EXAMPLE 15.1

Determine all support moments for the structure shown in the figure using the slope-deflection method. E and I are constant for this beam.

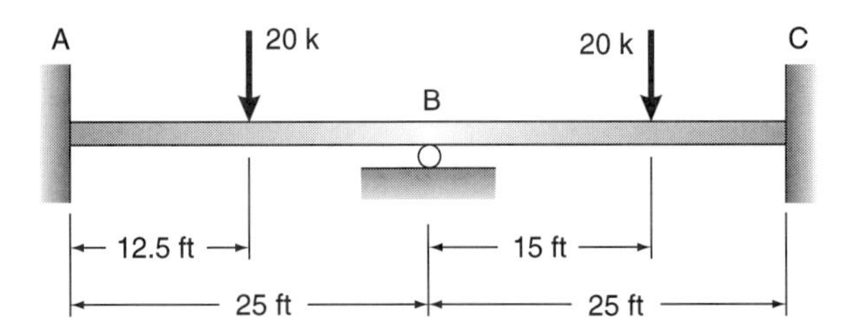

Solution. Begin by computing fixed-end moments with reference to Appendix D.

$$FEM_{AB} = -\frac{20(12.5)(12.5)^2}{25^2} = -62.5\,\text{k-ft}$$

$$FEM_{BA} = +\frac{20(12.5)(12.5)^2}{25^2} = +62.5\,\text{k-ft}$$

$$FEM_{BC} = -\frac{20(15)(10)^2}{25^2} = -48.0\,\text{k-ft}$$

$$FEM_{CB} = +\frac{20(10)(15)^2}{25^2} = +72.0\,\text{k-ft}$$

Because there is no settlement at the supports both chord rotation angles are zero. Further, because the supports at A and C are fixed supports, the rotations θ_A and θ_B are zero. Using Equation 15.4, we can write the equations for the moments at the ends of the members. These equations are:

$$\begin{aligned}
M_{AB} &= 2EK[2(0) + \theta_B - 3(0)] - 62.5 = 2EK\theta_B - 62.5\\
M_{BA} &= 2EK[0 + 2\theta_B - 3(0)] + 62.5 = 4EK\theta_B + 62.5\\
M_{BC} &= 2EK[2\theta_B + 0 - 3(0)] - 48.0 = 4EK\theta_B - 48.0\\
M_{CB} &= 2EK[\theta_B + 2(0) - 3(0)] + 72.0 = 2EK\theta_B + 72.0
\end{aligned}$$

We can next write the compatibility equation for moment at the middle support, which is

$$\begin{aligned}
\sum M_B &= 0 = M_{BA} + M_{BC}\\
0 &= 4EK\theta_B + 62.5 + 4EK\theta_B - 48.0 = EK\theta_B + 1.8125\\
\therefore\ EK\theta_B &= -1.8125
\end{aligned}$$

Lastly, by substituting this value of $EK\theta_B$ into the previously determined moment equations, we can compute the moments at each end of the members.

$$\begin{aligned}
M_{AB} &= 2EK\theta_B - 62.5 = 2(-1.8125) - 62.5 = 66.125\,\text{k-ft}\\
M_{BA} &= 4EK\theta_B + 62.5 = 4(-1.8125) - 62.5 = 55.25\,\text{k-ft}\\
M_{BC} &= 4EK\theta_B - 48.0 = 4(-1.8125) - 48.0 = -55.25\,\text{k-ft}\\
M_{CB} &= 2EK\theta_B + 72.0 = 2(-1.8125) + 72.0 = 68.375\,\text{k-ft}
\end{aligned}$$ ■

EXAMPLE 15.2

Find all end moments in the structure shown in the figure using slope deflection.

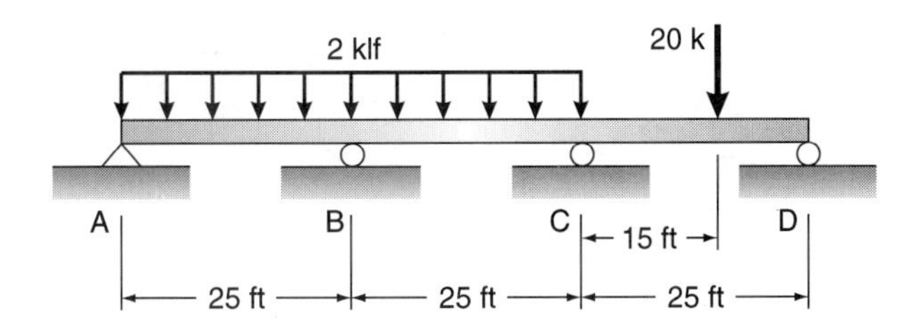

Solution. We begin evaluating the end moments by computing the fixed-end moments.

$$FEM_{AB} = FEM_{BC} = -\frac{w(25)^2}{12} = -104.16$$

$$FEM_{BA} = FEM_{CB} = +\frac{w(25)^2}{12} = +104.16$$

$$FEM_{CD} = -\frac{20(15)(10)^2}{25^2} = -48.0$$

$$FEM_{DC} = \frac{20(10)(15)^2}{25^2} = +72.0$$

Again, because there are no settlements at the supports, all chord rotation angles are zero. Further, because the supports at A and D are pinned supports, the moments M_{AB} and M_{DC} are zero. Using Equation 15.4, we can write the equations for the moments at the ends of the members. These equations are

$$M_{AB} = 2EK[2\theta_A + \theta_B - 3(0)] - 104.2 = 0$$
$$M_{BA} = 2EK[\theta_A + 2\theta_B - 3(0)] + 104.2$$
$$M_{BC} = 2EK[2\theta_B + \theta_C - 3(0)] - 104.2$$
$$M_{CB} = 2EK[\theta_B + 2\theta_C - 3(0)] + 104.2$$
$$M_{CD} = 2EK[2\theta_C + \theta_D - 3(0)] - 48$$
$$M_{DC} = 2EK[\theta_C + 2\theta_D - 3(0)] + 72 = 0$$
$$M_{BA} + M_{BC} = 0$$
$$M_{CB} + M_{CD} = 0$$

By solving these equations simultaneously, we find that

$$M_{AB} = 0 \qquad M_{BA} = 134.6\,\text{k-ft}$$
$$M_{BC} = -134.6\,\text{k-ft} \qquad M_{CB} = 86.47\,\text{k-ft}$$
$$M_{CD} = -86.47\,\text{k-ft} \qquad M_{DC} = 0 \quad \blacksquare$$

When a beam has simply supported ends, the moments at those ends must be equal to zero for equilibrium. Application of the usual slope-deflection equations will yield zero moments at these locations. However, it seems a waste of time to go through a process to determine the value of all of the moments in the beam when by inspection some of them obviously are equal to zero. The usual slope-deflection equations for a single span, as shown in Equation 15.4, are as follows:

$$M_{AB} = 2EK(2\theta_A + \theta_B - 3\psi) + FEM_{AB}$$
$$M_{AB} = 2EK(\theta_A + 2\theta_B - 3\psi) + FEM_{BA}$$

For this discussion, assume that end A is simply supported; the magnitude of M_{AB} is zero. Solving these two equations simultaneously by eliminating θ_A yields a simplified expression for M_{BA} that has only one unknown, θ_B. When performing hand calculations, the resulting simplified equation will expedite considerably the solution of continuous beams with simple ends. That equation is:

$$M_{BA} = 3EK(\theta_B - \psi) + FEM_{BA} - \tfrac{1}{2}FEM_{AB} \qquad \textbf{Eq. 15.5}$$

Using this equation, the number of equations to be solved is reduced by one for each simple support. When using computational tools, however, this may not be important.

EXAMPLE 15.3

Find the member-end moments in the beam shown in the figure. Support B settles 0.25 in. (0.0208 ft). For this beam, $I = 500$ in.^{4} and $E = 29{,}000$ ksi.

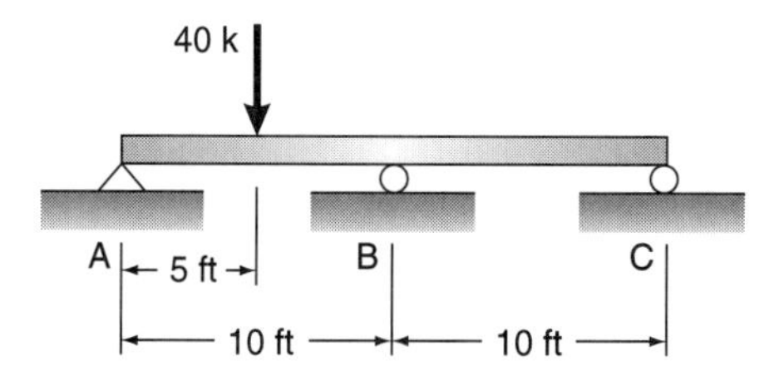

Solution. We will work this example in units of feet and kips. As such, the stiffness of the members is

$$EK = \frac{29{,}000(500)}{144(10)} = 10{,}070\,\text{k-ft}$$

Next, we will compute the values of the fixed-end moments:

$$FEM_{AB} = -\frac{50(5)(5)^2}{10^2} = -50\,\text{k-ft} \qquad FEM_{BC} = 0$$

$$FEM_{BA} = \frac{50(5)(5)^2}{10^2} = 50\,\text{k-ft} \qquad FEM_{CB} = 0$$

Because of the settlement at the middle support, the chord rotations in this example are not equal to zero. Rather, they are equal to

$$\psi_{AB} = \frac{0.25}{120} = \frac{0.02083}{10} = 0.002083\,\text{rad}$$

$$\psi_{BC} = -\frac{0.25}{120} = -\frac{0.02083}{10} = -0.002083\,\text{rad}$$

Using these terms, we are ready to write the four slope-deflection equations for this beam and the compatibility equation for equilibrium of moment at B. These equations are

$$M_{AB} = 2EK[2\theta_A + \theta_B - 3\psi_{AB}] - 50 = 0$$
$$M_{BA} = 2EK[\theta_A + 2\theta_B - 3\psi_{AB}] + 50$$

$$M_{BC} = 2EK[2\theta_B + \theta_C - 3\psi_{BC}] + 0$$
$$M_{CB} = 2EK[\theta_B + 2\theta_C - 3\psi_{BC}] + 0 = 0$$
$$M_{BA} + M_{BC} = 0$$

Upon solution of these equations and substitution of the values of EK and ψ, we find that the member-end moments are:

$$M_{AB} = 0 \qquad M_{BA} = -25.43\,\text{k-ft}$$
$$M_{BC} = 25.43\,\text{k-ft} \qquad M_{CB} = 0 \quad \blacksquare$$

EXAMPLE 15.4

Determine all the member-end moments for the beam shown in the figure. The settlement of the supports are A = 1.25 in, B = 2.40 in, C = 2.75 in, and D = 1.10 in. For this beam, I is equal to 8147.6 in.^{4} and E is equal to 29,000 ksi.

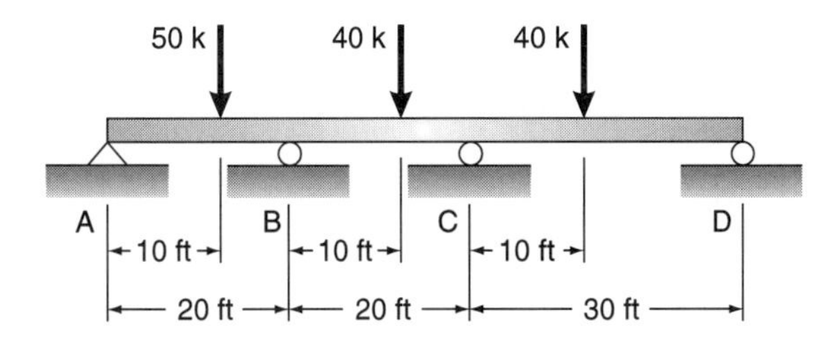

Solution. We begin this solution by determining the stiffness properties of the beam:

$$EK_{AB} = EK_{BC} = \frac{29{,}000(8147.6)}{144(20)} = 82{,}040\,\text{k-ft}$$

$$EK_{CD} = \frac{29{,}000(8147.6)}{144(30)} = 54{,}690\,\text{k-ft}$$

Next, we will compute the fixed-end moment in the beam members:

$$FEM_{AB} = -\frac{50(10)(10)^2}{20^2} = -125\,\text{k-ft} \qquad FEM_{BA} = \frac{50(10)(10)^2}{20^2} = 125\,\text{k-ft}$$

$$FEM_{BC} = -\frac{40(10)(10)^2}{20^2} = -100\,\text{k-ft} \qquad FEM_{CB} = \frac{40(10)(10)^2}{20^2} = 100\,\text{k-ft}$$

$$FEM_{CD} = -\frac{40(10)(20)^2}{30^2} = -177.8\,\text{k-ft} \qquad FEM_{DC} = \frac{40(20)(10)^2}{30^2} = 88.89\,\text{k-ft}$$

Because there is settlement at the supports, we need to compute the chord rotations that result from this settlement:

$$\psi_{AB} = \frac{2.4 - 1.25}{240} = 0.004792\,\text{rad}$$

$$\psi_{BC} = \frac{2.75 - 2.4}{240} = 0.001458\,\text{rad}$$

$$\psi_{CD} = \frac{1.1 - 2.75}{240} = -0.004583\,\text{rad}$$

The slope-deflection equations used for the analysis are

$$M_{AB} = 2EK_{AB}[2\theta_A + \theta_B - 3(0.004792)] - 125 = 0$$
$$M_{BA} = 2EK_{AB}[\theta_A + 2\theta_B - 3(0.004792)] + 125$$
$$M_{BC} = 2EK_{BC}[2\theta_B + \theta_C - 3(0.001458)] - 100$$
$$M_{CB} = 2EK_{BC}[\theta_B + 2\theta_C - 3(0.001458)] + 100$$
$$M_{CD} = 2EK_{CD}[2\theta_C + \theta_D - 3(-0.004583)] - 88.89$$
$$M_{DC} = 2EK_{CD}[\theta_C + 2\theta_D - 3(-0.004583)] + 88.89 = 0$$

In addition, the two necessary compatibility equations are

$$M_{BA} + M_{BC} = 0$$
$$M_{CB} + M_{CD} = 0$$

Upon evaluating the slope-deflection equations and the compatibility simultaneously for the unknown terms, we find that the member-end moments are

$$M_{AB} = 0 \qquad M_{BA} = -148.5\,\text{k-ft}$$
$$M_{BC} = 148.5\,\text{k-ft} \qquad M_{CB} = -371.8\,\text{k-ft}$$
$$M_{CD} = 371.8\,\text{k-ft} \qquad M_{DC} = 0$$ ■

15.4 ANALYSIS OF FRAMES WITH NO SIDESWAY

The slope-deflection equations may be applied to statically indeterminate frames in the same manner as they were to continuous beams if there is no possibility for the frames to lean or to deflect asymmetrically. Theoretically, a frame will not deflect to one side, or sway, if it is symmetrical and the loads applied to it are symmetrical, or if it is prevented from swaying by other parts of the structure. Neglecting axial deformation, the frame shown in Figure 15.4 cannot sway because the leftmost support restrains the structure against horizontal displacement. Illustrated in Example 15.5 is the analysis of a simple frame with no sidesway. When sidesway occurs, the joints of the frame move. This affects the values of θ at the joints and the values of ψ for the members. Frames with sidesway are discussed in Section 15.5.

Lehigh Shop Center (Courtesy Bethlehem Steel Corporation)

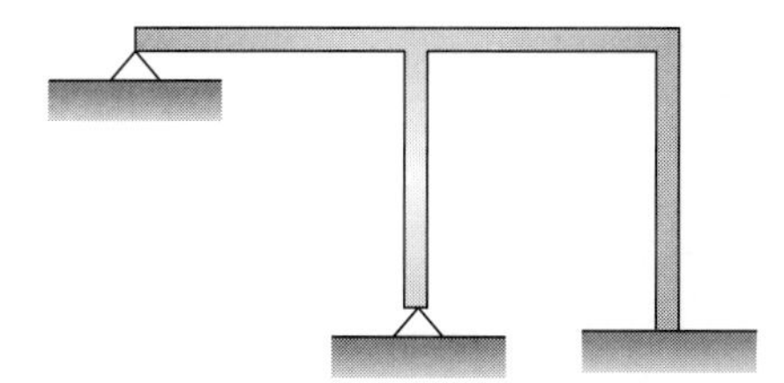

Figure 15.4 A frame that will sway under load

EXAMPLE 15.5

Find all moments for the frame shown in the figure. This frame has no sidesway because it is symmetric with respect to geometry and loads. EI is constant for all members and can be taken as unity.

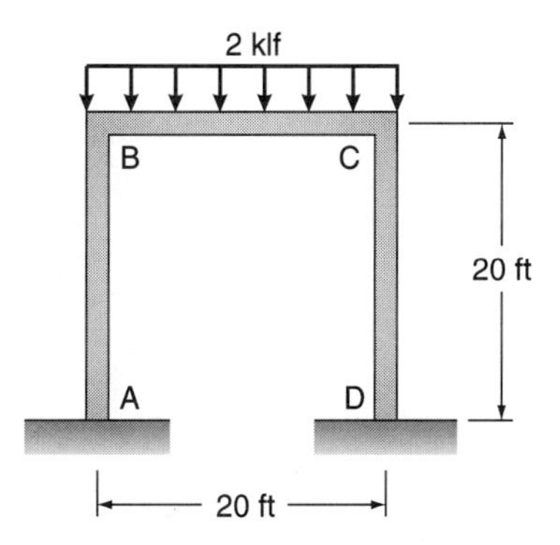

Solution. We begin the analysis of this frame by computing the fixed-end moments that occur in the members. The fixed-end moments are equal to zero for all members except BC. The fixed-end moments for that member are

$$M_{BC} = -\frac{2(20)^2}{12} = -66.67\,\text{k-ft}$$

$$M_{CB} = \frac{2(20)^2}{12} = 66.67\,\text{k-ft}$$

Because there is no sidesway in this frame, all of the chord rotations are equal to zero. The slope-deflection equations are written in the same manner that we did for beams. Because supports A and D are fixed supports, the rotations θ at those supports are equal to zero. EI and the lengths of the members are the same, so EK is the same for all members and will be taken as unity. The resulting slope-deflection equations are

$$M_{AB} = 2(1)[2(0) + \theta_B - 3(0)] + 0$$
$$M_{BA} = 2(1)[0 + 2\theta_B - 3(0)] + 0$$
$$M_{BC} = 2(1)[2\theta_B + \theta_C - 3(0)] - 66.67$$
$$M_{CB} = 2(1)[\theta_B + 2\theta_C - 3(0)] + 66.67$$
$$M_{CD} = 2(1)[2\theta_C + 0 - 3(0)] + 0$$
$$M_{DC} = 2(1)[\theta_C + 2(0) - 3(0)] + 0$$

As with beams, equilibrium at the interior joints must be satisfied. The equations of compatibility for rotational equilibrium to be used are

$$M_{BA} + M_{BC} = 0$$
$$M_{CB} + M_{CD} = 0$$

The unknown values in these equations are the member-end moments and the rotations of joints B and C. Upon evaluating the slope deflection and compatibility simultaneously for the unknown terms, we find that the member-end moments are

$$M_{AB} = 22.22\,\text{k-ft} \qquad M_{BA} = 44.44\,\text{k-ft}$$
$$M_{BC} = -44.44\,\text{k-ft} \qquad M_{CB} = 44.44\,\text{k-ft}$$
$$M_{CD} = -44.44\,\text{k-ft} \qquad M_{DC} = -22.22\,\text{k-ft}$$ ■

15.5 ANALYSIS OF FRAMES WITH SIDESWAY

The frame shown in Figure 15.5 is not symmetrical. Sidesway to one side, therefore, is possible as illustrated in the figure. Joints B and C deflect to the right, which causes chord rotations in members AB and CD. Theoretically, if axial deformation is neglected, there will be no rotation of member BC and joints B and C will displace the same horizontal distance, Δ.

For the frame shown in Figure 15.5, the chord rotations of members AB and CD, due to sidesway, can be seen to be equal to:

$$\psi_{AB} = \frac{\Delta}{\frac{2}{3}L} = \frac{3\Delta}{2L}$$
$$\psi_{CD} = \frac{\Delta}{L} \qquad \textbf{Eq. 15.6}$$

Observe for the same Δ that the shorter a member has a larger chord rotation and thus a larger the effect on the moments in the members. For this frame, ψ_{AB} is 1.5 times as large as ψ_{CD} because L_{AB} is only two-thirds as large as L_{CD}. Normally, to minimize the number of equations that must be used, the chord rotations in terms of Δ are used instead of the actual rotation.

Presented in Example 15.6 is the analysis of the frame shown in Figure 15.5. Because of the fixed ends, θ_A and θ_D are both equal to zero. As such, there are six unknown end moments, two unknown rotations (θ_A and θ_D) and an unknown displacement (Δ) in this frame. Nine equations are therefore necessary for solution. Six of the needed equations come from the usual slope-deflection equations for end moments. Two additional equations come from equilibrium of moment at joints B and C. The remaining equation

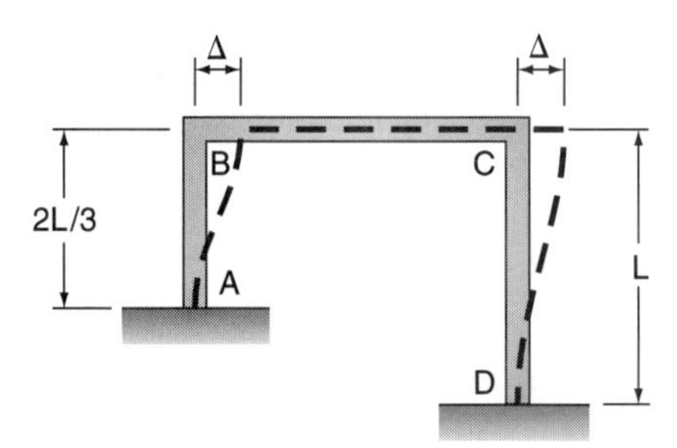

Figure 15.5 A frame that will sway sidesways under load

comes from equilibrium of forces in the horizontal direction. For the frame shown in the figure:

$$H_A = \frac{M_{AB} + M_{BA}}{L_{AB}}$$
$$H_D = \frac{M_{CD} + M_{DC}}{L_{CD}} \qquad \textbf{Eq. 15.7}$$

For equilibrium to occur in the horizontal direction, the following equation must be satisfied:

$$\sum F_H = 0 = H_A + H_D = \frac{M_{AB} + M_{BA}}{L_{AB}} + \frac{M_{CD} + M_{DC}}{L_{CD}} \qquad \textbf{Eq. 15.8}$$

This equation is the final equation needed for the analysis.

EXAMPLE 15.6

Determine all of the member-end moments for the frame shown in the figure. For this frame, EI is constant for all members and can be taken as 60 k-ft^2.

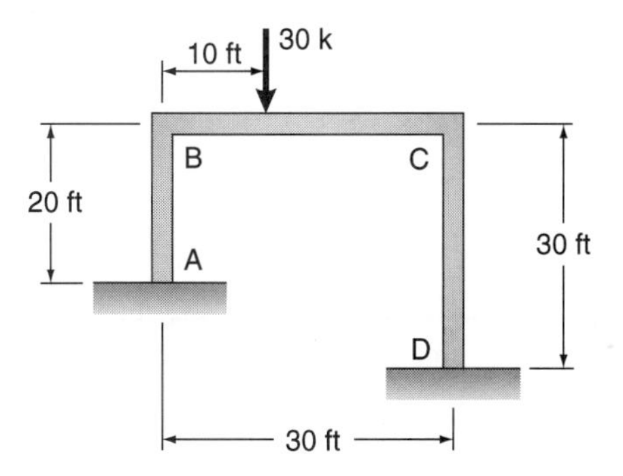

Solution. We will begin the solution by computing the stiffness terms for the slope-deflection equations:

$$EK_{AB} = \frac{EI}{L_{AB}} = \frac{60}{20} = 3\,\text{k-ft}$$

$$EK_{BC} = \frac{EI}{L_{BC}} = \frac{60}{30} = 2\,\text{k-ft}$$

$$EK_{CD} = \frac{EI}{L_{CD}} = \frac{60}{30} = 2\,\text{k-ft}$$

Next, we compute the fixed-end moments that occur in the members. The fixed-end moments are equal to zero for all members except BC. The fixed-end moments for that member are

$$M_{BC} = -\frac{30(10)(20)^2}{30^2} = -133.3\,\text{k-ft}$$

$$M_{CB} = \frac{30(20)(10)^2}{30^2} = 133.3\,\text{k-ft}$$

Because there is sidesway in this frame, all of the chord rotations are not equal to zero. Neglecting axial deformation, the chord rotation for member BC is equal to zero, but the chord rotations for the other two members are

$$\psi_{AB} = \frac{\Delta}{20}$$

$$\psi_{AB} = \frac{\Delta}{30}$$

We can now write the slope-deflection equations for this frame. They are written in the same manner as for the previous beams and frames. Because supports A and D are fixed supports, the rotations θ at those supports are equal to zero. The resulting slope-deflection equations are

$$M_{AB} = 2(3)\left[2(0) + \theta_B - 3\left(\frac{\Delta}{20}\right)\right] + 0$$

$$M_{BA} = 2(3)\left[0 + 2\theta_B - 3\left(\frac{\Delta}{20}\right)\right] + 0$$

$$M_{BC} = 2(2)[2\theta_B + \theta_C - 3(0)] - 133.3$$

$$M_{CB} = 2(2)[\theta_B + 2\theta_C - 3(0)] + 66.67$$

$$M_{CD} = 2(2)\left[2\theta_C + 0 - 3\left(\frac{\Delta}{30}\right)\right] + 0$$

$$M_{DC} = 2(2)\left[\theta_C + 2(0) - 3\left(\frac{\Delta}{30}\right)\right] + 0$$

As before, rotational equilibrium at the interior joints must be satisfied. The equations of rotational equilibrium to be used are

$$M_{BA} + M_{BC} = 0$$

$$M_{CB} + M_{CD} = 0$$

We must also satisfy equilibrium in the horizontal direction. That equation of equilibrium is

$$\sum F_H = 0 = \frac{M_{AB} + M_{BA}}{20} + \frac{M_{CD} + M_{DC}}{30}$$

The unknown values in these equations are the member-end moments, the rotations of joints B and C, and the lateral deflection of the frame, Δ. Upon evaluating the slope deflection and compatibility simultaneously for the unknown terms, we find that the member-end moments are

$$M_{AB} = 6.13\,\text{k-ft} \qquad M_{BA} = 69.59\,\text{k-ft}$$

$$M_{BC} = -69.59\,\text{k-ft} \qquad M_{CB} = 67.23\,\text{k-ft}$$

$$M_{CD} = -67.23\,\text{k-ft} \qquad M_{DC} = -46.35\,\text{k-ft} \quad \blacksquare$$

Slope deflection can be applied to frames with more than one condition of sidesway, such as the two-story frame shown in Figure 15.6. Hand analysis of frames of this type usually is handled more conveniently by the moment-distribution method, but knowledge of the slope-deflection solution is valuable in understanding the moment-distribution solution.

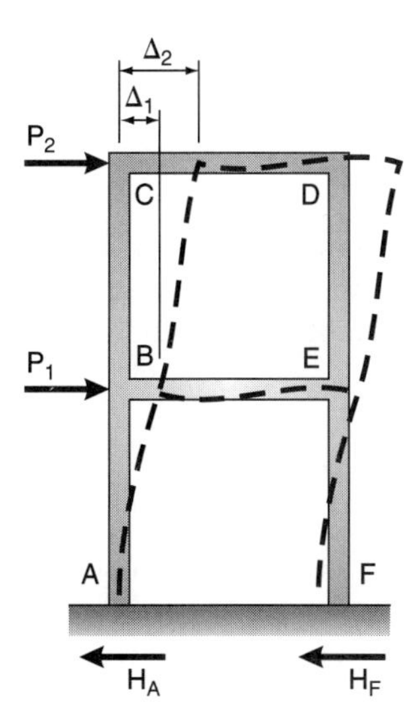

Figure 15.6 Sidesway in a two-story frame

The horizontal loads cause the structure to lean to the right; joints B and E to deflect horizontally Δ_1; and joints C and D to deflect horizontally Δ_2 as shown in Figure 15.6. For this structure, the chord rotations for the columns are:

$$\psi_{AB} = \psi_{EF} = \frac{\Delta_1}{L_{AB}}$$

$$\psi_{BC} = \psi_{DE} = \frac{\Delta_2 - \Delta_1}{L_{BC}} \qquad \textbf{Eq. 15.9}$$

Upon examination of these equations, we observe that the chord rotation is not dependent on the total deflection. Rather it is dependent upon the relative deflection of the ends of a member.

The slope-deflection equations may be written for the moment at each end of the six members. This results in 12 equations that contain 18 unknowns. As such, six additional equations are required for the solution. There are four compatibility equations for rotational equilibrium at the joints in the structure. These equations for this structure are

$$M_{BA} + M_{BC} + M_{BE} = 0$$

$$M_{CB} + M_{CD} = 0$$

$$M_{DC} + M_{DE} = 0$$

$$M_{ED} + M_{EB} + M_{EF} = 0$$

The remaining two equations can be obtained from equilibrium of the story shearing force. One equation can come from lateral equilibrium of the bottom story of the frame. That equation is

$$\sum F_H = P_1 + P_2 - H_A - H_F \qquad \textbf{Eq. 15.10}$$

In Section 15.5 we saw how to compute the shearing force at the ends of the members. For this frame, the values of H_A and H_F are computed from

$$H_A = -\frac{M_{AB} + M_{BA}}{L_{AB}}$$

$$H_F = -\frac{M_{EF} + M_{FE}}{L_{EF}} \qquad \textbf{Eq. 15.11}$$

As such, the equation of condition for horizontal equilibrium of the bottom story is

$$0 = \frac{M_{AD} + M_{BA}}{L_{AB}} + \frac{M_{EF} + M_{FE}}{L_{EF}} + P_1 + P_2 \qquad \textbf{Eq. 15.12}$$

A similar condition equation may be written for the top level if we cut a free-body just above member BE. The bending moments in the columns produce shearing forces that are equal and opposite to P_2. As such, the following equilibrium equation can be written for the second story:

$$\frac{M_{BC} + M_{CB}}{L_{BC}} + \frac{M_{DE} + M_{ED}}{L_{DE}} + P_2 = 0 \qquad \textbf{Eq. 15.13}$$

Six condition equations are available for determining the six remaining unknowns in the end-moment equations, and the problem may be solved, as illustrated in Example 15.7. No matter how many floors the building has, one shearing-force compatibility equation is available for each floor. The slope-deflection procedure is not very practical for multistory buildings. A six-story building that is four bays wide will require solution of 36 simultaneous equations since there are 36 unknown values. The solution is not quite as bad as it may seem, because each equation would contain only a few of the unknowns.

EXAMPLE 15.7

Find the moments for the frame shown in the figure using the slope-deflection method. For this frame, EI is constant and is assumed to be equal to 60 units.

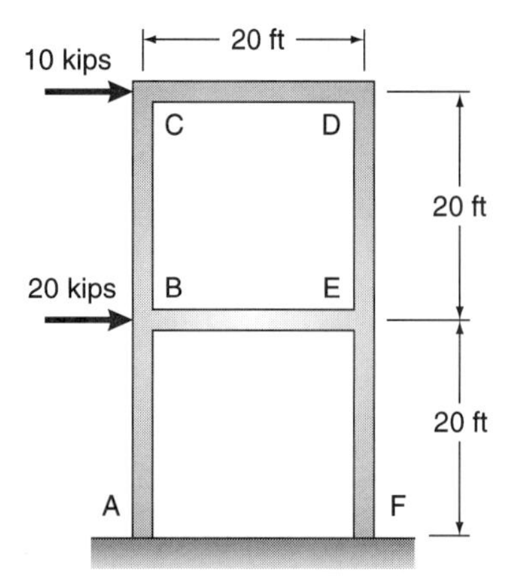

Solution. Since there are no loads applied to the members, the fixed-end moments in this frame are all equal to zero. As we saw previously, the chord rotations for the beams are equal to zero. Those for the columns are

$$\psi_{AB} = \psi_{EF} = \frac{\Delta_1}{L_{AB}}$$

$$\psi_{BC} = \psi_{DE} = \frac{\Delta_2 - \Delta_1}{L_{BC}}$$

The slope-deflection equations for this frame are then

$$M_{AB} = 2(3)\left[2(0) + \theta_B - 3\left(\frac{\Delta}{20}\right)\right] + 0 \qquad M_{BA} = 2(3)\left[0 + 2\theta_B - 3\left(\frac{\Delta}{20}\right)\right] + 0$$

$$M_{BC} = 2(3)\left[2\theta_B + \theta_C - 3\left(\frac{\Delta_2 - \Delta_1}{20}\right)\right] + 0 \qquad M_{CB} = 2(3)\left[\theta_B + 2\theta_C - 3\left(\frac{\Delta_2 - \Delta_1}{20}\right)\right] + 0$$

$$M_{CD} = 2(3)[2\theta_C + \theta_D - 3(0)] + 0 \qquad M_{CB} = 2(3)[\theta_C + 2\theta_D - 3(0)] + 0$$

$$M_{DE} = 2(3)\left[2\theta_D + \theta_E - 3\left(\frac{\Delta_2 - \Delta_1}{20}\right)\right] + 0 \qquad M_{ED} = 2(3)\left[\theta_D + 2\theta_E - 3\left(\frac{\Delta_2 - \Delta_1}{20}\right)\right] + 0$$

$$M_{EF} = 2(3)\left[2\theta_E + 0 - 3\left(\frac{\Delta}{20}\right)\right] + 0 \qquad M_{FE} = 2(3)\left[\theta_E + 2(0) - 3\left(\frac{\Delta}{20}\right)\right] + 0$$

$$M_{BE} = 2(3)[2\theta_B + \theta_E - 3(0)] + 0 \qquad M_{EB} = 2(3)[\theta_B + 2\theta_E - 3(0)] + 0$$

And the compatibility equations are

$$M_{BA} + M_{BC} + M_{BE} = 0$$
$$M_{CB} + M_{CD} = 0$$
$$M_{DC} + M_{DE} = 0$$
$$M_{ED} + M_{EB} + M_{EF} = 0$$
$$0 = \frac{M_{AB} + M_{BA}}{L_{AB}} + \frac{M_{EF} + M_{FE}}{L_{EF}} + P_1 + P_2$$
$$0 = \frac{M_{BC} + M_{CB}}{L_{BC}} + \frac{M_{DE} + M_{ED}}{L_{DE}} + P_2$$

The unknown values in these equations are the member-end moments; the rotations of joints B, C, D, and E; and the lateral deflections of the frame, Δ_1 and Δ_2. Upon evaluating the slope deflection and compatibility simultaneously for the unknown terms, we find that the member-end moments are

$$M_{AB} = M_{FE} = -176.4 \text{ k-ft} \qquad M_{BA} = M_{EF} = -123.6 \text{ k-ft}$$
$$M_{BC} = M_{ED} = -34.55 \text{ k-ft} \qquad M_{CB} = M_{DE} = -65.55 \text{ k-ft}$$
$$M_{CD} = M_{DC} = 64.55 \text{ k-ft} \qquad M_{BE} = M_{EB} = 158.2 \text{ k-ft}$$ ■

Transit Shed, Port of Long Beach, California (Courtesy of the American Institute of Steel Construction)

Alcoa Building, San Francisco. A dominant feature of the topped-out steel framework of the "tiara" of San Francisco's Golden Gateway project, the Alcoa Building, is the diamond-patterned seismic-resisting shear walls (Courtesy Bethlehem Steel Corporation)

15.6 ANALYSIS OF FRAMES WITH SLOPING LEGS

Space is not taken in this chapter to discuss the analysis of frames with sloping legs. A slope-deflection analysis of such a frame with hand calculations is tedious, and is therefore not likely to be used when other methods are available. Computer solutions using matrix methods and the moment-distribution procedure presented Chapters 17 and 21 are more commonly used today. Nevertheless, once the analysis of frames with sloping legs using moment distribution is understood, the analyst will be able to return to this chapter and use slope deflection for their analysis.

15.7 PROBLEMS FOR SOLUTION

For Problems 15.1 through 15.13, compute the end moments of the beams with the slope-deflection equations. Draw shearing force and bending moment diagrams. E is constant for all problems.

15.1

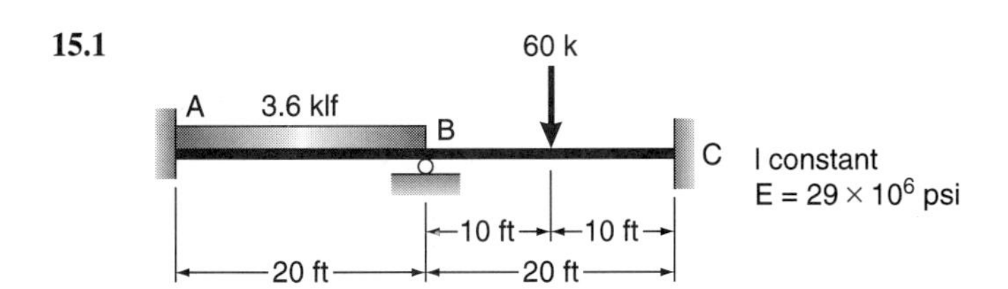

15.2

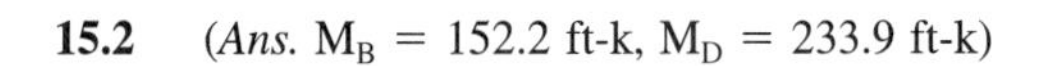
(*Ans.* $M_B = 152.2$ ft-k, $M_D = 233.9$ ft-k)

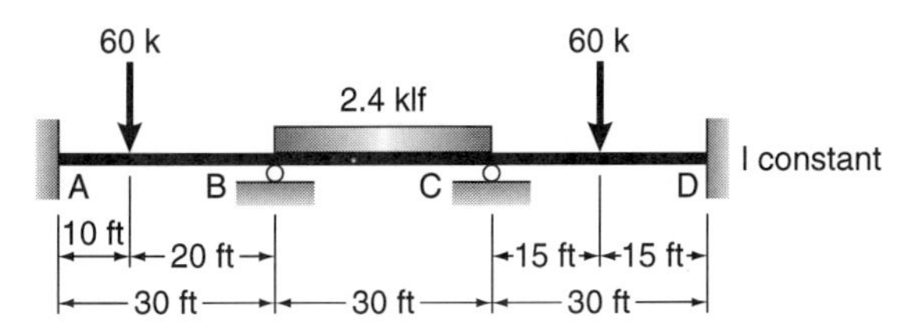

15.3

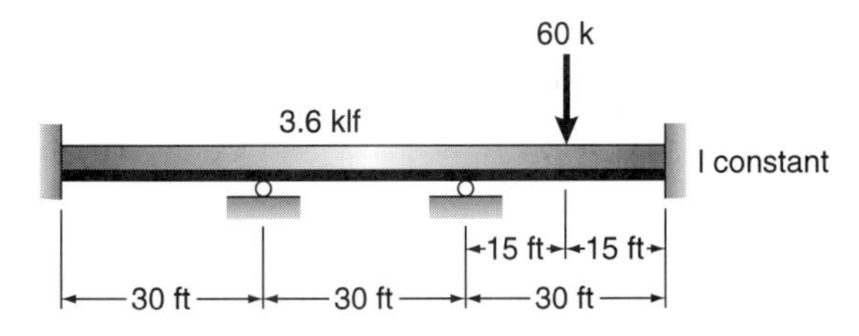

15.4 (*Ans.* $M_B = 560$ ft-k, $M_C = 655.4$ ft-k)

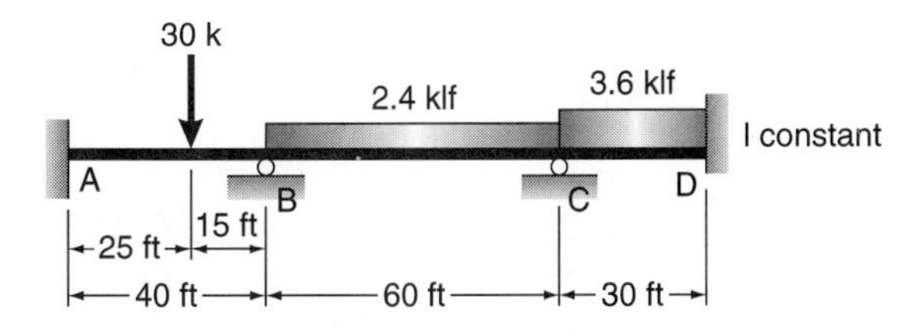

15.5

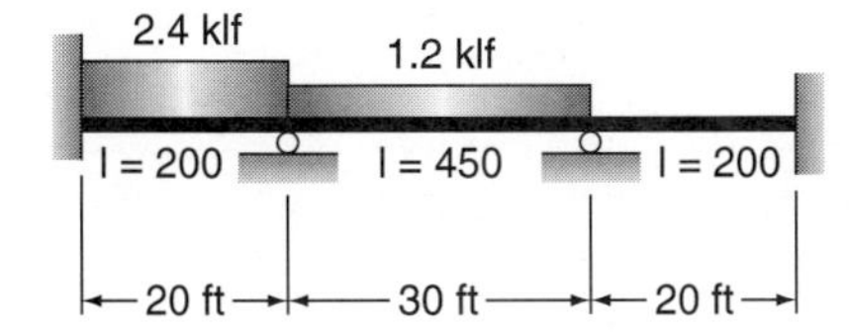

15.6 (*Ans.* $V_A = 67.5$ k ↑, $M_{AB} = 405$ ft-k)

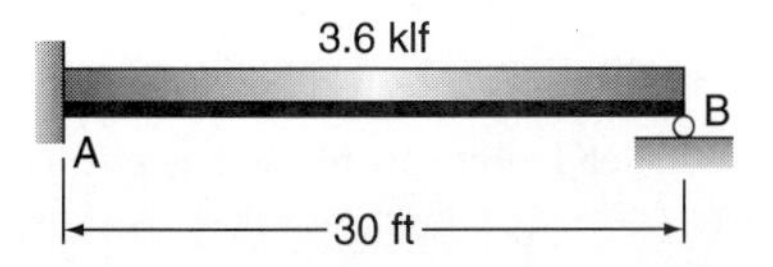

15.7

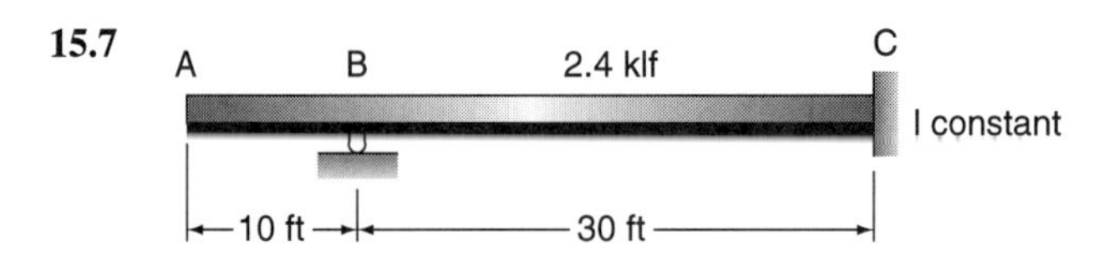

15.8 
(*Ans.* $M_A = 205$ kN-m, $M_B = 265$ kN-m)

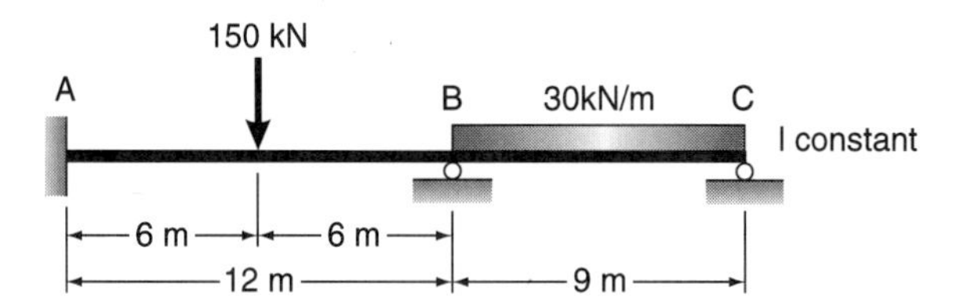

15.9

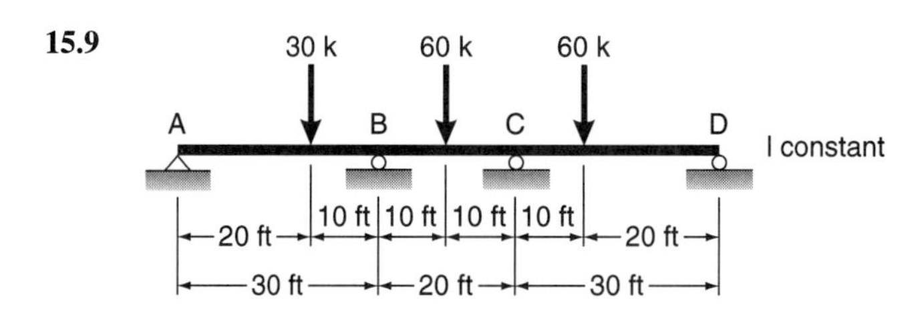

15.10 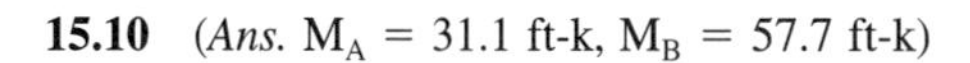
(*Ans.* $M_A = 31.1$ ft-k, $M_B = 57.7$ ft-k)

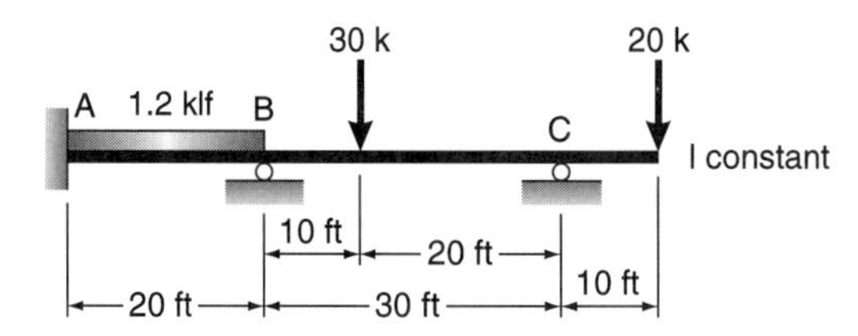

15.11 See Appendix D for fixed-end moments.

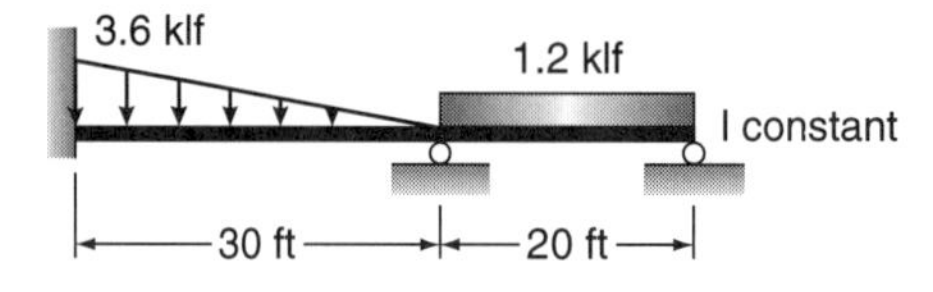

15.12 The beam of Problem 15.2 if both ends are simply supported (*Ans.* $M_B = 205.1$ ft-k, $M_C = 252.4$ ft-k)

15.13 The beam of Problem 15.4 if the right end is simply supported

For Problems 15.14 through 15.20, determine the end moments for the members of all of the structures by using the slope deflection equations.

15.14 (*Ans.* $M_A = 137$ ft-k, $M_B = 274.1$ ft-k)

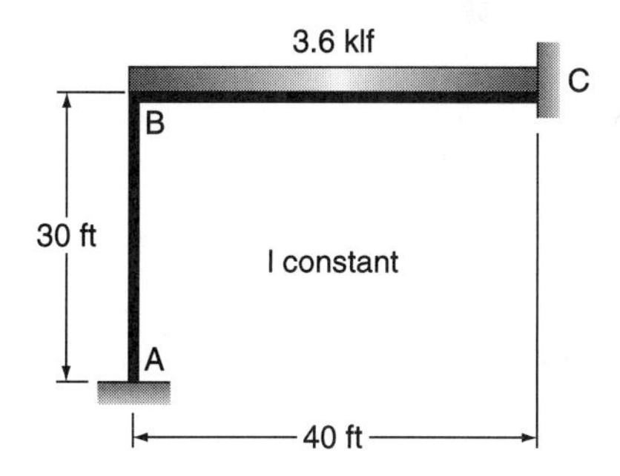

15.15

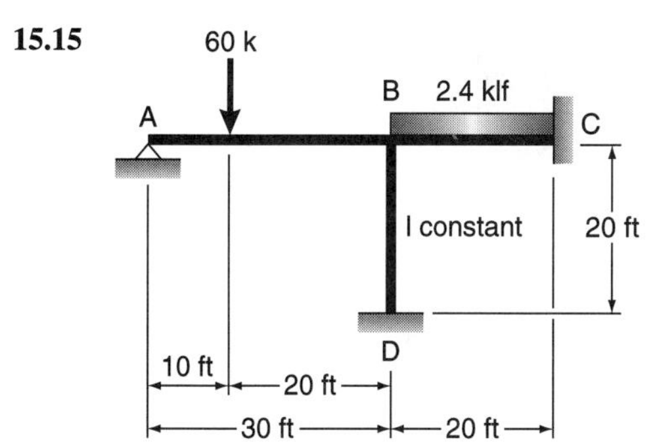

15.16 (*Ans.* $M_A = M_O = 70$ kN-m, $M_B = M_D = 380$ kN-m)

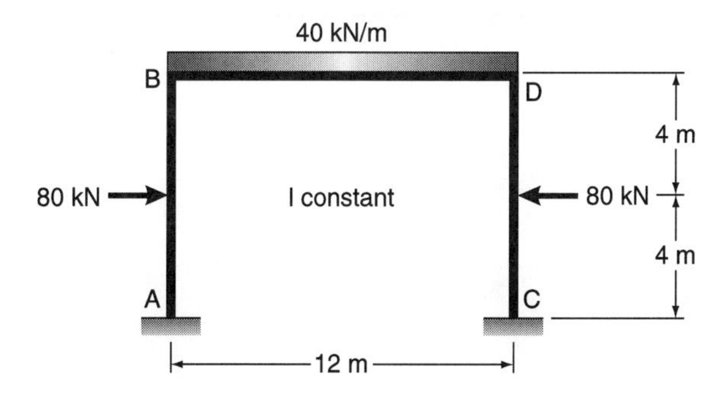

15.17

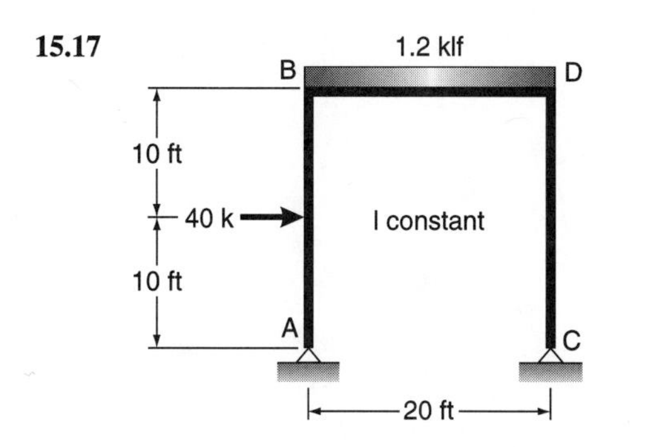

15.18 The frame of Problem 15.17 if the column bases are fixed (*Ans.* $M_A = 210.5$ ft-k, $M_B = 0.5$ ft-k)

15.19

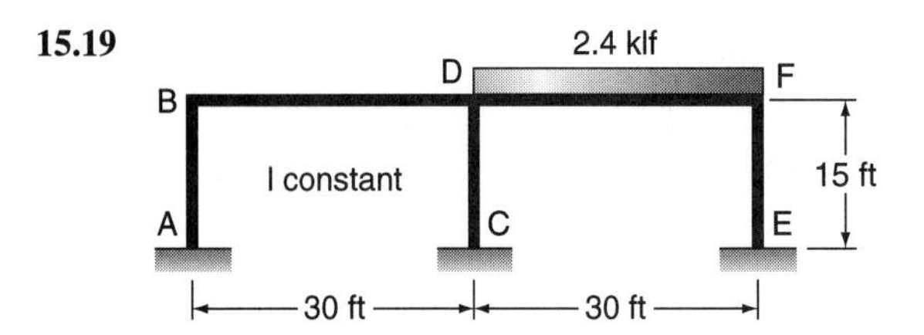

15.20 (*Ans.* $M_A = 138.5$ ft-k, $M_D = 192$ ft-k)

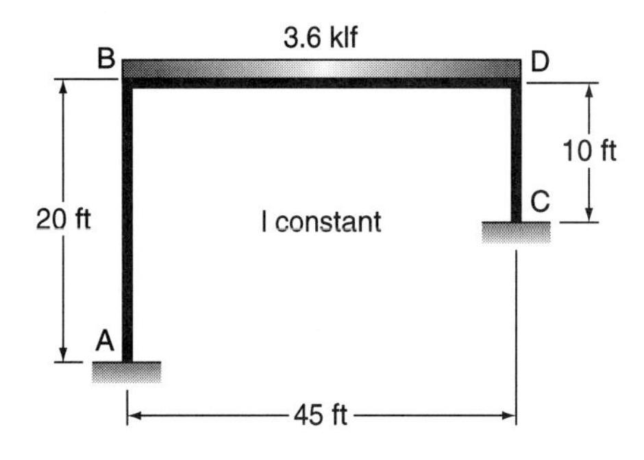

PART THREE

STATICALLY INDETERMINATE STRUCTURES

COMMON METHODS IN CURRENT PRACTICE

Chapter 16

Approximate Analysis of Statically Indeterminate Structures

16.1 INTRODUCTION

The approximate methods presented in this chapter for analyzing statically indeterminate structures could very well be designated as *classical methods*. The same designation could be made for the moment distribution method presented in Chapters 17 and 18. The methods discussed in this and the next several chapters will often be seen and used by an engineer in the course of everyday design—they are the methods of analysis commonly used in current engineering practice.

Statically indeterminate structures may be analyzed "exactly" or "approximately." Several "exact" methods, which are based on elastic distortions, are discussed in Chapters 13 through 15. Approximate methods, which involve the use of simplifying assumptions, are presented in this chapter. These methods have many practical applications such as the following:

1. When costs are being estimated for alternative structural configurations and design concepts, approximate analyses are often very helpful. Approximate analyses and approximate designs of the various alternatives can be made quickly and used for initial cost estimates.
2. To analyze a statically indeterminate structure, an estimate of the member sizes must be made before the structure can be analyzed using an "exact" method. This is necessary because the analysis of a statically indeterminate structure is based on the elastic properties of the members. An approximate analysis of the structure will yield forces from which reasonably good initial estimates can be made of member sizes.
3. Today, computers are available with which "exact" analyses and designs of highly indeterminate structures can be made quickly and economically. To make use of computer programs, preliminary estimates of the size of the members should be made. If an approximate analysis of the structure has been done, very reasonable estimates of member sizes are possible. The result will be appreciable savings of both computer time and design hours.
4. Approximate analyses are quite useful for checking computer solutions, which is a very important matter.

5. An "exact" analysis may be too expensive for small noncritical systems, particularly when preliminary designs are being made. An acceptable and applicable approximate method is very appropriate for such a situation.
6. An additional advantage of approximate methods is that they provide the analyst with an understanding for the actual behavior of structures under various loading conditions. This important ability probably will not be developed from computer solutions.

To make an "exact" analysis of a complicated statically indeterminate structure, a qualified analyst must model the structure, that is, the analyst must make certain assumptions about the behavior of the structure. For instance, the joints are assumed to be simple or to be semi-rigid. Characteristics of material behavior and loading conditions must be assumed, and so on. The result of these assumptions is that all analyses are approximate. We could say that we apply an "exact" analysis method to a structure that does not really exist. Furthermore, all analysis methods are approximate in the sense that every structure is constructed within certain tolerances—no structure is perfect—and its behavior cannot be determined precisely.

Many different methods are available for making approximate analyses. A few of the more common ones are presented here, with consideration being given to trusses, continuous beams, and building frames. The approximate methods described in this chapter hopefully will provide you with a general knowledge about a wide range of statically indeterminate structures. Not all types of statically indeterminate structures are considered in this chapter. However, based on the ideas presented, you should be able to make reasonable assumptions when other types of statically indeterminate structures are encountered.

To be able to analyze a structure using the equations of static equilibrium, there must be no more unknowns than there are available equations of static equilibrium. If a truss or frame has 10 more unknowns than equations of equilibrium, it is statically indeterminate to the 10th degree. To analyze it by an approximate method, one assumption for each degree of indeterminacy, a total of 10 assumptions must be made. Each assumption effectively provides another equation of equilibrium to use in the calculations.

16.2 TRUSSES WITH TWO DIAGONALS IN EACH PANEL

16.2.1 Diagonals Having Little Stiffness

The truss shown in Figure 16.1 has two diagonals in each panel. If one of the diagonal members were removed from each of the six panels, the truss would become statically determinate. Therefore the truss in Figure 16.1 is statically indeterminate to the sixth degree.

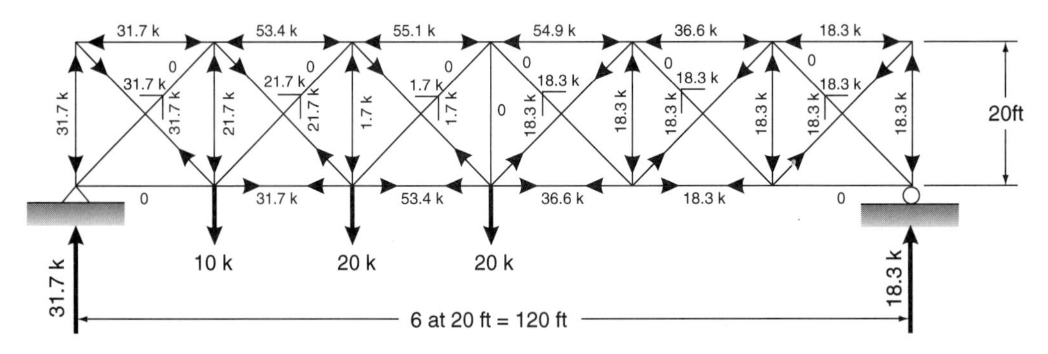

Figure 16.1 A truss analyzed assuming diagonals act only in tension

Frequently the diagonals in a truss are relatively long and slender, often being made of a pair of small steel angles. They can carry reasonably large tensile forces but have negligible capacity in compression. For this situation, it is logical to assume that the shearing force in each panel is carried entirely by the diagonal that would be in tension for that sense of the shearing force (positive or negative). The other diagonal is assumed to have no force. Making this assumption in each panel effectively provides six "equations" with which to evaluate the six redundants. The members in the remaining members can be evaluated with the equations of static equilibrium cast as the method of joints or the method of sections. The forces in Figure 16.1 were obtained on this basis.

16.2.2 Diagonals Having Considerable Stiffness

In some trusses, the diagonals are constructed with sufficient stiffness to resist significant compressive loads. In panels with two substantial diagonals, the shearing force is carried by both diagonals. The division of shear causes one diagonal to be in tension and the other to be in compression. The usual approximation made is that each diagonal carries 50% of the shearing force in the panel: other divisions of the shearing force are also possible. Another typical division is that one-third of shearing force is carried by the diagonal acting in compression and two-thirds is carried by the diagonal in tension.

The forces calculated for the truss in Figure 16.2 are based on a 50% division of the shearing force in each panel.

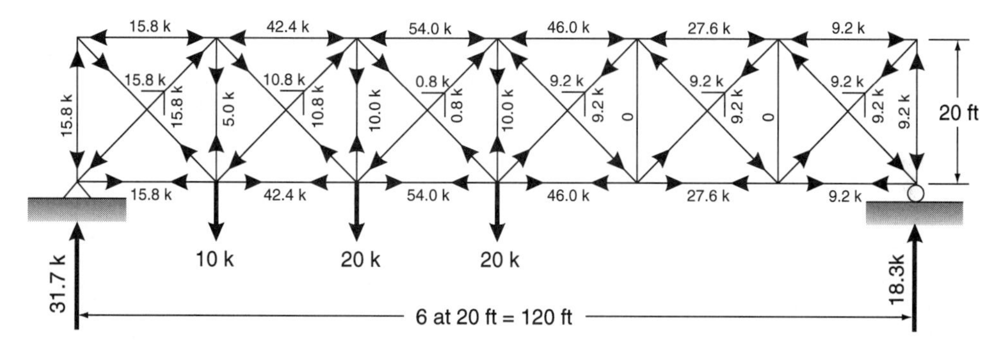

Figure 16.2 Approximate analysis of a truss assuming diagonals carry 50% of panel shearing force

16.3 CONTINUOUS BEAMS

Before beginning an "exact" analysis of a building frame, the sizes of the members in the frame must be estimated. Preliminary beam sizes can be obtained by considering their approximate moments. Frequently a portion of the building can be removed and analyzed separately from the rest of the structure. For instance, one or more beam spans may be taken out as a free body and assumptions made as to the moments in those spans. To facilitate such an analysis, moment diagrams are shown in Figure 16.3 for several different uniformly loaded beams.

It is obvious from the figure that the assumed types of supports can have a tremendous effect on the magnitude of the calculated moments. For instance, the uniformly loaded simple beam in Figure 16.3 will have a maximum moment equal to $wL^2/8$. On the other hand, the uniformly loaded single-span fixed-ended beam will have a maximum moment equal to $wL^2/12$. For a continuous uniformly loaded beam, the engineer may very well decide to estimate a maximum moment somewhere between the preceding values, at perhaps $wL^2/10$, and use that value for approximating the member size.

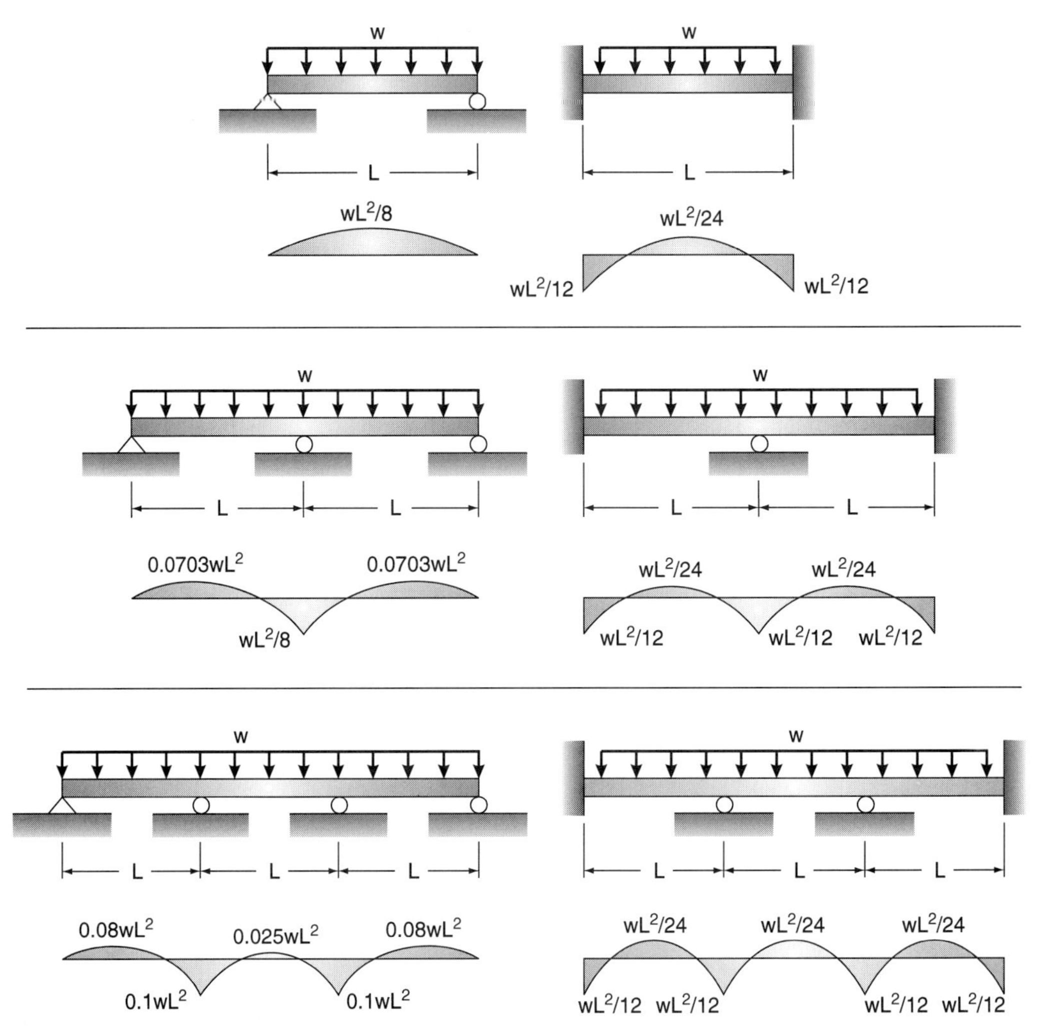

Figure 16.3 Moment diagrams for some typical beams

A very common method used for the approximate analysis of continuous reinforced-concrete structures involves the use of the American Concrete Institute bending moment and shearing force coefficients.[1] These coefficients, which are reproduced in Table 16.1, provide estimated maximum shearing forces and bending moments for buildings of normal proportions. The values calculated in this manner usually will be somewhat larger than those that would be obtained with an exact analysis. Consequently, appreciable economy can normally be obtained by taking the time or effort to make such an analysis. In this regard, the engineer should realize that these coefficients are considered to apply best to continuous frames having more than three or four continuous spans.

In developing the coefficients, the negative moment values were reduced to take into account the usual support widths and some moment redistribution before collapse. In addition, the positive moment values have been increased somewhat to account for the moment redistribution. It will also be noted that the coefficients account for the fact that

[1]*Building Code Requirements for Reinforced Concrete,* ACI 318-95 (Detroit: American Concrete Institute) Section 8.3.3, pp. 79–80.

TABLE 16.1 ACI MOMENT COEFFICIENTS*

Positive moment	
End spans	
If discontinuous end is restrained	$\frac{1}{11}wL_n^2$
If discontinuous end is integral with the support	$\frac{1}{14}wL_n^2$
Interior Spans	$\frac{1}{16}wL_n^2$
Negative moment at the exterior face of the first interior support	
Two spans	$\frac{1}{9}wL_n^2$
More than two spans	$\frac{1}{10}wL_n^2$
Negative moment at other faces of interior supports	$\frac{1}{11}wL_n^2$
Negative moment at face of all supports for (a) slabs with spans not exceeding 10 ft and (b) beams and girders where ratio of sum of column stiffness to beam stiffness exceeds 8 at each end of the span	$\frac{1}{12}wL_n^2$
Negative moment at interior faces of exterior supports for members built integrally with their supports	
Where the support is a spandrel beam or girder	$\frac{1}{24}wL_n^2$
Where the support is a column	$\frac{1}{16}wL_n^2$
Shear in end members at face of first interior support	$\frac{1.15wL_n}{2}$
Shear at face of all other supports	$\frac{wL_n}{2}$

*American Concrete Institute ACI 318-02

in monolithic construction, the supports are not simple and moments are present at end supports, such as where those supports are beams or columns.

In applying the coefficients, w is the design load per unit of length, while L_n is the clear span for calculating positive bending moments and the average of the adjacent clear spans for calculating negative bending moments. These values were developed for members with approximately equal spans—the larger of two adjacent spans does not exceed the smaller by more than 20%—and for cases where the ratio of the uniform service live load to the uniform service dead load is not greater than 3. In addition, the values are not applicable to prestressed-concrete members. Should these limitations not be met, a more precise method of analysis must be used.

For the design of a continuous beam or slab, the bending moment coefficients in effect provide two sets of moment diagrams for each span of the structure. One diagram is the result of placing the live loads so that they will cause maximum positive moment out in the span. The other is the result of placing the live loads to cause maximum negative moments at the supports. Actually, it is not possible to produce maximum negative moments at both ends of a span simultaneously. It takes one placement of the live loads to produce maximum negative moment at one end of the span and another placement to produce maximum negative moment at the other end. The assumption of both maximums occurring at the same time is on the safe side, however, because the resulting diagram will have greater critical values than are produced by either one of the two separate loading conditions.

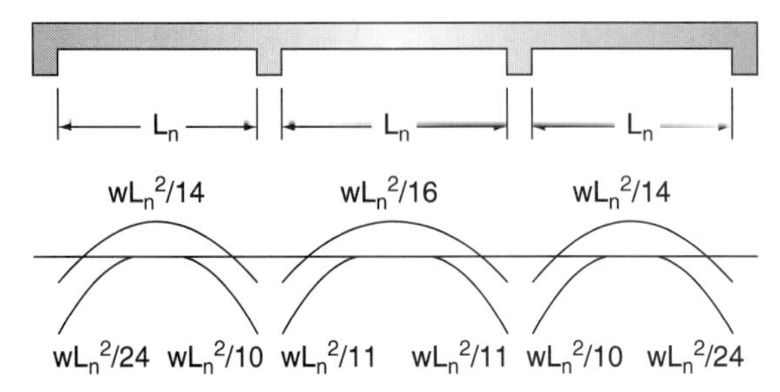

Figure 16.4 Moment envelopes for continuous slab constructed integrally with exterior supports that are spandrel girders

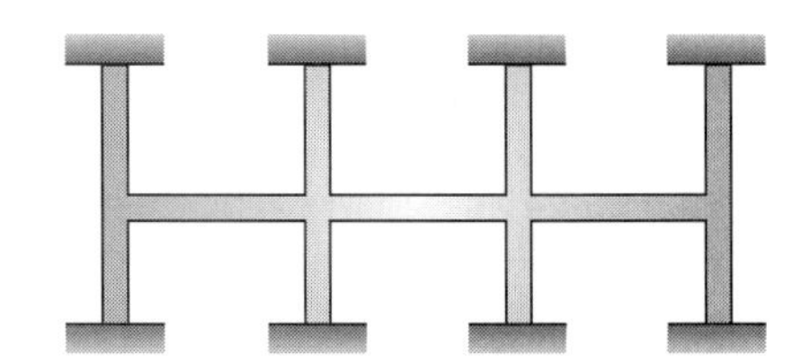

Figure 16.5 A portion of a building frame to be analyzed by the equivalent frame method

The ACI coefficients give maximum values for a bending moment envelope for each span of a continuous frame. Typical envelopes are shown in Figure 16.4 for a continuous slab that is constructed integrally with its exterior supports, which are spandrel girders.

On some occasions the analyst will take out a portion of a structure that includes not only the beams but also the columns for the floor above and the floor below, as shown in Figure 16.5. This procedure, usually called the *equivalent frame method,* is applicable only for gravity loads. The sizes of the members are estimated and an analysis is made using one of the exact methods of analysis we have discussed.

Multistory steel-frame building, showing structural skeleton; Hotel Sheraton, Philadelphia, Pennsylvania (Courtesy Bethlehem Steel Corporation)

16.4 ANALYSIS OF BUILDING FRAMES FOR VERTICAL LOADS

One approximate method for analyzing building frames for vertical loads involves the estimation of the location of the points of zero moment in the girders. These points, which occur where the moment changes from one sign to the other, are commonly called *points of inflection* (PIs) or points of contraflexure. One common practice is to assume the PIs are located at the one-tenth points from each girder end. In addition, the axial forces in the girders are assumed to be zero.[2]

These assumptions have the effect of creating a simple beam between the points of inflection, and the positive moments in the beam can be determined using simple equations of static equilibrium. Negative moments occur in the girders between their ends and the points of inflection. The negative moments are computed by treating the portion of the beam from the end to the point of inflection as a cantilever beam.

The shearing force at the end of the girders contributes to the axial forces in the columns. Similarly, the negative moments at the ends of the girders are transferred to the columns. At interior columns, the moments in the girders on each side oppose each other and may cancel. Exterior columns have moments caused by the girders framing into them on only one side; these moments need to be considered in design.

EXAMPLE 16.1

Determine the forces in beam AB of the building frame in the figure by assuming the points of inflection are at one-tenth points and the beam ends are fixed supports. The distributed loads acting on the frame are 3 klf each. The concentrated loads are acting at the third points on the beam.

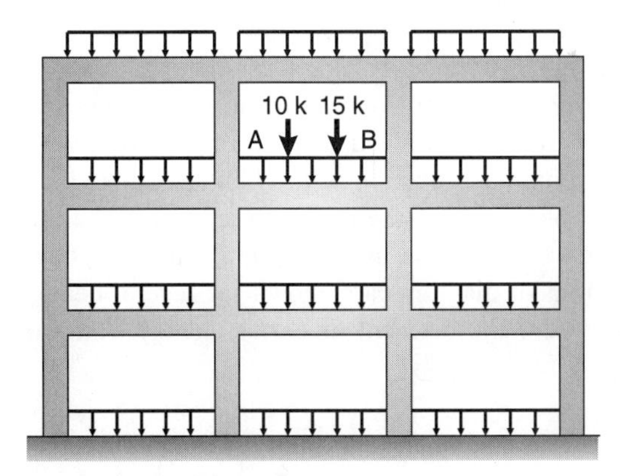

Solution. A detailed drawing of the beam and the location of the assumed points of inflection are shown in the next figure.

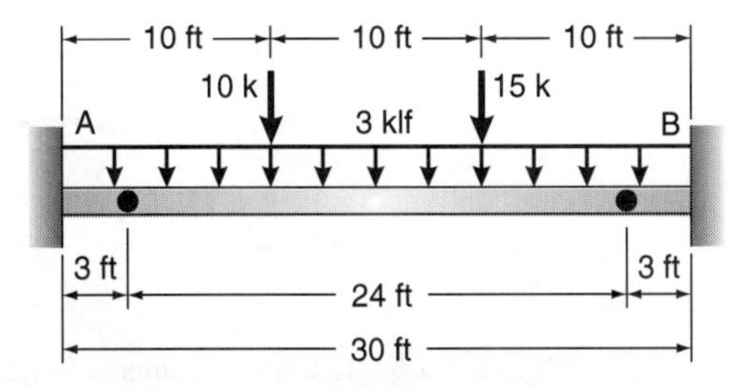

[2]C. H. Norris, J. B. Wilbur, and S. Utku, *Elementary Structural Analysis,* 3rd ed. (New York: McGraw-Hill, 1976), 200–201.

We can now split the beam into the "simple span" and the two "cantilever spans" and calculate the forces acting on each. After the forces are calculated the bending moment diagram can be drawn.

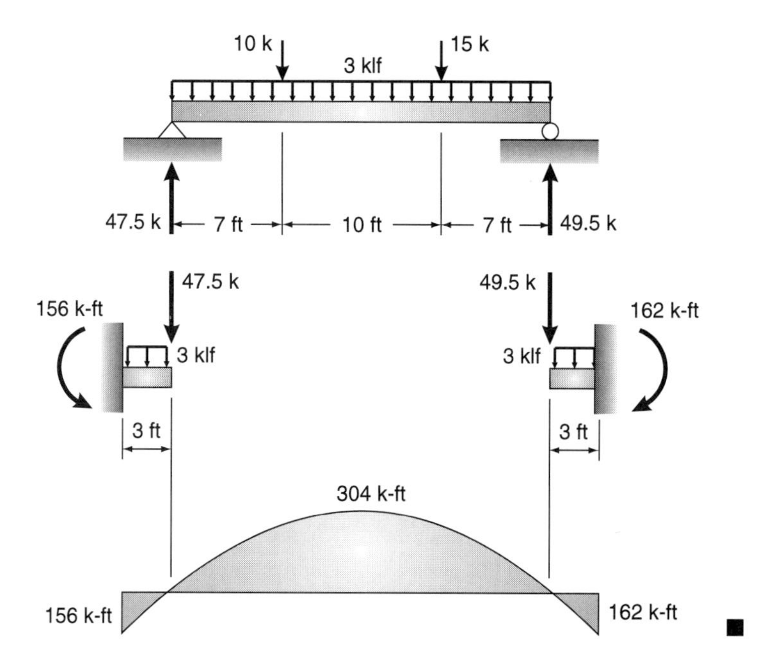

Sketching the deflected shape often helps to make reasonable estimates for the locations of the PIs. For illustration, a continuous beam is shown in Figure 16.6(a) and a sketch of its estimated deflected shape for the loads shown is given in part (b). From the sketch, an approximate location of the PIs is estimated.

It might be useful for the reader to see where PIs occur for a few types of statically indeterminate beams. These may be helpful in estimating where PIs will occur in other structures. Moment diagrams are shown for several beams in Figure 16.7. The PIs obviously occur where the moment diagrams change from positive to negative moments or vice versa.

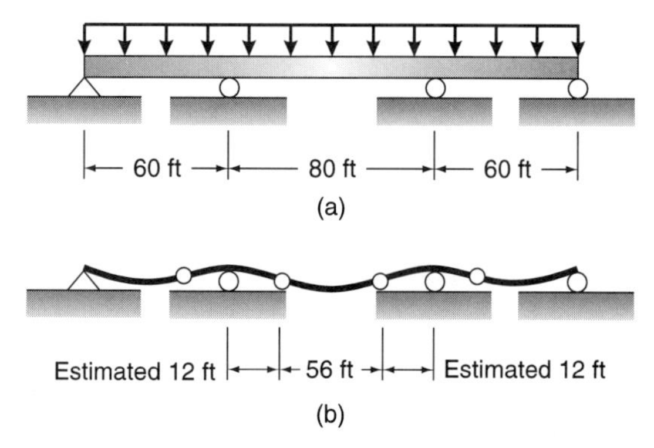

Figure 16.6 Estimated points of inflection in a continuous beam

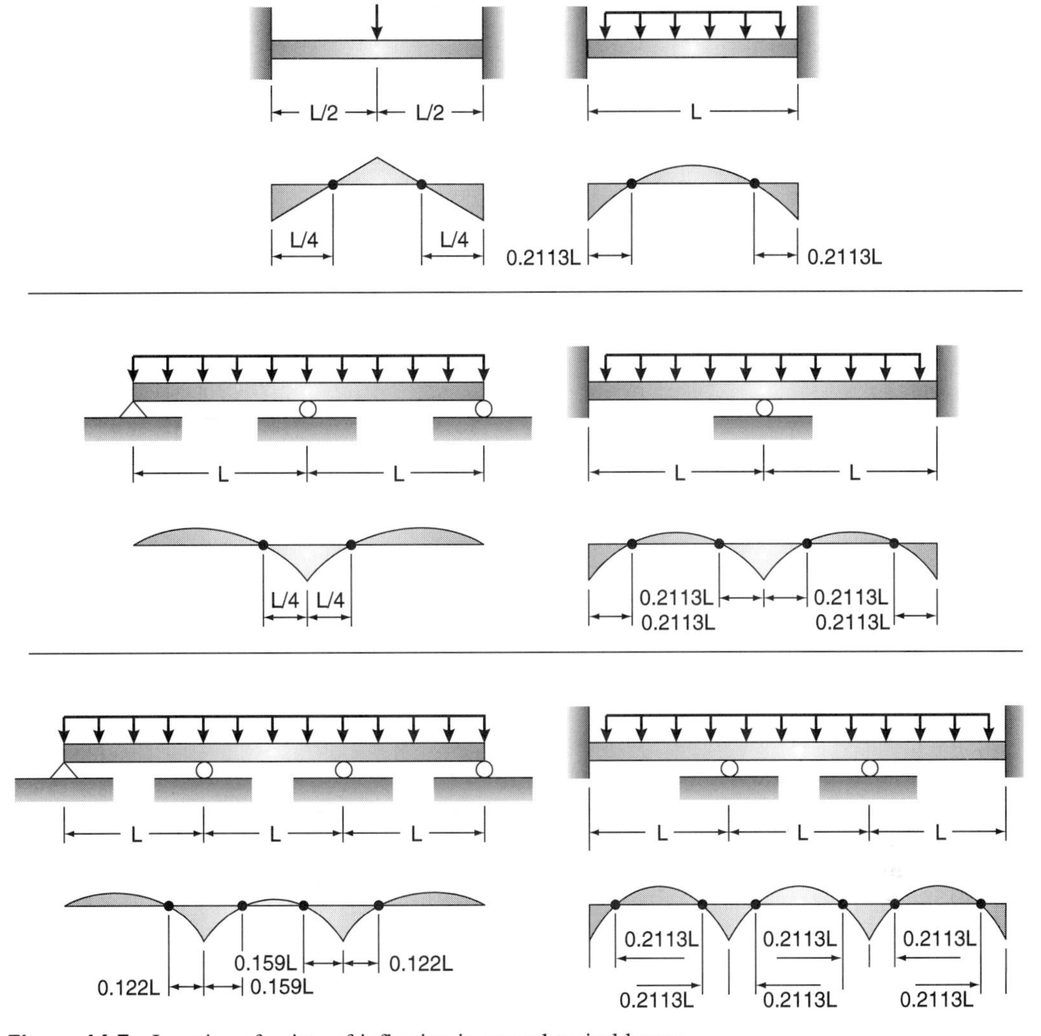

Figure 16.7 Location of points of inflection in several typical beams

16.5 ANALYSIS OF PORTAL FRAMES

Portal frames of the type shown in Figure 16.8 and Figure 16.9 may be fixed at their column bases, may be simply supported, or may be partially fixed. The columns of the frame of Figure 16.8(a) are assumed to have their bases fixed. Consequently, there will be three unknown forces of reaction at each support, for a total of six unknowns. The structure is statically indeterminate to the third degree, and to analyze it by an approximate method three assumptions must be made.

When a column is rigidly attached to its foundation, there can be no rotation at the base. Although the frame is subjected to wind loads causing the columns to bend laterally, a tangent to the column at the base will remain vertical. If the beam at the top of the columns is very stiff and rigidly fastened to the columns, the tangents to the columns at the junctions will remain vertical. A column rigidly fixed at the top and bottom will assume the shape of an S-curve when subjected to lateral loads, as shown in Figure 16.8(a).

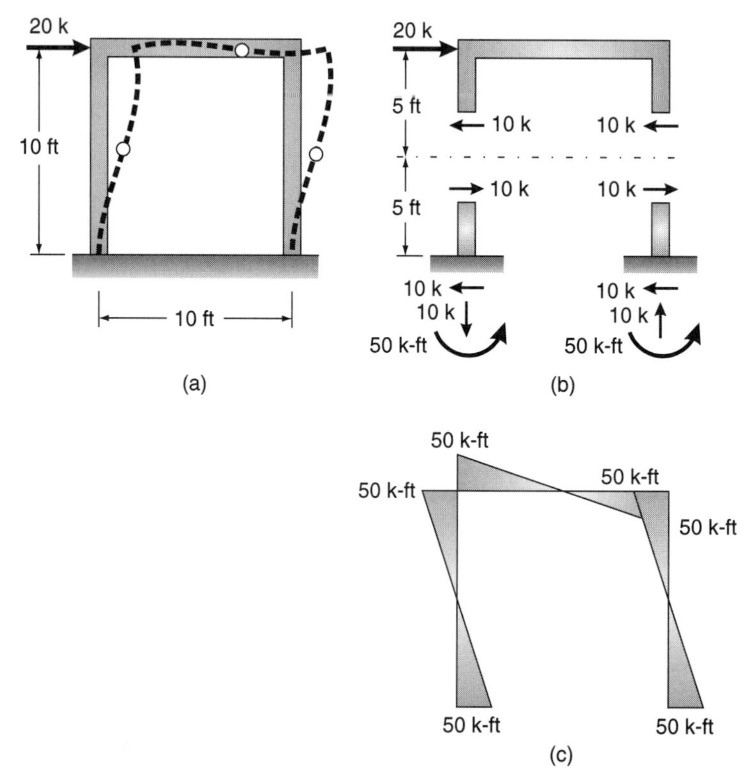

Figure 16.8 Approximate analysis of a portal frame with fixed column bases

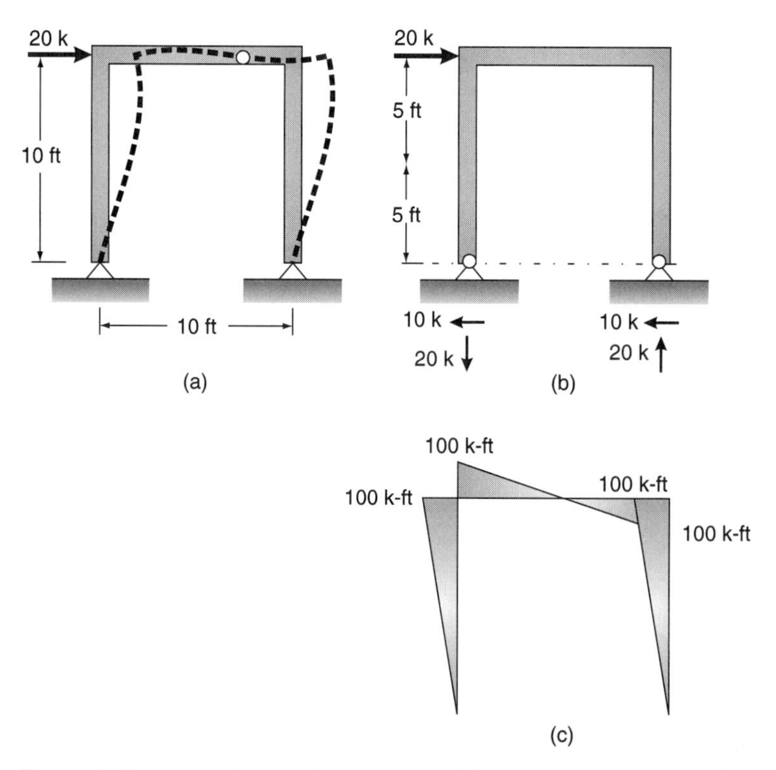

Figure 16.9 Approximate analysis of a portal frame with pinned column bases

At a point midway from the column base to the beam, the moment will be zero because it changes from a moment causing tension on one side of the column to a moment causing tension on the other side. If points of inflection are assumed in each column, two of the necessary three assumptions will have been made, that is, two equations for summation of moments are made available.

The third assumption usually made is that the horizontal shear is divided equally between the two columns at the plane of contraflexure. An "exact" analysis proves this is a very reasonable assumption as long as the columns are approximately the same size. If they are not similar in size, an assumption may be made that the shear is divided between them in a slightly different proportion, with the stiffer column carrying the larger proportional amount of the shear. A good assumption for such cases is to distribute the shear to the columns in proportion to their I/L^3 values.

To analyze the frame in Figure 16.8, the 20 kip lateral load is divided into 10 kip shearing forces at the column PIs as shown in Figure 16.8(b). The bending moment at the top and bottom of each column can be computed and is found to be equal to 10(5) = 50 k-ft. Moments then are taken about the left column PI to determine the right-hand vertical reaction as follows:

$$20(5) - 10V_R = 0$$

$$V_R = 10\text{ k}\uparrow \qquad \textbf{Eq. 16.1}$$

Finally, the moment diagrams are drawn with the results shown in Figure 16.9(d).

Wachovia Plaza, Charlotte, North Carolina (Courtesy Bethlehem Steel Corporation)

Should the column bases be assumed to be simply supported, that is, pinned or hinged, the points of contraflexure will occur at those hinges. Assuming the columns are the same size, the horizontal shear is split equally between the columns and the other values are determined as in Figure 16.9.

The principles that we have discussed here can be applied to other framed structures regardless of the number of bays or the number of stories. Except for preliminary analysis, most structures are not analyzed using approximate methods. The ready availability of structural analysis software and the rapidity with which it can be used have now caused most analysis to be conducted with computers.

16.6 MOMENT DISTRIBUTION

The moment distribution method described in the next two chapters for the analysis of statically indeterminate beams and frames involves successive cycles of computation, each cycle drawing closer to the "exact" answers. When the computations are carried out until the changes in the numbers are very small, it is considered an "exact" method of analysis. Should the number of cycles be limited, the method becomes a splendid approximate method.

16.7 ANALYSIS OF VIERENDEEL "TRUSSES"

In previous chapters a truss has been defined as a structure assembled with a group of ties and struts connected at their joints with frictionless pins so the members are subjected only to axial tension or axial compression. A special type of truss (although it is not really a truss by the preceding definition) is the Vierendeel "truss." A typical example of a Vierendeel truss is the pedestrian bridge shown in Figure 16.10. You can see that these

Figure 16.10 An Example of a Vierendeel truss (Clinic Inn Pedestrian Bridge Cleveland, Ohio, Courtesy of the American Institute of Steel Construction, Inc.)

trusses are actually rigid frames or, as some people say, they are girders with big holes in them. The Vierendeel truss was developed in 1896 by the Belgian engineer and builder Arthur Vierendeel. They are used rather frequently in Europe, but only occasionally in the United States.

A Vierendeel truss is usually constructed with reinforced concrete, but also may be fabricated with structural steel. The external loads are supported by means of the flexural resistance of the short heavy members. The continuous moment-resisting joints cause the "trusses" to be highly statically indeterminate. Vierendeel trusses are rather inefficient, but because of their aesthetics, they are used in situations in which their large clear openings are desirable. In addition, they are particularly convenient in buildings where they can be constructed with depths equal to the story heights.

These highly statically indeterminate structures can be analyzed approximately by the portal and cantilever methods described in the preceding section. Figure 16.11 illustrated a Vierendeel truss (which is very similar to the one shown in Figure 16.10) and the results obtained by applying the portal method. For this symmetrical truss, the cantilever method will yield the same results. To follow the calculations, turn the structure on its end because the shear being considered for the Vierendeel is vertical, whereas it was horizontal for the building frames previously considered. Because the truss is symmetric, the results are only shown for one half of the truss.

For many Vierendeel trusses—particularly those that are several stories in height—the lower horizontal members may be much larger and stiffer than the other horizontal members. To obtain better results with the portal or cantilever methods, the non-uniformity of sizes should be taken into account by assuming that more shearing force is carried by the stiffer members.

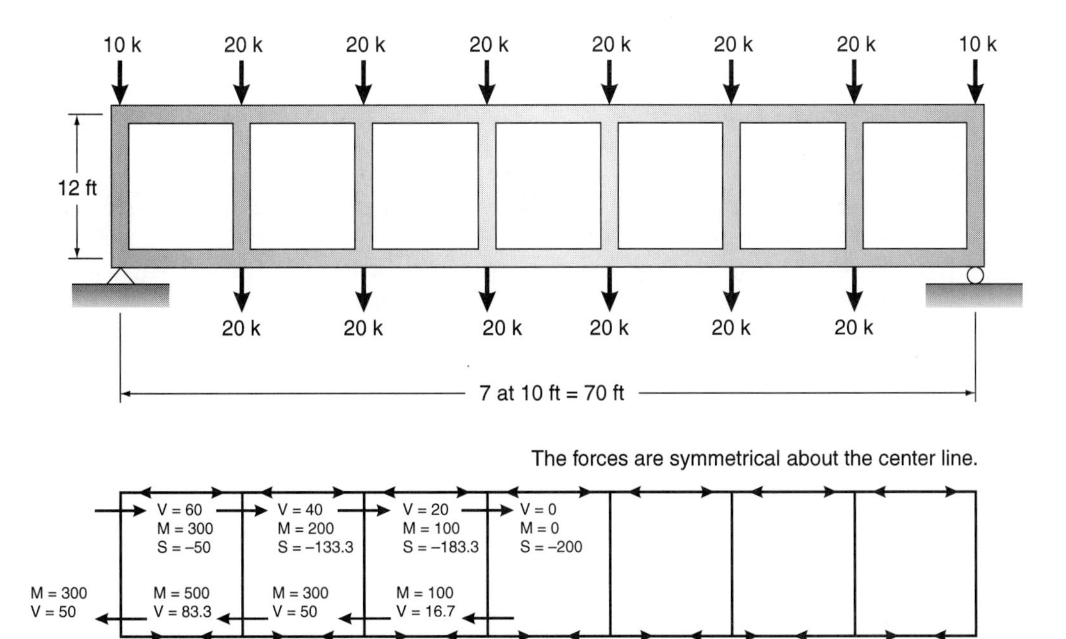

Figure 16.11 Analysis of a Vierendeel truss using the portal method

16.8 PROBLEMS FOR SOLUTION

16.1 Compute the forces in the members of the truss shown in the accompanying illustration if the diagonals are unable to carry compression.

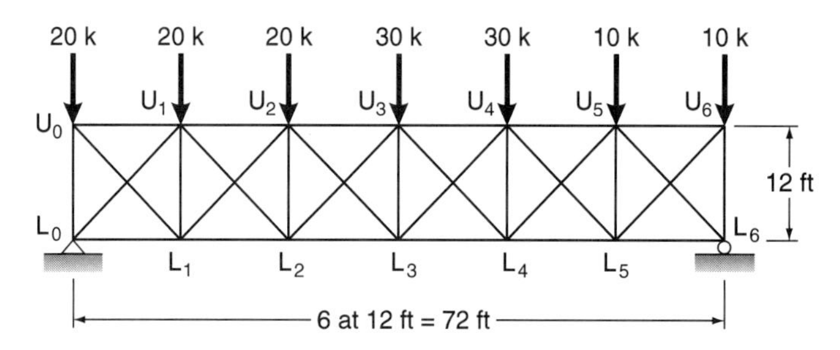

16.2 Repeat Problem 16.1 if the diagonals that theoretically are in compression can resist half of the shear in each panel. (*Ans.* $L_0L_1 = +28.33$ k, $L_1U_2 = -25.92$ k, $U_3U_4 = -103.35$ k)

16.3 Repeat Problem 16.1 if the diagonals that theoretically are in compression can resist one-third of the shear in each panel.

16.4 Compute the forces in the members of the cantilever truss shown if the diagonals are unable to carry any compression. (*Ans.* $U_0U_1 = +25$ k, $U_1L_1 = -60$ k, $U_1L_2 = +32.02$ k)

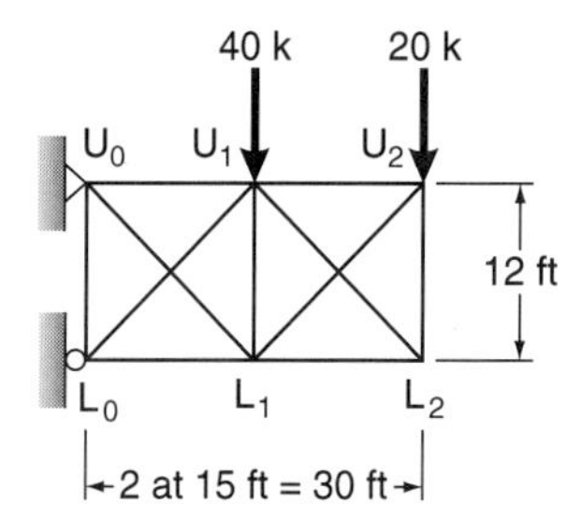

16.5 Repeat Problem 16.4 if the diagonals that would normally be in compression can resist only one-third of the shear in each panel.

16.6 Compute the forces in the members of the truss shown if the interior diagonals that would normally be in compression can resist no compression. (*Ans.* AF = +32.02 k, FG = −41.23 k, CG = −20 k)

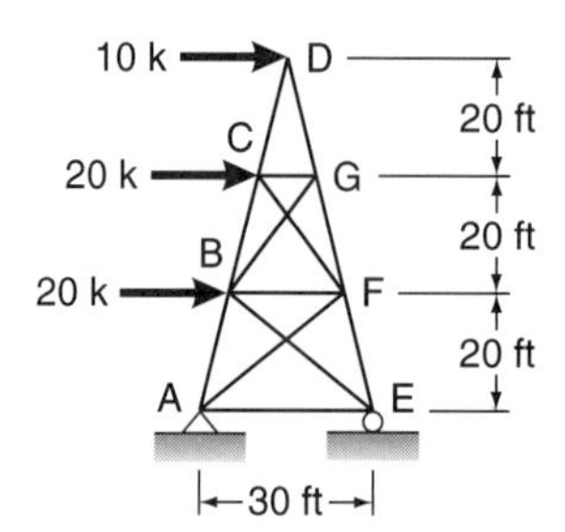

16.7 Prepare the shear and moment diagrams for the continuous beam shown using the ACI coefficients of Table 16.1. Assume the beam is constructed integrally with girders at all of its supports. The total load is to be 5 k/ft. Draw the moment diagrams as envelopes.

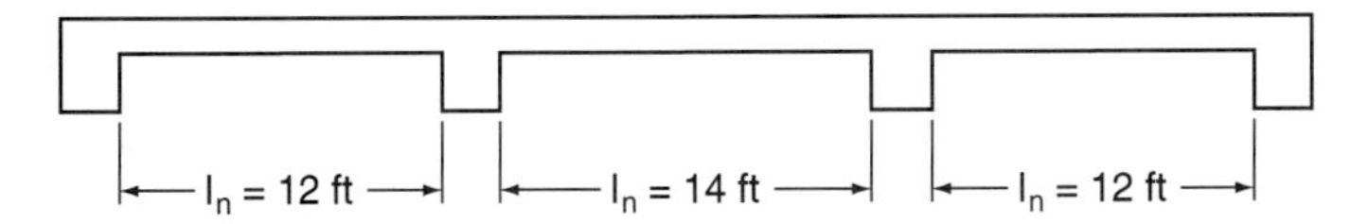

16.8 Analyze the portal shown if PIs are assumed to be located at column mid-depths. (*Ans.* Column moments top and bottom = 150 ft-k, right-hand vertical reaction is 10 k)

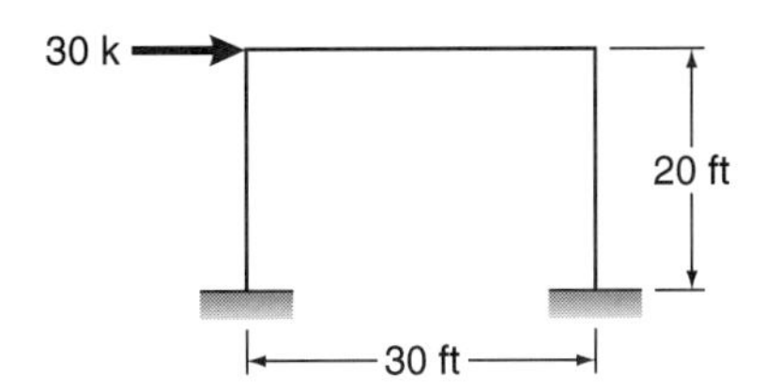

16.9 Repeat Problem 16.8 if the PIs are assumed to be located 8 ft above the column bases.

16.10 Repeat Problem 16.8 if the column bases are assumed to be pinned and the PIs located there. (*Ans.* Moment at top of columns = 300 ft-k, right-hand vertical reaction is 20 k)

16.11 Determine the forces for all of the members of the mill building truss shown in the accompanying illustration if points of inflection are assumed to be 12 ft above the column bases.

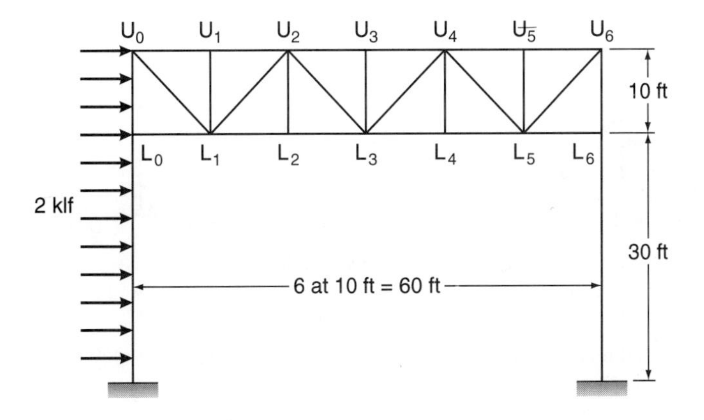

16.12 Repeat Problem 16.11 if points of inflection are assumed to be 15 ft above column bases. (*Ans.* $U_0U_1 = -14.58$ k, $L_2L_3 = -20.84$ k, $U_2L_3 = +14.74$ k)

16.13 Assuming points of inflection to be 10 ft from the column bases, determine the forces in all members of the structure shown in the accompanying illustration.

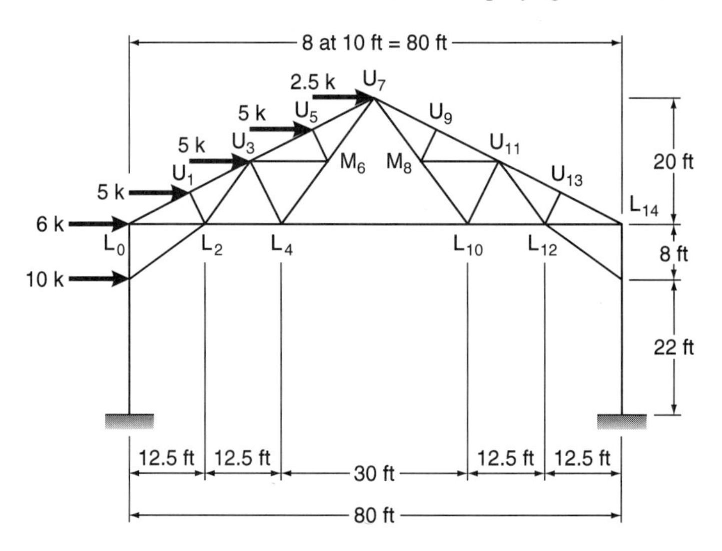

16.14 Rework Problem 16.13 with points of inflection assumed to be 6 ft above the column bases. (*Ans.* $L_0L_2 = -11.08$ k, $L_4L_{10} = -13.75$ k, $U_7U_9 = +10.13$ k)

For Problems 16.15 through 16.22, compute the bending moments, shear forces, and axial forces for all of the members of the frames shown (a) by using the portal method and (b) by using the cantilever method.

16.15

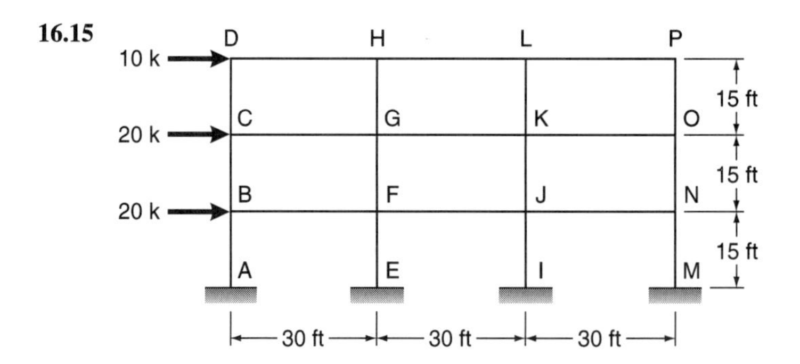

16.16 (*Ans.* Portal method for CD: V = 10 kN, M = 20 kN-m, S = +6.67 kN; Cantilever method: V = 6 kN, M = 12 kN-m, S = +4 kN)

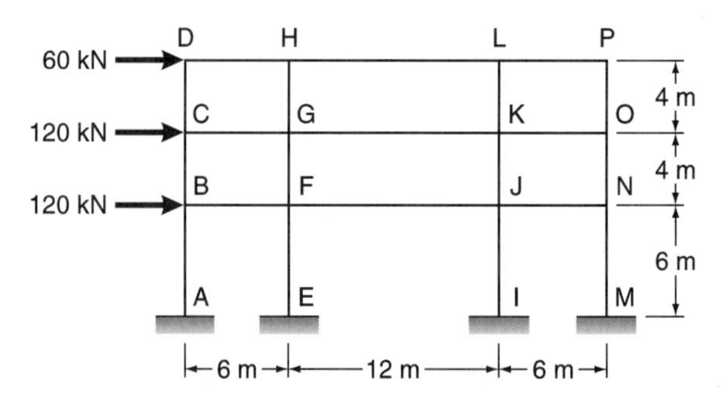

16.17

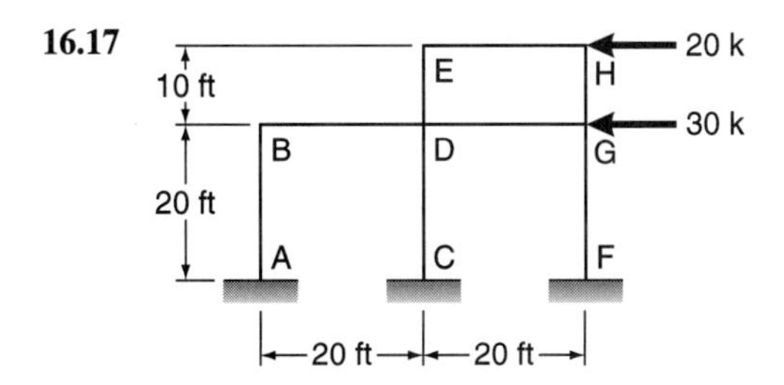

16.18 (*Ans.* Portal method for FG: V = 50 kN, M = 125 kN-m, S = −6.95 kN; Cantilever method: V = 41.52 kN, M = 103.8 kN-m, S = −3.66 kN)

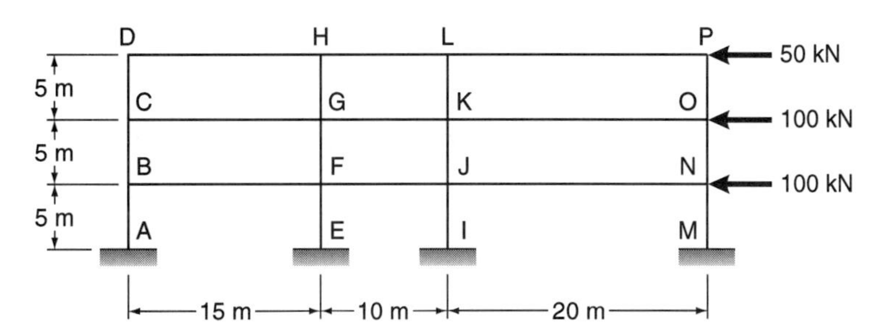

16.19

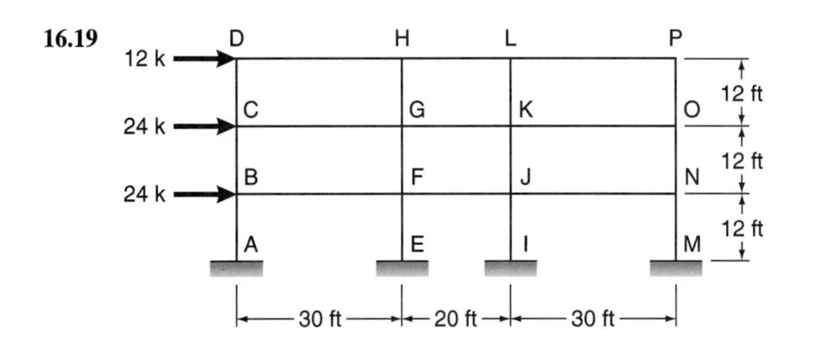

16.20 Rework Problem 16.19 if the column bases are assumed to be pinned. (*Ans.* for BC; Portal method: V = 6 k, M = 36 ft-k, S = +4.0 k; cantilever method: V = 6.35 k, M = 38.3 ft-k, S = +4.24 k)

16.21

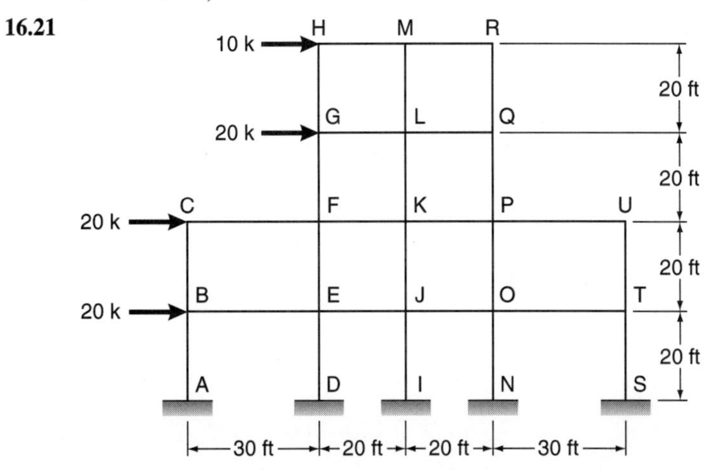

16.22 (*Ans.* Portal method for DH: V = 3.33 kN, M = 13.33 kN-m; Cantilever method: V = 1.81 kN, M = 7.24 kN-m)

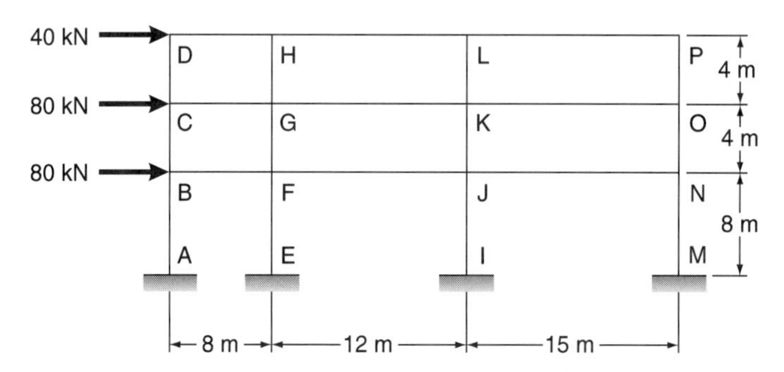

For Problems 16.23 and 16.24, compute the bending moments, shearing forces, and axial forces for all of the members of the Vierendeel trusses using the portal method.

16.23

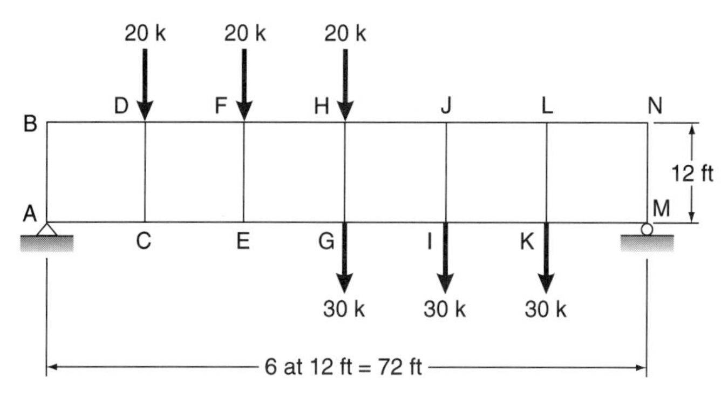

16.24 (*Ans.* for BC: V = 12.5 k, M = 75 ft-k, for GK: V = 15 k, M = 90 ft-k, S = 0)

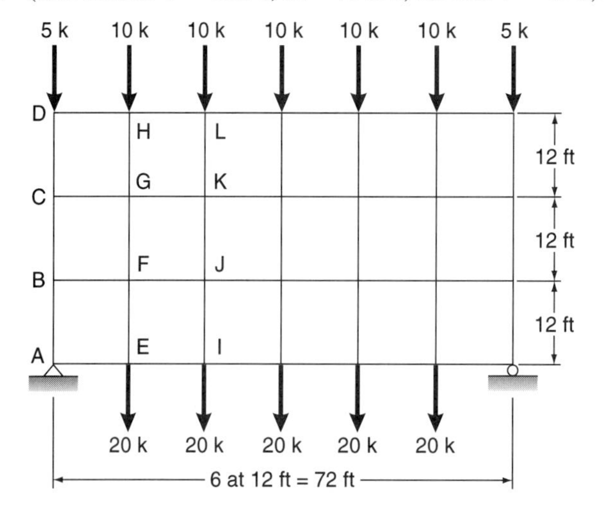

16.25 Using SABLE or SAP2000, analyze the Vierendeel truss shown in Figure 16.11. For the analysis, assume the members have constant areas and moments of inertia.

16.26 Using SABLE or SAP2000, repeat Problem 16.21 assuming the members have constant areas and moments of inertia.

16.27 Using SABLE or SAP2000, repeat Problem 16.22 assuming the members have constant areas and moments of inertia.

16.28 Using SABLE or SAP2000, repeat Problem 16.23 assuming the members have constant areas and moments of inertia.

Chapter 17

Moment Distribution for Beams

17.1 INTRODUCTION

The late Hardy Cross wrote papers about the moment-distribution method in 1929[1,2] and 1930[3] after having taught the subject to students at the University of Illinois since 1924. His papers began a new era in the analysis of statically indeterminate frames and gave added impetus to their use. The moment-distribution method of analyzing continuous beams and frames involves little more labor than the approximate methods but yields accuracy equivalent to that obtained from the infinitely more laborious "exact" methods.

The analysis of statically indeterminate structures in the preceding chapters frequently involved the solution of simultaneous equations. Simultaneous equations are not necessary in solutions by moment distribution, except in a few rare situations for complicated frames. The method that Cross developed involves successive cycles of computation, with each cycle drawing closer to the "exact" answers. The calculations may be stopped after two or three cycles, giving a very good approximate analysis, or they may be carried on to whatever degree of accuracy is desired. When these advantages are considered in light of the fact that the accuracy obtained by the lengthy "classical" methods is often questionable, the true worth of this quick and practical method is understood.

From the 1930s until the 1960s, moment distribution was the dominant method used for the analysis of continuous beams and frames. Since the 1960s, however, there has been a continually increasing use of computers for analyzing all types of structures. Computers are extremely efficient for solving the simultaneous equations that are generated by other methods of analysis. Generally, the software is developed from the matrix-analysis procedures described in Chapters 19 to 21 of this book.

Even with the computer software available, moment distribution continues to be an important hand-calculation method for analysis of continuous beams and frames. Structural engineers can use moment distribution to quickly make approximate analyses for

[1]Hardy Cross, "Continuity as a Factor in Reinforced Concrete Design," *Proceedings of the American Concrete Institute,* Vol. 25 (1929), pp. 669–708.

[2]Hardy Cross, "Simplified Rigid Frame Design," Report of Committee 301, *Proceedings of the American Concrete Institute,* Vol. 26, (1929), pp. 170–183.

[3]Hardy Cross, "Analysis of Continuous Frames by Distributing Fixed-End Moments," *Proceedings of the American Society of Civil Engineers,* Vol. 56, No. 5 (May 1930): pp. 919–928. Also, *Transactions of the American Society of Civil Engineers,* Vol. 96 (1932), pp. 1–10.

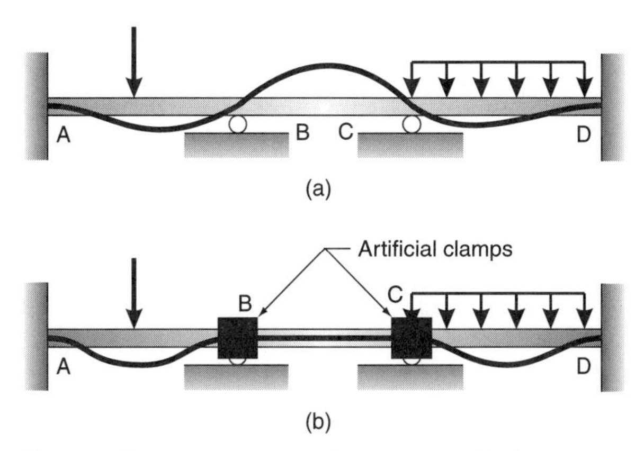

Figure 17.1 Development of moment distribution

preliminary designs and to check computer results, which is very important. In addition, moment distribution may be solely used for the analysis of small structures.

The beauty of moment distribution lies in the simplicity of its theory and application. Readers will be able to grasp quickly the principles involved and will clearly understand what they are doing and why they are doing it. In the discussion that follows, certain assumptions have been made. These assumptions are:

- The structures have members of constant cross section throughout their respective lengths. That is, the members are prismatic.
- The joints at which two or more members frame together do not translate.
- The joints to which members are connected can rotate, but the ends of all members connected to a joint rotate the same amount as the joint. At a joint, there is no rotation of the ends of members relative to each other or to the joint.
- Axial deformation of members is neglected.

Consider the continuous beam shown in Figure 17.1(a). Joints A and D are fixed against displacement. Joints B and C are not fixed: the loads on the structure will cause them to rotate. This fixity and rotation of the joints can be seen in the deformed shape of the beam.

Largest curved "horizontal skyscraper" in the United States, Boston, Massachusetts (Courtesy Bethlehem Steel Corporation)

If an imaginary clamp is placed at Joints B and C, fixing them so that they cannot rotate, the structure under load will have displaced the shape as shown in Figure 17.1(b). When the ends of all members are fixed against rotation in this manner, the fixed-end moments can be calculated with little difficulty. Tables of fixed-end moments have been prepared and are available from many sources. Solutions for common loading conditions are contained in Appendix D.

If the clamp at B is removed, the joint will rotate slightly, which will cause the ends of the members meeting there to rotate. This will cause a redistribution of the moments at the ends of the members. When a moment is applied to one end of a member and the other end is fixed, there is some effect or *carryover* to the fixed end. The changes in the moments at the B end of members AB and BC cause some effect at the other end of these members. There would be similar changes if the clamp at C were removed.

After the fixed-end moments are computed, the process of moment distribution may be stated briefly as consisting of the calculation of:

1. The moments caused in the B and C ends of the members by the rotation of joints B and C,
2. The magnitude of the moments carried over to the other ends of the members, and
3. The addition to or subtraction from the original fixed-end moments of these latter moments.

These steps can be simply written as being the fixed-end moments plus the moments due to the rotation of joints B and C. This process is represented by

$$M = M_{fixed} + M_{\theta_e} \qquad \textbf{Eq. 17.1}$$

17.2 BASIC RELATIONS

Two questions must be answered in order to apply the moment-distribution method to actual structures. They are

1. What is the moment developed or carried over to a fixed end of a member when the other end is subjected to a certain moment?
2. When a joint is unclamped and rotates, what is the distribution of the unbalanced moment to the members meeting at the joint, or how much resisting moment is supplied by each member?

17.2.1 Carryover Moment

To determine the carryover moment, the unloaded beam of constant cross section in Figure 17.2(a) is considered. If a moment M_1 is applied to the left end of the beam, a moment M_2 will be developed at the right end. The left end is at a joint that has been released and the moment M_1 causes it to rotate an amount θ_1. There will be, however, no deflection or translation of the left end with respect to the right end.

The second moment-area theorem may be used to determine the magnitude of M_2. The deflection of the tangent to the elastic curve of the beam at the left end with respect to the tangent at the right end (which remains horizontal) is equal to the moment of the area of the M/EI diagram taken about the left end and is also equal to zero. By drawing the M/EI diagram in Figure 17.2(b) and dividing it into two triangles to facilitate the area

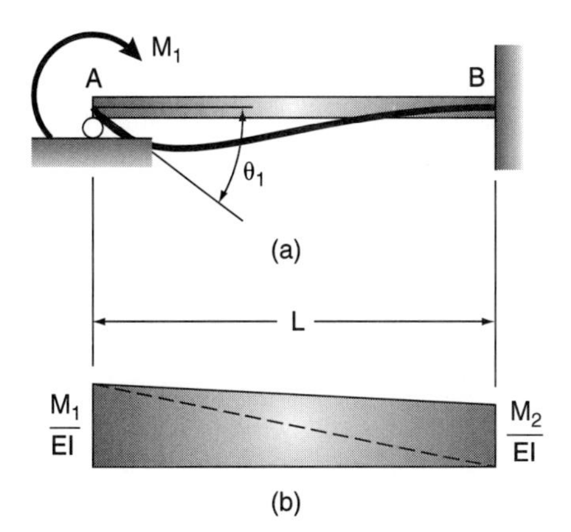

Figure 17.2 Concept of carryover moment

computations, the following expression may be written and solved for M_2:

$$\delta_A = \frac{\frac{1}{2}M_1L(\frac{1}{3}L) + \frac{1}{2}M_2L(\frac{2}{3}L)}{EI} = 0$$

$$\frac{M_1L^2}{6EI} + \frac{M_2L^2}{3EI} = 0 \qquad \textbf{Eq. 17.2}$$

$$M_2 = -\tfrac{1}{2}M_1$$

A moment applied at one end of a prismatic beam, the other end being fixed, will cause a moment half as large and of opposite sign at the fixed end. The carryover factor is −½. The minus sign refers to strength-of-materials sign convention: A distributed moment on one end causing tension in bottom fibers must be carried over so that it will cause tension in the top fibers of the other end. A study of Figures 17.2 and 17.3 shows that carrying over with a ½ value with the moment-distribution sign convention automatically takes care of the situation, and it is unnecessary to change signs with each carryover.

17.2.2 Distribution Factors

Usually members framed together at a joint have different flexural stiffness. When a joint is unclamped and begins to rotate under the unbalanced moment, the resistance to rotation varies from member to member. The problem is to determine how much of the unbalanced moment will be taken up by each of the members. A reasonable assumption is

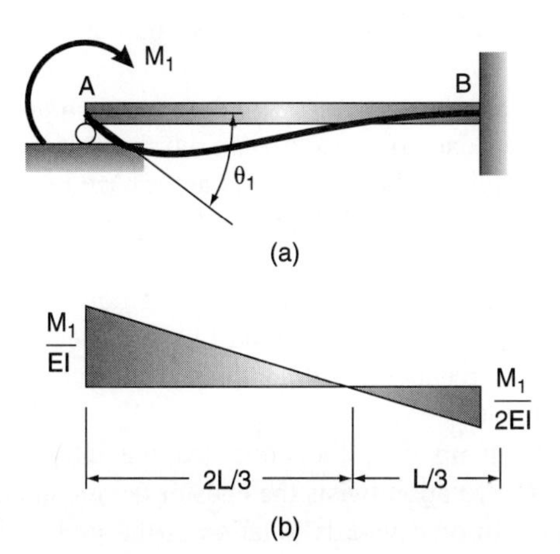

Figure 17.3 Moment Carry-over

that the unbalance will be resisted in direct relation to the respective resistance to end rotation of each member.

The beam and M/EI diagram of Figure 17.2 are redrawn in Figure 17.3, with the proper relationship between M_1 and M_2, and an expression is written for the amount of rotation caused by moment M_1.

Using the first moment-area theorem, the angle θ_1 may be represented by the area of the M/EI diagram between A and B, the tangent at B remaining horizontal:

$$\theta_1 = \frac{\frac{1}{2}M_1(\frac{2}{3})L - \frac{1}{2}(\frac{1}{2})M_1(\frac{1}{3})L}{EI}$$

$$\theta_1 = \frac{M_1 L}{4EI} \qquad \textbf{Eq. 17.3}$$

Assuming that all the members consist of the same material, having the same E values, the only variables in the foregoing equation affecting the amount of end rotation are the I and L values. The amount of rotation occurring at the end of a member obviously varies directly as the I/L value for the member. The larger the rotation of the member, the less moment it will carry. The moment resisted varies inversely as the amount of rotation or directly as the I/L value. This latter value is referred to as the stiffness factor K.

$$K = \frac{EI}{L} \qquad \textbf{Eq. 17.4}$$

To determine the unbalanced moment taken by each of the members at a joint, the stiffness factors at the joint are totaled, and each member is assumed to carry a proportion of the unbalanced moment equal to its K value divided by the sum of all the K values at the joint. These proportions of the total unbalanced moment carried by each of the members are the *distribution factors*.

$$DF_1 = \frac{K_1}{\sum K}$$

$$DF_2 = \frac{K_2}{\sum K} \qquad \textbf{Eq. 17.5}$$

17.3 DEFINITIONS

The following terms are constantly used in discussing moment distribution.

- **Fixed-End Moments:** When all of the joints of a structure are clamped to prevent any joint rotation, the external loads produce certain moments at the ends of the members to which they are applied. These moments are referred to as fixed-end moments. Appendix D presents fixed-end moments for various loading conditions.
- **Unbalanced Moments:** Initially the joints in a structure are considered to be clamped. When a joint is released, it rotates if the sum of the fixed-end moments at the joint is not zero. The difference between zero and the actual sum of the end moments is the unbalanced moment.
- **Distributed Moments:** After the clamp at a joint is released, the unbalanced moment causes the joint to rotate. The rotation twists the ends of the members at the joint and changes their moments. In other words, rotation of the joint is resisted

by the members and resisting moments are built up in the members as they are twisted. Rotation continues until equilibrium is reached—when the resisting moments equal the unbalanced moment—at which time the sum of the moments at the joint is equal to zero. The moments developed in the members resisting rotation are the distributed moments.

- **Carryover Moments:** The distributed moments in the ends of the members cause moments in the other ends, which are assumed fixed. These are the carryover moments.

17.4 SIGN CONVENTION

The moments at the end of a member are assumed to be negative when they tend to rotate the member end clockwise about the joint (the resisting moment of the joint would be counterclockwise). The continuous beam of Figure 17.4, with all joints assumed to be clamped, has clockwise (or −) moments on the left end of each span and counterclockwise (or +) moments on the right end of each span. The usual sign convention used in strength of materials shows fixed-ended beams to have negative moments on both ends for downward loads, because tension is caused in the top fibers of the beams at those points. It should be noted that this sign convention, to be used for moment distribution, is the same one used in Chapter 15 for slope deflection.

17.5 APPLICATION OF MOMENT DISTRIBUTION

The very few tools needed for applying moment distribution are now available, and the method of applying them is described (Figure 17.5).

A continuous beam with several loads applied to it is shown in Figure 17.5(a). In Figure 17.5(b), the interior joints B and C are assumed to be clamped, and the fixed-end moments are computed. At joint B, the unbalanced moment is computed and the clamp is removed, as seen in Figure 17.5(c). The joint rotates, thus distributing the unbalanced moment to the B ends of BA and BC in proportion to their distribution factors. The values of these distributed moments are carried over at the one-half rate to the other ends of the members. When equilibrium is reached, joint B is clamped in its new rotated position and joint C is released, as shown in Figure 17.5(d). Joint C rotates under its unbalanced moment until it reaches equilibrium; the rotation causes distributed moments in the C ends of members CB and CD and their resulting carryover moments. Joint C is now clamped and joint B is released, in Figure 17.5(e).

The same procedure is repeated again and again for joints B and C. The amount of unbalanced moment quickly diminishes until the release of a joint causes only negligible rotation. This process, in brief, is moment distribution.

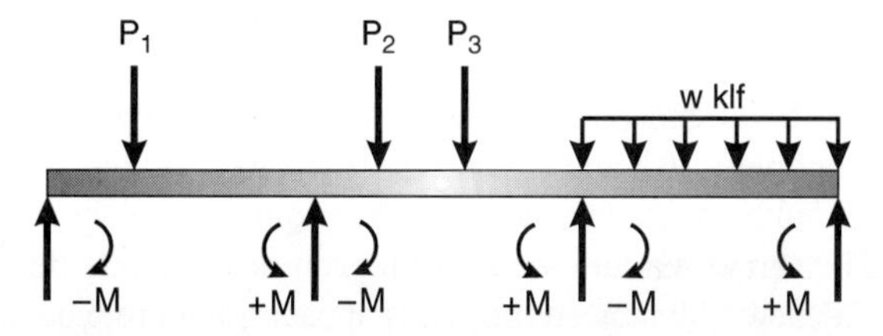

Figure 17.4 Sign convention for moment distribution

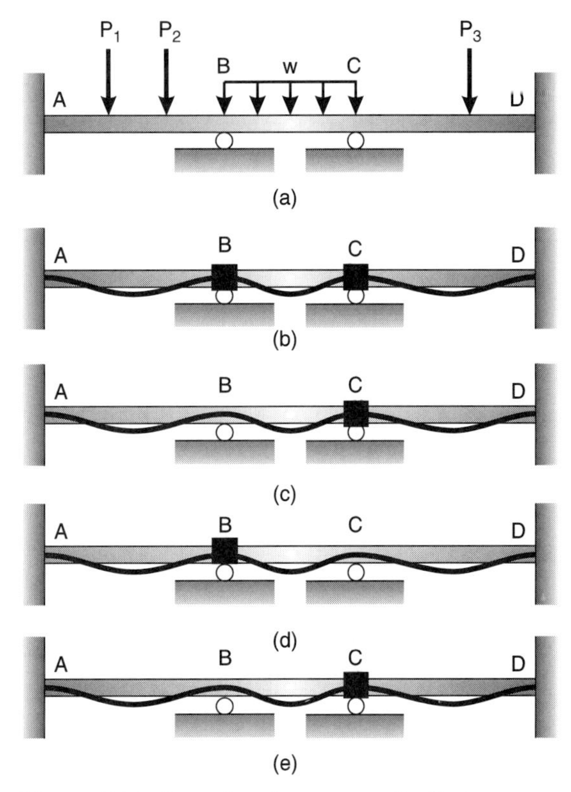

Figure 17.5 Procedure for moment distribution

Examples 17.1 to 17.3 illustrate the procedure used for analyzing relatively simple continuous beams. The stiffness factors and distribution factors are computed as follows for Example 17.1.

$$DF_{BA} = \frac{K_{BA}}{\sum K} = \frac{\frac{1}{20}}{\frac{1}{20} + \frac{1}{25}} = 0.43$$

Eq. 17.6

$$DF_{BA} = \frac{K_{BA}}{\sum K} = \frac{\frac{1}{15}}{\frac{1}{20} + \frac{1}{25}} = 0.57$$

A simple tabular form is used for Examples 17.1 and 17.2 to introduce the reader to moment distribution. For subsequent examples, a slightly varying but quicker solution much preferred by the authors is used. The tabular procedure may be summarized as follows:

1. The fixed-end moments are computed and recorded on one line (line FEM in Examples 17.1 and 17.2).
2. The unbalanced moments at each joint are balanced in the next line (Dist 1).
3. The carryovers are made from each of the joints on the next line (CO 1).
4. The new unbalanced moments at each joint are balanced (Dist 2), and so on. (As the beam of Example 17.1 has only one joint to be balanced, only one balancing cycle is necessary.)

When the distribution has reached the accuracy desired, a double line is drawn under each column of figures. The final moment in the end of a member equals the sum of the moments at its position in the table. Unless a joint is fixed, the sum of the final end moments in the ends of the members meeting at the joint must total zero.

EXAMPLE 17.1

Determine the end moments of the structure shown in the figure by moment distribution.

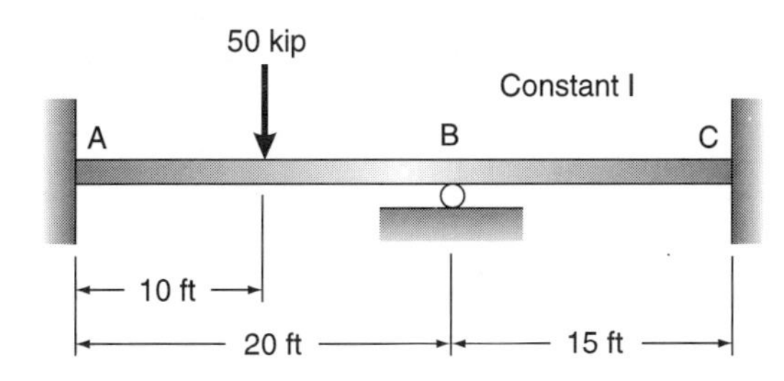

Solution

	0.43	0.57		
−125	+125	0.0	0.0	Fixed-End Moments
0.0	−53.8	−71.2	0.0	Distribution 1
−26.9			−35.9	Carry Over 1
−151.9	+71.2	−71.2	−35.9	Final Moments

■

EXAMPLE 17.2

Determine the end moments of the structure shown in the figure.

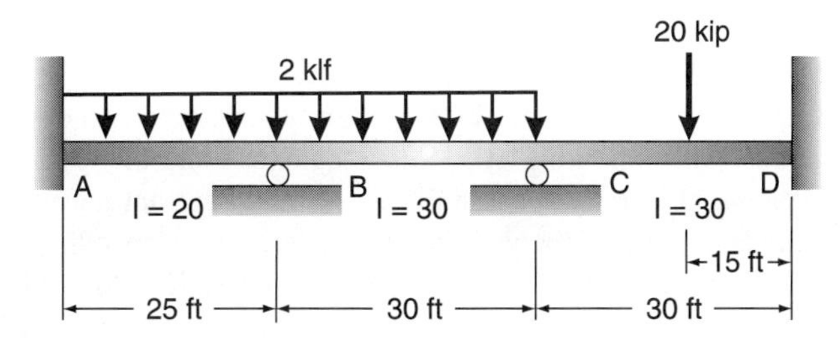

Solution

	0.44	0.56	0.5	0.5		
−104.2	+104.2	−150.0	+150.0	−75.0	+75.0	FEM's
	+20.2	+25.6	−37.5	−37.5		Dist 1
+10.1		−18.8	+12.8		−18.8	CO 1
	+8.3	+10.5	−6.4	−6.4		Dist 2
+4.2		−3.2	+5.3		−3.2	CO 2
	+1.4	+1.8	−2.7	−2.7		Dist 3
+0.7		−1.3	+0.9		−1.3	CO 3
	+0.6	+0.7	−0.4	−0.4		Dist 4
+0.3		−0.2	+0.4		−0.2	CO 4
	+0.1	+0.1	−0.2	−0.2		Dist 5
−88.9	+134.8	−134.8	+122.2	−122.2	+51.5	Final

■

TABLE 17.1 ACCURACY OF MOMENT DISTRIBUTION AFTER EACH CYCLE OF BALANCING FOR THE BEAM IN EXAMPLE 17.2

	Moments at Left Support		Moment at Support C	
Cycle	Moments at Cycle n	Ratio to Exact Moment	Moments at Support C	Ratio to Exact Moment
1	104.2	1.172	112.5	0.921
2	94.1	1.058	118.9	0.973
3	89.9	1.011	121.5	0.994
4	89.2	1.003	122.0	0.998
5	88.9	1.00	122.2	1.00

In Chapter 16, several methods for approximately analyzing statically indeterminate structures were introduced. Moment distribution is one of the "exact" methods of analysis if it's carried out until the moments to be distributed and carried over become quite small. However, it can be used as a superb approximate method for statically indeterminate structures if only a few cycles of distribution are made.

Consider the beam of Example 17.2. In this problem, each cycle of distribution is said to end when the unbalanced moments are balanced. Table 17.1 shows the ratios of the total moments up through each cycle at joints A and C to the final moments at those joints after all the cycles are completed. These ratios should give you an idea of how good partial moment distribution can be as an approximate method.

For many cases where the structure and/or the loads are very unsymmetrical there will be a few cycles of major adjustment of the moments throughout the structure. Until these adjustments are substantially made, moment distribution will not serve as a very good approximate method. It easily will be able, however, when the process can be stopped with good approximate results. This will occur when the unbalanced moments and carryover values become rather small as compared to the initial values.

Beginning with Example 17.3, a slightly different procedure is used for distributing the moments. Only one joint at a time is balanced and the required carryovers are made from that joint. Generally speaking it is desirable (but not necessary) to balance the joint that has the largest imbalance; make the carryovers; balance the joint with the largest imbalance, and so on, because such a process will result in the quickest convergence. This procedure is quicker than the tabular method used for Examples 17.1 and 17.2, and it follows along exactly with the description of the behavior of a continuous beam (with imaginary clamps) as pictured by the authors in Figure 17.5.

EXAMPLE 17.3

Compute the end moments in the beam shown in the figure.

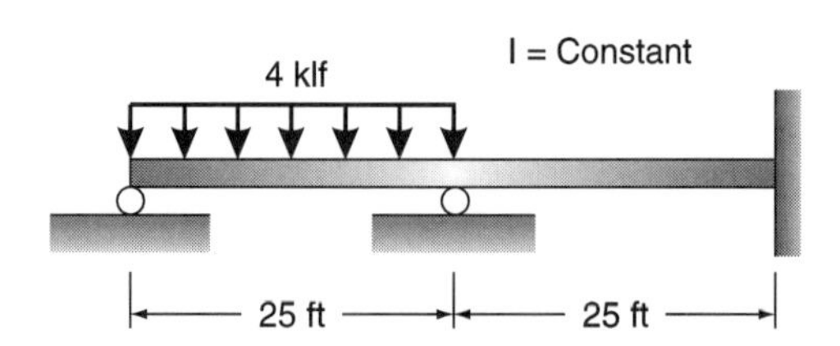

Solution

	0.5	0.5	
−208.3	+208.3		
+208.3	−104.2		
−78.1	−156.2	−156.2	−78.1
+78.1	+39.0		
−9.8	−19.5	−19.5	−9.8
+9.8	+4.9		
−1.2	−2.5	−2.5	−1.2
+1.2	+0.6		
−0.2	−0.3	−0.3	−0.2
+0.2	+178.5	−178.5	−89.3
0.0			

■

17.6 MODIFICATION OF STIFFNESS FOR SIMPLE ENDS

The carryover factor was developed for carrying over to fixed ends, but it is applicable to simply supported ends, which must have final moments of zero. The simple end of Example 17.3 was considered to be clamped; the carryover was made to the end; and the joint was freed and balanced back to zero. This procedure when repeated over and over is absolutely correct, but it involves a little unnecessary work, which may be eliminated by studying the stiffness of members with simply supported ends.

Shown in Figure 17.6(a) and (b) is a comparison of the relative stiffness of a member subjected to a moment M_1 when the far end is fixed and also when it is simply supported. In part (a), using principles of the moment-area method that were discussed in Chapter 10, the slope at the left end, which is represented by θ_1, is equal to $M_1L/4EI$.

For the simple end-supported beam shown in Figure 17.6(b), again using moment-area principles, the slope θ_1 is found to be $M_1L/3EI$. Therefore, the slope caused by the moment M_1 when the far end is fixed is only three-fourths as large ($M_1L/4EI \div M_1L/3EI = \frac{3}{4}$) when the far end is simply supported. The beam simply supported at the far end is only three-fourths as stiff as the one that is fixed. If the stiffness factors for end spans that are simply supported are modified by three-fourths, the simple end is initially balanced to zero,

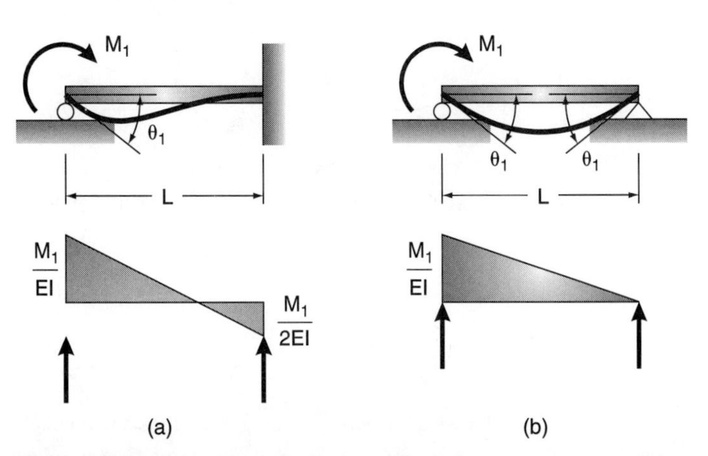

Figure 17.6 Comparison of relative stiffness for two support conditions

no carryovers are made to that end afterward, and the same results will be obtained. In Example 17.4, the stiffness modification is used for the beam that was previously evaluated in Example 17.3.

EXAMPLE 17.4

Determine the end moments of the structure shown in the figure by using the simple end-stiffness modification for the left end.

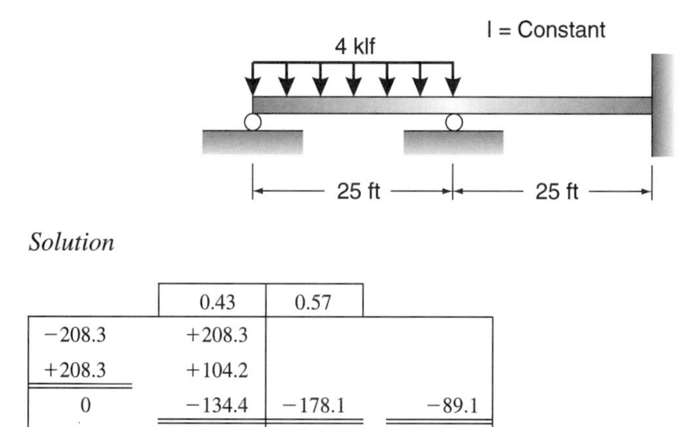

Solution

	0.43	0.57	
−208.3	+208.3		
+208.3	+104.2		
0	−134.4	−178.1	−89.1
	+178.1	−178.1	−89.1

■

17.7 SHEARING FORCE AND BENDING MOMENT DIAGRAMS

Drawing shear and moment diagrams is an excellent way to check the final moments computed by moment distribution and to obtain an overall picture of the stress condition in the structure. Before preparing the diagrams, it is necessary to consider a few points relating the shear and the moment diagram sign convention to the one used for moment distribution. The usual conventions for drawing the diagrams will be used—tension in bottom fibers of beam is positive moment and upward shear to the left is positive shear.

The relationship between the signs of the moments for the two conventions is shown with the beams of Figure 17.7. Part (a) of the figure illustrates a fixed-end beam for which the result of moment distribution is a negative moment. The clockwise moment bends the

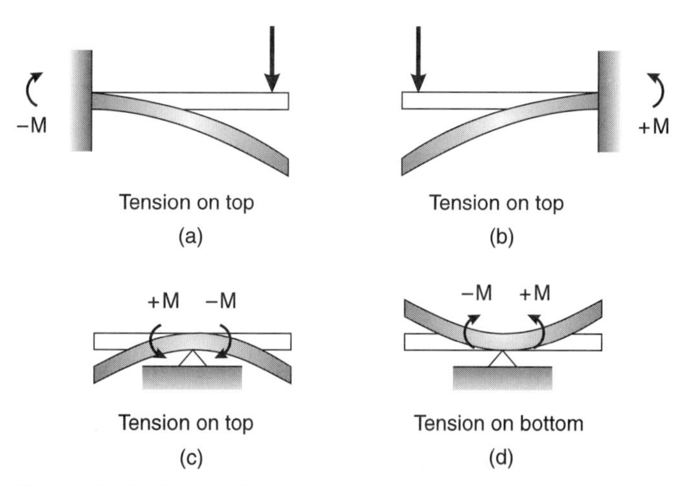

Figure 17.7 Interpreting sense of resulting moments

beam as shown, causing tension in the top fibers or a negative moment for the shear and moment diagram convention. In Figure 17.7(b) the result of moment distribution is a positive moment, but again the top beam fibers are in tension, indicating a negative moment for the moment diagram.

An interior simple support is represented by Figure 17.7(c) and (d). In part (c) moment distribution gives a negative moment to the right and a positive moment to the left, which causes tension in the top fibers. Part (d) shows the effect of moments of opposite sign at the same support considered in part (c).

From the preceding discussion it can be seen that the sign convention used herein for moment distribution for continuous beams agrees with the one used for drawing moment diagrams on the right-hand sides of supports, but disagrees on the left-hand sides.

To draw the diagrams for a vertical member, the right side is often considered to be the bottom side. Moments are distributed in the continuous beams in Examples 17.5 and 17.6, and the shearing force and bending moment diagrams are drawn.

The reactions shown in the solution of these problems were obtained by computing the reactions as though each span was simply supported and adding them to the reactions due to the moments at the beam supports.

EXAMPLE 17.5

Distribute moments and draw shear and moment diagrams for the structure shown in the figure.

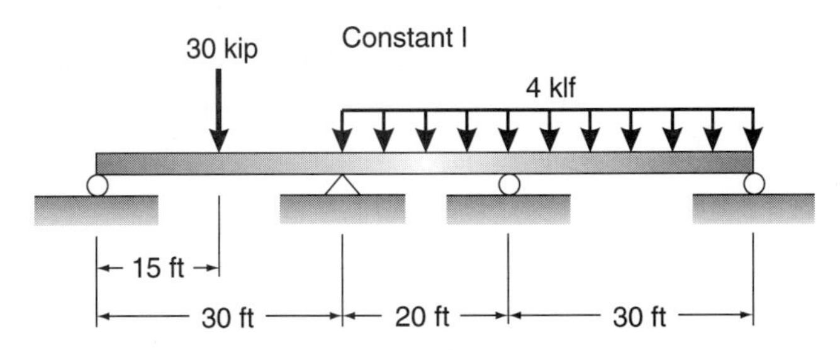

Solution

	0.33	0.67	0.67	0.33	
−112.5	+112.5	−133.3	+133.3	−300.0	+300.0
+112.5	+56.2			−150.0	−300.0
		+105.6	+211.1	+105.6	
	−47.0	−94.0	−47.0		
		+15.7	+31.3	+15.7	
	−5.2	−10.5	−5.2		
		+1.7	+3.5	+1.7	
	−0.6	−1.1	−0.6		
		+0.2	+0.4	+0.2	
	−0.1	−0.1			
0.0	+115.8	−115.8	+326.8	−326.8	0.0
Reaction Computations					
+15.00	+15.00	+40.00	+40.00	+60.00	+60.00
−3.86	+3.86	−10.55	+10.55	+10.89	−10.89
+11.14	+18.86	+29.45	+50.55	+70.89	+49.11

Shearing Force and Bending Moment Diagrams:

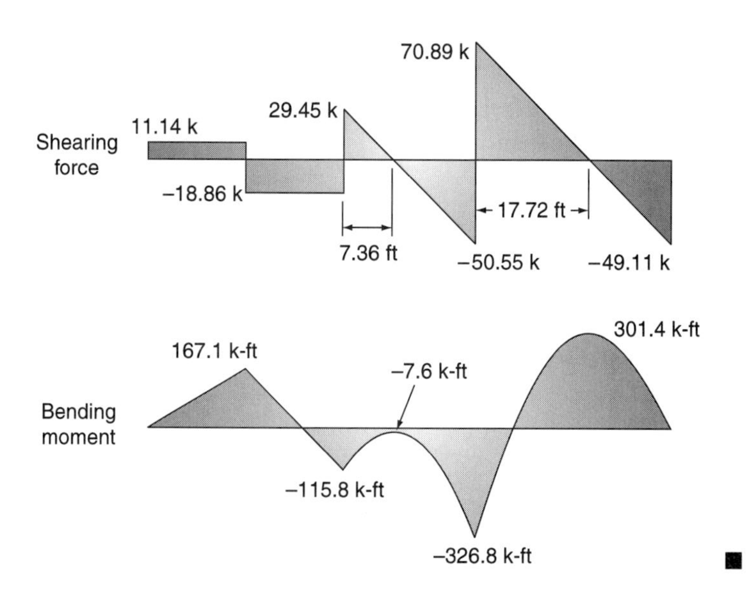

■

EXAMPLE 17.6

Distribute moments and draw shear and moment diagrams for the continuous beam shown in the figure.

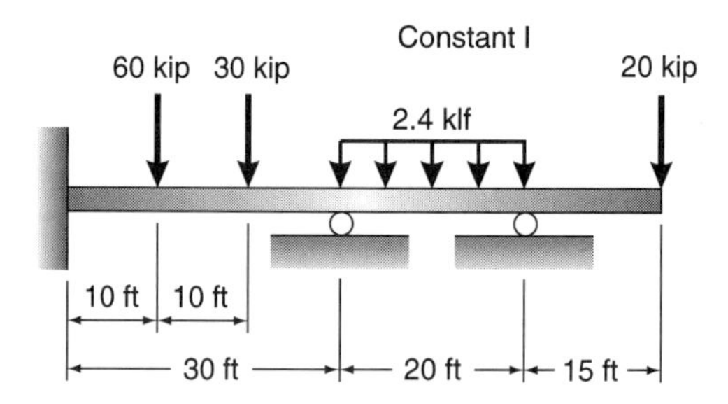

Solution

	0.471	0.529	1.00	0.00
−333.3	+266.7	−80.0	+80.0	−300.0
		+110.0	+220.0	
−69.9	−139.7	−157.0		
−403.2	+127.0	−127.0	+300	−300
Reaction Computations				
+50.00	+40.00	+24.0	+24.00	+20.00
+9.21	−9.21	−8.65	+8.65	
+59.21	+30.79	+15.35	+32.65	+20.00

Shearing Force and Bending Moment Diagrams:

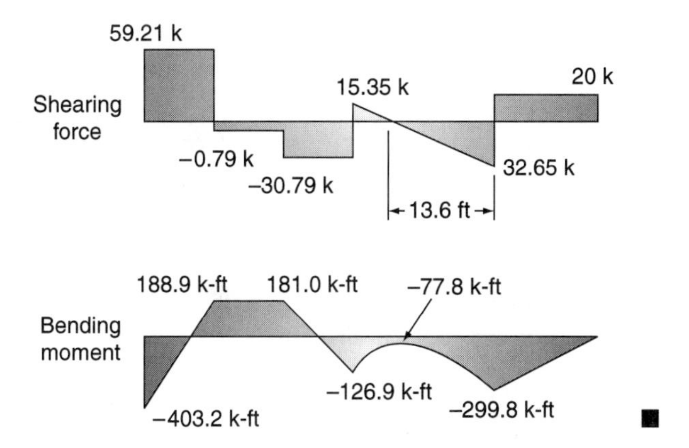

■

17.8 COMPUTER SOLUTIONS

The structural analysis software included with this book enables the reader to analyze very quickly all of the problems contained in this chapter. One such example follows.

EXAMPLE 17.7

Determine the final moments for the beam in the figure. This is the beam analyzed by moment distribution in Example 17.6.

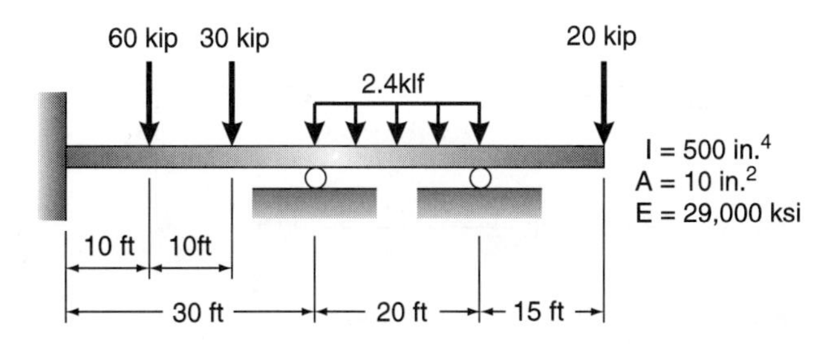

Solution. The structural model that we will use for the solution is shown next. The loads acting on the beam spans are a combination of point loads and uniformly distributed loads. The properties of the member are specified by creating a new section type, "general." The input data file for this example is on the CD-ROM contained in this book.

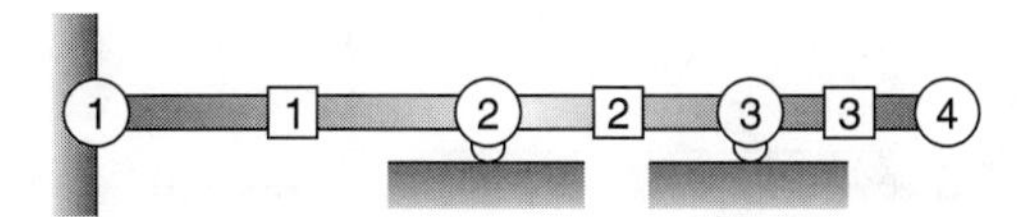

The results of the analysis for the beam members are in the table. Notice that these results are the same as we obtained using moment distribution.

		Beginning		End	
Member	F	M	V	M	V
1	0.0	−4835.9	−59.21	−1523.4	30.79
2	0.0	−1523.4	−15.35	−3600.0	32.65
3	0.0	−3600.0	−20.0	0.0	−20.0

17.9 PROBLEMS FOR SOLUTION

For Problems 17.1 through 17.22 analyze the structures by the moment-distribution method and draw shear and moment diagrams.

17.1

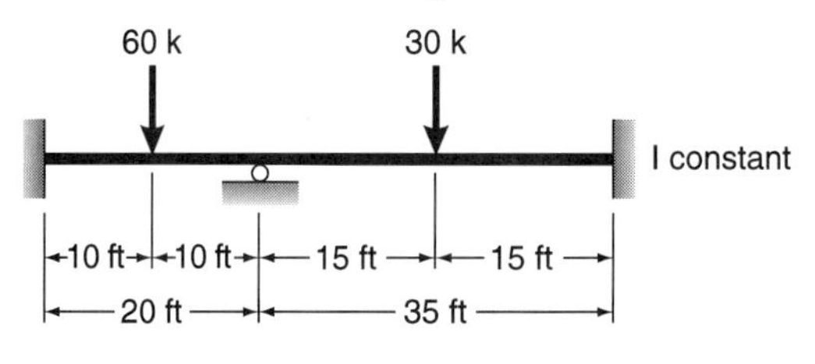

17.2 (*Ans.* M_A = 380 ft-k, M_B = 200 ft-k)

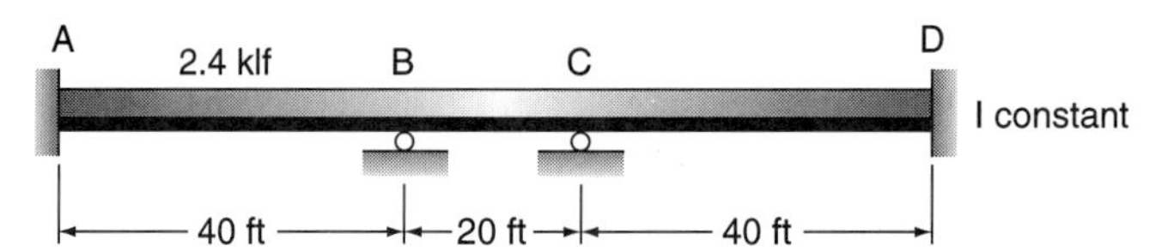

17.3

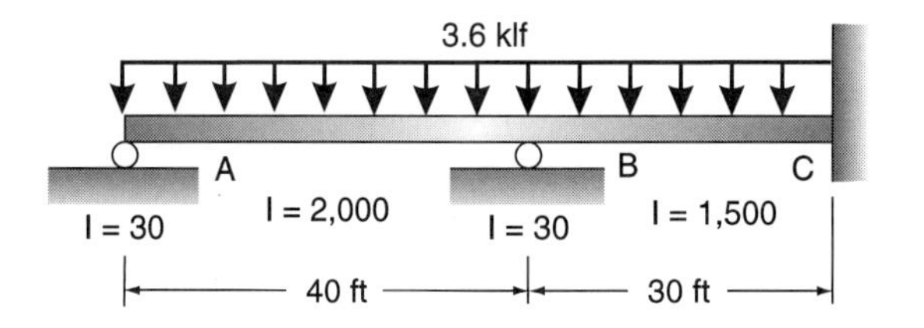

17.4 (*Ans.* M_B = 380 ft-k, M_C = 458 ft-k)

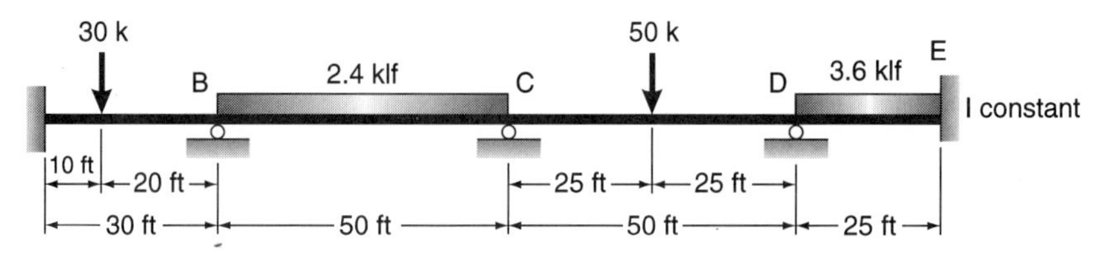

17.5

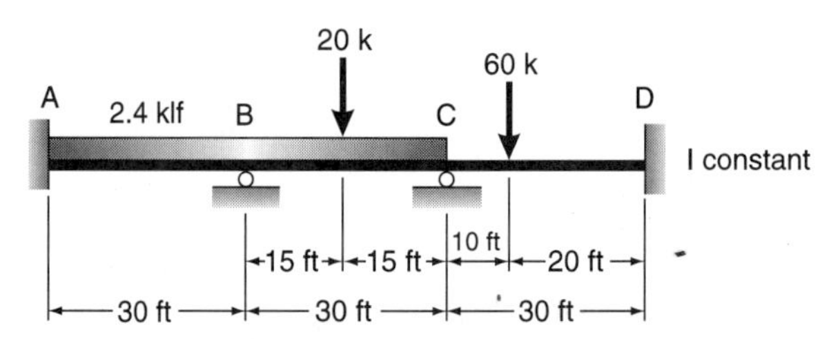

17.6 Rework Problem 17.3 if a 40-kip load is added at the centerline of each of the right two spans. (*Ans.* M_A = 135.5 ft-k, M_C = 685.5 ft-k)

17.7

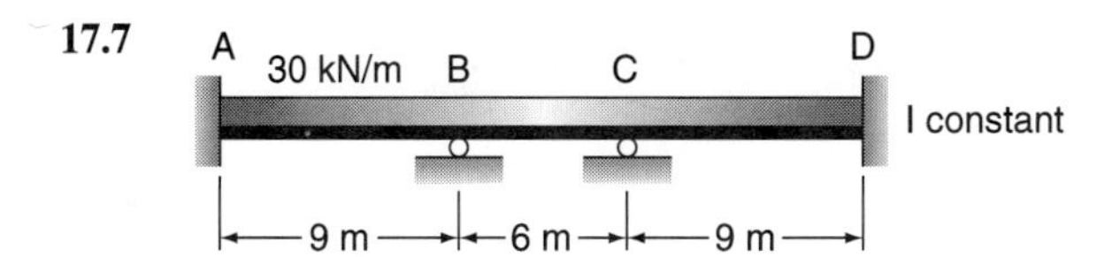

17.8 (*Ans.* M_B = 212.3 kN-m, M_D = 129.6 kN-m)

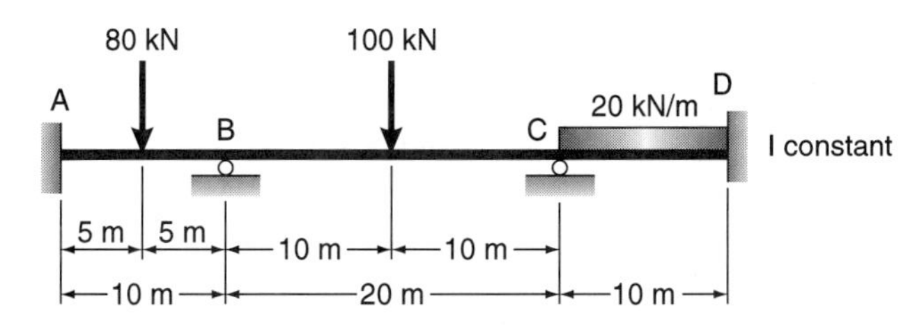

17.9

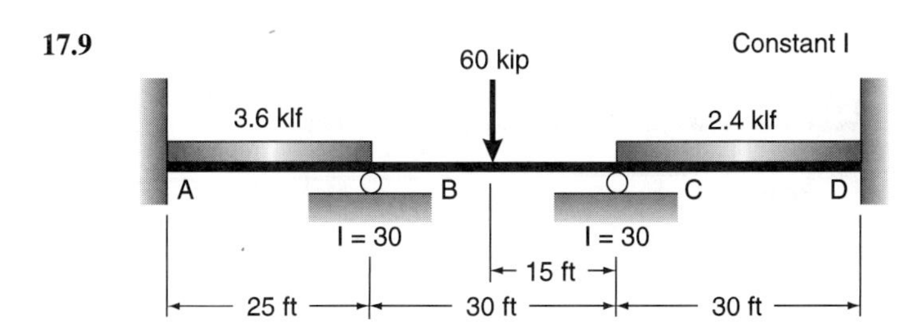

17.10 (*Ans.* M_A = 53.6 kN-m, M_B = 267.8 kN-m)

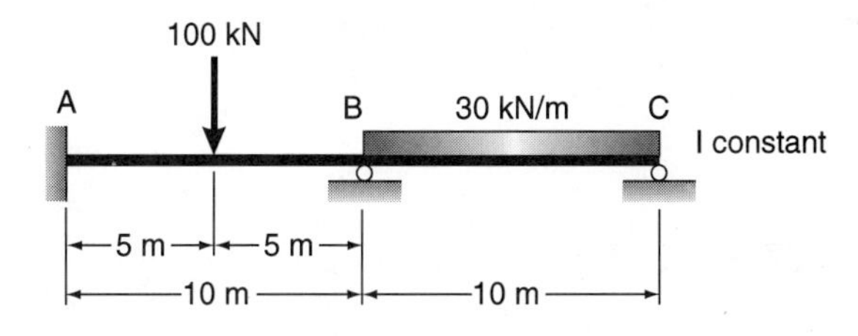

17.11

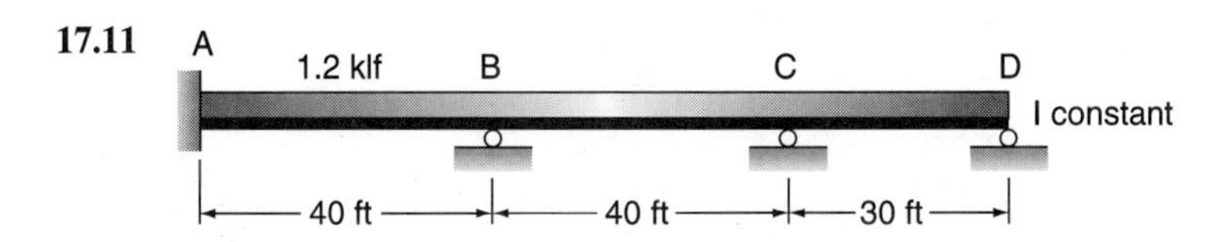

17.12 (*Ans.* $M_A = 150.3$ ft-k, $M_C = 95.6$ ft-k)

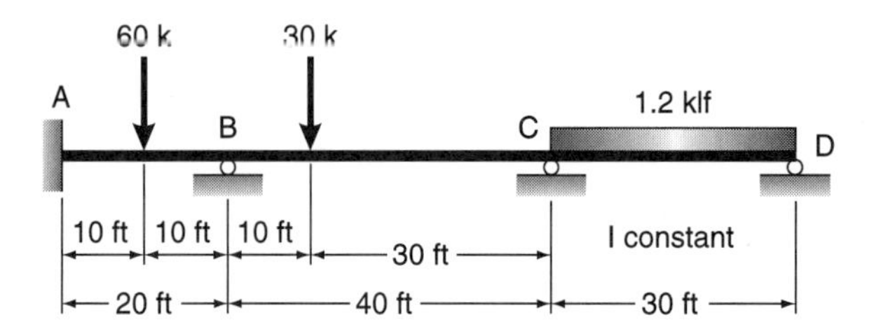

17.13

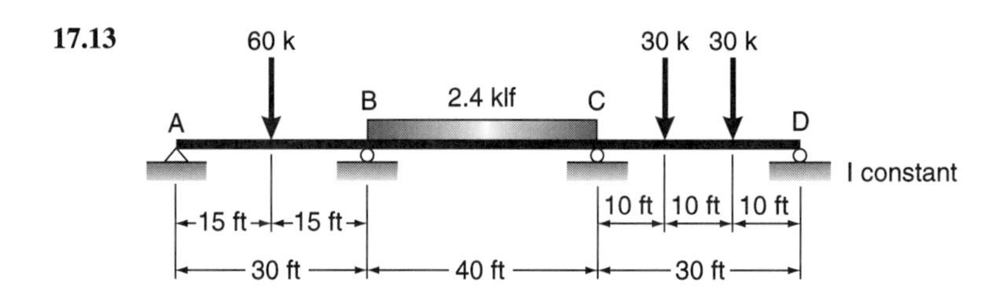

17.14 Repeat Problem 17.2 if the end supports A and D are made simple supports. (*Ans.* $M_B = M_C = 308.5$ ft-k)

17.15

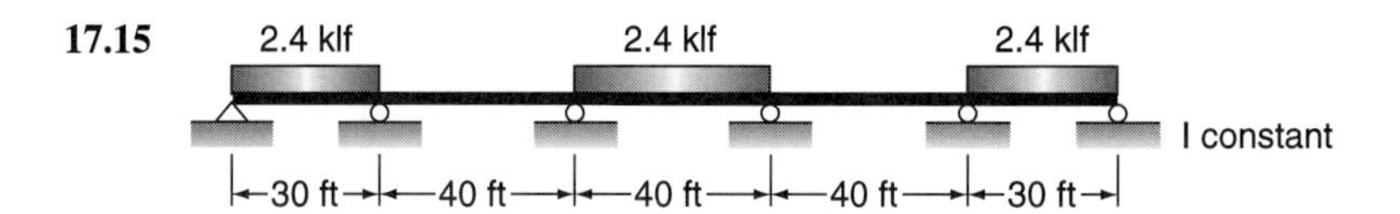

17.16 (*Ans.* $M_B = 499.8$ ft-k, $M_D = 267.2$ ft-k)

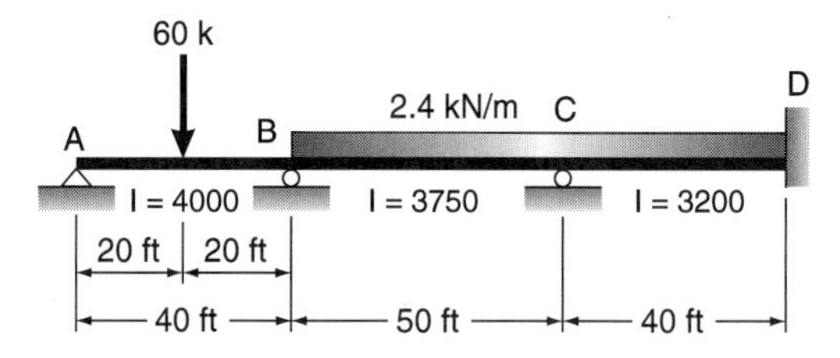

17.17

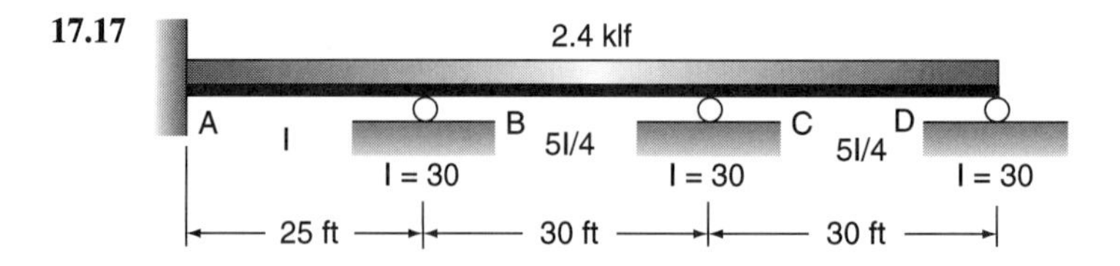

17.18 (*Ans.* $M_A = 151.1$ ft-k, $M_B = 258$ ft-k)

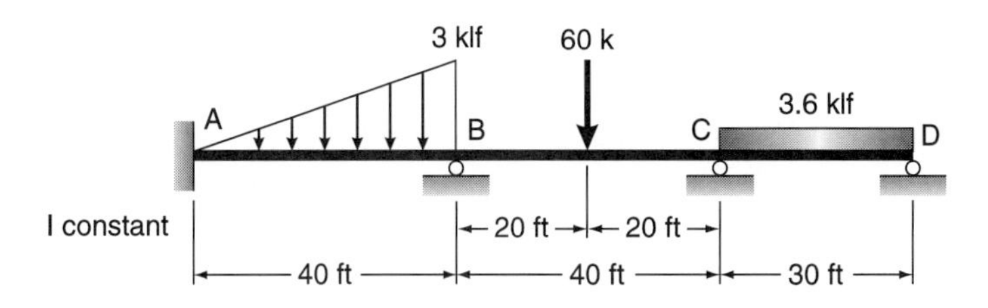

17.19

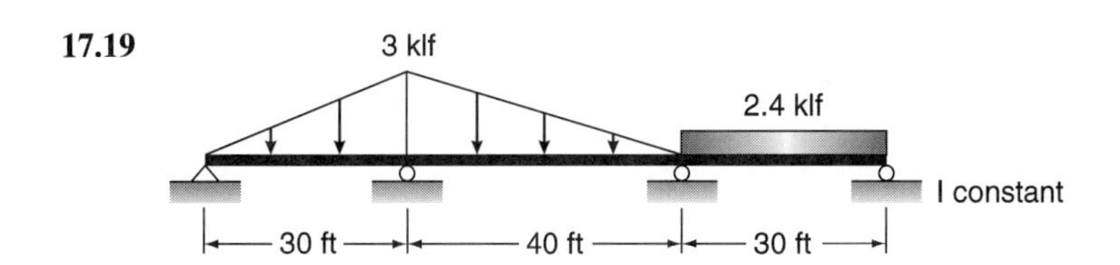

17.20 (*Ans.* $M_B = 277$ ft-k, $M_D = 362.4$ ft-k)

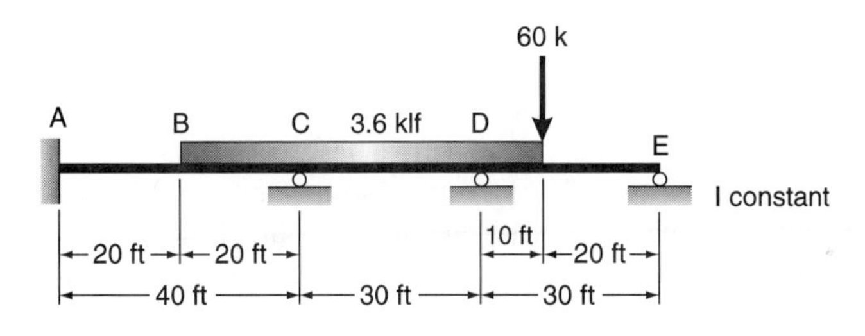

17.21

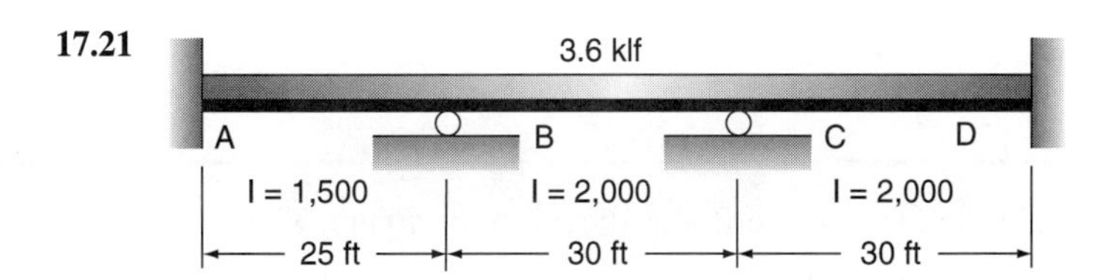

For Problems 17.22 through 17.25, rework the problems using SAP2000.

17.22 Problem 17.1

17.23 Problem 17.3 (*Ans.* $M_B = 193.7$ ft-k, $M_D = 169.1$ ft-k)

17.24 Problem 17.17

17.25 Problem 17.22 (*Ans.* $M_A = 176.4$ ft-k, $M_C = 362.4$ ft-k)

17.26 Go to the library and read the paper written by Hardy Cross about the moment distribution method. See the footnotes on the first page of this chapter for the citation.

Chapter 18

Moment Distribution for Frames

18.1 FRAMES WITH SIDESWAY PREVENTED

Moment distribution for frames is handled in the same manner as it is for beams when sidesway or lateral movement cannot occur. Analysis of frames without sidesway is illustrated by Examples 18.1 and 18.2. Where sidesway is possible, however, it must be taken into account because joint displacements cause rotations at the end of members connected to the joints, and therefore affect the moments in all the members.

As the structures being analyzed become more complex, a method of recording the calculations so they will not interfere with each other is necessary. In this chapter, a system is used for frames whereby the moments are recorded below beams on their left ends and above them on their right ends. For columns, the same system is used, the right sides being considered the bottom sides.

EXAMPLE 18.1

Determine the end moments for the frame shown in the figure. The relative value of I is shown for each member in the frame.

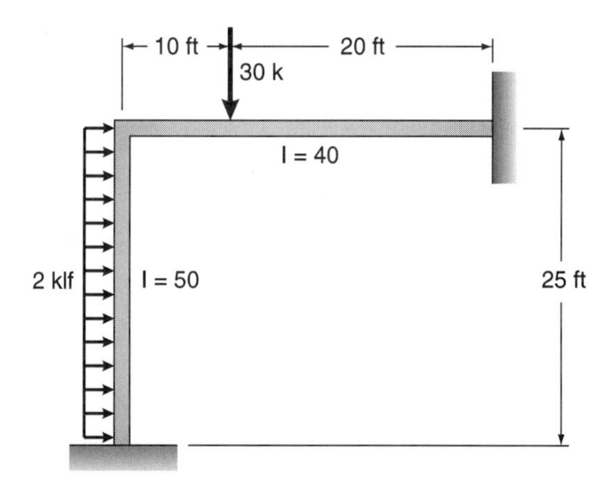

Solution

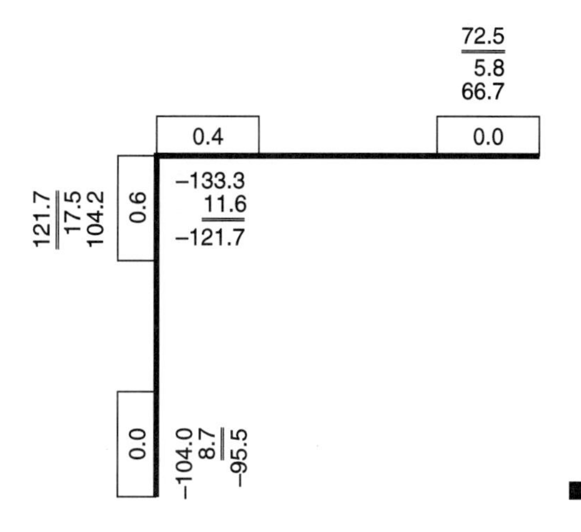

■

EXAMPLE 18.2

Compute the end moments of the structure shown in the figure. I is constant for all members in this structure.

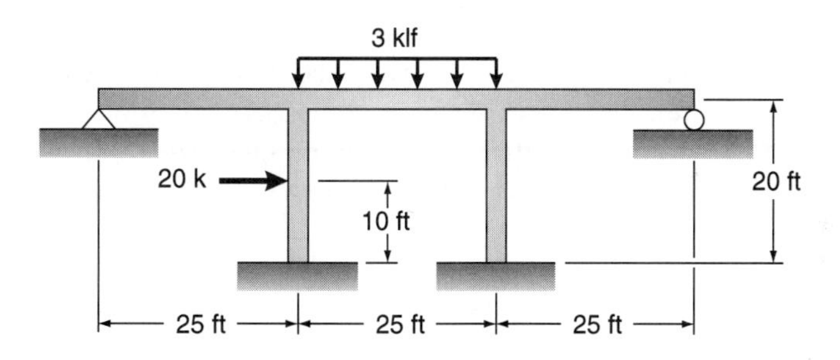

Solution

33.7
0.7
6.5
26.5

0.25 | 0.33

106.4
1.3
10.8
44.3
50.0

0.42

−156.2
35.4
−26.0
8.7
-3.0
1.0
−140.1

0.0

−50.0
22.2
5.4
0.6
−21.8

118.8
−1.5
4.4
−5.9
17.7
−52.1
156.2

0.33 | 0.25

−74.3
−1.8
−7.4
−65.1

0.42

−39.0
−4.4
−1.1
−44.5

0.0

−32.6
−3.7
−0.9
−37.2

■

18.2 FRAMES WITH SIDESWAY

Structural frames, similar to the one shown in Figure 18.1, are usually constructed so that they may possibly sway to one side or the other under load. Although the frame in this figure is symmetrical, it will tend to sway because the load is not centered on the structure—the structure is asymmetrical with respect to load. Analysis of the frame using moment distribution with the methods discussed so far will yield inconsistent results.

If the fixed-end moments are balanced using procedures we discussed in the previous section, the resulting member-end bending moments and shearing forces are as shown in Figure 18.2. To calculate the shearing forces, each member was taken as a free body and the equations of equilibrium were applied. Consider, for example, the right column. The bottom shearing force was calculated from the equation:

$$\sum M_{\text{Top of Column}} = 0$$

$$-35.5 - 17.8 + 30H = 0 \qquad \textbf{Eq. 18.1}$$

$$\therefore H = 1.78 \text{ kips} \leftarrow$$

The moments are those obtained from the moment-distribution analysis. The other shearing forces in the structure were calculated in a similar manner.

From the results obtained, we see that the sum of the horizontal forces on the entire structure is not equal to zero. The sum of the forces to the right is 0.89 kip to the right—the structure apparently is not in a state of static equilibrium.

The usual analysis does not yield consistent results because the structure actually sways, or deflects, to one side. Effectively, the structure that we analyzed is the one shown in Figure 18.3(a); there is an apparent support at the top of the structure. This artificial support is preventing the deflection shown in Figure 18.3(b) from occurring. This deflection causes bending moments that were not considered in the analysis. Hence, inconsistent results were obtained from the analysis.

One possible solution is to compute the deflections caused by applying a force of 0.89 kip acting to the right at the top of the bent. The moments caused by this force could be obtained for the computed deflections and added to the originally distributed fixed-end moments. This is a rather difficult and tedious approach.

A more convenient method is to assume the existence of an imaginary support that prevents the structure from swaying, as was shown in Figure 18.3(a). The fixed-end moments are distributed, and the force that the imaginary support must supply to hold the frame in place is computed. The support is then removed, which will allow the frame to

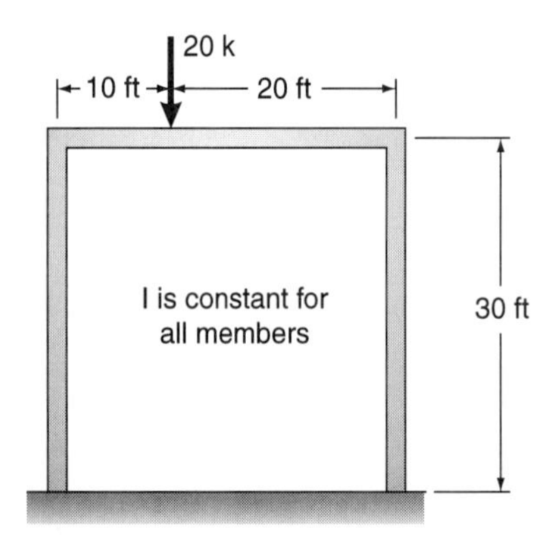

Figure 18.1 Frame subject to sidesway because of load

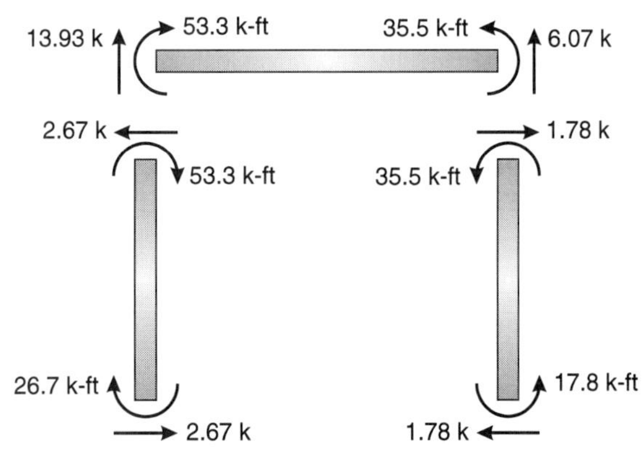

Figure 18.2 Results of analysis considering no sidesway

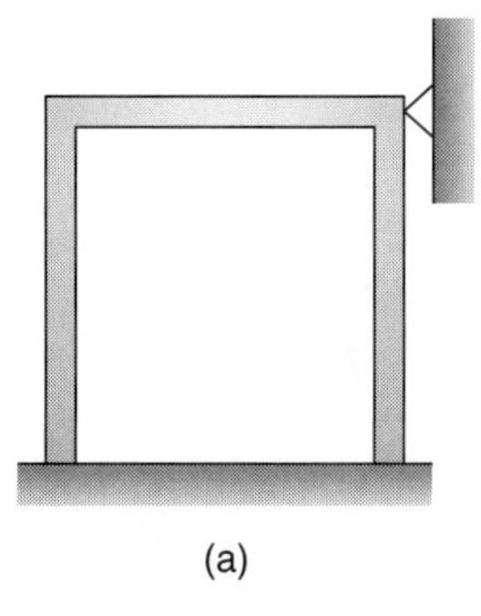

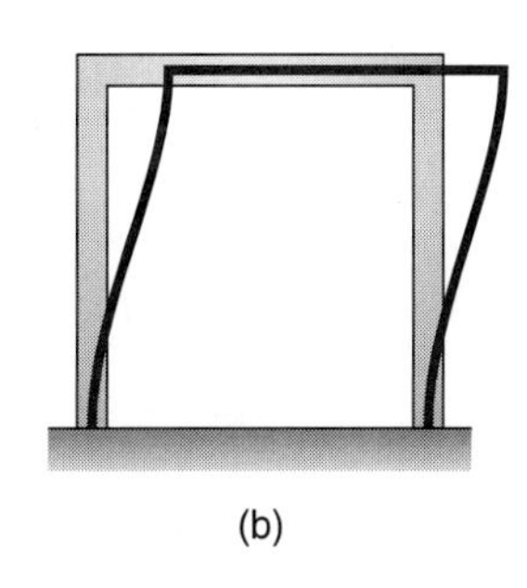

Figure 18.3 Virtual structure (a) when sidesway is not considered versus the actual behavior (b)

sway to the right. As the structure sways to the right, the joints are locked against rotation so moments are introduced into the ends of the columns. We assume values of the moments caused by the sidesway and then distribute those moments. After the moments are distributed, we can scale the results upward or downward until the resulting force offsets that in the imaginary support. The final forces in the members are equal to these scaled forces added to the results of the original analysis. In the next section, we will study the method for assuming a consistent set of moments caused by the sidesway and will demonstrate the procedure through an example.

18.3 SIDESWAY MOMENTS

Should all of the columns in a frame be the same length and have the same moments of inertia, the assumed sidesway moments will be the same for each column. However, should the columns have different lengths and/or moments of inertia, this will not be the case. You will see in the following paragraphs that the assumed sidesway moments will vary from column to column in proportion to their respective ratios of I/L^2.

If the frame of Figure 18.4(a) is pushed laterally an amount Δ by the lateral load P, it will have the deflected shape shown in Figure 18.4(b). Theoretically, both columns will become perfect S-curves, if the beam is considered to be rigid. At their mid-height, the deflection for both columns will equal $\Delta/2$. Mid-height of the columns may be considered points of contra-flexure; the bottom half of each column may be dealt with as though it was a simple cantilevered beam. The deflection of a cantilevered beam with a concentrated load at its end is

$$\Delta = \frac{PL^3}{3EI} \qquad \textbf{Eq. 18.2}$$

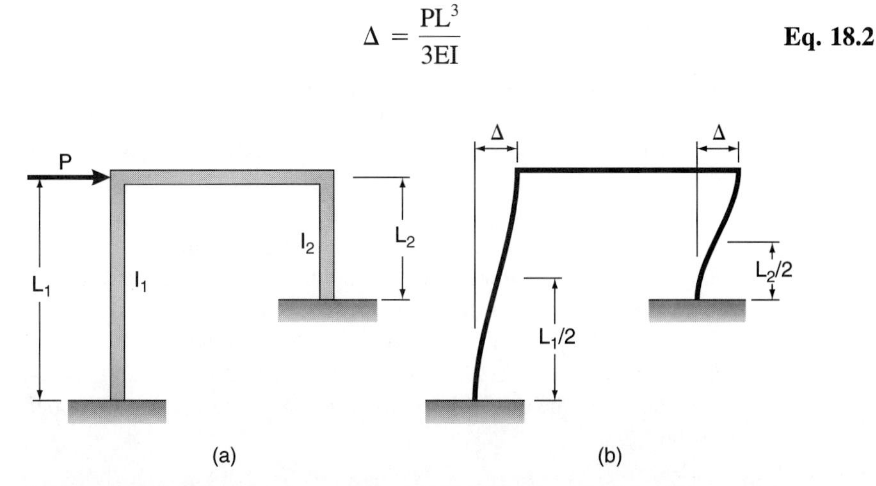

Figure 18.4 Simple frame for discuss of analysis considering sidesway

Since the deflections are the same for both columns, the following expressions may be written:

$$\frac{\Delta}{2} = \frac{P_1\left(\frac{L_1}{2}\right)^3}{3EI_1} = \frac{P_1L_1^3}{3EI_1}$$

$$\frac{\Delta}{2} = \frac{P_2\left(\frac{L_2}{2}\right)^3}{3EI_2} = \frac{P_2L_2^3}{3EI_2}$$

By solving these deflection expressions for P_1 and P_2, the forces pushing on the cantilevers, we have:

$$P_1 = \frac{12EI_1\Delta}{L_1^3}$$

$$P_2 = \frac{12EI_2\Delta}{L_2^3} \qquad \textbf{Eq. 18.3}$$

The moments caused by the two forces at the ends of their respective cantilevers are equal to the force times the cantilever length. These moments are written and the values of P_1 and P_2 are substituted in them:

$$M_1 = P_1\frac{L_1}{2} = \left(\frac{12EI_1\Delta}{L_1^3}\right)\frac{L_1}{2} = \frac{6EI_1\Delta}{L_1^2}$$

$$M_2 = P_2\frac{L_2}{2} = \left(\frac{12EI_2\Delta}{L_2^3}\right)\frac{L_2}{2} = \frac{6EI_2\Delta}{L_2^2} \qquad \textbf{Eq. 18.4}$$

From the expressions, a ratio of the moments may be written as follows because Δ is the same in each column:

$$\frac{M_1}{M_2} = \frac{6EI_1\Delta/L_1^2}{6EI_2\Delta/L_2^2} = \frac{I_1/L_1^2}{I_2/L_2^2} \qquad \textbf{Eq. 18.5}$$

This relationship must be used for assuming sidesway moments for the columns of a frame. Any convenient moments may be assumed, but they must be in proportion to each other as their I/L^2 values. Should their I and L values be equal, the assumed moments would be equal.

The procedure for applying the moment distribution method when sidesway is involved can be summarized as follows:

1. Determine the distribution factors for each member in the frame.
2. Calculate the fixed-end moments caused by the applied loads.
3. Distribute the fixed-end moments until convergence is achieved.
4. Compute the force imbalance, which is the force in the virtual support that is preventing sidesway from occurring.
5. Compute the value of I/L^2 for each of the columns.
6. Compute the assumed sidesway moments in proportion to the I/L^2 values of each of the columns and distribute these moments until convergence is achieved.
7. Compute the lateral force caused by these moments.
8. Add these latter final moments, times the ratio of lateral forces, to the moments obtained in the original distribution. These are the final member-end moments in the frame.

This procedure is demonstrated in the following two examples. In Example 18.3, we will apply a uniformly distributed load to a frame with unequal columns. Because the frame is not symetrical, it will sway laterally. In Example 18.4, we will apply both a uniformly distributed load and a lateral concentrated load to frame with equal length columns.

EXAMPLE 18.3

Using moment distribution, find all of the member-end moments in the structure shown in the figure.

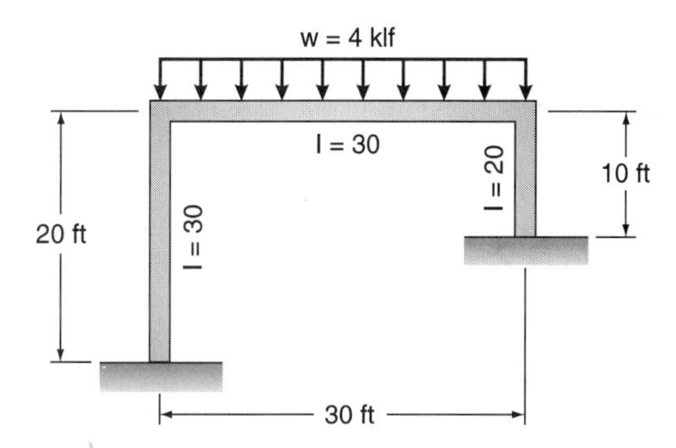

Solution. Distribute fixed-end moments and compute horizontal reactions at column bases.

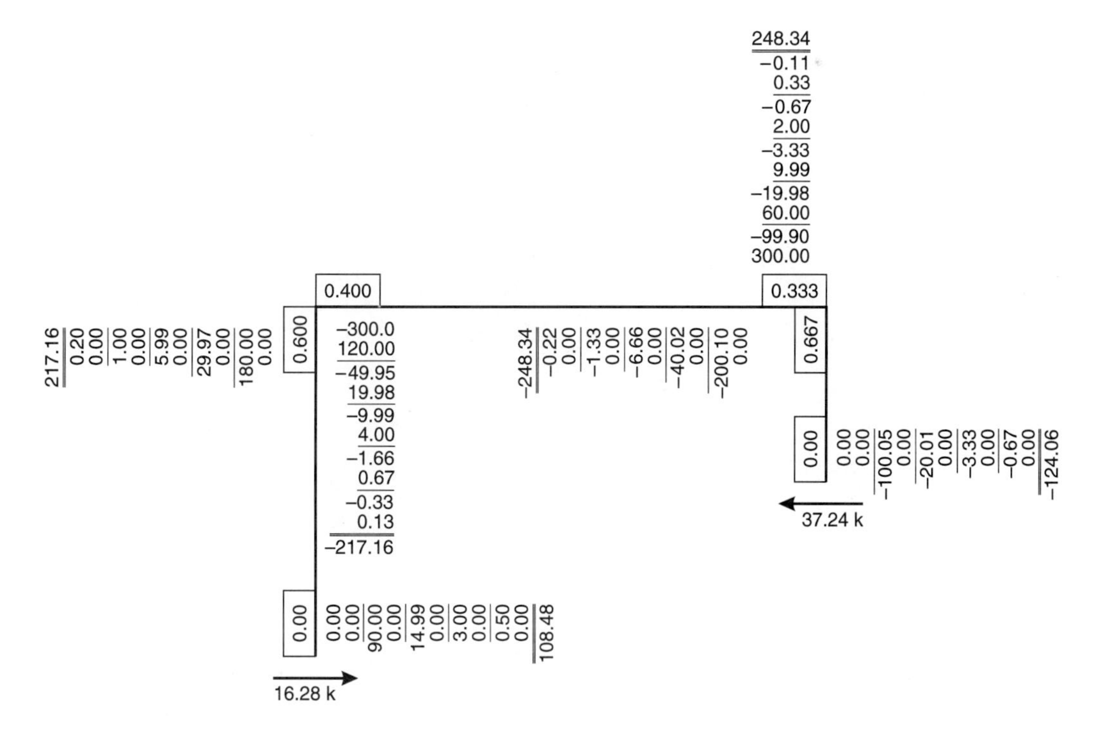

From these results we see that the imaginary support, the virtual support, exerts a force of 20.96 kips acting to the right. That is the force necessary for static equilibrium in the horizontal direction. When the structure sways to the right the assumed sidesway moments are negative and in the following proportion:

$$\frac{M_1}{M_2} = \frac{\frac{I_1}{L_1^2}}{\frac{I_2}{L_2^2}} = \frac{\frac{30}{20^2}}{\frac{20}{10^2}} = \frac{3}{8} \qquad \text{Let} \begin{cases} M_1 = 30 \text{ k-ft} \\ M_2 = 80 \text{ k-ft} \end{cases}$$

Next, distribute these assumed sidesway moments.

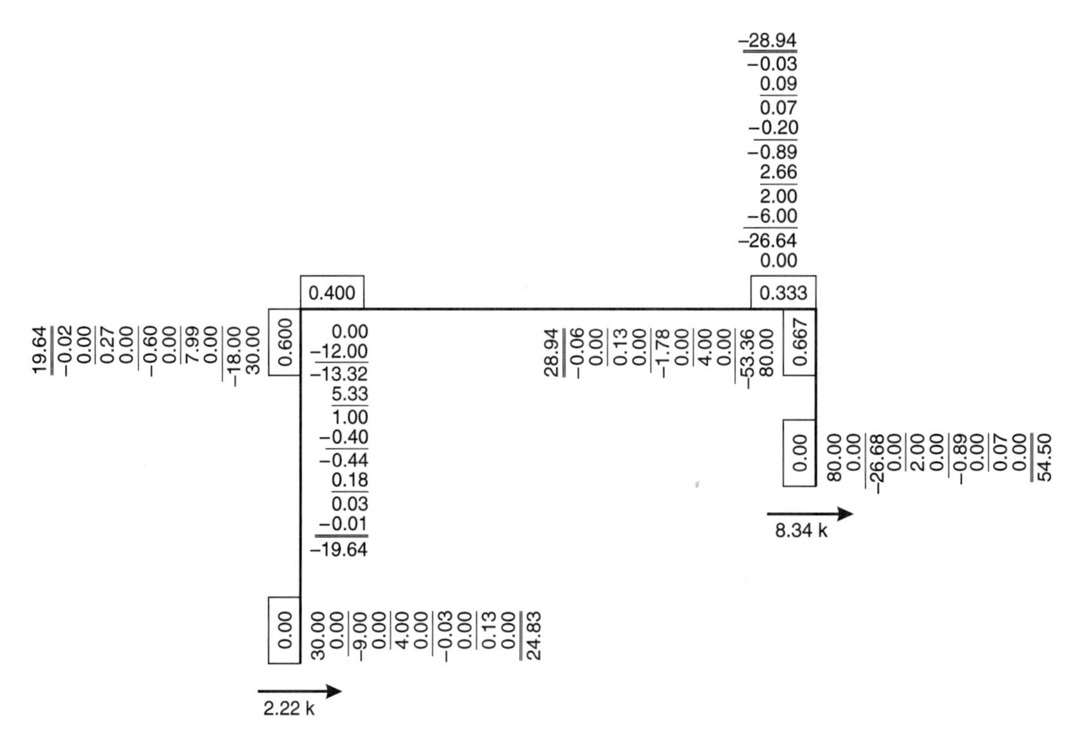

The factor by which the assumed sidesway moments are multiplied by is computed from

$$S = -\frac{16.28 - 37.24}{2.22 + 8.34} = 1.98$$

The final member-end moments in the frame are then

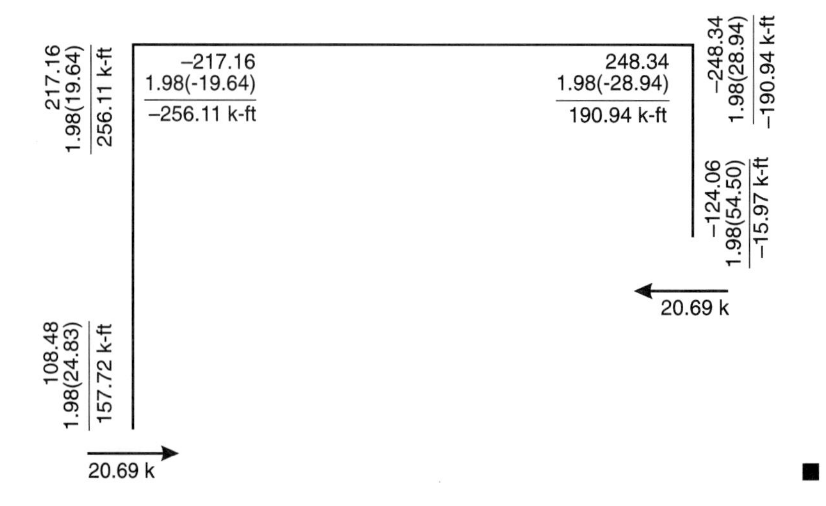

■

EXAMPLE 18.4

Determine the final member-end moments for the frame shown.

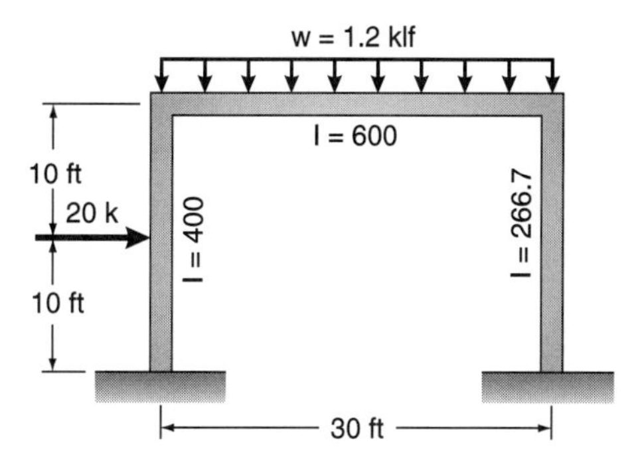

Solution. Distribute fixed-end moments and compute horizontal reactions at column bases.

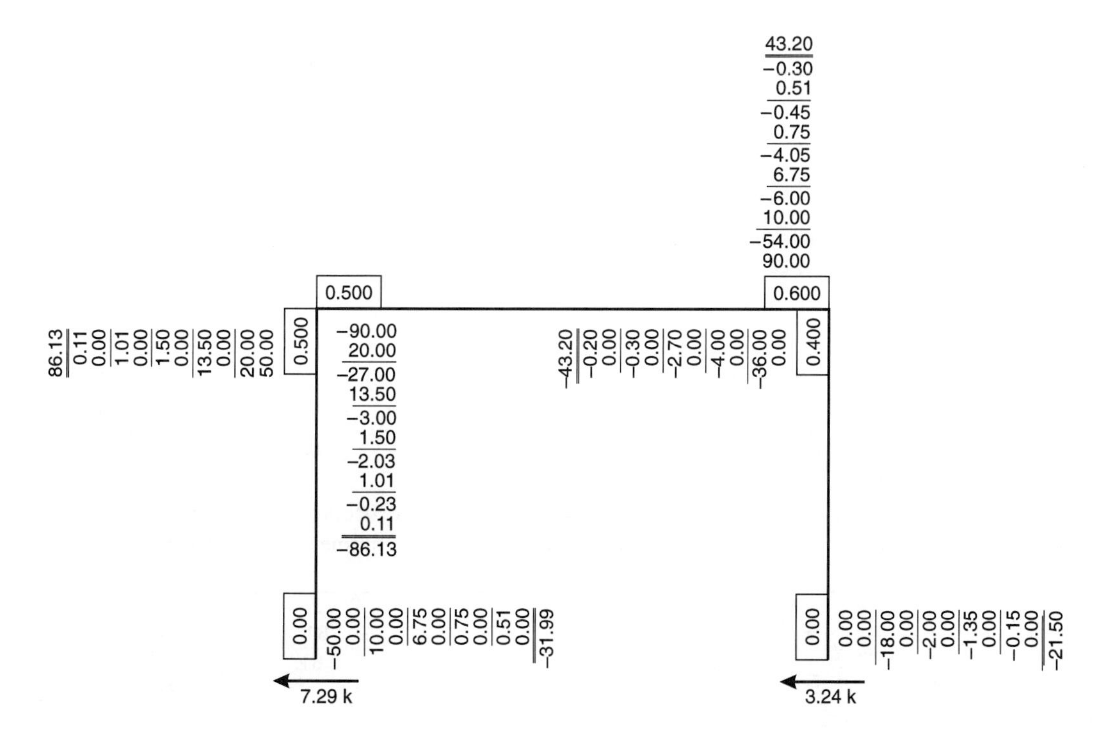

When calculating the shearing forces at the ends of the members, we must include any forces that are acting on the members. The left column has a concentrated load acting at midspan. The shearing force at the bottom of the left column, then, is

$$V_{\text{Left}} = \frac{-86.13 + 31.99 - 20(10)}{20} = 7.29 \text{ k} \leftarrow$$

From these results we see that the imaginary support, the virtual support, exerts a force of 10.53 kips acting to the left. That is the force necessary for static equilibrium in the horizontal direction, including the 20 kip lateral load. The assumed sidesway moments are determined as follows:

$$\frac{M_1}{M_2} = \frac{\frac{I_1}{L_1^2}}{\frac{I_2}{L_2^2}} = \frac{\frac{400}{20^2}}{\frac{266.7}{20^2}} = \frac{1.0}{0.667} \qquad \text{Let} \begin{cases} M_1 = 100 \text{ k-ft} \\ M_2 = 66.7 \text{ k-ft} \end{cases}$$

Next, distribute these assumed sidesway moments:

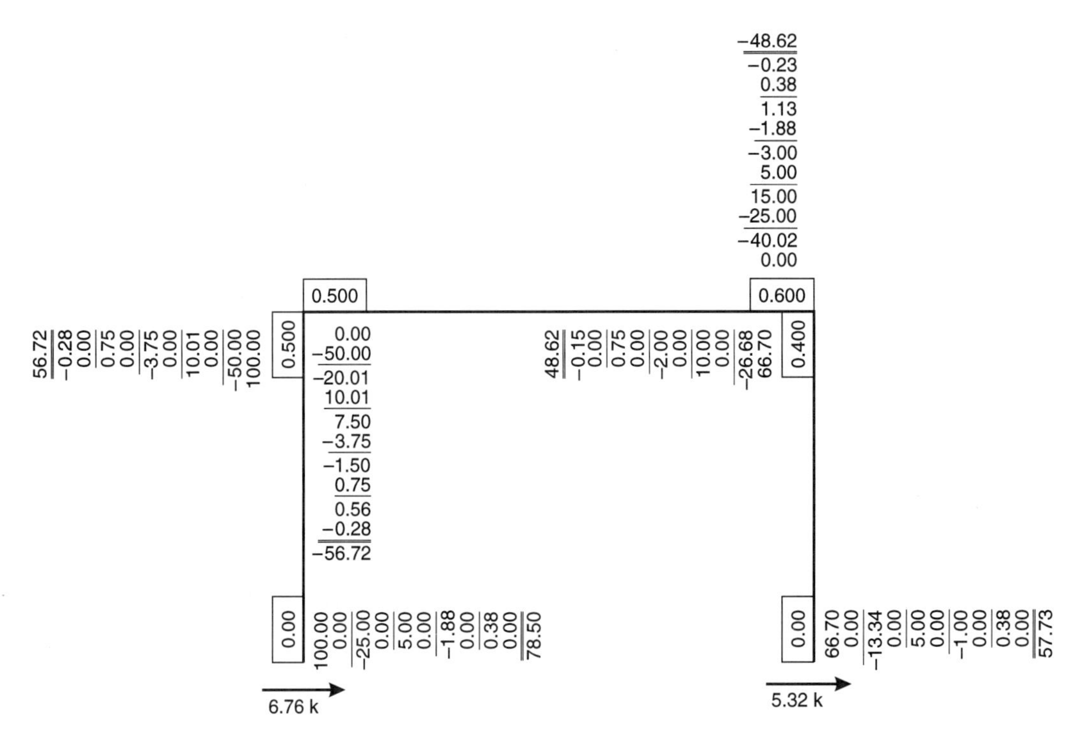

The factor by which the assumed sidesway moments are multiplied by is computed from:

$$S = -\frac{20 - 7.29 - 3.24}{6.76 + 5.32} = -0.784$$

The final member-end moments are equal to the distributed fixed-end moments minus 0.784 times the distributed assumed sidesway moments. The results are

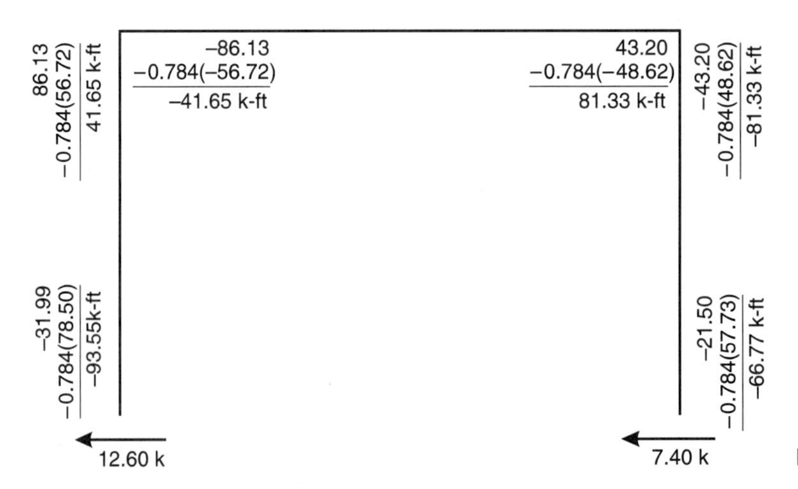

■

18.4 FRAMES WITH SLOPING LEGS

The frames considered up to this point have been made up of vertical and horizontal members. Earlier in this chapter, we proved that when sidesway occurs in such frames, it causes fixed-end moments in the columns proportional to their stiffness and the sidesway, namely $6EI\Delta/L^2$. Because $6E\Delta$ was constant for these frames, moments were assumed proportional to their value of I/l^2. Furthermore, lateral swaying did not produce fixed-end moments in the beams, in the horizontal members.

I-91 bridge, Lyndon, Vermont (Courtesy of the Vermont Agency of Transportation)

The sloping-leg frame of Figure 18.5 can be analyzed in much the same manner as the vertical-leg frames previously considered. The fixed-end moments due to the external loads are calculated and distributed, the horizontal reactions are computed, and the horizontal force imbalance is computed.

Under load, this frame deforms as indicated by the dashed line on Figure 18.5. The relative displacements of the joints in the frame were determined from the geometry of the frame, and by assuming there is no axial deformation and that the displacements are small relative to the frame. As before, the displacement of the joints causes

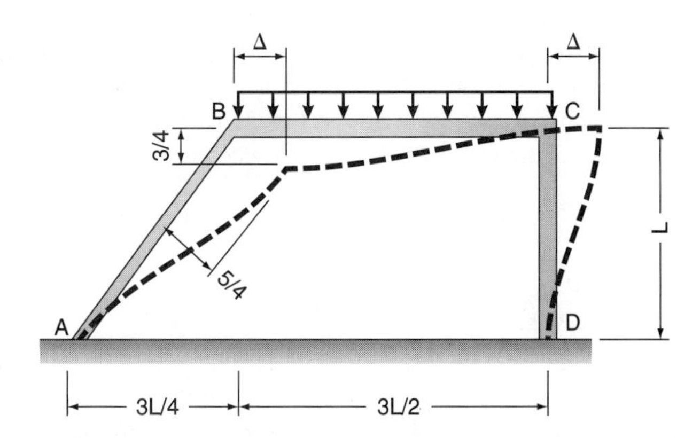

Figure 18.5 A frame with a sloping leg and its deflected shape

TABLE 18.1 COMPUTED RELATIVE SIDESWAY MOMENTS FOR EXAMPLE FRAME WITH SLOPING LEGS

Member	Relative Deformation	Relative Moment	Assumed Sidesway Moment
AB	$\frac{5\Delta}{4}$	$\left(\frac{5\Delta}{4}\right)\left(\frac{6EI}{L^2}\right)_{AB}$	$M_{CD}\left[\frac{\left(\frac{5}{4}\right)\left(\frac{EI}{L^2}\right)_{AB}}{\left(\frac{EI}{L^2}\right)_{CD}}\right]$
BC	$-\frac{3\Delta}{4}$	$\left(-\frac{3\Delta}{4}\right)\left(\frac{6EI}{L^2}\right)_{BC}$	$M_{CD}\left[\frac{\left(-\frac{3}{4}\right)\left(\frac{EI}{L^2}\right)_{AB}}{\left(\frac{EI}{L^2}\right)_{CD}}\right]$
CD	Δ	$\Delta\left(\frac{6EI}{L^2}\right)_{CD}$	M_{CD}

additional moments to be introduced into the frame. These sidesway moments, again, are proportional to the values of $6EI\Delta/L^2$ in each of the members. In sloping leg frames, the magnitude of Δ in each of the members is often unequal, as are the values of I and L. Therefore we must consider the entire value of $6EI\Delta/L^2$ when computing the assumed sidesway moments.

The assumed sidesway moments in sloping leg frames are usually established by assigning a value of sidesway moment to one of the members, and then computing the sidesway moments in Table 18.1 other members proportionally. This approach is illustrated in Figure 18.5. We begin by assuming the sidesway moment in member CD. The other assumed sidesway moments are computed so they are in proportion to this assumed moment. Other methods for establishing the assumed sidesway moments can be used, but the result must be the same.

These assumed moments are distributed, the horizontal reactions are computed, and the necessary moments needed for balancing are calculated and superimposed on the distributed fixed-end moments. This is done in exactly the same manner as was illustrated in the previous example.

One word of caution, however, is in order. The same procedure used for determining the horizontal reaction components at the bases of the sloping columns is used for the vertical columns. That is, a free-body diagram is drawn for each column and the shearing forces and axial forces are computed. When drawing the free-body diagrams for the members, be very careful to include all of the forces that are acting on the member. For vertical columns because the vertical reactions can be neglected as they pass through the points where moments are taken. This is not the case for sloping columns. Consider, for example, member AB of the frame in Figure 18.5. The free-body diagram for this member is shown in Figure 18.6. Observe that whether we sum moments about the top or the

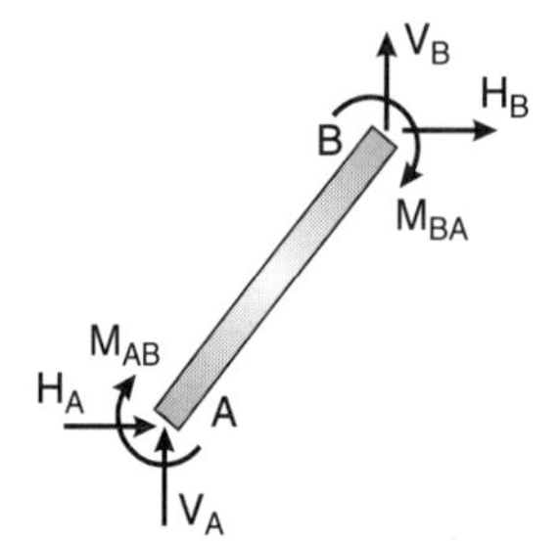

Figure 18.6 Forces acting at the ends of a general sloping member

bottom of the member, the horizontal and vertical forces at the other end will be included in the summation of moments equation; neither passes through the point about which moments are being summed.

From the analyses considered in this chapter it is obvious that frame members are subject to axial forces, and thus axial deformations, as well as to bending moments and shearing forces. The effects of axial deformations, although usually negligible, are neglected in the moment distribution procedures described herein.

18.5 MULTISTORY FRAMES

The frame shown in Figure 18.7(a), acted upon the loads P_1 and P_2, can sway in two fundamental ways, which are illustrated in Figure 18.7(b) and Figure 18.7(c). The loads P_1 and P_2 will cause both floors of the structure to displace to the right. How much displacement is going to occur in the top floor and how much will occur in the bottom floor is not known, however. Both of these sidesway conditions need to be considered. The actual lateral displacement of the frame is going to be some combination of these two fundamental sway modes.

Analyzing this frame using the usual sidesway procedure involves the following steps:

1. Sidesway moments are assumed for the top story for the sway condition in Figure 18.7(b). These moments are distributed through the entire frame.
2. Sidesway moments are assumed for the bottom story for the sway condition in Figure 18.7(c). These moments are distributed through the entire frame.

We then need two equations of horizontal equilibrium. One equation could be written for the story shear in the top floor:

$$K_1V_{1\ top} + K_2V_{2\ top} = P_1 \qquad \textbf{Eq. 18.6}$$

In this equation, $V_{1\ top}$ are the shearing forces in the upper columns caused by the moments obtained in step 1. Similarly, $V_{2\ top}$ are the shearing forces in the upper columns caused the moments obtained in step 2. A second equation could be written for the story shear in the lower floor:

$$K_1V_{1\ bottom} + K_2V_{2\ bottom} = P_1 + P_2 \qquad \textbf{Eq. 18.7}$$

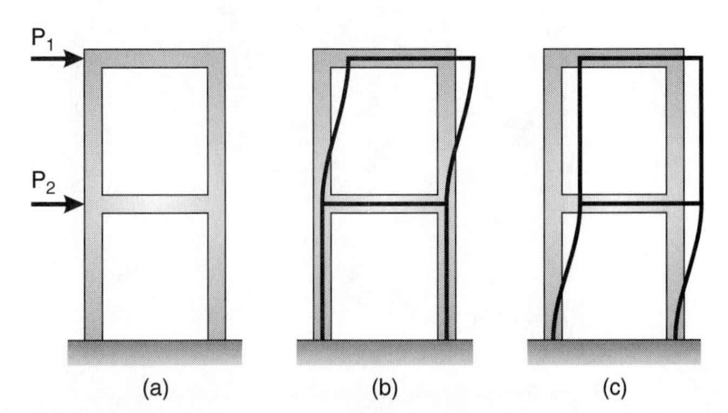

Figure 18.7 Story sway in a two story frame

Solving these two equations simultaneously yields the necessary values of K_1 and K_2. The final moments in the frame equal K_1 times the moments from step 1 plus K_2 times the moments from step 2.

The sidesway method is time-consuming, but not difficult to apply to a two-story frame. For multistory frames, however, it becomes unwieldy because each additional floor introduces another sidesway condition and another simultaneous equation. Since multistory frames will rarely be evaluated using moment distribution in engineering offices today, an example will not be worked.

18.6 ANALYSIS USING COMPUTERS

We continue to see that the amount of work necessary to analyze a structural system by hand is considerable. When using moment distribution with structures that can sway, two or more analyses must be performed. As we have said before, much analysis in the engineering office today is conducted with computers. The hand methods we have discussed, particularly moment distribution, can be used to perform quick checks on the computer solutions. In addition, as we shall see in the next chapters as we study matrix methods, the hand methods can form the basis to determine elemental characteristics for computer solutions.

Office building, 99 Park Avenue, New York City (Courtesy of the American Institute of Steel Construction, Inc.)

In the following example, we continue to develop our ability to perform correct computer analyses, and to demonstrate that the different methods provide the same—or similar—answers. We will analyze the sloping-leg frame that was shown in Figure 18.5. Observe that more information is directly obtained from the computer solutions. When using moment distribution, we only computed the end moments and then had to perform separate calculations to determine the shearing forces in the members. We never did compute the axial forces. In a computer solution, all of this information is obtained at the same time.

EXAMPLE 18.5

Determine the end moments, shearing forces and axial forces in the frame. This same frame was discussed in Section 18.4. As with the other computer examples, the input file is contained on the CD-ROM included with this book.

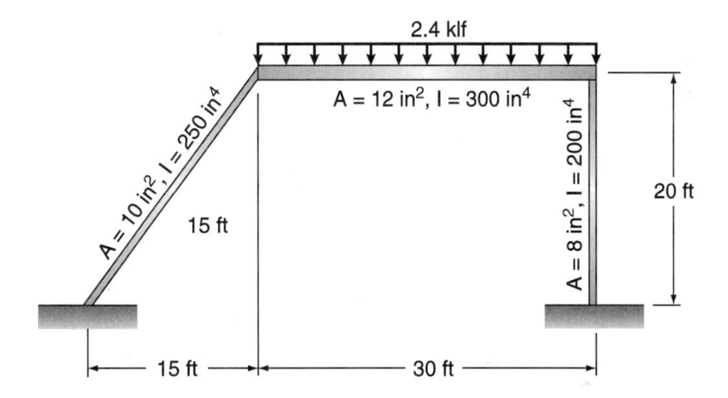

Solution. The computer model that we will use for the analysis is shown in the next figure. This is a rigid frame for which there are no releases at the ends of the members as we previously had with trusses.

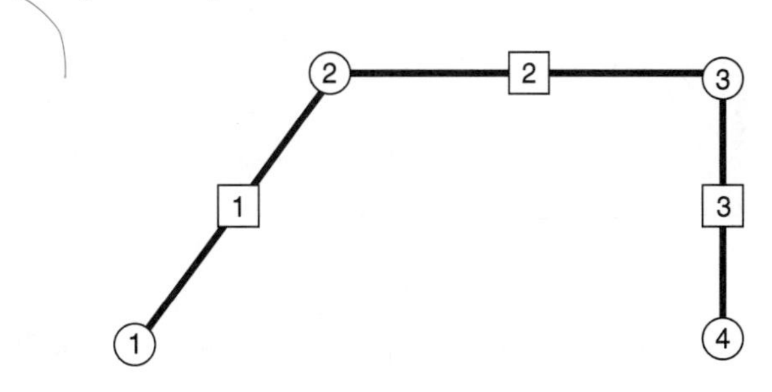

The results of the analysis are summarized in the following table:

		Beginning		End	
Member	P	M	V	M	V
1	−35.24	−667.07	−1.59	−201.41	−1.59
2	−19.88	−201.41	−29.14	−2670.04	42.86
3	−42.86	2099.98	19.88	−2067.04	19.88

■

18.7 PROBLEMS FOR SOLUTION

Balance moments and calculate horizontal reactions at bases of columns for Problems 18.1 through 18.6.

18.1

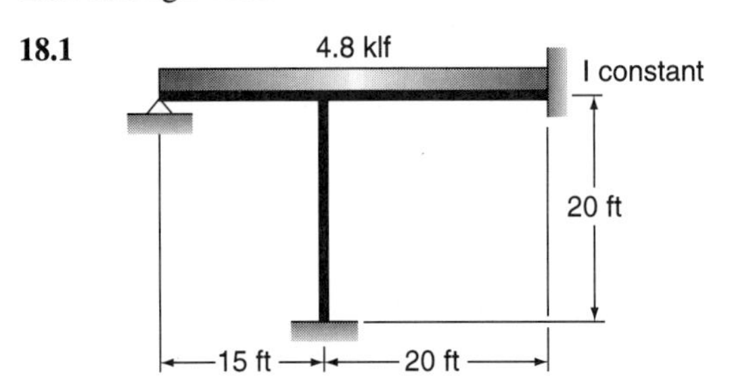

18.2 (*Ans.* $M_B = 14.44$ ft-k, $M_D = 15.12$ ft-k, $H_B = 2.2$ k)

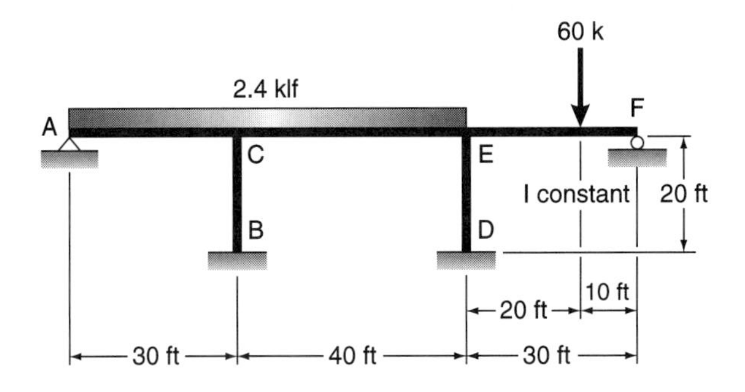

18.3

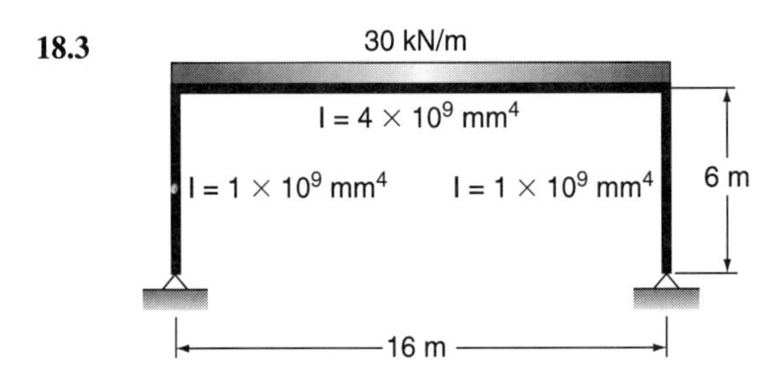

18.4 (*Ans.* $M_A = -172.4$ ft-k, $M_D = 105.1$ ft-k, $H_L = 33.37$ k, $H_R = 33.37$ k)

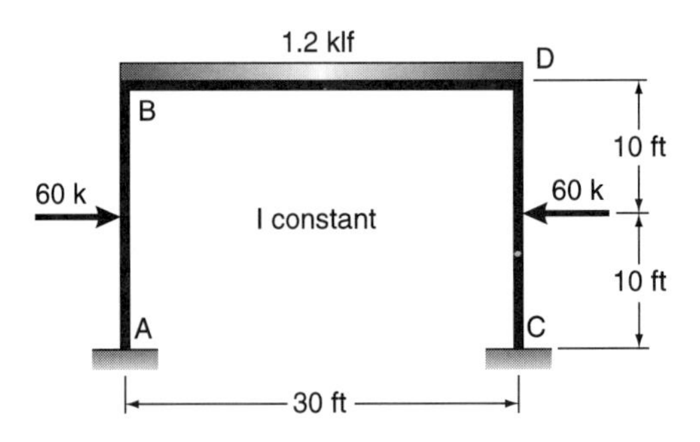

18.5

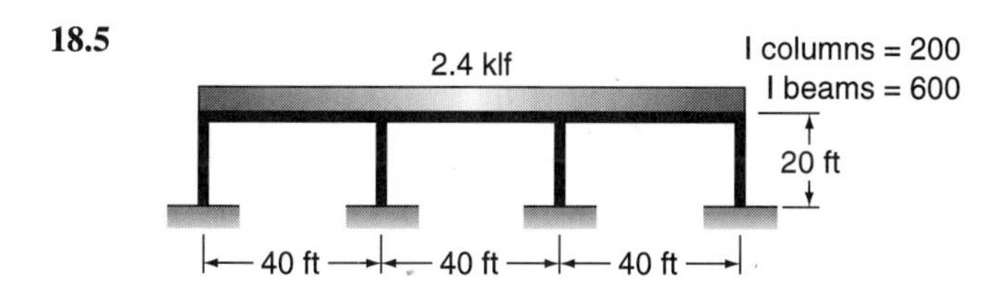

18.6 (*Ans.* $M_B = 389.7$ kN-m, $M_D = 537.6$ kN-m, $M_E = 39.6$ kN-m)

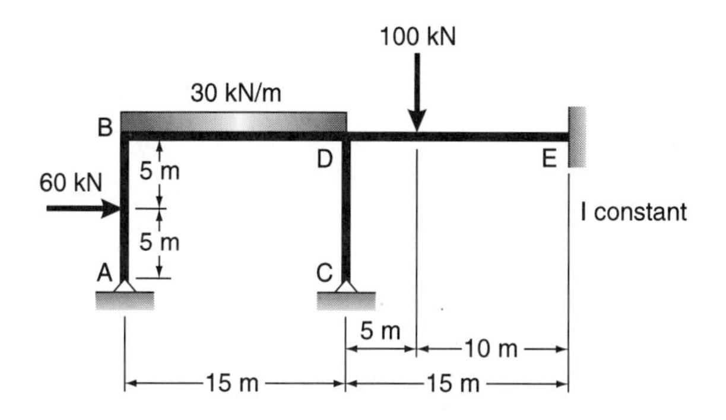

18.7

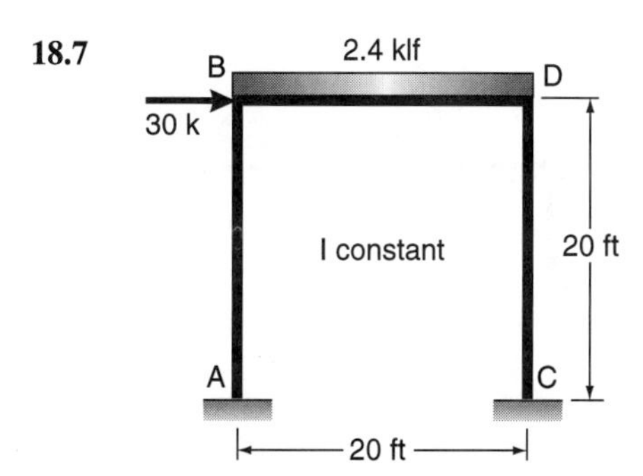

18.8 (*Ans.* $M_A = 101.3$ ft-k, $M_D = 179.9$ ft-k)

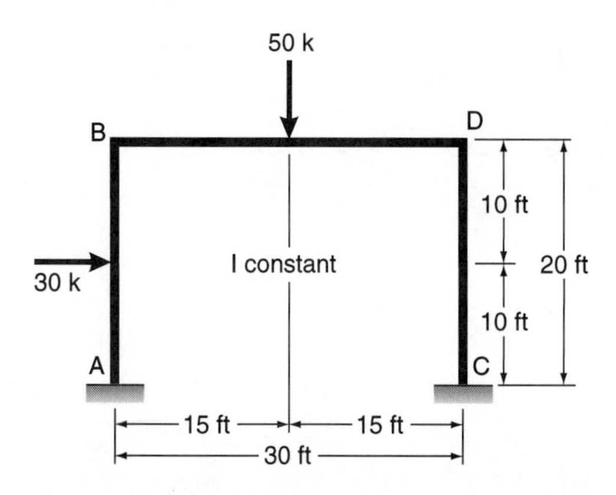

18.9

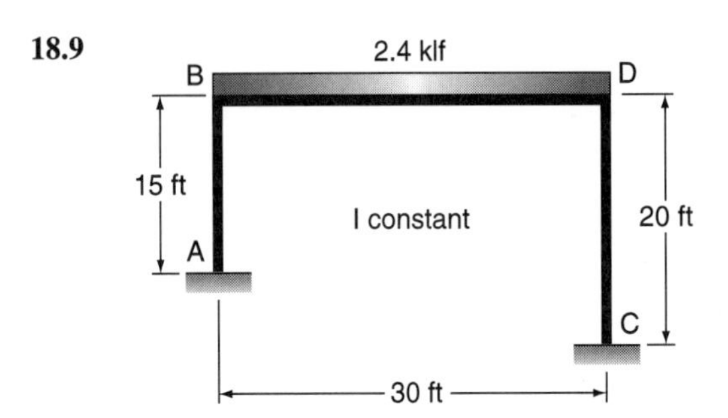

18.10 Rework Problem 19.8 if column bases are pinned. (*Ans.* $M_B = 2.9$ ft-k, $M_D = 297$ ft-k)

18.11 Rework Problem 19.7 if column CD is pinned at its base.

18.12 (*Ans.* $M_A = 22.9$ ft-k, $M_B = 305.0$ ft-k, $M_D = 315.6$ ft-k)

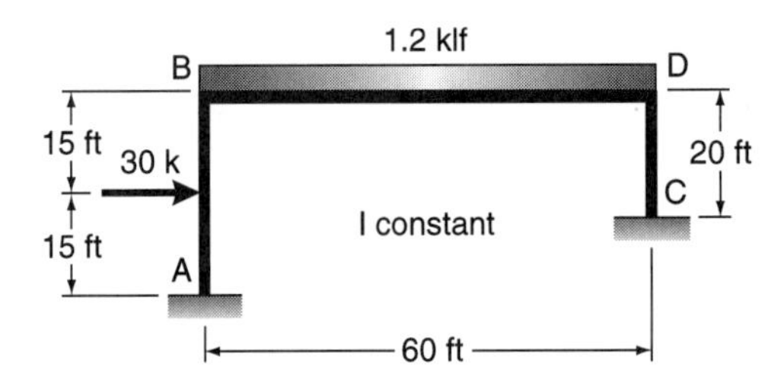

18.13

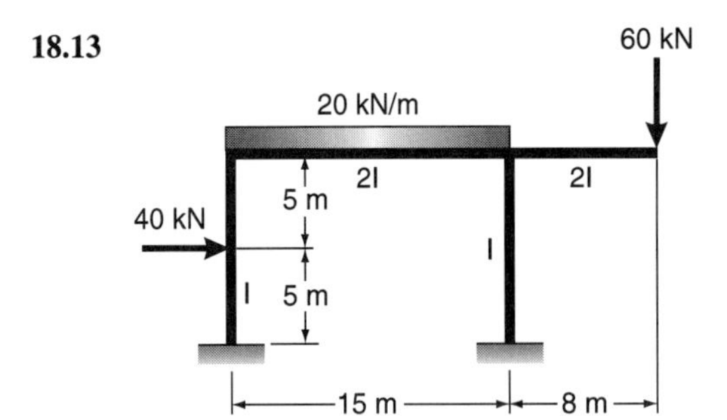

18.14 (*Ans.* $M_A = 221.2$ ft-k, $M_{DF} = 426.8$ ft-k, $H_A = 24.92$ k →).

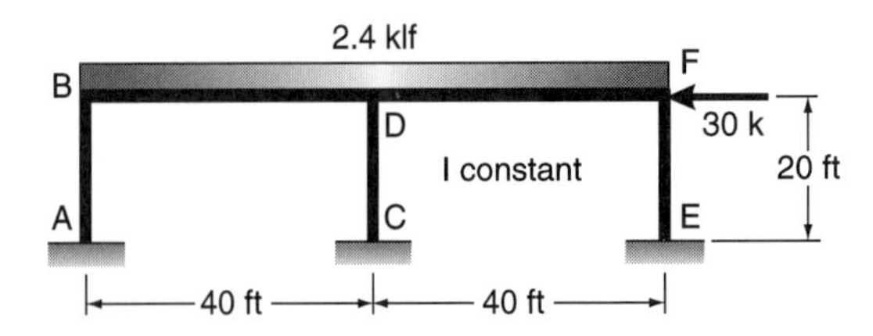

18.15

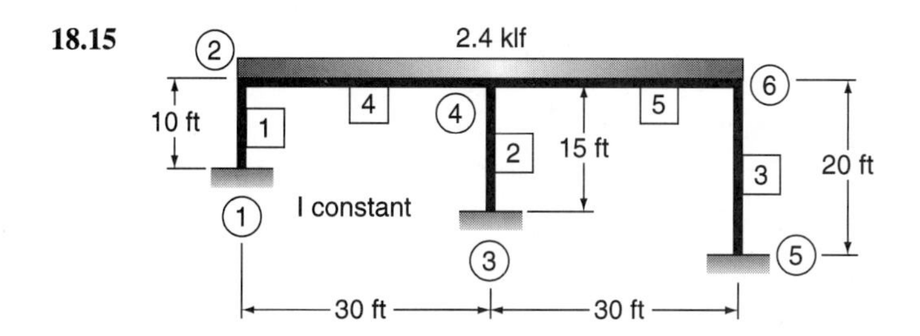

18.16 (*Ans*. M_A = 147.8 ft-k, M_{CD} = 83.5 ft-k, M_D = 133.3 ft-k, M_E = 102.8 ft-k)

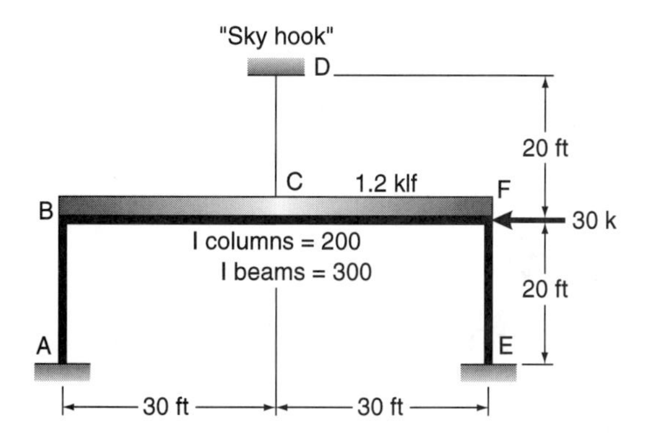

18.17

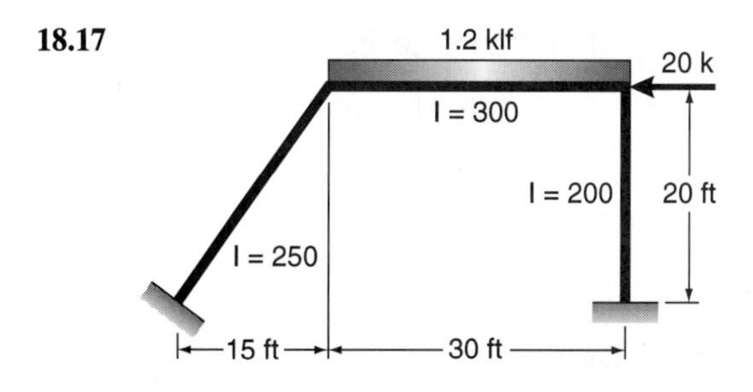

18.18

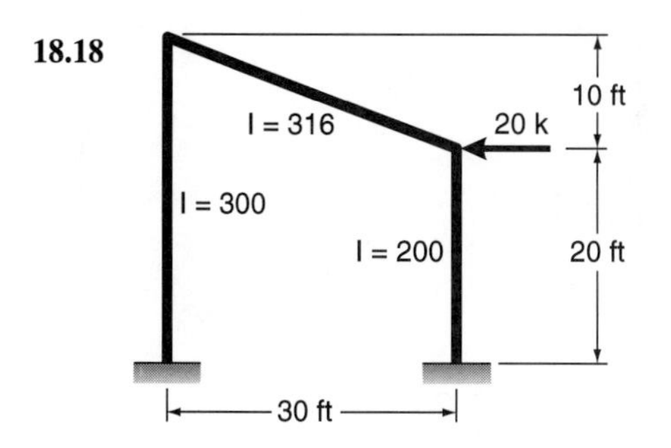

18.19 (*Ans.* $M_A = 1.5$ ft-k, $M_B = 7.6$ ft-k, $M_{DF} = 106.3$ ft-k, $M_F = 67.5$ ft-k)

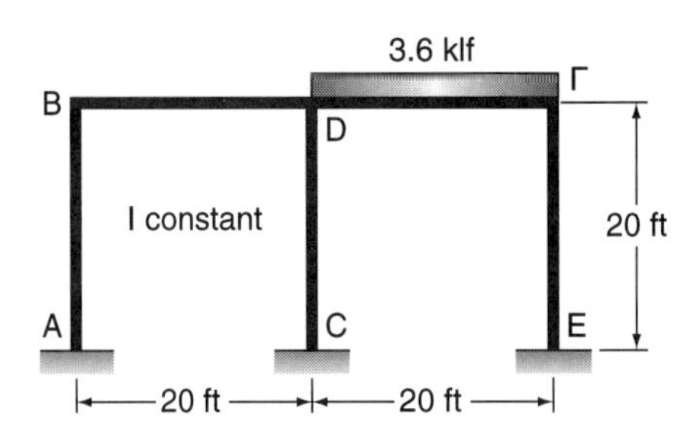

18.20 (*Ans.* $M_{BA} = 151$ ft-k, $M_{BE} = 222$ ft-k, $M_{CF} = 96$ ft-k, $M_{ED} = 94$ ft-k)

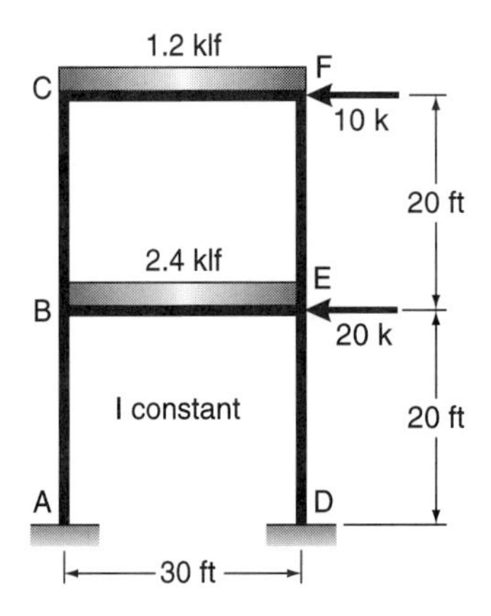

For Problems 18.21 through 18.26, rework the problems specified using SAP2000 or SABLE.

18.21 Problem 18.8 (*Ans.* $M_B = 120$ ft-k, $M_C = 180$ ft-k)

18.22 Problem 18.14

18.23 Problem 18.17

18.24 Problem 18.18

18.25 Problem 18.19

18.26 Assume a large constant cross-sectional area.

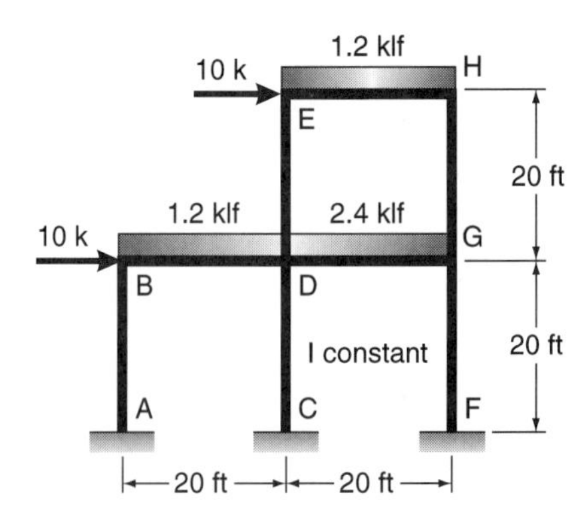

Chapter 19

Introduction to Matrix Methods

19.1 REASONS FOR MATRIX ANALYSIS

During the past three decades, there have been tremendous changes in the structural analysis methods used in engineering practice. These changes have occurred primarily because of the great developments made with high-speed digital computers and the increasing use of very complex structures. Matrix methods of structural analysis are a convenient mathematical representation of a structural system that is easily solved using digital computers. Prior to the advent of the computer, matrix methods were of little use because the equations were too cumbersome for solution using hand-calculation methods.

The change in analysis methods has accelerated during the past decade with the ready availability of very powerful personal computers and workstations. Today a majority of structural analysis is conducted using computer-based methods. Therefore, knowledge of the fundamental principles of matrix structural analysis and an appreciation of the strengths and weaknesses of the analysis procedure are very important for students of structural engineering.

Structural analysis using matrix methods involves no new concepts of structural engineering. We use principles learned in this book and principles learned in courses about mechanics of materials. We will simply organize the procedures and the resulting equations in a format conducive to analysis with a digital computer. The computer is capable of extraordinary feats of arithmetic, but it can do only those tasks that can be described with simple, precise, and unambiguous instructions. In the following chapters, we have attempted to explain the principles of matrix structural analysis and the application of these principles to the analysis of structural systems.

Knowledge of elementary matrix algebra is very important to an understanding of matrix structural analysis because most of the equations involved in the procedures are matrix equations. Although most engineers will have studied matrix algebra at some time, they may not have used it often and their memory of the principles may have become a little fuzzy. For this reason, an appendix about matrix algebra has been included in this textbook for review.

19.2 USE OF MATRIX METHODS

Structural engineers have been attempting to handle analysis problems for many years by applying the mathematical methods used in linear algebra. Although many structures could be analyzed with the resulting equations, the work was extremely tedious, at least until large-scale computers became available. In fact, the usual matrix equations are not manageable with hand-held calculators unless the most elementary structures are involved.

Today matrix analysis (using computers) has substantially replaced the classical methods of analysis in engineering offices. Therefore, engineering educators and writers of structural analysis textbooks are faced with a difficult decision. Should they require a thorough study of the classical methods followed by a study of modern matrix methods; should they require students to study both at the same time in an integrated approach; or should they present only a study of the modern methods? We feel that an initial study of some of the classical methods followed by a study of the matrix methods will result in an engineer who has a better understanding of structural behavior.

Erecting the world's largest radio telescope, Greenbank, West Virginia (Courtesy of Lincoln Electric Company)

Any method of analysis involving linear algebraic equations can be put into matrix notation, then matrix operations can be used to obtain their solution. The possibility of the application of matrix methods by the structural engineer is very important because all linearly elastic, statically determinate and indeterminate structures are governed by systems of linear equations.

The simple numerical examples presented in this chapter and the next could be solved more quickly by classical methods using a pocket calculator rather than by a matrix approach. However, as structures become more complex and as more loading patterns are considered, matrix methods using computers become increasingly useful.

19.3 FORCE AND DISPLACEMENT REPRESENTATIONS

The methods presented in earlier chapters for analyzing statically indeterminate structures can be placed in two general classes—force methods and displacement methods. Both of these methods have been developed to a stage where they can be applied to almost any structure—trusses, beams, frames, plates, shells, for example. The displacement procedures, however, are used more commonly today since they can be programmed more easily for solution by computers. These two methods of analysis, which were previously discussed in Section 12.5, are redefined here because the material presented in the last few chapters will enable the reader to better understand the definitions.

Force Method of Analysis With the force method, also called the *flexibility* or *compatibility method,* redundants are selected and removed from the structure so that a stable and statically determinate structure remains. An equation of deformation compatibility is written at each location where a redundant has been removed. These equations are written in terms of the redundants and the resulting equations are solved for the numerical values of the redundants. After the redundants are determined, equations of static equilibrium can be used to compute all other desired internal forces, moments, and so on. The method of consistent distortions discussed in Chapter 13 is a force method.

Displacement Method of Analysis With the displacement method of analysis, also called the *stiffness* or *equilibrium method,* the displacements of the joints necessary to describe fully the deformed shape of the structure are used in a set of simultaneous equations. When the equations are solved for these displacements, they are substituted into the force-deformation relations of each member to determine the various internal forces. The slope-deflection method discussed in Chapter 15 is a displacement method.

The number of unknowns in the displacement method is generally greater than the number of unknowns in flexibility methods. A greater number of simultaneous equations must therefore be evaluated. Despite this fact, displacement methods are the commonly used method for matrix analysis because representation of the structure is significantly easier. Furthermore, application of the method to statically determinate and statically indeterminate systems is formulated in the same manner. For this reason we will discuss only the displacement method for matrix analysis in this book. Excellent discussions of the flexibility method have been presented by Rubinstein[1] and Przemieniecki[2].

19.4 SOME NECESSARY DEFINITIONS

Before beginning a discussion of matrix methods, an understanding of the following basic terms is necessary.

- *Elements:* the pieces that comprise the structural system that is being represented. In the types of structures that have been analyzed in this book, the elements are the beams and columns.
- *Nodes:* the locations in the structure at which the elements are connected. In structures composed of beams and columns, the nodes usually are the joints.
- *Coordinate:* a possible displacement, a degree of freedom, in a structure at a node.[3] In planar structures there are three coordinates at each node. These possible displacements, or degrees of freedom, are two orthogonal translations and a rotation about an axis perpendicular to the plane defined by the translations. In a space structure, there are six coordinates at each node—three orthogonal translations and three rotations. Please note that a structural coordinate should not be confused with a Cartesian coordinate, which describes the location of a point in space.

[1]Rubinetein, Moshe F., *Matrix Computer Analysis of Structures,* Prentice-Hall, Incorporated, Englewood Cliffs, New Jersey, 1966.

[2]Przemieniecki, J. S., *Theory of Matrix Structural Analysis,* McGraw-Hill, Inc. New York, 1968.

[3]In the context of this book, a structural coordinate and a degree of freedom are really the same. In advanced courses, though, generalized coordinates often are used instead of degrees of freedom. For this reason, we will use the term *coordinate* here.

- *Force:* a general term that refers either to a force acting at a translational coordinate or to a moment acting at a rotational coordinate. There is no distinction as to whether the force is a known structural load or an unknown force of reaction.
- *Displacement:* a general term that refers to a translation at a translational coordinate or to a rotation at a rotational coordinate. There is no distinction as to whether the displacement is an unknown displacement at an unconstrained degree of freedom or a known displacement at a constrained degree of freedom.
- *Stiffness:* the force required to cause a unit deformation in an elastic material.

19.5 THE FUNDAMENTAL CONCEPT

In this book, we have been studying structural systems that behave in a linearly elastic manner. The structures considered also have undergone small displacements. Remember that the undeformed geometry can be used throughout the calculations in a structure undergoing small displacements. The behavior of such a structural system is described by Hooke's law[4], which we recall has the general form

$$F = kx \qquad \textbf{Eq. 19.1}$$

The fundamental equation used in matrix structural analysis, which is analogous to the basic expression of Hooke's law, is

$$\{F\} = [K]\{X\} \qquad \textbf{Eq. 19.2}$$

The basic difference is that now we are dealing with a system of simultaneous linear equations. The term {F} is the force vector and is analogous to the term F in Hooke's law. The force vector contains all of the forces that are acting at the structural coordinates. The vector {X} is the displacement vector and is analogous to the displacement in Hooke's law. It contains the displacements at each of the structural coordinates. Lastly, [K] is the stiffness matrix. It is analogous to the spring constant k in Hooke's law. The stiffness matrix is a representation of the ability of a structural system to resist load. In Section 19.8 we will see how the stiffness matrix for a structure is developed.

Before going on, the concept of stiffness deserves a little thought and discussion. By definition, the constant k for a spring—its stiffness—is the force required to cause a unit displacement. Consider for a moment the spring in Figure 19.1, which has stiffness $K = 50$ lb/in. If we prevent the left side of the spring from displacing and place a 50 lb force on the right side, the right end of the spring will displace one in. to the right. Conversely, to cause a unit displacement at the right, we must apply a force of 50 lbs at that point.

The coefficients in the stiffness matrix [K] follow the same general idea. Each of the coefficients in the matrix is identified by subscripts that indicate the row and column in which it is located. For example, K_{34} is the stiffness coefficient in third row and fourth column of the stiffness matrix. The definition of a stiffness coefficient is

> Stiffness coefficient K_{ij} is the force required at coordinate i to hold a unit displacement at coordinate j and zero displacement at all other coordinates.

[4]A common misconception is that Hooke's law is $\sigma = E\varepsilon$. Hooke was an English scientist who worked with springs. This relationship was derived from his observation for springs, which in Latin is *Ut Tensio sic Vis* (as the force, so the stretch). Nevertheless, $\sigma = E\varepsilon$ is often is referred to as Hooke's law.

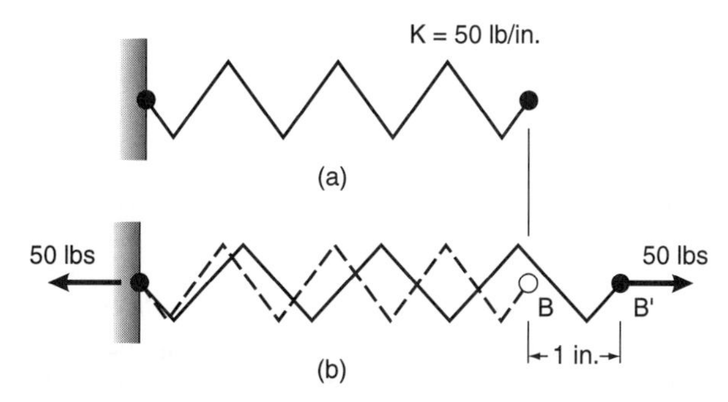

Figure 19.1 A simple spring and the forces necessary to maintain a unit displacement

This definition can be used to develop the stiffness matrix for a structural system. To see how this is done, let's return to the spring in Figure 19.1. We will consider two coordinates along the axis of the spring; one of the coordinates will be located at each end of the spring at the two possible displacement degrees of freedom, as illustrated in Figure 19.2(a). There is a row and column in the stiffness matrix for each coordinate that is being considered in the structure. The stiffness matrix for this spring will have two rows and two columns.

To determine the first column of the stiffness matrix, displace coordinate 1 one unit and prevent displacement at coordinate 2, as shown in Figure 19.2(b). Then determine the forces at all of the coordinates necessary to maintain this configuration. By inspection we see that a force of 50 lbs must be applied to the right at coordinate 1 and a force of 50 lbs must be applied to the left at coordinate 2 to maintain this configuration and satisfy the equations of equilibrium. We now have the first column of the stiffness matrix: K_{11} is the force at coordinate 1 maintain a unit displacement there and zero displacement elsewhere and is equal to 50 lbs/in.; and K_{21} is the force at coordinate 2 maintain a unit displacement at coordinate 1 and zero displacement elsewhere and is equal to -50 lbs/in.

This process is repeated for coordinate 2 in order to generate the second column of the stiffness matrix. If we displace it one unit, as shown in Figure 19.2(c), and hold coordinate 1 to zero displacement, it will be found that K_{12} is equal to -50 lbs/in. and K_{22} is equal to 50 lbs/in. The stiffness matrix, then, for this spring is

$$[k] = \begin{bmatrix} K_{11} & K_{12} \\ K_{21} & K_{22} \end{bmatrix} = \begin{bmatrix} 50 & -50 \\ -50 & 50 \end{bmatrix} \qquad \textbf{Eq. 19.3}$$

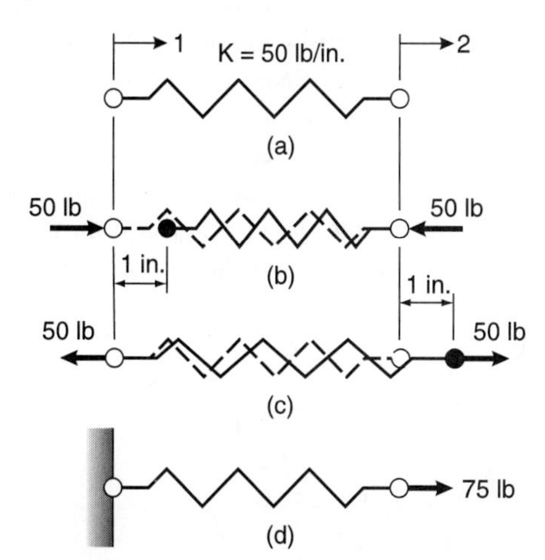

Figure 19.2 A simple spring with two coordinates and the forces necessary to hold a unit displacement at each

Now assume that the spring is used in a structure that is loaded and configured as shown in Figure 19.2(d). There is a force of 75 lbs applied to the right at coordinate 2 and coordinate 1 is constrained to zero displacement. The resulting system equation, in the form of Equation 19.2, becomes

$$\begin{bmatrix} 50 & -50 \\ -50 & 50 \end{bmatrix} \begin{Bmatrix} 0 \\ x_2 \end{Bmatrix} = \begin{Bmatrix} F_1 \\ 75 \end{Bmatrix} \qquad \textbf{Eq. 19.4}$$

Speaking in terms of structures, we would say that coordinate 1 is at a support (there is no displacement) and that the 75 lb force is an applied load. If the system of simultaneous linear equations implied by Equation 19.4 is solved for the unknowns x_2 and F_1 using whatever means is convenient, we find that F_1 is equal to -75 lbs and that x_2 is equal to 1.5 in. The unknown force of reaction is F_1 and x_2 is the unknown structural displacement.

19.6 SYSTEMS WITH SEVERAL ELEMENTS

Structures are rarely composed of single elements such as the single spring in Figure 19.2. There are usually many more elements, but the principles for developing the stiffness matrices are the same. Let's investigate a more complicated spring structure. Consider the three-spring structure shown in Figure 19.3. The system has three elements and four nodes. Each spring has the same stiffness, which is 50 lbs/in. The coordinates we will use are shown in the figure. Because a spring can direct force only along its axis, the only applicable coordinates are those parallel to the axis of the springs. For reasons that will become clear later, we have numbered the unconstrained coordinates first and the constrained coordinates last.

The structural stiffness matrix is developed one column at a time. To develop the stiffness coefficients for the first column, displace coordinate 1 one unit to the right and hold the displacement at all other coordinates to zero. To do this a force of 100 lbs must be applied at coordinate 1: a force of 50 lbs is required to move the right end of the left spring to the right one unit, and a force of 50 lbs is required to move the left end of the middle spring to the right one unit. Because Coordinate 1 was displaced one unit and the other coordinates were not displaced, there is a tensile force of 50 lbs in the left spring and a compressive force of 50 lbs in the middle spring. For equilibrium to be achieved there must be forces acting to the left at coordinates 2 and 3. The right spring did not deform because the coordinates to which this spring was attached did not displace. As such, there are no forces in the system as a result of deformation of the right spring. These forces are shown in Figure 19.3(b). The stiffness coefficients in the first column, then, are: $K_{11} = 100$ lbs/in., $K_{21} = -50$ lbs/in., $K_{31} = -50$ lbs/in., and $K_{41} = 0$.

The stiffness coefficients in the other columns are determined in the same manner as was the first column. The coordinate associated with each column is displaced 1 unit, the other coordinates are held to zero displacement, and the forces necessary to maintain this configuration are computed. After all of the stiffness coefficients have been found, the stiffness matrix for this three-spring structure is

$$[K] = \begin{bmatrix} 100 & -50 & -50 & 0 \\ -50 & 100 & 0 & -50 \\ -50 & 0 & 50 & 0 \\ 0 & -50 & 0 & 50 \end{bmatrix} \qquad \textbf{Eq. 19.5}$$

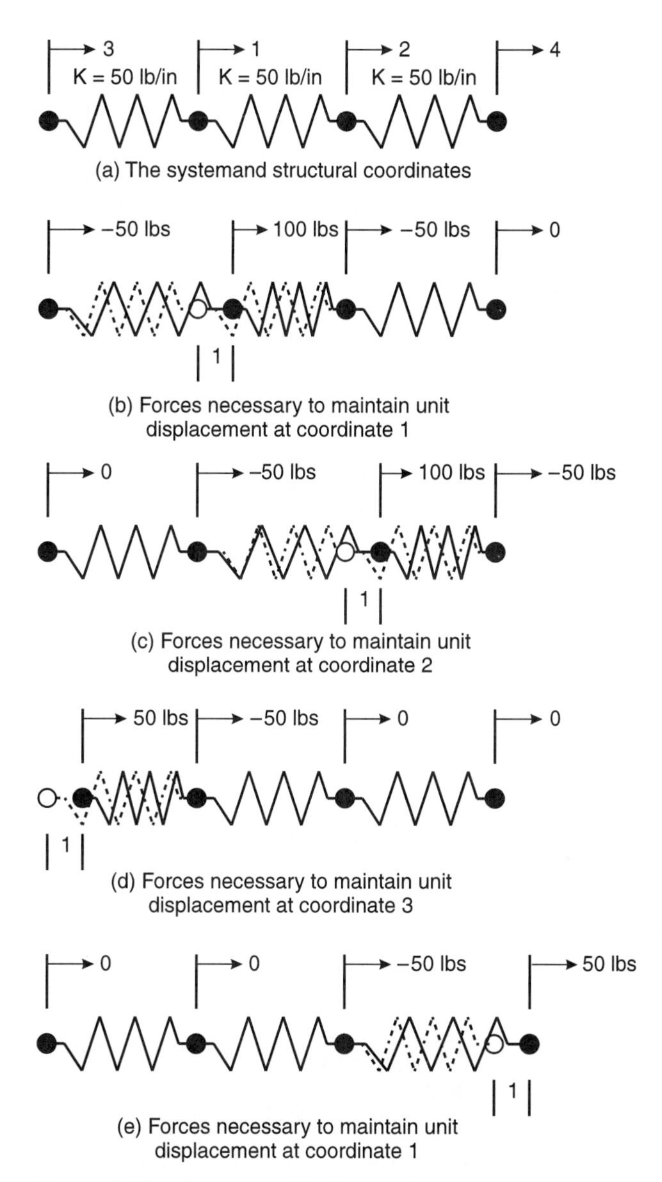

Figure 19.3 A system with three springs and four coordinates

Notice that the stiffness matrix is symmetrical about the main diagonal (the diagonal from upper left to lower right). Also observe that there are no zero terms on the main diagonal at rows or columns associated with unconstrained coordinates, and that the terms on the main diagonal are all positive. These are characteristics of the structural stiffness matrix that will always be true.

In structural analysis, the forces acting at constrained coordinates and the displacements at the unconstrained coordinates are the unknown quantities. The known displacements at constrained coordinates are the boundary conditions and the known forces at the unconstrained coordinates are the structural loads. Taking this example a little further, a force of 100 lbs is applied to the left at coordinate 2. The complete system equation becomes

$$\begin{bmatrix} 100 & -50 & -50 & 0 \\ -50 & 100 & 0 & -50 \\ -50 & 0 & 50 & 0 \\ 0 & -50 & 0 & 50 \end{bmatrix} \begin{Bmatrix} x_1 \\ x_2 \\ 0 \\ 0 \end{Bmatrix} = \begin{Bmatrix} 0 \\ -100 \\ F_3 \\ F_4 \end{Bmatrix} \qquad \textbf{Eq. 19.6}$$

The displacements at coordinates 1 and 2 are not known. The displacements at coordinates 3 and 4 are known because these are the structural supports. Likewise, the forces at coordinates 3 and 4 are not known because these are the structural reactions. There is no applied load at coordinate 1, so F_1 is zero. The force at coordinate 2 is negative because it is acting to the left, which is opposite to the assumed direction of the coordinate.

To solve for the unknowns in Equation 19.6, we need to partition the matrices. The partitioning will be between the constrained and unconstrained coordinates (this is why the unconstrained coordinates were numbered first and the constrained coordinates second). The partitioning lines are shown in Equation 19.7:

$$\left[\begin{array}{cc|cc} 100 & -50 & -50 & 0 \\ -50 & 100 & 0 & -50 \\ \hline -50 & 0 & 50 & 0 \\ 0 & -50 & 0 & 50 \end{array}\right] \left\{\begin{array}{c} x_1 \\ x_2 \\ \hline 0 \\ 0 \end{array}\right\} = \left\{\begin{array}{c} 0 \\ -100 \\ \hline F_3 \\ F_4 \end{array}\right\} \qquad \textbf{Eq. 19.7}$$

A partition of a matrix is another matrix. A partitioned matrix, then, is nothing more than a matrix of matrices. Therefore, Equation 19.7 is really

$$\begin{bmatrix} \begin{bmatrix} 100 & -50 \\ -50 & 100 \end{bmatrix} & \begin{bmatrix} -50 & 0 \\ 0 & -50 \end{bmatrix} \\ \begin{bmatrix} -50 & 0 \\ 0 & -50 \end{bmatrix} & \begin{bmatrix} 50 & 0 \\ 0 & 50 \end{bmatrix} \end{bmatrix} \begin{Bmatrix} \begin{Bmatrix} x_1 \\ x_2 \end{Bmatrix} \\ \begin{Bmatrix} 0 \\ 0 \end{Bmatrix} \end{Bmatrix} = \begin{Bmatrix} \begin{Bmatrix} 0 \\ -100 \end{Bmatrix} \\ \begin{Bmatrix} F_3 \\ F_4 \end{Bmatrix} \end{Bmatrix} \qquad \textbf{Eq. 19.8}$$

A shorthand way for representing this equation is

$$\begin{bmatrix} [K_{11}][K_{12}] \\ [K_{21}][K_{22}] \end{bmatrix} \begin{Bmatrix} \{X_1\} \\ \{X_2\} \end{Bmatrix} = \begin{Bmatrix} \{F_1\} \\ \{F_2\} \end{Bmatrix} \qquad \textbf{Eq. 19.9}$$

There are really two matrix equations represented by Equation 19.9. These equations are

$$[K_{11}]\{X_1\} + [K_{12}]\{X_2\} = \{F_1\} \qquad \textbf{Eq. 19.10}$$

and

$$[K_{21}]\{X_1\} + [K_{22}]\{X_2\} = \{F_2\} \qquad \textbf{Eq. 19.11}$$

The partition $\{X_2\}$ contains all zeros. Therefore, the second term on the left side of Equation 19.10 is equal to zero. The partition $\{X_1\}$ can be solved for from Equation 19.10 as

$$\{X_1\} = \begin{Bmatrix} x_1 \\ x_2 \end{Bmatrix} = [K_{11}]^{-1}\{F_1\} = \begin{Bmatrix} -0.667 \\ -1.333 \end{Bmatrix} \qquad \textbf{Eq. 19.12}$$

This result can be substituted into Equation 19.11 and the unknown forces, $\{F_2\}$, the forces of reaction, can be found to be

$$\{F_2\} = \begin{Bmatrix} F_3 \\ F_4 \end{Bmatrix} = \begin{Bmatrix} 33.33 \\ 66.67 \end{Bmatrix} \quad \textbf{Eq. 19.13}$$

The analysis for this more complicated structure is now complete. We determined values for all of the forces and displacements that were initially unknown.

19.7 BARS INSTEAD OF SPRINGS

In reality, structures are not composed of springs. They are composed of beams, bars, and columns. What would be the result if the system shown in Figure 19.3 contained bars instead of the springs? Such a system is shown in Figure 19.4.

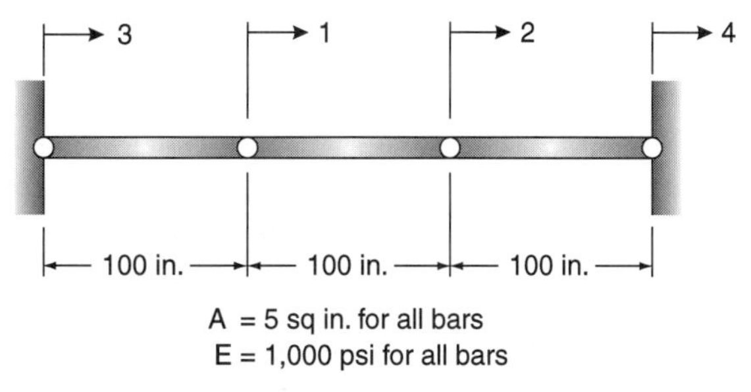

Figure 19.4 A simple structure composed of three bars

To evaluate this system, we need to determine the stiffness of each of the bars. This can be done using the equation for the deformation of a prismatic axially loaded bar, which we recall to be

$$\delta = \left(\frac{L}{AE}\right)P \quad \textbf{Eq. 19.14}$$

Equation 19.14 can be rearranged to resemble Equation 19.1.

$$\left(\frac{AE}{L}\right)\delta = P \quad \textbf{Eq. 19.15}$$

In this form, we observe that the term (AE/L) is analogous to the spring constant in Hooke's law. Indeed, (AE/L) is the stiffness for an axial bar. For the bars in Figure 19.4, we find that their stiffness is found to equal

$$k = \frac{5(1000)}{100} = 50 \text{ lbs/in.} \quad \textbf{Eq. 19.16}$$

This is the same value that we had used for the spring stiffness in Section 19.6. Because of that, the stiffness matrix, the equations used for analysis, and the results obtained are the same as we computed in that section.

19.8 SOLUTION FOR A TRUSS

Before finishing this chapter, let's evaluate a structure that looks a little more like the other structures we have analyzed. Consider the truss in Figure 19.5. This is a simple two-bar truss that has forces acting at the unconstrained node. The coordinates to be used in the

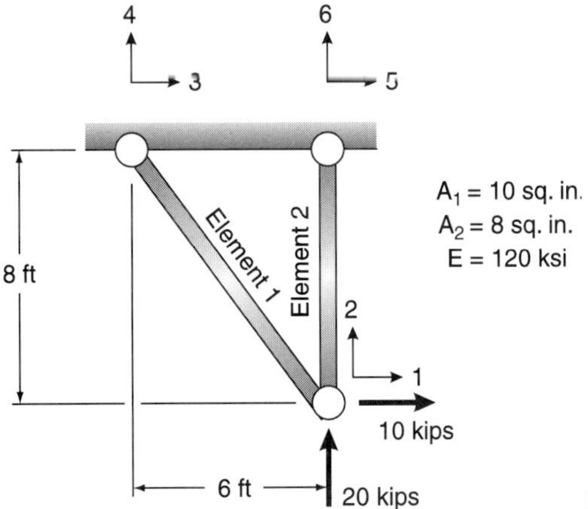

Figure 19.5 A two-bar truss

analysis and the member properties are shown in the figure. This is a pin-connected truss, so there are only two applicable coordinates at each node. The solution begins by first determining the stiffness of each member in the structure. From Equation 19.15, we find those to be

$$k_1 = \frac{10(120)}{10(12)} = 10 \text{ k/in.}$$
$$k_2 = \frac{8(120)}{8(12)} = 10 \text{ k/in.} \qquad \textbf{Eq. 19.17}$$

The procedures for building the stiffness matrix are a little more involved now than they were for the simpler structures in the previous sections. Regardless, the process for building the stiffness matrix is the same. We begin by displacing the first coordinate one unit and holding the other coordinates to zero displacement. This configuration is shown in Figure 19.6. Notice that element 1 has undergone a rigid body rotation and has elongated. The amount by which the bar has been elongated is 1 $\sin\theta$. Recall that we are considering small displacements so the undeformed geometry can be used in calculations with the deformed geometry. The force[5] in bar 1 caused by this elongation,

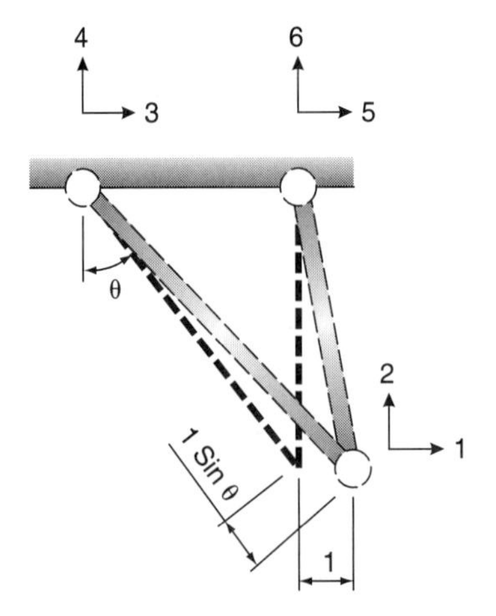

Figure 19.6 Geometry of two-bar truss when coordinate 1 is displaced one unit

[5]Here we use the notation ${}^{l}F_m$ where *m* indicates the coordinate that was displaced one unit and *l* indicates the bar being considered.

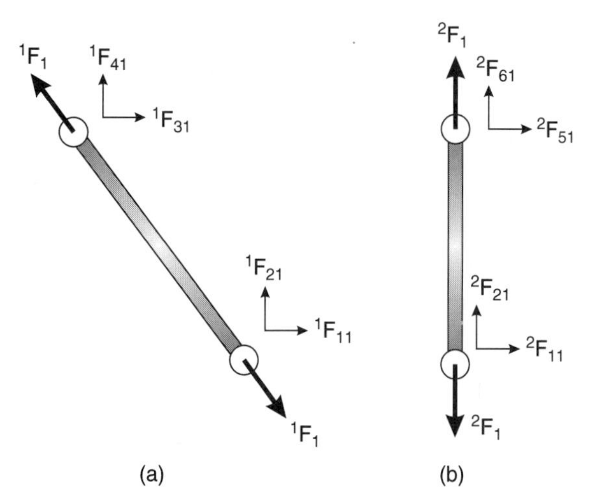

Figure 19.7 Contributions to stiffness at each coordinate when coordinate 1 is displaced one unit

then, is

$$^1F_1 = k_1 \sin\theta = 10(0.6) = 6 \text{ kips} \qquad \textbf{Eq. 19.18}$$

This is the force in the bar for every inch of displacement of coordinate 1. Components of this force are acting at each of the coordinates to which that bar is attached, as shown in Figure 19.7. The contribution to the stiffness of coordinate 1 from this bar, then, is as follows:

$$\begin{aligned}
^1F_{11} &= {^1F_1} \sin\theta = 6(0.6) = 3.6 \text{ k/in.} \\
^1F_{21} &= -{^1F_1} \cos\theta = 6(0.8) = -4.8 \text{ k/in.} \\
^1F_{31} &= -{^1F_1} \sin\theta = 6(0.6) = -3.6 \text{ k/in.} \\
^1F_{41} &= {^1F_1} \cos\theta = 6(0.8) = 4.8 \text{ k/in.}
\end{aligned} \qquad \textbf{Eq. 19.19}$$

Element 2 has only undergone a rigid body rotation; it has not changed length. As such, there is no force in the bar. Therefore element 2 does not contribute to the stiffness of coordinate 1. Column 1 of the stiffness matrix is simply the contribution from Element 1, which is:

$$\begin{aligned}
K_{11} &= {^1F_{11}} + {^2F_{11}} = 3.6 + 0 = 3.6 \text{ k/in.} \\
K_{21} &= {^1F_{21}} + {^2F_{21}} = -4.8 + 0 = -4.8 \text{ k/in.} \\
K_{31} &= {^1F_{41}} = -3.6 \text{ k/in.} \\
K_{41} &= {^1F_{41}} = 4.8 \text{ k/in.} \\
K_{51} &= {^2F_{51}} = 0 \text{ k/in.} \\
K_{61} &= {^2F_{61}} = 0 \text{ k/in.}
\end{aligned} \qquad \textbf{Eq. 19.20}$$

The second column of the stiffness matrix is determined by displacing only coordinate 2 one unit and determining the forces necessary to maintain this configuration. The geometry of this situation is shown in Figure 19.8. As before, the first bar has undergone a rigid body rotation and has become shorter by an amount 1 cos θ. The force in the bar is

$$^1F_2 = k_1 \cos\theta = 10(0.8) = 8 \text{ kips} \qquad \textbf{Eq. 19.21}$$

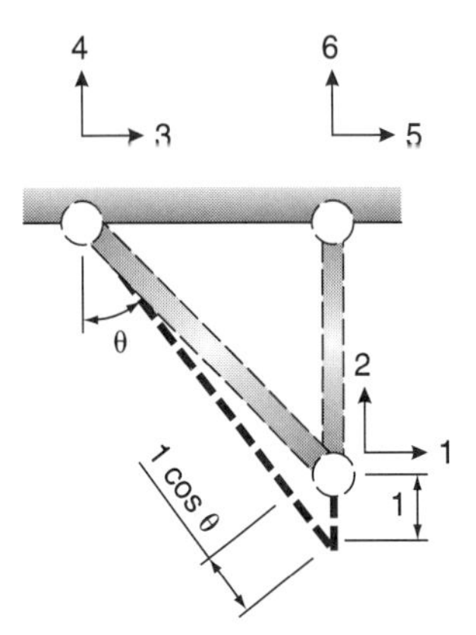

Figure 19.8 Geometry when coordinate 2 is displaced one unit

Again, this is the force in bar 1 for every unit of displacement at coordinate 2. As before, the forces at the coordinates to which the bar is attached are components of this compressive force, shown in Figure 19.9. The contribution to the stiffness of coordinate 2 caused by bar 1 is

$$
\begin{aligned}
{}^1K_{12} &= {}^1F_2 \sin\theta = -8(0.6) = -4.8 \text{ k/in.} \\
{}^1K_{22} &= {}^1F_2 \cos\theta = 8(0.8) = 6.4 \text{ k/in.} \\
{}^1K_{32} &= {}^1F_2 \sin\theta = 8(0.6) = 4.8 \text{ k/in.} \\
{}^1K_{42} &= -{}^1F_2 \cos\theta = 8(0.8) = -6.4 \text{ k/in.}
\end{aligned}
\qquad \textbf{Eq. 19.22}
$$

When coordinate 2 was displaced one unit, bar 2 became shorter by one unit. The force in bar 2, then, caused by the unit displacement of coordinate 2 is

$$ {}^2F_2 = 10(1) = 10 \text{ kips} \qquad \textbf{Eq. 19.23} $$

As with bar 1, this is the force in bar 2 for every unit of deformation of coordinate 2. The contributions to the stiffness at coordinate 2 from bar 2 are the components of this force in the coordinate directions, as shown on Figure 19.9. The components of

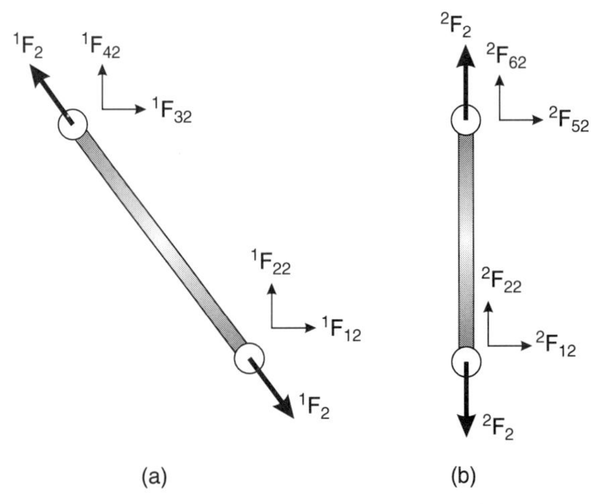

Figure 19.9 Contributions to stiffness at each coordinate when coordinate 2 is displaced one unit

force are

$$\begin{aligned} {}^2K_{12} &= 0 \text{ k/in.} \\ {}^2K_{22} &= {}^2F_2 = 10 \text{ k/in.} \\ {}^2K_{52} &= 0 \text{ k/in.} \\ {}^2K_{62} &= -{}^2F_2 = -10 \text{ k/in.} \end{aligned} \qquad \textbf{Eq. 19.24}$$

The total stiffness at coordinate 2 is the combination of the stiffness provided by all of the components. This result is

$$\begin{aligned} K_{12} &= -4.8 + 0 = -4.8 \text{ k/in.} \\ K_{22} &= 6.4 + 10 = 16.4 \text{ k/in.} \\ K_{32} &= 4.8 + 0 = 4.8 \text{ k/in.} \\ K_{42} &= -6.4 + 0 = -6.4 \text{ k/in.} \\ K_{52} &= 0 + 0 = 0 \text{ k/in.} \\ K_{62} &= 0 - 10 = -10 \text{ k/in.} \end{aligned} \qquad \textbf{Eq. 19.25}$$

The same procedures as these are followed for the other four coordinates. The resulting system equation is shown in Equation 19.26 along with the applied loads and the displacement boundary conditions.

$$\left[\begin{array}{cc|cccc} 3.6 & -4.8 & -3.6 & 4.8 & 0 & 0 \\ -4.8 & 16.4 & 4.8 & -16.4 & 0 & -10 \\ \hline -3.6 & 4.8 & 3.6 & -4.8 & 0 & 0 \\ 4.8 & -6.4 & -4.8 & 6.4 & 0 & 0 \\ 0 & 0 & 0 & 0 & 0 & 0 \\ 0 & -10 & 0 & 0 & 0 & -10 \end{array}\right] \left\{\begin{array}{c} x_1 \\ x_2 \\ \hline 0 \\ 0 \\ 0 \\ 0 \end{array}\right\} = \left\{\begin{array}{c} 10 \\ 20 \\ \hline F_3 \\ F_4 \\ F_5 \\ F_6 \end{array}\right\} \qquad \textbf{Eq. 19.26}$$

Using Equations 19.10 and 19.11, the unknown forces and displacements are found to be

$$\left\{\begin{array}{c} F_3 \\ F_4 \\ F_5 \\ F_6 \end{array}\right\} = \left\{\begin{array}{c} -9.97 \\ 13.29 \\ 0 \\ -33.3 \end{array}\right\} \quad \text{and} \quad \left\{\begin{array}{c} x_1 \\ x_2 \end{array}\right\} = \left\{\begin{array}{c} 7.21 \\ 3.33 \end{array}\right\} \qquad \textbf{Eq. 19.27}$$

Using principles learned previously in this textbook, the interested student could easily demonstrate that these are indeed the correct displacements and forces of reaction for this structure. The analysis of trusses containing more members would be handled in the same manner as the truss that we have just analyzed.

19.9 SYSTEM MATRICES USING STRAIN ENERGY

The system stiffness matrix could have been prepared through use of strain energy. Consider the single spring that was shown in Figure 19.2. The strain energy in the spring in terms of the displacement at the coordinates is

$$U = \frac{k}{2}(x_2 - x_1)^2 \qquad \textbf{Eq. 19.28}$$

Using Castigliano's first theorem we can show that

$$\frac{\partial U}{\partial x_1} = F_1 = \frac{k}{2}(2x_1 - 2x_2)$$
$$\frac{\partial U}{\partial x_2} = F_2 = \frac{k}{2}(2x_2 - 2x_1) \qquad \textbf{Eq. 19.29}$$

These two equations can be expressed in matrix form as

$$\begin{Bmatrix} F_1 \\ F_2 \end{Bmatrix} = \begin{bmatrix} k & -k \\ -k & k \end{bmatrix} \begin{Bmatrix} x_1 \\ x_2 \end{Bmatrix} \qquad \textbf{Eq. 19.30}$$

If the stiffness of the spring is 50 lbs/in. and the force at F_2 is 75 lbs, this is the same equation as given in Equation 19.4.

Strain energy and Castigliano's first theorem can be used for systems composed of multiple elements. Consider the spring system that was shown in Figure 19.3 as an example. The strain energy in this system in terms of displacements at the coordinates is

$$U = \frac{k_1}{2}(x_1 - x_3)^2 + \frac{k_2}{2}(x_2 - x_1)^2 + \frac{k_3}{2}(x_4 - x_2)^2 \qquad \textbf{Eq. 19.31}$$

Using Castigliano's first theorem, we can use this expression for strain energy in order to obtain the force at each of the structural coordinates. The result is the following four equations:

$$\frac{\partial U}{\partial x_1} = F_1 = \frac{k_1}{2}(2x_1 - 2x_3) + \frac{k_2}{2}(2x_1 - 2x_2)$$
$$\frac{\partial U}{\partial x_2} = F_2 = \frac{k_2}{2}(2x_2 - 2x_1) + \frac{k_3}{2}(2x_2 - 2x_4)$$
$$\frac{\partial U}{\partial x_3} = F_3 = \frac{k_1}{2}(2x_3 - 2x_1) \qquad \textbf{Eq. 19.32}$$
$$\frac{\partial U}{\partial x_4} = F_4 = \frac{k_3}{2}(2x_4 - 2x_2)$$

If these equations are expressed in matrix form, we obtain the system equation

$$\begin{Bmatrix} F_1 \\ F_2 \\ F_3 \\ F_4 \end{Bmatrix} = \begin{bmatrix} k_1 + k_2 & -k_2 & -k_1 & 0 \\ -k_2 & k_2 + k_3 & 0 & -k_3 \\ -k_1 & 0 & k_1 & 0 \\ 0 & -k_3 & 0 & k_3 \end{bmatrix} \begin{Bmatrix} x_1 \\ x_2 \\ x_3 \\ x_4 \end{Bmatrix} \qquad \textbf{Eq. 19.33}$$

Again, notice that this is the same result as obtained in Equation 19.6 if we substitute for the stiffness of the spring, the loads, and the boundary conditions.

Before leaving this discussion, there are two observations that should be made. First, as can be seen in Equation 19.31, each component in the system contributes to the total strain energy in the system. The first term on the right side of that equation is the contribution of the left spring, the second term is the contribution of the middle spring, and the last term is the contribution of the right spring.

Secondly, from Equations 19.32, and the representation of those equations in Equation 19.33, we see that each component connected to a structural coordinate contributes to the total stiffness of that coordinate. This observation will be used as we develop a generalized procedure to develop the system stiffness matrix in the next chapter.

19.10 LOOKING AHEAD

In this chapter, the basic concepts of matrix structural analysis have been introduced. When structural systems become very complicated and are composed of many elements, the direct procedures discussed in this chapter become very difficult and cumbersome to apply. In the next chapter, we will study a more general procedure for developing the system matrices. We also will look at how loads can be applied to the elements themselves. In that chapter, actual beams and bars instead of the springs used in this chapter will be considered. It is very interesting, though, that structural elements can be represented by springs.

19.11 PROBLEMS FOR SOLUTION

For problems 19.1 to 19.5, use the matrix methods presented in this chapter to determine the displacements at each joint in the truss and all of the forces of reaction.

19.1

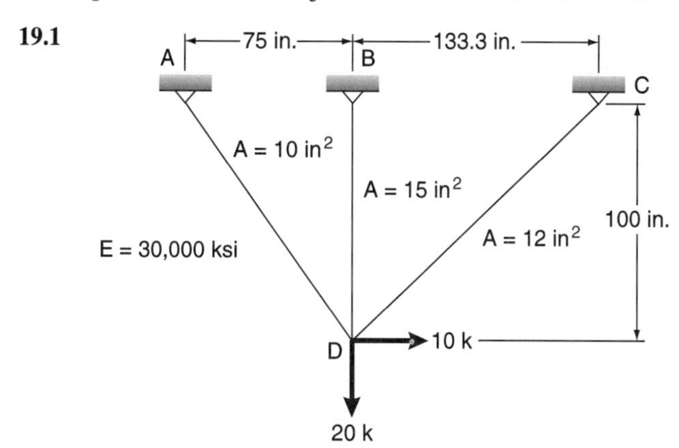

19.2

(*Ans.* $R_{AH} = -17.68$ kips, $R_{DV} = 17.68$ kips)

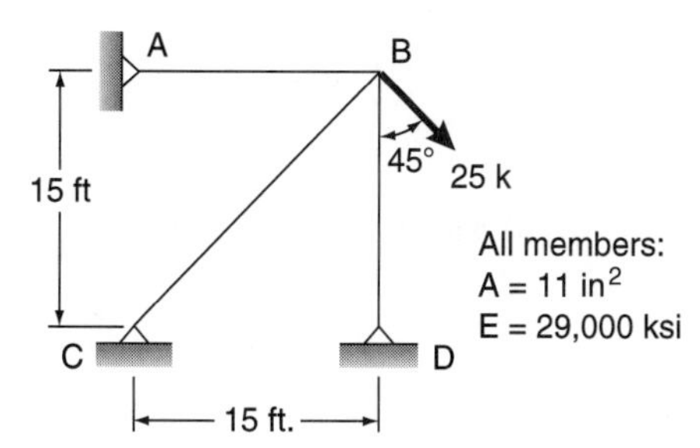

19.3

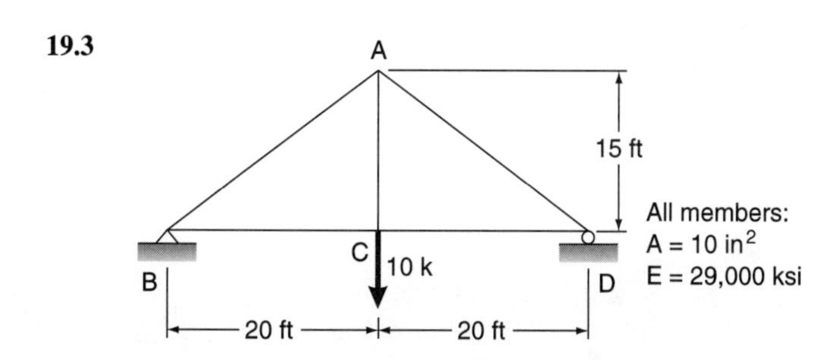

19.4 (*Ans.* R_{BH} = 11.13 kips, R_{BV} = 6.429 kips)

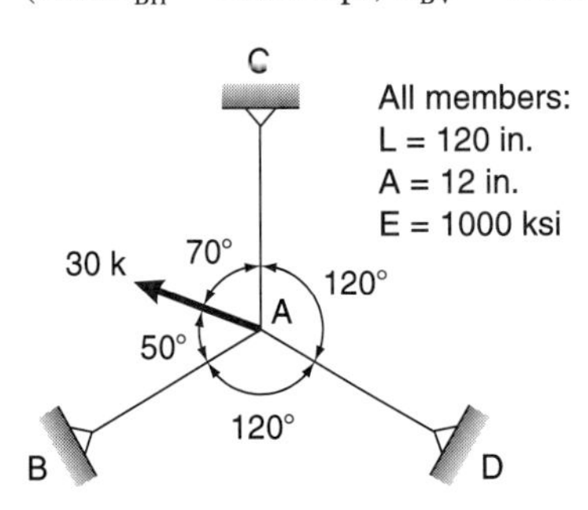

19.5

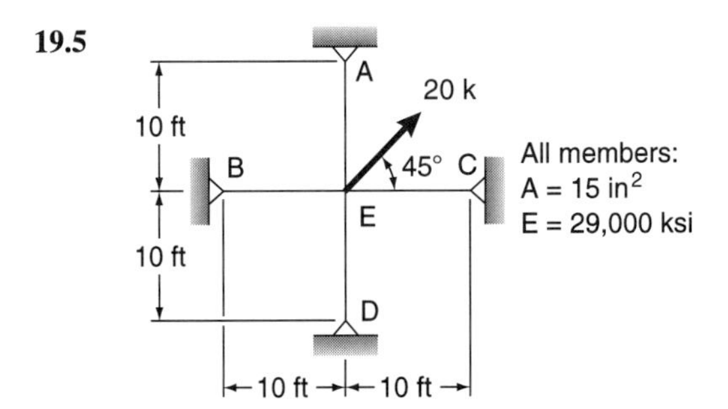

19.6 Repeat Problem 16.30. Assume for each member that A is 12 in^2 and that E is 29,999 ksi.

For Problems 19.7 to 19.11, use the computer program SABLE or SAP2000 to find the displacements at each joint in the structure, all of the forces of reaction, and the force in each bar. *Suggestion:* when using SABLE select the element type to be a PIN-PIN beam. Because such a beam cannot provide rotational stiffness, the rotation about the z axis will need to be constrained at each joint in the structure.

19.7 Repeat Problem 19.1 (*Ans.* δ_{DH} = 0.004305 in., δ_{DV} = −0.002863 in.)

19.8 Repeat Problem 19.2

19.9 Repeat Problem 19.3 (*Ans.* δ_{CV} = −0.02793 in.)

19.10 Repeat Problem 19.4

19.11 Repeat Problem 19.5 (*Ans.* δ_{EH} = 0.00196 in, δ_{EV} = 0.00195 in.)

19.12 Repeat Problem 19.6

Chapter 20

Generalizing Matrix Methods

20.1 INTRODUCTION

The fundamental concepts of matrix structural analysis were presented in the last chapter by employing very simple principles and applying them to the development of system matrices. By the time you finished studying that material, however, it probably was painfully clear to you that those methods would be very unwieldy for large structures. Furthermore, the preparation of system matrices in such a manner is not conducive to analysis using a computer.

U.S. Customs Court and Federal Office Building, New York City. The 41-story office building (rear) is a typical steel-framed tower, with columns supporting beams. The eight-story courthouse (foreground), designed for column-free courtrooms and a pedestrian concourse at street level, literally hangs from two parallel steel trusses above the roof level. (Courtesy Bethlehem Steel Corporation)

In this chapter, we will develop a general procedure that can be easily turned into algorithms solvable with computers. To do so, we will first develop the stiffness matrices for a general truss element and beam element in what we will call the *element coordinate system*. By changing the physical properties (area, length, etc.), the truss element can represent any member in a truss structure and the beam element can represent any element in a frame. We then go through a transformation from the element coordinate system to the *global coordinate system*—the coordinate system that is associated with the entire structure. Once the elemental stiffness matrices have been transformed to the global coordinate system, we can then assemble the system stiffness matrix and determine the forces and displacements in the structure.

The procedures developed in this chapter can be systematically applied to the analysis of an entire structure. Because the procedures are so systematic, they are ideally suited to analysis with a digital computer. These very same procedures have been used in SABLE as well as many commercial software packages that are available for structural analysis. These commercial software packages include SAP2000, STRUDL, NASTRAN, and ANSYS, among others.

20.2 DEFINITION OF COORDINATE SYSTEMS

In the previous chapter, we were working in what is called the global coordinate system. In this context, when we speak of coordinates recall that we are speaking of the structural coordinates. The global structural coordinates are used to define the displacements and forces acting on the entire structure. A coordinate is assigned to each relevant degree of freedom at each joint in the structure. Such coordinates are shown in Figure 20.1(a) for a plane truss and in Figure 20.1(b) for a plane frame. The joints in a truss can only translate so there are two translational coordinates at each joint. The joints in a plane frame, on the other hand, can translate and rotate. As such, there are three coordinates at each joint: two translations and a rotation. Generally, the global coordinates are aligned with the principal axes of the structure. In the case of frames and trusses, these axes usually are the horizontal and vertical axes of the structure.

The members in the structure each have a coordinate system associated with them. These element coordinate systems do not have to be oriented in the same direction as global coordinates. Typical elemental coordinates for a truss member and a frame member are shown in Figure 20.2(a) and (b), respectively. Notice that the element coordinates are oriented to be parallel and perpendicular to the longitudinal axis of the member. Also notice that the axial coordinates are directed from what has been named the "i" end of the member to what has been named the "j" end of the member. This establishes the positive direction of the element coordinates (which is important when interpreting the results) and the direction in which the angle θ is measured. The reference for the angle is the horizontal axis of the structure.

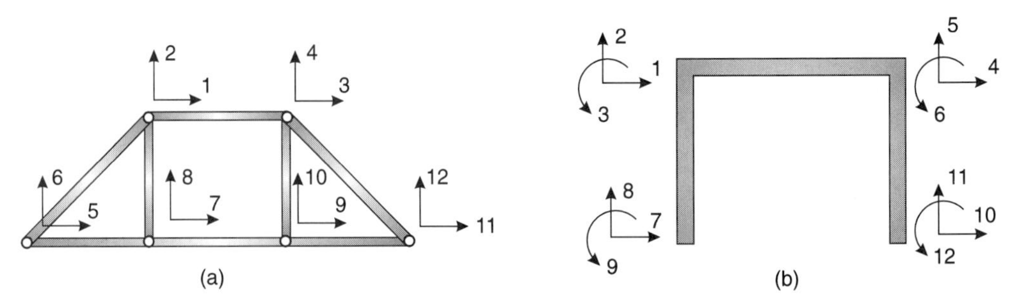

Figure 20.1 Global coordinates for (a) a truss and (b) a frame

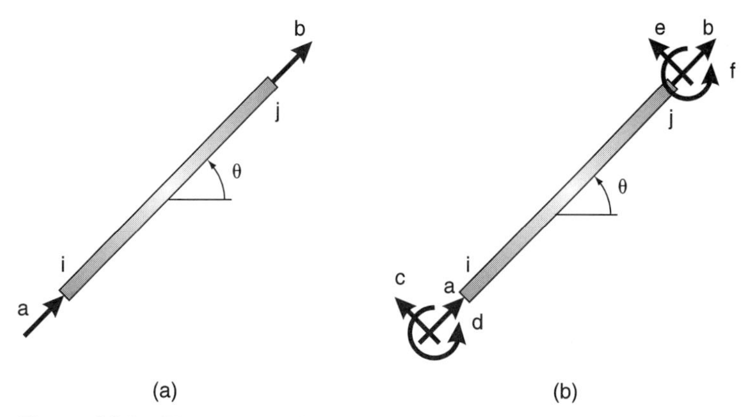

Figure 20.2 Element coordinates for (a) a truss element and (b) a frame element

20.3 THE ELEMENTAL STIFFNESS RELATIONSHIP

In Chapter 19 we presented the basic relationship between forces and displacements for the entire structure. That relationship was given in Equation 19.2. An analogous relationship exists when looking at the forces and displacements at the ends of a component of the structure. That relationship is:

$$\{f\} = [k]\{x\} \qquad \textbf{Eq. 20.1}$$

In Equation 20.1 {f} is the vector of forces, in local coordinates, acting at the ends of the member. The term {x} is the vector of displacements, also in local coordinates, of the end of the member. The matrix [k] is the elemental stiffness matrix in local coordinates. This relationship of Equation 20.1 for the member could have been expressed in terms of global coordinates as

$$\{f_g\} = [k_g]\{x_g\} \qquad \textbf{Eq. 20.2}$$

In Equation 20.2 each of the terms of the equation have the same significance as they did in Equation 20.1, except that the relationship is now expressed in terms of global coordinates—in reality, the global coordinates at the end of the member.

In the next two sections, we will develop the elemental stiffness matrix for a general truss element and a general beam element, respectively. In addition, we will develop the transformation matrix that is necessary to transform the quantities in local coordinates to global coordinates.

20.4 TRUSS ELEMENT MATRICES

When we develop the matrices for an element, two matrices need to be developed. These are the stiffness matrix and the transformation matrix. In Chapter 19, we have already developed the stiffness matrix for a truss element. As you recall from that discussion, the element coordinates are f_a and f_b as shown on Figure 20.3.

The stiffness matrix for a truss element is

$$[k] = \begin{bmatrix} \frac{AE}{L} & -\frac{AE}{L} \\ -\frac{AE}{L} & \frac{AE}{L} \end{bmatrix} \qquad \textbf{Eq. 20.3}$$

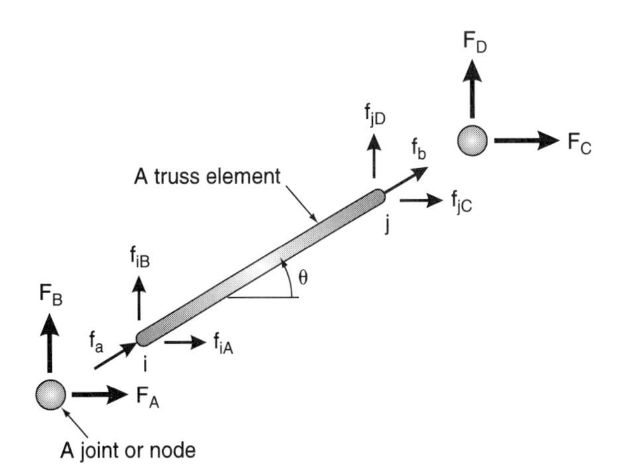

Figure 20.3 Relationship between global and elemental coordinates for a truss

This is nothing more than Equation 19.3 with the substitution in Equation 19.15 for the stiffness. Please notice that now we are using small letters to denote the element matrices and that we will continue to use large letters to denote the global matrices.

The transformation matrix for the truss element can be developed from principles of equilibrium. At each node on the element, the resultant forces acting at the node must be the same whether they are viewed in element coordinates (f_a and f_b) or in global coordinates (f_{1g}, f_{2g}, f_{3g}, and f_{4g}). As such the following four equations can be written by considering the geometry and coordinates shown in Figure 20.3:

$$\begin{aligned} f_a \cos\theta &= f_{1g} \\ f_a \sin\theta &= f_{2g} \\ f_b \cos\theta &= f_{3g} \\ f_b \sin\theta &= f_{4g} \end{aligned} \qquad \textbf{Eq. 20.4}$$

These four equations can be represented in matrix form as

$$\{f_g\} = \begin{Bmatrix} f_{1g} \\ f_{2g} \\ f_{3g} \\ f_{4g} \end{Bmatrix} = \begin{bmatrix} \cos\theta & 0 \\ \sin\theta & 0 \\ 0 & \cos\theta \\ 0 & \sin\theta \end{bmatrix} \begin{Bmatrix} f_a \\ f_b \end{Bmatrix} = [\beta]^T\{f\} \qquad \textbf{Eq. 20.5}$$

The left side of this equation is the vector of element end forces in global coordinates. The last term on the right-hand side is the vector of element end forces in element coordinates. The first term on the right-hand side is the matrix that maps the element coordinates onto the global coordinates. This matrix is the transpose of the transformation matrix [β] for the truss element. Similar equations could have been developed if we considered displacement instead of force. The transformation matrix maps forces and displacements from one coordinate system to the other; however, it does not map the stiffness of the element. The procedure to transform element stiffness from the local to the global coordinate system is discussed in Section 20.6.

20.5 BEAM ELEMENT MATRICES

The stiffness matrix for the beam element is composed of two parts. These parts are the axial force effects and the flexural force (shear and moment) effects. When displacements are small, which is one of the fundamental assumptions of the analysis procedure we are

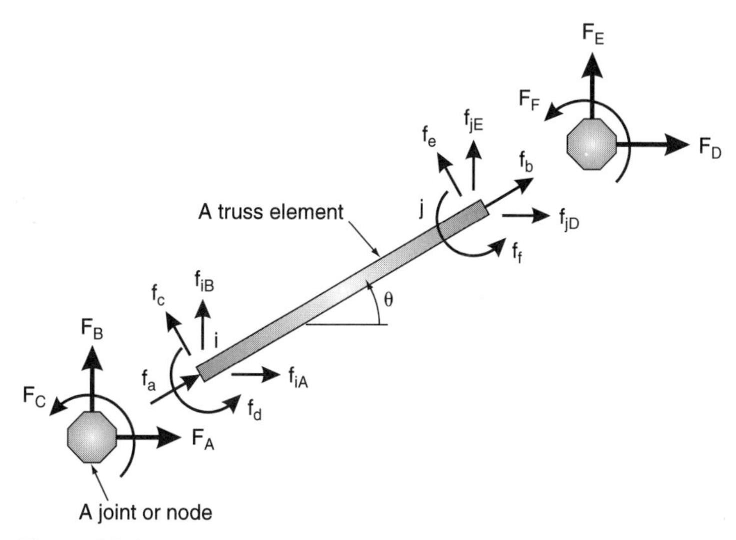

Figure 20.4 Relationship between global and elemental coordinates for a beam

developing, these two effects will not have an affect on each other; they are said to be uncoupled. Coupling of axial and flexural effects is a second-order effect that is negligible as long as displacements are small.

The stiffness matrix for the beam element has the general form:

$$[k] = \begin{bmatrix} [\text{Axial}] & [0] \\ [0] & [\text{Flexural}] \end{bmatrix} \qquad \textbf{Eq. 20.6}$$

If the elements coordinates shown in Figure 20.4 are used, the axial effects partition of the stiffness matrix is the same as that which we have been using for the truss element—that which is contained in Equation 20.3.

The flexural partition of the stiffness matrix is found in the same manner that we found the axial partition of the stiffness matrix. A unit displacement is imposed separately at each flexural coordinate; the displacement at the other flexural coordinates is constrained to zero; and the flexural forces necessary to hold this configuration are computed. These are the forces and displacements shown in Figure 20.5. These forces can be computed using any of the procedures discussed earlier in this book, for example, virtual work or slope deflection.

The forces necessary to maintain the displaced configuration are the stiffness coefficients in the column represented by the coordinate with the unit displacement. The flexural partition of the element stiffness matrix, then is

$$[\text{Flexural}] = \begin{bmatrix} \frac{12EI}{L^3} & \frac{6EI}{L^2} & -\frac{12EI}{L^3} & \frac{6EI}{L^2} \\ \frac{6EI}{L^2} & \frac{4EI}{L} & -\frac{6EI}{L^2} & \frac{2EI}{L} \\ -\frac{12EI}{L^3} & -\frac{6EI}{L^2} & \frac{12EI}{L^3} & -\frac{6EI}{L^2} \\ \frac{6EI}{L^2} & \frac{2EI}{L} & -\frac{6EI}{L^2} & \frac{4EI}{L} \end{bmatrix} \qquad \textbf{Eq. 20.7}$$

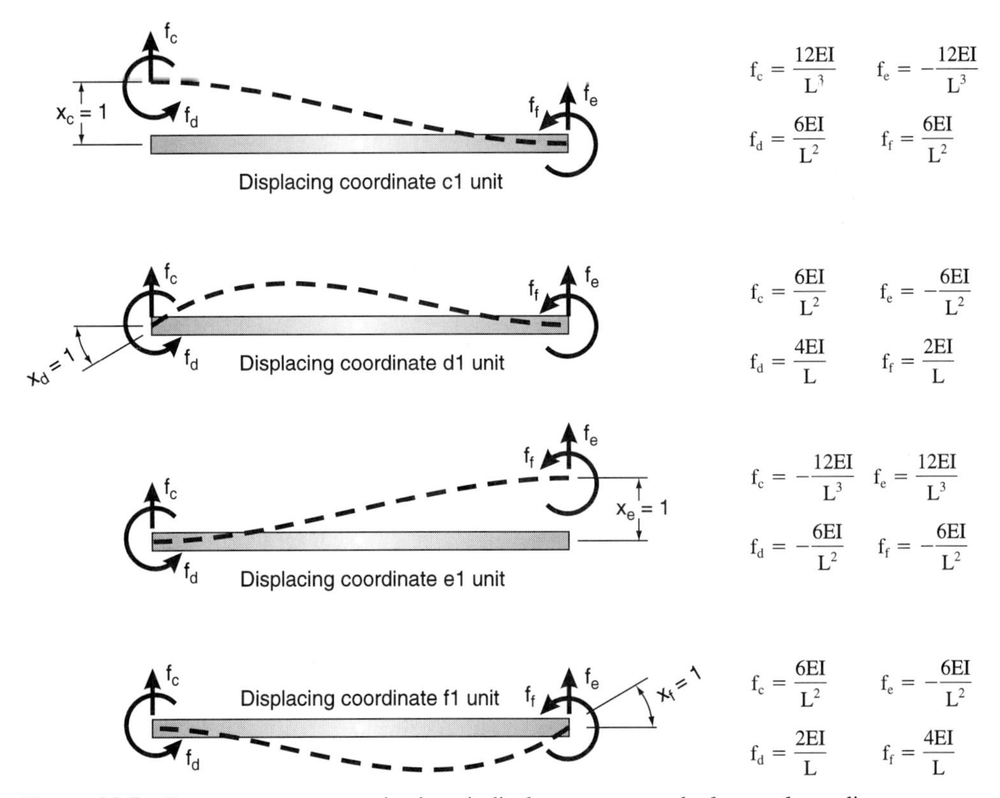

Figure 20.5 Forces necessary to maintain unit displacements at each elemental coordinate

and the final stiffness matrix for the beam element in local coordinates is then

$$[k] = \begin{bmatrix} \frac{AE}{L} & -\frac{AE}{L} & 0 & 0 & 0 & 0 \\ -\frac{AE}{L} & \frac{AE}{L} & 0 & 0 & 0 & 0 \\ 0 & 0 & \frac{12EI}{L^3} & \frac{6EI}{L^2} & -\frac{12EI}{L^3} & \frac{6EI}{L^2} \\ 0 & 0 & \frac{6EI}{L^2} & \frac{4EI}{L} & -\frac{6EI}{L^2} & \frac{2EI}{L} \\ 0 & 0 & -\frac{12EI}{L^3} & -\frac{6EI}{L^2} & \frac{12EI}{L^3} & -\frac{6EI}{L^2} \\ 0 & 0 & \frac{6EI}{L^2} & \frac{2EI}{L} & -\frac{6EI}{L^2} & \frac{4EI}{L} \end{bmatrix} \qquad \textbf{Eq. 20.8}$$

The transformation matrix for the beam element is developed in the same manner that the transformation matrix was developed for the truss element. The resultant force at each node must be the same whether the forces are expressed in terms of local coordinates or in terms of global coordinates. Because of this necessary consideration, the following six equations can be written:

$$\begin{aligned} f_a \cos\theta - f_c \sin\theta &= f_{1g} \\ f_a \sin\theta + f_c \cos\theta &= f_{2g} \\ f_d &= f_{3g} \\ f_b \cos\theta - f_e \sin\theta &= f_{4g} \\ f_b \sin\theta + f_e \cos\theta &= f_{5g} \\ f_f &= f_{6g} \end{aligned} \qquad \textbf{Eq. 20.9}$$

The six equations can be represented in matrix form as

$$\begin{Bmatrix} f_{1g} \\ f_{2g} \\ f_{3g} \\ f_{4g} \\ f_{5g} \\ f_{6g} \end{Bmatrix} = \begin{bmatrix} \cos\theta & 0 & -\sin\theta & 0 & 0 & 0 \\ \sin\theta & 0 & \cos\theta & 0 & 0 & 0 \\ 0 & 0 & 0 & 1 & 0 & 0 \\ 0 & \cos\theta & 0 & 0 & -\sin\theta & 0 \\ 0 & \sin\theta & 0 & 0 & \cos\theta & 0 \\ 0 & 0 & 0 & 0 & 0 & 1 \end{bmatrix} \begin{Bmatrix} f_a \\ f_b \\ f_c \\ f_d \\ f_e \\ f_f \end{Bmatrix} = [\beta]^T\{f\} \qquad \textbf{Eq. 20.10}$$

As with the transformation for the truss element, the left side of this equation is the vector of element end forces in global coordinates. The last term on the right-hand side is the vector of element end forces in element coordinates. The first term on the right hand side is the matrix that maps the element coordinates onto the global coordinates. This matrix is the transpose of the transformation matrix $[\beta]$ for the beam element. Again, this matrix maps the forces or displacements in local coordinates to global coordinates.

20.6 TRANSFORMATION TO GLOBAL COORDINATES

In the previous sections we developed the elemental stiffness and transformation matrices for the truss element and the beam element. Using these matrices, we can compute the elemental stiffness matrices in global coordinates. The complete transformation from elemental to global coordinates can be developed from energy principles.

The internal strain energy in the element in terms of local coordinates can be computed from

$$U = \{x\}^T\{f\} \qquad \textbf{Eq. 20.11}$$

and the strain energy in the element in terms of global coordinates can be written as

$$U = \{x_g\}^T\{f_g\} \qquad \textbf{Eq. 20.12}$$

The strain energy in the element must be the same whether it is written in terms of local coordinates or in terms of global coordinates. Therefore, the following must be true:

$$\{x\}^T\{f\} = \{x_g\}^T\{f_g\} \qquad \textbf{Eq. 20.13}$$

From the relationship developed in Equation 20.5 or Equation 20.10, we observe that

$$\{x\} = [\beta]\{x_g\} \qquad \textbf{Eq. 20.14}$$

Using Equation 20.14 and the law of the transpose of a product, we can show that

$$\{x\}^T = \{[\beta]\{x_g\}\}^T = \{x_g\}^T[\beta]^T \qquad \textbf{Eq. 20.15}$$

We can then substitute Equations 20.1, 20.2, and 20.15 into Equation 20.13 to obtain

$$\{x_g\}^T[\beta]^T[k][\beta]\{x_g\} = \{x_g\}^T[k_g]\{x_g\} \qquad \textbf{Eq. 20.16}$$

Observe in Equation 20.16 that both sides of the equation are premultiplied by $\{x_g\}^T$ and are post-multiplied by $\{x_g\}$. As such, both of these terms can be canceled from the equation. When doing so, Equation 20.16 reduces to

$$[\beta]^T[k][\beta] = [k_g] \qquad \textbf{Eq. 20.17}$$

which is the relationship that we wanted to develop. It is the transformation of the elemental stiffness in local coordinates to the elemental stiffness matrix in global coordinates.

20.7 ASSEMBLING THE GLOBAL STIFFNESS MATRIX

After we have developed the elemental stiffness for each component in the structure in terms of global coordinates, these matrices can be assembled to form the global stiffness matrix for the entire structure. Recall from our discussion in Section 19.9 that the total stiffness at a coordinate is the sum of the stiffness contributed to that coordinate by each element attached to that coordinate. As such, we can superimpose the element stiffness matrices to obtain the total global stiffness matrix for the structure. Unfortunately, all of the elements in the structure are not connected to all of the global coordinates. Generally, any one element is connected to only a few of the global coordinates. Because of this situation,

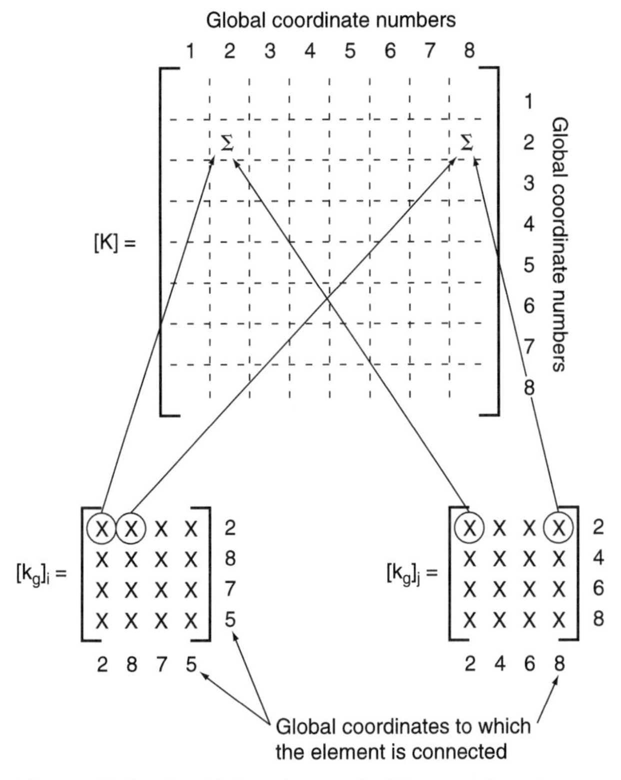

Figure 20.6 Combining elemental stiffness to form the global stiffness matrix

we cannot simply add all of the element stiffness matrices $[k_g]$ to obtain the global stiffness matrix; there is not a one-to-one correspondence between the stiffness coefficients.

To develop the global stiffness matrix from the elemental stiffness matrices, though, we can employ a scheme such as that shown in Figure 20.6. The elements in a structure are connected to a few of the global coordinates. Hypothetical coordinates for two elements are shown in the bottom part of the figure. The coefficients in the element stiffness matrices are added to the coefficients in the global stiffness matrix corresponding to the same coordinate rows and columns. This process is shown for a few of the coefficients in the two element matrices. The summation continues for all of the elements in the structure. After the summation has been performed for all of the elements in the structure, the global stiffness matrix for that structure will result.

20.8 LOADS ACTING ON THE SYSTEM

Loads that act on the structure can emanate from two sources. These are forces applied directly to the joints and forces applied to the members. The loads that act directly on the global coordinates are easy to deal with: they are the same forces that were dealt with in Chapter 19. For our discussion, those forces will be denoted as $\{F_j\}$. If all of the loads on the structure are acting on the joints, the global load vector is simply

$$\{F\} = \{F_j\} \qquad \textbf{Eq. 20.18}$$

However, all of the loads are not always acting only on the joints. Usually there also are forces acting on the members in the structure. These loads can include distributed loads and concentrated forces that act along the length of the member. To deal with the forces acting on the members, we must determine an equivalent set of global forces. This can be accomplished by again using principles of strain energy. In previous sections we said that the strain energy being computed in the element using local coordinates is equal to the strain energy being computed in the element using global coordinates. Similarly, the work done on the member by element forces in local coordinates is equal to the work done on the structure by the equivalent forces in global coordinates. This concept can be represented as

$$\tfrac{1}{2}\{x\}^T\{f^0\} = \tfrac{1}{2}\{x_g\}^T\{f_g^0\} \qquad \textbf{Eq. 20.19}$$

In Equation 20.19 we are using the notation $\{f^0\}$ to indicate the fixed-end forces from the member loads. We used these same fixed-end forces in our discussions about moment distribution and slope-deflection methods of analysis. Fixed-end forces for common member loads are shown in Appendix D. The term $\{f_g^0\}$ is the equivalent forces in global coordinates.

Recalling the relationship between local and global element displacements given by equation 20.14, we can show that:

$$\{x_g\}^T[\beta]^T\{f^0\} = \{x_g\}^T\{f_g^0\} \qquad \textbf{Eq. 20.20}$$

which reduces to

$$[\beta]^T\{f^0\} = \{f_g^0\} \qquad \textbf{Eq. 20.21}$$

This is the expression for the equivalent global forces caused by forces acting along the length of the member. By rearranging this equation to the form:

$$\{f_g^0\} - [\beta]^T\{f^0\} = 0 \qquad \textbf{Eq. 20.22}$$

observe that the fixed-end forces on the element caused by loads acting on the element are in equilibrium with their global effect on the structure. In other words, the fixed-end forces at the ends of the members act in the opposite direction when they are applied to the joints. The final global load vector is then

$$\{F\} = \{F_j\} - [\beta]\{f^0\} \qquad \textbf{Eq. 20.23}$$

This load vector will be used when solving the matrix equations. Note that when adding the element forces to the global force vector, the element forces are superimposed in the global force vector in the same way that the element stiffness matrices were superimposed to obtain the global stiffness matrices. This process was discussed in the previous section.

20.9 COMPUTING FINAL BEAM END FORCES

After computing the global forces and displacements using the procedures that were discussed in Chapter 19, we need to compute the forces that are acting at the ends of the

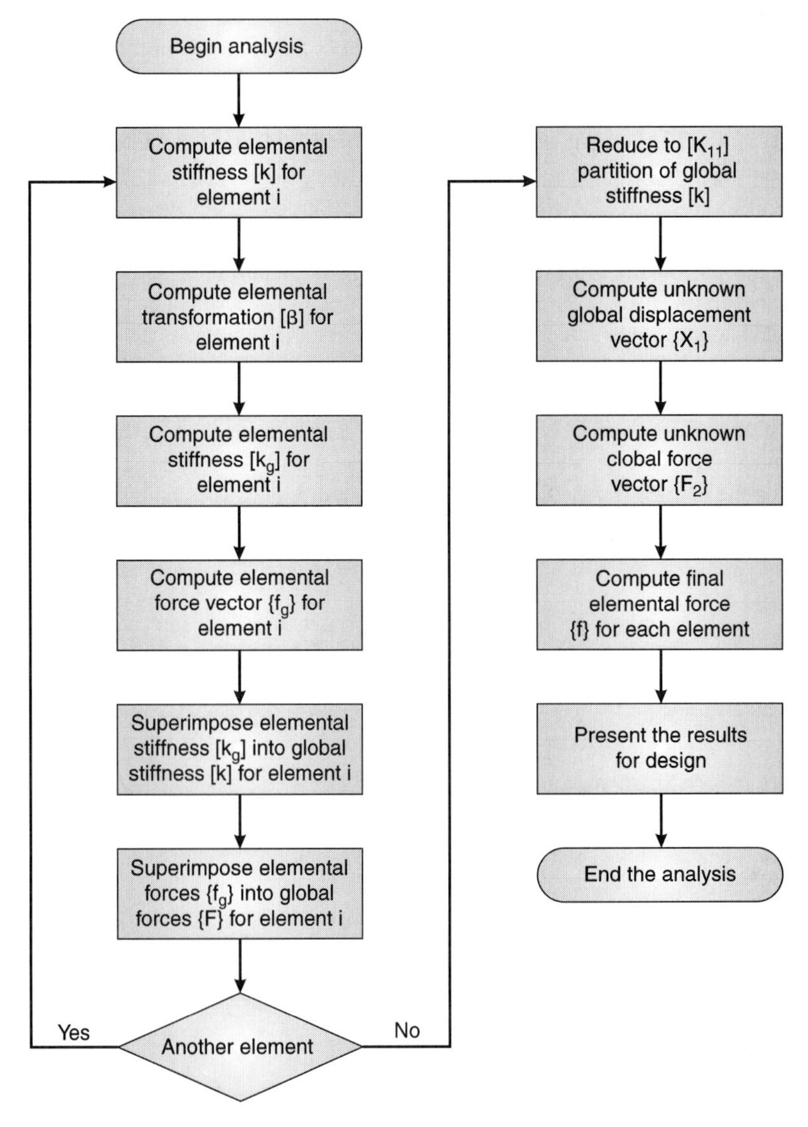

Figure 20.7 Flowchart of steps required for analysis of a structure

elements. There are two components to the final beam end forces: the forces caused by the response of the structure and the forces caused by loads applied directly to the element. The final beam end forces are

$$\{f\} = \{f_s\} + \{f^0\} = [k][\beta]\{x_g\} + \{f^0\} \qquad \textbf{Eq. 20.24}$$

These are the final forces acting on the end of the elements, in the coordinates of the element. The term $\{x_g\}$ are the global displacements at the global coordinates to which the member is connected.

20.10 PUTTING IT ALL TOGETHER

At this point, the systematic procedures may not be obvious and the entire process may be confusing and vague. To help your understanding of the necessary sequence of calculations, refer to the flowchart in Figure 20.7. This flowchart shows all of the steps necessary for a complete structural analysis using matrix methods or the analysis engine in a structural analysis computer program. We will use the principles that we have developed in this chapter for the analysis of a complete structure.

EXAMPLE 20.1

The structure to be analyzed is the T-Bar truss shown in the following figure. The two rods act in tension and stiffen the system. The post is rigidly connected to the beam.

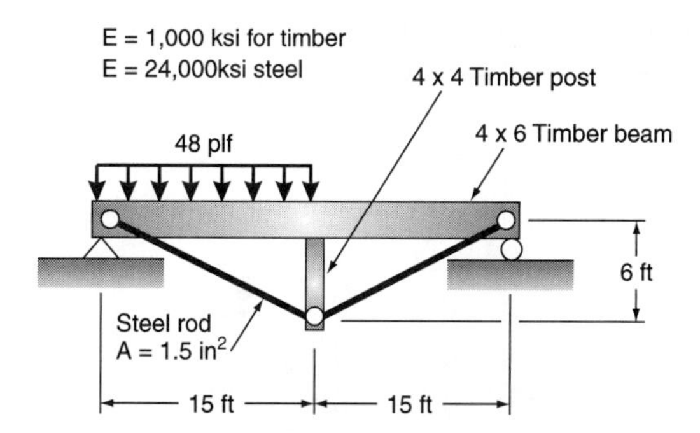

The global coordinates that we will use in the analysis are shown in the figure below. Notice that we have numbered the unconstrained degrees of freedom first and the constrained degrees of freedom last. This caused the unconstrained coordinates to be in the upper partitions of the matrices and the constrained coordinates to be in the lower partitions.

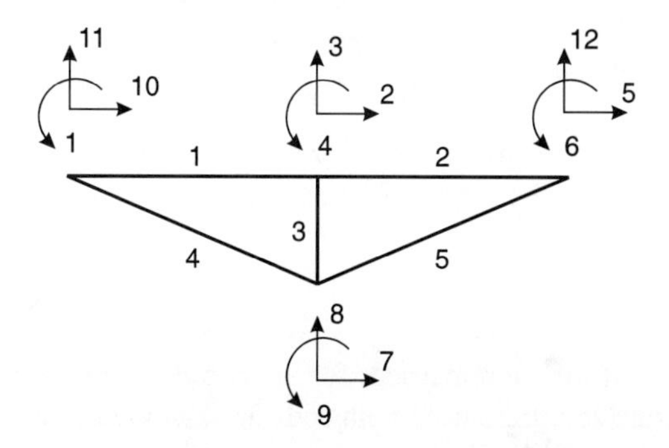

To begin the analysis, we must first compute the elemental stiffness and transformation matrices for each component in the structure. Because of the nature of this structure, some of these matrices are identical. The two horizontal components have the same stiffness and transformation matrices. The two rods have the same stiffness matrix.

Using Equations 20.8 and 20.10, the stiffness and transformation matrices, respectively, for the two horizontal beams are found to be

$$[k]_{1,2} = \begin{bmatrix} 133 & -133 & 0 & 0 & 0 & 0 \\ -133 & 133 & 0 & 0 & 0 & 0 \\ 0 & 0 & 0.148 & 13.3 & -0.148 & 13.3 \\ 0 & 0 & 13.3 & 1600 & -13.3 & 800 \\ 0 & 0 & -0.148 & -13.3 & 0.148 & -13.3 \\ 0 & 0 & 13.3 & 800 & -13.3 & 1600 \end{bmatrix}$$

$$[\beta]^T_{1,2} = \begin{bmatrix} 1.0 & 0 & 0 & 0 & 0 & 0 \\ 0 & 0 & 1.0 & 0 & 0 & 0 \\ 0 & 0 & 0 & 1.0 & 0 & 0 \\ 0 & 1.0 & 0 & 0 & 0 & 0 \\ 0 & 0 & 0 & 0 & 1.0 & 0 \\ 0 & 0 & 0 & 0 & 0 & 1.0 \end{bmatrix}$$

Each of these two components is taken to extend from the left to the right. The angle θ for these components is therefore equal to 0 degrees.

The post—the vertical member in the middle—is assumed to go from the bottom to the top. As such, this member is in the structure at an angle of 90 degrees. Its stiffness and transformation matrices are

$$[k]_3 = \begin{bmatrix} 222 & -222 & 0 & 0 & 0 & 0 \\ -222 & 222 & 0 & 0 & 0 & 0 \\ 0 & 0 & 0.686 & 24.7 & -0.686 & 24.7 \\ 0 & 0 & 24.7 & 1185 & -24.7 & 592 \\ 0 & 0 & -0.686 & -24.7 & 0.686 & -24.7 \\ 0 & 0 & 24.7 & 592 & -24.7 & 1185 \end{bmatrix}$$

$$[\beta]^T_3 = \begin{bmatrix} 0 & 0 & -1 & 0 & 0 & 0 \\ 1 & 0 & 0 & 0 & 0 & 0 \\ 0 & 0 & 0 & 1 & 0 & 0 \\ 0 & 0 & 0 & 0 & -1 & 0 \\ 0 & 1 & 0 & 0 & 0 & 0 \\ 0 & 0 & 0 & 0 & 0 & 1 \end{bmatrix}$$

The stiffness matrices for the two rods in the structure are determined using Equation 20.3. They are the same and are

$$[k]_{4,5} = \begin{bmatrix} 224 & -224 \\ -224 & 224 \end{bmatrix}$$

The transformation matrices for the two rods are different because their orientation in the structure is different. If both rods are assumed to extend from the left side to the right side, the orientation of the left rod is −20.8 degrees and that of the right rod is 20.8 degrees. The resulting transformation matrices, found using Equation 20.5, are

$$[\beta]_4^T = \begin{bmatrix} 0.928 & 0 \\ 0.371 & 0 \\ 0 & 0.928 \\ 0 & 0.371 \end{bmatrix} \qquad [\beta]_5^T = \begin{bmatrix} 0.928 & 0 \\ -0.371 & 0 \\ 0 & 0.928 \\ 0 & -0.371 \end{bmatrix}$$

Prior to assembling the system matrices, we need to compute the equivalent loads on the global structure caused by the uniformly distributed load on component 1. The equivalent loads, which can be computed using Equation 20.21, are

$$\{f_g^0\}_1 = \begin{bmatrix} 1 & 0 & 0 & 0 & 0 & 0 \\ 0 & 0 & 1 & 0 & 0 & 0 \\ 0 & 0 & 0 & 1 & 0 & 0 \\ 0 & 1 & 0 & 0 & 0 & 0 \\ 0 & 0 & 0 & 0 & 1 & 0 \\ 0 & 0 & 0 & 0 & 0 & 1 \end{bmatrix} \begin{Bmatrix} 0 \\ 0 \\ 0.36 \\ 10.8 \\ 0.36 \\ -10.8 \end{Bmatrix} = \begin{Bmatrix} 0 \\ 0.36 \\ 10.8 \\ 0 \\ 0.36 \\ -10.8 \end{Bmatrix}$$

The necessary information is now available to assemble the complete system equation in global coordinates. This is accomplished using the superposition procedure discussed in Section 20.7 for stiffness and in Section 20.8 for the loads. Using these procedures, the final stiffness equation for this structure is then

$$\left[\begin{array}{ccccccccc|ccc} 0.148 & 13.3 & -0.148 & 13.30 & 0 & 0 & 0 & 0 & 0 & 0 & 0 & 0 \\ 13.3 & 2419 & -146.3 & 824.7 & 0 & 0 & -0.686 & 0 & 24.7 & 0 & 0 & 0 \\ -0.148 & -146.3 & 355.1 & -13.3 & 0 & 0 & 0 & -222 & 0 & 0 & 0 & 0 \\ 13.3 & 824.7 & -13.3 & 2785 & 13.3 & 133.3 & -24.7 & 0 & 592 & 0 & 0 & -0.148 \\ 0 & 0 & 0 & 13.3 & 1793 & 800 & -192.9 & 77.12 & 0 & 0 & 0 & -90.42 \\ 0 & 0 & 0 & 13.3 & 800 & 1600 & 0 & 0 & 0 & 0 & 0 & -13.3 \\ 0 & -0.686 & 0 & -24.7 & -192.9 & 0 & 386.5 & 0 & -24.7 & -192.9 & -77.12 & 77.12 \\ 0 & 0 & -222 & 0 & 77.12 & 0 & 0 & 252.8 & 0 & -77.12 & -30.83 & -30.83 \\ 0 & 24.7 & 0 & 592 & 0 & 0 & -24.7 & 0 & 1185 & 0 & 0 & 0 \\ \hline 0 & 0 & 0 & 0 & 0 & 0 & -192.9 & -77.12 & 0 & 325.9 & -210.1 & 0 \\ 0 & 0 & 0 & 0 & 0 & 0 & -77.12 & -30.83 & 0 & -210.1 & 163.8 & 0 \\ 0 & 0 & 0 & -0.148 & -90.42 & -13.3 & 77.12 & -30.83 & 0 & 0 & 0 & 30.98 \end{array}\right] \begin{Bmatrix} X_1 \\ X_2 \\ X_3 \\ X_4 \\ X_5 \\ X_6 \\ X_7 \\ X_8 \\ X_9 \\ \hline 0 \\ 0 \\ 0 \end{Bmatrix} = \begin{Bmatrix} -10.8 \\ 0 \\ -0.36 \\ 10.8 \\ 0 \\ 0 \\ 0 \\ 0 \\ 0 \\ \hline F_{10} \\ F_{11} \\ F_{12} \end{Bmatrix}$$

The partitions in this equation are between the constrained and the unconstrained degrees of freedom. This equation can be solved using the procedures discussed in Chapter 19. The resulting unknown forces and displacements are

$$\begin{Bmatrix} F_{10} \\ F_{11} \\ F_{12} \end{Bmatrix} = \begin{Bmatrix} 0 \\ 0.54 \\ 0.18 \end{Bmatrix} \qquad \begin{Bmatrix} X_1 \\ X_2 \\ X_3 \\ X_4 \\ X_5 \\ X_6 \\ X_7 \\ X_8 \\ X_9 \end{Bmatrix} = \begin{Bmatrix} -0.00938 \\ -0.00443 \\ -0.0198 \\ 0.00493 \\ -0.00841 \\ -0.00230 \\ -0.00405 \\ -0.0178 \\ -0.00246 \end{Bmatrix}$$

The reaction force at coordinate 11 has to be treated a little differently. There is a force applied at this coordinate from the member. Because this coordinate is constrained to zero displacement, all force applied to this coordinate goes directly to the reaction. But there is also a reaction force at this coordinate resulting from analysis of Equation 20.31. The final reaction force at coordinate 11, which is shown in Equation 20.33, is 0.18 + 0.36 = 0.54. There is 0.18 kips reaction from the structure and the 0.36 kips reaction from the load applied by the left member.

The member end forces resulting from the analysis can be found using Equation 20.24. For the first member, these forces are computed from

$$\{f\}_1 = [k]_1[\beta]_1 = \begin{Bmatrix} 0 \\ 0 \\ -0.00938 \\ -0.00443 \\ -0.0198 \\ 0.00493 \end{Bmatrix} + \begin{Bmatrix} 0 \\ 0 \\ -0.360 \\ 10.8 \\ -0.360 \\ -10.8 \end{Bmatrix} = \begin{Bmatrix} 0.591 \\ -0.591 \\ 0.304 \\ 0 \\ 0.416 \\ -10.2 \end{Bmatrix}$$

Member end forces for the other components are computed in the same manner as the first member. The final member end forces for all of the components are shown in the following figure.

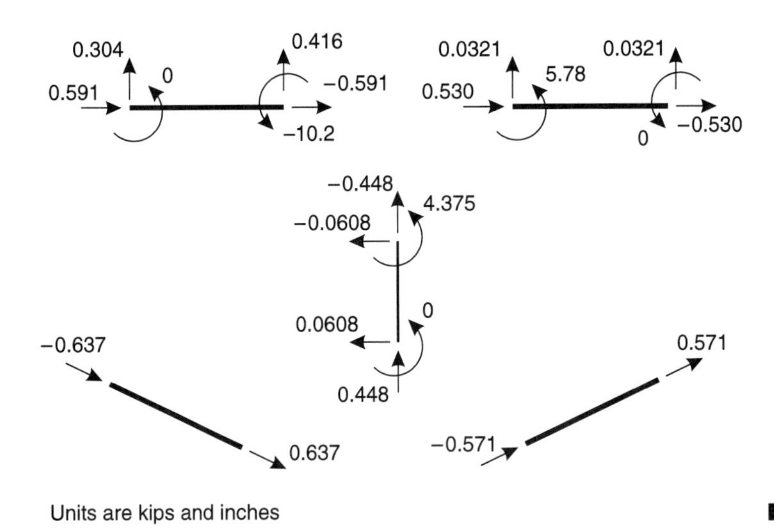

20.11 PROBLEMS FOR SOLUTION

For Problems 20.1 to 20.5, determine the displacements and forces of reaction for the structure, and the member end forces for each member.

20.1 Repeat Problem 19.1

20.2 Repeat Problem 19.2

20.3 Repeat Problem 19.3

20.4 Repeat Problem 19.4

20.5 Repeat Problem 19.5

For Problems 20.6 to 20.10, determine the displacements and forces of reaction for the structure, and the member end forces for each member. For these problems, if the moment of inertia is specified as being constant, assume that $I = 150 \text{ in}^4$. If the cross-sectional area of the member is not specified, assume that the area is equal to 10% of the moment of inertia.

20.6 Repeat Problem 15.2

20.7 Repeat Problem 15.4

20.8 Repeat Problem 15.7

20.9 Repeat Problem 15.14

20.10 Repeat Problem 18.8

20.11 Repeat Problem 18.20

20.12 Repeat Problem 18.26

Chapter 21

Additional Topics in Matrix Methods

21.1 INTRODUCTION

In the previous two chapters, we formulated the system matrices using fundamental principles of mechanics. In this chapter, we will continue that discussion using the principles of virtual work. This approach is the basis of the finite element method that you will study in other courses. In addition, we will look at members with different end conditions and the means to include the effects of support settlement, misfit, and temperature changes in the analysis.

21.2 THE TRUSS ELEMENT USING PRINCIPLES OF VIRTUAL WORK

The principles of virtual work can be used to formulate the elemental stiffness matrix for structural components. We first discussed the principles of virtual work in Chapter 11. Use of virtual work provides a significant generalization of the procedures that can be applied to simple truss elements as well as to plates and shells. In this book, however, we will limit our discussion to truss and beam elements.

Recall that the solution requires knowledge of the deflected shape of the system under load. The deflected shape is typically characterized using displacement functions, which can be formulated using polynomial, trigonometric, logarithmic, or other functions. The function selected, though, must be able to contain the rigid body displacement of the component. For simplicity when integrating and differentiating, polynomials are generally used. The rigid-body displacement is included through the constant term in the general polynomial.

Let us use virtual work to formulate the stiffness matrix for the truss element shown in Figure 21.1. This element has two nodes, one at the beginning of the element and one at the end. Because a truss element is an axial force element, it has only two possible displacements, there are two structural coordinates. These are a displacement at the left end of the element and a displacement at the right end of the element. Both of these coordinates are directed along the longitudinal axis of the element. To formulate the stiffness matrix for this element, we will assume that the displacement of any point along the bar can be represented by the function

$$u = a_0 + a_1 x \qquad \textbf{Eq. 21.1}$$

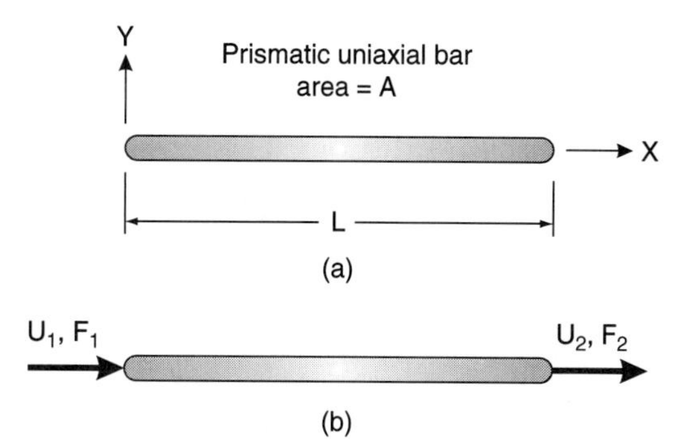

Figure 21.1 A truss element with two nodes

In this equation, the term a_0 will account for the rigid-body displacement as we will soon see, and the term a_1 will account for the deformation of the element. This equation can be represented in matrix form as

$$u = [1 \quad x]\begin{Bmatrix} a_0 \\ a_1 \end{Bmatrix} = [x][a] \qquad \textbf{Eq. 21.2}$$

The term [a] contains the constant coefficients. These coefficients are evaluated using the boundary conditions for the element. For the truss element that we are discussing, the boundary conditions are

$$\text{At} \quad x = 0 \qquad U = U_i$$
$$x = L \qquad U = U_j$$

If we substitute these boundary conditions into Equation 21.2, we obtain the following equation:

$$\begin{bmatrix} U_i \\ U_j \end{bmatrix} = \begin{bmatrix} 1 & 0 \\ 1 & L \end{bmatrix}\begin{bmatrix} a_0 \\ a_1 \end{bmatrix} = [L][a] \qquad \textbf{Eq. 21.3}$$

Solving this equation for [a], we find that

$$[a] = [L]^{-1}\{U\} \qquad \textbf{Eq. 21.4}$$

After evaluating $[L]^{-1}$ and substituting the resulting expression into Equation 21.2, we find that

$$u = [1 \quad 0]\left(\frac{1}{L}\right)\begin{bmatrix} L & 0 \\ -1 & 1 \end{bmatrix}\begin{Bmatrix} U_i \\ U_j \end{Bmatrix} \qquad \textbf{Eq. 21.5}$$

This equation further reduces to

$$u = \begin{bmatrix} 1 - \frac{x}{L} & \frac{x}{L} \end{bmatrix}\begin{Bmatrix} U_i \\ U_j \end{Bmatrix} \qquad \textbf{Eq. 21.6}$$

We can simplify this equation by representing it as

$$u = [N]\{U\} \qquad \textbf{Eq. 21.7}$$

by introducing the notation:

$$[N] = \begin{bmatrix} 1 - \frac{x}{L} & \frac{x}{L} \end{bmatrix} \qquad \textbf{Eq. 21.8}$$

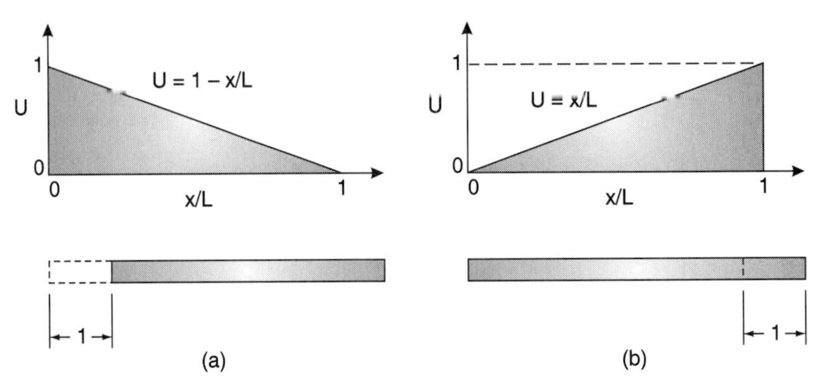

Figure 21.2 Displacement of a particle at x = X along the bar

The matrix [N] is called the matrix of shape functions. There is one shape function for each of the coordinates used to describe the element. The shape functions represent the shape of the element if a coordinate is displaced one unit and the displacement at the other coordinate is constrained to zero displacement. As such, the shape function $U = 1 - x/L$ represents axial deformation of the truss element if the coordinate 1, the left coordinate, is displaced one unit and the coordinate 2 is restrained to zero displacement. This relationship is shown in Figure 21.2(a). The deformation of any point along the bar in shown in Figure 21.2(b) if the right coordinate is displaced one unit and the left coordinate is constrained to zero displacement.

We can use the shape functions that we have developed to formulate the elemental stiffness matrix for a truss element. To begin, recall that

$$\varepsilon = \frac{\partial U}{\partial x} \qquad \textbf{Eq. 21.9}$$

As we go through the following derivation, please note that the local x-axis is directed along the member. It is in the local coordinate system that we will develop the stiffness matrix for the truss element. Using the relationships that we have previously developed, this relationship becomes

$$\varepsilon = \frac{\partial}{\partial x}[N]\{u\} = [B]\{u\} \qquad \textbf{Eq. 21.10}$$

where

$$[B] = \frac{\partial}{\partial x}[N] = \frac{1}{L}[-1 \quad 1] \qquad \textbf{Eq. 21.11}$$

The partial derivative of [N]{u} is equal to the partial derivative of [N] times {u} because {u} is not a function of x. The only term in the equation that is a function of x is [N].

The stress can also be expressed in terms of the shape functions. Recall that from Hooke's law $\sigma = E\varepsilon$. Substituting the relationships from Equations 21.10 and 21.11, we obtain the following expression for the stress in the truss element:

$$\sigma = E[B]\{u\} = \frac{E}{L}[-1 \quad 1]\begin{Bmatrix} U_i \\ U_j \end{Bmatrix} = \frac{E}{L}(U_j - U_i) \qquad \textbf{Eq. 21.12}$$

If we now apply a virtual displacement to the truss element, the virtual strain energy is

$$\delta\varepsilon = [B]\{\delta U\} \qquad \textbf{Eq. 21.13}$$

and the virtual strain energy is

$$\delta W_i = \int_{volume} (\delta\varepsilon)\sigma \, dV \qquad \textbf{Eq. 21.14}$$

If we substitute the expressions for virtual strain and stress in terms of strain we find that the virtual strain energy is

$$\delta W_i = \int_{vol} \{\delta u\}^T[B]^T E[B]\{u\} \, dV \qquad \textbf{Eq. 21.15}$$

If we take the constant terms outside of the integral, Equation 21.15 becomes

$$\delta W_i = \{\delta u\}^T\left(\int_{vol} [B]^T E[B] \, dV\right)\{u\} \qquad \textbf{Eq. 21.16}$$

As we continue with the solution, we can express the external virtual work in terms of the virtual displacements, namely

$$\delta W_e = \{\delta u\}^T\{p\} \qquad \textbf{Eq. 21.17}$$

We have seen this same expression previously. We know that the internal virtual work must be equal to the external virtual. Therefore, if we set Equation 21.16 equal to Equation 21.17 we obtain the equation

$$\{\delta u\}^T\left(\int_{vol} [B]^T E[B] \, dV\right)\{u\} = \{\delta u\}^T\{p\} \qquad \textbf{Eq. 21.18}$$

After rearranging the terms of this last equation, we obtain the equation:

$$\{\delta u\}^T\left[\{p\} - \left(\int_{vol} [B]^T E[B] \, dV\right)\{u\}\right] = 0 \qquad \textbf{Eq. 21.19}$$

For this equation to be true, the term in brackets must be equal to zero because, in general, the displacements are not equal to zero. As such we can show that

$$\left[\int_{vol} [B]^T E[B] \, dV\right]\{u\} = \{p\} \qquad \textbf{Eq. 21.20}$$

Let us examine this last equation for a moment. We have previously said that stiffness times displacement is equal to force. Upon examination of Equation 21.20, we observe that the term in brackets is multiplying displacement and the result is equal to the force. The term in brackets, then, must be equal to the stiffness of the bar, namely

$$[k] = \int_{vol} [B]^T E[B] \, dV \qquad \textbf{Eq. 21.21}$$

This last expression is a general expression for the elemental stiffness matrix in terms of shape functions. For the truss element with which we have been working, the matrix [B] was defined in Equation 21.11. If we substitute that equation into Equation 21.21, we obtain

$$[k] = \int_{vol} \left(\frac{1}{L}\right)\begin{bmatrix} -1 \\ 1 \end{bmatrix} E\left(\frac{1}{L}\right)[-1 \quad 1] \, dV \qquad \textbf{Eq. 21.22}$$

which reduces to

$$[k] = \frac{AE}{L}\begin{bmatrix} 1 & -1 \\ -1 & 1 \end{bmatrix} \qquad \textbf{Eq. 21.23}$$

We obtained this same elemental stiffness matrix in §20.4. In that section, we developed the matrix using fundamental principles of equilibrium and mechanics. In this section, we developed the stiffness matrix using principles of virtual work. The latter approach is more general and can be applied to a broader class of problems. In fact, this approach is used to develop generalized finite elements.

21.3 VIRTUAL WORK AND THE PRISMATIC BEAM ELEMENT

Let us continue using the principles of virtual work to develop the stiffness matrix for a prismatic beam element. The prismatic beam element is a long slender member that has a constant cross section. A long slender element is a member whose length is at least four times as long as any of its cross-sectional dimensions. These elements are typically subjected to transverse loads. By treating the element as a long slender beam, the predominant mode of deformation is flexural deformation. The stiffness matrix that we develop will be subject to the following assumptions:

- Small displacements,
- Linearly elastic deformation,
- Homogeneous and isotropic material, and,
- Infinite axial stiffness.

When developing this element, we will consider the four displacement coordinates and the related forces shown in Figure 21.3. At each end of the beam, the displacements considered are a translation and a rotation. As we begin developing the stiffness matrix for this element, we need to recall the moment-curvature relationship from mechanics. That relationship is

$$\frac{d^2v}{dx^2} = \frac{M}{EI} = \frac{1}{\rho} \qquad \textbf{Eq. 21.24}$$

We also need to recall the relationships between moment and displacement and between shearing force and displacement that we discussed in Chapter 5. Those relationships are

$$\frac{dM}{dx} = \frac{d^3v}{dx^3} = V$$
$$\frac{dV}{dx} = \frac{d^4v}{dx^4} = q \qquad \textbf{Eq. 21.25}$$

These are the basic equations from beam theory.

21.3.1 Developing the Shape Functions

To begin the solution we need to determine the shape functions for this element. Let us assume that the shape of the beam under load can be described by the equation:

$$v(x) = a_0 + a_1x + a_2x^2 + a_3x^3 \qquad \textbf{Eq. 21.26}$$

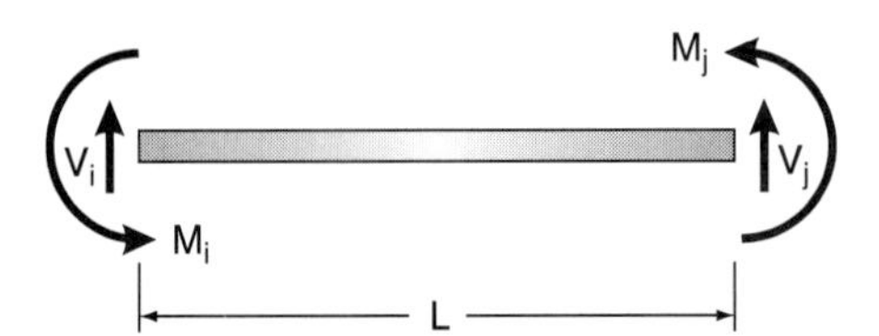

Figure 21.3 A prismatic beam element with four structural displacement coordinates

This approximation has sufficient constants for the number of degrees of freedom being considered. Further, it satisfies the constitutive differential equations. We can substitute the displacement boundary conditions into this equation to solve for the unknown constants. Because there are four unknown constants, we need four boundary conditions to evaluate all of the constants.

at $x = 0 \quad v = d_1$

$$v(0) = d_1 = a_0 + a_1(0) + a_2(0)^2 + a_3(0)^3 \qquad \textbf{Eq. 21.27}$$

$$a_0 = d_1$$

at $x = 0 \quad \dfrac{dv}{dx} = \theta_1$

$$\frac{dv}{dx}(0) = \theta_1 = a_1 + 2a_2(0) + 3a_3(0)^2 \qquad \textbf{Eq. 21.28}$$

$$a_1 = \theta_1$$

at $x = L \quad v = d_2$

$$v(L) = d_2 = d_1 + \theta_1(L) + a_2(L)^2 + a_3(L)^3 \qquad \textbf{Eq. 21.29}$$

at $x = L \quad \dfrac{dv}{dx} = \theta_2$

$$\frac{dv}{dx}(L) = \theta_2 = \theta_1 + 2a_2(L) + 3a_3(L)^2 \qquad \textbf{Eq. 21.30}$$

Equations 21.29 and 21.30 can be solved simultaneously for the constants a_2 and a_3. Doing so yields the equations:

$$\begin{aligned} a_2 &= \frac{-3}{L^2}(d_1 - d_2) - \frac{1}{L}(2\theta_1 + \theta_2) \\ a_3 &= \frac{2}{L^3}(d_1 - d_2) - \frac{1}{L^2}(\theta_1 + \theta_2) \end{aligned} \qquad \textbf{Eq. 21.31}$$

If we then substitute the constants obtained from Equations 21.27, 21.28, and 21.31 into Equation 21.26, we obtain the following function for displacement:

$$\begin{aligned} v(x) = d_1 + \theta_1 x &+ \left[\frac{-3}{L^2}(d_1 - d_2) - \frac{1}{L}(2\theta_1 + \theta_2)\right]x^2 \\ &+ \left[\frac{2}{L^3}(d_1 - d_2) + \frac{1}{L^2}(\theta_1 + \theta_2)\right]x^3 \end{aligned} \qquad \textbf{Eq. 21.32}$$

We can now group similar terms and express the result in matrix form. After doing so we obtain the equation:

$$v = [N]\{d\} = \begin{bmatrix} 2\left(\dfrac{x}{L}\right)^3 - 3\left(\dfrac{x}{L}\right)^2 + 1 \\ \dfrac{x^3}{L^2} - \dfrac{2x^2}{L} + x \\ -2\left(\dfrac{x}{L}\right)^3 + 3\left(\dfrac{x}{L}\right)^2 \\ \dfrac{x^3}{L^2} - \dfrac{x^2}{L} \end{bmatrix}^T \begin{Bmatrix} d_1 \\ \theta_1 \\ d_2 \\ \theta_2 \end{Bmatrix} \qquad \textbf{Eq. 21.33}$$

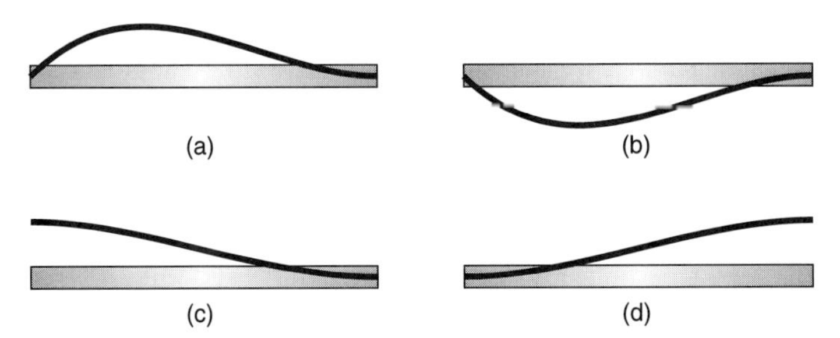

Figure 21.4 The shape functions for a prismatic beam

Observe again that the terms in the first matrix to the right of the second equal sign are the shape functions for the prismatic beam that we have been considering. These shape functions are shown graphically in Figure 21.4. Recall again that the shape functions are the deformed shape of the beam when one coordinate is displaced and the displacement at the other coordinates is constrained to zero. Figure 21.4(a) and (b) are the shape of the beam when the rotational coordinates are displaced. Figure 21.4(c) and (d) are the shape of the beam when the shearing coordinates are displaced. With these shape functions, we are now ready to formulate the stiffness matrix for this element using the principles of virtual work.

21.3.2 Applying Principles of Virtual Work

From our past study of mechanics, we remember that the relationship for stress at any point in a beam can be determined from the equation:

$$\sigma = \frac{My}{I} \qquad \textbf{Eq. 21.34}$$

Using the shape functions that we have just developed, this equation can be expanded in terms of the shape functions as

$$\sigma = \frac{My}{I} = -Ey\frac{d^2}{dx^2}[N]\{u\} = E[B]\{u\} \qquad \textbf{Eq. 21.35}$$

In Equation 21.35, the term [B] now has the form:

$$[B] = -y\frac{d^2}{dx^2}[N] \qquad \textbf{Eq. 21.36}$$

Accepting that strain and virtual strain in the beam can be represented, respectively, by

$$\begin{aligned} \varepsilon &= \frac{\sigma}{E} = [B]\{u\} \\ \delta\varepsilon &= [B]\{\delta u\} \end{aligned} \qquad \textbf{Eq. 21.37}$$

the virtual work equation for the beam then becomes

$$\begin{aligned} \{\delta u\}^T\{p\} &= \int_{vol} (\delta\varepsilon)\sigma\, dV \\ \{\delta u\}^T\{p\} &= \int_{vol} \{\delta u\}^T[B]^T E[B]\{u\}\, dV \end{aligned} \qquad \textbf{Eq. 21.38}$$

If we now move the constant terms to the outside of the integral, and if we drop similar terms on both sides of the equation, we obtain

$$\{p\} = \left[\int_{vol} [B]^T E [B]\, dV\right]\{u\} \qquad \textbf{Eq. 21.39}$$

which is the same as we previously obtained in Equation 21.20. The stiffness matrix for the beam element is obtained from

$$[k] = \int_{vol} [B]^T E [B]\, dV \qquad \textbf{Eq. 21.40}$$

This is the same form of the elemental stiffness matrix that we previously obtained for the truss element. The only difference is the form of the matrix [B]. After making the substitutions and performing the operations implied in Equation 21.40, we find that the elemental stiffness matrix for a prismatic beam element is

$$[k] = \begin{bmatrix} \frac{12EI}{L^3} & \frac{6EI}{L^2} & -\frac{12EI}{L^3} & \frac{6EI}{L^2} \\ \frac{6EI}{L^2} & \frac{4EI}{L} & -\frac{6EI}{L^2} & \frac{2EI}{L} \\ -\frac{12EI}{L^3} & -\frac{6EI}{L^2} & \frac{12EI}{L^3} & -\frac{6EI}{L^2} \\ \frac{6EI}{L^2} & \frac{2EI}{L} & -\frac{6EI}{L^2} & \frac{4EI}{L} \end{bmatrix} \qquad \textbf{Eq. 21.41}$$

21.4 SPECIAL MEMBER END CONDITIONS

Not all beams are connected to the joints with connections that transfer full moment and shear from the member to the joint, and vice versa. Sometimes members are connected to the joint with a pin so that shearing force, but not the moment, is transferred. We can develop the stiffness matrix for this element using the same procedures that we previously developed; all that changes is the boundary conditions. To understand the procedure, consider the beam element shown in Figure 21.5. This element is the same as the one with which we previously worked, except that now the moment at the left must be equal to zero because of the beam is connected to the joint with a pin.

The assumed displacement function for this element is the same as that shown in Equation 21.26, namely

$$v(x) = a_0 + a_1x + a_2x^2 + a_3x^3 \qquad \textbf{Eq. 21.42}$$

To evaluate the unknown constants in this displacement function we have available three displacement boundary conditions. These are the translational displacement at the left, the translational displacement at the right, and the rotational displacement at the right. The

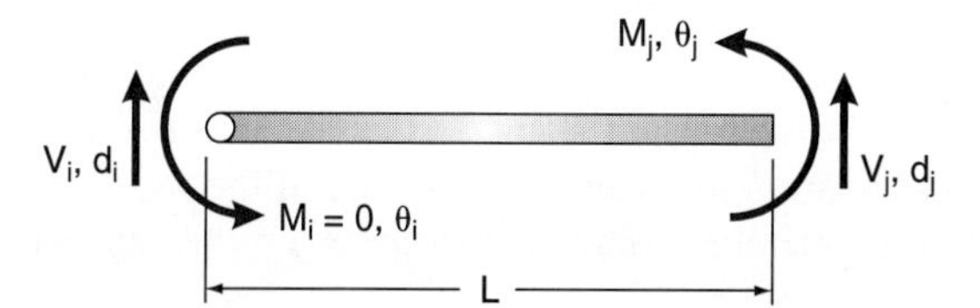

Figure 21.5 A prismatic beam element with a pinned end

fourth boundary condition is a force boundary condition; the rotational force at the left is equal to zero. Substituting these boundary conditions into Equation 21.42, we obtain

at $x = 0 \quad v = d_1$

$$v(0) = d_1 = a_0 + a_1(0) + a_2(0)^2 + a_3(0)^3 \qquad \textbf{Eq. 21.43}$$

$$a_0 = d_1$$

at $x = 0 \quad M_i = 0 = EI\dfrac{d^2v}{dx^2}$

$$0 = 2a_2 + 6a_3(0) \qquad \textbf{Eq. 21.44}$$

$$a_2 = 0$$

at $x = L \quad v = d_2$

$$v(L) = d_2 = d_1 + a_1(L) + a_3(L)^3 \qquad \textbf{Eq. 21.45}$$

at $x = L \quad \dfrac{dv}{dx} = \theta_2$

$$\frac{dv}{dx}(L) = \theta_2 = a_1 + 3a_3(L)^2 \qquad \textbf{Eq. 21.46}$$

Equations 21.45 and 21.46 can be solved simultaneously to determine the constants a_1 and a_2. Upon doing so we find that these constants are

$$a_1 = -\frac{\theta_2 L - 3d_2 + 3d_1}{2L}$$
$$a_2 = \frac{\theta_2 L - d_2 + d_1}{2L^3} \qquad \textbf{Eq. 21.47}$$

After substituting the constants a_0 through a_3 into Equation 21.42 we find that the equation describing the transverse displacement of the beam is

$$v(x) = d_1 - \left[\frac{\theta_2 L - 3d_2 + 3d_1}{2L}\right]x + \left[\frac{\theta_2 L - d_2 + d_1}{2L^3}\right]x^3 \qquad \textbf{Eq. 21.48}$$

As we did previously, we can group similar terms and express the result in matrix form. After doing so we obtain the equation:

$$v = [N]\{d\} = \begin{bmatrix} 1 - \dfrac{3x}{2L} + \dfrac{x^3}{2L^3} \\ 0 \\ \dfrac{3x}{2L} - \dfrac{x^3}{2L^3} \\ -\dfrac{x}{2} + \dfrac{x^3}{2L^2} \end{bmatrix}^T \begin{Bmatrix} d_1 \\ \theta_1 \\ d_2 \\ \theta_2 \end{Bmatrix} \qquad \textbf{Eq. 21.49}$$

Observe again that the terms in the first matrix to the right of the second equal sign are the shape functions for the prismatic with a pin at the left end. The zero in the second term of the matrix indicates that rotation of the joint at the end of the member does not have any effect on the forces in the member. That is consistent with the concept of a pin connection.

To obtain the elemental stiffness matrix for this member we apply Equations 21.36 and 21.40. When applying these equations recall that $\int_A y^2\,dA$ is equal to the moment of inertia of the beam. The resulting elemental stiffness matrix for a beam element with a pin at the left end is then

$$[k] = \begin{bmatrix} \frac{3EI}{L^3} & 0 & -\frac{3EI}{L^3} & \frac{3EI}{L^2} \\ 0 & 0 & 0 & 0 \\ -\frac{3EI}{L^3} & 0 & \frac{3EI}{L^3} & -\frac{3EI}{L^2} \\ \frac{3EI}{L^2} & 0 & -\frac{3EI}{L^2} & \frac{3EI}{L} \end{bmatrix} \quad \textbf{Eq. 21.50}$$

A similar result could have been derived for a pin located at the right end of the member.

21.5 CONSISTENT LOAD VECTORS

The principles of least work, which were developed in Chapter 13, will now be used to develop the consistent-load vectors. As before, these load vectors are the equivalent member end forces caused by loads acting along the member. They are called consistent load vectors because the same functions are used to develop the load vectors as were used to develop the stiffness matrices.

Recall from our previous discussions that

$$\delta W_e - \delta W_i = 0 \quad \textbf{Eq. 21.51}$$

The effects of all forces acting on the system, including those acting on the elements, are included in the energy terms. Let us more completely represent the internal and external work terms:

$$\delta W_e = \{\delta\bar{u}\}^T\{\bar{p}\} + \int_0^L \delta v(x)p(x)\,dx \quad \textbf{Eq. 21.52}$$

The first term to the right side of the equal sign is the energy caused by the forces acting at the ends of the member. The second term on the right-hand side is the energy caused by the forces acting along the length of the member. From previous discussion, we know that

$$\delta v = [N]\{\delta\bar{u}\} \quad \textbf{Eq. 21.53}$$

Equation 21.44 is then equal to

$$\delta W_e = \{\delta\bar{u}\}^T\{\bar{p}\} + \int_0^L \{\delta\bar{u}\}^T[N]^T p(x)\,dx \quad \textbf{Eq. 21.54}$$

This latter equation can be reduced to the following equation:

$$\delta W_e = \{\delta\bar{u}\}^T\left[\{\bar{p}\} + \int_0^L [N]^T p(x)\,dx\right] \quad \textbf{Eq. 21.55}$$

This equation can be represented as

$$\delta W_e = \{\delta\bar{u}\}^T[\{\bar{p}\} + \{\bar{p}\}^0] \quad \textbf{Eq. 21.56}$$

where

$$\{\bar{p}\}^0 = -\int_0^L [N]^T p(x)\, dx \qquad \text{Eq. 21.57}$$

This vector $\{\bar{p}\}^0$ is the consistent load vector. It is used in the same manner as we used equivalent member forces in the previous chapter. It is derived using the matrix of shape functions, [N], that we used to develop the stiffness matrix.

EXAMPLE 21.1

Determine the consistent load vector for the beam element shown in Figure 21.3 if it is subjected to a uniformly load w acting along its entire length. The load is acting downward, acting in the −y direction. That beam with load acting is shown in the following figure.

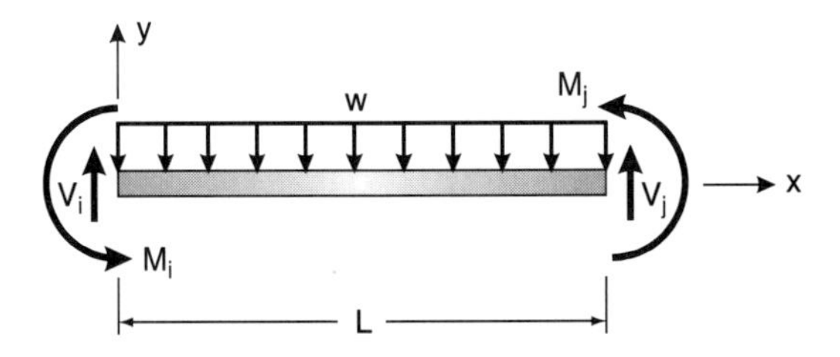

Solution. We solve for the consistent load vector by substituting the appropriate expressions into Equation 21.57. After doing so we find that

$$\{\bar{p}\}^0 = -\int_0^L \begin{bmatrix} 2\left(\dfrac{\bar{x}}{L}\right)^3 - 3\left(\dfrac{\bar{x}}{L}\right)^2 + 1 \\ \dfrac{\bar{x}^3}{L^2} - 2\left(\dfrac{\bar{x}^2}{L}\right) + \bar{x} \\ -2\left(\dfrac{\bar{x}}{L}\right)^3 + 3\left(\dfrac{\bar{x}}{L}\right)^2 \\ \dfrac{\bar{x}^3}{L^2} - \dfrac{\bar{x}^2}{L} \end{bmatrix} (-W)\, d\bar{x}$$

There is a negative sign on the load because it is acting in the direction opposite the y coordinate axis. After performing the necessary integration, we find that the consistent load vector is

$$\{\bar{p}\}^0 = \begin{Bmatrix} \dfrac{WL}{2} \\ \dfrac{WL^2}{12} \\ \dfrac{WL}{2} \\ -\dfrac{WL^2}{12} \end{Bmatrix} \quad \blacksquare$$

21.6 EFFECTS OF SUPPORT SETTLEMENT

As we learned in previous chapters, settlement of the structural supports can affect the forces in the components of a structural system. To develop a method for dealing with support settlement in matrix structural analysis, let us return to Equation 19.9, which was

$$\begin{bmatrix} [K_{11}][K_{12}] \\ [K_{21}][K_{22}] \end{bmatrix} \begin{Bmatrix} \{X_1\} \\ \{X_2\} \end{Bmatrix} = \begin{Bmatrix} \{F_1\} \\ \{F_2\} \end{Bmatrix} \qquad \textbf{Eq. 21.58}$$

This equation represents two matrix equations, namely:

$$\begin{aligned} [K_{11}]\{X_1\} + [K_{12}]\{X_2\} &= \{F_1\} \\ [K_{21}]\{X_1\} + [K_{22}]\{X_2\} &= \{F_2\} \end{aligned} \qquad \textbf{Eq. 21.59}$$

Recall that we partitioned the matrices between the free and constrained structural coordinates. The degrees of freedom at structural supports are always constrained structural coordinates, even if the support settles. The $\{X_1\}$ partition was the displacement at the unconstrained coordinates. The $\{X_2\}$ partition was the displacement at the constrained coordinates. Previously, all the coefficients in this partition were equal to zero. When this condition occurred, the displacements at the unconstrained structural coordinates and the forces of restraint at the constrained structural coordinates were calculated from

$$\begin{aligned} \{X_1\} &= [K_{11}]^{-1}\{F_1\} \\ \{F_2\} &= [K_{21}]\{X_1\} = [K_{21}][K_{11}]^{-1}\{F_1\} \end{aligned} \qquad \textbf{Eq. 21.60}$$

We used these same equations in Chapters 19 and 20.

When support settlement occurs, the displacements at the support may no longer be equal to zero. To include the effects of support settlement, the displacements at the constrained structural coordinates at the structural supports are included in the displacement vector $\{X_2\}$. When this occurs, the displacements at the unconstrained coordinates and the forces of restraint at the restrained coordinates are calculated from

$$\begin{aligned} \{X_1\} &= [K_{11}]^{-1}[\{F_1\} - [K_{12}]\{X_2\}] \\ \{F_2\} &= [K_{21}]\{X_1\} + [K_{22}]\{X_2\} \end{aligned} \qquad \textbf{Eq. 21.61}$$

The first of these equations was derived from the first equation in Equation 21.59. Notice that now the displacement vector $\{X_2\}$ is included in the equation. The second equation is identical to the second equation in Equation 21.59. When looking at the first equation, observe that the right term represents a modified set of forces acting on the structure.

21.7 PROBLEMS FOR SOLUTION

21.1 Determine the elemental stiffness matrix for the element shown in the figure. The element is pin-connected at the right end and can transfer moment at the left end. Plot the shape functions.

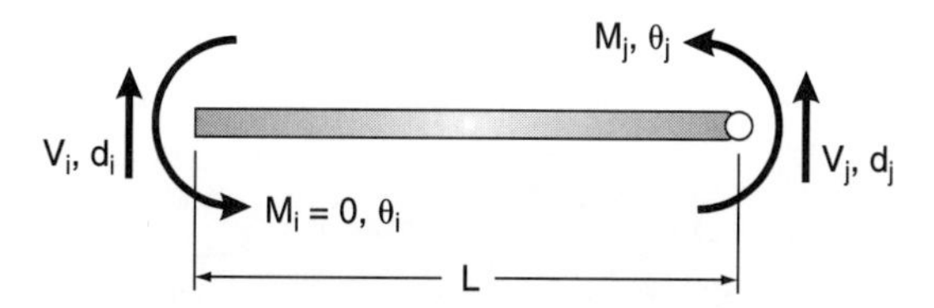

21.2 Determine the stiffness matrix for the three-degree-of-freedom truss element shown in the figure and plot the shape functions. The element has three degrees of freedom; it is not two truss elements connected together. *Hint:* Increase the displacement function previously used for the truss element by one order when computing the stiffness matrix.

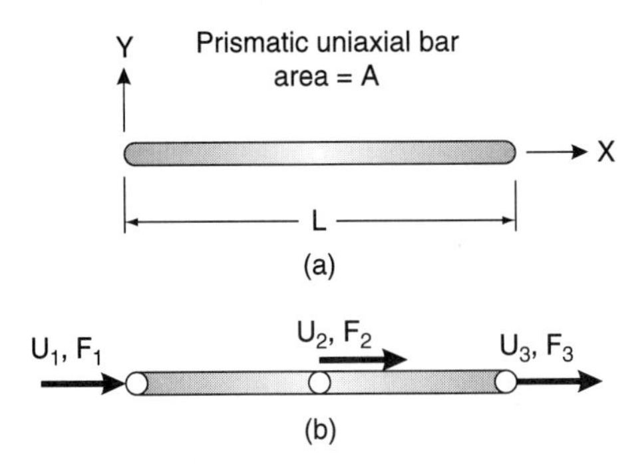

21.3 Determine the consistent load vector for the beam shown in Figure 21.5 if the beam is subjected to a uniform load acting downward.

21.4 Determine the consistent load vector for the beam shown in Figure 21.5 if the beam is subjected to a concentrated load at mid-span acting downward.

21.5 Determine the consistent load vector for the beam shown in Figure 21.3 if the beam is subjected to a concentrated load at mid-span acting downward.

21.6 Determine the consistent load vector for the beam shown in Problem 21.1 if the beam is subjected to a concentrated load at mid-span acting downward.

21.7 Determine the consistent load vector for the beam Problem 21.1 if the beam is subjected to a uniform load acting downward.

Appendix A

The Catenary Equation

Some civil engineering structures include cables as part of the load-carrying system. Among these are suspension bridges and offshore platform installations that include mooring lines, such as tension leg platforms. Because the cables are not weightless, as we often assume when designing guy-lines for small towers, the tension in the cable changes along its length and the cable sags along its length when in use. These cables assume the shape of a catenary. Developed in this appendix are the fundamental relationships necessary to evaluate the force in catenary cables.

The basic geometry of the catenary cable that we will be considering is shown in Figure A.1. Notice that the bottom of the cable is located at coordinate (0, 0) and that it is tangent to the horizontal coordinate axis at that point. The top of the cable is located at coordinate (X, Y).

To develop the governing equations for a cable, consider the segment of length Δs that is shown in Figure A.2. The equations of equilibrium for this segment of the cable are

$$\sum F_x = (T + \Delta T)\cos(\theta + \Delta\theta) - T\cos\theta = 0$$
$$\sum F_y = (T + \Delta T)\sin(\theta + \Delta\theta) - T\sin\theta - w\Delta s = 0 \qquad \textbf{Eq. A.1}$$

The term w in Eq. A.1 is the weight of the cable per unit length. For convenience in mathematical manipulation, let us introduce the function:

$$H(T, \theta) = T\cos\theta \qquad \textbf{Eq. A.2}$$

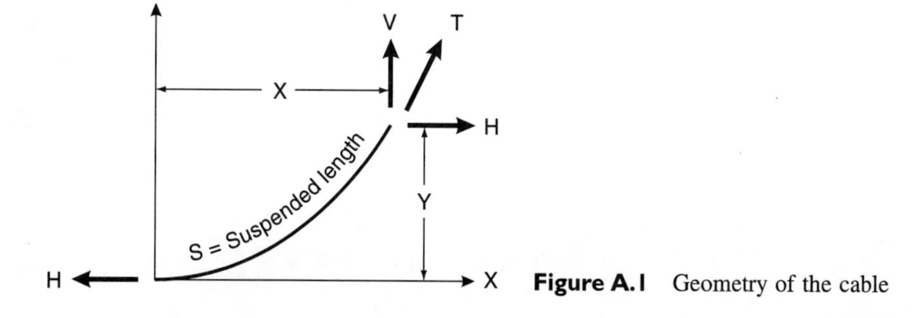

Figure A.1 Geometry of the cable

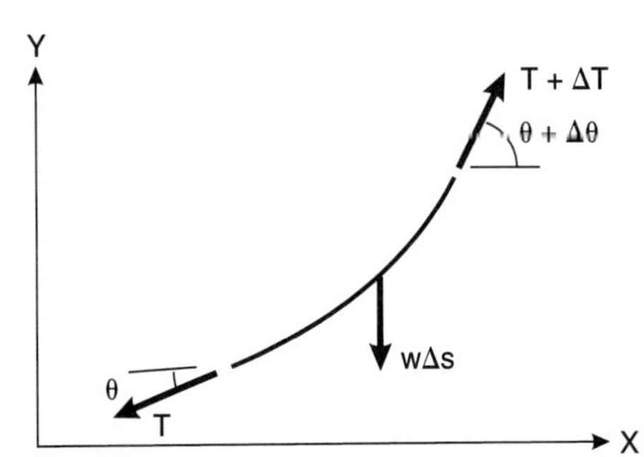

Figure A.2 A segment of the cable

This equation is the horizontal component of force in the cable. If we substitute Eq. A.2 into the first of the equations in Equation A.1, that equation becomes

$$H(T + \Delta T, \theta + \Delta\theta) - H(T, \theta) = \Delta H = 0 \qquad \textbf{Eq. A.3}$$

Along the segment Δs, then, the rate of change of the horizontal force is

$$\frac{dH}{ds} = \lim_{\Delta s \to 0}\left(\frac{\Delta H}{\Delta s}\right) = \lim_{\Delta s \to 0}\left(\frac{0}{\Delta s}\right) = 0 \qquad \textbf{Eq. A.4}$$

Because the function H, as defined in Equation A.2, is the horizontal component of force in the cable, we can conclude from Equation A.4 that the horizontal component of force is constant along the length of the cable, namely:

$$H = T\cos\theta \qquad \textbf{Eq. A.5}$$

From this equation, we can further conclude that for H to be constant the force in the cable, T, must change along its length.

Now let us examine the second of the equations in Equation A.1. If we rearrange that equation and divide both sides by $\Delta\theta$, we can obtain:

$$\frac{(T + \Delta T)\sin(\theta + \Delta\theta) - T\sin\theta}{\Delta\theta} = \frac{w\Delta s}{\Delta\theta} \qquad \textbf{Eq. A.6}$$

Taking the limit of this equation as $\Delta\theta$ approaches zero, we find that:

$$\lim_{\Delta\theta \to 0}\left[\frac{(T + \Delta T)\sin(\theta + \Delta\theta) - T\sin\theta}{\Delta\theta}\right] = \lim_{\Delta\theta \to 0}\left[\frac{w\Delta s}{\Delta\theta}\right] \qquad \textbf{Eq. A.7}$$

which becomes

$$\frac{d}{d\theta}[T\sin\theta] = w\frac{ds}{d\theta} \qquad \textbf{Eq. A.8}$$

But from Equation A.5 it is known that

$$T = \frac{H}{\cos\theta} \qquad \textbf{Eq. A.9}$$

so Equation A.8 becomes

$$\frac{d}{d\theta}[H\tan\theta] = H\sec^2\theta = w\frac{ds}{d\theta} \qquad \textbf{Eq. A.10}$$

and can be rearranged to the form

$$\frac{ds}{d\theta} = \frac{H}{w}\sec^2\theta \qquad \textbf{Eq. A.11}$$

Upon integrating Equation A.11, the suspended length of the cable is found to be

$$S = \int_0^S ds = \frac{H}{w}\int_0^\theta \sec^2\theta\, d\theta = \frac{H}{w}\tan\theta \qquad \textbf{Eq. A.12}$$

From calculus, it is known that

$$\cos\theta = \lim_{\Delta s\to 0}\left[\frac{\Delta x}{\Delta s}\right] = \frac{dx}{ds}$$
$$\sin\theta = \lim_{\Delta s\to 0}\left[\frac{\Delta y}{\Delta s}\right] = \frac{dy}{ds} \qquad \textbf{Eq. A.13}$$

By applying the chain rule with Equations A.11 and A.13, we obtain the equations:

$$\frac{dx}{d\theta} = \left[\frac{H}{w}\right]\sec^2\theta\cos\theta = \left[\frac{H}{w}\right]\sec\theta$$
$$\frac{dy}{d\theta} = \left[\frac{H}{w}\right]\sec^2\theta\sin\theta \qquad \textbf{Eq. A.14}$$

Integrating the first of these equations yields

$$X = \int_0^X dx = \frac{H}{w}\int_0^\theta \sec\theta\, d\theta \qquad \textbf{Eq. A.15}$$

which reduces to

$$X = \left[\frac{H}{w}\right]\ln(\sec\theta + \tan\theta) \qquad \textbf{Eq. A.16}$$

This is the X projection of the cable, the horizontal distance from one end to the other end. The Y projection can be found by working with the second equation in Equation A.14 and performing integration as was done to obtain the X projection. The result is

$$Y = \left[\frac{H}{w}\right](\sec\theta - 1) \qquad \textbf{Eq. A.17}$$

These last two equations are expressed in terms of the angle θ, which probably is not known. However, the angle can be expressed in terms of other parameters of cable.

From Equation A.16:

$$e^{wX/H} = \sec\theta + \tan\theta \qquad \textbf{Eq. A.18}$$

and from trigonometry we know that

$$\sec^2\theta - \tan^2\theta = (\sec\theta + \tan\theta)(\sec\theta - \tan\theta) = 1 \qquad \textbf{Eq. A.19}$$

If Equation A.18 is substituted into Equation A.19 the equation obtained is

$$\sec\theta - \tan\theta = e^{-wX/H} \qquad \textbf{Eq. A.20}$$

Then, by adding Equations A.18 and A.20, $\sec\theta$ is found to be

$$\sec\theta = \tfrac{1}{2}[e^{wX/H} + e^{-wX/H}] = \cosh\left(\frac{wX}{H}\right) \qquad \textbf{Eq. A.21}$$

and by subtracting Equation A.20 from Equation A.18 $\tan\theta$ is found to be

$$\tan\theta = \tfrac{1}{2}\left[e^{wX/H} - e^{-wX/H}\right] = \sinh\left(\frac{wX}{H}\right) \qquad \textbf{Eq. A.22}$$

The length of the cable is found by substituting Equation A.22 into Equation A.12. The result is

$$S = \left[\frac{H}{w}\right]\sinh\left(\frac{wX}{H}\right) \qquad \textbf{Eq. A.23}$$

By substituting Equation A.21 into Equation A.17, we find that the Y projection of the cable is

$$Y = \frac{H}{w}\left[\cosh\left(\frac{wX}{H}\right) - 1\right] \qquad \textbf{Eq. A.24}$$

Lastly, using Equations A.21, A.22, and A.24, the X projection of the cable is found to be

$$X = \frac{H}{w}\cosh^{-1}\left(\frac{Y + \frac{H}{w}}{\frac{H}{w}}\right) \qquad \textbf{Eq. A.25}$$

or if Equations A.21, A.22, and A.23 had been used, X would be of the form:

$$X = \frac{H}{w}\sinh^{-1}\left(\frac{Sw}{H}\right) \qquad \textbf{Eq. A.26}$$

The last quantity needed is the tension in the cable. This can be obtained by using Equations A.5, A.13, and A.26. The result is

$$T = \sqrt{(wS)^2 + H^2} \qquad \textbf{Eq. A.27}$$

From the form of this last equation, we recognize that T is computed using Pythagorean's theorem applied to the vertical and horizontal components of force in the cable. As such, the vertical component of force in the cable is

$$V = wS \qquad \textbf{Eq. A.28}$$

The equations developed are sufficient to determine the forces in a cable and its profile. They can be used to evaluate a cable that does not pass though the origin, as was shown in Figure A.1. To do so, a cable passing through the origin, which contains cable AB, is determined as shown in Figure A.3. The forces and projections at points A and B are then computed. They represent the forces and profile in the cable AB.

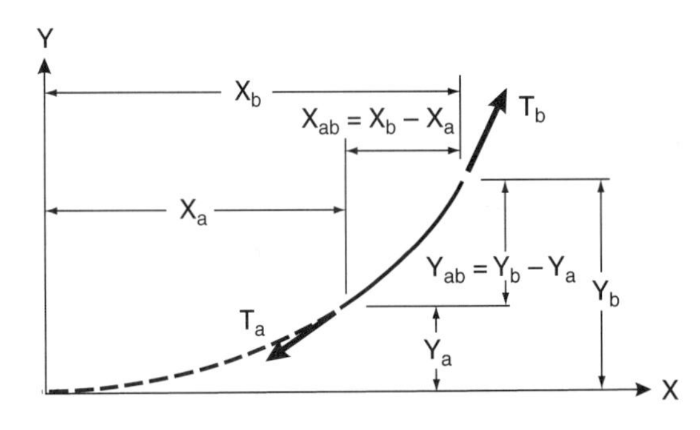

Figure A.3 Geometry of a cable segment not passing through the origin

EXAMPLE A.1

The anchor chain on a mobile offshore drilling unit is 2,000 ft long and has a submerged weight of 108 pounds per foot. It is tangent to the seafloor at the end with the anchor. The other end of the chain connects to the vessel at a point 800 ft above the seafloor. What is the tension in the chain and the distance from the vessel to the anchor?

By substituting Equation A.25 into Equation A.23 we obtain the equation:

$$S = \frac{H}{w}\sinh\left[\cosh^{-1}\left(\frac{Y + \frac{H}{w}}{\frac{H}{w}}\right)\right]$$

which can be solved for H, the horizontal component of force in the chain. The result is

$$H = \tfrac{1}{2}(S^2 + Y^2)\left(\frac{w}{Y}\right) = \tfrac{1}{2}(2{,}000^2 - 800^2)\left(\frac{108}{800}\right) = 226{,}800 \text{ lbs}$$

We then can substitute this value of H, and the other known values, into Equation A.25 to obtain X, the horizontal distance from the end of the chain to the anchor.

$$X = \frac{226{,}800}{108}\cosh^{-1}\left(\frac{800 + \frac{226{,}800}{108}}{\frac{226{,}800}{108}}\right) = 1{,}780 \text{ ft}$$

The tension in the chain can be computed using Equation A.27. The result is:

$$T = \sqrt{(108 \cdot 2000)^2 + (226{,}800)^2} = 313{,}200 \text{ lbs} \quad \blacksquare$$

Appendix B

Matrix Algebra

B.1 INTRODUCTION

An introduction to the basic rules of matrix algebra is presented here. The material is intended to provide introductory background of the subject. The focus of the material is on applications that may be of interest primarily to structural engineers. Other excellent references exist that will provide an interested reader with a more comprehensive treatment of the subject[1,2].

B.2 MATRIX DEFINITIONS AND PROPERTIES

A matrix is defined as an ordered arrangement of numbers in rows and columns, as:

$$[A] = \begin{bmatrix} a_{1,1} & a_{1,2} & \cdots & a_{1,n} \\ a_{2,1} & a_{2,2} & \cdots & a_{2,n} \\ \vdots & \vdots & \ddots & \vdots \\ a_{m,1} & a_{m,2} & \cdots & a_{m,n} \end{bmatrix} \qquad \textbf{Eq. B.1}$$

A representative matrix [A], such as shown in Eq. B.1, consists of m rows and n columns of numbers enclosed in brackets. Matrix [A] is said to be of order $m \times n$. The elements $a_{i,j}$ in the array are identified by two subscripts. The first subscript designates the row in which the element is and the second subscript designates the column. Thus $a_{3,5}$ is the element located at the intersection of the third row and the fifth column of the matrix. Elements with repeated subscripts, for example, $a_{i,i}$ are located on the main diagonal of the matrix.

Although the number of rows and columns of a matrix may vary from problem to problem, two special cases deserve mention. When m is equal to 1, the matrix consists of

[1] F. Ayres, *Theory and Problems of Matrices,* Schaum Outline Series (New York: McGraw-Hill, 1962).

[2] S. D. Conte and L. deBoor, *Elementary Numerical Analysis,* 2d ed. (New York: McGraw-Hill, 1972).

only one *row* of elements and is called a row matrix. It is written as

$$\lfloor A \rfloor = \lfloor a_1 \quad a_2 \; \dots \; a_n \rfloor \qquad \textbf{Eq. B.2}$$

When $n = 1$, the matrix consists of only one *column* and is called a column matrix or a vector. It is written as

$$\{A\} = \begin{Bmatrix} a_1 \\ a_2 \\ \vdots \\ a_n \end{Bmatrix} \qquad \textbf{Eq. B.3}$$

B.3 SPECIAL MATRIX TYPES

B.3.1 Square Matrix

When the number of rows, m, and the number of columns, n, are equal, the matrix is said to be *square*. As an example, if $m = n = 3$, the square matrix [A] may appear as

$$[A] = \begin{bmatrix} a_{11} & a_{12} & a_{13} \\ a_{21} & a_{22} & a_{23} \\ a_{31} & a_{32} & a_{33} \end{bmatrix} \qquad \textbf{Eq. B.4}$$

B.3.2 Symmetric Matrix

A symmetric matrix is one in which the off-diagonal terms are reflected about the main diagonal. As such:

$$a_{ij} = a_{ji} \qquad \text{for } i \neq j \qquad \textbf{Eq. B.5}$$

When the elements of a square matrix obey this rule, the matrix is said to be *symmetric* and the elements are arranged symmetrically about the main diagonal. An example of a symmetric 3×3 matrix is

$$[A] = \begin{bmatrix} 3 & -2 & 5 \\ -2 & 4 & 7 \\ 5 & 7 & 6 \end{bmatrix} \qquad \textbf{Eq. B.6}$$

Symmetric matrices occur frequently in structural theory and play an important role in the matrix manipulations used to develop the theory.

B.3.3 Identity Matrix

When all the elements on the main diagonal of a square matrix are equal to unity and all of the other elements are equal to zero, the matrix is called an *identity matrix*. Sometimes it is also called a *unit matrix*. Matrices of this type are identified with the symbol [I]. An example of a 3×3 identity matrix is

$$[I] = \begin{bmatrix} 1 & 0 & 0 \\ 0 & 1 & 0 \\ 0 & 0 & 1 \end{bmatrix} \qquad \textbf{Eq. B.7}$$

B.3.4 Transposed Matrix

When the elements of a given matrix are reordered so that the columns of the original matrix become the corresponding rows of the new matrix, the new matrix is said to be the *transpose* of the original matrix. In this book, the transpose of matrix [A] is given the symbol $[A]^T$. An example of a specific matrix and its transpose is

$$[A] = \begin{bmatrix} 1 & 3 & -2 \\ 5 & 4 & 7 \\ 8 & 2 & 6 \end{bmatrix}$$
$$[A]^T = \begin{bmatrix} 1 & 5 & 8 \\ 3 & 4 & 2 \\ -2 & 7 & 6 \end{bmatrix} \quad \textbf{Eq. B.8}$$

The reader should note that if the matrix [A] is a symmetric matrix, then

$$[A]^T = [A] \quad \textbf{Eq. B.9}$$

This property is used frequently in the development of structural theory using matrices. As special cases, the transpose of a row matrix becomes a column matrix and vice versa. Thus:

$$\lfloor A \rfloor^T = \{A\}$$
$$\{A\}^T = \lfloor A \rfloor \quad \textbf{Eq. B.10}$$

B.4 DETERMINANT OF A SQUARE MATRIX

A determinant of a square matrix [A] is given by the symbol $|A|$ and, in its expanded form, is written as:

$$|A| = \begin{vmatrix} a_{1,1} & a_{1,2} & \cdots & a_{1,n} \\ a_{2,1} & a_{2,2} & \cdots & a_{2,n} \\ \vdots & \vdots & \ddots & \vdots \\ a_{m,1} & a_{m,2} & \cdots & a_{m,n} \end{vmatrix} \quad \textbf{Eq. B.11}$$

This determinant is said to be of order m. Unlike matrix [A], which has no single value, the determinant $|A|$ does have a single numerical value. The value of $|A|$ is easily found for a 2×2 array of numbers, for example:

$$\begin{vmatrix} 2 & 1 \\ 4 & 5 \end{vmatrix} = 2 \cdot 5 - 1 \cdot 4 = 6 \quad \textbf{Eq. B.12}$$

The value of $|A|$ in the example was determined by multiplying the numbers on the main diagonal and subtracting from this product the product of the numbers on the other diagonal. Unfortunately, this simple procedure does not work for determinants of order greater than two.

A general procedure for finding the value of a determinant sometimes is called "expansion by minors." The first *minor* of the matrix [A], corresponding to the element $a_{i,j}$ is defined as the determinant of a reduced matrix obtained by eliminating the ith row and the jth column from matrix [A]. The minor is a specific number, like any other determinant.

As an illustration, several first minors are shown for a matrix [A]:

$$[A] = \begin{bmatrix} 1 & 2 & 3 \\ -2 & 3 & 4 \\ 1 & 5 & 2 \end{bmatrix}$$

$$\text{minor of } a_{1,1} = \begin{vmatrix} 3 & 4 \\ 5 & 2 \end{vmatrix} = 6 - 20 = -14 \qquad \textbf{Eq. B.13}$$

$$\text{minor of } a_{1,2} = \begin{vmatrix} -2 & 4 \\ 1 & 2 \end{vmatrix} = -4 - 4 = -8$$

$$\text{minor of } a_{2,3} = \begin{vmatrix} 1 & 2 \\ 1 & 5 \end{vmatrix} = 5 - 2 = 3$$

When the proper sign is attached to a minor, the result is called a *cofactor,* and is given the symbol $A_{i,j}$. The sign of a minor is determined by multiplying the minor by $(-1)^{i+j}$. Several cofactors for the example matrix [A] are

$$\begin{aligned} A_{1,1} &= (-1)^{1+1} \times \text{minor of } a_{1,1} = 1(-14) = -14 \\ A_{1,2} &= (-1)^{1+2} \times \text{minor of } a_{1,2} = (-1)(-8) = 8 \\ A_{2,3} &= (-1)^{2+3} \times \text{minor of } a_{2,3} = -1(3) = -3 \end{aligned} \qquad \textbf{Eq. B.14}$$

Now, to obtain the value of a general determinant, we can choose any arbitrary row i of matrix [A] and expand according to the relation:

$$|A| = \sum_{j=1}^{m} a_{i,j}A_{i,j} \qquad \textbf{Eq. B.15}$$

The value of a determinant can also be found by choosing an arbitrary column j and expanding according to the relation:

$$|A| = \sum_{i=1}^{m} a_{i,j}A_{i,j} \qquad \textbf{Eq. B.16}$$

If the order of the original determinant is large, the procedure described does not appear to produce a simple solution. For example, a 15th-order determinant will still have first minors that are of order 14. However, the 14th-order minors can be reduced to 13th-order minors by the expansion process. The process can be repeated until the resulting minors are of order 2. These minors can then be evaluated readily using the procedure described in the initial illustration of this section.

Although the procedure for evaluating determinants may appear long and tedious, computer algorithms can be written that will perform the necessary algebraic operations. Other simplifying procedures are available, which make use of special characteristics of determinants. These will not be discussed here, but interested readers may refer to books cited previously in footnotes 1 and 2 in this appendix.

B.5 ADJOINT MATRIX

A special matrix exists called an *adjoint* matrix and is given the symbol adj[A]. To find the adjoint matrix corresponding to an original matrix [A], first replace each element of [A] with its cofactor; the adjoint matrix is then the transpose of this resultant matrix.

Symbolically, adjoint matrix adj[A] is written as

$$\text{adj}[A] = \begin{vmatrix} A_{1,1} & A_{1,2} & \cdots & A_{1,m} \\ A_{2,1} & A_{2,2} & \cdots & A_{2,m} \\ \vdots & \vdots & \ddots & \vdots \\ A_{m,1} & A_{m,2} & \cdots & A_{m,m} \end{vmatrix} \qquad \textbf{Eq. B.17}$$

A specific numerical example of a matrix [A] and its adjoint matrix is

$$[A] = \begin{bmatrix} 1 & 2 & 3 \\ 2 & 3 & 4 \\ 1 & 5 & 3 \end{bmatrix}$$

$$\begin{aligned} A_{1,1} &= -11 & A_{1,2} &= -2 & A_{1,3} &= 7 \\ A_{2,1} &= 9 & A_{2,2} &= 0 & A_{2,3} &= -3 \\ A_{3,1} &= -1 & A_{3,2} &= 2 & A_{3,3} &= -1 \end{aligned} \qquad \textbf{Eq. B.18}$$

$$\text{adj}[A] = \begin{bmatrix} -11 & -2 & 7 \\ 9 & 0 & -3 \\ -1 & 2 & -1 \end{bmatrix}^T = \begin{bmatrix} -11 & 9 & -1 \\ -2 & 0 & 2 \\ 7 & -3 & -1 \end{bmatrix}$$

The adjoint matrix and the determinant both are used in the computation of the inverse of a matrix, a topic that is treated in a later section of this appendix.

B.6 MATRIX ARITHMETIC

B.6.1 Equality of Matrices

Two matrices are equal only if the corresponding elements of the two matrices are equal. Thus, equality of matrices can exist only between matrices of equal orders.

B.6.2 Addition and Subtraction of Matrices

Two matrices may be added or subtracted only if they have the same order. The addition of two matrices, [A] and [B], is performed as

$$[A] + [B] = [C] \qquad \textbf{Eq. B.19}$$

where the elements of matrix [C] are the sum of corresponding elements of [A] and [B], that is

$$c_{i,j} = a_{i,j} + b_{i,j} \qquad \textbf{Eq. B.20}$$

An example of the addition of two matrices is shown:

$$\begin{bmatrix} 1 & 2 \\ 3 & 4 \end{bmatrix} + \begin{bmatrix} 5 & 7 \\ 6 & 8 \end{bmatrix} = \begin{bmatrix} 6 & 8 \\ 10 & 12 \end{bmatrix} \qquad \textbf{Eq. B.21}$$

Subtraction of two matrices is performed similarly by subtracting corresponding elements.

Both the commutative and the associative laws hold for the addition and subtraction of matrices. Thus:

$$[A] + [B] = [B] + [A] \qquad \textbf{Eq. B.22}$$

and

$$[A] + ([B] + [C]) = ([A] + [B]) + [C] \qquad \textbf{Eq. B.23}$$

B.6.3 Scalar Multiplication of Matrices

To multiply a matrix by a scalar, *each element* of the matrix is multiplied by the scalar. Thus:

$$\alpha[A] = \begin{bmatrix} \alpha a_{1,1} & \alpha a_{1,2} & \cdots & \alpha a_{1,n} \\ \alpha a_{2,1} & \alpha a_{2,2} & \cdots & \alpha a_{2,n} \\ \vdots & \vdots & \ddots & \vdots \\ \alpha a_{m,1} & \alpha a_{m,2} & \cdots & \alpha a_{m,n} \end{bmatrix} \quad \textbf{Eq. B.24}$$

B.6.4 Multiplication of Matrices

The product of two matrices exists only if the matrices are *conformable*. For the matrix product [A][B], conformability means that the number of columns of [A] equals the number of rows of [B]. The two matrices shown are conformable (in the order shown) and may be multiplied.

$$[A] = \begin{bmatrix} 1 & 3 \\ 2 & 5 \end{bmatrix} \qquad [B] = \begin{bmatrix} 2 & 6 & 5 \\ 1 & 3 & -4 \end{bmatrix} \quad \textbf{Eq. B.25}$$

The number of columns in [A] is equal to the number of rows in [B]. However, if the order of the matrix multiplication is reversed, [B][A], the matrices are *not* conformable and the matrix product does not exist. Furthermore, even if the matrices [A] and [B] are square, and thus conformable in the order [A][B] and [B][A], the two matrix products are generally not the same. In general,

$$[A][B] \neq [B][A] \quad \textbf{Eq. B.26}$$

The formal definition of a matrix product between matrices that are conformable is given as follows:

$$[A]_{m\times l}[B]_{l\times n} = [C]_{m\times n} \quad \textbf{Eq. B.27}$$

where

$$c_{i,j} = \sum_{k=1}^{l} a_{i,k}b_{k,j} \quad \textbf{Eq. B.28}$$

Note that matrix [A] is not of the same order as [B], but the two matrices are conformable since the number of columns of [A] equals the number of rows of [B]. The order of the product matrix [C] is $m \times n$.

A simple illustration of a matrix product is shown:

$$\begin{bmatrix} 1 & 2 \\ -3 & 2 \end{bmatrix}\begin{bmatrix} 1 & 3 & 2 \\ 4 & 5 & 3 \end{bmatrix} = \begin{bmatrix} 9 & 13 & 8 \\ 5 & 1 & 0 \end{bmatrix} \quad \textbf{Eq. B.29}$$

In this matrix multiplication the terms in the matrix [C] are computed as follows:

$$\begin{aligned} c_{1,1} &= (1)(1) + (2)(4) = 9 \\ c_{1,2} &= (1)(3) + (2)(5) = 13 \\ c_{1,3} &= (1)(2) + (2)(3) = 8 \\ c_{2,1} &= (-3)(1) + (2)(4) = 5 \\ c_{2,2} &= (-3)(3) + (2)(5) = 1 \\ c_{2,3} &= (-3)(2) + (2)(3) = 0 \end{aligned} \quad \textbf{Eq. B.30}$$

Although the order in which two matrices are multiplied may not be reversed, in general, without obtaining different results, both the associative and the distributive laws are valid for matrix products. Thus

$$[A][B][C] = ([A][B])[C] = [A]([B][C]) \qquad \textbf{Eq. B.31}$$

and

$$[A]([B] + [C]) = [A][B] + [A][C] \qquad \textbf{Eq. B.32}$$

If a matrix [A] is multiplied by the identity matrix [I] (assuming that the matrices are conformable) the matrix [A] remains unchanged. Thus

$$\begin{bmatrix} 1 & 2 & 3 \\ -2 & 4 & 6 \\ 3 & 5 & 2 \end{bmatrix}\begin{bmatrix} 1 & 0 & 0 \\ 0 & 1 & 0 \\ 0 & 0 & 1 \end{bmatrix} = \begin{bmatrix} 1 & 2 & 3 \\ -2 & 4 & 6 \\ 3 & 5 & 2 \end{bmatrix} \qquad \textbf{Eq. B.33}$$

In general,

$$\begin{aligned} [A][I] &= [A] \\ [I][A] &= [A] \end{aligned} \qquad \textbf{Eq. B.34}$$

B.6.5 Transpose of a Product

If a matrix product is transposed, the result is the reverse product of transpose of the individual matrices. The result is shown symbolically as:

$$([A][B])^T = [B]^T[A]^T \qquad \textbf{Eq. B.35}$$

The transpose of a triple matrix product may be found by using Equation B.35 and the associative law in several stages, as shown

$$\begin{aligned} ([A][B][C])^T &= \{[A]([B][C])\}^T \\ &= ([B][C])^T[A]^T \\ &= [C]^T[B]^T[A]^T \end{aligned} \qquad \textbf{Eq. B.36}$$

Note that in finding the transpose of a matrix product the order of multiplication changes.

B.6.6 Matrix Inverse

Although matrix addition, subtraction, and multiplication have been defined in the preceding sections, no mention has been made of matrix division. In fact, division of matrices in the form [A]/[B] does not exist. However, a matrix operation does exist that closely parallels algebraic division. This operation makes use of a matrix *inverse*.

The inverse of a square matrix [A] is given the symbol $[A]^{-1}$. It is defined such that:

$$[A][A]^{-1} = [I] \qquad \textbf{Eq. B.37}$$

Many techniques exist with which a matrix inverse may be determined. One formal technique is described by the following relationship:

$$[A]^{-1} = \frac{\text{adj}[A]}{|A|} \qquad \textbf{Eq. B.38}$$

As an example, consider the matrix [A] given by Equation B.18. The adjoint matrix adj[A] is also shown in Equation B.18. The determinant $|A|$ may be found by using the elements and first minors of the first column of [A]:

$$\begin{aligned} |A| &= a_{1,1}A_{1,1} + a_{2,1}A_{2,1} + a_{3,1}A_{3,1} \\ &= 1(-11) + 2(9) + 1(-1) \\ &= 6 \end{aligned} \qquad \textbf{Eq. B.39}$$

Thus:

$$[A]^{-1} = \frac{1}{6}\begin{bmatrix} -11 & 9 & -1 \\ -2 & 0 & 2 \\ 7 & -3 & -1 \end{bmatrix} \qquad \textbf{Eq. B.40}$$

The correctness of the values given for the coefficients of $[A]^{-1}$ may be verified by forming the matrix product $[A][A]^{-1}$ and checking to see if the result is the identity matrix. For the example given:

$$[A][A]^{-1} = \begin{bmatrix} 1 & 2 & 3 \\ 2 & 3 & 4 \\ 1 & 5 & 3 \end{bmatrix}\left(\frac{1}{6}\right)\begin{bmatrix} -11 & 9 & -1 \\ -2 & 0 & 2 \\ 7 & -3 & -1 \end{bmatrix} = \frac{1}{6}\begin{bmatrix} 6 & 0 & 0 \\ 0 & 6 & 0 \\ 0 & 0 & 6 \end{bmatrix} = [I] \qquad \textbf{Eq. B.41}$$

A special situation involving the inverse of a matrix deserves attention. If the determinant of a matrix [A] is equal to zero ($|A| = 0$), then the division operation indicated by Equation B.38 cannot be performed. Under these circumstances, the inverse of matrix [A] does not exist and matrix [A] is said to be singular. Singular matrices occur frequently in structural theory and a reader should be aware of the meaning of this term. An example of a singular matrix is shown:

$$[A] = \begin{bmatrix} 1 & 2 & 4 \\ -2 & 3 & 2 \\ 3 & 6 & 12 \end{bmatrix} \qquad \textbf{Eq. B.42}$$

and

$$|A| = a_{1,1}A_{1,1} + a_{2,1}A_{2,1} + a_{3,1}A_{3,1} = 24 + 0 - 24 = 0 \qquad \textbf{Eq. B.43}$$

The inverse of a matrix product may be found using rules that are very similar to those used in finding the transpose of a matrix product as shown in Equation B.35. Specifically,

$$([A][B])^{-1} = [B]^{-1}[A]^{-1} \qquad \textbf{Eq. B.44}$$

and

$$\begin{aligned} ([A][B][C])^{-1} &= ([B][C])^{-1}[A]^{-1} \\ &= [C]^{-1}[B]^{-1}[A]^{-1} \end{aligned} \qquad \textbf{Eq. B.45}$$

B.6.7 Application of the Matrix Inverse

Consider a set of algebraic equations, each of which contains a number of unknown quantities, x_i, namely:

$$\begin{aligned} a_{1,1}x_1 + a_{1,2}x_2 + a_{1,3}x_3 &= b_1 \\ a_{2,1}x_1 + a_{2,2}x_2 + a_{2,3}x_3 &= b_2 \\ a_{3,1}x_1 + a_{3,2}x_2 + a_{3,3}x_3 &= b_3 \end{aligned} \qquad \textbf{Eq. B.46}$$

The set of algebraic equations may be cast in matrix form as follows:

$$\begin{bmatrix} a_{1,1} & a_{1,2} & a_{1,3} \\ a_{2,1} & a_{2,2} & a_{2,3} \\ a_{3,1} & a_{3,2} & a_{3,3} \end{bmatrix} \begin{Bmatrix} x_1 \\ x_2 \\ x_3 \end{Bmatrix} = \begin{Bmatrix} b_1 \\ b_2 \\ b_3 \end{Bmatrix} \qquad \textbf{Eq. B.47}$$

or, symbolically, as

$$[A]\{X\} = \{B\} \qquad \textbf{Eq. B.48}$$

The solution for the unknown quantities, x_1, x_2, and x_3, may be found by premultiplying both sides of Equation B.48 by $[A]^{-1}$, namely:

$$\begin{aligned} [A]^{-1}[A]\{X\} &= [A]^{-1}\{B\} \\ [I]\{X\} &= [A]^{-1}\{B\} \\ \{X\} &= [A]^{-1}\{B\} \end{aligned} \qquad \textbf{Eq. B.49}$$

Therefore, if $[A]^{-1}$ is known or can be computed, the values of x_i can be determined from a simple matrix product.

As a numerical example, consider the following algebraic equations. The coefficients in these equations are the same as those shown in matrix [A] in Equation B.18, namely:

$$\begin{aligned} 1x_1 + 2x_2 + 3x_3 &= 13 \\ 2x_1 + 3x_2 + 4x_3 &= 19 \\ 1x_1 + 5x_2 + 3x_3 &= 22 \end{aligned} \qquad \textbf{Eq. B.50}$$

The solution for x_1, x_2, and x_3 is

$$\begin{Bmatrix} x_1 \\ x_2 \\ x_3 \end{Bmatrix} = \begin{bmatrix} 1 & 2 & 3 \\ 2 & 3 & 4 \\ 1 & 5 & 3 \end{bmatrix}^{-1} \begin{Bmatrix} 13 \\ 19 \\ 20 \end{Bmatrix} = \frac{1}{6} \begin{bmatrix} -11 & 9 & -1 \\ -2 & 0 & 2 \\ 7 & -3 & -1 \end{bmatrix} \begin{Bmatrix} 13 \\ 19 \\ 20 \end{Bmatrix} = \begin{Bmatrix} 1 \\ 2 \\ 3 \end{Bmatrix} \qquad \textbf{Eq. B.51}$$

Many other techniques exist for solving algebraic equations simultaneously. The use of the matrix inverse is a special technique that may be used at the option of the analyst.

B.7 GAUSS'S METHOD FOR SOLVING SIMULTANEOUS EQUATIONS

One of the most widely used methods for solving linear, algebraic equations simultaneously is the Gauss method. This method, or some variation of it, is used in many of the currently available computer programs that deal with structural problems. Interestingly, the method is also well adapted to hand calculations. The fundamentals of this method are illustrated next.

Consider a set of three algebraic equations written in terms of three unknowns x_1, x_2, and x_3:

$$\begin{aligned} 3x_1 + 1x_2 - 1x_3 &= 2 \\ 1x_1 + 4x_2 + 1x_3 &= 12 \\ 2x_1 + 1x_2 + 2x_3 &= 10 \end{aligned} \qquad \textbf{Eq. B.52}$$

To solve for the unknown terms, first divide each equation of the set by its leading coefficient, so that the leading coefficient becomes unity, namely:

$$\begin{aligned} x_1 + \tfrac{1}{3}x_2 - \tfrac{1}{3}x_3 &= \frac{2}{3} \\ x_1 + 4x_2 + x_3 &= 12 \\ x_1 + \tfrac{1}{2}x_2 + x_3 &= 5 \end{aligned} \qquad \textbf{Eq. B.53}$$

Next, subtract the first equation of the resultant set from each of the other equations so that the leading coefficients of the second and third equations are equal to zero, namely:

$$\begin{aligned} x_1 + 0.3333x_2 - 0.3333x_3 &= 0.6667 \\ 3.6667x_2 + 1.3333x_3 &= 11.3333 \\ 0.1667x_2 + 1.3333x_3 &= 4.3333 \end{aligned} \qquad \textbf{Eq. B.54}$$

Repeat the two operations described, but now start with the second equation; divide the second and third equations by their leading coefficient so that the leading coefficients become unity:

$$\begin{aligned} x_1 + 0.3333x_2 - 0.3333x_3 &= 0.6667 \\ x_2 + 0.3636x_3 &= 3.0909 \\ x_2 + 7.9996x_3 &= 25.9946 \end{aligned} \qquad \textbf{Eq. B.55}$$

Subtract the second equation from the third equation so that the leading coefficient of the third equation becomes zero:

$$\begin{aligned} x_1 + 0.3333x_2 - 0.3333x_3 &= 0.6667 \\ x_2 + 0.3636x_3 &= 3.0909 \\ 7.6360x_3 &= 22.9037 \end{aligned} \qquad \textbf{Eq. B.56}$$

Divide the third equation by its leading coefficient. This operation yields a solution for x_3:

$$x_3 = \frac{22.9037}{7.6360} = 2.9994 \qquad \textbf{Eq. B.57}$$

Now substitute the derived value of x_3 into the latest form of the second equation and solve for x_2:

$$\begin{aligned} x_2 + 0.3636x_3 = x_2 + 0.3636(2.9994) &= 3.0909 \\ x_2 &= 2.0003 \end{aligned} \qquad \textbf{Eq. B.58}$$

Finally, substitute the derived value of x_2 and x_3 into the first equation to solve for x_1:

$$\begin{aligned} x_1 + 0.3333x_2 - 0.3333x_3 = x_1 + 0.3333(2.0003) - 0.3333(2.9994) &= 0.6667 \\ x_1 &= 0.9997 \end{aligned} \qquad \textbf{Eq. B.59}$$

The solution, then, obtained using Gauss's method is

$$\begin{Bmatrix} x_1 \\ x_2 \\ x_3 \end{Bmatrix} = \begin{Bmatrix} 0.9997 \\ 2.0003 \\ 2.9994 \end{Bmatrix} \qquad \textbf{Eq. B.60}$$

This solution compares closely to the exact solution, which is

$$\begin{Bmatrix} x_1 \\ x_2 \\ x_3 \end{Bmatrix} = \begin{Bmatrix} 1 \\ 2 \\ 3 \end{Bmatrix} \qquad \textbf{Eq. B.61}$$

The inexactness of the solution shown is a function of round-off errors. Improved accuracy is attained by carrying a larger number of significant figures in the solution.

B.8 SPECIAL TOPICS

B.8.1 Matrix Partitioning

The manipulation of matrix equations is frequently made simpler by dividing the matrices into smaller matrices, called *partitions*. Partitioning is indicated in this book by horizontal and vertical lines between the rows and the columns of the matrices. Illustrations of matrices that have been partitioned are as follows:

$$[A] = \left[\begin{array}{cc|c} a_{1,1} & a_{1,2} & a_{1,3} \\ a_{2,1} & a_{2,2} & a_{2,3} \\ \hline a_{3,1} & a_{3,2} & a_{3,3} \end{array}\right]$$

$$\lfloor A \rfloor = \lfloor a_1 \quad a_2 \mid a_3 \rfloor \qquad \textbf{Eq. B.62}$$

$$\{A\} = \left\{\begin{array}{c} a_1 \\ a_2 \\ \hline a_3 \end{array}\right\}$$

Partitioning of a matrix equation is

$$[A]\{X\} = \{B\}$$

$$\left[\begin{array}{cc|c} a_{1,1} & a_{1,2} & a_{1,3} \\ a_{2,1} & a_{2,2} & a_{2,3} \\ \hline a_{3,1} & a_{3,2} & a_{3,3} \end{array}\right] \left\{\begin{array}{c} x_1 \\ x_2 \\ \hline x_3 \end{array}\right\} = \left\{\begin{array}{c} b_1 \\ b_2 \\ \hline b_3 \end{array}\right\} \qquad \textbf{Eq. B.63}$$

Note that the horizontal partition lines in Equation B.63 extend between the same two rows for each matrix in the equation. Furthermore, the column numbers that define the vertical partition line for the square matrix are the same as the row numbers that define the horizontal partition lines for the complete equation. Thus, if the horizontal partition lines run between the second and third rows, then the vertical partition line runs between the second and third column of the square matrix.

As an illustration of the use of partitioning, consider the same matrix equation as was described in Equation B.50:

$$\left[\begin{array}{cc|c} 1 & 2 & 3 \\ 2 & 3 & 4 \\ \hline 1 & 5 & 3 \end{array}\right] \left\{\begin{array}{c} x_1 \\ x_2 \\ \hline x_3 \end{array}\right\} = \left\{\begin{array}{c} 13 \\ 19 \\ \hline 22 \end{array}\right\} \qquad \textbf{Eq. B.64}$$

In Equation B.64, partition lines have been drawn between the second and third rows of each matrix, and between the second and third columns of matrix [A]. When the matrix is partitioned, the partitions, are generally still are matrices and are manipulated as matrices. In this example, though, one of the partitions is a 1×1 matrix, which can be treated as a scalar. The original matrix equation may now be written as two matrix equations, namely:

$$\begin{bmatrix} 1 & 2 \\ 2 & 3 \end{bmatrix} \begin{Bmatrix} x_1 \\ x_2 \end{Bmatrix} + \begin{Bmatrix} 3 \\ 4 \end{Bmatrix} x_3 = \begin{Bmatrix} 13 \\ 19 \end{Bmatrix} \qquad \textbf{Eq. B.65}$$

and

$$\lfloor 1 \quad 5 \rfloor \begin{Bmatrix} x_1 \\ x_2 \end{Bmatrix} + 3x_3 = 22 \qquad \textbf{Eq. B.66}$$

Equation B.66 may be solved for x_3 in terms of x_1 and x_2 as

$$3x_3 = 22 - \lfloor 1 \quad 5 \rfloor \begin{Bmatrix} x_1 \\ x_2 \end{Bmatrix}$$

$$x_3 = \frac{22}{3} - \frac{1}{3}\lfloor 1 \quad 5 \rfloor \begin{Bmatrix} x_1 \\ x_2 \end{Bmatrix} \qquad \textbf{Eq. B.67}$$

The value of x_3 from Equation B.67 is substituted into Equation B.65, and the resultant equation is solved for x_1 and x_2:

$$\begin{bmatrix} 1 & 2 \\ 2 & 3 \end{bmatrix} \begin{Bmatrix} x_1 \\ x_2 \end{Bmatrix} + \begin{Bmatrix} 3 \\ 4 \end{Bmatrix} \left(\frac{22}{3} - \frac{1}{3}\lfloor 1 \quad 5 \rfloor \begin{Bmatrix} x_1 \\ x_2 \end{Bmatrix} \right) = \begin{Bmatrix} 13 \\ 19 \end{Bmatrix}$$

$$\begin{bmatrix} 1 & 2 \\ 2 & 3 \end{bmatrix} \begin{Bmatrix} x_1 \\ x_2 \end{Bmatrix} - \frac{1}{3}\begin{bmatrix} 3 & 15 \\ 4 & 20 \end{bmatrix} \begin{Bmatrix} x_1 \\ x_2 \end{Bmatrix} = \begin{Bmatrix} 13 \\ 19 \end{Bmatrix} - \frac{1}{3}\begin{Bmatrix} 66 \\ 88 \end{Bmatrix}$$

$$\frac{1}{3}\begin{bmatrix} 0 & -9 \\ 2 & -11 \end{bmatrix} \begin{Bmatrix} x_1 \\ x_2 \end{Bmatrix} = \frac{1}{3}\begin{Bmatrix} -27 \\ -31 \end{Bmatrix} \qquad \textbf{Eq. B.68}$$

$$\begin{Bmatrix} x_1 \\ x_2 \end{Bmatrix} = \begin{bmatrix} 0 & -9 \\ 2 & -11 \end{bmatrix}^{-1} \begin{Bmatrix} -27 \\ -31 \end{Bmatrix} = \begin{Bmatrix} 1 \\ 3 \end{Bmatrix}$$

The values of x_1 and x_2 from Equation B.68 may now be substituted in Equation B.67 to obtain the value of x_3, namely:

$$x_3 = \frac{22}{3} - \frac{1}{3}\lfloor 1 \quad 5 \rfloor \begin{Bmatrix} 1 \\ 3 \end{Bmatrix} = \frac{1}{3}(22 - 16) = 2 \qquad \textbf{Eq. B.69}$$

One obvious advantage of partitioning is that the order of the matrices for which inverses must be found is reduced. However, this advantage is offset somewhat by the fact that additional algebraic manipulations are required when using partitioning schemes. Nonetheless, partitioning of matrix equations is used extensively in developing computer solutions for structural problems.

B.8.2 Differentiating and Integrating a Matrix

Previously in this appendix a matrix was defined as an ordered arrangement of numbers in rows and columns. Although the elements of a matrix generally are thought of as constants, they may also be variables, as shown by the following example:

$$[A] = \begin{bmatrix} 3x & -x^2 & 2x^4 \\ -x^2 & 5x^3 & 7x \\ 2x^4 & 7x & 2x^2 \end{bmatrix} \qquad \textbf{Eq. B.70}$$

Differentiating matrix [A] with respect to x is performed by differentiating each element in the matrix with respect to x. The result for the example is shown as follows:

$$\frac{d}{dx}[A] = \begin{bmatrix} 3 & -2x & 8x^3 \\ -2x & 15x^2 & 7 \\ 8x^3 & 7 & 4x \end{bmatrix} \qquad \textbf{Eq. B.71}$$

The elements of [A] are functions of more than one variable, say, x and y. Partial differentiation of matrix [A] with respect to either x or y is performed similarly, by partially differentiating each element of the matrix with respect to that variable.

In a similar way, integration of a matrix is performed by integrating each element of the matrix. Thus, for the matrix [A] given by Equation B.70, integration produces the following results:

$$\int_a^b [A]\,dx = \begin{bmatrix} \frac{3x^2}{2} & \frac{-x^3}{3} & \frac{2x^5}{5} \\ \frac{-x^3}{3} & \frac{5x^4}{4} & \frac{7x^2}{2} \\ \frac{2x^5}{5} & \frac{7x^2}{2} & \frac{2x^3}{3} \end{bmatrix}_a^b = \begin{bmatrix} \frac{3(b^2 - a^2)}{2} & \frac{-(b^3 - a^3)}{3} & \frac{2(b^5 - a^5)}{5} \\ \frac{-(b^3 - a^3)}{3} & \frac{5(b^4 - a^4)}{4} & \frac{7(b^2 - a^2)}{2} \\ \frac{2(b^5 - a^5)}{5} & \frac{7(b^2 - a^2)}{2} & \frac{2(b^3 - a^3)}{3} \end{bmatrix} \quad \textbf{Eq. B.72}$$

A special application of matrix differentiation is worth noting because it frequently appears in the development of structural theory. Assume that a scalar variable U is defined in terms of a matrix triple product:

$$U = \tfrac{1}{2}\lfloor x \rfloor [A]\{x\} \quad \textbf{Eq. B.73}$$

where $\{x\}$ is a column matrix consisting of n variables $(x_1, x_2, \ldots, x_n)$. The square matrix [A] is a symmetric matrix. If U is differentiated successively with respect to x_1, x_2, ... x_n, and the results arranged in a column matrix, the result is remarkably simple, namely:

$$\begin{Bmatrix} \frac{\partial U}{\partial x_1} \\ \frac{\partial U}{\partial x_2} \\ \vdots \\ \frac{\partial U}{\partial x_n} \end{Bmatrix} = [A]\{x\} \quad \textbf{Eq. B.74}$$

A further differentiation yields:

$$\frac{\partial^2 U}{\partial x_i \partial x_j} = a_{i,j} \quad \textbf{Eq. B.75}$$

where $a_{i,j}$ are the elements of the original matrix [A]. Although the proof of Equations B.74 and B.75 is not given here, a simple example will verify their correctness. Consider the following matrix triple product:

$$U = \tfrac{1}{2}\lfloor x_1 \quad x_2 \rfloor \begin{bmatrix} 2 & 4 \\ 4 & 3 \end{bmatrix} \begin{Bmatrix} x_1 \\ x_2 \end{Bmatrix} \quad \textbf{Eq. B.76}$$

When U is expanded the result is

$$U = \tfrac{1}{2}(2x_1^2 + 8x_1x_2 + 3x_2^2) \quad \textbf{Eq. B.77}$$

Differentiating U with respect to x_1 and with respect to x_2 yields

$$\frac{\partial U}{\partial x_1} = 2x_1 + 4x_2$$
$$\frac{\partial U}{\partial x_2} = 4x_1 + 3x_2 \qquad \textbf{Eq. B.78}$$

Arrangement of these results in matrix form produces

$$\begin{Bmatrix} \dfrac{\partial U}{\partial x_1} \\ \dfrac{\partial U}{\partial x_2} \end{Bmatrix} = \begin{bmatrix} 2 & 4 \\ 4 & 3 \end{bmatrix} \begin{Bmatrix} x_1 \\ x_2 \end{Bmatrix} \qquad \textbf{Eq. B.79}$$

This result corresponds to Equation B.74. Further differentiation produces

$$\frac{\partial^2 U}{\partial x_1^2} = 2 \qquad \frac{\partial^2 U}{\partial x_1 x_2} = 4$$
$$\frac{\partial^2 U}{\partial x_2 x_1} = 4 \qquad \frac{\partial^2 U}{\partial x_2^2} = 3 \qquad \textbf{Eq. B.80}$$

These results agree with Equation B.75.

The resultant form of the matrix triple product given by Equation B.73 is sometimes called a *quadratic form* (because of the second-order appearance of the variables in the matrix product). Quadratic forms occur frequently in structural theory when strain energy is used to help derive the stiffness matrix. Although quadratic forms have not been used in this book, the topic is important in more advanced treatment of structural theory and is included here for future reference for interested readers.

Appendix C

Wind and Snow Load Tables and Figures

TABLE C.1 CLASSIFICATION OF BUILDINGS AND OTHER STRUCTURES FOR FLOOD, WIND, SNOW, AND EARTHQUAKE LOADS

Nature of Occupancy	Category
Buildings and other structures that represent a low hazard to human life in the event of failure including, but not limited to: ▪ Agricultural facilities ▪ Certain temporary facilities ▪ Minor storage facilities	I
All buildings and other structures except those listed in Categories I, III and IV	II
Buildings and other structures that represent a substantial hazard to human life in the event of failure including, but not limited to: ▪ Buildings and other structures where more than 300 people congregate in one area ▪ Buildings and other structures with day-care facilities with capacity greater than 150 ▪ Buildings and other structures with elementary or secondary school facilities with capacity greater than 150 ▪ Buildings and other structures with a capacity greater than 500 for colleges or adult education facilities ▪ Health-care facilities with a capacity of 50 or more resident patients but not having surgery or emergency treatment facilities ▪ Jails and detention facilities ▪ Power generating stations and other public utility facilities not included in Category IV Buildings and other structures containing sufficient quantities of toxic, explosive or other hazardous substances to be dangerous to the public if released including, but not limited to: ▪ Petrochemical facilities ▪ Fuel storage facilities ▪ Manufacturing or storage facilities for hazardous chemicals ▪ Manufacturing or storage facilities for explosives	III
Buildings and other structures that are equipped with secondary containment of toxic, explosive or other hazardous substances (including, but not limited to double-wall tank, dike of sufficient size to contain a spill, or other means to contain a spill or a blast within the property boundary of the facility and prevent release of harmful quantities of contaminants to the air, soil, ground water, or surface water) or atmosphere (where appropriate) shall be eligible for classification as a Category II structure. In hurricane-prone regions, buildings and other structures that contain toxic, explosive, or other hazardous substances and do not qualify as Category IV structures shall be eligible for classification as Category II structures for wind loads if these structures are operated in accordance with mandatory procedures that are acceptable to the authority having jurisdiction and which effectively diminish the effects of wind on critical structural elements or which alternatively protect against harmful releases during and after hurricanes. Buildings and other structures designated as essential facilities including, but not limited to: ▪ Hospitals and other health care facilities having surgery or emergency treatment facilities ▪ Fire, rescue, and police stations and emergency vehicle garages ▪ Designated earthquake, hurricane, or other emergency shelters ▪ Communications centers and other facilities required for emergency response ▪ Power generating stations and other public utility facilities required in an emergency ▪ Ancillary structures (including, but not limited to communication towers, fuel storage tanks, cooling towers, electrical substation structures, fire water storage tanks or other structures housing or supporting water or other fire-suppression material or equipment) required for operation of Category IV structures during an emergency ▪ Aviation control towers, air traffic control centers and emergency aircraft hangars ▪ Water storage facilities and pump structures required to maintain water pressure for fire suppression ▪ Buildings and other structures having critical national defense functions	IV

Source: American Society of Civil Engineers, Minimum Design Loads for Buildings and Other Structures, ASCE 7-98, 2000. Reproduced with permission of the American Society of Civil Engineers.

TABLE C.2 VELOCITY PRESSURE EXPOSURE COEFFICIENTS, K_h AND K_z

Height above ground level, z		Exposure (Note 1)					
		A		B		C	D
ft	(m)	Case 1	Case 2	Case 1	Case 2	Cases 1 & 2	Cases 1 & 2
0–15	(0–4.6)	0.68	0.32	0.70	0.57	0.85	1.03
20	(6.1)	0.68	0.36	0.70	0.62	0.90	1.08
25	(7.6)	0.68	0.39	0.70	0.66	0.94	1.12
30	(9.1)	0.68	0.42	0.70	0.70	0.98	1.16
40	(12.2)	0.68	0.47	0.76	0.76	1.04	1.22
50	(15.2)	0.68	0.52	0.81	0.81	1.09	1.27
60	(18)	0.68	0.55	0.85	0.85	1.13	1.31
70	(21.3)	0.68	0.59	0.89	0.89	1.17	1.34
80	(24.4)	0.68	0.62	0.93	0.93	1.21	1.38
90	(27.4)	0.68	0.65	0.96	0.96	1.24	1.40
100	(30.5)	0.68	0.68	0.99	0.99	1.26	1.43
120	(36.6)	0.73	0.73	1.04	1.04	1.31	1.48
140	(42.7)	0.78	0.78	1.09	1.09	1.36	1.52
160	(48.8)	0.82	0.82	1.13	1.13	1.39	1.55
180	(54.9)	0.86	0.86	1.17	1.17	1.43	1.58
200	(61.0)	0.90	0.90	1.20	1.20	1.46	1.61
250	(76.2)	0.98	0.98	1.28	1.28	1.53	1.68
300	(91.4)	1.05	1.05	1.35	1.35	1.59	1.73
350	(106.7)	1.12	1.12	1.41	1.41	1.64	1.78
400	(121.9)	1.18	1.18	1.47	1.47	1.69	1.82
450	(137.2)	1.24	1.24	1.52	1.52	1.73	1.86
500	(152.4)	1.29	1.29	1.56	1.56	1.77	1.89

Notes:

1. **Case 1:** a. All components and cladding.
   b. Main wind-force resisting system in low-rise buildings designed using Figure C.2.

   Case 2: a. All main wind-force resisting systems in buildings except those in low-rise buildings designed using Figure C.2.
   b. All main wind-force resisting systems in other structures.
2. The velocity pressure exposure coefficient K_z may be determined from the following formula:

   For $15\text{ ft} \leq z \leq z_g$ For $z < 15\text{ ft}$

   $K_z = 2.01(z/z_g)^{2/\alpha}$ $K_z = 2.01(15/z_g)^{2/\alpha}$

   Note: z shall not be taken less than 100 ft for Case 1 in exposure A or less than 30 ft for Case 1 in exposure B.
3. α and z_g are tabulated in ASCE 7-98 Table 6.4.
4. Linear interpolation for intermediate values of height z is acceptable.
5. Exposure categories are defined in Section 2.9.3.

Source: American Society of Civil Engineers, Minimum Design Loads for Buildings and Other Structures, ASCE 7-98, 2000.

TABLE C.3 INTERNAL PRESSURE COEFFICIENTS FOR BUILDINGS, GC_{pi}

Enclosure Classification	GC_{pi}
Open Buildings	0.00
Partially Enclosed Buildings	+0.55 −0.55
Enclosed Buildings	+0.18 −0.18

Notes:

1. Plus and minus signs signify pressures acting toward and away from the internal surfaces.

2. Values of GC_{pi} shall be used with q_z or q_h as specified in Section 2.9.5.

3. Two cases shall be considered to determine the critical load requirements for the appropriate condition:
(i) a positive value of GC_{pi} applied to all internal surfaces
(ii) a negative value of GC_{pi} applied to all internal surfaces

Source: American Society of Civil Engineers, Minimum Design Loads for Buildings and Other Structures, ASCE 7-98, 2000.

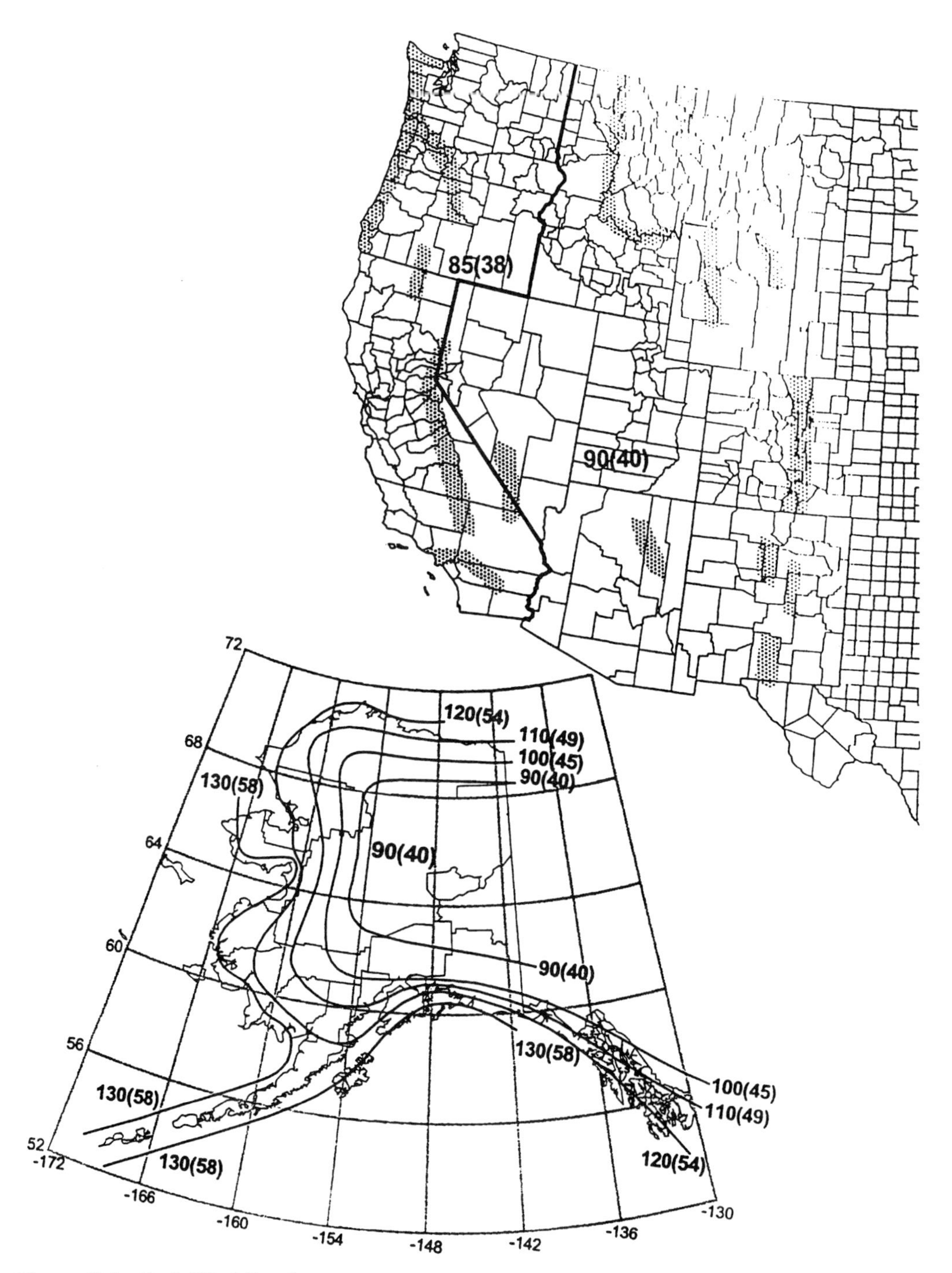

Figure C.1 Basic Wind Speed

Source: American Society of Civil Engineers, Minimum Design Loads for Buildings and Other Structures, ASCE 7-98, 2000. Reproduced with permission of the American Society of Civil Engineers.

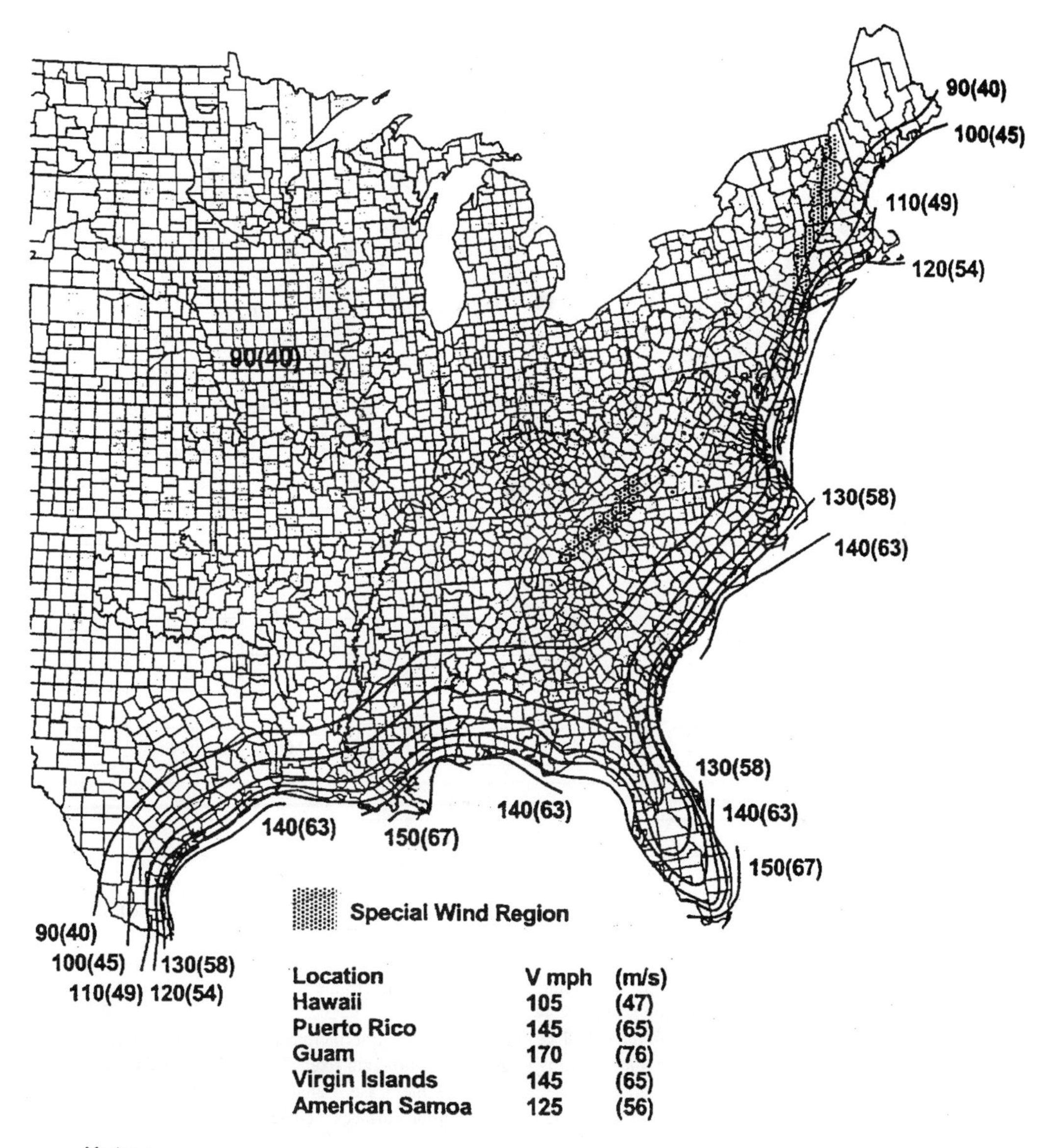

Location	V mph	(m/s)
Hawaii	105	(47)
Puerto Rico	145	(65)
Guam	170	(76)
Virgin Islands	145	(65)
American Samoa	125	(56)

Notes:

1. **Values are nominal design 3-second gust wind speeds in miles per hour (m/s) at 33 ft (10 m) above ground for Exposure C category.**
2. **Linear interpolation between wind contours is permitted.**
3. **Islands and coastal areas outside the last contour shall use the last wind speed contour of the coastal area.**
4. **Mountainous terrain, gorges, ocean promontories, and special wind regions shall be examined for unusual wind conditions.**

Figure C. I *(Continued)*

Source: American Society of Civil Engineers, Minimum Design Loads for Buildings and Other Structures, ASCE 7-98, 2000. Reproduced with permission of the American Society of Civil Engineers.

Figure C.2 Main Wind Force Resisting System, $h \leq 60$ ft; External Pressure Coefficients, GC_{pf}, Enclosed and Partially Enclosed Buildings; Walls & Gable Roof

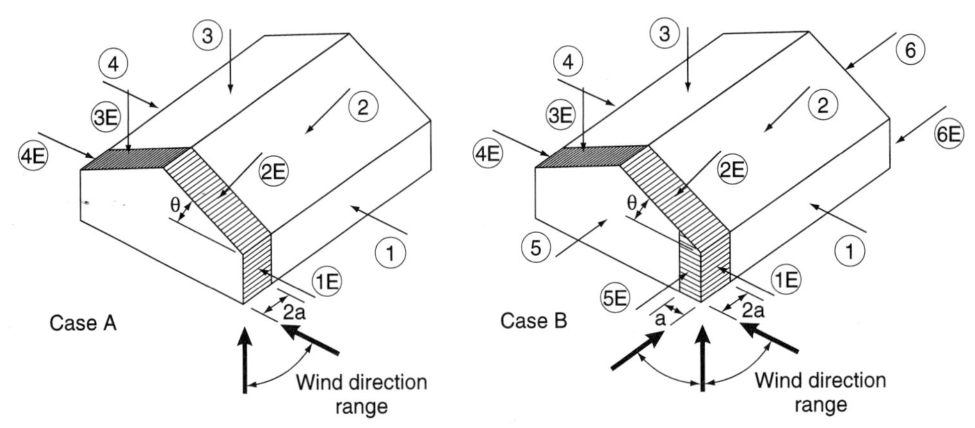

Roof Angle θ (degrees)	CASE A							
	Building Surface							
	1	2	3	4	1E	2E	3E	4E
0–5	0.40	−0.69	−0.37	−0.29	0.61	−1.07	−0.53	−0.43
20	0.53	−0.69	−0.48	−0.43	0.80	−1.07	−0.69	−0.64
30–45	0.56	0.21	−0.43	−0.37	0.69	0.27	−0.53	−0.48
90	0.56	0.56	−0.37	−0.37	0.69	0.69	−0.48	−0.48

Roof Angle θ (degrees)	CASE B											
	Building Surface											
	1	2	3	4	5	6	1E	2E	3E	4E	5E	6E
0–90	−0.45	−0.69	−0.37	−0.45	0.40	−0.29	−0.48	−1.07	−0.53	−0.48	0.61	−0.43

Notes:

1. Case A and Case B are required as two separate loading conditions to generate the wind actions, including torsion, to be resisted by the main wind-force resisting system.
2. To obtain the critical wind actions, the building shall be rotated in 90° degree increments so that each corner in turn becomes the windward corner while the loading patterns in the sketches remain fixed. For the design of structural systems providing lateral resistance in the direction parallel to the ridge line, Case A shall be based on $\theta = 0°$.
3. Plus and minus signs signify pressures acting toward and away from the surfaces, respectively.
4. For Case A loading the following restrictions apply:
   a. The roof pressure coefficient GC_{pf}, when negative in Zone 2, shall be applied Zone 2 for a distance from the edge of roof equal to 0.5 times the horizontal dimensions of the building measured perpendicular to the eave line or 2.5h, whichever is less; the remainder of Zone 2 extending to the ridge line shall use the pressure coefficient GC_{pf} for Zone 3.
   b. Except for moment-resisting frames, the total horizontal shear shall not be less than that determined by neglecting wind forces on roof surfaces.
5. Combinations of external and internal pressures (see Table C.3) shall be evaluated as required to obtain the most severe loadings.
6. For values of θ other than those shown, linear interpolation is permitted.
7. Notation:
   a: 10 percent of least horizontal dimension or 0.4h, whichever is smaller, but not less than either 4% of least horizontal dimension of 3 ft (1 m).
   h: Mean roof height, in feet (meters), except that eave height shall be used for $\theta \leq 10°$.
   θ: Angle of plane of roof from horizontal, in degrees.

Source: American Society of Civil Engineers, Minimum Design Loads for Buildings and Other Structures, ASCE 7-98, 2000. Reproduced with permission of the American Society of Civil Engineers.

Figure C.3a Components and Cladding; h ≤ 60 ft; External Pressure Coefficients, GC_p; Enclosed and Partially Enclosed Buildings; Walls

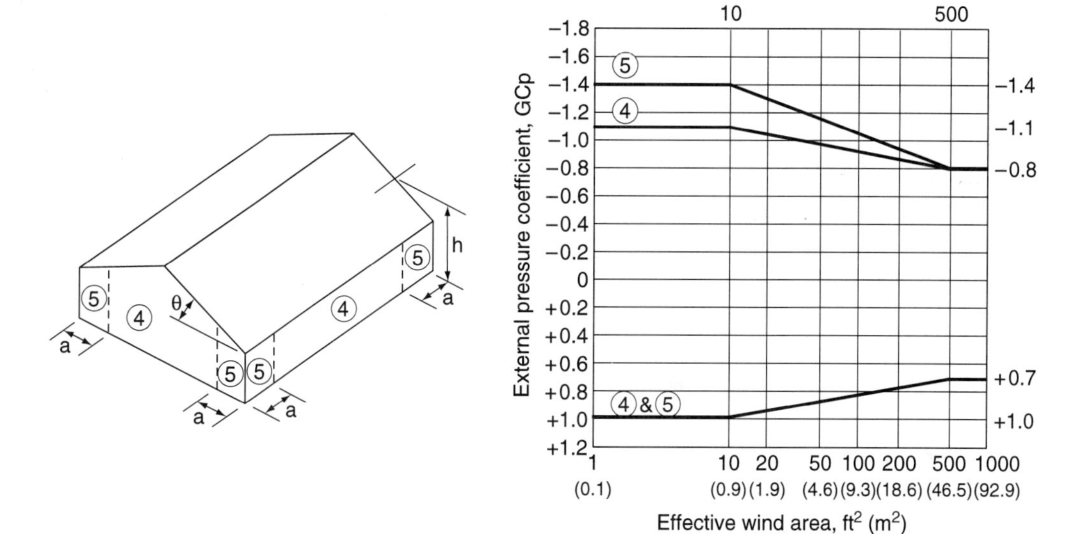

Notes:

1. Vertical scale denotes GC_p to be used with q_h.
2. Horizontal scale denotes effective wind area, in square feet (square meters).
3. Plus and minus signs signify pressures acting toward and away from the surfaces, respectively
4. Each component shall be designed for maximum positive and negative pressures.
5. Values of GC_p for walls shall be reduced by 10% when $\theta \leq 10°$.
6. Notation:
   a: 10 percent of least horizontal dimension or 0.4h, whichever is smaller, but not less than either 4% of least horizontal dimension or 3 ft (1 m).
   h: Mean roof height, in ft (ms), except that eave height shall be used for $\theta \leq 10°$.
   θ: Angle of plane of roof from horizontal, in degrees.

Source: American Society of Civil Engineers, Minimum Design Loads for Buildings and Other Structures, ASCE 7-98, 2000. Reproduced with permission of the American Society of Civil Engineers.

Figure C.3b Components and Cladding; h ≤ 60 ft; External Pressure Coefficients, GC_p; Enclosed and Partially Enclosed Buildings; Gable Roofs θ ≤ 10°

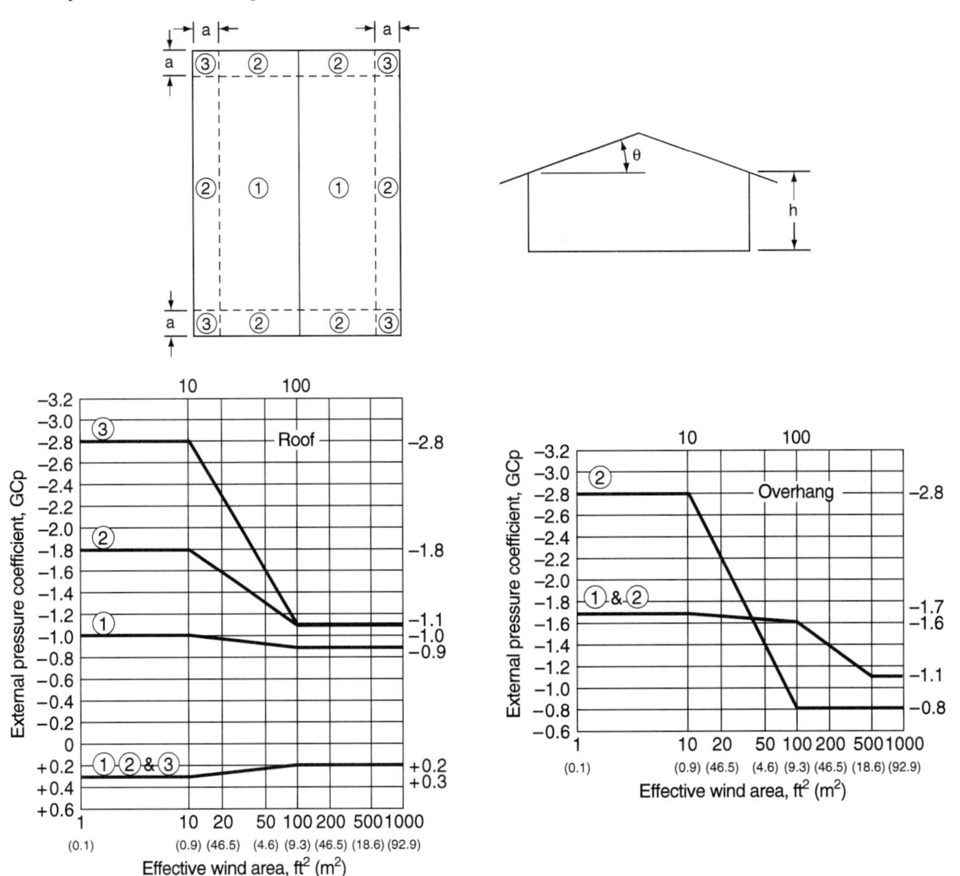

Notes:

1. Vertical scale denotes GC_p to be used with q_h.
2. Horizontal scale denotes effective wind area, in square feet (square meters).
3. Plus and minus signs signify pressures acting toward and away from the surfaces, respectively
4. Each component shall be designed for maximum positive and negative pressures.
5. If a parapet equal to or higher than 3 ft (1 m) is provided around the perimeter of the roof with θ ≤ 10°, Zone 3 shall be treated as Zone 2.
6. Values of GC_p for roof overhangs include pressure contributions from both upper and lower surfaces.
7. Notation:
   - a: 10 percent of least horizontal dimension or 0.4h, whichever is smaller, but not less than either 4% of least horizontal dimension or 3 ft (1 m).
   - h: Eave height shall be used for θ ≤ 10°.
   - θ: Angle of plane of roof from horizontal, in degrees.

Source: American Society of Civil Engineers, Minimum Design Loads for Buildings and Other Structures, ASCE 7-98, 2000. Reproduced with permission of the American Society of Civil Engineers.

Figure C.3c Components and Cladding; $h \leq 60$ ft; External Pressure Coefficients, GC_p; Enclosed and Partially Enclosed Buildings; Stepped Roofs

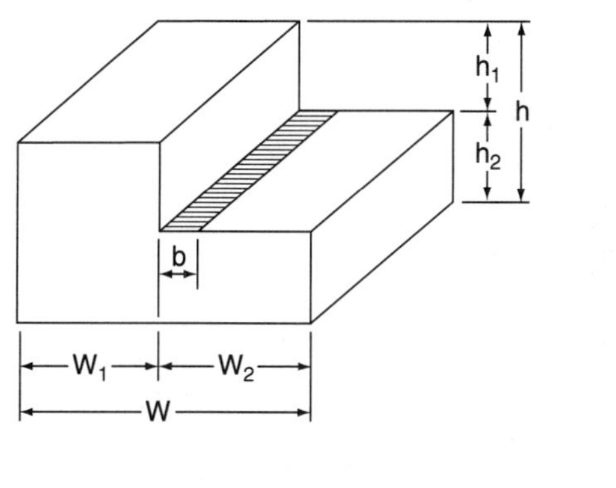

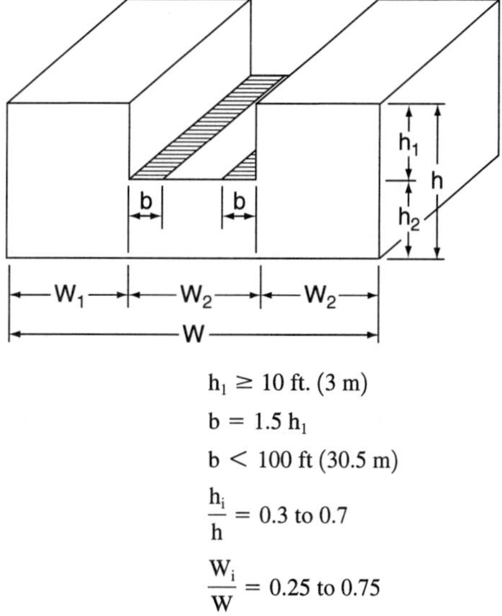

$h_1 \geq 10$ ft. (3 m)

$b = 1.5\ h_1$

$b < 100$ ft (30.5 m)

$\frac{h_i}{h} = 0.3$ to 0.7

$\frac{W_i}{W} = 0.25$ to 0.75

Notes:

1. On the lower level of flat, stepped roofs shown in Fig. C.3c, the zone designations and pressure coefficients shown in Fig. C.3b ($\theta \leq 10°$) shall apply, except that at the roof-upper wall intersection(s), Zone 3 shall be treated as Zone 2 and Zone 2 shall be treated as Zone 1. Positive values of GC_p equal to those for walls in Fig. C.3a shall apply on the cross-hatched areas shown in Fig. C.3c.
2. Notation:

   b: $1.5h_1$ in Fig. C.3c, but not greater than 100 ft (30.5 m).

   h: Mean roof height, in feet (meters).

   h_i: h_1 or h_2 in Fig. C.3c; $h = h_1 + h_2$; $h_1 \geq 10$ ft (3.1 m); $h_i/h = 0.3$ to 0.7.

   W: Building width in Fig. C.3c.

   W_i: W_1 or W_2 or W_3 in Fig. C.3c. $W = W_1 + W_2$ or $W_1 + W_2 + W_3$; $W_i/W = 0.25$ to 0.75.

   θ: Angle of plane of roof from horizontal, in degrees.

Source: American Society of Civil Engineers, Minimum Design Loads for Buildings and Other Structures, ASCE 7-98, 2000. Reproduced with permission of the American Society of Civil Engineers.

Figure C. 4 Components and Cladding; $h \leq 60$ ft; External Pressure Coefficients, GC_p; Enclosed and Partially Enclosed Buildings; Monoslope Roofs $3° < \theta \leq 10°$

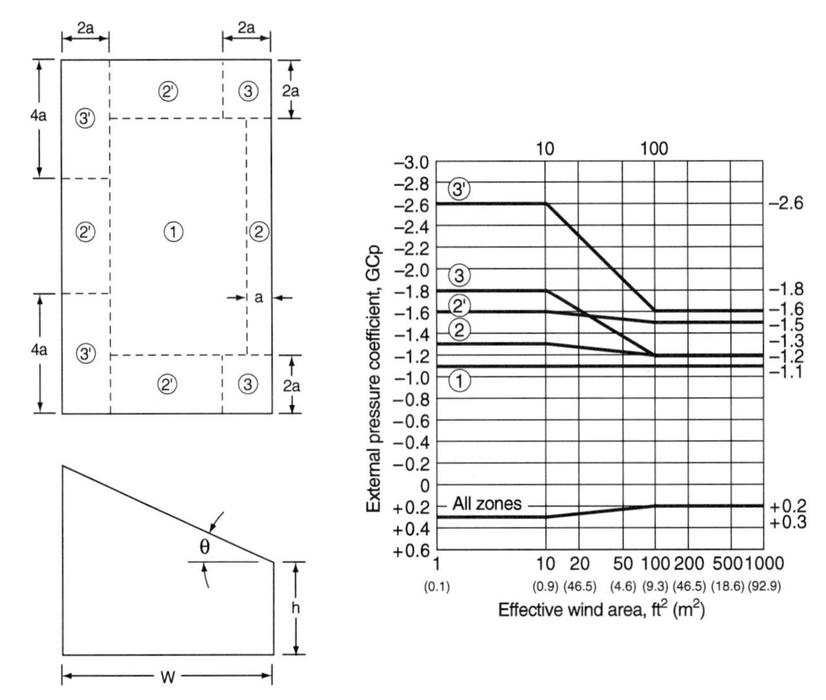

Notes:

1. Vertical scale denotes GC_p to be used with q_h.
2. Horizontal scale denotes effective wind area A, in square ft (square ms).
3. Plus and minus signs signify pressures acting toward and away from the surfaces, respectively
4. Each component shall be designed for maximum positive and negative pressures.
5. For $\theta \leq 3°$, values of GC_p from Fig. C.3b shall be used.
6. Notation:
   - a: 10% of least horizontal dimension or 0.4h, whichever is smaller, but not less than either 4% of least horizontal dimension or 3 ft (1 m).
   - h: Eave height shall be used for $\theta \leq 10°$.
   - W: Building width, in ft (ms).
   - θ: Angle of plane of roof from horizontal, in degrees.

Source: American Society of Civil Engineers, Minimum Design Loads for Buildings and Other Structures, ASCE 7-98, 2000. Reproduced with permission of the American Society of Civil Engineers.

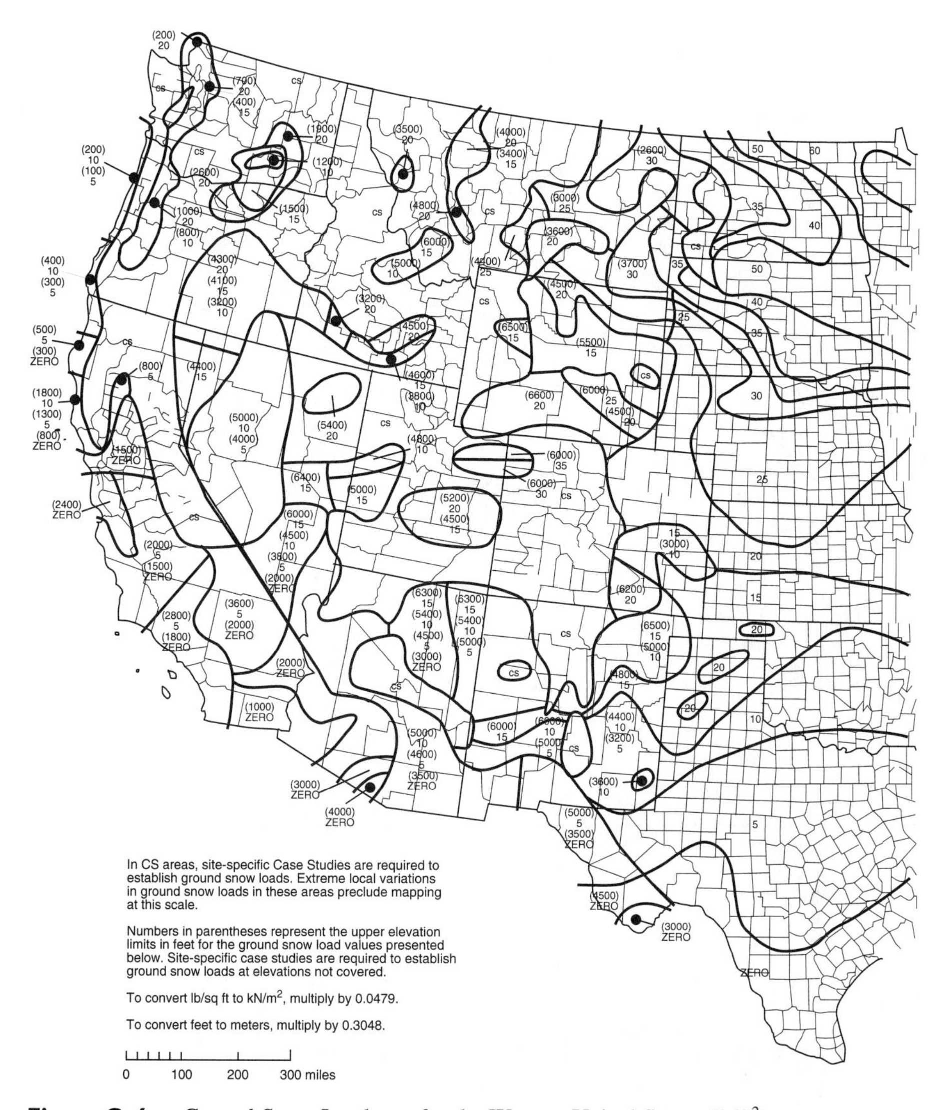

Figure C. 6a Ground Snow Loads, p_g for the Western United States (lb/ft^2)

Source: American Society of Civil Engineers, Minimum Design Loads for Buildings and Other Structures, ASCE 7-98, 2000. Reproduced with permission of the American Society of Civil Engineers.

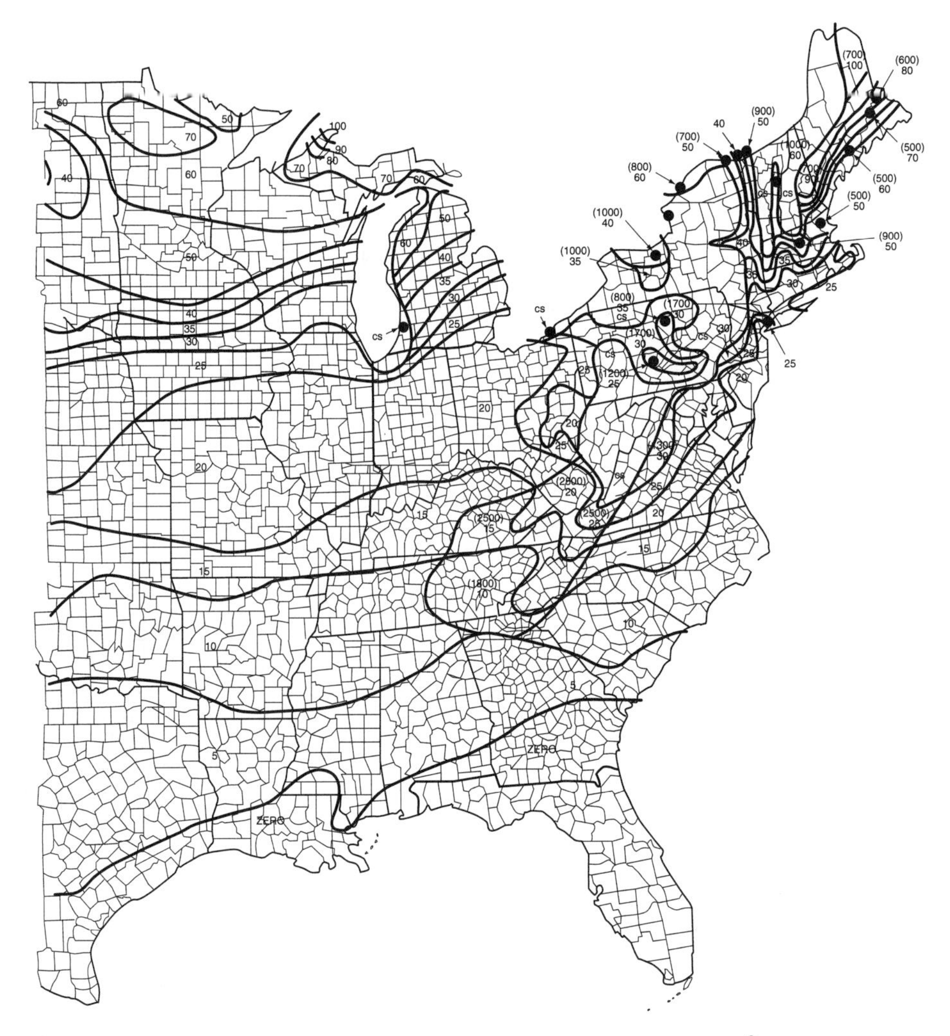

Figure C. 6b Ground Snow Loads, p_g for the Eastern United States (lb/ft^2).

Source: American Society of Civil Engineers, Minimum Design Loads for Buildings and Other Structures, ASCE 7-98, 2000. Reproduced with permission of the American Society of Civil Engineers.

Appendix D

Beam Fixed-End Moments

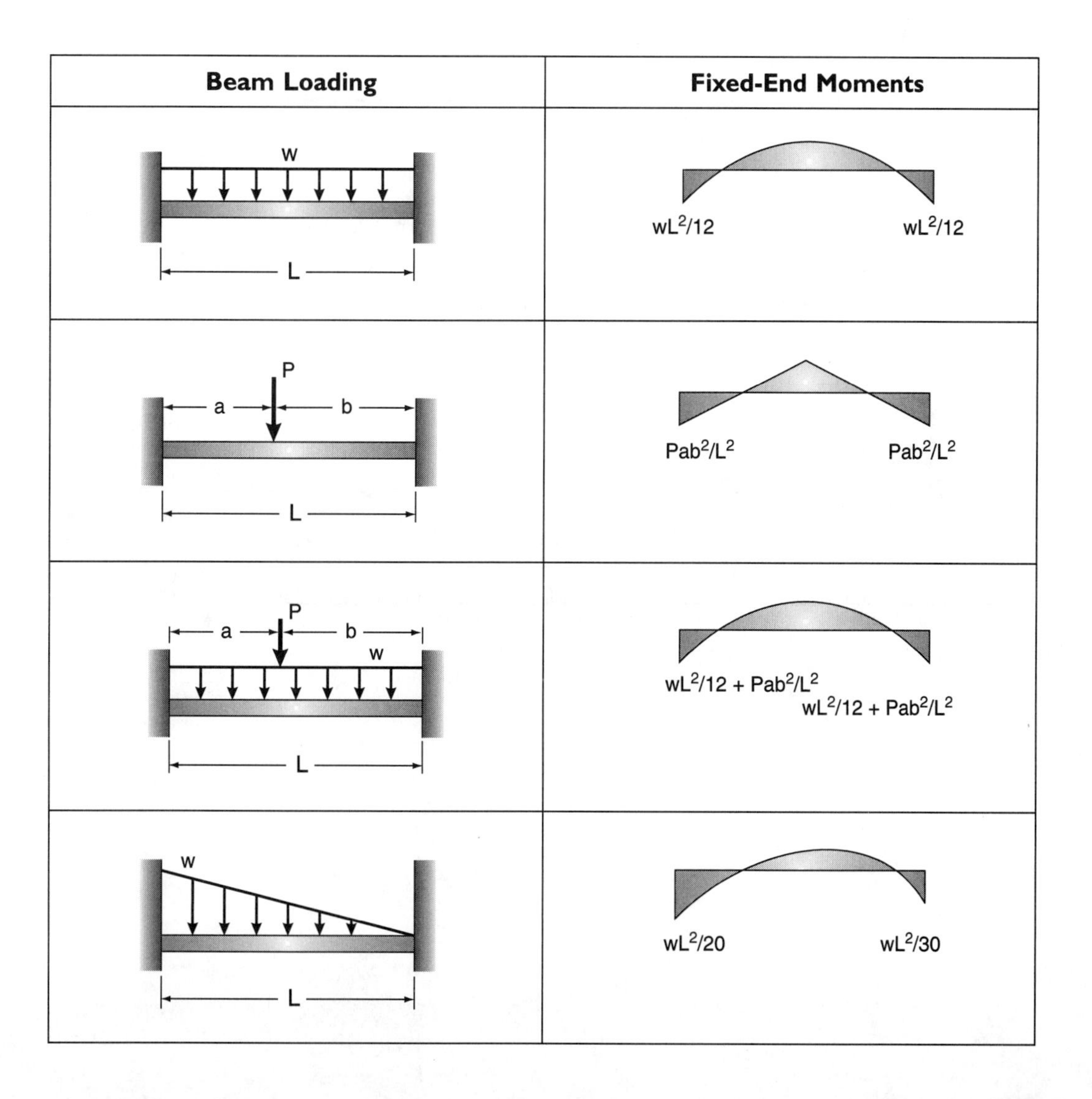

Beam Loading	Fixed-End Moments
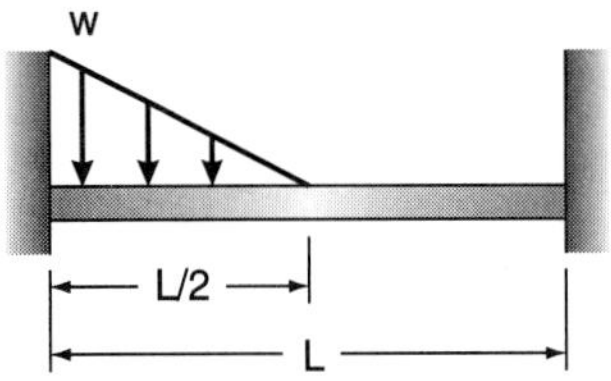 w; L/2; L	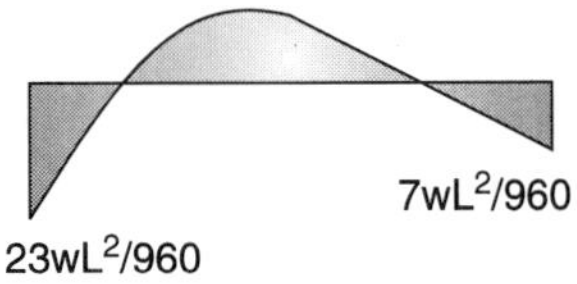 $23wL^2/960$; $7wL^2/960$
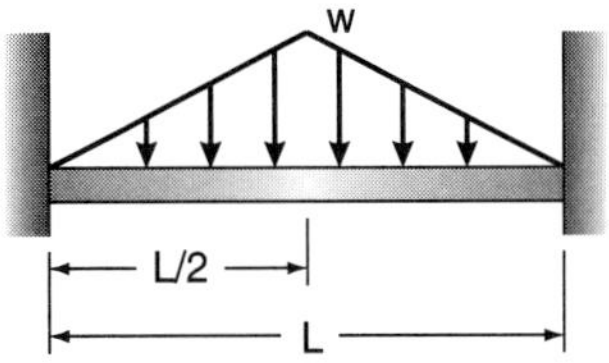 w; L/2; L	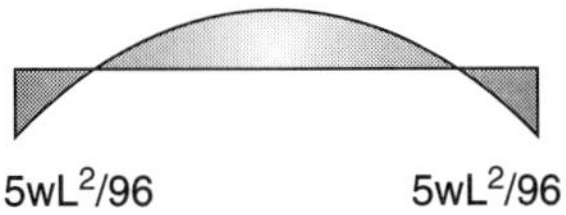 $5wL^2/96$; $5wL^2/96$
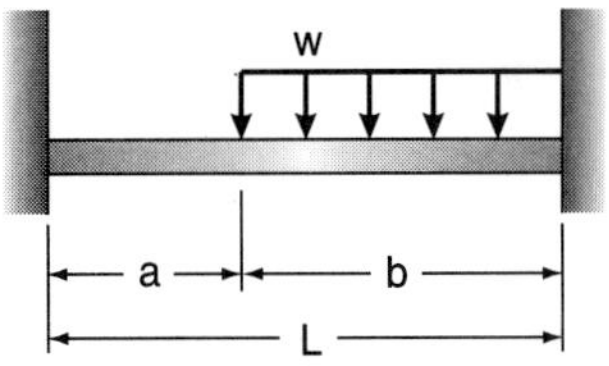 w; a; b; L	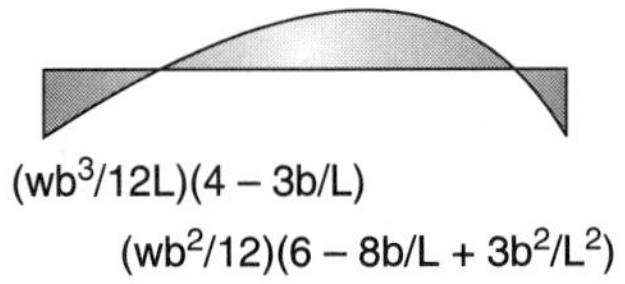 $(wb^3/12L)(4 - 3b/L)$; $(wb^2/12)(6 - 8b/L + 3b^2/L^2)$
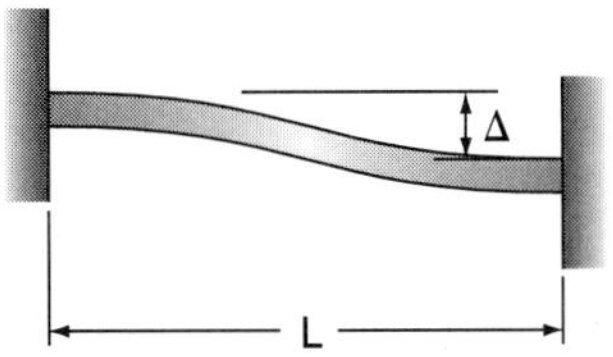 Δ; L	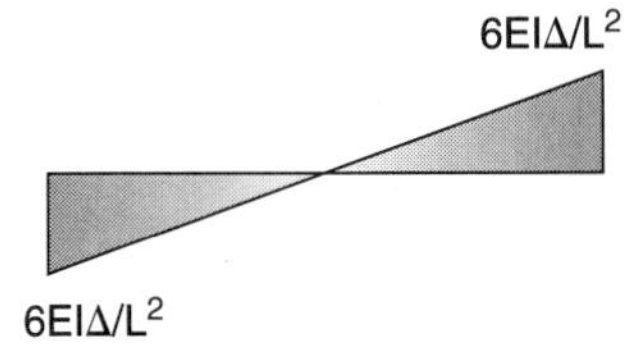 $6EI\Delta/L^2$; $6EI\Delta/L^2$

Appendix E

Properties of Commonly Used Areas

Geometric Area	Area Properties
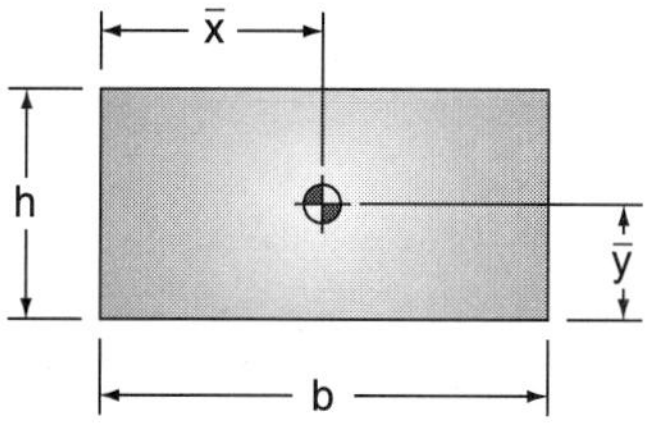 h, b, $\bar{x}$, $\bar{y}$	$A = bh$ $\bar{x} = b/2$ $\bar{y} = h/2$
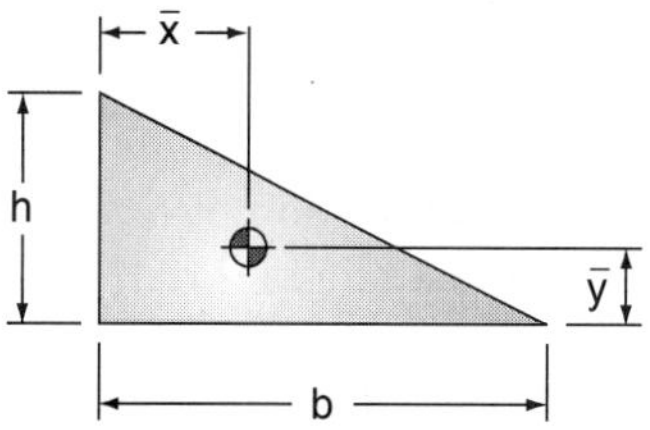 h, b, $\bar{x}$, $\bar{y}$	$A = bh/2$ $\bar{x} = b/3$ $\bar{y} = h/3$
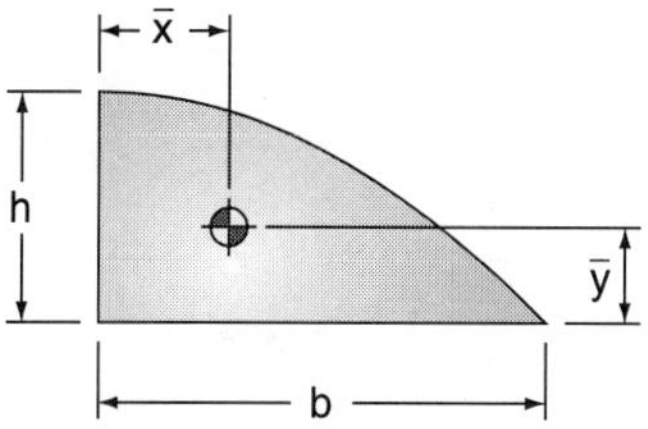 h, b, $\bar{x}$, $\bar{y}$	$A = 2bh/3$ $\bar{x} = 3b/8$ $\bar{y} = 2h/5$
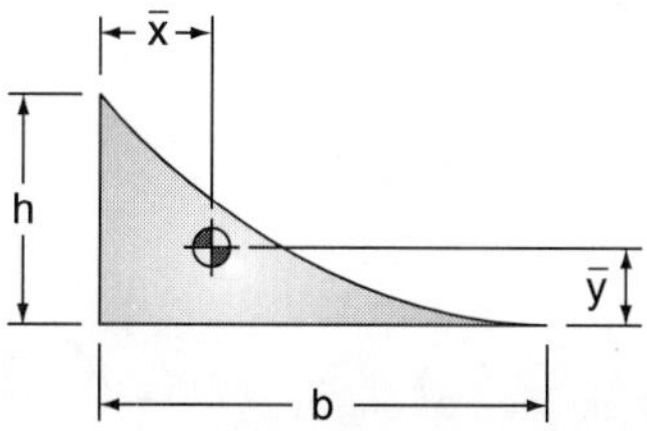 h, b, $\bar{x}$, $\bar{y}$	$A = bh/3$ $\bar{x} = b/4$ $\bar{y} = 3h/10$

Appendix F

Elastic Weight and Conjugate Beam Methods

F.1 THE METHOD OF ELASTIC WEIGHTS

A careful study of the procedure used in applying the area-moment theorems will reveal a simpler and more practical method of computing slopes and deflections for most beams. In reviewing this procedure the beam and M/EI diagram of Figure F.1 are considered.

By letting A equal the area of the M/EI diagram, the deflection of the tangent at R from the tangent at L equals Ay, and the change in slope between the two tangents is A.

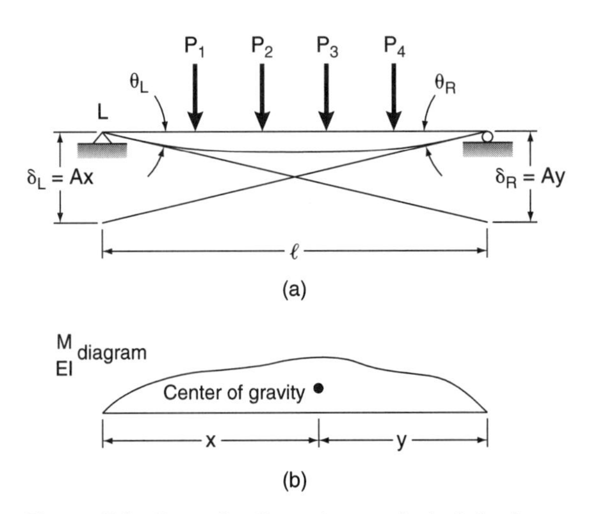

Figure F.1 Beam for discussing method of elastic weights

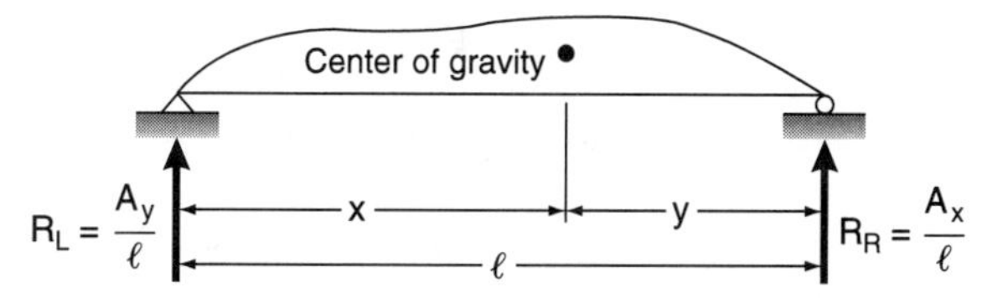

Figure F.2 Reactions for beam loaded with M/EI diagram

An imaginary beam is loaded with the M/EI diagram, as shown in Figure F.2, and the reactions R_L and R_R are determined. They equal Ay/ℓ and Ax/ℓ, respectively.

In Figure F.1 the slopes of the tangents to the elastic curve at each end of the beam (θ_L and θ_R) are equal to the deflections between the tangents at each end divided by the span length, as follows:

$$\theta_L = \frac{\delta_R}{\ell} \qquad \theta_R = \frac{\delta_L}{\ell}$$

The values of δ_L and δ_R have previously been found to equal A_x and A_y, respectively, and may be substituted in these expressions:

$$\theta_L = \frac{A_y}{\ell} \qquad \theta_R = \frac{A_x}{\ell}$$

The end slopes are exactly the same as the reactions for the beam in Figure F.2. At either end of the fictitious beam the shear equals the reaction and thus the slope in the actual beam. Further experiments will show that the shear at any point in the beam loaded with the M/EI diagram equals the slope at that point in the actual beam.

A similar argument can be made concerning the computation of deflections, and it will be found that the deflection at any point in the actual beam equals the moment at that point in the fictitious beam. In detail, the two theorems of elastic weights may be stated as follows:

1. The slope of the elastic curve of a simple beam at a point, measured with respect to a chord between the supports, equals the shear at that point if the beam is loaded with the M/EI diagram.
2. The deflection of the elastic curve of a simple beam at a point, measured with respect to a chord between the supports, equals the moment at that point if the beam is loaded with the M/EI diagram.

F.2 APPLICATION OF THE METHOD OF ELASTIC WEIGHTS

The method of elastic weights in its present form is applicable only to beams simply supported at each end. It will be found in using the method that maximum deflections in the actual beam occur at points of zero shear in the imaginary beam. The reasoning is the same as that presented for shear and moment diagrams in Chapter 5, where maximum moments were found to occur at points of zero shear.

Consideration has not been given to the subject of sign conventions for either the elastic-weight methods. Study of the shears and moments on the fictitious beam reveals the directions of slopes and deflections. A positive shear in the fictitious beam shows the left side is being pushed up with respect to the right side, or the beam is sloping downward from left to right. Similarly, a positive moment indicates downward deflection.

Examples F.1 and F.2 illustrate the application of elastic weights.

EXAMPLE F.1

Determine the deflection at the centerline of the beam shown in the figure.

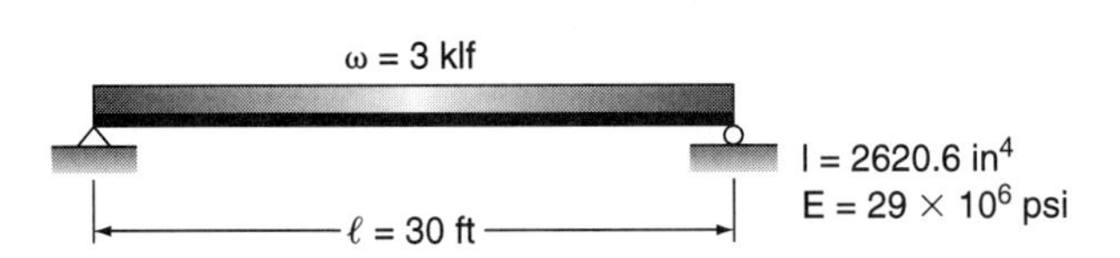

Solution

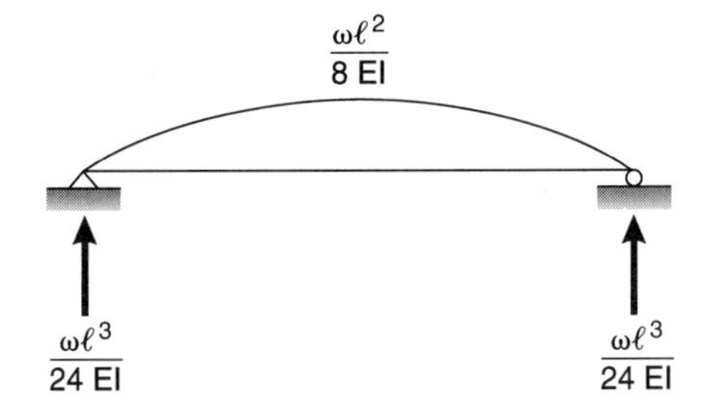

Deflection at centerline:

$$\delta_{\text{℄}} = \text{moment}_{\text{℄}}\left(\frac{w\ell^3}{24EI}\right)\left(\frac{\ell}{2}\right) - \left(\frac{2}{3}\right)\left(\frac{\ell}{2}\right)\left(\frac{w\ell^2}{8EI}\right)\left(\frac{3}{8}\frac{\ell}{2}\right)$$

$$= \frac{5w\ell^4}{384EI} = \frac{(5)(3000/12)(30 \times 12)^4}{(384)(29 \times 10^6)(2620.6)} = 0.719 \text{ in.} \quad \blacksquare$$

EXAMPLE F.2

Determine the slope and deflection at the centerline of the beam shown in the figure.

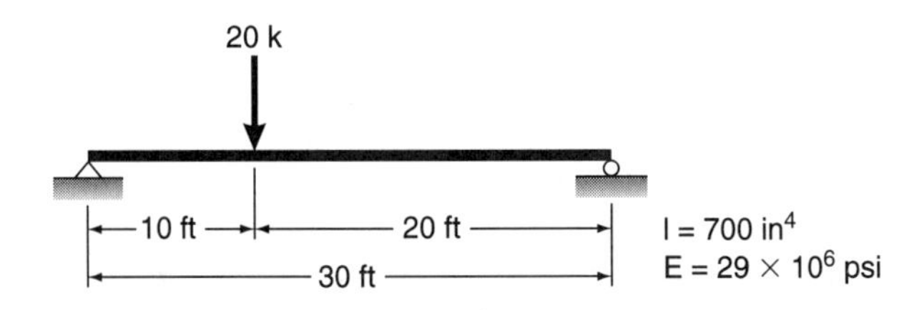

Solution. Loading the fictitious beam with the M/EI diagram and computing the reactions.

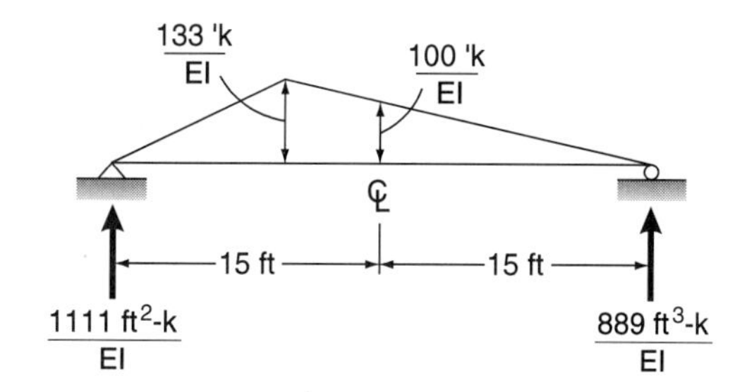

Deflection

$$\delta_{℄} = \frac{(889)(15) - (\frac{1}{2})(100)(15)(5)}{EI} = \frac{9585 \text{ ft}^3\text{-k}}{EI}$$

$$= \frac{(9585)(1728)(1000)}{(29 \times 10^6)(700)} = 0.816 \text{ in.}$$

Slope

$$\theta_{℄} = \frac{-889 + (\frac{1}{2})(100)(15)}{EI} = -\frac{139 \text{ ft}^2\text{-k}}{EI}$$

$$= -\frac{(139)(144)(1000)}{(29 \times 10^6)(700)} = -0.000986 \text{ rad}$$

$$= -0.056^6 \text{ (negative slope /)}$$ ■

EXAMPLE F.3

Compute the maximum deflection for the beam shown in the figure.

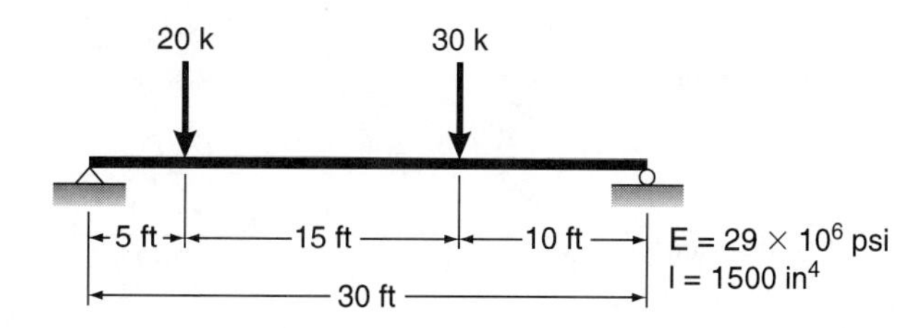

Solution

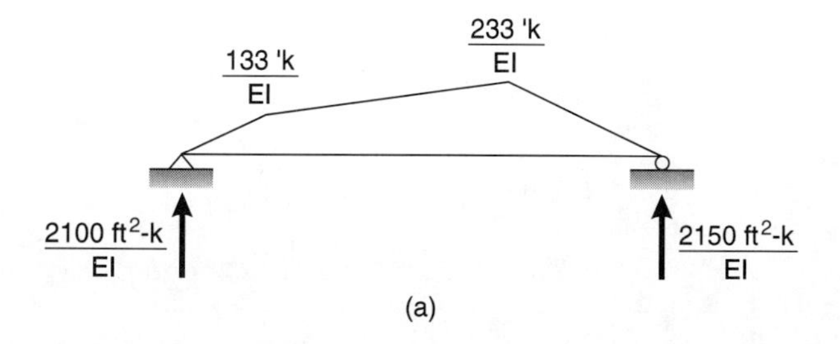

(a)

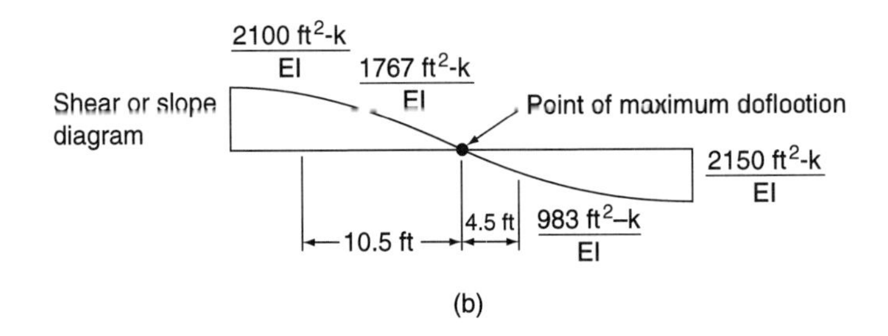

(b)

$$\delta_{max} = \frac{(2100)(15.5) - (\frac{1}{2})(5)(133)(12.17) - (10.5)(133)(5.25) - (\frac{1}{2})(10.5)(70)(3.5)}{EI}$$

$$= \frac{19{,}890 \text{ ft}^3\text{-k}}{EI} = \frac{(19{,}890)(1728)(1000)}{(29 \times 10^6)(1500)} = 0.790 \text{ in.} \quad \blacksquare$$

EXAMPLE F.4

Compute the centerline deflection for the simple beam shown in the figure.

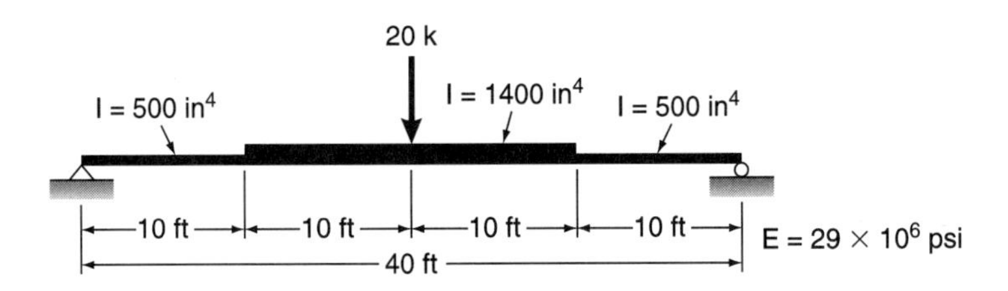

Solution

200 'k
100 'k
100 'k
M diagram
0.2 'k / E
0.0715 'k / E
0.143 'k / E
0.0715 'k / E
0.2 'k / E
M/EI diagram
2.07 ft²-k / E
2.07 ft²-k / E

$$\delta_{℄} = \left(\frac{2.07}{E}\right)(20) - \left(\frac{1}{2}\right)(10)\left(\frac{0.2}{E}\right)(13.33) - (10)\left(\frac{0.0715}{E}\right)(5)$$

$$- \left(\frac{1}{2}\right)(10)\left(\frac{0.0715}{E}\right)(3.33)$$

$$= \frac{23.3}{E} = \frac{(23.3)(1728)(1000)}{29 \times 10^6} = 1.39 \text{ in.} \quad \blacksquare$$

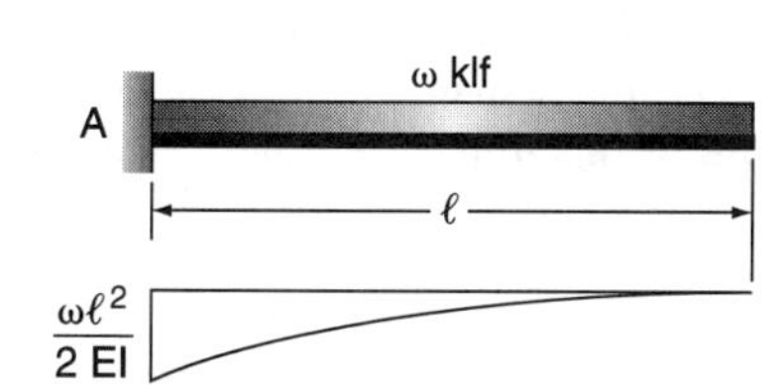

Figure F.3 Cantilever beam loaded with M/EI diagram for a uniform load

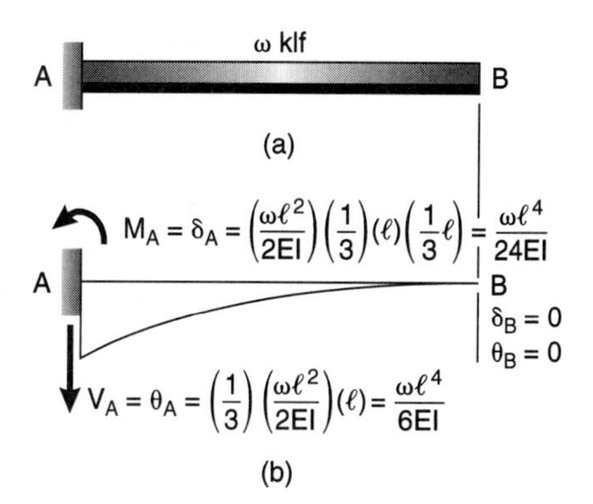

Figure F.4 Slope and deflection found using method of elastic weights

F.3 LIMITATIONS OF THE ELASTIC-WEIGHT METHOD

The method of elastic weights was developed for simple beams, and in its present form will not work for cantilevered beams, overhanging beams, fixed-ended beams, and continuous beams. The moment-area theorems are used to determine the correct slope and deflection at the free end of the uniformly loaded cantilevered beam of Figure F.3.

$$\theta_B = \left(\frac{1}{3}\right)(\ell)\left(\frac{w\ell^2}{2EI}\right) = \frac{w\ell^3}{6EI}$$

$$\delta_B = \left(\frac{1}{3}\right)(\ell)\left(\frac{w\ell^2}{2EI}\right)\left(\frac{3}{4}\ell\right) = \frac{w\ell^4}{8EI}$$

If the elastic-weight method was used in an attempt to find the slope and deflection at the ends of the same beam, the result would be slopes and deflections of zero at the free end and $w\ell^3/6EI$ and $w\ell^4/24EI$ at the fixed end, as shown in Figure F.4.

The slope and deflection at the fixed end A must be zero; however, application of elastic weights to the beam results in both shear and moment, falsely indicating slope and deflection.

If the fixed end of the beam was moved to the free end and the resulting beam loaded with the M/EI diagram, the shears and moments would correspond exactly to the slopes and deflections on the actual beam as found by the moment-area method.

F.4 CONJUGATE-BEAM METHOD

The conjugate-beam method makes use of an analogous or "conjugate" beam to be handled by elastic weights in place of the actual beam, to which it cannot be correctly applied. The shear and moment in the imaginary beam, loaded with the M/EI diagram, must correspond exactly with the slope and deflection of the actual beam.

The correct mathematical relationship is obtained for a beam simply supported if it is loaded "as is" with the M/EI diagram. If the elastic-weight method is applied to other types of beams, the largest moments due to the M/EI loading occur at the supports, incorrectly indicating that the largest deflections occur at those points. For elastic weights to be applied correctly, use must be made of substitute beams, of conjugate beams, that have the supports changed so the correct relationships are obtained.

The loads and properties of the true beam have no effect on the manner in which the conjugate beam is supported. The only factors affecting the supports of the imaginary beam are the supports of the actual beam. The lengths of the two beams are equal. We will discuss what the various types of beam support must become in the conjugate beam so that the elastic-weight method will apply. The mathematical proof of these relationships is explained in detail in books on strength of materials. The relationships are summarized in Figure F.5.

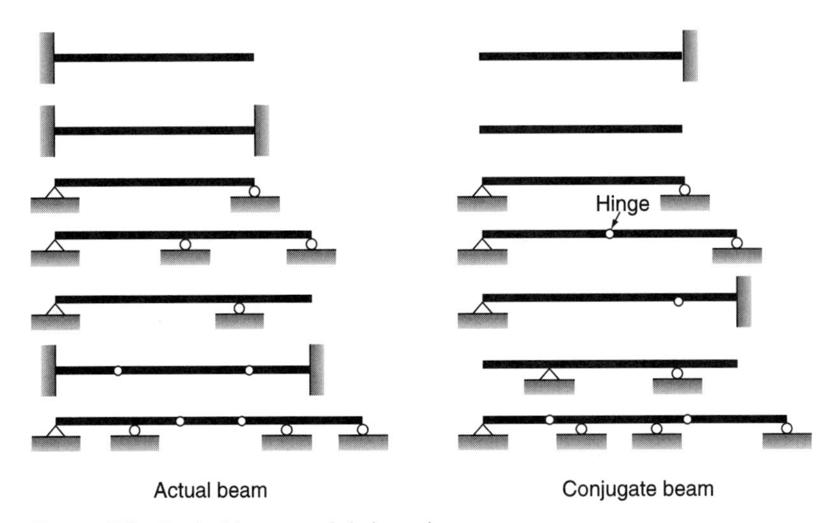

Figure F.5 Typical beams and their conjugates

Free End

The free end of a beam slopes and deflects when the beam is loaded. The conjugate beam must have both shear and moment at that end when it is loaded with the M/EI diagram. The only type of end support having both shear and moment is the fixed end. *A free end in the actual beam becomes a fixed end in the conjugate beam.*

Fixed End

A similar discussion in reverse order can be made for a fixed end. No slope or deflection can occur at a fixed end, and there must not be any shear or moment in the conjugate beam at that point. *A fixed end in the actual beam becomes a free end in the conjugate beam.*

Simple End Support

A simple end slopes but does not deflect when the beam is loaded. The imaginary beam will have shear but no moment at that point, a situation that can occur only at a simple support. *A simple end support in the actual beam remains a simple end support in the conjugate beam.*

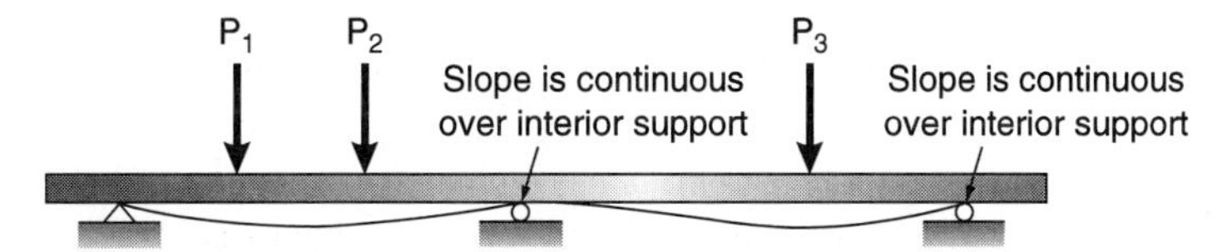

Figure F.6 Qualitative deflections for a beam continuous over one support

Simple Interior Support

There is no deflection at either a simple interior or a simple end support. Both types may slope when the beam is loaded, but the situations are somewhat different. The slope at a simple interior support is continuous across the support; that is, no sudden change of slope occurs. This condition is not present at a simple end support where the slope suddenly begins, as shown in the deflection curve for the beam of Figure F.6. If there is no change of slope at a simple interior support, there can be no change of shear at the corresponding support in the conjugate beam. Any type of external support at this point would cause a change in the shear; therefore, an internal pin (or unsupported hinge) is required. *A simple interior support in the actual beam becomes an unsupported internal hinge in the conjugate beam.*

Internal Hinge

At an unsupported internal hinge there is both slope and deflection, which means that the corresponding support in the conjugate beam must have shear and moment. *An internal hinge in the actual beam becomes a simple support in the conjugate beam.*

The reactions, moments, and shears of the conjugate beam are easily computed by static equilibrium because the conjugate beam is always statically determinate, even though the real beam may be statically indeterminate. Sometimes the conjugate beam may appear to be completely unstable. The most conspicuous example is the conjugate beam for the fixed-end beam as shown in Figure F.7, which has no supports whatsoever. On second glance,

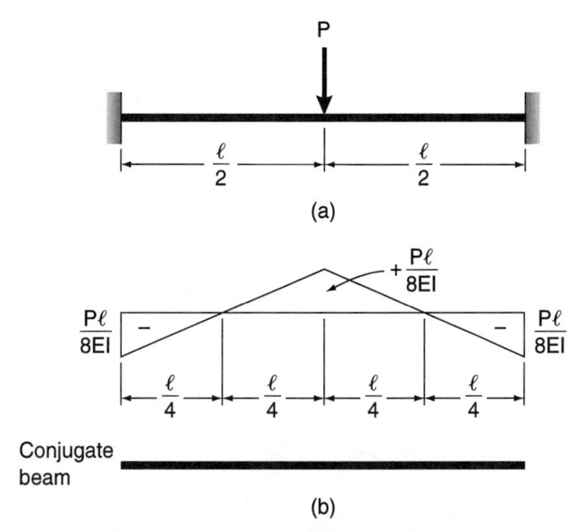

Figure F.7 A fixed end beam whose conjugate appears unstable, but is geometrically stable

the areas of the M/EI diagram are seen to be so precisely balanced between downward and upward loads (positive and negative areas of the diagram, respectively) as to require no supports. Any supports seemingly required would have zero reactions, and the proper shears and moments are supplied to coincide with the true slopes and deflections. Even a real beam continuous over several simple supports has a conjugate that is simply end-supported.

F.5 SUMMARY OF BEAM RELATIONS

A brief summary of the relations that exist between loads, shears, moments, slope changes, slopes, and deflections is presented in Figure F.8. The relations are shown for a uniformly loaded beam but are applicable to any type of loading. For the two sets of curves shown, the ordinate on one curve equals the slope at that point on the following curve. It is obvious from these figures that the same mathematical relations that exist between load, shear, and moment hold for M/EI loading, slope, and deflection.

F.6 APPLICATION OF THE CONJUGATE METHOD TO BEAMS

Examples F.5 and F.6 illustrate the conjugate method of calculating slopes and deflections for beams. The procedure as to symbols and units used in applying the method is in general the same as that used for the moment-area and elastic-weight methods.

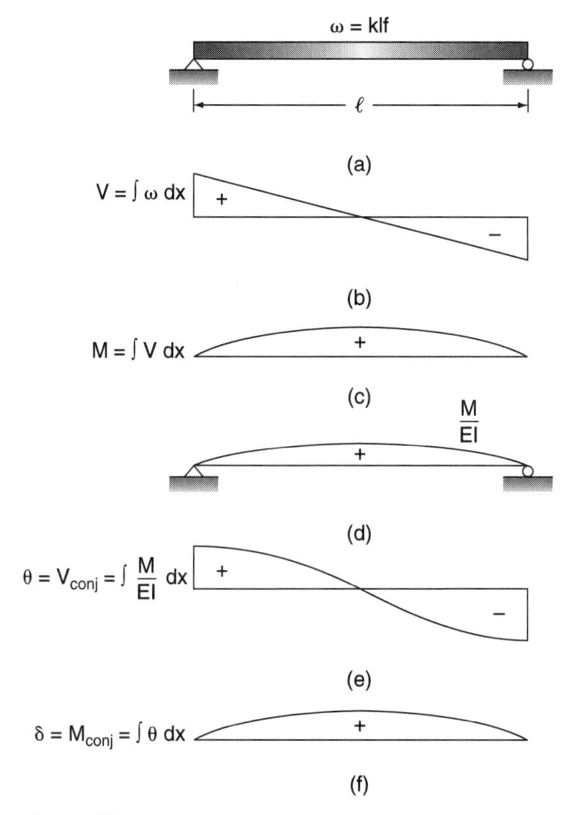

Figure F.8 Summary of relations for conjugate beam method

Maximum deflections occur at points of zero shear on the conjugate structure. For example, the point of zero shear in the beam of Figure F.7 is the centerline. The deflection is as follows:

$$\begin{aligned}\delta_{\mathbb{C}} &= \text{moment}_{\mathbb{C}} \\ &= \left(\frac{1}{2}\right)\left(\frac{\ell}{4}\right)\left(\frac{P\ell}{8EI}\right)\left(\frac{5}{12}\ell\right) - \left(\frac{1}{2}\right)\left(\frac{\ell}{4}\right)\left(\frac{P\ell}{8EI}\right)\left(\frac{\ell}{12}\right) \\ &= \frac{P\ell^3}{192EI}\end{aligned}$$

EXAMPLE F.5

Determine the slope and deflection of point A in the figure.

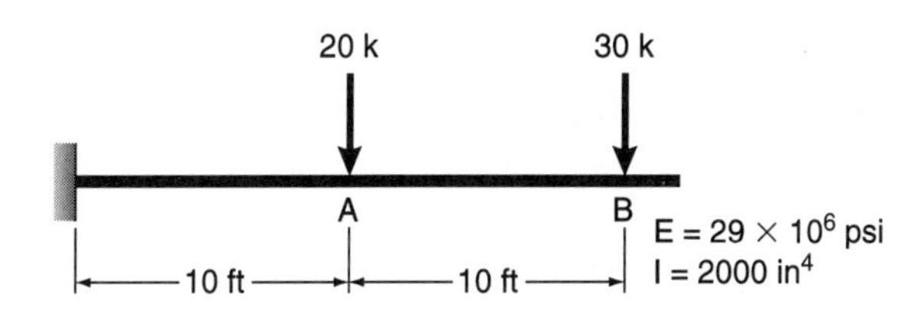

Solution

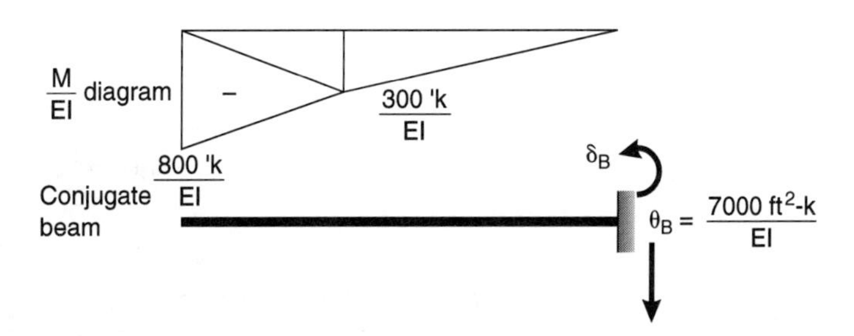

Slope

$$\begin{aligned}\theta_A &= \frac{(\frac{1}{2})(300)(10) + (\frac{1}{2})(800)(10)}{EI} = \frac{5500 \text{ ft}^2\text{-k}}{EI} \\ &= \frac{(5500)(144)(1000)}{(29 \times 10^6)(2000)} = 0.0136 \text{ rad} = 0.78°\backslash\end{aligned}$$

Deflection

$$\begin{aligned}\delta_A &= \frac{(\frac{1}{2})(300)(10)(3.33) + (\frac{1}{2})(800)(10)(6.67)}{EI} = \frac{31{,}667 \text{ ft}^2\text{-k}}{EI} \\ &= \frac{(31{,}667)(1728)(1000)}{(29 \times 10^6)(2000)} = 0.943 \text{ in.} \downarrow \quad \blacksquare\end{aligned}$$

EXAMPLE F.6

Determine deflections at points A and B in the overhanging beam shown in the figure.

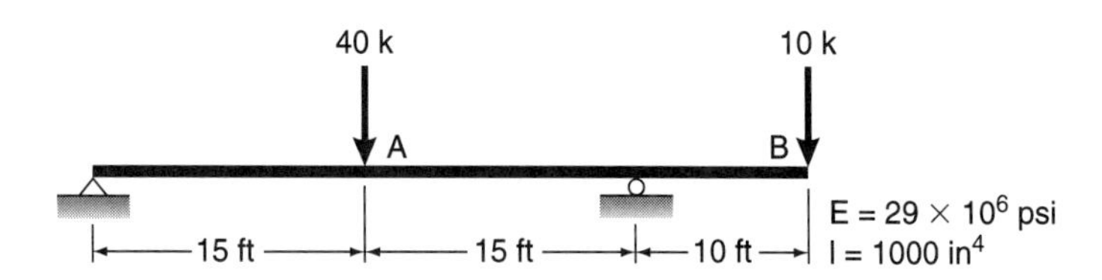

Solution. The M/EI diagram is drawn and placed on the conjugate beam, which has an interior hinge. The reactions are determined as they were for the cantilever-type structures of Chapter 4. The portion of the beam to the left of the hinge is considered a simple beam, and its reactions are determined. The reaction at the hinge is applied as a concentrated load acting at the end of the cantilever to the right of the hinge in the opposite direction, and the reactions at the fixed end are determined. To simplify the mathematics, a separate moment diagram is drawn for each of the concentrated loads.

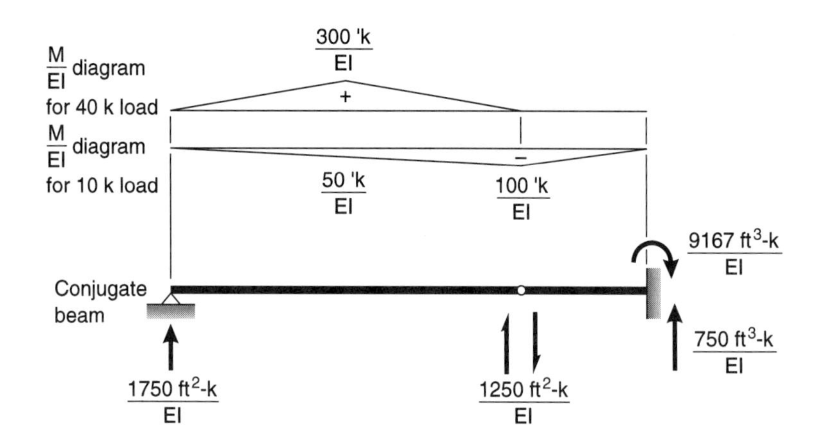

$$\delta_A = \frac{(1750)(15) - (\tfrac{1}{2})(15)(300)(5) + (\tfrac{1}{2})(15)(50)(5)}{EI} = \frac{16{,}875 \text{ ft}^3\text{-k}}{EI} = 1.01 \text{ in.} \downarrow$$

$$\delta_B = \frac{-(1250)(10) + (\tfrac{1}{2})(10)(100)(6.67)}{EI} = -\frac{9167 \text{ ft}^3\text{-k}}{EI} = 0.546 \text{ in.} \uparrow \quad \blacksquare$$

F.7 PROBLEMS FOR SOLUTION

Use the conjugate-beam method for solving Problems F.1 through F.10.

F.1 θ_A, δ_A: $E = 29 \times 10^6$ psi. $I = 1800$ in^4 (*Ans.* 0.00123 rad, 0.294 in. ↓)

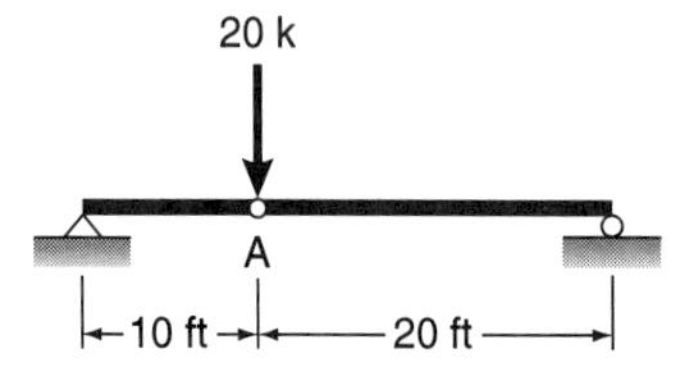

F.2 Determine the maximum deflection in the beam of Problem F.1.

F.3 θ_A, θ_B, δ_A, δ_B: $E = 29 \times 10^6$ psi. $I = 4000$ in.^{4} (*Ans.* 0.000310 rad, 0.00590 rad, 1.27 in. ↓, 0.906 in. ↓)

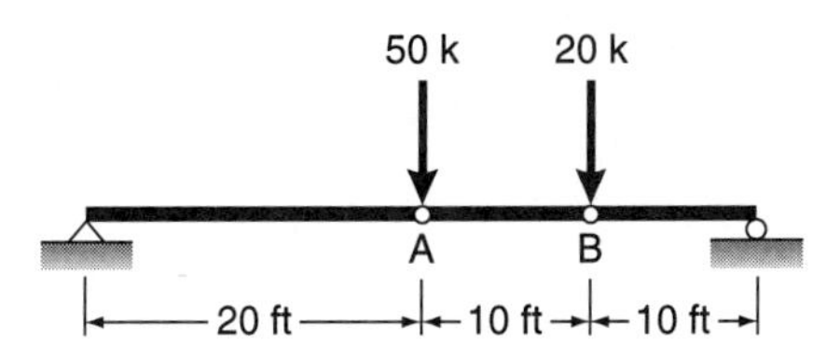

F.4 Determine the maximum deflection in the beam of Problem F.3.

F.5 Determine the slope and deflection at points A and B for the beam. (*Ans.* $\theta_A = 0.0034$ rad, $\theta_B = 0.00306$ rad, $\delta_A = 0.631$ in. ↓, $\delta_B = 0.673$ in. ↓)

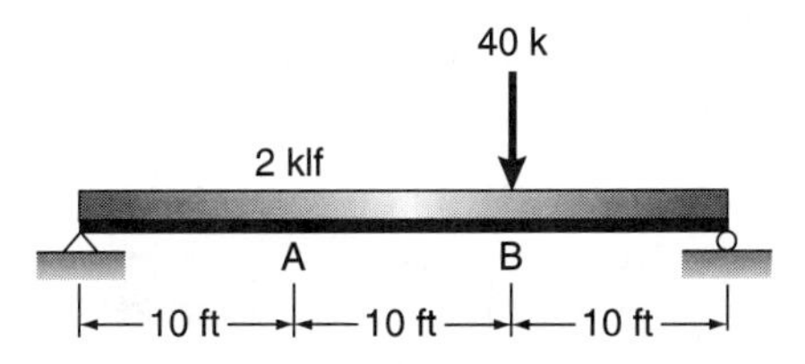

F.6 θ_A, δ_A: $E = 29 \times 10^6$ psi. $I = 2370$ in.^{4}

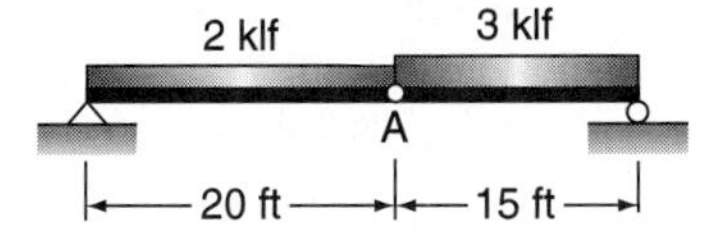

Problem F.6

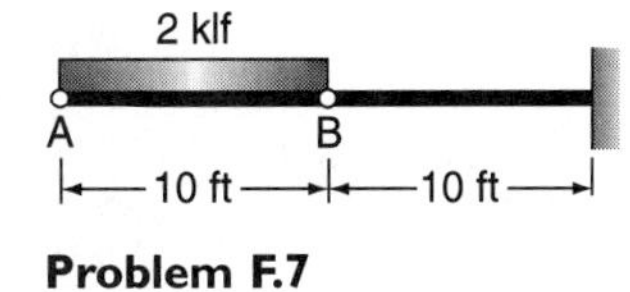

Problem F.7

F.7 θ_A, θ_B, δ_A, δ_B: $E = 29 \times 10^6$ psi. $I = 1820$ in^4 (*Ans.* 0.00637 rad, 0.00546 rad, 1.12 in. ↓, 0.382 in. ↓)

F.8 θ_A, δ_A: $E = 200{,}000$ MPa

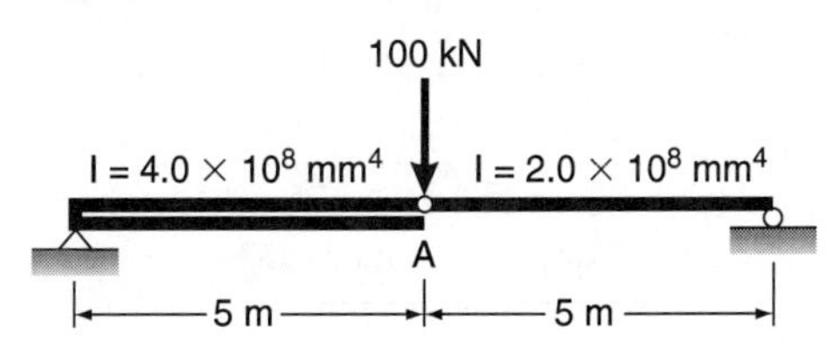

F.9 θ_A, δ_A: $E = 29 \times 10^6$ psi. $I = 1600$ in.^{4} (*Ans.* 0.00414 rad, 0.795 in. ↓)

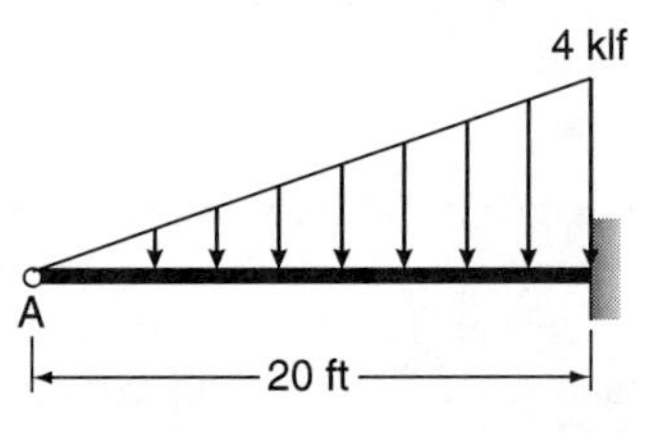

Problem F.9

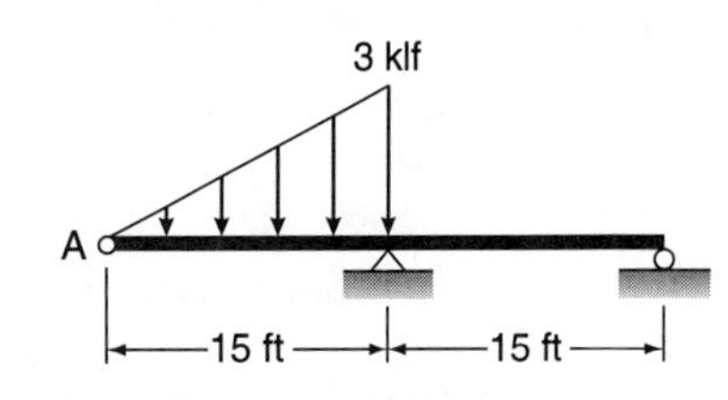

Problem F.10

F.10 θ_A, δ_A: $E = 29 \times 10^4$ psi. $I = 513$ in.^{4}

Glossary

Approximate structural analysis Analysis of structures making use of certain simplifying assumptions or "reasonable approximations."

Beam A member that supports loads that are acting transverse to the member's axis.

Bending moment Algebraic sum of the moments of all of the external forces to one side or the other of a particular section in a member. The moments are taken about an axis through the centroid of the section.

Braced frame A frame that has resistance to lateral loads supplied by some type of auxiliary bracing.

Camber The construction of a member bent or arched in one direction so that it won't look bad when the loads bend it in the opposite direction.

Cantilever A projecting or overhanging beam.

Cantilever construction Two simple beams, each with overhanging or cantilevered ends with another simple beam in between supported by the cantilevered ends.

Castigliano's theorems Energy methods for computing deformations and for analyzing statically indeterminate structures.

Cladding The exterior covering of the structural parts of a building.

Column A structural member whose primary function is to support compressive loads.

Concrete A mixture of sand, gravel, crushed rock, or other aggregates held together in a rocklike mass with a paste of cement and water.

Conjugate beam An imaginary beam that has the same length as a real beam being analyzed, and that has a set of boundary and internal continuity conditions such that the slopes and deflections in the real beam equal the shear and moment in the fictitious beam when it is loaded with the M/EI diagram.

Conservation of energy See Law of conservation of energy.

Dead loads Loads of constant magnitude that remain in one position. Examples are weights of walls, floors, roofs, fixtures, structural frames, and so on.

Diaphragms Structural components that are flat plates.

Effective length The distance between points of zero moments in a column, that is, the distance between its inflection points.

Elastic behavior When the external forces are removed, an elastic member will return to its original length.

Environmental loads The loads caused by the environment in which the structure is located. Included are snow, wind, rain, and earthquakes. Strictly speaking, these also are live loads.

"Exact" structural analysis Theoretical analysis of structures.

Fixed-end moments The moments at the ends of loaded members when the member joints are clamped to prevent rotation.

Floor beams The larger beams in many bridge floors that are perpendicular to the roadway of the bridge and that are used to transfer the floor loads from the stringers to the supporting girders or trusses.

Geometric instability A situation existing when a structure has a number of reaction components equal to or greater than the number of equilibrium equations available, and yet still is unstable.

Girder A rather loosely used term usually indicating a large beam and perhaps one into which smaller beams are framed.

Hooke's law A statement of the linear relationship existing between force and deformation in elastic members.

Impact loads The difference between the magnitudes of live loads actually caused and the magnitudes of those loads had they been dead loads.

Influence area The floor area of a building that *directly influences* the forces in a particular member.

Influence line A diagram whose ordinates show the magnitude and character of some function of a structure (shear, moment, deflection, etc.) as a unit load moves across the structure.

Joists The closely spaced beams supporting the floors and roofs of buildings.

Law of conservation of energy When a set of external loads is applied to a structure, the work performed by those loads equals the work performed in the elements of the structure by the internal forces.

Least work, principle of The internal work accomplished by each member or each portion of a structure subjected to a set of external loads is the least possible amount necessary to maintain equilibrium in supporting the loads.

Live loads Loads that change position and magnitude: They move or are moved. Examples are trucks, people, warehouse, materials, furniture, and so on.

Load and resistance factor design A method of design in which the loads are multiplied by certain load or over-capacity factors (larger than 1.0) and the members are designed to have design strengths sufficient to resist these so-called factored loads.

Matrix An ordered arrangement of numbers in rows and columns.

Maxwell's law of reciprocal deflections The deflection at one point A in a structure due to a load applied at another point B is exactly the same as the deflection at B if the same load is applied at A. The law is applicable to members consisting of materials that follow Hooke's law.

Moment distribution A successive correction or iteration method of analysis whereby fixed end and/or sidesway moments are balanced by a series of corrections.

Muller–Breslau's principle The deflected shape of a structure represents to some scale the influence line for a function of the structure such as shear, moment, deflection, among others, if the function in question is allowed to act through a unit displacement.

Nodes The locations in a structure where the elements are connected. In structures composed of beams and columns, the nodes usually are the joints.

Nominal strength Theoretical strength.

Open-web joist A small parallel chord truss whose members often are made from bars (hence the common name *bar joist*) or small angles or other shapes. These joists are very commonly used to support floor and roof slabs.

Plane frame A frame that for purposes of analysis and design is assumed to lie in a single (or two-dimensional) plane.

Point of contraflexure *See* Point of inflection.

Point of inflection (PI) A point of zero moment. Also called *point of contraflexure.*

Ponding A situation in which water accumulates on a roof faster than it runs off.

Principle of superposition If a structure is linearly elastic, the forces acting on the structure may be separated or divided in any convenient fashion and the structure analyzed for the separate cases. The final results can be obtained by adding together the individual parts.

Purlins Roof beams that span between trusses.

Qualitative influence line A sketch of an influence line in which no numerical values are given.

Quantitative influence line An influence line that shows numerical values.

Reinforced concrete A combination of concrete and steel reinforcing wherein the steel provides the tensile strength lacking in the concrete. The steel reinforcing also can be used to help the concrete resist compressive forces.

Scuppers Large holes or tubes in walls or parapets that enable water above a certain depth to quickly drain from roofs.

Seismic Of or having to do with an earthquake.

Service loads The actual loads that are assumed to be applied to a structure when it is in service (also called *working loads*).

Shear The algebraic summation of the external forces in a member to one side or the other of a particular section that are perpendicular to the axis of the member.

Sidesway The lateral movement of a structure caused by lateral or unsymmetrical loads and/or by an unsymmetrical arrangement of the members of the structure.

Skeleton construction Building construction in which the loads are transferred from each floor by beams to columns and thence to the foundation.

Slenderness ratio The ratio of the effective length of a member to its radius of gyration, both values pertaining to the same axis of bending.

Slope deflection A classical method of analyzing statically indeterminate structures in which the moments at the ends of the members are expressed in terms of the rotations (or slopes) and deflections of the joints.

Space truss A three-dimensional truss.

Statically determinate structures Structures for which the equations of equilibrium are sufficient to compute all of the external reactions and internal forces.

Statically indeterminate structures Structures for which the equations of equilibrium are insufficient for computing the external reactions and internal forces.

Steel An alloy consisting almost entirely of iron (usually over 98%). It also contains small quantities of carbon, silicon, manganese, sulfur, phosphorus, and other elements.

Stringers The beams in bridge floors that run parallel to the roadway.

Structural analysis The computation of the forces and deformations of structures under load.

Struts Structural members that are subjected only to axial compression forces.

Superposition principle *See* Principle of superposition.

Tension coefficient The force in a truss member divided by its length.

Three-moment theorem A classical theorem that presents the relationship between the moments in the different supports of a continuous beam.

Ties Structural members that are subjected only to axial tension forces.

Tributary area The loaded area of a structure that *directly contributes* to the load applied to a particular member.

Truss A structure formed by a group of members arranged in the shape of one or more triangles.

Unbraced frame A frame whose resistance to lateral forces is provided by its members and their connections.

Unstable equilibrium A support situation whereby a structure is stable under one arrangement of loads but is not stable under other load arrangements.

Vierendeel "truss" Though not really a truss by our usual definition, it is considered a special type of truss whose members are arranged in the shape of a set of rectangles. It requires moment-resisting joints.

Virtual displacement A fictitious displacement imposed on a structure.

Virtual work The work performed by a set of real forces during a virtual displacement.

Voussoirs The truncated wedge-shaped parts of a stone arch that are pushed together in compression.

Wichert truss A continuous statically determinate truss formerly patented by E. M. Wichert.

Working loads *See* Service loads.

Yield stress The stress at which there is a decided increase in the elongation or strain in a member without a corresponding increase in stress.

Zero-load test A procedure in which one member of a truss subjected to no external loads is given a force and the forces in the other members are computed. If all the joints balance or are in equilibrium, the structure is unstable.

Index

Limited Use License Agreement